Computational Methods in Engineering

Computational Methods in Engineering: Finite Difference, Finite Volume, Finite Element, and Dual Mesh Control Domain Methods provides readers with the information necessary to choose appropriate numerical methods to solve a variety of engineering problems. Explaining common numerical methods in an accessible yet rigorous manner, the book details the finite element method (FEM), finite volume method (FVM), and importantly, a new numerical approach, dual mesh control domain method (DMCDM).

Numerical methods are crucial to everyday engineering. The book begins by introducing the various methods and their applications, with example problems from a range of engineering disciplines including heat transfer, solid and structural mechanics, and fluid mechanics. It highlights the strengths of FEM, with its systematic procedure and modular steps, and then goes on to explain the uses of FVM. It explains how DMCDM embodies useful parts of both FEM and FVM, particularly in its use of the control domain method and how it can provide a comprehensive computational approach. The final chapters look at ways to use different numerical methods, primarily FEM and DMCDM, to solve typical problems of bending of beams, axisymmetric circular plates, and other nonlinear problems.

This book is a useful guide to numerical methods for professionals and students in all areas of engineering and engineering mathematics.

APPLIED AND COMPUTATIONAL MECHANICS

A Series of Textbooks and Reference Books

Founding Editor

J. N. Reddy, Distinguished Professor and the Holder of the O'Donnell Foundation Chair in the Department of Mechanical Engineering at Texas A&M University, is a highly-cited researcher, author of 24 textbooks and over 800 journal papers, and a leader in the applied and computational mechanics fields for nearly 50 years.

Advanced Thermodynamics Engineering, Second Edition
Kalyan Annamalai, Ishwar K. Puri, Miland Jog

Applied Functional Analysis
J. Tinsley Oden, Leszek F. Demkowicz

Classical Continuum Mechanics, Second Edition
Karan S. Surana

Combustion Science and Engineering
Kalyan Annamalai, Ishwar K. Puri

Computational Methods in Engineering: Finite Difference, Finite Volume, Finite Element, and Dual Mesh Control Domain Methods
J.N. Reddy

Computational Modeling of Polymer Composites: A Study of Creep and Environmental Effects
Samit Roy and J.N. Reddy

Continuum Mechanics for Engineers, Fourth Edition
G. Thomas Mase, Ronald E. Smelser, Jenn Stroud Rossmann

Dynamics in Engineering Practice, Eleventh Edition
Dara W. Childs, Andrew P. Conkey

Exact Solutions for Buckling of Structural Members
C.M. Wang, C.Y. Wang, J.N. Reddy

Failure Analysis of Composite Materials with Manufacturing Defects
Ramesh Talreja

The Finite Element Method for Boundary Value Problems: Mathematics and Computations
Karan S. Surana, J. N. Reddy

The Finite Element Method in Heat Transfer and Fluid Dynamics, Third Edition
J.N. Reddy, D.K. Gartling

Mechanics of Laminated Composite Plates and Shells: Theory and Analysis, Second Edition
J.N. Reddy

Mechanics of Materials
Clarence W. de Silva

Mechanics of Solids and Structures, Second Edition
Roger T. Fenner, J.N. Reddy

Microchemical Analysis and Multi-Scale Modeling Using the Voronoi Cell Finite Element Method
Somnath Ghosh

Numerical and Analytical Methods with MATLAB® for Electrical Engineers
William Bober, Andrew Stevens

Numerical and Analytical Methods with MATLAB®
William Bober, Chi-Tay Tsai, Oren Masory

Physical Components of Tensors
Wolf Altman, Antonio Marmo De Oliveira

Plates and Shells: Theory and Analysis, Fourth Edition
Ansel C. Ugural

Practical Analysis of Composite Laminates
J.N. Reddy, Antonio Miravete

Solving Ordinary and Partial Boundary Value Problems in Science and Engineering
Karel Rektorys

Theories and Analyses of Beams and Axisymmetric Circular Plates
J.N. Reddy

Computational Methods in Engineering

Finite Difference, Finite Volume, Finite Element, and Dual Mesh Control Domain Methods

J. N. Reddy
J. Mike Walker '66 Department of Mechanical Engineering
Texas A&M University, College Station, TX

CRC Press
Taylor & Francis Group
Boca Raton London New York

CRC Press is an imprint of the
Taylor & Francis Group, an **informa** business

Designed cover image: J. N. Reddy

First edition published 2024
by CRC Press
2385 NW Executive Center Drive, Suite 320, Boca Raton FL 33431

and by CRC Press
4 Park Square, Milton Park, Abingdon, Oxon, OX14 4RN

CRC Press is an imprint of Taylor & Francis Group, LLC

ISBN: 978-1-032-46637-8 (hbk)
ISBN: 978-1-032-46681-1 (pbk)
ISBN: 978-1-003-38281-2 (ebk)

DOI: 10.1201/9781003382812

Typeset in CMR10
by KnowledgeWorks Global Ltd.

Dedicated to the loving memory of
my teacher, mentor, and friend

John Tinsley Oden
(25 Dec 1936–27 Aug 2023)

Professor J. T. Oden was a pioneer of the finite element method (especially, the mathematical foundations of the FEM) and its applications to a host of problems in engineering and applied sciences. His legacy as an engineering educator, researcher, scholar, mentor to many researchers around the world, and the founder of the Oden Institute for Computational Engineering & Sciences at the University of Texas at Austin will last forever.

As my Ph.D. advisor (I was his second PhD student and the first one to go into academics), I knew Dr. Oden as a kind and gentle person, great teacher, mentor, and friend for all my life. I may not understand why he left this world so soon - before I was ready to say good-bye to him - but I know that his life gave me memories too beautiful to forget. He is alive in my heart through the kind words he said to me, the caring deeds he did for me, and the professional mentor and role model that he was to me.

The life of the dead is placed in the memory of the living.

– Marcus Tullius Cicero

Contents

Preface

Over the last several decades, computational methods made it possible, with the help of sophisticated physics-based mathematical models, numerical methods, and high-speed computers, to analyze many practical problems of engineering for analysis, design, and manufacturing. Most of the practicing engineers may end up using a commercial software package or an open source code. But by having a background in most commonly used numerical methods, namely, the finite difference method (FDM), finite volume method (FVM), and finite element method (FEM), practicing engineers and scientists can benefit at work in judiciously selecting the suitable numerical method for the solution of the problem at hand. To date, there does not exist a single place where an interested engineer or scientist can find all these popular numerical methods explained in a single volume in a fashion they can readily understand. The present book is aimed at fulfilling this gap in the literature, while introducing a variant of the FEM and FVM, namely, the *Dual Mesh Control Domain Method* (DMCDM).

This book is a result of a study of the FVM during the COVID-19 period by the author, who ventured out to understand the inner workings of the FVM. During this study, it became clear to him that, while the FVM has a physically desirable basis (i.e., satisfying the differential equations in the integral sense), the method is presented and practiced by various authors through an introduction of ad-hoc assumptions and approximations. To overcome and eliminate the arbitrariness in selecting approximations of derivatives of the dependent variables, the DMCDM was conceived by the author as a better alternative to FVM. While retaining FVM's positive attributes, the author has brought into play the polynomial approximation of the dependent variables and the concept of duality. In comparison with the FVM, the DMCDM has a unique set of steps, free of ad-hoc assumptions and approximations. During the last two years, the author and his colleagues have tested the DMCDM for a variety of problems in one and two dimensions and for linear and nonlinear problems. Like the FEM, which was introduced originally to solve linear problems of solid and structural mechanics and later generalized and endowed with a strong mathematical basis, the author felt the DMCDM should be introduced to others so that it can be advanced both in theory and application to a variety of practical problems described by differential equations. In this context, the author felt that including a review of other competing methods such as the finite difference method, FVM, and FEM is needed so that the readers have the benefit of finding all popular numerical methods in one place to compare and understand their distinguishing features.

This book describes in detail, in a manner easily understandable to beginning researchers, the finite difference, finite volume, and finite element methods along with the dual mesh control domain method, which is based on the main characteristic features of two popular methods, namely, the FEM and the FVM. The DMCDM brings ideas of interpolation and duality from the FEM and discretization of governing equations using the control volume idea from the FVM. While it will take some time for others to embrace and advance the DMCDM, the author felt that by describing the method and illustrating its applications to a variety of science and engineering problems, others in the computational mechanics community would see the benefits of the new method and advance it by providing sound mathematical basis for the method and finding applications in a number of subject areas of engineering and applied science.

As stated at the outset, introductions to finite difference, finite volume, finite element, and dual mesh control domain methods and their applications in the solution of differential equations arising in applied science and engineering are presented in this book. Each of the four methods are introduced using second-order differential equations in one and two dimensions, and their applications are illustrated using a number of problems from heat transfer, solid and structural mechanics, and fluid mechanics. While the FEM has a systematic procedure and steps (i.e., modular) beginning with a set of differential equations to the final step of solving the resulting (discrete) algebraic equations among the nodal values of the duality pairs, the FVM appears to have the same issues as the traditional finite difference method in obtaining the final algebraic equations. These issues are largely connected with the approximation of derivatives, evaluation of domain integrals, and imposition of gradient boundary conditions. The DMCDM makes use of the desirable features of the FEM and FVM.

Chapter 1 provides the necessary background and mathematical and mechanics preliminaries needed for the subsequent chapters. These include: types of differential equations and problems (i.e., boundary and initial value problems), Taylor's series and elements of matrix theory, variational formulations and methods, interpolation theory, numerical integration, and solution of linear equations. A number of finite difference methods are discussed and their use in the solution of ordinary and partial differential equations is presented in Chapter 2. Chapter 3 is dedicated to the introduction of the finite volume method, while Chapter 4 is concerned with the finite element method. The dual mesh control domain method is introduced in Chapter 5 for one- and two-dimensional problems in the context of solving a Poisson equation. Chapter 6 contains finite element and dual mesh control domain formulations of nonlinear one- and two-dimensional problems in a single variable. Here the iterative methods for the solution of nonlinear algebraic equations are also discussed. The remaining chapters, Chapter 7 through 10, deal with the FEM and DMCDM as applied to analyses of beams, axisymmetric circular plates, plane elasticity and viscous incompressible flows, and plate bending problems, respectively. In all chapters, the governing equations are reviewed, the FEM and DMCDM models are derived, and numerical examples are presented. In Chapter 9, the DMCDM is extended to the use of arbitrary meshes of triangles and quadrilaterals (using the isoparametric concept) in connection with plane elasticity. Applications of the DMCDM to vibration, buckling, and coupled and nonlinear problems of engineering physics are underway.

The author extends his sincere thanks to his colleagues, Professors N. K. Anand and Arun Srinivasa, for their interest and time in discussing the workings of the FVM and DMCDM. Thanks are also to Tanmaye Heblekar for his careful reading of and valuable comments on the manuscript of this book and for his help with Section 9.9. The author realizes that a book of this mathematical and computational complexity is bound to have typos, errors, and incomplete information about specific examples included herein. The author sincerely requests the readers to send any errors they find and queries to his attention.

J. N. Reddy
College Station, Texas

List of symbols

Symbol	Meaning
a_{ij}	Coefficients of matrix $[A] = \mathbf{A}$
ds	Line or surface elements
$\mathbf{D}$	Symmetric part of the velocity gradient tensor; that is, $\mathbf{D} = \frac{1}{2}\left[(\boldsymbol{\nabla}\mathbf{v})^{\mathrm{T}} + \boldsymbol{\nabla}\mathbf{v}\right]$
$\mathbf{e}_i$	Basis vector in the x_i-direction
$(\hat{\mathbf{e}}_r, \hat{\mathbf{e}}_\theta, \hat{\mathbf{e}}_z)$	Basis vectors in the (r, θ, z) system
$(\hat{\mathbf{e}}_x, \hat{\mathbf{e}}_y, \hat{\mathbf{e}}_z)$	Basis vectors in the (x, y, z) system
$\mathbf{f}$	Body force vector
f_x, f_y, f_z	Body force components in the x, y, and z directions
g	Acceleration due to gravity; internal heat generation per unit volume
J	Determinant of $\mathbf{J}$ (Jacobian)
k	Spring constant; thermal conductivity
$\mathbf{k}$	Thermal conductivity tensor
$\hat{\mathbf{n}}$	Unit normal vector
(n_x, n_y, n_z)	Components of the unit normal vector $\hat{\mathbf{n}}$
P	Hydrostatic pressure; perimeter
q_n	Heat flux normal to the boundary, $q_n = -k\hat{\mathbf{n}} \cdot \boldsymbol{\nabla}T$
$\mathbf{q}$	Heat flux vector; diffusion flux
r	Radial coordinate in the cylindrical polar system; $r = \lvert\mathbf{r}\rvert$
$\mathbf{r}$	Position vector in cylindrical coordinates, $\mathbf{x}$
(r, θ, z)	Cylindrical coordinate system
$\mathbf{t}$	Stress vector; traction vector
$\mathbf{t}_i$	Stress vector on x_i-plane, $\mathbf{t}_i = \sigma_{ij}\hat{\mathbf{e}}_j$
T	Temperature
(u_x, u_y, u_z)	Components of displacement vector $\mathbf{u}$ in (x, y, z) system
(v_x, v_y, v_z)	Components of velocity vector $\mathbf{v}$ in (x, y, z) system
(x, y, z)	Rectangular Cartesian coordinates

Greek symbols

Symbol	Meaning
α	Angle; coefficient of thermal expansion; kinetic energy coefficient
β	Heat transfer coefficient
γ	Penalty parameter
Γ	Total boundary
δ	Dirac delta; variational symbol
δ_{ij}	Components of the unit tensor, $\mathbf{I}$ (Kronecker delta)
ϵ	Tolerance specified for steady-state solution;
ζ	Natural (normalized) coordinate
η	Natural (normalized) coordinate
θ	Angular coordinate in the cylindrical and spherical coordinate systems; angle; absolute temperature
λ	Lamé constant; Lagrange multiplier
μ	Lamé constant; viscosity
ξ	Natural (normalized) coordinate
ρ	Mass density
$\boldsymbol{\sigma}$	Stress tensor
σ_{ij}	Components of the stress tensor in the rectangular coordinate system (x_1, x_2, x_3)
$\sigma_{rr}, \sigma_{\theta\theta}, \sigma_{r\theta}, \cdots$	Components of the stress tensor $\boldsymbol{\sigma}$ in the cylindrical coordinate system (r, θ, z)
τ	Shear stress
$\boldsymbol{\tau}$	Viscous stress tensor
ϕ	A typical scalar function; velocity potential; angular coordinate in the spherical coordinate system
φ_i	Hermite cubic interpolation functions
ψ_i^e	Lagrange interpolation functions of a finite element
Ω	Domain of a problem
$\boldsymbol{\nabla}$	Gradient operator with respect to $\mathbf{x}$
∇^2	Laplace operator, $\nabla^2 = \boldsymbol{\nabla} \cdot \boldsymbol{\nabla}$
∇^4	Biharmonic operator, $\nabla^4 = \nabla^2 \nabla^2$

About the Author

J. N. Reddy, O'Donnell Foundation Chair IV Professor in the J. Mike Walker '66 Department of Mechanical Engineering at Texas A&M University, is a highly cited researcher and a leader in the applied mechanics field for more than 50 years. He is well-known worldwide for his significant contributions to the field of applied and computational mechanics through the authorship of widely used textbooks. His pioneering works on the development of shear deformation theories of beams, plates, and shells (that bear his name in the literature as *the Reddy third-order plate theory* and *the Reddy layerwise theory*), nonlocal and nonclassical continuum mechanics have had a major impact and have led to new research developments and applications.

Recent honors and awards include: 2023 Leonardo da Vinci Award from the European Academy of Sciences, 2023 Michael P. Païdoussis Medal from the Royal Society of Canada, 2022 IACM Congress (Gauss-Newton) Medal from the International Association of Computational Mechanics, 2019 Timoshenko Medal from the American Society of Mechanical Engineers, 2018 Theodore von Kármán Medal from the American Society of Civil Engineers, 2017 John von Neumann Medal from the U.S. Association of Computational Mechanics, 2016 Prager Medal from the Society of Engineering Science, and 2016 ASME Medal from the American Society of Mechanical Engineers. He is a member of the US National Academy of Engineering and foreign fellow of the Canadian Academy of Engineering, the Chinese Academy of Engineering, the Brazilian National Academy of Engineering, the Indian National Academy of Engineering, the Royal Academy of Engineering of Spain, the European Academy of Sciences (Honorary Member), and the Academia Scientiarum et Artium Europaea (the European Academy of Sciences and Arts). For additional information about the author, visit **http://mechanics.tamu.edu/**.

There is no "complete" mathematical model of anything we study. We only try to "improve" on what we already know. Junuthula Narasimha (J.N.) Reddy

1
Introduction and Preliminaries

1.1 Background

Study and analysis of all physical processes require three major components (or pillars): (1) mathematical models of the process[1], (2) experimental characterization of certain process parameters that enter the mathematical models, and (3) a suitable numerical method (i.e., *algorithm* or procedure) to *compute* the solution. All three components involve approximation and inherent errors. Errors in the development of mathematical model come from an incomplete understanding of the process, resulting in mathematical models that do not account exactly for everything the process has. Characterization of physical (or material) properties through experiments is also approximate. Finally, by definition, numerical methods introduce errors at the algorithm level as well as at the computational level (e.g., truncation and round-off errors).

There exist a variety of numerical methods for the solution of differential equations. Typically, all approximate methods are endowed with a procedure to convert a system of coupled differential equations, $\mathbf{Au} = \mathbf{f}$, among a set of variables $\mathbf{u}$ into a set of algebraic equations, which can be expressed in matrix form as $\mathbf{KU} = \mathbf{F}$. Here $\mathbf{U}$ denotes the vector of discrete values of the variables $\mathbf{u}$ and $\mathbf{F}$ is the vector of discrete values of the dual variable $\mathbf{f}$ at a selected number of points (called mesh points or nodes) inside the domain and on its boundary. The *duality pairs* $(\mathbf{U}, \mathbf{F})$ (or effect and cause; e.g., velocity and force; temperature and heat; displacement and force) exist in all fields of science and engineering (the dual variables involve the derivatives of the primary unknowns). A physical process may have more than one duality pair, depending on the number of equations and unknowns. These duality pairs are unique and no polygamy exists among them; that is, each element of a pair is dual only to the other element of the same pair but not to an element in the other pairs. The resulting algebraic equations are solved after imposition of boundary conditions on the elements of the duality pairs (and only one element of the duality pair is necessarily known, or a relation between the pair is known at every point of the domain and its boundary).

[1]Derivation of mathematical equations from physical principles involves making assumptions and placing restrictions concerning the process, both of which introduce error into the mathematical model. In fact, there is no such thing as "exact" mathematical model of any natural process we study. We can only improve what is currently available.

The actual process that results in the final matrix equation $\mathbf{KU} = \mathbf{F}$ differs from one method to another. Most commonly used methods are finite difference methods (FDMs), the finite volume method (FVM), and the finite element method (FEM); of course, there are a host of other numerical methods out there in the literature that have not received the same level of attention across the computational mechanics community as these three methods. A brief description of the three most popular methods as well as the recently proposed method, namely, the dual mesh control domain method, as applied to the numerical solution of differential equations is presented here; a detailed review of these methods is considered in the coming chapters.

1.2 Popular Numerical Methods

1.2.1 Finite Difference Method

In the *finite difference method*, the derivatives of the variable of a differential equation are replaced by difference quotients (or the solution variables are expanded in a Taylor series) that involve the values of the variables at a finite number points of the domain (see Fig. 1.2.1), leading to a formula, called the finite difference stencil whose repeated application yields the final set of algebraic equations among the values of the unknowns at the mesh points (see, e.g., Carnahan, Luther, and Wilkes [1]). The finite difference method, as commonly practiced, suffers from several disadvantages or restrictions, such as difficulty in applying the method to nonrectangular domains, representing gradient boundary conditions, using nonuniform and non-rectangular meshes, and making problem-dependent approximations. In addition, the concept of duality and the explicit approximation formulas (only implicit or implied forms exist) for the variables being approximated are not readily available. These disadvantages, although some of them may have been overcome in recent years and in specific contexts, precluded the development of commercially successful robust, general-purpose, computational software.

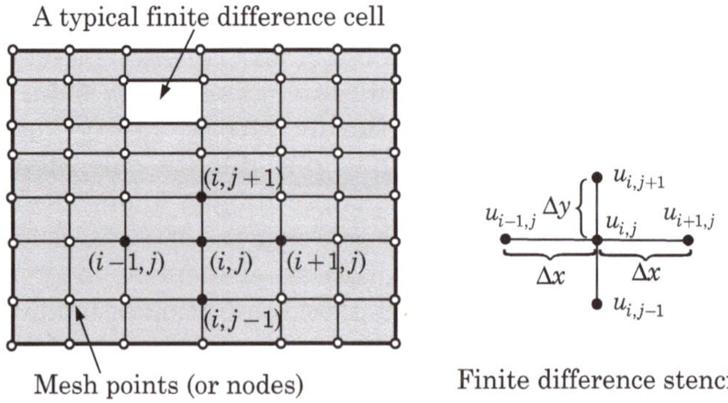

Fig. 1.2.1 Representation of a rectangular domain with a mesh of finite difference cells, each of size $\Delta x \times \Delta y$. A typical finite difference (e.g., central difference) scheme results in a stencil relating the discrete values of the unknown u at five mesh points.

1.2.2 Finite Volume Method

In the *finite volume method* (FVM), one represents a given domain as a collection of nonoverlapping domains, called *control volumes* (see [2–7]). Then an integral (not weighted-integral) statement of the governing equation (after invoking the Green–Gauss theorem to convert the domain integral to the boundary integral for second-order equations) is used over a typical control volume to derive the algebraic equations among the values of the variables at mesh points of the domain. The algebraic equations derived using a control volume consist of nodal values from four neighboring control volumes, as shown in Fig. 1.2.2 (a notable difference from FEM). In the FVM, at the center of each control volume (for uniform meshes) lies a computational node, and the derivatives of the dependent variables at the control volume interfaces are calculated in terms of the values of the dependent variables at the nodes using the Taylor series approximations. The resulting algebraic equations resemble a finite difference stencil, that is, a relation among values of the unknowns at mesh points on the left, right, top, and bottom (also front and back in three-dimensional problems) of the mesh point of interest. Then the algebraic equation is evaluated at all nodes of the mesh, except where the nodal value is specified, to obtain the required algebraic equations of the problem. Thus, in the FVM there is no explicit assembly of elements. The imposition of gradient type boundary conditions involves, sometimes, fictitious points from inside and outside the domain, and there is no unique methodology that seems to exist for the computation of domain integrals and the imposition of boundary conditions in the FVM. Since a vast majority developers of the FVM came from the traditional finite difference community, they tend to borrow ideas like upwinding, artificial viscosity, and other ad-hoc approaches to make the numerical schemes to perform to their satisfaction. Again, there are no explicit and unique forms of approximations used, and no concept of duality exploited in the FVM.

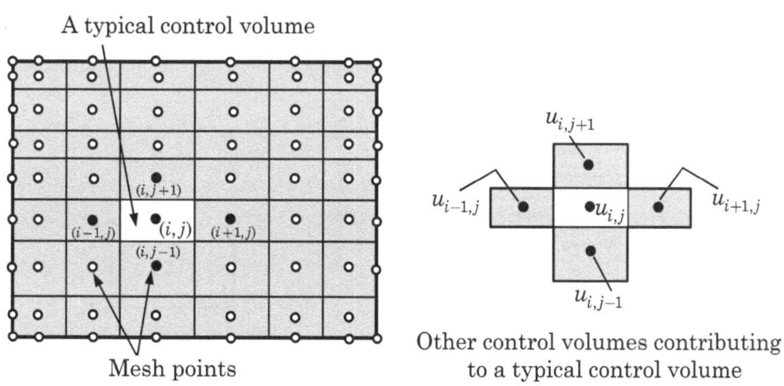

Fig. 1.2.2 Representation of a rectangular domain with a mesh of control volumes, each control volume centered around a mesh point. The governing differential equation is satisfied in an integral sense over each control volume, which brings the discrete values at mesh points of other neighboring control volumes.

1.2.3 Finite Element Method

The *finite element method* is based on the following three-fold idea (see [8–15] for details of the method and applications in heat transfer, fluid dynamics, and structural mechanics):

(1) the total domain Ω can be represented as a collection of a finite number of nonoverlapping but interconnected (at the boundaries of the) subdomains, called *finite elements*, Ω^e; the elements are of a particular geometry that allows the construction of approximation (or interpolation) functions;

(2) over each element Ω^e, the dependent unknown u is interpolated through a set of points (nodes) of the element as $\mathbf{u} \approx \sum \mathbf{u}_j \psi_j$, $\mathbf{u}_j$ being the value of $\mathbf{u}$ at the jth node (see Fig. 1.2.3) and ψ_j are suitable approximation functions, and the governing equation is converted to a set of algebraic equations $\mathbf{K}^e\mathbf{u}^e = \mathbf{F}^e$ (called *finite element model*) using a method of approximation (e.g., weak-form Galerkin or Ritz, subdomain, least-squares, and so on); all of which, except for the subdomain method, are a form of weighted-integral statements; the element equations contain nodal variables from only the element under consideration; and

(3) the element equations from all elements are put together (element assembly) using balance and continuity conditions at element interfaces to obtain a set of algebraic equations for the whole mesh, $\mathbf{KU} = \mathbf{F}$, which are then solved after applying the boundary conditions at the nodes on the boundary only.

We note that in the FEM, the domain (i.e., element) used for the approximation of dependent variable(s) as well as for the satisfaction of the governing equation(s) is the same. Of course, the mesh used for the approximation of the geometry can be different from that used for the variables (e.g., in the isoparametric formulations, both meshes are the same). The major disadvantage of the

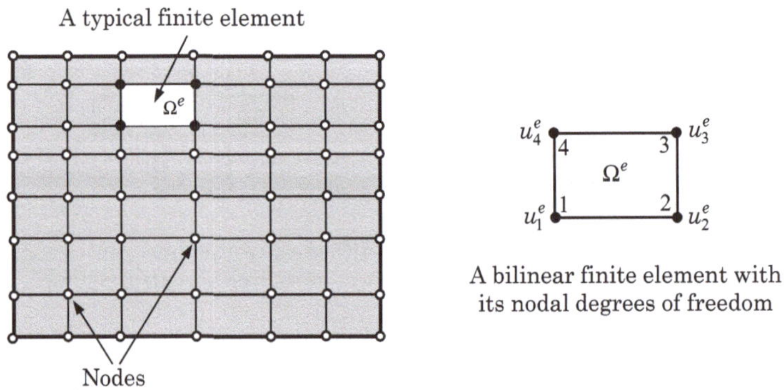

Fig. 1.2.3 Representation of a rectangular domain with a mesh of bilinear finite elements, each finite element having four nodes which are interfaced with neighboring elements. The governing differential equation is satisfied in an integral sense over each finite element, which brings the discrete values at mesh points of other neighboring elements.

FEM is that it introduces discontinuity in the dual variables, although the dual variables are "balanced" in the assembly process, at the element interfaces in the post-computation. The method also introduces discontinuity of the derivatives of the function across interelement boundaries unless the approximations are C^1-continuous. The common FEM formulations are based on weighted-integral statements, with weight functions being the same as the approximation functions. This has the effect of smoothing of the solution.

1.2.4 Dual Mesh Control Domain Method

The *dual mesh control domain method* (DMCDM), as introduced by Reddy [16], has three basic features: (1) The DMCDM makes use of the concept of duality. (2) The dependent unknowns, called *primary variables*, are interpolated over, what is called, the *primal mesh.* (3) The discretized equations are obtained by satisfying the governing equations over the domains of the *dual mesh.* The primal mesh is a set of finite elements with their nodes and interpolation functions, and the dual mesh is a set of subdomains, called *control domains*, each of which includes a node of the primal mesh; the interior nodes will be inside of the control domains, and the boundary nodes will be on one edge or at the intersection of two edges of the control domains (see Fig. 1.2.4). The discretized equations resulting from the interior control domains are algebraic relations among the nodal values of the primary variables, and the nodal values of all finite elements connected to the control domain appear in the equations. The discretized equations of control domains on the boundary contain the nodal values of both the primary and secondary variables.

Clearly, the first two features are the same as those of the FEM and the third feature is one of the FVM. As a result, there is no assembly of discretized equations, solution gradients on the boundary are not replaced in terms of the nodal values but expressed in terms of the dual variables. Then the governing equation

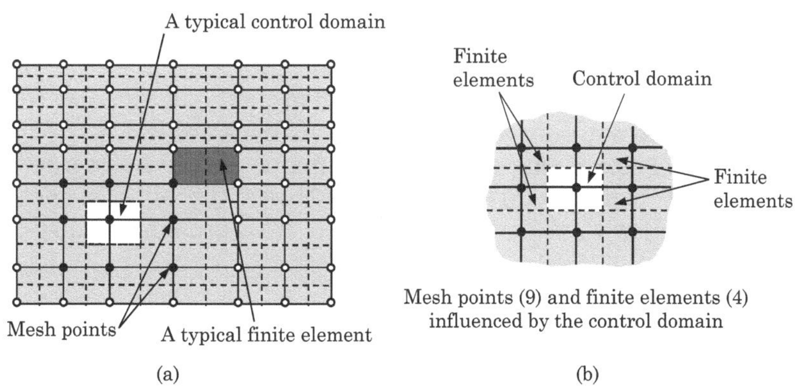

Fig. 1.2.4 (a) Representation of a rectangular domain with a primal mesh (shown with broken lines) of *bilinear* rectangular finite elements and a dual mesh of rectangular control domains (shown with solid lines which bisect the broken lines) such that the dual mesh contains the mesh points (or nodes) of the primal mesh in the interior of the control domains. For a uniform primal mesh, the dual mesh will be uniform and the nodes of the primal mesh will be at the center of the dual mesh. (b) Nodes and finite elements influenced by the control domain.

is required to be satisfied in an integral sense (*not* in a weighted-integral sense [17]) over the finite domain. The second-order terms in the differential equation are integrated by parts and expressed as dual variables on the interfaces of the dual mesh. When the interfaces fall on the boundary, either the dual variables or their counterparts (i.e., primary variables) are known and thereby the derivatives are not replaced at the boundary nodes, eliminating the need for the so-called zero-thickness control volumes or fictitious control volumes.

The DMCDM brings the best features of the FEM, namely, the interpolation of the variables, duality, and imposition of physical boundary conditions, and of the FVM in satisfying the actual balance equations over the control domain. The major merits of the DMCDM are that the method inherits the desirable features of the FVM (in satisfying the global form of the governing equations over the finite domains and avoiding assembly of elements as well as evaluation of boundary integrals), and it overcomes the disadvantage of the discontinuity of the secondary variables at the interfaces of the finite elements by calculating them at the boundaries of the control domains, where they are continuous (i.e., uniquely defined).

1.3 Common Features of the Numerical Methods

All four numerical methods have the following three major features (the features were illustrated in Figs. 1.2.1–1.2.4 using two-dimensional rectangular domains):

1. The domain (i.e., a geometric region on which the governing equations are valid) is discretized into a set of connected but nonoverlapping subdomains. These subdomains are called *cells* in the FDM, *control volumes* in the FVM and *finite elements* in the FEM and DMCDM. The collection of subdomains is often called a *grid* or *mesh*. The subdomains are polygons (often, rectangular in the FDM and FVM), and the corner points where the sides of a subdomain meet are called mesh/grid points or nodes. In the FVM, each control volume is centered around a mesh point (of course, near the boundary, only partial control volumes are used). In the FEM, only those polygonal shapes that allow unique derivation of interpolation functions are considered. The DMCDM uses the finite elements (geometric shapes) with their shape functions to approximate the dependent variables.

2. The discretized (algebraic) equations from the governing equations are derived using a method of approximation, either by representing the function as a linear combination of nodal values of the unknown dependent variables and known interpolation functions (in the FEM and DMCDM) or by representing the derivatives of the unknowns in terms of their values at the mesh points using a Taylor series approximation (in the FDM and FVM). The method of approximation in an FDM is to use a Taylor series to replace the derivatives of the unknowns in the differential equations at a given generic mesh point (i, j) in terms of their values at the generic

mesh point and mesh points that are neighbors to the generic mesh point, as dictated by the order of the Taylor series approximation. The resulting algebraic equation is called a *finite difference stencil,* which can be applied to each and every mesh point in the discretized domain to obtain the required number of algebraic equations (i.e., the same number of equations as there are unknown nodal values). Obviously, this process of applying the stencil to mesh points that are on or near to the boundary of the domain may bring the values of the unknowns at (fictitious) mesh points outside the domain, unless the type of approximation is changed there to avoid the fictitious points. In the FVM, the discretization is based on the satisfaction of the governing differential equations in the integral sense (i.e., global form of the differential equations as derived in continuum mechanics) over a typical control volume and replacing the derivatives again as in the FDM. This process also leads to a "stencil" valid for a control volume, and the stencil contains values of the unknowns from the neighboring control volumes. In the FEM, the governing equations are satisfied in a weighted-integral sense (actually, in a weak-form sense), and the unknowns are replaced in terms of their nodal values of the element (see Fig. 1.2.3). Thus, in the FEM, the discretized equations over an element contain the nodal values of the element only (requiring "assembly" of element equations). In the DMCDM, the dependent variables are approximated using finite element interpolation functions (as in the FEM), and the governing equation is satisfied in an integral sense over a typical control domain of the dual mesh (as in the FVM). The DMCDM also results in a stencil-like discrete equation among the values of the mesh points.

3. The discretized equations are obtained in terms of the unknowns at mesh points by evaluating the stencil at all mesh points in the FDM, FVM, and DMCDM or by the assembly of element equations in the FEM, apply suitable boundary conditions, and solve the discrete equations to determine all the unknowns at the mesh points/nodes.

We note that, independent of the numerical method and the method of approximation, all numerical methods ultimately result in a system of algebraic equations in terms of the unknowns at the mesh points. Thus, the mesh of subdomains in all numerical methods can ultimately be viewed as a mesh of points or nodes (i.e., the cell or element identity is lost after discretization) at which the solution is determined. The way the FDM and FVM are practiced, all of the data (i.e., coefficients in the differential equations and the source terms) are either treated as constant or averaged over each cell. This is not the case in the FEM or DMCDM. In addition, in the FVM and FDM, the dual (or secondary) variables (which contain derivatives of the dependent variables, and they are often quantities of significant practical interest) do not enter the discretized equations explicitly; as a consequence, their specification on the boundary is implemented only in terms of the primary variables (i.e., the dependent variables of the differential equations) by expressing the derivatives in terms of the primary variables at mesh points on the boundary and in the interior of the domain. On the other hand, in the FEM and DMCDM, the dual variables are preserved without replacing them in terms of the dependent variables. In particular, the DMCDM makes use of (a) the concept of duality in a problem, (b) a discretization procedure based on the *explicit* approximation of dependent variables, and (c) the satisfaction of governing equations over the dual mesh of control domains.

1.4 Present Study

In this book, we present basic introductions to the FDM, FVM, FEM, and DMCDM in detail and illustrate their applications to one- and two-dimensional problems of engineering (e.g., heat transfer, solid and structural mechanics, and fluid dynamics). Therefore, the reader will get a good understanding of relative efforts involved in each method in solving differential equations.

Following this introduction, some essential preliminaries regarding the Taylor series, interpolation theory, and variational methods are presented in the coming sections of this chapter. Chapters 2-5 are dedicated to the discussion of the main ideas behind the FDM, FVM, FEM, and DMCDM, respectively. All four methods are introduced for second-order differential equations in a single variable in one and two dimensions (typical of heat transfer problems). Chapter 6 is devoted to one- and two-dimensional nonlinear problems in a single unknown. Chapters 7-12 deal with the FEM and DMCDM for the bending of straight beams (Chapter 7), axisymmetric bending of circular plates (Chapter 8), plane elasticity and viscous incompressible flows in two dimensions (Chapter 9), and plate bending (Chapter 10).

1.5 Types of Differential Equations and Problems

1.5.1 Preliminary Comments

The numerical methods to be discussed in this book require a background in Taylor's series, variational formulations and methods, elements of the interpolation theory and approximation, numerical integration, solution of linear equations, and types of errors present in a numerical solution. Therefore, this chapter is dedicated to the discussion of these topics in order to prepare the reader with the required background. Readers familiar with these topics may skip this chapter and go straight to Chapter 2.

1.5.2 Order and Types of Differential Equations

A differential equation of nth order in a single dependent variable u in a two-dimensional domain Ω with closed boundary Γ is of the form

$$F\left(x, y, u, \frac{\partial u}{\partial x}, \frac{\partial u}{\partial y}, \frac{\partial^2 u}{\partial x^2}, \frac{\partial^2 u}{\partial x \partial y}, \frac{\partial^2 u}{\partial y^2}, \ldots, \frac{\partial^n u}{\partial x^n}, \ldots, \frac{\partial^n u}{\partial x^{n-i} \partial y^i}, \ldots, \frac{\partial^n u}{\partial y^n}\right) = 0 \tag{1.5.1}$$

for $i = 1, 2, \ldots, n$. An equation of type in Eq. (1.5.1) is called nth order because the highest derivative appearing in the equation is n. It is a *partial differential equation* (PDE) because there are two independent coordinates, namely, x and y. Equation (1.5.1) reduces to an *ordinary differential equation* (ODE) when we remove all terms involving y coordinate:

$$F\left(x, u, \frac{du}{dx}, \frac{d^2u}{dx^2}, \ldots, \frac{d^nu}{dx^n}\right) = 0. \tag{1.5.2}$$

Equations (1.5.1) and (1.5.2) can also be expressed in the alternative forms as

$$A\left(u, \frac{\partial u}{\partial x}, \frac{\partial u}{\partial y}, \frac{\partial^2 u}{\partial x^2}, \frac{\partial^2 u}{\partial x \partial y}, \frac{\partial^2 u}{\partial y^2}, \ldots, \frac{\partial^n u}{\partial x^n}, \ldots, \frac{\partial^n u}{\partial x^{n-i}\partial y^i}, \ldots, \frac{\partial^n u}{\partial y^n}\right) = f(x,y) \tag{1.5.3}$$

for $i = 1, 2, \ldots, n$, and

$$A\left(u, \frac{du}{dx}, \frac{d^2u}{dx^2}, \ldots, \frac{d^nu}{dx^n}\right) = f(x). \tag{1.5.4}$$

When the operator A is a function of u and/or its derivatives, the equations are called *nonlinear* ODEs or PDEs. In particular, if α is a real number, the operator A is said to be linear if and only if the following condition holds:

$$A(\alpha\, u) = \alpha\, A(u). \tag{1.5.5}$$

Examples of an ODE are provided by

$$\begin{aligned}
a\frac{du}{dx} + bu &= f(x),\\
-\frac{d}{dx}\left(a\frac{du}{dx}\right) + b\frac{du}{dx} + cu &= f(x),\\
\frac{d^2}{dx^2}\left(a\frac{d^2u}{dx^2}\right) + b\frac{d^2u}{dx^2} + cu &= f(x).
\end{aligned} \tag{1.5.6}$$

In Eq. (1.5.6), the coefficients a, b, c, and f are known functions of x and possibly the solution u and/or its derivatives. When the coefficients are only functions of x, the ODEs in Eq. (1.5.5) are linear ODEs; otherwise, they are nonlinear ODEs. Clearly, Eq. $(1.5.6)_1$ [i.e., the first equation in Eq. (1.5.6)] is a first-order ODE, Eq. $(1.5.6)_2$ is a second-order ODE, and Eq. $(1.5.6)_3$ is a fourth-order ODE.

Examples of a PDE are provided by

$$\begin{aligned}
m\frac{\partial u}{\partial t} - \frac{\partial}{\partial x}\left(a\frac{\partial u}{\partial x}\right) + cu &= f(x,t),\\
m_0\frac{\partial u}{\partial t} - m_1\frac{\partial^3 u}{\partial x^2 \partial t} + \frac{\partial}{\partial t}\left(m_2\frac{\partial u}{\partial t}\right) - \frac{\partial}{\partial x}\left(a\frac{\partial u}{\partial x}\right) + cu &= f(x,t),\\
-\frac{\partial}{\partial x}\left(a_{11}\frac{\partial u}{\partial x}\right) - \frac{\partial}{\partial y}\left(a_{22}\frac{\partial u}{\partial y}\right) + a_{10}\frac{\partial u}{\partial x} + a_{02}\frac{\partial u}{\partial y} + a_{00}u &= f(x,y),\\
\frac{\partial^2}{\partial x^2}\left(a\frac{\partial^2 u}{\partial x^2}\right) + \frac{\partial^2}{\partial x \partial y}\left(b\frac{\partial^2 u}{\partial x \partial y}\right) + \frac{\partial^2}{\partial y^2}\left(c\frac{\partial^2 u}{\partial y^2}\right) + d\,u &= f(x,y).
\end{aligned} \tag{1.5.7}$$

The first three in (1.5.7) are second-order PDEs while the fourth is a fourth-order PDE in spatial coordinates.

1.5.3 Types of Problems Described by Differential Equations

A function $u(x, y)$ that satisfies Eq. (1.5.1), implying that it is at least n times differentiable with respect to x and y, is said to be a *solution* of the equation. In general, there are many functions $u(x, y)$ that satisfy Eq. (1.5.1). To obtain a *unique solution* to Eq. (1.5.1), it is necessary to provide additional conditions on u and its derivatives up to and including order $n - 1$ at some specific values of (x, y) in Ω and/or on Γ. An nth order differential equation requires n integrations resulting in n constants of integration whose determination requires n conditions.

A nth order ordinary differential equation (1.5.2) or a partial differential equation (1.5.1) with all n conditions specified at the same value of x (or y), then the problem is termed an *initial-value problem* (IVP) . When the additional conditions are specified at more than one value of x or y, the problem is called a *boundary-value problem* (BVP). All single first-order ODEs, needing only one condition to solve them, are necessarily IVPs. The second-order ODE in Eq. $(1.5.6)_2$ is an IVP if two conditions on $u(x)$ and its derivative are specified for the same $x = x_0$, and it is a BVP if the conditions are specified for two different values of x. All time-dependent problems described by ODEs are IVPs. However, not all IVPs are time-dependent problems because one may rewrite a higher-order ODE as a set of first-order ODEs.

A PDE can be a BVP or both IVP and BVP, depending on the specified conditions. A PDE in the dependent variable $u(x, t)$, with n derivatives in x and m derivatives in t requires n integrations with respect to x and m integrations with respect to t to determine $u(x, t)$. If the constants that appear due to integration with respect to x are determined using conditions on u and its derivatives at different values of x, then the equation describes a BVP with respect to x, and the specified conditions are called the *boundary conditions.* If the constants appearing due to integration with respect to t are determined using conditions on u and its derivatives at one fixed value (say, at $t = t_0$), the equation describes an IVP with respect to t, and the conditions are termed as *initial conditions.* When the constants of integration with respect to both coordinates are determined using conditions on u and its derivatives with respect to both coordinates at different values of the two coordinates, the PDE describes a boundary value problem with respect to both coordinates. Thus, all time-dependent problems described by PDEs are both IVPs and BVPs. For example, Eqs. $(1.5.7)_3$ and $(1.5.7)_4$ are BVPs if the conditions on u and its derivatives are specified at different points (x, y) on the boundary Γ, whereas Eq. $(1.5.7)_2$ is both an IVP and BVP when conditions on u and its derivatives with respect to t are specified at a fixed t (say, $t = 0$) and conditions on u and its derivatives with respect to x are specified at different values of x.

1.5.4 Homogeneous and Inhomogeneous Equations

All nth order differential equations, ODEs or PDEs, have two types of terms: (1) terms that contain the dependent variable (u) and its derivatives of order n; these terms may also contain coefficients that are functions of the variable (in the nonlinear case) and the coordinates (x, y, and t); and (2) terms that contain only functions of the coordinates. The terms that contain only the coordinates (i.e., x, y, and/or t) are known as the *source terms.* For example, f is the source term in Eqs. (1.5.6) and (1.5.7).

When the source term in a differential equation is zero, the equation is termed a *homogeneous differential equation*; otherwise (i.e., when the source term is nonzero), the equation is termed a nonhomogeneous equation. Similarly, when a specified initial value or boundary value is nonzero, it is said to be a nonhomogeneous initial or boundary condition. Otherwise, the initial or boundary condition is called a homogeneous initial or boundary condition.

The phrase *general solution* to a differential equation means it is the solution of the associated homogeneous differential equation. The part of the total solution to a differential equation that satisfies the nonhomogeneous differential equation is known as the *particular solution.*

1.5.5 Examples of IVPs and BVPs

Here we list three examples of IVPs and BVPs that the readers may be familiar with or those considered in this book in the later chapters. Of course, not all IVPs and BVPs considered in this book are visited in this section. Only well-defined one-dimensional problems are considered here. The primary reason for introducing them here is to make the reader know the types of initial and boundary conditions these IVPs and BVPs have to satisfy in order to have unique solutions.

Engineering is a problem-solving discipline. When one is presented with analyzing a real-world problem[2] (with a specific goal of determining the "solution" of the problem), following three major phases of engineering analysis can be identified (see Fig. 1.5.1):

(1) *Formulation.* This step involves converting a real-world system to an engineering problem that can be "solved." The main components of this step include identifying: (a) the domain with its boundary, (b) constitutive properties, (c) the source terms (or "loads" or stimuli) that drive the solution, (d) the governing differential equations, and (e) the boundary and/or initial conditions. In identifying these elements, we make assumptions concerning the geometry, material behavior, "load" representation, and the nature of the solution (e.g., linear or nonlinear), and make use of the physical laws (e.g., conservation of mass, balance of linear and angular momenta, and balance of energy) governing the phenomena experienced

[2]There are so many types of problems in engineering that it is not possible to be abstract to include a general discussion that fits all of them; instead, we focus on problems that are of interest to the present study (mostly from heat transfer, fluid mechanics, and solid and structural mechanics).

by the system. Thus, formulating a problem means setting up the problem as an IVP, a BVP, or an IVP-BVP that can be solved using tools of analysis.

(2) *Solution.* Here we determine the solution to the given IVP or BVP using a method of solution. All solutions are approximate because of the simplifying assumptions made in the formulation (i.e., there is no such thing

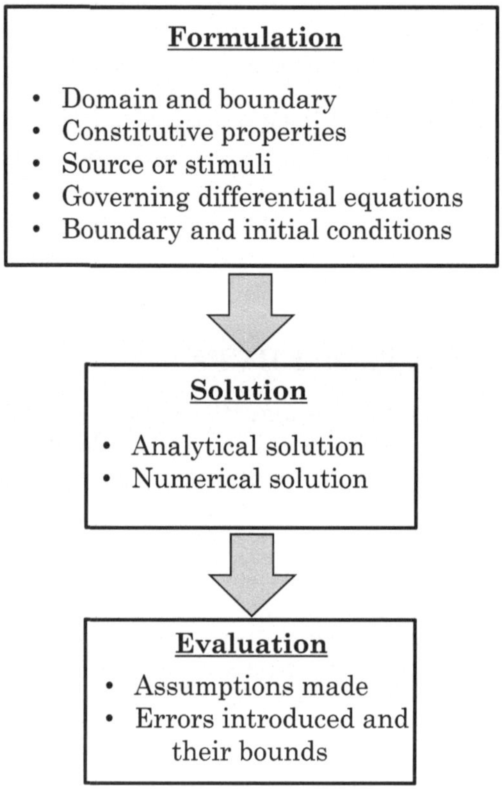

Fig. 1.5.1 The three phases of engineering problem solving.

as "exact solution" to a real-world problem). The solution methods can be broadly classified as analytical and numerical. An analytical solution approach makes use of mathematical tools such as direct integration of the differential equations, separation of variables technique, series method, and host of others (see, e.g., Kreyszig [18] and Pipes and Harvill [19]). Analytical solutions can be a closed-form solution (i.e., finite number of terms in the solution) or a series solution with an infinite number of terms, which needs to be truncated after a certain number of terms. The phrase "exact solution" of a problem refers to the analytical solution of the identified governing differential equation with its coefficients, which are only approximate because of the assumptions made in formulating the problem. A numerical solution generally refers to an approximate solution obtained using a (numerical) method of approximation (e.g., classical variational methods such as the Ritz, Galerkin, and weighted-residual methods, FEM, FDM, and FVM, among many other methods). Sometimes (especially for simple problems), depending on the nature of the equations being solved, the numerical solution may coincide with the analytical solution.

(3) *Evaluation of the results.* The obtained analytical or numerical results must be interpreted in the context of their intended use, often to help the designer to make a decision based on degree to which the functionality requirement is met and the cost (by selecting geometry, materials, types of loads, and so on). The designer must know of the assumptions made in the formulation and errors (and their bounds) in the solution obtained.

1.5.5.1 Simple pendulum

Consider a bob of constant mass m (kg) attached to one end of a string of length ℓ (m) and the other end is pivoted to a fixed point O (without friction), as shown in Fig. 1.5.2. The bob is assumed to be rigid (i.e., not deformable) and the string is massless and inextensible. We wish to determine the motion of the pendulum by determining the angular motion of the bob.

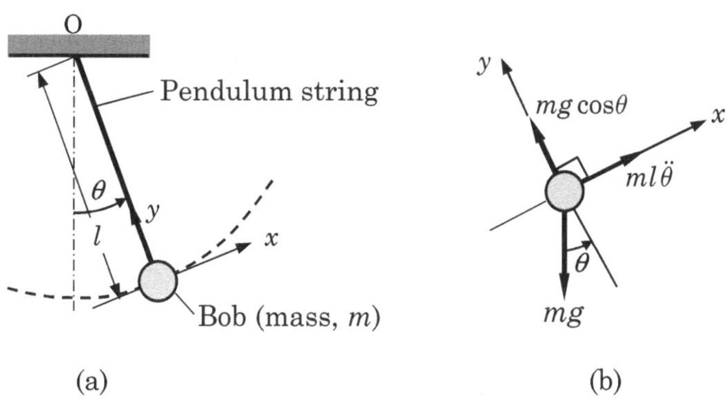

Fig. 1.5.2 Simple pendulum.

The equation governing the motion of a simple pendulum can be derived using the law of balance of linear momentum, alternatively known as Newton's second law of motion, which states that the vector sum of externally applied forces on a system is equal to the time rate of change of the linear momentum (mass times the velocity) of the system:

$$m\frac{d^2\theta}{dt^2} + \frac{mg}{\ell}\sin\theta = 0 \tag{1.5.8}$$

with the conditions

$$\theta(0) = \theta_0, \quad \dot{\theta}(0) = v_0, \tag{1.5.9}$$

where θ is the angular displacement (radians), g is the acceleration due to gravity (m/s^2), t denotes time (s), the superposed dot indicates the derivative with respect to t, and θ_0 and v_0 denote the initial angular displacement and velocity, respectively. The conditions in Eq. (1.5.9) are the initial conditions.

Clearly, the problem described by Eqs. (1.5.8) and (1.5.9) is an initial value problem (with $t = t_0 = 0$). The IVP is homogeneous because $f(t) = 0$; it is nonlinear because

$$A(\theta) \equiv m\frac{d^2\theta}{dt^2} + \frac{mg}{\ell}\sin\theta \quad \text{and} \quad A(\alpha\theta) = m\alpha\frac{d^2\theta}{dt^2} + \frac{mg}{\ell}\sin(\alpha\theta) \neq \alpha A(\theta).$$

Although α is factored out in the first term (hence, the first term is linear), it cannot be factored out in the second term (hence, the second term is nonlinear).

Under the assumption of small angular motions, $\sin\theta$ can be approximated as $\sin\theta \approx \theta$, and then the IVP becomes a homogeneous, second-order, linear ODE (the independent coordinate being t):

$$m\frac{d^2\theta}{dt^2} + \frac{mg}{\ell}\theta = 0. \tag{1.5.10}$$

The general solution to the second-order linear homogeneous differential equation in Eq. (1.5.10) is obtained using the standard solution procedure[3]:

$$\theta(t) = c_1 \cos\alpha t + c_2 \sin\alpha t, \quad \alpha^2 \equiv \frac{g}{\ell}, \tag{1.5.11}$$

where c_1 and c_2 are constants to be determined using the initial conditions. We have

$$\theta(0) = c_1 = \theta_0, \quad \dot\theta(0) = -\alpha\, c_2 = v_0 \;\rightarrow\; c_2 = -\frac{v_0}{\alpha} \tag{1.5.12}$$

and the analytical solution becomes

$$\theta(t) = \theta_0 \cos\alpha t - \frac{v_0}{\alpha}\sin\alpha t, \quad \alpha^2 \equiv \frac{g}{\ell}. \tag{1.5.13}$$

If we were to solve the nonlinear equation, Eq. (1.5.8), subject to the conditions in Eq. (1.5.9), we may consider using a numerical method because it is not possible to solve the nonlinear problem analytically when θ is large.

1.5.5.2 One-dimensional heat flow in a long rod

Here we consider a long rod whose length L is very large compared to its diameter d (i.e., $L >> d$, so that heat flow by conduction in the radial direction is assumed to be negligible, and the problem becomes one dimensional; see Fig. 1.5.3). If the rod is not insulated, heat transfer by convection can take place between the surface of the rod and the surrounding. The governing equation in this case is obtained by using the law of balance of energy (which states that the time rate of change of the internal energy is equal to the heat input to the system; since this is a steady-state problem, there is no time rate involved):

---

[3]The homogeneous second-order equation at hand can be written as $(D^2+\alpha^2)\theta = 0$ with $\alpha^2 = g/\ell$. The solution is $D = \pm i\,\alpha$, $i = \sqrt{-1}$, giving $u(x) = Ae^{-i\alpha x} + Be^{i\alpha x} = c_1\cos\alpha x + c_2\sin\alpha x$.

$$-\frac{d}{dx}\left(Ak\frac{dT}{dx}\right) + \beta P(T - T_\infty) = g(x), \tag{1.5.14}$$

where $T(x)$ denotes temperature (°C) above certain reference temperature, $A = \pi d^2/4$ is the area of cross-section (m^2) of the rod, $P = \pi d$ is the perimeter (m), k is the conductivity [W/(m·°C)], β is the convective heat transfer coefficient [W/(m^2·°C)], and $g(x)$ is the internal heat generation per unit length (W/m).

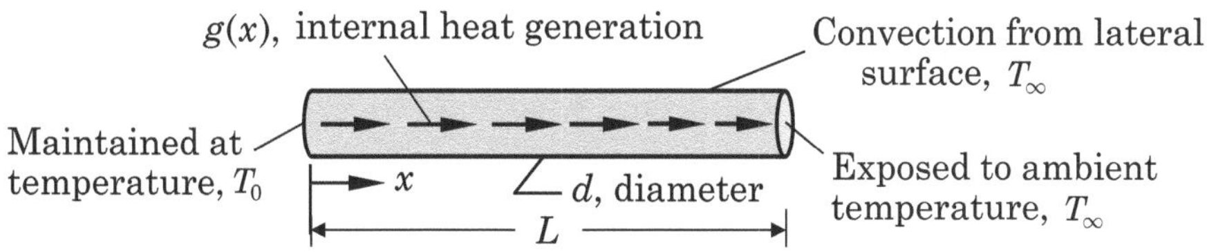

Fig. 1.5.3 One-dimensional heat flow in a rod.

The first term in Eq. (1.5.14) denotes the transfer of energy due to conduction, and the second term is the energy transfer due to convection through the surface of the rod.

Equation (1.5.14) can be solved with conditions on either temperature T at two different points, or heat $Q = -kA(dT/dx)$ at two different points, or T at one point and Q at a different point (typically, the points are at the two ends of the bar). To be specific, let us consider the following two sets of boundary conditions:

$$\textbf{Set 1:} \quad T(0) = T_0, \quad T(L) = T_L, \tag{1.5.15a}$$

$$\textbf{Set 2:} \quad T(0) = T_0, \quad \left[kA\frac{dT}{dx} + \beta A(T - T_\infty)\right]_{x=L} = Q_0, \tag{1.5.15b}$$

where T_0 and T_L are the specified temperatures and Q_0 is the specified heat. These conditions, being specified at two different values of the coordinate x, are clearly the boundary conditions. The conditions in Eq. (1.5.15a) are called the *Dirichlet* boundary conditions. The condition in Eq. (1.5.15b)$_2$ represents the balance of heat due to conduction $[kA(dT/dx)]$ and convection $[\beta A(T - T_\infty)]$ at $x = L$, and it is known as the *Newton*, *mixed*, or *Robin* boundary condition. In heat transfer literature, it is also known as *convective* boundary condition. As a special case, the boundary condition when the end $x = L$ is insulated is given by setting $\beta = 0$ and $Q_0 = 0$ in Eq. (1.5.15b)$_2$. The case where $\beta = 0$ is known as the *Neumann* boundary condition.

Equation (1.5.14) is a second-order ODE because it contains highest derivative of two in one independent coordinate, namely, x. Also, it is a *nonhomogeneous differential equation* because the source term (g) is nonzero. We also note that the boundary conditions in Eqs. (1.5.15a) and (1.5.15b) are nonhomogeneous, as long as at least one of the three quantities (T_0, T_∞, Q_0) is nonzero.

The problem described by Eqs. (1.5.14) with boundary conditions in Eq. (1.5.15a) or Eq. (1.5.15b) is a BVP because the conditions are specified at two different points (i.e., $x = 0$ and $x = L$). Quantities k, A, β, P, T_∞, T_0, Q_0, and f are called the *data* of the problem because they are prescribed quantities of the problem.

The problem defined by Eqs. (1.5.14) and (1.5.15a) or Eqs. (1.5.14) and (1.5.15b) can be solved analytically for constant values of $Ak > 0$ and $P\beta > 0$. Equation (1.5.11) can be expressed in terms of $u \equiv T - T_\infty$ as

$$-\frac{d^2u}{dx^2} + m^2 u = f, \quad m^2 = \frac{P\beta}{Ak}, \quad f = \frac{g}{Ak}, \tag{1.5.16a}$$

$$\textbf{Set 1:} \quad u(0) = T(0) - T_\infty \equiv u_0, \quad u(L) = T(L) - T_\infty \equiv u_L. \tag{1.5.16b}$$

$$\textbf{Set 2:} \quad u(0) = T(0) - T_\infty \equiv u_0, \quad \left[\frac{du}{dx} + \frac{\beta}{k}u\right]_{x=L} = Q_0. \tag{1.5.16c}$$

The homogeneous (or complementary) solution of Eq. (1.5.16a) is obtained as (see the previous footnote; $(-D^2 + m^2)u = 0 \;\rightarrow\; D = \pm m$)

$$u_h(x) = A\,e^{-mx} + B\,e^{mx} = c_1 \cosh mx + c_2 \sinh mx. \tag{1.5.17}$$

The particular solution for a constant source $f = f_0$ is given by (see [19])

$$u_p(x) = \frac{1}{2m}\left[-e^{mx}\int e^{-mx} f_0\,dx + e^{-mx}\int e^{mx} f_0\,dx\right] = \frac{f_0}{m^2} \tag{1.5.18}$$

so that the complete solution is

$$u(x) = u_h(x) + u_p(x) = c_1 \cosh mx + c_2 \sinh mx + \frac{f_0}{m^2}. \tag{1.5.19}$$

The solution $u(x)$ in Eq. (1.5.19) satisfies the nonhomogeneous differential equation (1.5.16a). The constants c_1 and c_2 are determined using the boundary conditions in Eq.(1.5.16b) or Eq.(1.5.16c).

For Set 1 boundary conditions, we have [$\cosh(0) = 1$ and $\sinh(0) = 0$]

$$\begin{aligned} c_1 = u_0 - \frac{f_0}{m^2}, \quad c_2 &= \frac{\hat{u}_L - \hat{u}_0 \cosh mL}{\sinh mL}, \\ \hat{u}_0 \equiv u_0 - \frac{f_0}{m^2}, \quad \hat{u}_L &\equiv u_L - \frac{f_0}{m^2}. \end{aligned} \tag{1.5.20a}$$

The solution for Set 1 boundary conditions is

$$u(x) = \hat{u}_0 \frac{\sinh m(L-x)}{\sinh mL} + \hat{u}_L \frac{\sinh mx}{\sinh mL}. \tag{1.5.20b}$$

For Set 2 boundary condition, the constants c_1 and c_2 become

$$c_1 = u_0 - \frac{f_0}{m^2}, \quad c_2 = c_1 \left(\frac{m \sinh mL + (\beta/k)\cosh mL}{m \cosh mL + (\beta/k)\sinh mL}\right) \tag{1.5.21a}$$

and the solution for Set 2 boundary conditions is

$$u(x) = c_1 \frac{\cosh m(L-x) + (\beta/mk)\sinh m(L-x)}{\cosh mL + (\beta/mk)\sinh mL}. \tag{1.5.21b}$$

1.5.5.3 Deformation of a bar with an axial load

This last example is concerned with the time-dependent axial deformation (i.e., change in the geometry) of a bar subjected to an axial force. In particular, we consider a straight bar of length L, area of cross-section A, and made of linear elastic material with modulus of elasticity E, as shown in Fig. 1.5.4. In general, the product EA, called axial stiffness, can be a function of the axial coordinate x (e.g., functionally graded material).

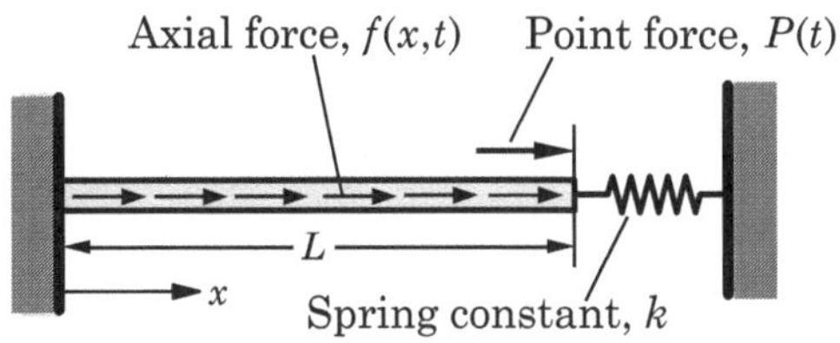

Fig. 1.5.4 Axial deformation and motion of a bar fixed at $x = 0$ and axially-supported by a linear elastic spring at $x = L$, and subjected to axial distributed load $f(x)$ and point load at $x = L$; the bar is initially at rest [i.e., $u(x, 0) = 0$ and $\dot{u}(x, 0) = 0$].

The governing equation of motion for this problem is obtained using the principle of the balance of linear momentum, and it is given by

$$\rho A \frac{\partial^2 u}{\partial t^2} - \frac{\partial}{\partial x}\left(EA\frac{\partial u}{\partial x}\right) = f(x,t), \tag{1.5.22}$$

where $u(x,t)$ denotes axial displacement (m), A is the area of cross-section (m^2) of the bar, E is the modulus of elasticity ($\mathrm{N/m}^2$), ρ is the mass density ($\mathrm{kg/m}^3$), and $f(x,t)$ is the body force per unit length (N/m). The first term in Eq. (1.5.22) denotes the time rate of change of linear momentum and the remaining terms are forces (internal and external) acting on the bar.

Equation (1.5.22) is subjected to the following conditions:

$$u(0,t) = 0, \quad \left[EA\frac{\partial u}{\partial x} + ku\right]_{x=L} = P(t), \tag{1.5.23}$$

where $P(t)$ is the specified point force. These conditions, being specified at two different values of the coordinate x, are clearly the boundary conditions.

In addition, we have the following conditions on $u(x,t)$ at $t = 0$:

$$u(x,0) = u_0(x), \quad \dot{u}(x,0) = v_0(x), \tag{1.5.24}$$

where $u_0(x)$ and $v_0(x)$ are the specified displacements and velocities, respectively, at time $t = 0$. These conditions, being specified at the same value of the coordinate t, are clearly the initial conditions. Thus, the problem described by Eqs. (1.5.23) and (1.5.24) is an IVP with respect to t and a BVP with respect to x. The problem is linear and nonhomogeneous (because $f \neq 0$ and $P \neq 0$).

The analytical solution for the problem can be determined for the case in which $f = 0$ and $P = 0$ and the parameters ρ, A, and E are constant, using the method of *separation of variables*[4] (see [18]). Here we consider the case in which $f = 0$. In the separation of variables method, we assume that $u(x,t)$ can be expressed as a product of two separate functions, one of x and one of t:

$$u(x,t) = F(x)G(t), \quad \frac{\partial^2 u}{\partial x^2} = \frac{d^2 F}{dx^2}G, \quad \frac{\partial^2 u}{\partial t^2} = F\frac{d^2 G}{dt^2}. \tag{1.5.25}$$

Substituting these expressions into the equation of motion (1.5.16a), we obtain

$$\rho A F\,\frac{d^2G}{dt^2} - EA\,\frac{d^2F}{dx^2}\,G = 0 \tag{1.5.26}$$

or, dividing with FG,

$$\frac{1}{G}\frac{d^2G}{dt^2} = \frac{E}{\rho}\frac{1}{F}\frac{d^2F}{dx^2}. \tag{1.5.27}$$

The only way that two functions of independent variables x and t can be equal to each other is for them to be equal to a constant. This constant is conveniently denoted as $-\lambda^2$ (this is the only choice that results in a meaningful solution):

$$\frac{1}{G}\frac{d^2G}{dt^2} = -\lambda^2, \quad \frac{E}{\rho}\frac{1}{F}\frac{d^2F}{dx^2} = -\lambda^2. \tag{1.5.28}$$

The solutions of these two differential equations are

$$F(x) = A\,\cos\alpha x + B\,\sin\alpha x, \quad G(t) = C\,\cos\lambda t + D\,\sin\lambda t, \quad \alpha^2 = \lambda^2\frac{\rho}{E}, \tag{1.5.29}$$

where A, B, C, and D are constants to be determined. The complete solution is

$$u(x,t) = (A\,\cos\alpha x + B\,\sin\alpha x)\,(C\,\cos\lambda t + D\,\sin\lambda t)\,. \tag{1.5.30}$$

The boundary condition $u(0,t) = 0$ gives $A = 0$. The second boundary condition $(1.5.23)_2$ yields

$$B\Big(-\alpha\,\cos\alpha L + \frac{k}{EA}\sin\alpha L\Big)\,(C\,\cos\lambda t + D\,\sin\lambda t) = 0, \tag{1.5.31}$$

which implies, for arbitrary t and nontrivial solution (i.e., $B \neq 0$), the requirement

$$-\alpha\,\cos\alpha L + \frac{k}{EA}\sin\alpha L = 0 \;\rightarrow\; \frac{kL}{EA}\tan\alpha L = \alpha\,L. \tag{1.5.32}$$

This transcendental equation provides an infinite number of values of $\alpha\,L$ for which Eq. (1.5.32) holds. The first two roots, for $kL/EA = 1$, are $\alpha_1 L = 4.493$ and $\alpha_2 L = 7.725$.

Now the complete solution can be expressed as

$$u(x,t) = \sum_{n=1}^{\infty}(C_n^*\,\cos\lambda_n t + D_n^*\,\sin\lambda_n t)\sin\alpha_n x, \quad \lambda_n = \sqrt{E/\rho}\alpha_n. \tag{1.5.33}$$

[4]It takes considerably more effort to find the analytical solution, if one exists, for the case in which f and P are nonzero.

The initial conditions $u(x,0) = u_0(x)$ and $\dot{u}(x,0) = v_0(x)$ give the relations

$$\sum_{n=1}^{\infty} C_n^* \sin \alpha_n x = u_0(x), \quad \sum_{n=1}^{\infty} D_n^* \lambda_n \sin \alpha_n x = v_0(x), \tag{1.5.34}$$

which can be used to determine the constants C_n^* and D_n^* $(n = 1, 2, \ldots)$ as

$$C_n^* = \frac{2}{L} \int_0^L u_0(x) \sin \alpha_n x \, dx, \quad D_n^* \lambda_n = \frac{2}{L} \int_0^L v_0(x) \sin \alpha_n x \, dx. \tag{1.5.35}$$

Another example of an IVP-BVP is provided by the equation governing transient heat transfer in a rod:

$$\rho c A \frac{\partial T}{\partial t} - \frac{\partial}{\partial x}\left(kA \frac{\partial T}{\partial x} \right) = g(x,t), \tag{1.5.36}$$

where c denotes the specific heat (a material property).

1.6 Taylor's Series and Elements of Matrix Theory

1.6.1 Introduction

Taylor's series is of fundamental importance in both the finite difference method and the finite volume method. Therefore, it is useful to review it here. It is also useful to review the elements of the matrix theory because all numerical methods applied to the solution of differential equations ultimately end up as the solution of linear equations among undetermined coefficients of the approximation. These equations are expressed in the form of matrices because the matrix representations facilitate algorithm development and their execution using computers.

1.6.2 Taylor's Series and Taylor's Formula

The basic premise of the Taylor series is that every analytic function[5] $f(x)$ can be represented by a power series, called *Taylor series* of $f(x)$. The Taylor series of an analytic function $f(x)$ about x_0 is given by

$$\begin{aligned} f(x) &= \sum_{n=0}^{\infty} (x - x_0)^n \frac{f^{(n)}(x_0)}{n!} \\ &= f(x_0) + \frac{(x - x_0)}{1!} f'(x_0) + \frac{(x - x_0)^2}{2!} f''(x_0) + \cdots \end{aligned} \tag{1.6.1}$$

where $f' = df/dx$ and $f^{(n)} = d^n f/dx^n$. The particular case in which $x_0 = 0$ is called *MaClaurin* series of $f(x)$. When the series is truncated and the remainder

[5]A function $f(x)$ is said to be *analytic* at a point $x = x_0$ if it can be represented by a power series in powers of $(x - x_0)$ with radius of convergence $R > 0$.

$R_n(x)$ is added, the finite power series is known as *Taylor's formula:*

$$\begin{aligned} f(x) &= f(x_0) + \frac{(x-x_0)}{1!}f'(x_0) + \frac{(x-x_0)^2}{2!}f''(x_0) + \cdots \\ &\quad + \frac{(x-x_0)^n}{n!}f^{(n)}(x_0) + R_n(x) \\ &= p_n(x) + R_n(x) \end{aligned} \tag{1.6.2}$$

where p_n is the polynomial approximation of $f(x)$ and R_n is the remainder

$$R_n(x) = (x-x_0)^{n+1}\frac{f^{(n+1)}(\xi)}{(n+1)!} \tag{1.6.3}$$

for $x_0 < \xi < x$ when $x > x_0$ and $x < \xi < x_o$ when $x < x_0$. The series converges if R_n satisfies the condition

$$\lim_{n\to\infty} R_n(x) = 0 \tag{1.6.4}$$

For example, the Taylor series of the function $f(x) = 1/(1-x)$ for $|x| < 1$ is ($f' = df/dx$, $f'' = d^2f/dx^2$, and $f^{(n)} = d^nf/dx^n$)

$$\frac{1}{1-x} = \sum_{n=0}^{\infty} x^n = 1 + x + x^2 + x^3 + \cdots + \tag{1.6.5}$$

The function $f(x)$ is singular at $x = 1$. Similarly, we have

$$e^x = \sum_{n=0}^{\infty}\frac{x^n}{n!} = 1 + x + \frac{x^2}{2!} + \frac{x^3}{3!} + \cdots \tag{1.6.6}$$

$$\cos x = \sum_{n=0}^{\infty}(-1)^n\frac{x^{2n}}{(2n)!} = 1 - \frac{x^2}{2!} + \frac{x^4}{4!} - \cdots \tag{1.6.7}$$

$$\sin x = \sum_{n=0}^{\infty}(-1)^n\frac{x^{2n+1}}{(2n+1)!} = 1 - \frac{x^3}{3!} + \frac{x^5}{5!} - \cdots, \tag{1.6.8}$$

$$\cosh x = \sum_{n=0}^{\infty}\frac{x^{2n}}{(2n)!} = 1 + \frac{x^2}{2!} + \frac{x^4}{4!} + \cdots \tag{1.6.9}$$

$$\sinh x = \sum_{n=0}^{\infty}\frac{x^{2n+1}}{(2n+1)!} = x + \frac{x^3}{3!} + \frac{x^5}{5!} + \cdots \tag{1.6.10}$$

$$\log(1+x) = \sum_{n=0}^{\infty}(-1)^n\frac{x^{n+1}}{(n+1)!} = x - \frac{x^2}{2!} + \frac{x^3}{3!} - \cdots. \tag{1.6.11}$$

If a continuous function $f(x)$ has continuous derivatives up to and including $(n+1)$ everywhere in the interval $[x_0, x]$, it can be represented by a finite power series

$$f(x) = f(x_0) + (x - x_0)f'(x_0) + \frac{(x - x_0)^2}{2!}f''(x_0) + \cdots + \frac{(x - x_0)^n}{n!}f^{(n)}(x_0) + R_n(x), \tag{1.6.12}$$

where $f^{(n)}$ denotes the nth derivative with respect to x.

For example, for $x = x_0+h$, for any real number $h > 0$, Eq. (1.6.12) becomes

$$f(x_0 + h) = f(x_0) + hf'(x_0) + \frac{h^2}{2!}f''(x_0) + \frac{h^3}{3!}f^{(3)}(x_0) + \cdots + \frac{h^n}{n!}f^{(n)}(x_0) + R_n(x). \tag{1.6.13}$$

For $x = x_0 - h$, Eq. (1.6.12) becomes

$$f(x_0 - h) = f(x_0) - hf'(x_0) + \frac{h^2}{2!}f''(x_0) - \frac{h^3}{3!}f^{(3)}(x_0) + \cdots + (-1)^n\frac{h^n}{n!}f^{(n)}(x_0) + R_n(x) \tag{1.6.14}$$

If we truncate the Taylor series in Eq. (1.6.13) after the second term on the right-hand side and solving for $f'(x_0)$, we obtain

$$\left.\frac{df}{dx}\right|_{x=x_0} = f'(x_0) \approx \frac{f(x_0 + h) - f(x_0)}{h} + O(h). \tag{1.6.15}$$

Here 'O' refers to the "order of" the remainder. Equation (1.6.15) implies that the slope of the function $f(x)$ at $x = x_0$ can be approximated using the values of $f(x)$ at $x = x_0$ and $x = x_0 + h$. Equation (1.6.15) is known as the *forward difference formula*. Clearly, the forward difference approximation of a derivative has error (truncation error) of the order h, as this is the highest order term that is dropped for the approximation. Typically, h is a mesh size that is measured in relation to the domain length, the value of $h/L < 1$ and $h > h^2 > h^3, \ldots$. So (1.6.15) is called first-order accurate forward difference formula.

Similarly, truncating the series in Eq. (1.6.14) we obtain

$$f'(x_0) \approx \frac{f(x_0) - f(x_0 - h)}{h} + O(h), \tag{1.6.16}$$

which represents the slope based on the function values at $x = x_0 - h$ and $x = x_0$, and it is known as the *backward difference formula* (see Fig. 1.6.1).

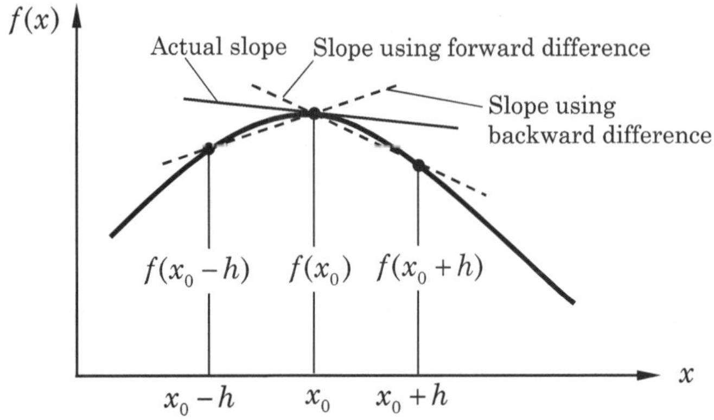

Fig. 1.6.1 Slopes (first derivative of a function) approximated using the first-order forward and backward difference formulas.

The truncated Taylor series can be used to approximate the slope (i.e., first-order derivative) with second-order accuracy. By subtracting (1.6.14) from (1.6.13), we obtain

$$\left.\frac{df}{dx}\right|_{x=x_0} = f'(x_0) \approx \frac{f(x_0+h)-f(x_0-h)}{2h} + O(h^2). \tag{1.6.17}$$

Equation (1.6.17) is known as the central difference formula for the first derivative.

The truncated Taylor series can also be used to find approximation of higher-order derivatives with varying degree of accuracy in terms of the function values. For example, by adding Eqs. (1.6.13) and (1.6.14), and solving for the second derivative $f''(x_0)$, we obtain

$$f''(x_0) \approx \frac{f(x_0-h)-2f(x_0)+f(x_0+h)}{2h^2} + O(h^2), \tag{1.6.18}$$

which is known as the *second-order central difference formula* for the approximation of the second derivative of the function $f(x)$. Equations (1.6.15)–(1.6.17) are only a few examples of many possible approximations of first and second derivatives with different orders of errors. Some examples are provided here.

Forward difference formulas with second-order accuracy:

$$f'(x_0) = \frac{-3f(x_0)+4f(x_0+h)-f(x_0+2h)}{2h}, \tag{1.6.19}$$

$$f''(x_0) = \frac{2f(x_0)-5f(x_0+h)+4f(x_0+2h)-f(x_0+3h)}{h^3}. \tag{1.6.20}$$

Backward difference formulas with second-order accuracy:

$$f'(x_0) = \frac{3f(x_0)-4f(x_0-h)+f(x_0-2h)}{2h}, \tag{1.6.21}$$

$$f''(x_0) = \frac{2f(x_0)-5f(x_0-h)+4f(x_0-2h)-f(x_0-3h)}{h^3}. \tag{1.6.22}$$

Before we embark on the discussion of various direct and iterative methods of solution of linear equations, we review some useful concepts and formulas from matrix algebra because the linear equations are often expressed in the form of matrix equations.

1.6.3 Theory of Matrices

1.6.3.1 Definition of a matrix

Consider the linear equations

$$\begin{aligned} a_{11}\,x_1 + a_{12}\,x_2 + \cdots + a_{1n}\,x_n &= b_1, \\ a_{21}\,x_1 + a_{22}\,x_2 + \cdots + a_{2n}\,x_n &= b_2, \\ &\vdots, \\ a_{n1}\,x_1 + a_{n2}\,x_2 + \cdots + a_{nn}\,x_n &= b_n. \end{aligned} \tag{1.6.23}$$

There are $n \times n = n^2$ coefficients, a_{ij} $(i,j = 1,2,\ldots,n)$, relating x_i to b_i. The form of these equations suggests writing the scalars a_{ij} (jth component of the ith equation) in the rectangular array

$$[A] = \begin{bmatrix} a_{11} & a_{12} & \cdots & a_{1n} \\ a_{21} & a_{22} & \cdots & a_{2n} \\ \vdots & \vdots & \vdots & \vdots \\ a_{m1} & a_{m2} & \cdots & a_{mn} \end{bmatrix}. \tag{1.6.24}$$

The rectangular (in the present case, it is a square) array $[A]$ of elements a_{ij} is called a *matrix.*

A matrix is denoted in this book either by bold symbol such as $\mathbf{A}$ or $[A]$. A matrix with m rows and n columns is termed a $m \times n$ matrix, with the number of rows listed first. The element in the ith row and jth column of a matrix $[A]$ is denoted by a_{ij}. A matrix with only one row is called a row vector (not in the sense of having a magnitude and direction) and a matrix with only one column is called column vector. Typically, we denote the elements of a row or column vector with a single subscript:

$$\{X\} = \begin{Bmatrix} x_1 \\ x_2 \\ \vdots \\ x_{n-1} \\ x_n \end{Bmatrix}, \quad \{Y\} = \{\, y_1 \;\; y_2 \;\; \cdots \;\; y_{m-1} \;\; y_m \,\}.$$

Row and column vectors can be used to denote the components of a vector quantity. For example, consider a vector $\mathbf{V} = v_1\,\hat{\mathbf{e}}_1 + v_2\,\hat{\mathbf{e}}_2 + v_3\,\hat{\mathbf{e}}_3$ in a rectangular Cartesian coordinate system $(x_1, x_2, x_3) = (x, y, z)$ shown in Fig. 1.6.2, where v_i are components of the vector and $\hat{\mathbf{e}}_i$ are the unit basis vectors. We can represent vector $\mathbf{V}$ as a *product* of a row matrix with a column matrix as

$$\mathbf{V} = \{v_1 \;\; v_2 \;\; v_3\} \begin{Bmatrix} \hat{\mathbf{e}}_1 \\ \hat{\mathbf{e}}_2 \\ \hat{\mathbf{e}}_3 \end{Bmatrix}.$$

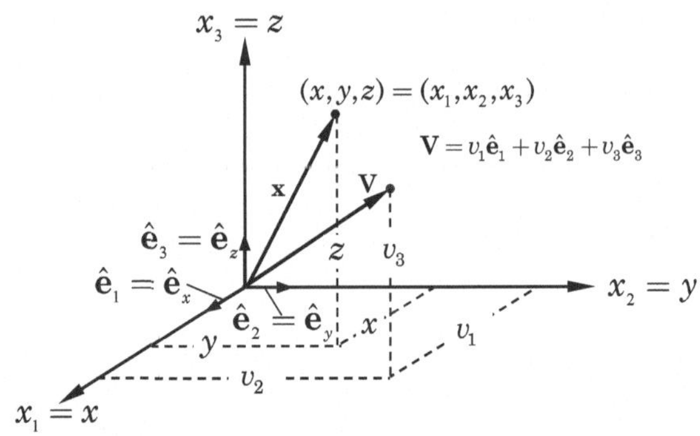

Fig. 1.6.2 The rectangular Cartesian coordinate system $(x_1, x_2, x_3) = (x, y, z)$, with unit basis vectors $\hat{\mathbf{e}}_i$.

Note that the vector **A** is obtained by multiplying the ith element in the row matrix with the ith element in the column matrix and adding them.

A square matrix is one that has the same number of rows as columns. The elements of a square matrix for which the row number and the column number are the same, a_{ii}, are called the *diagonal elements*. A square matrix is said to be a diagonal matrix if all of its off-diagonal elements are zeros. An identity matrix, denoted by **I** or $[I]$, is a diagonal matrix whose elements are all 1's:

$$\begin{bmatrix} 1 & 0 & 0 & 0 & \cdots & 0 \\ 0 & 1 & 0 & 0 & \cdots & 0 \\ 0 & 0 & 1 & 0 & \cdots & 0 \\ \vdots & \vdots & \vdots & \vdots & & \\ 0 & 0 & 0 & \cdots & 0 & 1 \end{bmatrix}.$$

The sum of the diagonal elements of a square matrix is called the *trace* of the matrix.

1.6.3.2 Addition of matrices and transpose of a matrix

The *sum* of two matrices of the same size is defined to be a matrix of the same size obtained by simply adding the corresponding elements. If $[A] = [a_{ij}]$ is an $m \times n$ matrix and $[B] = [b_{ij}]$ is an $m \times n$ matrix, their sum is an $m \times n$ matrix, $[C] = [c_{ij}]$, with

$$c_{ij} = a_{ij} + b_{ij} \quad \text{for all}\, i, j. \tag{1.6.25}$$

If $[A]$ is an $m \times n$ matrix, then the $n \times m$ matrix obtained by interchanging its rows and columns is called the *transpose* of $[A]$ and is denoted by $[A]^{\mathrm{T}}$ or $\mathbf{A}^{\mathrm{T}}$. A square matrix **A** is said to be *symmetric* if $\mathbf{A} = \mathbf{A}^{\mathrm{T}}$ holds (i.e., $a_{ij} = a_{ji}$). A *anti-symmetric* square matrix is one for which the condition $\mathbf{A}^{\mathrm{T}} = -\mathbf{A}$ or $a_{ij} = -a_{ji}$ holds for all $i, j = 1, 2, \ldots, n$. Obviously, the diagonal elements of an anti-symmetric matrix are necessarily zero. Every arbitrary square matrix **A** can be represented as the sum of a symmetric matrix and an anti-symmetric matrix:

$$\mathbf{A} = \tfrac{1}{2}\left(\mathbf{A} + \mathbf{A}^{\mathrm{T}}\right) + \tfrac{1}{2}\left(\mathbf{A} - \mathbf{A}^{\mathrm{T}}\right). \tag{1.6.26}$$

1.6.3.3 Matrix multiplication

Let $[A]$ be $m \times n$ and $[B]$ be $n \times p$ matrices (note that the number of columns of $[A]$ must be the same as the number of rows of $[B]$ to define this product), the product $[A]\,[B]$ is defined to be the $m \times p$ matrix $[C]$ with coefficients

$$c_{ij} = \{i\text{th row of } [\mathbf{A}]\}\left\{\begin{matrix} j\text{th} \\ \text{column} \\ \text{of } [B] \end{matrix}\right\} = \{a_{i1}, a_{i2}, \ldots, a_{in}\}\left\{\begin{matrix} b_{1j} \\ b_{2j} \\ \vdots \\ b_{nj} \end{matrix}\right\}$$

$$= a_{i1}b_{1j} + a_{i2}b_{2j} + \cdots + a_{in}b_{nj} = \sum_{k=1}^{n} a_{ik}\, b_{kj}. \tag{1.6.27}$$

Thus, the linear equations in Eq. (1.6.23) can written in matrix form as

$$\begin{bmatrix} a_{11} & a_{12} & \cdots & a_{1n} \\ a_{21} & a_{22} & \cdots & a_{2n} \\ \vdots & \vdots & \vdots & \vdots \\ a_{m1} & a_{m2} & \cdots & a_{mn} \end{bmatrix} \left\{\begin{matrix} x_1 \\ x_2 \\ \vdots \\ x_n \end{matrix}\right\} = \left\{\begin{matrix} b_1 \\ b_2 \\ \vdots \\ b_n \end{matrix}\right\} \quad \text{or} \quad \mathbf{AX} = \mathbf{B}. \tag{1.6.28}$$

A constant multiple of a matrix is equal to the matrix obtained by multiplying all of the elements by the constant.

For any nonzero vector (i.e., at least one component of the vector is nonzero) $\mathbf{x}$, the product $\mathbf{x}^T\mathbf{A}\mathbf{x}$ is said to be the *quadratic form* (which is a scalar expression) associated with a square matrix $\mathbf{A}$. A square matrix $\mathbf{A}$ is said to be *positive-definite* if its quadratic form is positive for all nonzero vectors $\mathbf{x}$:

$$\mathbf{x}^{\mathrm{T}}\mathbf{A}\mathbf{x} > 0. \tag{1.6.29}$$

1.6.3.4 Determinant of a matrix

The *determinant* of the matrix $[A]$ is defined to be the scalar $\det[A] = |\mathbf{A}|$ computed according to

$$\det[A] \equiv |\mathbf{A}| = \sum_{i=1}^{n}(-1)^{i+1} a_{i1}\, |A_{i1}|, \tag{1.6.30}$$

where $|A_{i1}|$ is the determinant of the $(n-1) \times (n-1)$ matrix that remains on deleting out the ith row and the first column of matrix $[A]$. For 1×1 matrices, the determinant is defined according to $|a_{11}| = a_{11}$. For convenience we define the determinant of a zeroth-order matrix to be unity. In the above definition, special attention is given to the first column of the matrix $[A]$. We call it the expansion of $|\mathbf{A}|$ according to the first column of $[A]$. One can expand $|\mathbf{A}|$ according to any column or row:

$$|\mathbf{A}| = \sum_{i=1}^{n}(-1)^{i+j} a_{ij}\, |A_{ij}| \quad \text{for any fixed value of } j = 1, 2, \ldots, \text{ or } n, \tag{1.6.31}$$

where $|A_{ij}|$ is the determinant of the matrix obtained by deleting the ith row and jth column of matrix $[A]$.

Example 1.6.1

Find the determinant of the matrix

$$[A] = \begin{bmatrix} 2 & 5 & -1 \\ 1 & 4 & 3 \\ 2 & -3 & 5 \end{bmatrix}.$$

Solution: Using the definition in Eq. (1.6.31), we have

$$\begin{aligned}
|\mathbf{A}| &= \sum_{i=1}^{3} (-1)^{i+1} a_{i1} \, |A_{i1}| \\
&= (-1)^2 a_{11} \begin{vmatrix} 4 & 3 \\ -3 & 5 \end{vmatrix} + (-1)^3 a_{21} \begin{vmatrix} 5 & -1 \\ -3 & 5 \end{vmatrix} + (-1)^4 a_{31} \begin{vmatrix} 5 & -1 \\ 4 & 3 \end{vmatrix} \\
&= 2[(-1)^2(4)(5) + (-1)^3(-3)3] - 1[(-1)^2(5)(5) + (-1)^3(-3)(-1)] \\
&\quad + 2[(-1)^2(5)(3) + (-1)^3(4)(-1)] \\
&= 2(20 + 9) - (25 - 3) + 2(15 + 4) = 74.
\end{aligned}$$

The following properties of determinants should be noted:

(1) det $([A][B]) = \det[A] \cdot \det[B]$.
(2) det $[A]^{\mathrm{T}} = \det[A]$.
(3) det $(\alpha[A]) = \alpha^n \det[A]$, where α is a scalar and $[A]$ is a $n \times n$ matrix.
(4) If $[A']$ is a matrix obtained from $[A]$ by multiplying a row (or column) of $[A]$ by a scalar α, then det $[A'] = \alpha \det[A]$.
(5) If $[A']$ is the matrix obtained from $[A]$ by interchanging any two rows (or columns) of $[A]$, then det $[A'] = -\det[A]$.
(6) If $[A]$ has two rows (or columns) one of which is a scalar multiple of another (i.e., linearly dependent), det $[A] = 0$.
(7) If $[A']$ is the matrix obtained from $[A]$ by adding a multiple of one row (or column) to another, then det $[A'] = \det[A]$.

For an $n \times n$ matrix $[A]$, the determinant of the $(n-1) \times (n-1)$ submatrix of $[A]$ obtained by deleting row i and column j of $[A]$ is called the *minor* of a_{ij} and is denoted by $|A_{ij}|$. The quantity $\text{cof}_{ij}(A) \equiv (-1)^{i+j}|A_{ij}|$ is called the *cofactor* of a_{ij}. The determinant of $[A]$ can be cast in terms of the minor and cofactor of a_{ij}:

$$\det[A] = \sum_{i=1}^{n} a_{ij} \, \text{cof}_{ij} = \sum_{i=1}^{n} (-1)^{i+j} \, a_{ij} \, |A_{ij}| \tag{1.6.32}$$

for any fixed value of j. One can also verify that

$$\sum_{i=1}^{n} (-1)^{i+j} \, a_{ip} \, |A_{iq}| = 0 \text{ when } p \neq q. \tag{1.6.33}$$

In other words, the following identity holds:

$$\sum_{i=1}^{n}(-1)^{i+j}\, a_{ip}\, |A_{iq}| = \det[A]\, \delta_{pq}. \tag{1.6.34}$$

The *adjunct* (also called *adjoint*) of a matrix $[A]$ is the transpose of the matrix obtained from $[A]$ by replacing each element by its cofactor. The adjunct of $[A]$ is denoted by $\mathrm{Adj}(\mathbf{A})$.

1.6.3.5 Matrix inverse

If $[A]$ is an $n \times n$ matrix and $[B]$ is any $n \times n$ matrix such that $[A][B] = [B][A] = [I]$, then $[B]$ is called an *inverse* of $[A]$. If it exists, the inverse of a matrix is unique. To verify this, let both $[B]$ and $[C]$ be inverses for $[A]$; then by definition,

$$[A][B] = [B][A] = [A][C] = [C][A] = [I].$$

This shows that $[B] = [C]$, and the inverse is unique. The inverse of $[A]$ is denoted by $[A]^{-1}$. If $[A]$ is nonsingular (i.e., $\det[A] \neq 0$), the inverse $[A]^{-1}$ of $[A]$ can be computed according to

$$[A]^{-1} = \frac{1}{\det[A]}\,\mathrm{Adj}(\mathbf{A}). \tag{1.6.35}$$

Example 1.6.2

Consider the matrix of **Example 1.6.1**. Determine the minors, cofactors, and the inverse.

Solution: The minors and cofactors of the matrix are

$$|A_{11}| = \begin{vmatrix} 4 & 3 \\ -3 & 5 \end{vmatrix}, \quad |A_{12}| = \begin{vmatrix} 1 & 3 \\ 2 & 5 \end{vmatrix}, \quad |A_{13}| = \begin{vmatrix} 1 & 4 \\ 2 & -3 \end{vmatrix},$$

$$\begin{aligned}
\mathrm{cof}_{11}(A) &= (-1)^2\,|A_{11}| = 4 \times 5 - (-3)3 = 29,\\
\mathrm{cof}_{12}(A) &= (-1)^3\,|A_{12}| = -(1 \times 5 - 3 \times 2) = 1,\\
\mathrm{cof}_{13}(A) &= (-1)^4\,|A_{13}| = 1 \times (-3) - 2 \times 4 = -11.
\end{aligned}$$

Then the $\mathrm{Adj}(\mathbf{A})$ is given by

$$\mathrm{Adj}(\mathbf{A}) = \begin{bmatrix} \mathrm{cof}_{11}(A) & \mathrm{cof}_{12}(A) & \mathrm{cof}_{13}(A) \\ \mathrm{cof}_{21}(A) & \mathrm{cof}_{22}(A) & \mathrm{cof}_{23}(A) \\ \mathrm{cof}_{31}(A) & \mathrm{cof}_{32}(A) & \mathrm{cof}_{33}(A) \end{bmatrix}^{\mathrm{T}} = \begin{bmatrix} 29 & -22 & 19 \\ 1 & 12 & 19 \\ -11 & 16 & 3 \end{bmatrix}.$$

The determinant is given by (expanding by the first row)

$$|\mathbf{A}| = 2(29) + 5(1) + (-1)(-11) = 74.$$

The inverse of $[A]$ can be now computed using Eq. (1.6.35),

$$[A]^{-1} = \mathbf{A}^{-1} = \frac{1}{74}\begin{bmatrix} -29 & -22 & 19 \\ 1 & 12 & -7 \\ -11 & 16 & 3 \end{bmatrix}.$$

It can be easily verified that $\mathbf{A}\mathbf{A}^{-1} = \mathbf{I}$.

The following additional properties of matrix addition, transpose, and multiplications should be noted.

(1) Matrix addition is commutative: $[A] + [B] = [B] + [A]$.

(2) Matrix addition is associative: $[A] + ([B] + [C]) = ([A] + [B]) + [C]$.

(3) There exists a unique matrix $[0]$, such that $[A] + [0] = [0] + [A] = [A]$. The matrix $[0]$ is called *zero matrix*; it contains all zeros.

(4) Matrix addition is distributive with respect to multiplication by a scalar α: $\alpha([A] + [B]) = \alpha[A] + \alpha[B]$; matrix addition is also distributive with respect to matrix multiplication: $([A] + [B])[C] = [A][C] + [B][C]$.

(5) If $[A][B]$ is defined, $[B][A]$ may or may not be defined. If both $[A][B]$ and $[B][A]$ are defined, it is not necessary that they be of the same size.

(6) Matrix multiplication is associative: $([A][B])[C] = [A]([B][C])$.

(7) $([A]^{\mathrm{T}})^{\mathrm{T}} = [A]$; $([A] + [B])^{\mathrm{T}} = [A]^{\mathrm{T}} + [B]^{\mathrm{T}}$.

(8) $([A][B])^{\mathrm{T}} = [B]^{\mathrm{T}}[A]^{\mathrm{T}}$ (note the order).

(9) If $[A]$ is nonsingular, $\left([A]^{-1}\right)^{\mathrm{T}} = \left([A]^{\mathrm{T}}\right)^{-1}$.

1.7 Interpolation Theory

1.7.1 Introduction

All numerical methods developed to solve differential equations have approximation of the problem variables (or their derivatives) by suitable functions. These functions, as we will see in the case of the Ritz method discussed in Section 1.11.3.6, are required to satisfy certain differentiability requirements and interpolation properties (i.e., attain certain values at some selective points of the domain). Interpolation and approximation of functions are closely related.

We approximate a function $u(x)$ with another "suitable" function $u_h(x)$ for two main reasons: (1) The original function $u(x)$ is difficult to evaluate [e.g., differential or integral of $u(x)$] compared to the chosen function $u_h(x)$. Functions such as $\log x$, $\sin x$, and e^x provide examples of functions that cannot be evaluated by strictly arithmetic operations without first finding approximation functions such as finite power series [see Eqs. (1.6.6)–(1.6.11)]. (2) The function $u(x)$ is only known for a finite number of values of x, say x_i, $i = 1, 2, \ldots, n$ (e.g., experimental data). We wish to construct a function $u_h(x)$ that will allow an estimation of the value of $u(x)$ for $x \neq x_i$, $i = 1, 2, \ldots, n$. Once u_h is presumed to be a good approximation of $u(x)$, then one can analytically differentiate and integrate u_h with respect to x. The second reason is equally useful when the function $u(x)$ is unknown and we seek its values $u_i = u(x_i)$ at finite number of points x_i, $i = 1, 2, \ldots, n$ by constructing a function $u_h(x)$ such that $u_h(x_i) = u_i$. Then $u_h(x)$ is said to be an *interpolant* of $u(x)$. In this section, we discuss the derivation of polynomial interpolation functions, which can be used as the approximating functions.

1.7.2 Interpolating Polynomials

Here we consider polynomial approximation of a function $f(x)$ in the form

$$f(x) \approx f_n(x) = \sum_{i=1}^{n+1} c_i\, x^{i-1}, \tag{1.7.1}$$

where c_i $(i = 1, 2, \ldots, n+1)$ are constants which are determined such that $f_n(x)$ is *exactly* equal to $f(x_1)$, $f(x_2)$, ..., $f(x_{n+1})$ at $n+1$ points $(x_1, x_2, \ldots, x_{n+1})$, called the *base points*. Thus, given the pairs of values $(x_i, f(x_i))$, $i = 1, 2, \ldots, n+1$), we require that

$$f_n(x_i) = f(x_i), \quad i = 1, 2, \ldots, n+1. \tag{1.7.2}$$

According to a fundamental theorem of algebra, there is one and only one polynomial of degree n or less which assumes specified values for $n+1$ distinct points (not necessarily equally spaced). This polynomial $f_n(x)$ is termed the nth degree *interpolating polynomial*, as illustrated in Fig. 1.7.1.

By definition, the coefficients c_i are the solution of the following linear equations:

$$\begin{aligned}
c_1 + c_2\, x_1 + c_3\,(x_1)^2 + \cdots + c_{n+1}\,(x_1)^n &= f(x_1)\\
c_1 + c_2\, x_2 + c_3\,(x_2)^2 + \cdots + c_{n+1}\,(x_2)^n &= f(x_2)\\
c_1 + c_2\, x_3 + c_3\,(x_3)^2 + \cdots + c_{n+1}\,(x_3)^n &= f(x_3)\\
&\vdots\\
c_1 + c_2\, x_{n+1} + c_3\,(x_{n+1})^2 + \cdots + c_{n+1}\,(x_{n+1})^n &= f(x_{n+1}).
\end{aligned} \tag{1.7.3}$$

The determinant of the $(n+1) \times (n+1)$ matrix of coefficients of the linear equations in Eq. (1.7.3), namely,

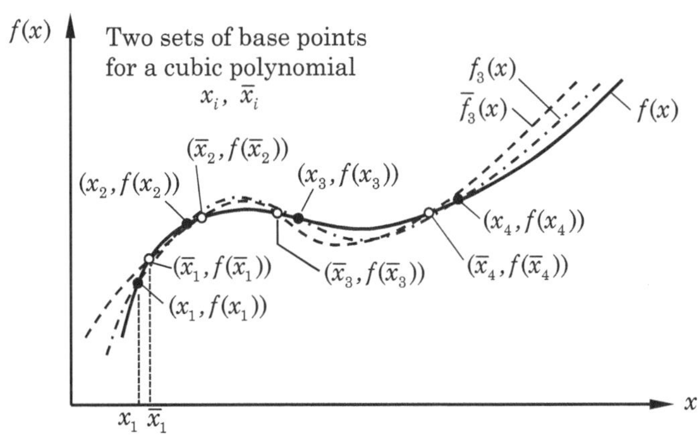

Fig. 1.7.1 Approximation of a function by an interpolating polynomial.

$$
\begin{vmatrix}
1 & x_1 & x_1^2 & \cdots & x_1^n \\
1 & x_2 & x_2^2 & \cdots & x_2^n \\
1 & x_3 & x_3^2 & \cdots & x_3^n \\
\vdots & & & & \\
1 & x_{n+1} & x_{n+1}^2 & \cdots & x_{n+1}^n
\end{vmatrix}
\tag{1.7.4}
$$

is known as the *Vandermonde* determinant and is nonzero if $x_i \neq x_j$, $i \neq j$ (i.e., duplicate base points are not allowed). Consequently, the system of linear equations in Eq. (1.7.3) has a unique solution for the c_i and thus a unique interpolating polynomial $f_n(x)$ which exactly reproduces $f(x)$ at the base points x_i $(i = 1, 2, 3, \ldots, n+1)$.

Solving the set of linear equations, especially for large values of n, is computationally expensive; therefore, one can use the fact that $f_n(x)$ has n roots (i.e., $f_n(x)$ vanishes at $n+1$ base points) and write (here Π symbol is used for multiplicative "summation")

$$
\begin{aligned}
\psi_i(x) &= \frac{x - x_1}{x_i - x_1}\,\frac{x - x_2}{x_i - x_2}\cdots\frac{x - x_{i-1}}{x_i - x_{i-1}}\,\frac{x - x_{i+1}}{x_i - x_{i+1}}\cdots\frac{x - x_n}{x_i - x_n} \\
&= \Pi_{j=1, j\neq i}^{n+1}\,\frac{x - x_j}{x_i - x_j}
\end{aligned}
\tag{1.7.5}
$$

for $i = 1, 2, \ldots, n+1$. Thus, the nth degree interpolation polynomial is given by

$$
\begin{aligned}
f_n(x) &= \sum_{i=1}^{n+1} f(x_i)\,\psi_i(x), \\
\psi_i(x_j) &= \delta_{ij} = \begin{cases} 1, & \text{if } i = j \\ 0, & \text{if } i \neq j \end{cases}.
\end{aligned}
\tag{1.7.6}
$$

We note that the interpolation functions $\psi_i(x)$ only depend on the base points x_i and the distance between them, but not on the function values. Approximating functions that interpolate only the functions values – not their derivative values – are termed the *Lagrange interpolation functions.*

Approximating polynomials may be derived using the derivative values as well as the function values. For example, a cubic polynomial ($n = 3$) that reproduces the function values $f(x)$ and its first derivative values $f^{'} = df/dx$ at $x = x_1$ and $x = x_2$ [i.e., four values, $f(x_1)$, $f(x_2)$, $f^{'}(x_1)$, and $f^{'}(x_2)$ at two base points] can be derived from the equations

$$
\begin{aligned}
c_1 + c_2\,x_1 + c_3\,x_1^2 + c_4\,x_1^3 &= f(x_1) \\
c_1 + c_2\,x_2 + c_3\,x_2^2 + c_4\,x_2^3 &= f(x_2) \\
c_2 + 2c_3\,x_1 + 3c_4\,x_1^2 &= f^{'}(x_1) \\
c_2 + 2c_3\,x_2 + 3c_4\,x_2^2 &= f^{'}(x_2).
\end{aligned}
\tag{1.7.7}
$$

The determinant of these equations is

$$\begin{vmatrix} 1 & x_1 & x_1^2 & x_1^3 \\ 1 & x_2 & x_2^2 & x_2^3 \\ 0 & 1 & 2x_1 & 3x_1^2 \\ 0 & 1 & 2x_2 & 3x_2^2 \end{vmatrix}, \tag{1.7.8}$$

which is not zero for distinct points $x_1 \neq x_2$. Interpolation functions derived using values of the function and values of the derivatives of the function are termed *Hermite family* of approximation functions.

We close this section with an example of the Lagrange interpolation of a function. It is important to note that there is no guaranty as to which degree of interpolation would yield the best results. The accuracy of approximation, for a given n, also depends on the choice of the base points.

Example 1.7.1

Construct linear and higher degree interpolating polynomials, with suitable equidistant points, of the function $f(x) = \cos \pi x + 2x - 1$ in the interval $[0, 1]$.

Solution: For a linear polynomial ($n = 1$), we need two ($n + 1$) points. If we use the end points, we have $f_1 = f(x_1) = f(0) = 0.0$ and $f_2 = f(x_2) = f(1) = 0.0$ and the interpolant is a trivial polynomial. On the other hand, if we select $x_1 = 0.25$ and $x_2 = 0.75$, we have $f_1 = f(0.25) = 0.2071$ and $f_2 = f(0.75) = -0.2071$, and functions $\psi_i(x)$ $(i = 1, 2)$ are

$$\begin{aligned} \psi_1(x) &= \frac{x - 0.75}{0.25 - 0.75} = (1.5 - 2x), \\ \psi_2(x) &= \frac{x - 0.25}{0.75 - 0.25} = (2x - 0.5) \end{aligned} \tag{1}$$

and

$$f_h(x) = 0.2071(1.5 - 2x) + 0.2071(0.5 - 2x) = 0.4142(1 - 2x). \tag{2}$$

For a quadratic interpolation, we need three points. We choose $x_1 = 0.25$, $x_2 = 0.5$, and $x_3 = 0.75$, and obtain $f_1 = 0.2071$, $f_2 = 0$, and $f_3 = -0.2071$. The $\psi_i(x)$ $(i = 1, 2, 3)$ are

$$\begin{aligned} \psi_1(x) &= \frac{x - 0.5}{0.25 - 0.5}\frac{x - 0.75}{0.25 - 0.75} = (2 - 4x)(1.5 - 2x) \\ \psi_2(x) &= \frac{x - 0.25}{0.5 - 0.25}\frac{x - 0.75}{0.5 - 0.75} = (1 - 4x)(3 - 4x) \\ \psi_3(x) &= \frac{x - 0.25}{0.75 - 0.25}\frac{x - 0.5}{0.75 - 0.5} = (0.5 - 2x)(2 - 4x) \end{aligned} \tag{3}$$

and the interpolating polynomial is

$$f_h(x) = 0.2071(3 - 10x + 8x^2) - 0.2071(1 - 6x + 8x^2) = 0.4142(1 - 2x). \tag{4}$$

For a cubic interpolation, we choose the following four points: $x_1 = 0.2$, $x_2 = 0.4$, $x_3 = 0.6$, and $x_4 = 0.8$, and obtain $f_1 = 0.209$, $f_2 = 0.109$, $f_3 = -0.109$, and $f_4 = -0.209$. The polynomials functions $\psi_i(x)$ $(i = 1, 2, 3, 4)$ are

$$\begin{aligned}
\psi_1(x) &= \frac{x-0.4}{0.2-0.4}\,\frac{x-0.6}{0.2-0.6}\,\frac{x-0.8}{0.2-0.8} = \tfrac{1}{3}(2-5x)(15-2.5x)(4-5x)\\
\psi_2(x) &= \frac{x-0.2}{0.4-0.2}\,\frac{x-0.6}{0.4-0.6}\,\frac{x-0.8}{0.4-0.8} = (1-5x)(3-5x)(2-2.5x)\\
\psi_3(x) &= \frac{x-0.2}{0.6-0.2}\,\frac{x-0.4}{0.6-0.4}\,\frac{x-0.8}{0.6-0.8} = (0.5-2.5x)(2-5x)(4-5x)\\
\psi_4(x) &= \frac{x-0.2}{0.8-0.2}\,\frac{x-0.4}{0.8-0.4}\,\frac{x-0.6}{0.8-0.6} = -\tfrac{1}{3}(1-5x)(3-5x)
\end{aligned} \tag{5}$$

and the interpolating polynomial is

$$\begin{aligned}
f_h(x) = &\ 0.209\,\tfrac{1}{3}(2-5x)(15-2.5x)(4-5x) + 0.109(1-5x)(3-5x)(2-2.5x)\\
&- 0.109(0.5-2.5x)(2-5x)(4-5x) + 0.209\,\tfrac{1}{3}(1-5x)(3-5x).
\end{aligned} \tag{6}$$

Figure 1.7.2 contains plots of the function $f(x)$ along with its interpolation functions as functions of the coordinate x for $n = 2$ to $n = 5$. Clearly, as the value of n is increased, the interpolation function closely approximates a transcendental function with good accuracy.

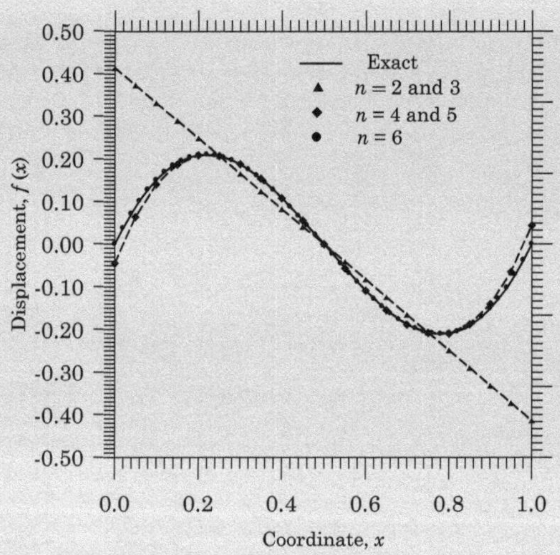

Fig. 1.7.2 Approximation of a function by various degree interpolation polynomials.

1.8 Numerical Integration

1.8.1 Preliminary Comments

This section is devoted to the numerical evaluation of integrals of the type

$$\int_{x_a}^{x_b} f(x)\,dx, \qquad \int_{\Omega} F(x,y)\,dxdy,$$

when $f(x)$ and $F(x, y)$ are not simple functions or the geometry of the domain Ω is complicated. Such integral evaluations are common in all numerical methods. Here we review few commonly used numerical integration methods, especially numerical integration using the Gauss numerical integration (called *Gauss quadrature*). The readers may consult [1, 8] for additional details on numerical integration.

1.8.2 Trapezoidal and Simpson's Formulas

Consider the integral,

$$I = \int_{x_a}^{x_b} f(x)\, dx, \tag{1.8.1}$$

where $f(x)$ is a function x. We choose to interpolate the function $f(x)$ in $[x_a, x_b]$ using a polynomial and evaluate it:

$$f(x) \approx \sum_{I=1}^{N+1} f_I\, \psi_I(x), \tag{1.8.2}$$

where f_I denotes the value of $f(x)$ at the Ith point x_I of the interval $[x_a, x_b]$ and $\psi_I(x)$ are polynomials of degree N in the interval. Substitution of Eq. (1.8.2) into Eq. (1.8.1) and evaluation of the integral gives an approximate value of the integral I. For example, suppose that we choose linear interpolation of $f(x)$, as shown in Fig. 1.8.1(a). For $N = 1$, we have

$$\psi_1(x) = \frac{x - x_b}{x_a - x_b}, \quad \psi_2(x) = \frac{x - x_a}{x_b - x_a}, \quad \int_{x_a}^{x_b} \psi_i(x)\, dx = \frac{x_b - x_a}{2} \tag{1.8.3a}$$

and

$$I = \frac{x_b - x_a}{2}\Big(f_1 + f_2\Big), \quad f_1 = f(x_a), \quad f_2 = f(x_b). \tag{1.8.3b}$$

Thus, the value of the integral is given by the area of a trapezoid used to approximate the area under the function $f(x)$ between x_a and x_b [see Fig. 1.8.1(a)]. Equation (1.8.3b) is known as the *trapezoidal rule* of numerical integration. The error in the evaluation of an integral by the trapezoidal rule is of the order $O(h^3)$.

If we use the quadratic interpolation of $f(x)$ [see Fig. 1.8.1(b)], with

$$\begin{aligned}
\psi_1(x) &= \left(\frac{x - x_\alpha}{x_a - x_\alpha}\right)\left(\frac{x - x_b}{x_a - x_b}\right), \\
\psi_2(x) &= \left(\frac{x - x_a}{x_\alpha - x_a}\right)\left(\frac{x - x_b}{x_\alpha - x_b}\right), \\
\psi_3(x) &= \left(\frac{x - x_a}{x_b - x_a}\right)\left(\frac{x - x_\alpha}{x_b - x_\alpha}\right),
\end{aligned} \tag{1.8.4a}$$

where x_α is a point between x_a and x_b (i.e., $x_a < x_\alpha < x_b$).

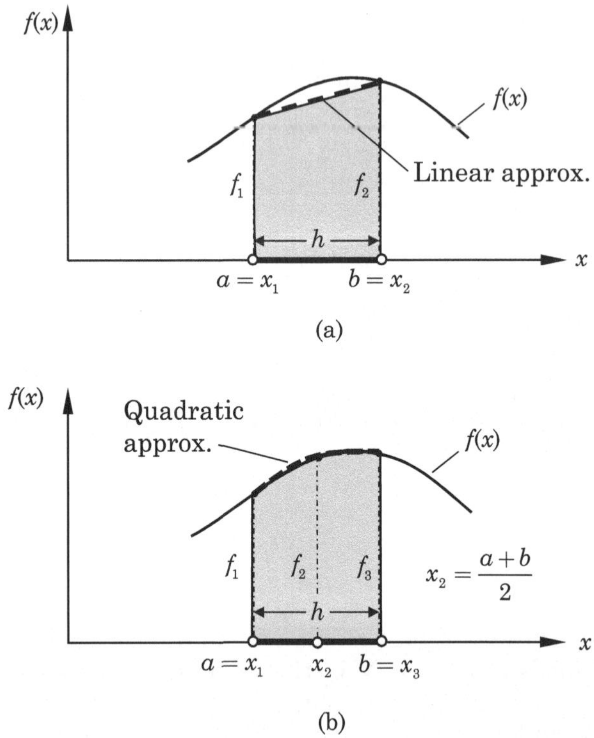

Fig. 1.8.1 Approximate evaluation of an integral using the (a) trapezoidal rule (linear interpolation) and (b) Simpson one-third rule (quadratic interpolation).

For uniform spacing of the base points, we take $x_\alpha = x_a + 0.5(x_b - x_a)$. For this case, we have

$$\int_{x_a}^{x_b} \psi_1(x)\,dx = \int_{x_a}^{x_b} \psi_3(x)\,dx = \frac{x_b - x_a}{6}, \quad \int_{x_a}^{x_b} \psi_2(x)\,dx = 4\frac{x_b - x_a}{6}, \quad (1.8.4b)$$

and the integral I is approximated to

$$I = \frac{h}{3}(f_1 + 4f_2 + f_3), \quad (1.8.4c)$$

where $h = 0.5(x_b - x_a)$ is the distance between the base points and

$$f_1 = f(x_a), \quad f_2 = f(x_a + 0.5(x_b - x_a)), \quad f_3 = f(x_b). \quad (1.8.4d)$$

Equation (1.8.4c) is known as *Simpson's one-third rule.* The error in the evaluation of an integral by the one-third Simpson rule is of the order $O(h^5)$.

From Eqs. (1.8.3b) and (1.8.4c), the form of the numerical integration formula is:

$$I = \int_{x_a}^{x_b} f(x)\,dx \approx (b-a)\sum_{I=1}^{N} f(x_I)\,w_I, \quad (1.8.5)$$

where x_I are called the *quadrature points*, W_I are the *quadrature weights*, and N is the number of equally-spaced quadrature (or base) points. The formula in Eq. (1.8.5), known as the *Newton–Cotes closed integration formula*, only requires functional evaluations, multiplications, and additions to obtain the numerical value of an integral of a function. They yield exact values of the integral whenever $f(x)$ is a polynomial of order N.

The weighting coefficients for $N = 1, 2, \ldots, 7$ are given in Table 1.8.1. Note that $\sum_{I=1}^{N} w_I = 1$. The base point location for $N = 1$ is $x_1 = \frac{1}{2}(a+b)$. For $N > 1$, the base point locations are

$$x_1 = a, \quad x_2 = a + \Delta x, \ \ldots, \ x_N = a + (N-1)\Delta x = b$$

and $\Delta x = (b-a)/(N-1)$. We note that when N is *even* (i.e., when there is an even number of intervals or an odd number of base points), the formula is exact when $f(x)$ is a polynomial of degree N or less; when N is *odd*, the formula is exact for a polynomial of degree $N-1$ or less. Odd-point formulas are frequently used because of their high order of accuracy (see Carnahan, Luther and Wilkes [1]).

Table 1.8.1 Weighting coefficients for the Newton–Cotes formula.

N	w_1	w_2	w_3	w_4	w_5	w_6	w_7
1	1						
2	$\frac{1}{2}$	$\frac{1}{2}$					
3	$\frac{1}{6}$	$\frac{4}{6}$	$\frac{1}{6}$				
4	$\frac{1}{8}$	$\frac{3}{8}$	$\frac{3}{8}$	$\frac{1}{8}$			
5	$\frac{7}{90}$	$\frac{32}{90}$	$\frac{12}{90}$	$\frac{32}{90}$	$\frac{7}{90}$		
6	$\frac{19}{288}$	$\frac{75}{288}$	$\frac{50}{288}$	$\frac{50}{288}$	$\frac{75}{288}$	$\frac{19}{288}$	
7	$\frac{41}{840}$	$\frac{216}{840}$	$\frac{27}{840}$	$\frac{272}{840}$	$\frac{27}{840}$	$\frac{216}{840}$	$\frac{41}{840}$

1.8.3 Gauss Quadrature Formula

The Gauss quadrature formula in one dimension is of the form

$$I = \int_{-1}^{+1} f(\xi)\, d\xi \approx \sum_{I=1}^{N} f(\xi_I)\, w_I, \tag{1.8.6}$$

where ξ_I and w_I are the *Gauss points* and the *Gauss weights*, respectively. A Gauss quadrature with N points can exactly evaluate an integral of a polynomial of degree $(2N-1)$.

The main idea of the Gauss quadrature is to determine the points ξ_I and weights W_I such that the integrals of all polynomials of degree $2N-1$ are exact. Thus, the exact evaluation of a linear ($2N-1 = 1$ or $N = 1$) polynomial requires one quadrature point (ξ_1) and one weight (W_1); in other words, we need two conditions to determine the two unknowns. These are provided by

$$\int_{-1}^{1} f(\xi)\, d\xi \text{ is exact for } f(\xi) = 1 \text{ and } f(\xi) = \xi \tag{1.8.7}$$

In other words, we have

$$\int_{-1}^{1} 1\, d\xi = 2 = f(\xi_1)\, w_1 = 1 \cdot w_1, \quad \int_{-1}^{1} \xi\, d\xi = 0 = f(\xi_1)\, w_1 = \xi_1\, w_1, \tag{1.8.8}$$

which yield $w_1 = 2$ and $\xi_1 = 0$.

The two-point Gauss quadrature, $N = 2$ (or $2N-1 = 3$) can exactly evaluate integrals of polynomial degree 3. The two-point quadrature has four unknowns, ξ_1, ξ_2, w_1, and w_2. We require

$$\int_{-1}^{1} f(\xi)\, d\xi \text{ is exact for } f(\xi) = 1,\ \xi,\ \xi^2, \text{ and } \xi^3 \tag{1.8.9}$$

Obviously, a polynomial of degree 2 will be exactly evaluated by a two-point quadrature. Thus, we have

$$\begin{aligned}
\int_{-1}^{1} 1\, d\xi &= 2 = w_1 + w_2, \\
\int_{-1}^{1} \xi\, d\xi &= 0 = \xi_1\, w_1 + \xi_2\, w_2, \\
\int_{-1}^{1} \xi^2\, d\xi &= \frac{2}{3} = \xi_1^2\, w_1 + \xi_2^2\, w_2, \\
\int_{-1}^{1} \xi^3\, d\xi &= 0 = \xi_1^3\, w_1 + \xi_2^3\, w_2.
\end{aligned} \tag{1.8.10}$$

The solution of these equations gives $\xi_1 = -\xi_2 = -1/\sqrt{3} = -0.57735\ldots$ and $w_1 = w_2 = 1$.

Thus, for any given N, there are $2N$ conditions to determine the Gauss points and Gauss weights, as illustrated for $N = 1$ and $N = 2$. These points and weights have nothing to do with the function whose integral is being evaluated. Table 1.8.2 contains the Gauss points and Gauss weights for $N = 1, 2, \ldots, 8$.

Since the Gauss quadrature is developed for $-1 \leq \xi \leq 1$, it is necessary to transform the integral written in terms of x [see Eq. (1.8.1)] to one in terms of ξ. This requires a coordinate transformation from $x_a \leq x \leq x_b$ to $-1 \leq \xi \leq 1$. This relationship is linear, and it can be established with the help of the conditions

$$\text{when } x = x_a,\ \xi = -1; \text{ and when } x = x_b,\ \xi = 1. \tag{1.8.11a}$$

This transformation between x and ξ can be represented by the linear "stretch" mapping

$$x = c_1 + c_2\, \xi \text{ for } x \text{ in } [x_a, x_b], \tag{1.8.11b}$$

where c_1 and c_2 are to be determined such that conditions in Eq. (1.8.11a) hold:

$$x_a = c_1 + c_2(-1), \quad x_b = c_1 + c_2\,(1).$$

Table 1.8.2 Weights (positive) and Gauss points for the Gauss quadrature (included up to 15 significant figures*).

$$\int_{-1}^{1} f(\xi)\, d\xi = \sum_{I=1}^{N} f(\xi_I) W_I.$$

ξ_I	N	W_I
0.00000 00000 00000	One-point formula	2.00000 00000 00000
± 0.57735 02691 89626	Two-point formula	1.00000 00000 00000
± 0.77459 66692 41483	Three-point formula	0.55555 55555 55556
0.00000 00000 00000		0.88888 88888 88889
± 0.86113 63115 94053	Four-point formula	0.34785 48451 37454
± 0.33998 10435 84856		0.65214 51548 62546
± 0.90617 98459 38664	Five-point formula	0.23692 68850 56189
± 0.53846 93101 05683		0.47862 86704 99366
0.00000 00000 00000		0.56888 88888 88889
± 0.93246 95142 03152	Six-point formula	0.17132 44923 79170
± 0.66120 93864 66265		0.36076 15730 48139
± 0.23861 91860 83197		0.46791 39345 72691
± 0.94910 79123 42759	Seven-point formula	0.12948 49661 68870
± 0.74153 11855 99394		0.27970 53914 89277
± 0.40584 51513 77397		0.38183 00505 05119
0.00000 00000 00000		0.41795 91836 73469
± 0.96028 98564 97536	Eight-point formula	0.10122 85362 90376
± 0.79666 64774 13627		0.22238 10344 53374
± 0.52553 24099 16329		0.31370 66458 77887
± 0.18343 46424 95650		0.36268 37833 78362

* 0.57735... $= 1/\sqrt{3}$, 0.77459... $= \sqrt{3/5}$, 0.888... $= 8/9$, 0.555... $= 5/9$.

Solving for c_1 and c_2, we obtain

$$c_2 = \tfrac{1}{2}(x_b - x_a) = \tfrac{1}{2}h, \quad c_1 = \tfrac{1}{2}(x_b + x_a) = x_a + \tfrac{1}{2}h, \quad h = x_b - x_a.$$

Hence, the transformation between x and ξ becomes ($x_b = x_a + h$)

$$\begin{aligned} x(\xi) &= \tfrac{1}{2}(x_b + x_a) + \tfrac{1}{2}h\,\xi \\ &= x_a\,\tfrac{1}{2}(1-\xi) + x_b\,\tfrac{1}{2}(1+\xi) && (1.8.12a) \\ &= x_a + \tfrac{h}{2}(1+\xi). && (1.8.12b) \end{aligned}$$

The local coordinate ξ is called the *normal coordinate* or *natural coordinate*, and its values always lie between -1 and 1, with its origin at the center of the domain.

The derivation of the Lagrange family of interpolation functions in terms of the natural coordinate ξ is made easy by the interpolation property [see Eq. (1.7.6)]:

$$\psi_i(\xi_j) = \begin{cases} 1, & \text{if } i = j \\ 0, & \text{if } i \neq j \end{cases} \tag{1.8.13}$$

where ξ_j is the ξ-coordinate of the jth base point in the one-dimensional domain. The $(n-1)$st degree Lagrange interpolation functions ψ_i $(i = 1, 2, \ldots, n)$ are given by:

$$\psi_i(\xi) = \frac{(\xi - \xi_1)(\xi - \xi_2)\cdots(\xi - \xi_{i-1})(\xi - \xi_{i+1})\cdots(\xi - \xi_n)}{(\xi_i - \xi_1)(\xi_i - \xi_2)\cdots(\xi_i - \xi_{i-1})(\xi_i - \xi_{i+1})\cdots(\xi_i - \xi_n)}. \tag{1.8.14}$$

The transformation in Eq. (1.8.12b) allows us to rewrite integrals involving x as those in terms of ξ:

$$\int_{x_a}^{x_b} f(x)\, dx = \int_{-1}^{1} \hat{f}(\xi)\, d\xi, \quad \hat{f}(\xi)\, d\xi = f(\xi)\, J\, d\xi \tag{1.8.15}$$

so that the Gauss quadrature can be used to evaluate the integral over $[-1, 1]$. The differential element dx in the global coordinate system x is related to the differential element $d\xi$ in the natural coordinate system ξ by

$$dx = \frac{dx}{d\xi} d\xi = J\, d\xi, \tag{1.8.16}$$

where J is called the *Jacobian* of the transformation, and its value is

$$J = \frac{dx}{d\xi} = \sum_{i=1}^{2} x_i \frac{d\psi_i}{d\xi} = \tfrac{1}{2}(x_2 - x_1) = \tfrac{h}{2}. \tag{1.8.17}$$

The result in Eq. (1.8.17) holds whenever the domain is a straight line (i.e., not a curve).

Example 1.8.1

Evaluate the integral of the following polynomial using the (a) Newton–Cotes and (b) Gauss integration rules.

$$f(x) = 5 + 3x + 2x^2 - x^3 - 4x^4, \quad -1 \leq x \leq 1. \tag{1}$$

Solution: The exact value of the integral of $f(x)$ is

$$\int_{-1}^{1} f(x)\, dx = \int_{-1}^{1} (5+3x+2x^2-x^3-4x^4) dx = \left[5x + 3\frac{x^2}{2} + 2\frac{x^3}{3} - \frac{x^4}{4} - 4\frac{x^5}{5}\right]_{-1}^{1} = \frac{146}{15}. \tag{2}$$

(a) The Newton–Cotes formula is exact for polynomial degree of $N - 1 = 4$ (or $N = 5$). Let us check first what we obtain with the trapezoidal and Simpson's rules.

Using the trapezoidal rule ($N = 2$) in Eq. (1.8.3b), we obtain ($h = b - a = 2$, $x_1 = -1$, $x_2 = 1$, $f(x_1) = 1$, and $f(x_2) = 5$)

$$\int_{-1}^{1} f(x)\, dx \approx \frac{h}{2}[f(x_1) + f(x_2)] = \frac{2}{2}(1 + 5) = 6. \tag{3}$$

Using the one-third Simpson rule ($N = 3$) in Eq. (1.8.4c), we obtain ($b - a = 2$, $x_1 = -1$, $x_2 = 0$, $x_3 = 1$, $f(x_1) = 1$, $f(x_2) = 5$, and $f(x_3) = 5$)

$$\int_{-1}^{1} f(x)\,dx \approx \frac{2}{6}\left[f(x_1) + 4f(x_2) + f(x_3)\right] = \frac{1}{3}(1 + 4 \times 5 + 5) = \frac{26}{3}. \tag{4}$$

Using $N = 5$ from Table 1.8.1 in Eq. (1.8.5), we obtain ($x_1 = -1$, $x_2 = -0.5$, $x_3 = 0$, $x_4 = 0.5$, $x_5 = 1$, $f(x_1) = 1$, $f(x_2) = 31/8$, $f(x_3) = 5$, $f(x_4) = 53/8$, and $f(x_5) = 5$)

$$\int_{-1}^{1} f(x)\,dx \approx \frac{2}{90}\left[7 + \frac{31}{8}32 + 5 \times 12 + \frac{53}{8}32 + 5 \times 7\right] = \frac{1}{45}\left[102 + 4(31 + 53)\right] = \frac{146}{15}, \tag{5}$$

which coincides with the exact value.

(b) Since the integrand is a polynomial of degree 4, the number of Gauss points N needed to exactly evaluate the integral is $2N - 1 = 4 \rightarrow N = 3$. Let us check what values we obtain with one- and two-point Gauss rules. Since $-1 \leq x \leq 1$, there is no need to transform x to ξ (i.e., $\xi = x$).

First we evaluate $f(x)$ at various Gauss points ($f(0) = 5$):

$$\begin{aligned}
f(-1/\sqrt{3}) &= 5 + 3\left(-1/\sqrt{3}\right) + 2\left(-1/\sqrt{3}\right)^2 - \left(-1/\sqrt{3}\right)^3 - 4\left(-1/\sqrt{3}\right)^4 \\
&= \frac{17}{3} - \frac{4}{9} - \frac{8}{3\sqrt{3}} \\
f(1/\sqrt{3}) &= 5 + 3\left(1/\sqrt{3}\right) + 2\left(1/\sqrt{3}\right)^2 - \left(1/\sqrt{3}\right)^3 - 4\left(1/\sqrt{3}\right)^4) \\
&= \frac{17}{3} - \frac{4}{9} + \frac{8}{3\sqrt{3}} \\
f(-\sqrt{3/5}) &= 5 + 3\left(-\sqrt{3/5}\right) + 2\left(-\sqrt{3/5}\right)^2 - \left(-\sqrt{3/5}\right)^3 - 4\left(-\sqrt{3/5}\right)^4 \\
&= \frac{155}{25} - \frac{36}{25} - 12\sqrt{\frac{3}{5}} \\
f(\sqrt{3/5}) &= 5 + 3\left(\sqrt{3/5}\right) + 2\left(\sqrt{3/5}\right)^2 - \left(\sqrt{3/5}\right)^3 - 4\left(\sqrt{3/5}\right)^4 \\
&= \frac{31}{5} - \frac{36}{25} + 12\sqrt{\frac{3}{5}}.
\end{aligned} \tag{6}$$

Using the one-point Gauss rule [$x_1 = 0.0$, $W_1 = 2$, and $f(x_1) = 5$], we obtain:

$$\int_{-1}^{1} f(x)\,dx = f(\xi_1)w_1 = 10 \quad (2.74\%\ \text{error}). \tag{7}$$

The two-point Gauss rule [$x_1 = -1/\sqrt{3}$, $x_2 = 1/\sqrt{3}$, $f(x_1) = (47/9) - 8/3\sqrt{3}$, $f(x_2) = (47/9) + 8/3\sqrt{3}$, $w_1 = 1$, and $w_2 = 1$] gives

$$\int_{-1}^{1} f(\xi)\,d\xi = f(\xi_1)W_1 + f(\xi_2)w_2 = \frac{94}{9} \quad (7.31\%\ \text{error}). \tag{8}$$

The three-point Gauss rule ($\xi_1 = -\sqrt{0.6}$, $w_1 = 5/9$, $\xi_2 = 0.0$, $w_2 = 8/9$, $\xi_3 = \sqrt{0.6}$, and $w_3 = 5/9$) gives the exact result:

$$\begin{aligned}
\int_{-1}^{1} f(\xi)\,d\xi &= f(\xi_1)w_1 + f(\xi_2)w_2 + f(\xi_3)w_3 \\
&= \left(\frac{119}{25} - 12\sqrt{\frac{3}{5}}\right)\frac{5}{9} + 5 \times \frac{8}{9} + \left(\frac{119}{25} + 12\sqrt{\frac{3}{5}}\right)\frac{5}{9} = \frac{146}{15}.
\end{aligned} \tag{9}$$

1.8.4 Extension to Two Dimensions

The numerical integration formulas developed in Section 1.8.3 can be extended to two dimensions. For example, the Gauss quadrature for the evaluation of functions defined over a square domain $\hat{\Omega} = [-1, 1] \times [-1, 1]$, termed the *master element*, can be expressed as

$$\int_{\hat{\Omega}} F(\xi, \eta)\, d\xi d\eta = \int_{-1}^{1} \int_{-1}^{1} F(\xi, \eta)\, d\xi d\eta \approx \sum_{I=1}^{M} \sum_{J=1}^{N} F(\xi_I, \eta_J)\, w_I\, w_J. \tag{1.8.18}$$

To apply the Gauss quadrature to integral of a function $F(x, y)$ defined on an arbitrary two-dimensional domain Ω (presumably a quadrilateral),

$$I = \int_{\Omega} F(x, y)\, dxdy \tag{1.8.19}$$

we must transform Ω to $\hat{\Omega}$ by a coordinate transformation between the normalized coordinates (ξ, η) to the (x, y) coordinates as

$$x = \sum_{j=1}^{m} x_j \hat{\psi}_j(\xi, \eta), \quad y = \sum_{j=1}^{m} y_j \hat{\psi}_j(\xi, \eta), \tag{1.8.20}$$

where $\hat{\psi}_j(\xi, \eta)$ are the functions used to interpolate the geometry over the master domain $\hat{\Omega}$ and (x_i, y_i) are points that define the geometry of Ω. As an example, consider Ω to be a quadrilateral with straight sides[6] as shown in Fig. 1.8.2. Since the four corner points of the quadrilateral define the geometry, (x_i, y_i) are the coordinates of these four corner nodes. The transformation in Eq. (1.8.20) maps a point (ξ, η) in the master domain $\hat{\Omega}$ into a point (x, y) in the actual domain Ω and vice versa; if the Jacobian J, the determinant of the matrix $\mathbf{J}$, of the transformation is positive-definite, $J > 0$ [the two-dimensional version of Eq. (1.8.16)], then:

$$\begin{Bmatrix} \frac{\partial}{\partial \xi} \\ \frac{\partial}{\partial \eta} \end{Bmatrix} = \begin{bmatrix} \frac{\partial x}{\partial \xi} & \frac{\partial y}{\partial \xi} \\ \frac{\partial x}{\partial \eta} & \frac{\partial y}{\partial \eta} \end{bmatrix} \begin{Bmatrix} \frac{\partial}{\partial x} \\ \frac{\partial}{\partial y} \end{Bmatrix} \equiv \mathbf{J} \begin{Bmatrix} \frac{\partial}{\partial x} \\ \frac{\partial}{\partial y} \end{Bmatrix}, \quad \mathbf{J} = \begin{bmatrix} \frac{\partial x}{\partial \xi} & \frac{\partial y}{\partial \xi} \\ \frac{\partial x}{\partial \eta} & \frac{\partial y}{\partial \eta} \end{bmatrix}. \tag{1.8.21}$$

The inverse of Eq. (1.8.21) is what is needed to express the derivatives of functions with respect to (x, y) to those with respect to (ξ, η), which requires the inverse of $\mathbf{J}$ (and $J \neq 0$).

[6]When a domain is geometrically complex, it can be represented as a collection of quadrilaterals, and then the Gauss quadrature can be used on each quadrilateral. When the quadrilateral has curved sides, quadratic or higher-order approximations (which requires use of additional base points in the quadrilateral) can be used.

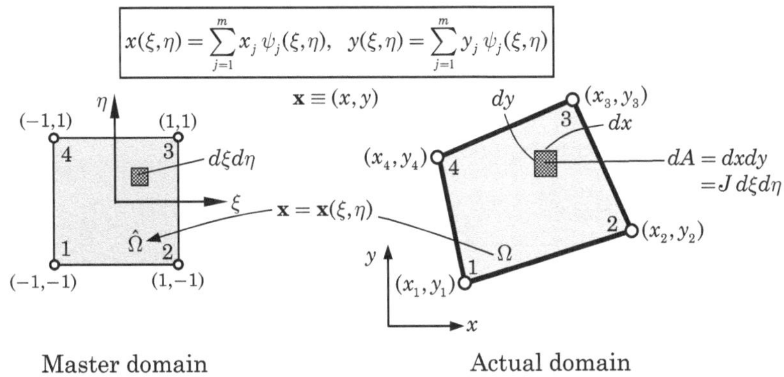

Fig. 1.8.2 Transformation of a function $f(x, y)$ from an arbitrary quadrilateral domain Ω to a function $F(x(\xi, \eta), y(\xi, \eta)) = F(\xi, \eta)$ in the master domain $\hat{\Omega}$. The transformation from Ω to $\hat{\Omega}$ is to facilitate numerical evaluation of the integral of $F(x, y)$ using the Gauss quadrature.

The integral in Eq. (1.8.19) is evaluated using the Gauss quadrature as

$$\begin{aligned} I &= \int_{\Omega} F(x, y)\, dxdy \approx \int_{\hat{\Omega}} F(x(\xi, \eta), y(\xi, \eta))\, J\, d\xi d\eta \\ &= \int_{-1}^{1} \int_{-1}^{1} \hat{F}(\xi, \eta)\, d\xi d\eta \\ &\approx \sum_{I=1}^{M} \sum_{J=1}^{N} \hat{F}(\xi_I, \eta_J)\, w_I\, w_J, \end{aligned} \tag{1.8.22}$$

where

$$\hat{F}(\xi_I, \eta_J) \equiv F(x(\xi, \eta), y(\xi, \eta))\, J(\xi, \eta). \tag{1.8.23}$$

It should be noted that the transformation of a given domain Ω to the master domain $\hat{\Omega}$ is solely for the purpose of numerically evaluating the integrals of the form in Eq. (1.8.19) (i.e., there is no transformation of the physical domain). Also, when one needs to evaluate an integral inexactly, numerical integration allows one to use reduced number of Gauss points. For example, if one desires to evaluate the integral of a quadratic polynomial as that of a linear polynomial, use of the one-point Gauss rule can evaluate it as an integral of a linear polynomial. Such a technique is known in the literature as *reduced integration*, which is used to avoid "locking" experienced by discretized equations of shear deformation beam, plate, and shell theories (see, e.g., Chapter 7 for additional discussion).

The numerical integration ideas discussed in this section are very useful in the evaluation of integrals involving finite element and dual mesh control domain formulations. Numerical integration is necessary when the coefficients are functions of unknown dependent variables, as in nonlinear problems.

1.9 Solution of Linear Algebraic Equations

1.9.1 Introduction

All numerical methods used to solve differential equations ultimately result in a system of linear algebraic equations of the form

$$\left.\begin{aligned} a_{11}\,x_1 + a_{12}\,x_2 + \cdots + a_{1n}\,x_n &= b_1,\\ a_{21}\,x_1 + a_{22}\,x_2 + \cdots + a_{2n}\,x_n &= b_2,\\ &\ \vdots\\ a_{n1}\,x_1 + a_{n2}\,x_2 + \cdots + a_{nn}\,x_n &= b_n, \end{aligned}\right\} \rightarrow\ \mathbf{AX} = \mathbf{B}, \tag{1.9.1}$$

where a_{ij} and b_i are known coefficients, and x_i are unknowns to be determined. Typically, x_i are the unknown parameters of the approximation in a numerical method (e.g., $x_i = c_i$ in the Ritz method).

In Eq. (1.9.1), when all b_i are zero, then the system is called homogeneous. If any one of the right-hand side elements is nonzero (i.e., not all b_i are zero), then the system is called nonhomogeneous.

We wish to find answers to the following questions concerning the solution of Eq. (1.9.1):

(1) Does the system possess a solution?
(2) If a solution exists, is it unique?
(3) How is (are) the solution(s) to be determined?

The existence of solutions and their uniqueness can be explained in geometric terms using a pair of linear algebraic equations (see Reddy and Rasmussen [20] and Reddy [21] for additional discussion):

$$3x_1 - 2x_2 = 4,\quad 2x_1 + x_2 = 5\ \ \rightarrow\ \ \begin{bmatrix} 3 & -2\\ 2 & 1\end{bmatrix}\begin{Bmatrix} x_1\\ x_2\end{Bmatrix} = \begin{Bmatrix} 4\\ 5\end{Bmatrix}. \tag{1.9.2}$$

The determinant of the coefficient matrix $\mathbf{A}$ is $A = 7$. Hence, the solution is

$$\begin{Bmatrix} x_1\\ x_2\end{Bmatrix} = \frac{1}{7}\begin{bmatrix} 1 & 2\\ -2 & 3\end{bmatrix}\begin{Bmatrix} 4\\ 5\end{Bmatrix} = \begin{Bmatrix} 2\\ 1\end{Bmatrix}. \tag{1.9.3}$$

In this case, the two lines represented by the equations intersect at the point (2,1), as can be seen from Fig. 1.9.1(a).

Next, consider the linear equations

$$6x_1 + 4x_2 = 4,\quad 3x_1 + 2x_2 = 2\ \ \rightarrow\ \ \begin{bmatrix} 6 & 4\\ 3 & 2\end{bmatrix}\begin{Bmatrix} x_1\\ x_2\end{Bmatrix} = \begin{Bmatrix} 4\\ 2\end{Bmatrix}. \tag{1.9.4}$$

In this case, we have $A = 0$. We note that the first equation is a multiple (twice that) of the second equation. In other words, there is only one relation between the two unknowns. Therefore, there are infinitely many solutions, for example, $(x_1, x_2) = (2, -2)$, $(4, -5)$, $(-2, 4)$, and so on. Geometrically, this case corresponds to two lines on top of each other [i.e., every point on one line corresponds to a point on the other line; see Fig. 1.9.1(b)].

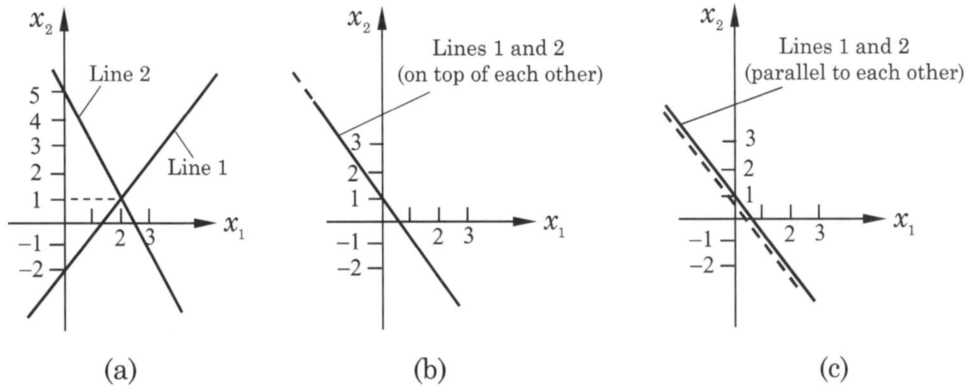

Fig. 1.9.1 Geometric interpretation of the solutions of two simultaneous algebraic equations.

Finally, consider the pair of equations

$$6x_1 + 4x_2 = 3, \quad 3x_1 + 2x_2 = 2$$

or, in matrix form, we have

$$\begin{bmatrix} 6 & 4 \\ 3 & 2 \end{bmatrix} \begin{Bmatrix} x_1 \\ x_2 \end{Bmatrix} = \begin{Bmatrix} 3 \\ 2 \end{Bmatrix}. \tag{1.9.5}$$

We find that the determinant of the coefficient matrix is zero, $A = 0$, because the row 1 of the matrix is twice that of the second row (i.e., $R_1 = 2R_2$) but $b_1 \neq 2b_2$. Hence the equations are inconsistent, and thus there is no solution of the equations. Geometrically, the two equations are parallel [i.e., they never intersect each other; see Fig. 1.9.1(c)].

Concerning how the solutions to linear equations are determined, there are two classes of methods: (1) direct methods and (2) iterative methods. The direct methods involve a formal inversion of the coefficient matrix **A** and multiplication of the right-hand side with the inverse. These methods have a fixed number (a function of n) of arithmetic operations to obtain the solution. The solution is exact within the truncation and round-off errors in the computer. On the other hand, iterative methods are based on certain assumptions and approximations, resulting in an approximate solution within an error tolerance prescribed by the user. Depending on the algorithm and error tolerance, the number of operations can vary with n, the error tolerance, and the particular iterative method used. Here we discuss few commonly used direct and iterative methods.

1.9.2 Direct Methods

The following direct methods are discussed here:

(1) Cramer's rule
(2) Gaussian elimination method
(3) Gauss–Jordan method

1.9.2.1 Cramer's rule

Recall from Section 1.6.2 that the determinant $|\mathbf{A}|$ of a square matrix $\mathbf{A}$ can be defined in terms of its cofactors, cof_{ij}, as [see Eq. (1.6.32)]

$$|\mathbf{A}| = \sum_{i=1}^{n} a_{ij}\,\mathrm{cof}_{ij} \tag{1.9.6}$$

for any fixed value of j ($j = 1, 2, \ldots,$ or n).

Using Eq. (1.6.34), we can solve the linear equations

$$\sum_{j=1}^{n} a_{kj}\, x_j = b_k. \tag{1.9.7}$$

Using the summation convention[7], Eq. (1.6.34) can be expressed as

$$|\mathbf{A}|\,\delta_{pq} = a_{pk}\,\mathrm{cof}_{qk} \quad \text{(sum on } k \text{ over the range of 1 to } n \text{ is implied).} \tag{1.9.8}$$

Similarly, Eq. (1.9.7) can be expressed using the summation convention as

$$a_{kj}\, x_j = b_k \quad \text{(sum on } j \text{ over the range of 1 to } n \text{ is implied).} \tag{1.9.9}$$

Multiplying Eq. (1.9.9) with cof_{kp}, we obtain (note that $a_{jk}\,\mathrm{cof}_{pk} = a_{kj}\,\mathrm{cof}_{kp}$)

$$a_{kj}\,\mathrm{cof}_{kp}\, x_j = b_k\,\mathrm{cof}_{kp} \text{ (sum on } j \text{ and } k \text{ is implied).} \tag{1.9.10}$$

Using the identity in Eq. (1.9.8), we can write

$$|\mathbf{A}|\,\delta_{jp}\, x_j = b_k\,\mathrm{cof}_{kp}$$

or

$$x_p = \frac{1}{|\mathbf{A}|}\, b_k\,\mathrm{cof}_{kp}, \quad |\mathbf{A}| \neq 0. \tag{1.9.11}$$

This result is known as *Cramer's rule* for solving a set of n linear equations.

We now specialize the result to a system of $n = 3$ equations among (x_1, x_2, x_3):

$$\begin{aligned} a_{11}x_1 + a_{12}x_2 + a_{13}x_3 &= b_1, \\ a_{21}x_1 + a_{22}x_2 + a_{23}x_3 &= b_2, \\ a_{31}x_1 + a_{32}x_2 + a_{33}x_3 &= b_3. \end{aligned} \tag{1.9.12}$$

For this case, Cramer's rule gives

$$x_1 = \frac{1}{|\mathbf{A}|}\begin{vmatrix} b_1 & a_{12} & a_{13} \\ b_2 & a_{22} & a_{23} \\ b_3 & a_{32} & a_{33} \end{vmatrix}, \quad x_2 = \frac{1}{|\mathbf{A}|}\begin{vmatrix} a_{11} & b_1 & a_{13} \\ a_{21} & b_2 & a_{23} \\ a_{31} & b_3 & a_{33} \end{vmatrix}, \quad x_3 = \frac{1}{|\mathbf{A}|}\begin{vmatrix} a_{11} & a_{12} & b_1 \\ a_{12} & a_{22} & b_2 \\ a_{13} & a_{32} & b_3 \end{vmatrix}, \tag{1.9.13}$$

[7] In summation convention, the sum $\sum_{i=1}^{n} A_i\, B_i$ is replaced with $A_i\, B_i$ by omitting the summation convention and with the understanding that the repeated index i implies summation over the range of i. The repeated index is called *dummy index* and thus can be replaced by *any other symbol that has not already been used in the expression.*

where $|\mathbf{A}|$ is the determinant of matrix $[A]$

$$|\mathbf{A}| = \begin{vmatrix} a_{11} & a_{12} & a_{13} \\ a_{21} & a_{21} & a_{23} \\ a_{31} & a_{32} & a_{33} \end{vmatrix}. \tag{1.9.14}$$

When $|\mathbf{A}| \neq 0$, we obtain a unique solution. For a homogeneous system, the right-hand side of Eq. (1.9.12) is zero (i.e., all b_i are zero). Then the numerator of Eq. (1.9.13) is zero, giving a trivial solution, $x_1 = x_2 = x_3 = 0$.

Example 1.9.1

Determine the solution to the following set of linear equations using Cramer's rule:

$$\begin{aligned} x_1 + x_2 + x_3 &= 1 \\ x_1 + x_2 - 3x_3 &= 2 \\ 3x_1 + x_2 - x_3 &= 3. \end{aligned}$$

Solution: First, we note that the determinant of the coefficient matrix is

$$|\mathbf{A}| = \begin{vmatrix} 1 & 1 & 1 \\ 1 & 1 & -3 \\ 3 & 1 & -1 \end{vmatrix} = [(1)(-1) - (1)(-3)] - [(1)(-1) - (3)(-3)] + [(1)(1) - (3)(1)] = -8.$$

The solution by Cramer's rule is given by

$$x_1 = -\frac{1}{8}\begin{vmatrix} 1 & 1 & 1 \\ 2 & 1 & -3 \\ 3 & 1 & -1 \end{vmatrix} = -\frac{1}{8}\{[(1)(-1) - (1)(-3)] - [(2)(-1) - (3)(-3)] + [(2)(1) - (3)(1)]\} = \frac{3}{4},$$

$$x_2 = -\frac{1}{8}\begin{vmatrix} 1 & 1 & 1 \\ 1 & 2 & -3 \\ 3 & 3 & -1 \end{vmatrix} = -\frac{1}{8}\{[(2)(-1) - (3)(-3)] - [(1)(-1) - (3)(-3)] + [(1)(3) - (3)(2)]\} = \frac{1}{2},$$

$$x_3 = -\frac{1}{8}\begin{vmatrix} 1 & 1 & 1 \\ 1 & 1 & 2 \\ 3 & 1 & 3 \end{vmatrix} = -\frac{1}{8}\{[(1)(3) - (1)(2)] - [(1)(3) - (3)(2)] + (1)(1) - (1)(3)]\} = -\frac{1}{4}.$$

1.9.2.2 Gaussian elimination method

In the Gaussian elimination method, elementary operations are used to reduce the coefficient matrix of the linear equations to upper diagonal form (i.e., all elements below the diagonal are made zero through elementary operations, while making sure that all diagonal elements are nonzero). An elementary row (column) operation on a matrix is an operation of one of the following types: (1) interchange of two rows (columns), (2) multiplication of the elements of a row (column) by a scalar, and (3) addition to the elements of a row (column), of k times the corresponding elements of another row (column).

For a $n \times n$ matrix, the upper-diagonal form (through forward elimination) is of the form

$$\begin{bmatrix} a'_{11} & a'_{12} & a'_{13} & \dots & a'_{1n} \\ 0 & a'_{21} & a'_{23} & \dots & a'_{2n} \\ 0 & 0 & a'_{33} & \dots & a'_{3n} \\ \vdots & \vdots & \vdots & \vdots & \vdots \\ 0 & \dots & 0 & a'_{(n-1)(n-2)} & a'_{(n-1)(n-1)} \\ 0 & \dots & 0 & 0 & a'_{nn} \end{bmatrix} \begin{Bmatrix} x_1 \\ x_2 \\ x_3 \\ \vdots \\ x_{n-1} \\ x_n \end{Bmatrix} = \begin{Bmatrix} b'_1 \\ b'_2 \\ b'_3 \\ \vdots \\ b'_{n-1} \\ b'_n \end{Bmatrix}, \tag{1.9.15}$$

where a'_{ij} and b'_i are the modified coefficients resulting from the elementary operations to arrive at the upper diagonal form. Once the forward elimination is completed, then the unknowns are calculated by a process of *backward substitution.* Thus, we have

$$\begin{aligned} x_n &= b'_n / a'_{33}, \\ x_{n-1} &= \left(b'_{(n-1)} - a'_{(n-1)n}\, x_n \right) / a'_{(n-1)(n-1)}, \\ x_{n-2} &= \left(b'_{n-2} - a'_{(n-2)(n-1)}\, x_{(n-1)} - a_{(n-2)n}\, x_n \right) / a'_{(n-2)(n-2)} \\ &\vdots \\ x_i &= \frac{1}{a'_{ii}} \left(b'_i - \sum_{j=i+1}^{n} a'_{2j}\, x_j \right), \qquad (1.9.16) \\ &\vdots \\ x_1 &= \frac{1}{a'_{22}} \left(b'_2 - \sum_{j=2}^{n} a'_{2j}\, x_j \right). \qquad (1.9.17) \end{aligned}$$

This process gives the values of all x_i, $i = 1, 2, \dots, n$. We illustrate the method using an example.

Example 1.9.2

Solve the same system of linear equations in **Example 1.9.1** using the Gaussian elimination method.

Solution: We have

$$\begin{bmatrix} 1 & 1 & 1 \\ 1 & 1 & -3 \\ 3 & 1 & -1 \end{bmatrix} \begin{Bmatrix} x_1 \\ x_2 \\ x_3 \end{Bmatrix} = \begin{Bmatrix} 1 \\ 2 \\ 3 \end{Bmatrix}.$$

We perform elementary operations to obtain the upper diagonal form. We begin with multiplying row 2 (R_2) with 3:

$$\begin{bmatrix} 1 & 1 & 1 \\ 3 & 3 & -9 \\ 3 & 1 & -1 \end{bmatrix} \begin{Bmatrix} x_1 \\ x_2 \\ x_3 \end{Bmatrix} = \begin{Bmatrix} 1 \\ 6 \\ 3 \end{Bmatrix}.$$

Then subtracting row 3 (R_3) from the modified row 2 (R_2'), we obtain

$$\begin{bmatrix} 1 & 1 & 1 \\ 0 & 2 & -8 \\ 3 & 1 & -1 \end{bmatrix} \begin{Bmatrix} x_1 \\ x_2 \\ x_3 \end{Bmatrix} = \begin{Bmatrix} 1 \\ 3 \\ 3 \end{Bmatrix}.$$

Now we subtract 3 times row 1 (R_1) from row 3 R_3 and obtain

$$\begin{bmatrix} 1 & 1 & 1 \\ 0 & 2 & -8 \\ 0 & -2 & -4 \end{bmatrix} \begin{Bmatrix} x_1 \\ x_2 \\ x_3 \end{Bmatrix} = \begin{Bmatrix} 1 \\ 3 \\ 0 \end{Bmatrix}.$$

Next, add the modified R_2' to modified R_3' and arrive at the upper diagonal form:

$$\begin{bmatrix} 1 & 1 & 1 \\ 0 & 2 & -8 \\ 0 & 0 & -12 \end{bmatrix} \begin{Bmatrix} x_1 \\ x_2 \\ x_3 \end{Bmatrix} = \begin{Bmatrix} 1 \\ 3 \\ 3 \end{Bmatrix}.$$

Now starting from R_3'', we will begin the process of backward substitution and obtain $x_3 = -3/12 = -1/4$. Using R_2', we obtain

$$x_2 = \frac{1}{2}\left[3 - (-8)x_3\right] = \frac{1}{2}.$$

Finally, using R_1:

$$x_1 = (1 - x_2 - x_3) = \frac{3}{4}.$$

Thus the solution of the system is $(x_1, x_2, x_3) = 0.25\,(3, -2, 1)$, which coincides with the solution of **Example 1.9.1**.

A Fortran listing of the Gauss elimination procedure for banded symmetric coefficient matrices as well as for banded unsymmetric matrices can be found at the author's website (**http://mechanics.tamu.edu**). Readers can possibly convert them to MATLAB® (The Math Works, Inc.) or Python codes.

1.9.2.3 Gauss–Jordan elimination

In the Gauss–Jordan method, the coefficient matrix is augmented with the right-hand column vector and the identity matrix, and then the off-diagonal elements are eliminated in *all* rows as opposed to eliminating off-diagonal elements in subsequent rows as in the Gaussian elimination method. A matrix $[A]$ is said to be *augmented* by another matrix $[B]$ having the same number of rows as matrix $[A]$, if $[A]$ and $[B]$ are placed column-wise in the same matrix: $[a_{ij}|b_{ij}]$.

Through algebraic operations, when the coefficient matrix is reduced to the identity matrix, the augmented identity matrix becomes the inverse matrix ($[A]^{-1}$), and the right-hand side becomes the solution vector ($\{X\}$). Next, we will illustrate these steps using the same system of equations as in the previous two examples.

Example 1.9.3

Solve the linear equations of **Example 1.9.1** using the Gauss–Jordan method.

Solution: The matrix $[A]$ augmented with $\{B\}$ and $[I]$, we have

$$\left[\begin{array}{ccc|c|ccc} 1 & 1 & 1 & 1 & 1 & 0 & 0 \\ 1 & 1 & -3 & 2 & 0 & 1 & 0 \\ 3 & 1 & -1 & 3 & 0 & 0 & 1 \end{array}\right].$$

Subtracting R_3 from R_1, and dividing the resulting row R_1' with -2, we obtain

$$\left[\begin{array}{ccc|c|ccc} 1 & 0 & -1 & 1 & -0.5 & 0 & 0.5 \\ 1 & 1 & -3 & 2 & 0.0 & 1 & 0.0 \\ 3 & 1 & -1 & 3 & 0.0 & 0 & 1.0 \end{array}\right].$$

Subtracting R_1' from R_2, we obtain

$$\left[\begin{array}{ccc|c|ccc} 1 & 0 & -1 & 1 & -0.5 & 0 & 0.5 \\ 0 & 1 & -2 & 1 & 0.5 & 1 & -0.5 \\ 3 & 1 & -1 & 3 & 0.0 & 0 & 1.0 \end{array}\right].$$

The operation $R_3 - 3R_1' \rightarrow R_3'$ gives

$$\left[\begin{array}{ccc|c|ccc} 1 & 0 & -1 & 1 & -0.5 & 0 & 0.5 \\ 0 & 1 & -2 & 1 & 0.5 & 1 & -0.5 \\ 0 & 1 & 2 & 0 & 1.5 & 0 & -0.5 \end{array}\right].$$

The operation $R_2' + R_3' \rightarrow R_2''$ gives

$$\left[\begin{array}{ccc|c|ccc} 1 & 0 & -1 & 1 & -0.5 & 0 & 0.5 \\ 0 & 2 & 0 & 1 & 2.0 & 1 & -1.0 \\ 0 & 1 & 2 & 0 & 1.5 & 0 & -0.5 \end{array}\right].$$

Normalizing R_2'' (called pivoting)

$$\left[\begin{array}{ccc|c|ccc} 1 & 0 & -1 & 1.0 & -0.5 & 0.0 & 0.5 \\ 0 & 1 & 0 & 0.5 & 1.0 & 0.5 & -0.5 \\ 0 & 1 & 2 & 0.0 & 1.5 & 0.0 & -0.5 \end{array}\right].$$

The operation $R_3' - R_2'' \rightarrow R_3''$ and normalizing R_3'' gives

$$\left[\begin{array}{ccc|c|ccc} 1 & 0 & -1 & 1.00 & -0.50 & 0.00 & 0.5 \\ 0 & 1 & 0 & 0.50 & 1.00 & 0.50 & -0.5 \\ 0 & 0 & 1 & -0.25 & 0.25 & -0.25 & 0.0 \end{array}\right].$$

Finally, adding row 3 to row 1 in the above augmented matrix, we obtain

$$\left[\begin{array}{ccc|c|ccc} 1 & 0 & 0 & 0.75 & -0.25 & -0.25 & 0.5 \\ 0 & 1 & 0 & 0.50 & 1.00 & 0.50 & -0.5 \\ 0 & 0 & 1 & -0.25 & 0.25 & -0.25 & 0.0 \end{array}\right].$$

Thus, the solution and inverse are

$$\begin{Bmatrix} x_1 \\ x_2 \\ x_3 \end{Bmatrix} = \begin{Bmatrix} 0.75 \\ 0.50 \\ -0.25 \end{Bmatrix}, \quad [A]^{-1} = \begin{bmatrix} -0.25 & -0.25 & 0.5 \\ 1.00 & 0.50 & -0.5 \\ 0.25 & -0.25 & 0.0 \end{bmatrix}.$$

Obviously, the number of elementary operations in the Gauss–Jordan method are higher than those in the Gaussian elimination method. This also means there are more round-off errors in the Gauss–Jordan method compared to the Gaussian elimination method. But the advantage is that, along with the solution, one obtains the inverse $[A]^{-1}$. Once a matrix inverse for a stiffness matrix is determined ($\mathbf{A}^{-1}$), then response to different right-hand sides (forcing function) can be obtained by simple matrix manipulation, $\mathbf{X} = \mathbf{A}^{-1}\mathbf{B}$.

A Fortran listing of the Gauss–Jordan procedure for general (i.e., symmetric or not and banded or not) coefficient matrices is available from the author's website (**http://mechanics.tamu.edu**). Readers can possibly convert it to a MATLAB or Python code.

1.9.3 Iterative Methods

1.9.3.1 Preliminary comments

Numerical solutions of most real-world problems involve the solution of a large number of equations. Solution of such large systems of equations by direct methods is computationally intensive (i.e., large amounts of CPU time and memory). Iterative methods, as the phrase suggests, are concerned with solving the linear equations without inverting a matrix. Typically, we start with an initial approximation of the solution (or guess solution vector) and solve for a particular unknown x_i using the known b_i and x_j ($j = 1, 2, \ldots, n; j \neq i$). One may go (or sweep) from $i = 1$ to n and then sweep backward or start again from $i = 1$. At the end of each iteration (or sweep), the solution vector from the current iteration, $\mathbf{X}^r$, is compared with the solution vector from the previous iteration, $\mathbf{X}^{r-1}$, to determine the difference between them (root-mean-square value of the difference normalized with the previous solution norm). When the normalized difference is less than a certain prescribed tolerance ϵ, convergence of the iterative process is declared (i.e., iteration stopping criterion):

$$\text{Error} = \sqrt{\frac{\sum_{j=1}^{n}\left(x_j^r - x_j^{r-1}\right)^2}{\sum_{j=1}^{n} x_j^r\, x_j^r}} \leq \epsilon \tag{1.9.18}$$

and $\mathbf{x}^r$ is taken as the solution to the system of linear equations, $\mathbf{AX} = \mathbf{B}$, if the process converges within a prescribed number of iterations (say, *itmax*).

The convergence of the iterative process depends on the conditioning (e.g., diagonal dominance) of the system of equations as well as the initial guess vector. The system of equations

$$\sum_{j=1}^{n} a_{ij} x_j = b_i \tag{1.9.19a}$$

is said to be *diagonally dominant* if

$$|a_{ii}| > \sum_{j=1, j\neq i}^{n} |a_{ij}|. \tag{1.9.19b}$$

The diagonal dominance is a sufficient condition for convergence. For example, the linear system of equations considered in **Example 1.9.1** is *not* diagonally dominant because $|a_{11}| > |a_{12}|+|a_{13}|$, $|a_{22}| > |a_{21}|+|a_{23}|$, and $|a_{33}| > |a_{31}|+|a_{32}|$ are not satisfied. Of course, the system will converge if the guess vector is very close to the actual solution (which is a very rare thing to happen when one does not know the solution).

In this section, we discuss the following two widely used iterative techniques:

(1) Jacobi iteration
(2) Gauss–Seidel iteration

These methods differ from each other only slightly. In the Gauss–Seidel iteration, the latest known components of the solution vector are used in computing the next component of the vector. But this difference makes the Gauss–Seidel iteration to converge faster (if it converges at all).

1.9.3.2 Jacobi iteration

In the Jacobi iteration method, the ith components of the solution vector at the rth iteration is computed using:

$$x_i^r = \frac{1}{a_{ii}}\left[b_i - \sum_{j=1,j\neq i}^{n} a_{ij}\, x_j^{(r-1)}\right]. \tag{1.9.20}$$

For $r = 1$, the solution vector $\mathbf{X}^{(0)}$ is the initial guess vector. We note that in the Jacobi iteration method the ith component of the solution at rth iteration, x_i^r, is calculated using the values of all other components of the solution vector from the $(r-1)$st iteration, although x_j^r for $j = 1, 2, \ldots, i-1$ are known from the current iteration. The Gauss–Seidel iteration method makes use of these latest known components.

When $n = 3$, Eq. (1.9.20) takes the form

$$\begin{aligned}
x_1^r &= \frac{1}{a_{11}}\left[b_1 - a_{12}\, x_2^{(r-1)} - a_{13}\, x_3^{(r-1)}\right],\\
x_2^r &= \frac{1}{a_{22}}\left[b_2 - a_{21} x_1^{(r-1)} - a_{23} x_3^{(r-1)}\right],\\
x_3^r &= \frac{1}{a_{33}}\left[b_3 - a_{31} x_1^{(r-1)} - a_{32} x_2^{(r-1)}\right].
\end{aligned}$$

We now consider an example using a diagonally–dominant system.

Example 1.9.4

Solve the following system of linear equations using the Jacobi iteration method:

$$\begin{bmatrix} 4 & 2 & 1 \\ -1 & 2 & 0 \\ 2 & 1 & 4 \end{bmatrix} \begin{Bmatrix} x_1 \\ x_2 \\ x_3 \end{Bmatrix} = \begin{Bmatrix} 11 \\ 3 \\ 16 \end{Bmatrix}.$$

The actual solution is $(x_1, x_2, x_3) = (1, 2, 3)$.

Solution: We begin with the guess vector $\mathbf{X}^{(0)} = \mathbf{0}$. Then we have

$$x_1^1 = \frac{1}{a_{11}}\left(b_1 - a_{12}\,x_2^0 - a_{13}\,x_3^0\right) = \frac{1}{4}\left(11 - 2\times 0 - 1\times 0\right) = \frac{11}{4} = 2.75,$$
$$x_2^1 = \frac{1}{a_{22}}\left(b_2 - a_{21}\,x_1^0 - a_{23}\,x_3^0\right) = \frac{1}{2}\left[3 - (-1)\times 0 - 0\times 0\right] = \frac{3}{2} = 1.50,$$
$$x_3^1 = \frac{1}{a_{33}}\left(b_3 - a_{31}\,x_1^0 - a_{32}\,x_2^0\right) = \frac{1}{4}\left(16 - 2\times 0 - 1\times 0\right) = \frac{16}{4} = 4.00.$$

The second iteration yields

$$x_1^2 = \frac{1}{4}\left(11 - 2\times 1.5 - 1\times 4\right) = 1.000,$$
$$x_2^2 = \frac{1}{2}\left[3 - (-1)\times 2.75 - 0\times 4\right] = 2.875,$$
$$x_3^2 = \frac{1}{4}\left(16 - 2\times 2.75 - 1\times 1.5\right) = 2.250.$$

The numerical solutions obtained at various iterations are shown in Table 1.9.1. The solution obtained with the Jacobi iteration for the choice of the initial guess (0, 0, 0) converges to the exact solution up to four decimal points (i.e., the normalized root-mean value of the difference between two consecutive iterations is less than $\epsilon = 10^{-4}$) in twelve iterations. Other choices of guess vector will yield the same results but may take slightly fewer or more iterations.

Table 1.9.1 Solution obtained with the Jacobi iteration method.

Iter. no.	x_1	x_2	x_3	Error
0	0.0000	0.0000	0.0000	– – –
1	2.7500	1.5000	4.0000	1.0000
2	1.0000	2.8750	2.2500	0.7479
3	0.7500	2.0000	2.7812	0.3005
4	1.0547	1.8750	3.1250	0.1255
5	1.0312	2.0273	3.0039	0.0520
6	0.9853	2.0156	2.9775	0.0145
7	0.9978	1.9927	3.0034	0.0098
8	1.0028	1.9989	3.0029	0.0021
9	0.9998	2,0014	2.9989	0.0015
10	0.9996	1.9999	2.9997	0.0005
11	1.0001	1.9998	3.0002	0.0002
12	1.0000	2.0001	3.0000	0.0001
13	1.0000	2.0000	3.0000	0.0000
14	1.0000	2.0000	3.0000	0.0000

1.9.3.3 Gauss–Seidel iteration

As stated earlier, the Gauss–Seidel iteration method is the same as the Jacobi iteration scheme except that in the Gauss–Seidel iteration the latest computed values of the unknowns (in the rth iteration) are used in iteration r. Thus, the ith component in the rth iteration is computed using the formula

$$x_i^r = \frac{1}{a_{ii}}\Big(b_i - \sum_{j=1}^{i-1} a_{ij}\, x_j^{(r)} - \sum_{j=i+1}^{n} a_{ij}\, x_j^{(r-1)}\Big). \tag{1.9.21}$$

When $n = 3$, we have

$$\begin{aligned} x_1^r &= \frac{1}{a_{11}}\Big(b_1 - a_{12}\, x_2^{(r-1)} - a_{13}\, x_3^{(r-1)}\Big), \\ x_2^r &= \frac{1}{a_{22}}\Big(b_2 - a_{21}\, x_1^r - a_{23}\, x_3^{(r-1)}\Big), \\ x_3^r &= \frac{1}{a_{33}}\Big(b_3 - a_{31}\, x_1^r - a_{32}\, x_2^r\Big). \end{aligned} \tag{1.9.22}$$

Example 1.9.5

Solve the following system of linear equations using the Gauss–Seidel iteration method:

$$\begin{bmatrix} 4 & 2 & 1 \\ -1 & 2 & 0 \\ 2 & 1 & 4 \end{bmatrix} \begin{Bmatrix} x_1 \\ x_2 \\ x_3 \end{Bmatrix} = \begin{Bmatrix} 11 \\ 3 \\ 16 \end{Bmatrix}.$$

Solution: Again, we use the guess vector to be $\mathbf{X}^{(0)} = \mathbf{0}$. Then we have

$$\begin{aligned} x_1^1 &= \frac{1}{4}\Big(11 - 2\times 0 - 1\times 0\Big) = 2.750, \\ x_2^1 &= \frac{1}{2}\Big[3 - (-1)\times 2.75 - 0\times 0\Big] = 2.875, \\ x_3^1 &= \frac{1}{4}\Big(16 - 2\times 2.75 - 1\times 2.875\Big) = 1.90625. \end{aligned}$$

The second iteration yields

$$\begin{aligned} x_1^2 &= \frac{1}{4}\Big(11 - 2\times 2.875 - 1\times 1.90625\Big) = 0.8359375, \\ x_2^2 &= \frac{1}{2}\Big[3 - (-1)\times 0.8359375 - 0\times 1.90625\Big] = 1.9179687, \\ x_3^2 &= \frac{1}{4}\Big(16 - 2\times 0.8359375 - 1\times 1.91796875\Big) = 3.1025391. \end{aligned}$$

The numerical solution obtained with the Gauss–Seidel iteration method for various iterations is shown in Table 1.9.2. From Table 1.9.2, it is clear that the solution has converged to its exact value (up to four decimal places) in six iterations, as opposed to twelve iterations in the Jacobi iteration method. This is expected because the latest known values of the solution vector are used in the Gauss–Seidel iteration method.

Table 1.9.2 Solution obtained with the Gauss–Seidel iteration method.

Iter. no.	x_1	x_2	x_3	Error
0	0.0000	0.0000	0.0000	– – –
1	2.7500	2.8750	1.9062	1.0000
2	0.8359	1.9180	3.1025	0.6552
3	1.0154	2.0077	2.9904	0.0614
4	0.9985	1.9993	3.0009	0.0057
5	1.0001	2.0001	2.9999	0.0005
6	1.0000	2.0000	3.0000	0.0001
7	1.0000	2.0000	3.0000	0.0000

1.9.4 Iterative Methods for Nonlinear Equations

1.9.4.1 Introduction

In the previous section, attention was focused on methods for the solution of linear algebraic equations of the form $[A]\{X\} = \{B\}$, in which the $n \times n$ matrix $[A]$ is independent of the vector $\{X\}$; that is, $[A]$ is a linear matrix operator acting on the vector of unknowns $\{X\}$ producing the vector[8] $\{B\}$. If the operator is a function of some or all unknowns $x_1, x_2, \ldots x_n$, the resulting equations are said to be nonlinear. An example of nonlinear algebraic equations is

$$\begin{aligned} 5\,x_1\,x_3 + x_2^2 + 3\,x_1\,x_3 &= 16, \\ x_1^2 + 4\,x_2^3 + x_3^2 + x_4 &= 11, \\ -x_1^2 + 2\,x_4\,x_2^2 + 6\,x_3^2 - 2\,x_4 &= 23, \\ x_1 - x_1\,x_2^2 - x_2\,x_3 + 4\,x_4 &= -2. \end{aligned} \tag{1.9.23}$$

In this case, the 4×4 matrix operator $[A]$ (not unique because of the product terms involving x_i and x_j, $i \neq j$) is a function of x_i:

$$[A] = \begin{bmatrix} 5\,x_3 & x_2 & 3\,x_1 & 0 \\ x_1 & 4\,x_2^2 & x_3 & 1 \\ -x_1 & 2\,x_2\,x_4 & 6\,x_3 & -2 \\ 1 & -x_1\,x_2 & -x_2 & 4 \end{bmatrix} \tag{1.9.24}$$

acting on the vector (clearly, $[A]\{\alpha\,X\} \neq \alpha[A]\{X\}$)

$$\{X\} = \begin{Bmatrix} x_1 \\ x_2 \\ x_3 \\ x_4 \end{Bmatrix}. \tag{1.9.25}$$

When we are faced with the solution of such nonlinear algebraic equations, we *linearize* the equations first (i.e., $[A]$ is made linear) and then use a direct or iterative method discussed in the preceding sections to solve the linearized equations (giving an approximate solution to the nonlinear equations). For example, if we use a guess vector $\mathbf{X}^0$ to evaluate the matrix $[A]$ in Eq. (1.9.24), we obtain

$$[A] = \begin{bmatrix} 5\,x_3^0 & x_2^0 & 3\,x_1^0 & 0 \\ x_1^0 & 4(x_2^0)^2 & x_3^0 & 1 \\ -x_1^0 & 2\,x_2^0\,x_4^0 & 6\,x_3^0 & -2 \\ 1 & -x_1^0\,x_2^0 & -x_2^0 & 4 \end{bmatrix} \tag{1.9.26}$$

[8]An operator A maps a set of elements $X = \{x_i\}$ to another set of elements $Y = \{y_i\}$, and it is stated as $A : X \to Y$. Operationally, we use the notation $A(x_i) = y_i$. The operator A is said to be *linear* if and only if $A(\alpha\,x_1 + \beta\,x_2) = \alpha\,A(x_1) + \beta\,A(x_2)$ for any scalars α and β; otherwise, the operator A is said to be *nonlinear*. Such operators can be of the algebraic, differential, or integral type. The linearity of the operator $A(\cdot)$ can be checked only when it operates on an element $(\cdot)$.

then $[A]$ is no longer dependent on the unknowns x_i; that is, $[A]$ is now a linear matrix operator.

It is obvious that such a linearization requires another loop of iteration within possibly an iterative solver. The inner iteration, called *nonlinear iteration,* is to update the matrix $[A]$ with the latest known solution vector $\{X\}$ before solving the equations again. This iteration continues till the difference, measured in appropriate norm, between two consecutive iterations is less than a prescribed tolerance.

In this section, we discuss two commonly used linearization methods for nonlinear iteration:

1. Picard iteration (or direct iteration)
2. Newton iteration

1.9.4.2 Picard iteration

In the Picard iteration, also known as the *direct iteration* or the method of successive substitutions, we begin with an initial guess for $\mathbf{X}$, say $\mathbf{X}^{(0)}$, and determine a first approximation of $\mathbf{X}$ by solving the equation

$$\mathbf{X}^{(1)} = [\mathbf{A}_0]^{-1}\,\mathbf{B}, \quad \mathbf{A}_0 \equiv \mathbf{A}(\mathbf{X}^{(0)}), \tag{1.9.27}$$

where the notation $\mathbf{A}_0$ is used to indicate that it is evaluated using the vector $\mathbf{X}^{(0)}$. The second approximation for $\mathbf{X}$ is sought by using the latest known solution $\mathbf{X}^{(1)}$ to evaluate the coefficient matrix $\mathbf{A}$:

$$\mathbf{X}^{(2)} = [\mathbf{A}_1]^{-1}\,\mathbf{B}. \tag{1.9.28}$$

Thus, at the rth iteration we have

$$\mathbf{X}^{(r)} = [\mathbf{A}_{r-1}]^{-1}\,\mathbf{B}, \quad \mathbf{A}_{r-1} \equiv \mathbf{A}(\mathbf{X}^{(r-1)}). \tag{1.9.29}$$

This procedure is continued until the difference between two consecutive approximations of $\mathbf{X}$, namely $\mathbf{X}^{(r-1)}$ and $\mathbf{X}^{(r)}$, differ by a pre-selected value. Thus, the algorithm and criterion for nonlinear convergence may be expressed as

$$\text{Algorithm:} \qquad \mathbf{X}^{(r)} = \left[\mathbf{A}_{(k-1)}\right]^{-1}\mathbf{B} \tag{1.9.30}$$

$$\text{Convergence criterion:} \qquad \sqrt{\frac{\sum_{i=1}^{n}\left[x_i^{(r)} - x_i^{(r-1)}\right]^2}{\sum_{i=1}^{n}[x_i^{(r)}]^2}} < \epsilon_{NL}, \tag{1.9.31}$$

where ϵ_{NL} denotes the *nonlinear convergence tolerance.*

A geometric interpretation of the direct iteration procedure for a single nonlinear equation, $Ax = b$, is illustrated in Fig. 1.9.2 for an initial guess of $x^{(0)} = 0$. At the beginning of the rth iteration, the secant of the curve $R(x) = A(x)x - b = 0$ is found at the point $x = x^{(r-1)}$ and the solution $x^{(r)}$ is computed using Eq. (1.9.29). Figure 1.9.2 shows the convergence to the actual solution x_c. The nonlinear convergence of the algorithm depends on the nature of the curve $A(x)x - b = 0$, the initial guess, and the external source b.

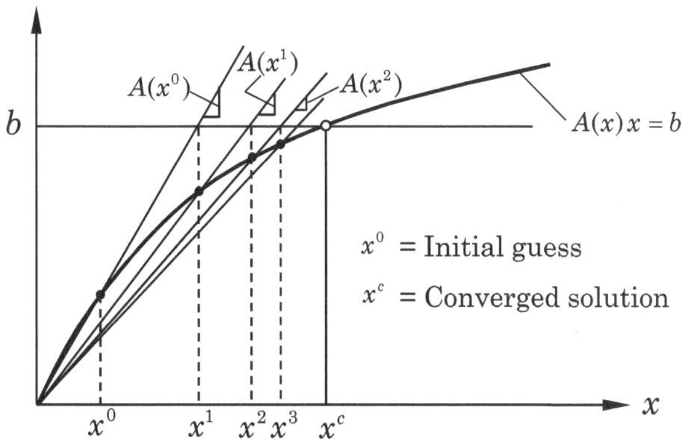

Fig. 1.9.2 Direct iteration scheme (the case of convergence is illustrated).

The rate of convergence of the iterative procedure can be accelerated, in certain types of nonlinear behavior, by a *relaxation* procedure in which a weighted average of the solutions from the last two iterations is used to evaluate $\mathbf{A}(\bar{\mathbf{x}})$, where $\bar{\mathbf{x}} = \beta \mathbf{x}^{(r-2)} + (1-\beta)\mathbf{x}^{(r-1)}$, and β $(0 \leq \beta \leq 1)$ is called the relaxation or acceleration parameter. The actual value of β varies from problem to problem.

1.9.4.3 Newton iteration

The Newton iteration technique is based on finite Taylor series expansion of the residual $\mathbf{R}(\mathbf{X}) \equiv \mathbf{A}(\mathbf{X})\mathbf{X} - \mathbf{B}$. To formulate the Newton iterative scheme, suppose that we know solution $\mathbf{X}^{(r-1)}$ of $\mathbf{AX} = \mathbf{B}$ at the beginning of the rth iteration. Expanding $\mathbf{R}(\mathbf{X})$ about the known solution $\mathbf{X}^{(r-1)}$ in a Taylor series, we have

$$\begin{aligned}\mathbf{R}(\mathbf{X}) = \mathbf{R}(\mathbf{X}^{(r-1)}) &+ \left(\frac{\partial \mathbf{R}}{\partial \mathbf{X}}\right)\Bigg|_{\mathbf{X}^{(r-1)}} \left(\mathbf{X}^{(r)} - \mathbf{X}^{(r-1)}\right) \\ &+ \frac{1}{2!}\left(\frac{\partial^2 \mathbf{R}}{\partial \mathbf{X}^2}\right)\Bigg|_{\mathbf{X}^{(r-1)}} \left(\mathbf{X}^{(r)} - \mathbf{X}^{(r-1)}\right)^2 + \cdots . \end{aligned} \tag{1.9.32}$$

Since we are interested in determining $\mathbf{X}^{(r)}$, we truncate the Taylor series after the linear term in $\mathbf{X}^{(r)} - \mathbf{X}^{(r-1)}$ and force the residual to zero:

$$\mathbf{X}^{(r)} - \mathbf{X}^{(r-1)} = -\left[\mathbf{T}(\mathbf{X}^{(r-1)})\right]^{-1} \mathbf{R}(\mathbf{X}^{(r-1)}), \tag{1.9.33a}$$

$$\mathbf{X}^{(r)} = \mathbf{X}^{(r-1)} + \left[\mathbf{T}(\mathbf{X}^{(r-1)})\right]^{-1} \left(\mathbf{B} - \mathbf{A}(\mathbf{X}^{(r-1)})\,\mathbf{X}^{(r-1)}\right), \tag{1.9.33b}$$

where $\mathbf{T}$ is the matrix, known as the *tangent coefficient matrix*

$$\mathbf{T} \equiv \frac{\partial \mathbf{R}}{\partial \mathbf{X}}\Bigg|_{\mathbf{X}^{(r-1)}}. \tag{1.9.34}$$

The tangent coefficient matrix components t_{ij} can be computed as explained next. First note that the ith component of the residual vector is

$$R_i = \sum_{k=1}^{n} a_{ik} x_k - b_i, \quad i = 1, 2, 3, \ldots, n. \tag{1.9.35}$$

Hence, we have

$$t_{ij} = \frac{\partial R_i}{\partial x_j} = \sum_{k=1}^{n} \left(\frac{\partial a_{ik}}{\partial x_j} x_k + a_{ik} \frac{\partial x_k}{\partial x_j} \right) - \frac{\partial b_i}{\partial x_j} = a_{ij} + \sum_{k=1}^{n} \frac{\partial a_{ik}}{\partial x_j} x_k, \tag{1.9.36}$$

where we have used the following identities in arriving at the last line of Eq. (1.9.36):

$$\frac{\partial x_k}{\partial x_j} = \delta_{kj}, \quad \sum_{k=1}^{n} a_{ik}\, \delta_{kj} = a_{ij}. \tag{1.9.37}$$

For example, the second term on the right-hand side of Eq. (1.9.36) can be computed for the specific coefficients a_{ij} in Eq. (1.9.24) as follows:

$$\begin{aligned}
&\frac{\partial a_{13}}{\partial x_1} = 3, \quad \frac{\partial a_{1j}}{\partial x_1} = 0, \quad \text{for } j = 1, 3, 4;\\
&\frac{\partial a_{21}}{\partial x_1} = 1, \quad \frac{\partial a_{2j}}{\partial x_1} = 0, \quad \text{for } j = 2, 3, 4;\\
&\frac{\partial a_{31}}{\partial x_1} = -1, \quad \frac{\partial a_{3j}}{\partial x_1} = 0, \quad \text{for } j = 2, 3, 4; \quad \frac{\partial a_{42}}{\partial x_1} = -x_2,\\
&\frac{\partial a_{4j}}{\partial x_1} = 0, \quad \text{for } j = 1, 3, 4; \quad \frac{\partial a_{12}}{\partial x_2} = 1, \quad \frac{\partial a_{1j}}{\partial x_2} = 0, \quad \text{for } j = 1, 3, 4;\\
&\frac{\partial a_{22}}{\partial x_2} = 8\, x_2, \quad \frac{\partial a_{2j}}{\partial x_2} = 0, \quad \text{for } j = 1, 3, 4; \quad \frac{\partial a_{31}}{\partial x_2} = 2\, x_4,\\
&\frac{\partial a_{3j}}{\partial x_2} = 0, \quad \text{for } j = 1, 3, 4; \quad \frac{\partial a_{42}}{\partial x_2} = -x_1, \quad \frac{\partial a_{43}}{\partial x_2} = -1,\\
&\frac{\partial a_{11}}{\partial x_3} = 5, \quad \frac{\partial a_{1j}}{\partial x_3} = 0, \quad \text{for } j = 2, 3, 4;\\
&\frac{\partial a_{23}}{\partial x_3} = 1, \quad \frac{\partial a_{2j}}{\partial x_3} = 0, \quad \text{for } j = 1, 2, 4; \quad \frac{\partial a_{33}}{\partial x_3} = 6,\\
&\frac{\partial a_{3j}}{\partial x_3} = 0, \quad \text{for } j = 1, 2, 4; \quad \frac{\partial a_{4j}}{\partial x_3} = 0, \quad \text{for } j = 1, 2, 3, 4;\\
&\frac{\partial a_{1j}}{\partial x_4} = 0, \quad \text{for } j = 1, 2, 3, 4; \quad \frac{\partial a_{2j}}{\partial x_4} = 0, \quad \text{for } j = 1, 2, 3, 4;\\
&\frac{\partial a_{32}}{\partial x_4} = 2\, x_2, \quad \frac{\partial a_{3j}}{\partial x_4} = 0, \quad \text{for } j = 1, 3, 4;\\
&\frac{\partial a_{4j}}{\partial x_4} = 0, \quad \text{for } j = 1, 2, 3, 4.
\end{aligned} \tag{1.9.38}$$

Thus, the tangent stiffness matrix is $[t_{ij} = a_{ij} + \sum_{k=1}^{n}(\partial a_{ik}/\partial x_j)x_k]$

$$[T] = \begin{bmatrix} 5\,x_3 & x_2 & 3\,x_1 & 0 \\ x_1 & 4\,x_2^2 & x_3 & 1 \\ -x_1 & 2\,x_2\,x_4 & 6\,x_3 & -2 \\ 1 & -x_1\,x_2 & -x_2 & 4 \end{bmatrix} + \begin{bmatrix} 3\,x_3 & x_2 & 5\,x_1 & 0 \\ x_1 & 8x_2^2 & x_3 & 0 \\ -x_1 & 2\,x_2\,x_4 & 6\,x_3 & 2\,x_2^2 \\ -x_2^2 & -x_1\,x_2 - x_3 & 0 & 0 \end{bmatrix}. \tag{1.9.39}$$

The residual or *imbalance source*, $\mathbf{R}(\mathbf{X}^{(r-1)})$, is reduced to zero with each iteration, if the procedure converges. Equation (1.9.33b) gives the solution increment $\Delta\mathbf{X}^{(r)}$ at the rth iteration by solving the equation

$$\Delta\mathbf{X}^{(r)} = -\left[\mathbf{T}(\mathbf{X}^{(r-1)})\right]^{-1}\mathbf{R} \tag{1.9.40a}$$

and updating the solution

$$\mathbf{X}^{(r)} = \mathbf{X}^{(r-1)} + \Delta\mathbf{X}^{(r)}. \tag{1.9.40b}$$

The iteration is continued until a convergence criterion, say Eq. (1.9.31), is satisfied. Other convergence criteria include checking the magnitude of the imbalance force $\mathbf{R}$.

A geometrical interpretation of the Newton procedure using a single nonlinear equation $A(x)x = b$ (or $R = A(x)x - b = 0$) is shown in Fig. 1.9.3. For most problems, the method has faster convergence characteristics.

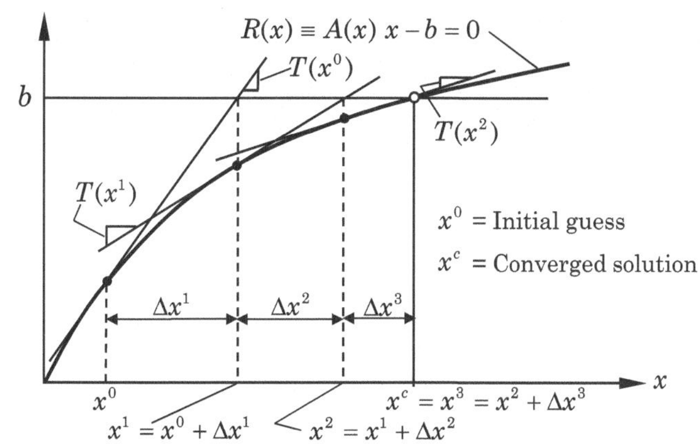

Fig. 1.9.3 Newton iteration scheme (the case of convergence is illustrated).

1.10 Method of Manufactured Solutions

When a new numerical method is introduced, its accuracy is often verified through known analytical solutions. The availability of exact solutions of PDEs is limited due to a combination of geometry, boundary conditions, source, and material properties. Analytical solutions based on conventional methods like direct integration, series method, and separation of variables method are not always possible. Due to such limitations, one can actually "manufacture" analytical solutions (see [22]). In this method, one assumes a solution that satisfies a set of boundary conditions and varies in a desired manner with respect to the independent coordinates. Of course, this task itself is not easy; it requires the knowledge of the physics of the problem governed by the PDE. Substitution of the assumed solution into the PDE gives rise to a function that is identified as the source function.

To fix the thoughts, consider the following linear operator equation posed on a domain Ω with a closed boundary Γ:

$$A(u) = f \ \text{ in } \ \Omega; \quad B(u) = q \ \text{ on } \ \Gamma, \tag{1.10.1}$$

where A and B are linear operators, u is the dependent variable, f is the source term, and q is the boundary value. We assume that the given problem is well-posed and has a solution. Now suppose that u_m is the assumed solution that has the following properties: (1) u_m is continuous and its derivatives of the order up to and including those that appear in the operator A exist and are continuous within the domain and (2) satisfies the boundary conditions. The solution must be valid both inside the domain (i.e., satisfy the governing differential equation) and on the boundary (i.e., satisfy the specified boundary conditions). If some prescribed boundary conditions are not met by the assumed solution, they can be identified during the process of developing the manufactured solution. When the assumed solution is substituted into $A(u)$, it yields $A(u_m)$, which will not, in most cases, be a zero. In order to satisfy the governing equation $A(u_m) = f$, the source term f is identified as $f = A(u_m)$. If the assumed solution does not satisfy the specified boundary conditions, one can identify boundary conditions as well by evaluating u_m and/or its derivatives on the boundary. Then the original problem has a manufactured solution u_m that satisfies the operator equation for $f = A(u_m)$ and the boundary conditions, $B(u_m) = q$. We consider two examples: a problem with a known form of the solution and another problem without any such information.

Example 1.10.1

Consider the Poisson equation over a rectangular domain of dimensions $3a \times 2a$ (see Fig. 1.10.1):

$$-k\left(\frac{\partial^2 u}{\partial x^2} + \frac{\partial^2 u}{\partial y^2}\right) = f(x, y). \tag{1.10.2}$$

Determine the analytical solution to Eq. (1.10.2) with the following boundary conditions:

$$-k\frac{\partial u}{\partial y}\bigg|_{y=0} = 0, \quad -k\frac{\partial u}{\partial x}\bigg|_{x=0} = 0, \quad u(3a, y) = 0, \quad u(x, 2a) = u_0 \cos\frac{\pi x}{6a} \tag{1.10.3}$$

where k and u_0 are constants.

Solution: Using the boundary conditions, we assume (knowing the solution form from other sources!) a manufactured solution of the form

$$u_m(x, y) = \cos\frac{\pi x}{6a} \cosh\frac{\pi y}{6a}. \tag{1.10.4}$$

The assumed solution is continuous with all its derivatives in the domain.

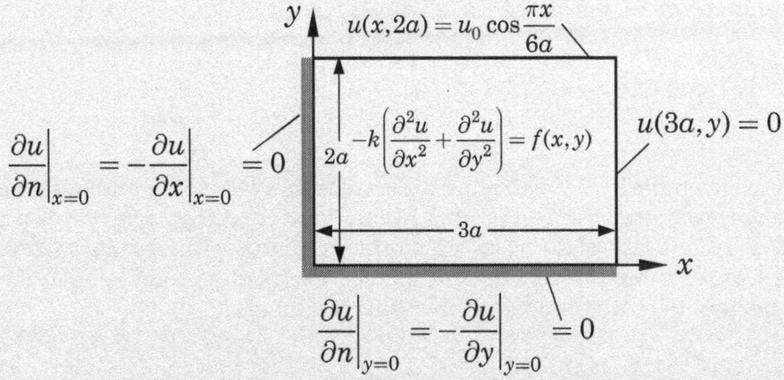

Fig. 1.10.1 Domain and boundary conditions for a second-order PDE in two dimensions.

Note that the solution when evaluated on the boundary segments has the values

$$
\begin{aligned}
&-k\left.\frac{\partial u_m}{\partial y}\right|_{y=0} = 0, \quad -k\left.\frac{\partial u_m}{\partial x}\right|_{x=0} = 0, \\
&u_m(3a, y) = 0, \quad u_m(x, 2a) = \cos\frac{\pi x}{6a}\cosh\frac{\pi}{3}.
\end{aligned} \tag{1.10.5}
$$

All boundary conditions in Eq. (1.10.3), except at $y = 2a$, are satisfied exactly. To satisfy the boundary condition at $y = 2a$, we must modify our assumed solution to

$$
u_m(x, y) = u_0 \frac{\cos\frac{\pi x}{6a}\cosh\frac{\pi y}{6b}}{\cosh(\pi/3)}. \tag{1.10.6}
$$

Next, we substitute the assumed solution from Eq. (1.10.6) into the governing differential equation to determine the source term

$$
\begin{aligned}
f(x, y) &= -k\left(\frac{\partial^2 u_m}{\partial x^2} + \frac{\partial^2 u_m}{\partial y^2}\right) \\
&= -\frac{ku_0}{\cosh(\pi/3)}\left[-\frac{\pi^2}{36a^2}\cos\frac{\pi x}{6a}\cosh\frac{\pi y}{6a} + \frac{\pi^2}{36a^2}\cos\frac{\pi x}{6a}\cosh\frac{\pi y}{6a}\right] = 0
\end{aligned} \tag{1.10.7}
$$

Thus, the assumed manufactured solution in Eq. (1.10.6) exactly satisfies the governing equation in Eq. (1.10.1) and boundary conditions in Eq. (1.10.5) for $f = 0$ and arbitrary k. Thus, the manufactured solution $u_m(x, y)$ is the exact solution of Eqs. (1.10.2) and (1.10.3). We note that the chosen boundary conditions must be physically meaningful (e.g., one may not specify both u and its derivatives on the same boundary). We consider another example of the manufactured solution.

Example 1.10.2

When the cylinder is of finite length and material properties, boundary conditions, and applied loads vary along the length of the cylinder but independent of the circumferential coordinate θ, we can use any typical plane (a wedge) of the domain [see Fig. 1.10.2(a)] as the domain of the problem. The equation governing a diffusion process (e.g., heat flow) in this case is of the form

$$
-\frac{k}{r}\left[\frac{\partial}{\partial r}\left(r\frac{\partial u}{\partial r}\right) + \frac{\partial}{\partial z}\left(r\frac{\partial u}{\partial z}\right)\right] = f(r, z). \tag{1.10.8}
$$

Determine a manufactured solution of Eq. (1.10.6) for the boundary conditions shown in Fig. 1.10.2(b),

$$u(r,b) = g_1(r), \quad u(a,z) = g_2(z), \quad \left.\frac{\partial u}{\partial r}\right|_{r=0} = 0, \quad \left.\frac{\partial u}{\partial z}\right|_{x=0} = 0. \tag{1.10.9}$$

Solution: Before we embark on developing a manufactured solution and defining the problem for which it is valid, we note that the gradient boundary condition at $r = 0$ is a mathematical requirement on a symmetry plane (i.e., any other boundary condition to replace it is not meaningful). All other boundary conditions can be changed as long as one does not specify both u and its normal derivative along a boundary segment.

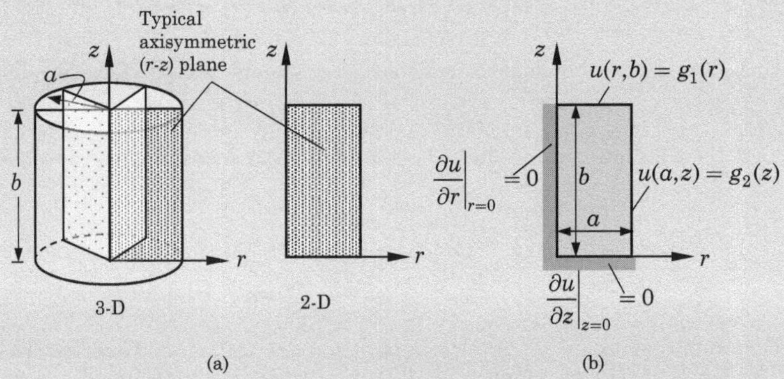

Fig. 1.10.2 Domain and boundary conditions for an axisymmetric problem.

One logical way to construct the manufactured solution is to seek the solution in the form

$$u_m(r,z) = g_1(r)\, g_2(z). \tag{1.10.10}$$

The boundary conditions in Eq. (1.10.8) require $g_1(r)$ and $g_2(z)$ to satisfy the following relations:

$$\left.\frac{dg_1}{dr}\right|_{r=0} = 0, \quad \left.\frac{dg_2}{dz}\right|_{z=0} = 0, \quad g_1(a) = 1, \quad g_2(b) = 1. \tag{1.10.11}$$

The lowest-order polynomials that we can choose for g_1 and g_2 are $g_1(r) = c_1 + c_2 r + c_3 r^2$ and $g_2(z) = d_1 + d_2 z + d_3 z^2$. Using the relations in Eq. (1.10.11), we find that ($c_2 = 0$, $c_1 = 1 - a^2 c_3$, $d_2 = 0$, and $d_1 = 1 - b^2 d_3$)

$$g_1(r) = 1 + c_3(-a^2 + r^2), \quad g_2(z) = 1 + d_3(-b^2 + z^2), \tag{1.10.12}$$

where c_3 and d_3 are arbitrary nonzero constants. To avoid $g_1(r)$ and $g_2(z)$ being negative (from physical considerations), we choose

$$c_3 = -\frac{1}{2a^2}, \quad d_3 = -\frac{1}{2b^2} \tag{1.10.13}$$

so that the manufactured solution is [and making certain that the boundary conditions on boundaries $u_m(r,b) = g_1(r)$ and $u_m(a,z) = g_2(z)$ are satisfied]:

$$u_m(r,z) = \tfrac{1}{4}\left(3 - \frac{r^2}{a^2}\right)\left(3 - \frac{z^2}{b^2}\right), \quad g_1(r) = \tfrac{1}{2}\left(3 - \frac{r^2}{a^2}\right), \quad g_2(z) = \tfrac{1}{2}\left(3 - \frac{z^2}{b^2}\right). \tag{1.10.14}$$

One can verify that the boundary conditions in Eq. (1.10.9) are satisfied.

Next, we determine $f(r,z)$ by substituting u_m from Eq. (1.10.14) into the governing equation (1.10.8). First, we compute the various derivatives required in Eq. (1.10.8):

$$\frac{\partial u_m}{\partial r} = -\frac{1}{2a^2} r \left(3 - \frac{z^2}{b^2}\right),$$

$$\frac{\partial u_m}{\partial z} = -\frac{1}{2b^2} \left(3 - \frac{r^2}{a^2}\right) z$$

$$\frac{\partial}{\partial r}\left(r\frac{\partial u_m}{\partial r}\right) = -\frac{r}{a^2}\left(3 - \frac{z^2}{b^2}\right),$$

$$\frac{\partial^2 u_m}{\partial z^2} = -\frac{1}{2b^2}\left(3 - \frac{r^2}{a^2}\right).$$

Then we have

$$f(r,z) = -k\left[\frac{1}{r}\frac{\partial}{\partial r}\left(r\frac{\partial u_m}{\partial r}\right) + \frac{\partial}{\partial z}\left(\frac{\partial u_m}{\partial z}\right)\right]$$

$$= k\left[\frac{1}{a^2}\left(3 - \frac{z^2}{b^2}\right) + \frac{1}{2b^2}\left(3 - \frac{r^2}{a^2}\right)\right]. \quad (1.10.15)$$

Thus, the manufactured solution of Eq. (1.10.7) subject to boundary conditions in Eq. (1.10.8) and the source term $f(r,z)$ of Eq. (1.10.15) is given by u_m of Eq. (1.10.14).

Plots of the solution $u(r,z)$ versus r for three different values of $z = 0.0, 1.0$, and 1.5 are presented in Fig. 1.10.3.

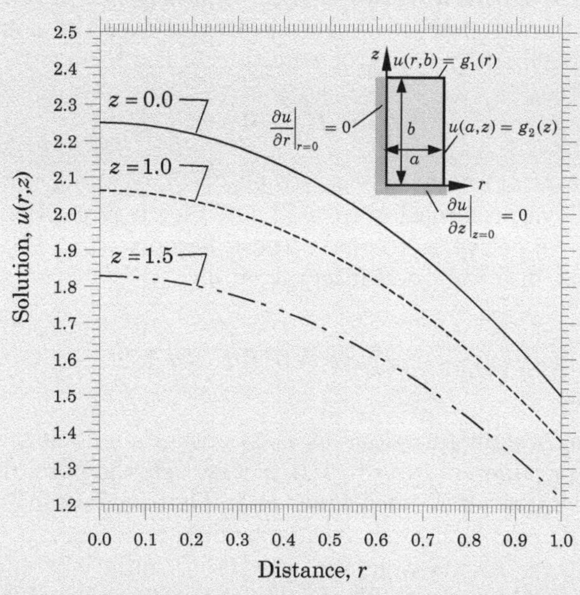

Fig. 1.10.3 Plots of $u(r,z)$ versus r for $z = 0.0, 1.0$, and 1.5.

1.11 Variational Formulations and Methods

1.11.1 Background

The classical use of the phrase "variational formulations" refers to the construction of a functional (whose meaning will be made clear shortly) or a variational principle that is equivalent to the governing equations of the problem. The principles of virtual work and the principle of minimum total potential energy in solid and structural mechanics (see Reddy [17]) provide good examples of variational formulations. The modern use of the phrase *variational problem* refers to the formulation in which the governing equations (and some boundary conditions) are translated into equivalent integral statements.

In almost all approximate methods used to determine the solution of differential equations, we seek solution in the form

$$u(x) \approx u_N(x) = \sum_{j=1}^{N} c_j \phi_j(x), \tag{1.11.1}$$

where u represents the solution of a particular differential equation and associated boundary conditions, and u_N is its approximation that is represented as a linear combination of unknown parameters c_j and known functions $\phi_j(x)$, called *approximation functions.* We shall discuss the conditions on the approximation functions ϕ_j in the sequel (see Section 1.11.5.2). The parameters c_j are determined by requiring u_N to satisfy the governing differential equation in some sense. One of the ways we choose to satisfy the governing differential equation is by requiring that its weighted-integral over the domain be zero. If the problem is described by operator equation of the form

$$A(u(x)) = f(x) \quad 0 < x < L \tag{1.11.2}$$

where A is a differential operator [e.g., see Eq. (1.5.5)] and f is a given function. Substitution of the approximation from (1.11.1) for u gives $A(u_N)$, which is not equal to f. The objective is to make the difference $R = A(u_N) - f$, called *residual,* go to zero in a weighted-integral sense:

$$\int_0^L w_i \left[A(u_N(x)) - f(x)\right] dx = 0 \quad \text{for} \quad i = 1, 2, \ldots, N. \tag{1.11.3}$$

The need for using weight functions w_i is to obtain a sufficient number of relations among the parameters c_j, $j = 1, 2, \ldots, N$. This is possible by selecting a linearly independent set of N such functions. That is, for each choice of w_i, we obtain an algebraic equation among the parameters $c_1, c_2, \ldots, c_N$.

In this section, we discuss weighted-integral formulations and various classical variational methods of approximation to determine the solution to the original equation (1.11.2). Toward this end, we first list the gradient and divergence theorems, which are useful in developing weak forms from weighted-integral statements.

1.11.2 Integral Identities

The integral statement in Eq. (1.11.3) requires approximation functions $\phi_j(x)$ [see Eq. (1.11.1)] that have continuous derivatives of the same order as those appearing in the operator equation (1.11.2). Thus, if A is a differential operator of $2m$th order, $\phi_j(x)$ must be functions that have $2m$th order nonzero derivatives. One may integrate by parts the expression $w_i A(u_N)$ such that m derivatives from u_N are transferred to w_i. This "weakening" of the differentiability required of u_N (and thus ϕ_j) results in an integral form known as a *weak form*. Such weak forms are used in the Ritz method and finite element models that use the Ritz method. Therefore, integration by parts and its generalization to multi-dimensions[9] are the topics of this subsection.

1.11.2.1 Integration by parts formulas

Let p, q, u, and w be functions of x that are sufficiently differentiable (i.e., as required in the integrand of the integral) with respect to the coordinate x. Then the following integration by parts formulas hold:

$$\begin{aligned}\int_a^b \frac{d}{dx}\left(p\frac{du}{dx}\right) dx &= \left[p\frac{du}{dx}\right]_{x=a}^{x=b} \\ &= p(x_b)\left.\frac{du}{dx}\right|_{x=b} - p(a)\left.\frac{du}{dx}\right|_{x=a},\end{aligned} \tag{1.11.4}$$

$$\int_a^b w\left[\frac{d}{dx}\left(p\frac{du}{dx}\right)\right] dx = -\int_a^b p\frac{dw}{dx}\frac{du}{dx}dx + \left[w(x)\cdot\left(p\frac{du}{dx}\right)\right]_{x=a}^{x=b} \tag{1.11.5}$$

$$\begin{aligned}\int_a^b w\left[\frac{d^2}{dx^2}\left(q\frac{d^2u}{dx^2}\right)\right] dx &= \int_a^b q\frac{d^2w}{dx^2}\frac{d^2u}{dx^2}\,dx + \left[w(x)\cdot\frac{d}{dx}\left(q\frac{d^2u}{dx^2}\right)\right]_{x=a}^{x=b} \\ &\quad + \left[-\frac{dw}{dx}\cdot\left(q\frac{d^2u}{dx^2}\right)\right]_{x=a}^{x=b}.\end{aligned} \tag{1.11.6}$$

To establish the result in Eq. (1.11.5), we begin with the identity (using the product rule of differentiation)

$$\frac{d}{dx}\left(w\cdot p\frac{du}{dx}\right) = \frac{dw}{dx}\cdot p\frac{du}{dx} + w\cdot\frac{d}{dx}\left(p\frac{du}{dx}\right). \tag{1.11.7}$$

Integrating both sides of the identity in Eq. (1.11.7) and rearranging the terms, we arrive at

$$\begin{aligned}\int_a^b w\frac{d}{dx}\left(p\frac{du}{dx}\right)dx &= \int_a^b \frac{d}{dx}\left(w\cdot p\frac{du}{dx}\right)dx - \int_a^b p\frac{dw}{dx}\frac{du}{dx}\,dx \\ &= \left[w(x)\cdot\left(p\frac{du}{dx}\right)\right]_{x=a}^{x=b} - \int_a^b p\frac{dw}{dx}\frac{du}{dx}\,dx.\end{aligned} \tag{1.11.8}$$

[9]In two- and three-dimensional cases, integration by parts amounts to using the gradient and divergence theorems.

To establish the relation in Eq. (1.11.6), we begin with the identity

$$\frac{d}{dx}\left[w \cdot \frac{d}{dx}\left(q\frac{d^2u}{dx^2}\right)\right] = \frac{dw}{dx}\frac{d}{dx}\left(q\frac{d^2u}{dx^2}\right) + w\frac{d^2}{dx^2}\left(q\frac{d^2u}{dx^2}\right). \tag{1.11.9}$$

Integrating both sides and rearranging the terms, we obtain

$$\begin{aligned}\int_a^b w\frac{d^2}{dx^2}\left(q\frac{d^2u}{dx^2}\right)dx &= \int_a^b \frac{d}{dx}\left[w \cdot \frac{d}{dx}\left(q\frac{d^2u}{dx^2}\right)\right]dx - \int_a^b \frac{dw}{dx}\frac{d}{dx}\left(q\frac{d^2u}{dx^2}\right)dx\\ &= \left[w \cdot \frac{d}{dx}\left(q\frac{d^2u}{dx^2}\right)\right]_a^b - \int_a^b \frac{dw}{dx}\frac{d}{dx}\left(q\frac{d^2u}{dx^2}\right)dx. \end{aligned} \tag{1.11.10}$$

Next, replace p with q, w with dw/dx, and u with du/dx in Eq. (1.11.8):

$$\int_a^b \frac{dw}{dx}\frac{d}{dx}\left(q\frac{d^2u}{dx^2}\right)dx = \left[\frac{dw}{dx} \cdot q\frac{d^2u}{dx^2}\right]_a^b - \int_a^b q\frac{d^2w}{dx^2}\frac{d^2u}{dx^2}\,dx. \tag{1.11.11}$$

Substituting Eq. (1.11.11) into the right-hand side of Eq. (1.11.10), we obtain

$$\begin{aligned}\int_a^b w\frac{d^2}{dx^2}\left(q\frac{d^2u}{dx^2}\right)dx &= \left[w \cdot \frac{d}{dx}\left(q\frac{d^2w}{dx^2}\right)\right]_a^b - \int_a^b \frac{dw}{dx}\frac{d}{dx}\left(q\frac{d^2u}{dx^2}\right)dx\\ &= \left[w \cdot \frac{d}{dx}\left(q\frac{d^2u}{dx^2}\right)\right]_a^b - \left[\frac{dw}{dx} \cdot q\frac{d^2u}{dx^2}\right]_a^b + \int_a^b q\frac{d^2w}{dx^2}\frac{d^2u}{dx^2}\,dx.\end{aligned}$$

1.11.2.2 Gradient and divergence theorems

The operator $\boldsymbol{\nabla}$ is a vector differential operator which is used to define the gradient, divergence, and curl operations that are introduced in the calculus of vectors (see Reddy [23]). The operator $\boldsymbol{\nabla}$ takes different forms in different coordinate systems. In rectangular Cartesian system $(x_1, x_2, x_3) = (x, y, z)$ [see Fig. 1.11.1(a)], it has the form

$$\boldsymbol{\nabla} = \hat{\mathbf{e}}_1\frac{\partial}{\partial x_1} + \hat{\mathbf{e}}_2\frac{\partial}{\partial x_2} + \hat{\mathbf{e}}_3\frac{\partial}{\partial x_3} = \hat{\mathbf{e}}_x\frac{\partial}{\partial x} + \hat{\mathbf{e}}_y\frac{\partial}{\partial y} + \hat{\mathbf{e}}_z\frac{\partial}{\partial z}, \tag{1.11.12}$$

whereas in a cylindrical coordinate system (r, θ, z) [see Fig. 1.11.1(b)], $\boldsymbol{\nabla}$ is expressed as

$$\boldsymbol{\nabla} = \hat{\mathbf{e}}_r\frac{\partial}{\partial r} + \hat{\mathbf{e}}_\theta\frac{1}{r}\frac{\partial}{\partial \theta} + \hat{\mathbf{e}}_z\frac{\partial}{\partial z}. \tag{1.11.13}$$

The gradient of a function (whether it is a scalar, vector, or tensor) F is denoted as $\boldsymbol{\nabla}F$, the divergence of a vector or a tensor $\mathbf{T}$ is denoted as $\boldsymbol{\nabla} \cdot \mathbf{T}$, and the curl of a vector or tensor $\mathbf{S}$ is denoted with $\boldsymbol{\nabla} \times \mathbf{S}$. The divergence operation with itself is known as the Laplace operator, denoted with $\nabla^2 \equiv \boldsymbol{\nabla} \cdot \boldsymbol{\nabla}$.

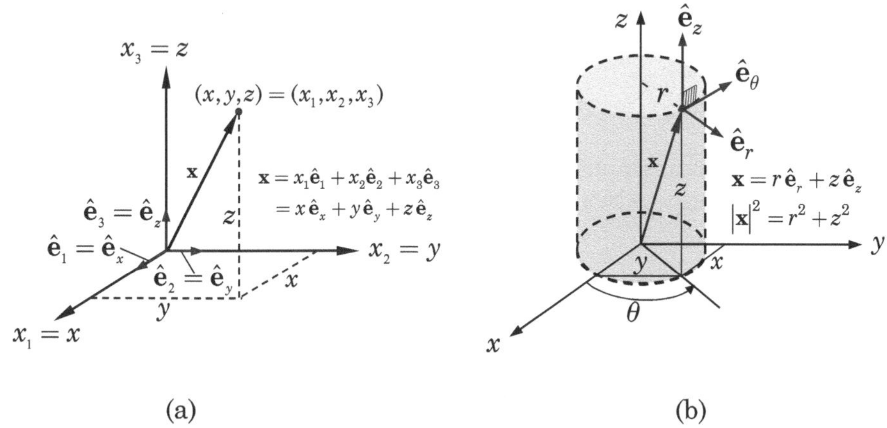

Fig. 1.11.1 (a) The rectangular Cartesian coordinate system (x, y, z). (b) The cylindrical coordinate system (r, θ, z).

The Laplace operator has the following form in the rectangular Cartesian and cylindrical coordinate systems:

$$\nabla^2(\cdot) = \left(\frac{\partial^2}{\partial x^2} + \frac{\partial^2}{\partial y^2} + \frac{\partial^2}{\partial z^2}\right)(\cdot), \tag{1.11.14}$$

$$\nabla^2(\cdot) = \frac{1}{r}\left[\frac{\partial}{\partial r}\left(r\frac{\partial}{\partial r}\right) + \frac{1}{r}\frac{\partial^2}{\partial r^2} + r\frac{\partial^2}{\partial z^2}\right](\cdot). \tag{1.11.15}$$

Let $F(\mathbf{x})$ and $\mathbf{G}(\mathbf{x})$ be continuous scalar and vector functions, respectively, in a three-dimensional domain Ω with closed boundary Γ, where $\mathbf{x}$ denotes the coordinates (i.e., position vector) of a generic point in Ω. The following theorems are well known in the literature, which can be specialized to two dimensions by simply restricting the index i on x_i to 1 and 2. *Gradient theorem*

$$\int_\Omega \boldsymbol{\nabla} F \, dv = \oint_\Gamma \hat{\mathbf{n}} \, F \, ds, \tag{1.11.16}$$

where $\hat{\mathbf{n}}$ denotes the unit vector normal to the surface Γ of the domain Ω, the circle on the boundary integral indicates that the integration is taken over the entire closed boundary Γ, dv is a volume element, and ds is a surface element. The rectangular Cartesian component form of the above equation is

$$\int_\Omega \frac{\partial F}{\partial x_i} \, dv = \oint_\Gamma n_i F \, ds, \tag{1.11.17}$$

where n_i denotes the ith rectangular component of the unit normal vector $\hat{\mathbf{n}}$, that is, $n_i = \cos(x_i, \hat{\mathbf{n}})$. As a special case, Eq. (1.11.17) yields the following statements in two dimensions:

$$\int_\Omega \frac{\partial F}{\partial x} \, dxdy = \oint_\Gamma n_x \, F \, ds, \quad \int_\Omega \frac{\partial F}{\partial y} \, dxdy = \oint_\Gamma n_y \, F \, ds, \tag{1.11.18}$$

where now ds denotes an element of the curve enclosing the two-dimensional domain. The direction cosines n_x and n_y of the unit vector $\hat{\mathbf{n}}$ can be written as

$$n_x = \cos(x, \hat{\mathbf{n}}) = \hat{\mathbf{e}}_x \cdot \hat{\mathbf{n}}, \quad n_y = \cos(y, \hat{\mathbf{n}}) = \hat{\mathbf{e}}_y \cdot \hat{\mathbf{n}}, \tag{1.11.19}$$

where $\cos(x, \hat{\mathbf{n}})$, for example, is the cosine of the angle between the positive x direction and the unit vector $\hat{\mathbf{n}}$.

Divergence theorem

$$\int_\Omega \boldsymbol{\nabla} \cdot \mathbf{G}\, dv = \oint_\Gamma \hat{\mathbf{n}} \cdot \mathbf{G}\, ds. \tag{1.11.20}$$

Here the dot between vectors denotes the scalar (or dot) product of the vectors. In component form ($\mathbf{G} = G_x\hat{\mathbf{e}}_x + G_y\hat{\mathbf{e}}_y + G_z\hat{\mathbf{e}}_y$ and $\hat{\mathbf{n}} = n_x\hat{\mathbf{e}}_x + n_y\hat{\mathbf{e}}_y + n_z\hat{\mathbf{e}}_z$), we have

$$\int_\Omega \left(\frac{\partial G_x}{\partial x} + \frac{\partial G_y}{\partial y} + \frac{\partial G_z}{\partial z} \right) dv = \oint_\Gamma (n_x G_x + n_y G_y + n_z G_z)\, ds. \tag{1.11.21}$$

For a two-dimensional case, Eq. (1.11.21) becomes

$$\int_\Omega \left(\frac{\partial G_x}{\partial x} + \frac{\partial G_y}{\partial y} \right) dxdy = \oint_\Gamma (n_x G_x + n_y G_y)\, ds. \tag{1.11.22}$$

The gradient and divergence theorems can be used to establish some useful integral identities. Let $w(\mathbf{x})$ and $u(\mathbf{x})$ be continuous functions with continuous derivatives defined in a three-dimensional domain Ω. Then, noting the identity $\boldsymbol{\nabla}(wu) = (\boldsymbol{\nabla} w)\, u + w\, (\boldsymbol{\nabla} u)]$, we can establish the following identity:

$$\int_\Omega w(\boldsymbol{\nabla} u)\, d\Omega + \int_\Omega (\boldsymbol{\nabla} w)\, u\, d\Omega = \oint_\Gamma \hat{\mathbf{n}}\, w\, u\, d\Gamma. \tag{1.11.23}$$

The component form of Eq. (1.11.23) in two dimensions is given by (by equating the components on two sides of the equality)

$$\int_\Omega w\, \frac{\partial u}{\partial x}\, dxdy + \int_\Omega \frac{\partial w}{\partial x}\, u\, dxdy = \oint_\Gamma n_x\, w\, u\, ds, \tag{1.11.24}$$

$$\int_\Omega w\, \frac{\partial u}{\partial y}\, dxdy + \int_\Omega \frac{\partial w}{\partial y}\, u\, dxdy = \oint_\Gamma n_y\, w\, u\, ds. \tag{1.11.25}$$

Similarly, using the identity $\boldsymbol{\nabla} \cdot (\boldsymbol{\nabla} u\, w) = w\, \nabla^2 u + \boldsymbol{\nabla} u \cdot \boldsymbol{\nabla} w$ we can write

$$\int_\Omega w(\nabla^2 u)\, dv + \int_\Omega \boldsymbol{\nabla} w \cdot \boldsymbol{\nabla} u\, dv = \oint_\Gamma w\, \frac{\partial u}{\partial n}\, ds, \tag{1.11.26}$$

where ∇^2 is the Laplace operator defined in Eq. (1.11.14), $\partial/\partial n$ denotes the normal derivative operator

$$\frac{\partial}{\partial n} \equiv \hat{\mathbf{n}} \cdot \boldsymbol{\nabla} = n_x\, \frac{\partial}{\partial x} + n_y\, \frac{\partial}{\partial y} + n_z\, \frac{\partial}{\partial z}. \tag{1.11.27}$$

The component form of Eq. (1.11.26) is given by

$$\int_{\Omega} w\Big(\frac{\partial^2 u}{\partial x^2}+\frac{\partial^2 u}{\partial y^2}\Big)dxdy+\int_{\Omega}\Big(\frac{\partial u}{\partial x}\frac{\partial w}{\partial x}+\frac{\partial u}{\partial y}\frac{\partial w}{\partial y}\Big)dxdy$$
$$=\oint_{\Gamma} w\Big(n_x\frac{\partial u}{\partial x}+n_y\frac{\partial u}{\partial y}\Big)ds. \qquad (1.11.28)$$

1.11.3 Integral Formulations and Methods of Approximation

As discussed earlier, the motivation for the use of weighted-integral statements of differential equations comes from the fact that we wish to have a means to determine the unknown parameters c_j in the approximate solution $u_N = \sum_{j=1}^{N} c_j\phi_j(x)$ [see Eq. (1.11.1)]. The variational methods of approximation, e.g., the Ritz, Galerkin, least-squares, collocation, subdomain, or, in general, weighted-residual methods to be discussed in this section are based on weighted-integral statements of the governing equations.

In this section, our primary objective is to construct weak forms of differential equations from the weighted-integral statements. As a part of the development, the notion of duality and identification of variables of a problem into primary type and secondary type as well as the form of the boundary conditions associated with the equation(s) are established. In solid mechanics, the weak forms are equivalent to the principle of virtual displacements or the principle of minimum total potential energy (see Reddy [17]).

1.11.3.1 Residual of approximation

We begin with a typical second-order ordinary differential equation of the form

$$-\frac{d}{dx}\Big[a(x)\frac{du}{dx}\Big]+cu=f(x) \quad \text{for} \quad 0<x<L, \qquad (1.11.29)$$

which is subject to the boundary conditions of the type

$$u(0)=u_0, \quad \Big[a\frac{du}{dx}+\beta(u-u_\infty)\Big]_{x=L}=Q_L. \qquad (1.11.30)$$

Here $a(x)$, $c(x)$, and $f(x)$ are known functions of the coordinate x, u_0, u_∞, β, and Q_L are known values, and L is the size of the one-dimensional domain. The boundary conditions in Eq. (1.11.30) are nonhomogeneous; the homogeneous form of the boundary condition $u(0) = u_0$ is $u(0) = 0$, and the homogeneous form of the boundary condition $[a(du/dx)+\beta(u-u_\infty)]_{x=L} = Q_L$ is $[a(du/dx)+\beta(u-u_\infty)]_{x=L} = 0$.

Equations of the type in Eq. (1.11.29) arise, for example, in the study of one-dimensional heat flow in a rod with surface convection [see Eq. (1.5.14)]. Another example is provided by axial deformation of a bar. The operator A in Eq. (1.11.2) in this case can be identified as

$$A(\cdot)=-\frac{d}{dx}\Big[a(x)\frac{d(\cdot)}{dx}\Big]+c(\cdot). \qquad (1.11.31)$$

Returning to the weighted-residual integral statement in Eq. (1.11.3), the expression

$$0 \neq -\frac{d}{dx}\left[a(x)\frac{du_n}{dx}\right] + cu_n - f(x) \equiv \mathbb{R}(x, c_1, c_2, \ldots, c_n) \quad \text{for} \quad 0 < x < L \tag{1.11.32}$$

is called the *residual* of approximation in the differential equation. It is a function of x and $c_1, c_2, \ldots$, and c_n. The objective of any approximate method is to make the residual to be zero in some acceptable sense. Here we discuss several alternatives.

1.11.3.2 Collocation method

One possible acceptable sense in which $\mathbb{R}$ can be made zero is to require $\mathbb{R}$ to vanish at N selected points (so that there are N equations among the constants, c_i) of the domain

$$\mathbb{R}(x_i, c_1, c_2, \ldots, c_N) = 0 \quad \text{for} \quad i = 1, 2, \ldots, N. \tag{1.11.33}$$

This particular method of making $\mathbb{R}$ equal to zero at selective N points is known as the *collocation method* and the points x_i are called the *collocation points.* We note that the resulting solution u_N makes R to be zero at the collocation points, but $\mathbb{R}$ is nonzero at other points.

1.11.3.3 Subdomain method

Another way to make $\mathbb{R}$ zero is to make the integral of the residual to be zero in N different subintervals of the total domain. That is, we divide the domain $(0, L)$ into N nonoverlapping but connected subintervals (x_a^i, x_b^i):

$$\int_{x_a^i}^{x_b^i} \mathbb{R}(x, c_1, c_2, \ldots, c_N)\, dx = 0 \quad \text{for} \quad i = 1, 2, \ldots, N. \tag{1.11.34}$$

Here x_a^i and x_b^i are the x-coordinates of the left and right ends of the domain, respectively. In this case, the residual is made zero, in the integral sense, in each interval. However, $\mathbb{R}$ is not zero point-wise.

1.11.3.4 Least-squares method

Yet another way to make $\mathbb{R}$ zero is to minimize the integral of the square of the residual (the squaring of $\mathbb{R}$ is to make it positive; otherwise, there is a possibility of positive and negative errors canceling each other over the domain) with respect to c_i:

$$\text{minimize } I \equiv \int_0^L \mathbb{R}^2\, dx \quad \text{or} \quad \frac{\partial}{\partial c_i}\int_0^L \mathbb{R}^2\, dx = 2\int_0^L \frac{\partial \mathbb{R}}{\partial c_i}\mathbb{R}\, dx = 0 \tag{1.11.35}$$

for $i = 1, 2, \ldots, n$. The method based on Eq. (1.11.35) is called the *least-squares method.* When the operator A is linear, we have $\partial\mathbb{R}/\partial c_i = A(\phi_i)$.

1.11.3.5 Weighted-residual methods

The least-squares method in Eq. (1.11.35) gives another idea, namely, weighting the residual with a linearly independent set of functions and setting it to zero. That is, determine the c_j by requiring $\mathbb{R}$ to vanish in a "weighted-residual" sense (i.e., making the residual orthogonal to a set of weight functions):

$$\int_0^L w_i(x)\,\mathbb{R}(x, c_1, c_2, \cdots, c_n)\, dx = 0, \quad i = 1, 2, \ldots, n, \tag{1.11.36}$$

where $\{w_i(x)\}$ are a set of linearly independent functions, called *weight functions*, which, in general, can be different from the set of approximation functions $\{\phi_i(x)\}$. This method is known as the *weighted-residual method*. Indeed, the weighted-residual statement in Eq. (1.11.36) includes, as special cases, the collocation method of Eq. (1.11.33) as well as the least-squares method of Eq. (1.11.35). When $w_i = \delta(x - x_i)$, we obtain the result in Eq. (1.11.33), and when we set $w_i = (\partial\mathbb{R}/\partial c_i)$ we obtain the result in Eq. (1.11.35). Various known special cases of Eq. (1.11.36) are listed below.

$$\begin{array}{ll} \textit{Petrov–Galerkin method:} & w_i = \psi_i \neq \phi_i \\ \textit{Galerkin's method:} & w_i = \phi_i \\ \textit{Least-squares method:} & w_i = A(\phi_i), \quad A = -\frac{d}{dx}\left[a(x)\frac{d}{dx}\right] + c \\ \textit{Collocation method:} & w_i = \delta(x - x_i), \end{array} \tag{1.11.37}$$

where x_i denotes the ith collocation point of the domain of the problem, and $\delta(\cdot)$ is the Dirac delta "function," which is defined by

$$\delta(x - x_0) = 0 \quad \text{when} \quad x \neq x_0, \tag{1.11.38a}$$

and

$$\int_{-\infty}^{\infty} f(x)\delta(x - x_0)\, dx = f(x_0). \tag{1.11.38b}$$

Due to the different choices of the weight function w_i, the system of algebraic equations will have different characteristics in different methods. For linear differential equations of any order, only the least-squares method (among the weighted-residual methods) yields a symmetric system of matrix equations.

1.11.3.6 Ritz method

The Ritz method is originally presented as one in which a quadratic functional, which is equivalent to the governing equation and certain boundary conditions, is minimized with respect to the parameters $(c_1, c_2, \ldots, c_N)$ of the approximation. As discussed by Reddy [17], the method is applicable to any integral statement that is equivalent to the governing differential equations and the natural boundary conditions of the problem. Such integral statement is known as a *weak form*, which is the topic of the next section. We shall return to the Ritz method in Section 1.11.5. The method of deriving finite element equations is often based on the Ritz method[10].

[10] In most finite element models, the idea of the Ritz method is used, although it is never acknowledged as such; instead, they have been termed incorrectly as Galerkin finite element models.

1.11.4 Weak (Integral) Forms

The main idea behind the weak form development is to "weaken" the differentiability requirement on u_N (equivalently, on ϕ_i). There are three steps in the development of the weak form of any differential equation. These steps are illustrated by means of the model differential equation, Eq. (1.11.29), and boundary conditions in Eq. (1.11.30). Weak forms can be developed for any differential equation of order 2 and higher.

Step 1: *Weighted-integral statement.* This step is the same as the weighted-residual statement of a differential equation. Move all terms of the differential equation to one side (so that it reads $\ldots = 0$), multiply the entire equation with an arbitrary function $w_i(x)$, and integrate over the domain $\Omega = (0, L)$ of the problem:

$$0 = \int_0^L w_i \left[-\frac{d}{dx}\left(a\frac{du_N}{dx} \right) + cu_N - f \right] dx. \tag{1.11.39}$$

Recall that the expression in the square brackets is not identically zero since u is replaced by its approximation, u_n.

Mathematically, Eq. (1.11.39) is a statement that the error in the differential equation (due to the approximation of the solution) is zero in the weighted-integral sense. The integral statement in Eq. (1.11.39) yields one algebraic equation among the parameters $c_1, c_2, \cdots, c_N$ for each choice of w_i. By choosing N linearly independent functions for w_i, we obtain N equations for $c_1, c_2, \ldots, c_N$ from Eq. (1.11.39).

Note that the weighted-integral statement of any – first-order or higher-order – differential equation can be readily written. The weighted-integral statement is equivalent only to the differential equation, and it does not include any boundary conditions. The set of weight functions $\{w_i\}$ in Eq. (1.11.39) can be any linearly independent set of integrable functions. Equation (1.11.39) is the basis of all weighted-residual methods listed in Eq. (1.11.3.5).

Step 2: *Integration by parts to distribute differentiation equally between the dependent variables and the weight function and the identification of primary and secondary variables.* While the weighted-integral statement in Eq. (1.11.39) allows us to obtain the necessary number (N) of algebraic relations among c_j for N different choices of the weight function w_i, it requires that the approximation functions $\{\phi_i\}$ be such that u_N [see Eq. (1.11.32)] is differentiable as many times as called for in the original differential equation, Eq. (1.11.29), and it satisfies the specified boundary conditions.

If we plan to use the approximation functions ϕ_i for w_i (as in the Galerkin method), it makes sense to move half of the derivatives from u_N to w_i in the weighted-integral statement in Eq. (1.11.39) so that both w_i and u_N are differentiated equally (this comment applies for even order differential equations), and we have lesser (or weaker) continuity requirements on ϕ_i, $i = 1, 2, \ldots, N$. The resulting integral form is known, for obvious reasons, as the *weak form.* Of course, weakening the differentiability of u_N (and hence ϕ_i) is purely a mathematical (and perhaps computational) consideration. As will be seen shortly, the weak formulation has two desirable characteristics. First, it requires weaker, as already indicated, continuity of the dependent variable, and for even order differential equations (as is the case with problems studied in this book), it always results in a symmetric coefficient matrix. Second, the boundary conditions on the derivative of u_N of the problem are included in the weak form, and therefore, the approximate solution u_N is required to satisfy boundary conditions only on u_N. These two features of a weak form play an important role in the Ritz method.

Returning to the integral statement in Eq. (1.11.39), we integrate the first term of the expression by parts to obtain

$$\begin{aligned} 0 &= \int_0^L \left\{ w_i \left[-\frac{d}{dx}\left(a\frac{du_N}{dx} \right) \right] + cw_i u_N - w_i\, f \right\} dx \\ &= \int_0^L \left(a\frac{dw_i}{dx}\frac{du_N}{dx} + cw_i u_N - w_i\, f \right) dx - \left[w_i \cdot a\frac{du_N}{dx} \right]_0^L, \end{aligned} \tag{1.11.40}$$

where the integration by parts formula [see Eq. (1.11.5)] is used with $p = a$ and $w = w_i$ on the first term to arrive at the second line of Eq. (1.11.40). Note that now the weight function w_i is required to be differentiable at least once. But this is not a concern because we plan to use $\{\phi_i\}$ for $\{w_i\}$ (in the Galerkin and Ritz methods).

An important part of Step 2 is to identify the duality pair (cause and effect). We shall identify the variable appearing in the differential equation, namely u, as the *primary variable.* After trading differentiation between the weight function w_i and the variable u_N of the equation, examine the boundary expression(s) resulting from the integration by parts. The boundary expression(s) will involve both the weight function w_i and derivatives of the dependent variable u_N. Coefficients of the weight function (and possibly its derivatives for higher-order equations) in the boundary expression(s) are termed the *secondary variable(s).*

Remark. A word of caution regarding trading the differentiation between the weight function and the dependent variable is in order. The trading of differentiation is also dictated, in addition to the weakening of the continuity requirements on ϕ_i, by the need to include physically meaningful boundary terms into the weak form, regardless of the effect on the continuity requirements. Therefore, *trading of differentiation from the dependent variable to the weight function should not be performed if it results in boundary terms that are not physically meaningful.*

The specification of the primary variable u constitutes the *essential boundary condition* (EBC), whereas the specification of the secondary variable constitutes the *natural boundary condition* (NBC). For example, for the problem at hand, the coefficient of the weight function w_i in the boundary expression is $a(du_N/dx)$; hence, $a(du_N/dx)$ is the secondary variable of the formulation, and its specification constitutes the NBC.

The secondary variables always have physical meaning and are often quantities of interest; if they do not have physical meaning to be able to specify them, one should not have carried out integration by parts (see the Remark above). In the case of heat transfer problems, temperature is the primary variable and the secondary variable represents heat, $kA\,(dT/dx)$. In the axial deformation of bars, the displacement u is the primary variable and $a(du/dx)$, which represents the axial force, is the secondary variable. We shall denote the secondary variable by

$$Q = n_x \left(a \frac{du_N}{dx} \right) \to Q(0) = -a \frac{du_N}{dx} \Big|_{x=0}, \quad Q(L) = a \frac{du_N}{dx} \Big|_{x=L}, \tag{1.11.41}$$

where n_x denotes the direction cosine (i.e., cosine of the angle between the positive x-axis and the normal to the boundary). For one-dimensional problems, the normal at the boundary points is always along the length of the domain. Thus, we have $n_x = -1$ at the left end and $n_x = 1$ at the right end of the domain.

It should be noted that the number and form of the primary and secondary variables depend on the order of the differential equation. The number of primary and secondary variables is always the same, and with each primary variable there is an associated secondary variable; that is, they always appear in pairs (e.g., displacement and force, temperature and heat, and so on). Only one member of the pair, either the primary variable or the secondary variable, may be specified at a point. Thus, a given problem can have its specified boundary conditions in one of three categories: (1) all specified boundary conditions are of the essential type; (2) some of the specified boundary conditions are of the essential type and the remaining are of the natural type; or (3) all specified boundary conditions are of the natural type. For a single second-order equation, as in the present case, there is one primary variable u and one secondary variable Q, and only one element of the pair (u, Q) can be specified at a point. For a fourth-order equation, such as that for the classical (i.e., Euler–Bernoulli) theory of beams, there are two of each kind (i.e., two primary variables and two secondary variables), as will be seen later. In general, a $2m$th-order differential equation requires m integration-by-parts to transfer m derivatives from u_N to w_i and, therefore, there will be m boundary terms, involving m primary variables and m secondary variables (with derivatives up to the order $m-1$); that is, there are m pairs of primary and secondary variables.

Returning to Eq. (1.11.40), we rewrite it using the notation of Eq. (1.11.41):

$$\begin{aligned} 0 &= \int_0^L \left(a \frac{dw_i}{dx} \frac{du_N}{dx} + c w_i u_N - w_i f \right) dx - \left[w_i \, a \frac{du_N}{dx} \right]_0^L \\ &= \int_0^L \left(a \frac{dw_i}{dx} \frac{du_N}{dx} + c w_i u_N - w_i f \right) dx - w_i(0) Q(0) - w_i(L) Q(L). \end{aligned} \tag{1.11.42}$$

Equation (1.11.42) is called the *weak form* of Eq. (1.11.29). The word "weak" refers to the reduced differentiability required of u_N in Eq. (1.11.40) when compared to the u_N in the weighted-integral statement in Eq. (1.11.39) or the differential equation in Eq. (1.11.29), where u_N is required to be twice-differentiable but only once-differentiable in Eq. (1.11.40).

Step 3: *Replace the expression for the secondary variables by their specified values and finalize the weak form.* The third and last step of the weak form development is to impose the actual boundary conditions of the problem under consideration. It is here that we require the weight function w_i to vanish at boundary points where the essential boundary conditions are specified; that is, w_i is required to satisfy the *homogeneous form* of the specified essential boundary conditions of the problem. That is, the weight function w_i is treated as a *virtual* change (or *variation*) of the primary variable u, $w_i \sim \delta u$. When a primary variable is specified at a point, the virtual change there must be zero. For the problem at hand, the boundary conditions are given by Eq. (1.11.30). By the rules of classification of boundary conditions, $u = u_0$ is the essential boundary condition and $(adu/dx)|_{x=L} = Q_L$ is the natural boundary condition. Thus, the weight function w_i, for $i = 1, 2, \cdots, N$, is required to satisfy

$$w_i(0) = 0, \quad \text{because} \quad u(0) = u_0. \tag{1.11.43}$$

Since $w_i(0) = 0$ and

$$Q(L) = \left(a\frac{du_N}{dx}n_x\right)\bigg|_{x=L} = \left(a\frac{du_N}{dx}\right)\bigg|_{x=L} = Q_L - \beta[u_N(L) - u_\infty], \tag{1.11.44}$$

Eq.(1.11.42) reduces to the expression

$$\begin{aligned} 0 = &\int_0^L \left(a\frac{dw_i}{dx}\frac{du_N}{dx} + cw_iu_N - w_i f\right)dx \\ &+ \beta w_i(L)u_N(L) - w_i(L)Q_L - \beta w_i(L)u_\infty, \end{aligned} \tag{1.11.45}$$

which is the weak form equivalent to the differential equation in Eq. (1.11.29) and the natural boundary condition in Eq. (1.11.30). This completes the steps involved in the development of the weak form of a differential equation.

It should be recalled that the primary purpose of developing a weighted-integral statement or the weak form of a differential equation is to obtain as many algebraic equations as there are unknown coefficients c_i in the approximation of the dependent variable u of the equation. For different choices of the weight function w_i, different sets of algebraic equations are obtained. However, because of the restrictions placed on the weight function w_i in Step 3 of the weak form development, w_i belongs to the same vector space of functions as the approximation functions (i.e., $w_i \in \{\phi_i\}$). The resulting discrete model is known as the *Ritz model*, which is discussed in Section 1.11.5. Here we consider couple of examples to illustrate the development of weak forms.

Example 1.11.1

The equations governing the pure bending of a beam using the Euler–Bernoulli beam theory, under the Euler-Bernoulli hypothesis that plane sections perpendicular to the axis of the beam before deformation remain (a) plane, (b) inextensible, and (c) perpendicular to the bent axis after deformation (see Chapter 7 and Reddy [17] for details), are given by

$$-\frac{dV}{dx} + kw - q = 0, \quad \frac{dM}{dx} - V = 0, \quad 0 < x < L, \tag{1.11.46a}$$

where $w(x)$ denotes the transverse deflection of the beam (along the z-axis), $V(x)$ is the shear force, $M(x)$ is the bending moment, $k(x)$ is the foundation modulus (if any), $q(x)$ is the distributed transverse load, and L is the length of the beam, as shown in Fig. 1.11.2. The two equations in Eq. (1.11.46a) can be combined into a single equation to yield

$$-\frac{d^2M}{dx^2} + kw - q = 0, \quad 0 < x < L. \tag{1.11.46b}$$

Using the kinematics and Hooke's law, the bending moment and shear force in an isotropic beam can be related to the deflection w by

$$M(x) = -EI\frac{d^2w}{dx^2}, \quad V(x) = \frac{dM}{dx} = -\frac{d}{dx}\left(EI\frac{d^2w}{dx^2}\right), \tag{1.11.47}$$

where $E(x)I(x) > 0$ is the flexural rigidity of the beam, E being the modulus of elasticity and I the second moment of area about the y-axis (into the plane of the page). Replacing M in terms of w using Eq. $(1.11.47)_1$, we obtain the equation governing the deflection $w(x)$ as

$$\frac{d^2}{dx^2}\left(EI\frac{d^2w}{dx^2}\right) + kw - q = 0, \quad 0 < x < L. \tag{1.11.48}$$

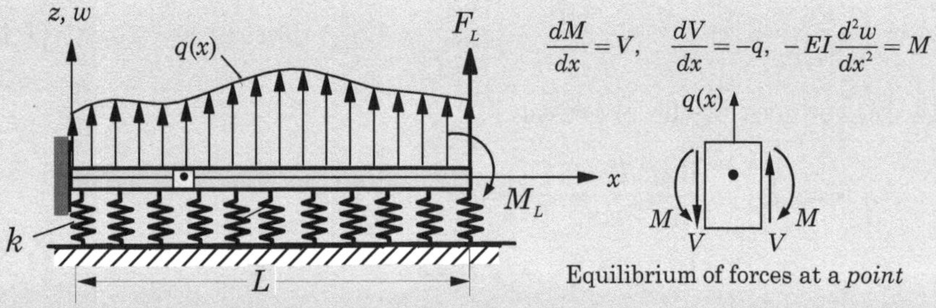

Fig. 1.11.2 A beam on elastic foundation and fixed at the left end, $x = 0$, subjected to distributed transverse load $q(x)$, and point force F_L and moment M_L at the right end, $x = L$.

If the beam is clamped at the left end and subjected to a transverse point load F_0 and bending moment M_0 at $x = L$, as indicated in Fig. 1.11.2, develop the weak form of the problem.

Solution: Since the equation contains a fourth-order derivative, the weak-form development requires transferring two derivatives from w to the weight function, $v(x)$, which must be twice differentiable and satisfy the homogeneous form of essential boundary conditions.

Step 1: Multiplying Eq. (1.11.48) by v, and integrating the first term by parts twice with respect to x, we obtain [see Eq. (1.11.6)]

$$0 = \int_0^L v\left[\frac{d^2}{dx^2}\left(EI\frac{d^2w}{dx^2}\right) + kw - q\right] dx. \tag{1.11.49a}$$

Step 2: Trade differentiation between the weight function v and the dependent variable w:

$$0 = \int_0^L \left(EI\,\frac{d^2v}{dx^2}\,\frac{d^2w}{dx^2} + kvw - vq\right) dx$$
$$+ \left[v\,\frac{d}{dx}\left(EI\frac{d^2w}{dx^2}\right) - \frac{dv}{dx}\,EI\frac{d^2w}{dx^2}\right]_0^L. \tag{1.11.49b}$$

From Eq. (1.11.49b), it follows (dictated by the form in which the weight function v appears in the boundary terms) that w and dw/dx are the primary variables and V and M are the secondary variables . Thus, the duality pairs associated with the weak form are: (w, V) and (θ, M). Specification of w and/or $\theta = -dw/dx$ constitutes an essential (geometric) boundary condition, and the specification of

$$V(x) \equiv -\frac{d}{dx}\left(EI\frac{d^2w}{dx^2}\right) \quad \text{(shear force)} \tag{1.11.50a}$$

and

$$M(x) \equiv -EI\frac{d^2w}{dx^2} \quad \text{(bending moment)} \tag{1.11.50b}$$

constitutes the natural boundary conditions for the Euler–Bernoulli beam theory.

The boundary conditions can be expressed in terms of the deflection w as

$$\begin{aligned} &w(0) = 0, \quad \theta(0) \equiv \left(-\frac{dw}{dx}\right)\bigg|_{x=0} = 0, \\ &\left(-EI\frac{d^2w}{dx^2}\right)\bigg|_{x=L} = M_L, \quad \left[-\frac{d}{dx}\left(EI\frac{d^2w}{dx^2}\right)\right]_{x=L} = F_L, \end{aligned} \tag{1.11.51}$$

where M_L is the applied bending moment and F_L is the applied transverse point load at $x = L$. Since w and $\theta = -dw/dx$ (both are primary variables) are specified at $x = 0$, we require the weight function v and its derivative dv/dx to be zero there (i.e., at $x = 0$):

$$v(0) = \left(\frac{dv}{dx}\right)\bigg|_{x=0} = 0. \tag{1.11.52}$$

The remaining two boundary conditions in Eq. (1.11.51) are natural boundary conditions, which place no restrictions on v and its derivatives.

Step 3: Thus, Eq. (1.11.49b) becomes

$$0 = \int_0^L \left(EI\frac{d^2v}{dx^2}\frac{d^2w}{dx^2} + kvw - vq\right) dx - v(L)F_L + \left(\frac{dv}{dx}\right)\bigg|_{x=L} M_L, \tag{1.11.53}$$

which is the weak form equivalent to the fourth-order differential equation (1.11.48) and the natural boundary conditions in Eq. $(1.11.51)_2$ for the Euler–Bernoulli beam theory.

The variational problem in this case can be written as

$$B(v, w) = \ell(v), \tag{1.11.54a}$$

where

$$\begin{aligned} B(v, w) &= \int_0^L \left(EI\frac{d^2v}{dx^2}\frac{d^2w}{dx^2} + kvw\right) dx, \\ \ell(v) &= \int_0^L v\,q\,dx + v(L)F_L - \left(\frac{dv}{dx}\right)\bigg|_{x=L} M_L. \end{aligned} \tag{1.11.54b}$$

Note that for the fourth-order equation, the essential boundary conditions involve not only the dependent variable but also its first derivative. As pointed out earlier, at any boundary point, only one of the two boundary conditions (essential or natural) can be specified. For example, since the transverse deflection w is specified (to be zero or not) at $x = 0$, then one cannot specify the shear force $V(0)$ (and vice versa). Similarly, since the slope $\theta = -dw/dx$ is specified at $x = 0$, the bending moment $M(0)$ cannot be specified. Both $V(0)$ and $M(0)$ are reactions to be determined. Similar comments apply to the end $x = L$.

The next example is concerned with the weak-form development of a pair of coupled second-order differential equations in one dimension. The equations arise in connection with the modeling of a straight beam using the Timoshenko beam theory.

Example 1.11.2

Consider the following pair of coupled differential equations governing the equilibrium of straight beams according to the Timoshenko beam theory, in which the normality condition (c) of the Euler–Bernoulli beam hypothesis (see **Example 1.11.1**) is removed (i.e., plane cross-sections normal to the longitudinal axis do not remain perpendicular after deformation; consequently, the transverse shear strain is accounted). The governing equations of the Timoshenko beam theory are (see Section 7.2.2 for details)

$$-\frac{d}{dx}\left[S\left(\frac{dw}{dx}+\phi_x\right)\right]+kw=q, \tag{1.11.55a}$$

$$-\frac{d}{dx}\left(D\frac{d\phi_x}{dx}\right)+S\left(\frac{dw}{dx}+\phi_x\right)=0. \tag{1.11.55b}$$

where w is the transverse deflection, ϕ_x is the rotation of a transverse normal line about the y-axis, S is the shear stiffness $S = K_sGA$ (K_s being the shear correction coefficient), G is the shear modulus, A is the area of cross-section, $D = EI$ is the bending stiffness, k is the foundation modulus, and q is the distributed transverse load. Develop the weak form of the pair of equations in (1.11.55a) and (1.11.55b), using the three-step procedure and identify the duality pairs and form of the boundary conditions.

Solution: We use the three-step procedure to each of the two differential equations to develop the weak forms.

Step 1a: Multiply the first equation with weight function v_1 and integrate over the length of the beam:

$$0=\int_0^L v_1\left\{-\frac{d}{dx}\left[S\left(\frac{dw}{dx}+\phi_x\right)\right]+kw-q\right\}dx. \tag{1.11.56a}$$

Step 2a: Trade the differentiation between v_1 and w and arrive at

$$\begin{aligned}0&=\int_0^L\left[\frac{dv_1}{dx}S\left(\frac{dw}{dx}+\phi_x\right)+kv_1w-v_1q\right]dx-\left[v_1\,S\left(\frac{dw}{dx}+\phi_x\right)\right]_0^L\\&=\int_0^L\left[\frac{dv_1}{dx}S\left(\frac{dw}{dx}+\phi_x\right)+kv_1w-v_1q\right]dx-v_1(L)\left[S\left(\frac{dw}{dx}+\phi_x\right)\right]_{x=L}\\&\quad+v_1(0)\left[S\left(\frac{dw}{dx}+\phi_x\right)\right]_{x=0}\end{aligned} \tag{1.11.56b}$$

Step 1b: Multiply the second equation with weight function v_2 and integrate over the length of the beam:

$$0=\int_0^L v_2\left[-\frac{d}{dx}\left(D\frac{d\phi_x}{dx}\right)+S\left(\frac{dw}{dx}+\phi_x\right)\right]dx \tag{1.11.57a}$$

Step 2b: Trade the differentiation between v_2 and ϕ_x:

$$\begin{aligned}0&=\int_0^L\left[D\frac{dv_2}{dx}\frac{d\phi_x}{dx}+v_2\,S\left(\frac{dw}{dx}+\phi_x\right)\right]dx-\left[v_2\,D\frac{d\phi_x}{dx}\right]_0^L\\&=\int_0^L\left[D\frac{dv_2}{dx}\frac{d\phi_x}{dx}+v_2\,S\left(\frac{dw}{dx}+\phi_x\right)\right]dx-v_2(L)\left[D\frac{d\phi_x}{dx}\right]_{x=L}+v_2(0)\left[D\frac{d\phi_x}{dx}\right]_{x=0}\end{aligned} \tag{1.11.57b}$$

Note that integration by parts was used such that the expression $w_{,x}+\phi_x$ is preserved, because the secondary variable appearing in the boundary term is indeed a physically meaningful quantity, namely, the shear force. Such considerations can only be used by knowing the physics of the problem at hand. Also, note that the pair of weight functions (v_1, v_2) satisfy the homogeneous form of specified essential boundary conditions associated with the pair (w, ϕ_x) (with the correspondence $v_1 \sim w$ and $v_2 \sim \phi_x$).

Step 3: An examination of the boundary terms shows that $w \sim v_1$ and $\phi_x \sim v_2$ are the primary variables, and the secondary variables are given by

$$V(x) \equiv S\left(\frac{dw}{dx} + \phi_x\right) \quad \text{(shear force)}, \qquad M(x) \equiv D\frac{d\phi_x}{dx} \quad \text{(bending moment)}. \tag{1.11.58}$$

Thus, a beam problem modeled using the Timoshenko beam theory admits boundary conditions on the duality pairs: (w, V) and (ϕ_x, M).

To finalize the weak forms, we must take care of the boundary terms by considering a specific beam problem. Using the beam of Fig. 1.11.2, we see that $v_1(0) = 0$ (because w is specified at $x = 0$) and $v_2(0) = 0$ (because ϕ_x is specified at $x = 0$), and

$$V(L) = \left[S\left(\frac{dw}{dx} + \phi_x\right)\right]_{x=L} = F_L, \quad M(L) = \left[D\frac{d\phi_x}{dx}\right]_{x=L} = M_L.$$

Consequently, the weak forms in Eqs. (1.11.56b) and (1.11.57b) become

$$0 = \int_0^L \left[S\,\frac{dv_1}{dx}\left(\frac{dw}{dx} + \phi_x\right) + kv_1 w - v_1 q\right] dx - v_1(L)\,F_L, \tag{1.11.59a}$$

$$0 = \int_0^L \left[D\,\frac{dv_2}{dx}\,\frac{d\phi_x}{dx} + v_2\,S\left(\frac{dw}{dx} + \phi_x\right)\right] dx - v_2(L)\,M_L. \tag{1.11.59b}$$

To identify the variational problem of the Timoshenko beam problem, we must combine the two weak forms into a single expression

$$\begin{aligned} 0 = &\int_0^L \left[S\left(\frac{dv_1}{dx} + v_2\right)\left(\frac{dw}{dx} + \phi_x\right) + D\,\frac{dv_2}{dx}\,\frac{d\phi_x}{dx} + kv_1 w - v_1 q\right] dx \\ &- v_1(L)\,F_L - v_2(L)\,M_L. \end{aligned} \tag{1.11.60}$$

Then the variational problem is to seek the pair (w, ϕ_x) such that

$$B((v_1, v_2), (w, \phi_x)) = \ell((v_1, v_2)) \tag{1.11.61}$$

holds for all (v_1, v_2). Here, the bilinear and linear forms of the problem are given by

$$\begin{aligned} B((v_1, v_2), (w, \phi_x)) &= \int_0^L \left[S\left(\frac{dv_1}{dx} + v_2\right)\left(\frac{dw}{dx} + \phi_x\right) + D\frac{dv_2}{dx}\frac{d\phi_x}{dx} + kv_1 w\right] dx, \\ \ell((v_1, v_2)) &= \int_0^L v_1\,q\;dx + v_1(L)\,F_0 + v_2(L)\,M_0. \end{aligned} \tag{1.11.62}$$

Clearly, $B((v_1, v_2), (w, \phi_x))$ is symmetric in its arguments (i.e., interchange of v_1 with w and v_2 with ϕ_x yields the same expression).

The last example of this section is concerned with a second-order differential equation in two dimensions. The equation arises in a number of fields, including heat transfer, transverse deflections of a membrane, and torsion of cylindrical members [see, e.g., Reddy [8, 17]].

Example 1.11.3

Consider the problem of determining the solution $u(x, y)$ of the partial differential equation,

$$-\frac{\partial}{\partial x}\left(a_{xx}\frac{\partial u}{\partial x}\right) - \frac{\partial}{\partial y}\left(a_{yy}\frac{\partial u}{\partial y}\right) = f(x, y) \tag{1.11.63}$$

in a closed, bounded, two-dimensional domain Ω with boundary Γ, as shown in Fig. 1.11.3. Here a_{xx}, a_{yy}, and f are known functions of position (x, y) in Ω. The function $u(x, y)$ is required to satisfy, in addition to the differential equation (1.11.63), the following boundary conditions on the boundary Γ of Ω:

$$u = \hat{u} \;\text{ on } \Gamma_u, \quad a_{xx}\frac{\partial u}{\partial x}n_x + a_{yy}\frac{\partial u}{\partial y}n_y = \hat{q} \;\text{ on } \Gamma_q. \tag{1.11.64}$$

where the portions Γ_u and Γ_q such that $\Gamma = \Gamma_u \cup \Gamma_q$ and $\Gamma = \Gamma_u \cap \Gamma_q = \emptyset$, as shown in Fig. 1.11.3. Develop the weak form and the associated variational problem.

Solution: The three-step procedure applied to Eq. (1.11.63) yields:

Step 1:
$$0 = \int_\Omega w\left[-\frac{\partial}{\partial x}\left(a_{xx}\frac{\partial u}{\partial x}\right) - \frac{\partial}{\partial y}\left(a_{yy}\frac{\partial u}{\partial y}\right) - f\right] dxdy. \tag{1.11.65a}$$

Step 2:
$$\begin{aligned} 0 = &\int_\Omega \left(a_{xx}\frac{\partial w}{\partial x}\frac{\partial u}{\partial x} + a_{yy}\frac{\partial w}{\partial y}\frac{\partial u}{\partial y} - wf\right) dxdy \\ &- \oint_\Gamma w\left(a_{xx}\frac{\partial u}{\partial x}n_x + a_{yy}\frac{\partial u}{\partial y}n_y\right) ds, \end{aligned} \tag{1.11.65b}$$

where we used integration by parts [see Eqs. (1.11.24)–(1.11.28)] to transfer the differentiation from u to w so that both u and w have the same order derivatives with respect to both x and y. The boundary term shows that u is the primary variable while

$$q_n \equiv a_{xx}\frac{\partial u}{\partial x}n_x + a_{yy}\frac{\partial u}{\partial y}n_y \tag{1.11.66}$$

is the secondary variable. One can interpret q_n as the flux normal to the boundary, $q_n = \hat{\mathbf{n}} \cdot \mathbf{q}$, where $\mathbf{q}$ is the flux vector

$$\mathbf{q} \equiv a_{xx}\frac{\partial u}{\partial x}\,\hat{\mathbf{e}}_x + a_{yy}\frac{\partial u}{\partial y}\,\hat{\mathbf{e}}_y. \tag{1.11.67}$$

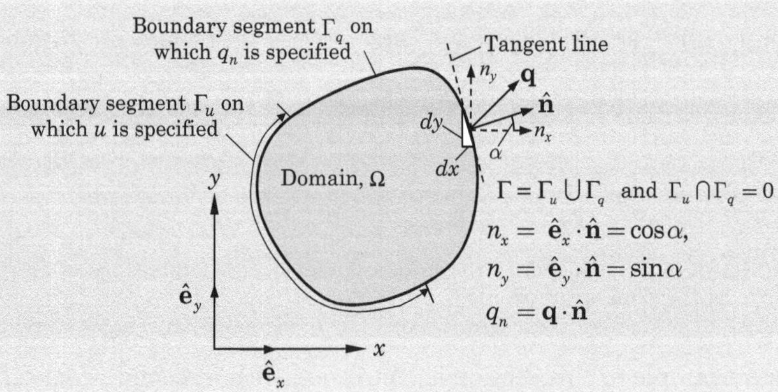

Fig. 1.11.3 A two-dimensional bounded domain Ω with closed boundary Γ.

Step 3: The last step in the procedure is to impose the specified boundary conditions in Eq. (1.11.64). Since u is specified on Γ_u, then the weight function w is zero on Γ_u and arbitrary on Γ_q. Consequently, Eq. (1.11.65b) simplifies to

$$0 = \int_\Omega \left(a_{xx} \frac{\partial w}{\partial x} \frac{\partial u}{\partial x} + a_{yy} \frac{\partial w}{\partial y} \frac{\partial u}{\partial y} - wf \right) dxdy - \int_{\Gamma_q} w\hat{q}\, ds. \tag{1.11.68}$$

The weak form in Eq. (1.11.68) can be expressed as $B(w,u) = \ell(w)$, where the bilinear form and linear form are

$$B(w,u) = \int_\Omega \left(a_{xx} \frac{\partial w}{\partial x} \frac{\partial u}{\partial x} + a_{yy} \frac{\partial w}{\partial y} \frac{\partial u}{\partial y} \right) dxdy \tag{1.11.69}$$

$$\ell(w) = \int_\Omega wf\, dxdy + \int_{\Gamma_q} w\hat{q}\, ds. \tag{1.11.70}$$

The next section is devoted to the discussion of the Ritz method and its applications, where we make use of the weak forms developed here.

1.11.5 The Ritz Method of Approximation

1.11.5.1 Basic idea

In this section, we introduce the Ritz method of approximation because it is closely connected with the development of finite element models. In the Ritz method, we seek an approximate solution in the form of a linear combination of suitable approximation functions $\phi_j(x)$ and undetermined parameters c_j: $\sum_j c_j \phi_j$, as given in Eq. (1.11.1). In the Ritz method, the coefficients c_j are determined using the weak form[11], which is equivalent to the equation being solved as well as its specified natural boundary conditions.

In the Ritz method, the choice of weight functions is restricted to the approximation functions, $w_i(x) = \phi_i(x)$ which, in addition to being sufficiently differentiable as dictated by the weak form, satisfy the homogeneous form of the specified essential boundary conditions. The method is described here for a linear variational problem (which is the same as the weak form).

Consider the variational problem of finding the solution u such that

$$B(w,u) = \ell(w) \tag{1.11.71}$$

for all sufficiently differentiable functions w_i that satisfy the homogeneous form of specified essential boundary conditions on u. In general, $B(\cdot,\cdot)$ can be unsymmetric in w and u, and it can be even nonlinear in u [however, $B(\cdot,\cdot)$ is always linear in w]. The discrete problem consists of finding an approximate solution u_N such that

$$B(w_i, u_N) = \ell(w_i), \quad i = 1, 2, \ldots, n. \tag{1.11.72}$$

[11]The weighted-residual methods also use Eq. (1.11.1), but the properties of ϕ_j are different; they are usually higher-order than those used in the Ritz method because in the Ritz method, which uses the weak form, ϕ_j satisfy lower differentiability requirements and satisfy the homogeneous form of the specified essential boundary conditions (see, e.g., Reddy [8, 17]).

The form of the approximate solution to Eq. (1.11.71) is slightly modified from that in Eq. (1.11.1) in order to separate the homogeneous solution from the particular solution:

$$u_N(x) = \sum_{j=1}^{N} c_j\, \phi_j(x) + \phi_0(x), \tag{1.11.73}$$

where the constants $c_1, c_2, \ldots, c_N$, called the *Ritz coefficients*, are determined such that Eq. (1.11.72) holds for N different choices $\{w_i\}$, so that N independent algebraic relations among $c_1, c_2, \ldots, c_N$ are obtained. The functions $\{\phi_i\}_{i=0}^{N}$ and ϕ_0 are chosen such that u_N satisfies the specified essential boundary conditions [recall that the specified natural boundary conditions are already included in the variational problem]. The ith algebraic equation is obtained from Eq. (1.11.72) by substituting $w_i = \phi_i$ and u_N from Eq. (1.11.73)

$$B\Big(\phi_i, \sum_{j=1}^{N} c_j\phi_j + \phi_0\Big) = \ell(\phi_i) \;\Rightarrow\; \sum_{j=1}^{N} B(\phi_i, \phi_j)c_j + B(\phi_i, \phi_0) = \ell(\phi_i) \tag{1.11.74}$$

or

$$\sum_{j=1}^{N} R_{ij}c_j = b_i, \quad i = 1, 2, \ldots, N, \tag{1.11.75a}$$

where

$$R_{ij} = B(\phi_i, \phi_j), \quad b_i = \ell(\phi_i) - B(\phi_i, \phi_0). \tag{1.11.75b}$$

The algebraic equations in Eq. (1.11.75a) can be expressed in matrix form as

$$\mathbf{Rc} = \mathbf{b}. \tag{1.11.76}$$

1.11.5.2 Properties of the approximation functions

The set of approximation functions $\{\phi_i\}$ and ϕ_0 used in the N-parameter Ritz solution, Eq. (1.11.73), are required to satisfy certain properties. The particular form chosen in Eq. (1.11.73) facilitates the satisfaction of nonhomogeneous EBCs to be satisfied by $\phi_0(x)$ while the remaining part provides the solution to the homogeneous case. At points where the EBCs are specified, $\phi_0(x)$ is equal to the specified value at that point, requiring the remaining part of Eq. (1.11.73) be zero there. Since c_j are arbitrary, we choose ϕ_j to vanish identically at the points where the EBCs are specified (zero or not). To see this, suppose that u_N is specified to be u_0 at $x = x_0$. If we select $\phi_0(x)$ such that $\phi_0(x_0) = u_0$, then

$$u_N(x_0) = \sum_{j=1}^{N} c_j\phi_j(x_0) + \phi_0(x_0)$$

$$u_0 = \sum_{j=1}^{N} c_j\phi_j(x_0) + u_0 \;\rightarrow\; \sum_{j=1}^{N} c_j\phi_j(x_0) = 0,$$

which is satisfied, for arbitrary c_j, by choosing $\phi_j(x_0) = 0$.

If all specified essential boundary conditions are homogeneous (i.e., the specified value u_0 is zero), then ϕ_0 is taken to be zero and ϕ_j must still satisfy the homogeneous form of specified essential boundary conditions, $\phi_j(x_0) = 0$, $j = 1, 2, \ldots, N$. Note that the requirement that w_i be zero at the boundary points where the essential boundary conditions are specified is satisfied by the choice $w_i = \phi_i(x)$.

In summary, the approximation functions $\phi_i(x)$ and $\phi_0(x)$ are required to satisfy the following conditions:

(1) (a) $\{\phi_i\}_{i=1}^N$ must be such that $B(\phi_i, \phi_j)$ is defined and nonzero [i.e., ϕ_i are sufficiently differentiable and integrable as required in the evaluation of $B(\phi_i, \phi_j)$]. (b) ϕ_i must satisfy the homogeneous form of the specified essential boundary conditions of the problem.

(2) For any N, the set $\{\phi_i\}_{i=1}^N$ along with the columns (and rows) of $B(\phi_i, \phi_j)$ must be *linearly independent.*

(3) The set $\{\phi_i\}$ must be *complete.* For example, when ϕ_i are algebraic polynomials, completeness requires that the set $\{\phi_i\}$ contain all terms of the lowest order admissible, and up to the highest order desired.

(4) The only requirement on ϕ_0 is that it satisfies the specified essential boundary conditions. When the specified essential boundary conditions are zero, then ϕ_0 is identically zero. Also, for completeness reasons, ϕ_0 must be the lowest-order function that satisfies the specified essential boundary conditions.

Next, we consider a few examples of the application of the Ritz method. Since most finite element models in the literature, including all commercial codes, are based on the element-wise application of the Ritz method. Therefore, the reader should follow the steps closely in arriving at the algebraic equations in terms of the coefficients, c_j.

Example 1.11.4

Consider the differential equation [cf. Eq. (1.5.16a)]

$$-\frac{d^2u}{dx^2} + \alpha\, u = 0 \quad \text{for} \quad 0 < x < L. \tag{1.11.77}$$

Determine the N-parameter Ritz solution using algebraic polynomials for the following two sets of boundary conditions:

$$\textbf{Set 1:} \quad u(0) = u_0, \quad u(L) = u_L \tag{1.11.78}$$

$$\textbf{Set 2:} \quad u(0) = u_0, \quad \left[\frac{du}{dx} + \beta_L u\right]_{x=L} = 0. \tag{1.11.79}$$

Numerically evaluate the solutions for $N = 1, 2,$ and 3 using the following data:

$$L = 0.05 \text{ m}, \quad \alpha = 400, \quad u_0 = 300, \quad u_L = 0.0. \tag{1.11.80}$$

Solution for Set 1 boundary conditions: The bilinear form and the linear functional associated with Eqs. (1.11.77) and (1.11.78) are

$$B(w,u) = \int_0^L \left(\frac{dw}{dx}\frac{du}{dx} + \alpha\, wu \right) dx, \quad \ell(w) = 0. \tag{1.11.81}$$

Since both of the specified boundary conditions in this case are of the essential type and nonhomogeneous, we have $\phi_0 \neq 0$. The algebraic equations for this case are given by

$$\sum_{j=1}^{n} R_{ij} c_j = b_i, \quad i = 1, 2, \ldots, n \tag{1.11.82a}$$

$$R_{ij} = B(\phi_i, \phi_j) = \int_0^L \left(\frac{d\phi_i}{dx}\frac{d\phi_j}{dx} - \phi_i \phi_j \right) dx, \tag{1.11.82b}$$

$$b_i = \ell(\phi_i) - B(\phi_i, \phi_0) = -B(\phi_i, \phi_0). \tag{1.11.82c}$$

Clearly, for this problem, $R_{ij} = R_{ji}$ (i.e., the coefficient matrix $\mathbf{R}$ is symmetric).

Next we discuss the choice of $\phi_0(x)$ and ϕ_i, which are required to satisfy the conditions

$$\phi_0(0) = u_0, \quad \phi_0(L) = u_L; \quad \phi_i(0) = \phi_i(L) = 0.$$

Clearly, the choice $\phi_0(x) = u_0 + (u_L - u_0)(x/L)$ satisfies the required conditions. Then we work with $\phi_1(x)$. The polynomial $\phi_1(x) = (0 - x)(L - x)$ vanishes at $x = 0$ and $x = 1$, and its first derivative is nonzero. Hence, we take $\phi_1 = x(L - x)$. The next function in the sequence of complete functions is obviously $\phi_2 = x^2(L - x)$ [or $\phi_2 = x(L - x)^2$]. Thus, the following set of functions are admissible:

$$\phi_0(x) = u_0 + (u_L - u_0)\frac{x}{L}, \quad \phi_1(x) = x(L - x), \quad \phi_2(x) = x^2(L - x), \ldots,$$
$$\phi_i(x) = x^i(L - x), \ldots, \; \phi_N(x) = x^N(L - x). \tag{1.11.83}$$

Then the N-parameter Ritz approximation is

$$u_N = c_1\, x(L - x) + c_2\, x^2(L - x) + \cdots + c_N\, x^N(L - x) + \left[u_0 + (u_L - u_0)\frac{x}{L} \right]. \tag{1.11.84}$$

As a rule (to achieve convergence of the solution with an increasing number of terms in the sequence), one must start with the lowest-order admissible function and include all admissible higher-order functions up to the desired degree.

For the choice of approximation functions in Eq. (1.11.83), the matrix coefficients R_{ij} and vector coefficients b_i of Eqs. (1.11.82b) and (1.11.82c) can be computed as follows:

$$\begin{aligned} R_{ij} &= \int_0^L \left\{ [ix^{i-1}L - (i+1)x^i][jx^{j-1}L - (j+1)x^j] + \alpha(Lx^i - x^{i+1})(Lx^j - x^{j+1}) \right\} dx \\ &= (L)^{i+j+1} \left[\frac{ij}{i+j-1} - \frac{2ij + i + j}{i+j} + \frac{(i+1)(j+1)}{i+j+1} \right] \\ &\quad + \alpha(L)^{i+j+3} \left[\frac{1}{i+j+1} - \frac{2}{i+j+2} + \frac{1}{i+j+3} \right], \end{aligned} \tag{1.11.85a}$$

$$\begin{aligned} b_i &= -\int_0^L \left\{ [ix^{i-1}L - (i+1)x^i] \frac{u_L - u_0}{L} + \alpha(x^i L - x^{i+1}) \left[u_0 + (u_L - u_0)\frac{x}{L} \right] \right\} dx \\ &= -\alpha(L)^{i+2} \left[\frac{u_0}{(i+1)(i+2)} + \frac{u_L - u_0}{(i+2)(i+3)} \right]. \end{aligned} \tag{1.11.85b}$$

for $i, j = 1, 2, \ldots, N$. We note that the coefficients R_{ij} and b_i computed for the previous value of N remain unchanged for the higher values of N. For example, for $N = 3$, one only needs to compute $R_{i3} = R_{3i}$ for $i = 1, 2, 3$ and b_3.

Next, we consider the one-, two-, and three-parameter approximations to illustrate how the Ritz solution converges to the exact solution of the problem as we increase the value of N. The exact solution of Eq. (1.11.77) subject to the boundary conditions in Eq. (1.11.78) is [see Eq. (1.5.20b) with $\alpha = m^2$ and $f_0 = 0$]

$$u(x) = u_L \frac{\sinh mx}{\sinh mL} + u_0 \frac{\sinh m(L-x)}{\sinh mL}, \quad m = \sqrt{\alpha}. \tag{1.11.86}$$

Now we specialize the equations for $N = 1, 2,$ and 3 using the data $\alpha = m^2 = 400$, $L = 0.05$, $u_0 = 300$, and $u_L = 0$. For $N = 1$, we have

$$R_{11} = 0.458333 \times 10^{-4}, \quad b_1 = -1.25 \rightarrow c_1 = -\frac{b_1}{R_{11}} = -27271.11.$$

The one-parameter Ritz solution is given by

$$u_1(x) = c_1\phi_1(x) + \phi_0 = -27272.7(Lx - x^2) + u_0\left(1 - \frac{x}{L}\right).$$

For $N = 2$, we have

$$10^{-7}\begin{bmatrix} 458.333 & 11.458 \\ 11.458 & 0.446 \end{bmatrix}\begin{Bmatrix} c_1 \\ c_2 \end{Bmatrix} = -\begin{Bmatrix} 1.250 \\ 0.025 \end{Bmatrix}.$$

Solving the linear equations, we obtain

$$c_1 = -37040.2, \quad c_2 = 390698.0.$$

The two-parameter Ritz solution is given by

$$\begin{aligned} u_2(x) &= c_1\phi_1(x) + c_2\phi_2(x) + \phi_0 \\ &= -37040.2(Lx - x^2) + 390698(Lx^2 - x^3) + u_0\left(1 - \frac{x}{L}\right). \end{aligned}$$

For $N = 3$, we have

$$10^{-7}\begin{bmatrix} 458.333 & 11.458 & 0.3423 \\ 11.458 & 0.446 & 0.0165 \\ 0.342 & 0.017 & 0.0007 \end{bmatrix}\begin{Bmatrix} c_1 \\ c_2 \\ c_3 \end{Bmatrix} = -\begin{Bmatrix} -1.250 \\ -0.025 \\ -0.001 \end{Bmatrix}.$$

The solution of the above equations is

$$c_1 = -37493, \quad c_2 = 435658, \quad c_3 = -899197.$$

The three-parameter Ritz solution is given by

$$\begin{aligned} u_3(x) &= c_1\phi_1(x) + c_2\phi_2(x) + c_3\phi_3(x) + \phi_0 \\ &= -37493(Lx - x^2) + 435658(Lx^2 - x^3) - 899197(Lx^3 - x^4) + u_0\left(1 - \frac{x}{L}\right). \end{aligned}$$

A comparison of the Ritz solutions for $N = 1, 2, 4,$ and 8 with the exact solution in Eq. (1.11.86) (with $u_0 = 300$, $u_L = 0$, $m = 20$, and $L = 0.05$) is presented in Table 1.11.1. The Ritz solution converges to the exact solution for $N \geq 4$.

Table 1.11.1 Comparison of the Ritz solution with the exact solution of $-\frac{d^2u}{dx^2} + 400u = 0,\quad 0 < x < 0.05;\quad u(0) = 300,\quad u(L) = 0$

	Exact	Ritz Solution, u_N			
x	Solution	$N=1$	$N=2$	$N=4$	$N=8$
0.000	300.000	300.000	300.000	300.000	300.000
0.005	262.045	263.864	262.105	262.045	262.045
0.010	226.712	229.091	226.747	226.712	226.712
0.015	193.648	195.682	193.631	193.648	193.648
0.020	162.522	163.636	162.464	162.522	162.522
0.025	133.023	132.955	132.955	133.023	133.023
0.030	104.855	103.636	104.808	104.855	104.855
0.035	77.737	75.682	77.733	77.736	77.737
0.040	51.396	49.091	51.435	51.396	51.396
0.045	25.570	23.864	25.622	25.570	25.570
0.050	0.000	0.000	0.000	0.000	0.000

Solution for Set 2 boundary conditions: For the second set of boundary conditions in Eq. (1.11.79), the bilinear and linear forms become [see Eq. (1.11.45)]

$$B(w,u) = \int_0^L \left(\frac{dw}{dx}\frac{du}{dx} + \alpha\, wu \right) dx + \beta_L\, w(L)\, u(L), \quad \ell(w) = 0. \tag{1.11.87}$$

The approximation function ϕ_0 is required to satisfy only $\phi_0(0) = u_0$. Therefore, $\phi_0(x) = u_0$ is the lowest order function that satisfies the condition. Functions $\phi_i(x)$ must be selected to satisfy the homogeneous boundary condition $\phi_i(0) = 0$. Clearly, $\phi_1(x) = x$, $\phi_2(x) = x^2$, $\cdots$, $\phi_i(x) = x^i$ meet the requirement. Thus, we have

$$u_N(x) = c_1 x + c_2 x^2 + \cdots + c_i x^i + \cdots + c_N x^N. \tag{1.11.88}$$

The coefficients R_{ij} and b_i can be evaluated as

$$\begin{aligned} R_{ij} = B(\phi_i, \phi_j) &= \int_0^L \left(ij x^{i+j-2} + \alpha\, x^{i+j} \right) dx + \beta_L\, L^{i+j} \\ &= L^{i+j-1} \left(\frac{ij}{i+j-1} + \alpha\, \frac{L^2}{i+j+1} + \beta_L\, L \right) \end{aligned} \tag{1.11.89a}$$

$$b_i = -\alpha\, u_0 \int_0^L x^i\, dx - \beta_L\, L^i\, u_0 = -u_0 L^i \left(\alpha\, \frac{L}{i+1} + \beta_L \right). \tag{1.11.89b}$$

For example, for $N = 1$ we have

$$R_{11} = L\left(1 + \tfrac{1}{3}\alpha L^2 + \beta_L L\right), \quad b_1 = -u_0\left(\tfrac{1}{2}\alpha L + \beta_L\right).$$

For the choice of the data $L = 0.5$, $\alpha = 400$, $u_0 = 300$, $k = 50$, $\beta = 100$, and $\beta_L = \beta/k = 2.0$, we obtain

$$c_1 = b_1/R_{11} = -180/0.071667 = -2511.63.$$

The one-parameter Ritz solution becomes

$$u_1(x) = u_0 + c_1\, x = 300 - 2511.63\, x.$$

For $N = 2$ we have

$$10^{-3}\begin{bmatrix} 71.6667 & 3.3750 \\ 3.3750 & 0.2042 \end{bmatrix}\begin{Bmatrix} c_1 \\ c_2 \end{Bmatrix} = -\begin{Bmatrix} -180.0 \\ -6.5 \end{Bmatrix}.$$

The solution of these equations is

$$c_1 = -4569.90, \quad c_2 = 43706.5.$$

Them the two-parameter Ritz solution is given by

$$u_2 = c_1\phi_1 + c_2\phi_2 + \phi_0 = 300 - 4569.9\,x + 43706.5\,x^2.$$

For $N = 3$, the Ritz equations are

$$10^{-3}\begin{bmatrix} 71.6667 & 3.3750 & 0.1625 \\ 3.3750 & 0.2042 & 0.0110 \\ 0.1625 & 0.0110 & 0.0006 \end{bmatrix}\begin{Bmatrix} c_1 \\ c_2 \\ c_3 \end{Bmatrix} = -\begin{Bmatrix} 180.0000 \\ 6.5000 \\ 0.2625 \end{Bmatrix}.$$

The solution of these equations is

$$c_1 = -4765.95, \quad c_2 = 55352.9, \quad c_3 = -155424.$$

The three-parameter Ritz solution is

$$u_2 = c_1\phi_1 + c_2\phi_2 + c_3\phi_3 + \phi_0 = 300 - 4765.95\,x + 55352.9\,x^2 + -155424\,x^3.$$

The exact solution of Eqs. (1.11.77) and (1.11.79) is [see (1.5.21b) with $f_0 = 0$, $\beta_L = \beta/k$, and $m = \sqrt{\alpha}$]

$$u(x) = u_0\,\frac{\cosh m(L-x) + (\beta/mk)\sinh m(L-x)}{\cosh mL + (\beta/mk)\sinh mL}. \tag{1.11.90}$$

A comparison of the Ritz solutions for $N = 1, 2, 4$ and 8 with the exact solution is presented in Table 1.11.2. Clearly, the Ritz solution coincides with the exact solution for $N \geq 4$.

Table 1.11.2 Comparison of the Ritz solution with the exact solution.

	Exact	Ritz solution, u_N			
x	Solution	$N=1$	$N=2$	$N=4$	$N=8$
0.000	300.000	300.000	300.000	300.000	300.000
0.005	277.443	287.442	278.243	277.445	277.443
0.010	257.662	274.884	258.672	257.661	257.662
0.015	240.460	262.326	241.285	240.457	240.460
0.020	225.665	249.767	226.085	225.663	225.665
0.025	213.128	237.209	213.069	213.128	213.128
0.030	202.724	224.651	202.239	202.727	202.724
0.035	194.350	212.093	193.594	194.352	194.350
0.040	187.920	199.535	187.134	187.921	187.920
0.045	183.371	186.977	182.860	183.369	183.371
0.050	180.658	174.419	180.771	180.658	180.658

The next example deals with axial natural vibration of an elastic bar (see Reddy [8, 17]).

Example 1.11.5

Consider an elastic bar of uniform cross-section (A), length L, and modulus E. The bar is fixed at the left end and connected to a rigid support via a linear elastic spring (with spring constant k), as shown in Fig. 1.11.4. Set up the N-parameter Ritz approximation using algebraic polynomials to determine the axial natural frequencies, ω, and determine, as a special case, the first two natural frequencies. The governing differential equation and boundary conditions associated with the bar shown in Fig. 1.11.4 are:

$$-\frac{d}{dx}(EA\frac{du}{dx}) - \rho A\omega^2 u = 0, \quad 0 < x < L \tag{1.11.91}$$

$$u(0) = 0, \quad EA\frac{du}{dx} + ku = 0 \quad \text{at } x = L, \tag{1.11.92}$$

where ω is the frequency of natural vibration and $u(x)$ is the mode shape. Use $kL = EA$ in the numerical evaluation to find the frequencies.

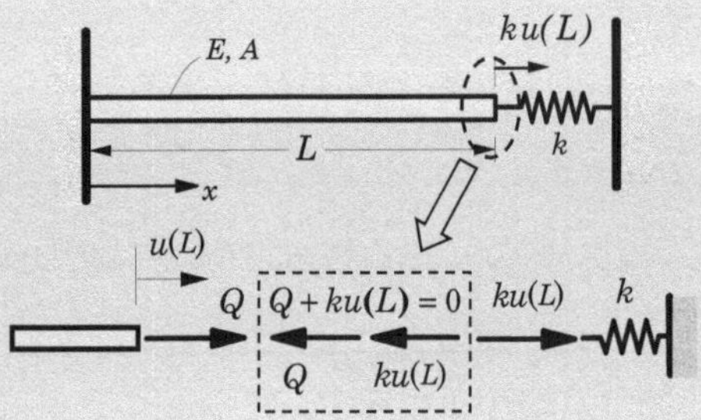

Fig. 1.11.4 A uniform elastic bar fixed at the left end and supported axially with a linear elastic spring.

Solution: First, we note that a continuous system has infinite number of frequencies, while a discretized problem (e.g., the Ritz solution) will have a finite number of frequencies. The number of frequencies is equal to the number of unconstrained degrees of freedom; in the Ritz method, the number of natural frequencies is equal to the number of parameters (N). Second, we also note that this second-order ODE is a special case of Eq. (1.11.29) with c replaced by $c = -\rho A\omega^2$, which contains the unknown (ω) to be determined, requiring us to keep the first term and the second term in the ODE separate. Also, the first boundary conditions, $u(0) = 0$ is the essential boundary condition and the second one is a natural (mixed type) boundary condition.

From the three-step procedure discussed for the weak form development [see Eqs. 1.11.39–(1.11.45)], we have the following weak form of the problem:

$$0 = \int_0^L \left(EA\frac{dw}{dx}\frac{du}{dx} - \rho A\omega^2 wu \right) dx + kw(L)u(L), \tag{1.11.93}$$

where w is the weight function.

The N-parameter Ritz approximation is

$$u(x) \approx u_N(x) = \sum_{j=1}^{N} c_j\phi_j(x) + \phi_0(x), \tag{1.11.94}$$

where $\phi_0 = 0$ (because the only specified essential boundary condition is homogeneous) and $\phi_j(x)$ must be selected to satisfy the condition $\phi_j(0) = 0$ for $j = 1, 2, \ldots, N$ and be differentiable at least once with respect to x. Substituting Eq. (1.11.94) (with $\phi_0 = 0$) into Eq. (1.11.93), we obtain

$$0 = \sum_{j=1}^{N} \left[-\lambda \int_0^L \rho A \phi_i \phi_j \, dx + \left(\int_0^L EA \frac{d\phi_i}{dx} \frac{d\phi_j}{dx} \, dx + k\phi_i(L)\phi_j(L) \right) \right] c_j, \tag{1.11.95}$$

where λ is the square of the frequency, $\lambda = \omega^2$. Equation (1.11.95) is the ith equation of the following set of N equations expressed in matrix form:

$$(\mathbf{R} - \lambda \mathbf{M})\, \mathbf{c} = \mathbf{0}, \tag{1.11.96}$$

where

$$R_{ij} = \int_0^L EA \frac{d\phi_i}{dx} \frac{d\phi_j}{dx} \, dx + k\phi_i(L)\phi_j(L), \quad M_{ij} = \int_0^L \rho A \phi_i \phi_j \, dx. \tag{1.11.97}$$

Equation (1.11.96) represents a matrix eigenvalue problem to determine the eigenvalues λ_i and modes shapes $u_i(x)$ for $i = 1, 2, 3, \ldots, N$.

The obvious algebraic function for $\phi_i(x)$ is

$$\phi_i(x) = \left(\frac{x}{L} \right)^i. \tag{1.11.98}$$

Substituting ϕ_i from Eq. (1.11.98) into Eq. (1.11.97), we obtain

$$R_{ij} = \int_0^L EA \frac{d\phi_i}{dx} \frac{d\phi_j}{dx} \, dx + k\phi_i(L)\phi_j(L) = \frac{EI}{L} \left(\frac{ij}{i + j - 1} \right) + k \tag{1.11.99a}$$

$$M_{ij} = \int_0^L \rho A \phi_i \phi_j \, dx = \rho AL \left(\frac{1}{i + j + 1} \right). \tag{1.11.99b}$$

As a specific case, we take $N = 2$ (to obtain *two* natural frequencies) and obtain

$$R_{11} = \frac{EA}{L} + k, \quad R_{21} = R_{12} = \frac{EA}{L} + k, \quad R_{22} = \frac{4EI}{3L} + k,$$
$$M_{11} = \frac{\rho AL}{3}, \quad M_{12} = \frac{\rho AL}{4}, \quad M_{22} = \frac{\rho AL}{5}$$

and the matrix eigenvalue problem in Eq. (1.11.96) becomes

$$\left(\frac{EA}{3L} \begin{bmatrix} 3 + 3\alpha & 3 + 3\alpha \\ 3 + 3\alpha & 4 + 3\alpha \end{bmatrix} - \lambda \frac{\rho AL}{60} \begin{bmatrix} 20 & 15 \\ 15 & 12 \end{bmatrix} \right) \begin{Bmatrix} c_1 \\ c_2 \end{Bmatrix} = \begin{Bmatrix} 0 \\ 0 \end{Bmatrix}, \tag{1.11.100}$$

where $\alpha = kL/EA$. The algebraic eigenvalue problem in Eq. (1.11.100) must be solved for $\lambda = \omega^2$ and c_i (hence, for the mode shape $u_N(x) = \sum_i^N c_i \phi_i(x)$).

We carry out the remaining calculations with $\alpha = kL/EA = 1$. For a nontrivial solution (i.e., $c_1 \neq 0, \;\; c_2 \neq 0$), we set the determinant of the coefficient matrix in Eq. (1.11.100) to zero:

$$\begin{vmatrix} 2 - \frac{\bar\lambda}{3} & 2 - \frac{\bar\lambda}{4} \\ 2 - \frac{\bar\lambda}{4} & \frac{7}{3} - \frac{\bar\lambda}{5} \end{vmatrix} = 0, \quad \bar\lambda = \frac{\lambda \rho L^2}{E} = \frac{\omega^2 \rho L^2}{E}$$

or

$$15\bar\lambda^2 - 640\bar\lambda + 2400 = 0.$$

The quadratic equation has two roots

$$\bar{\lambda}_1 = 4.1545, \quad \bar{\lambda}_2 = 38.512 \;\rightarrow\; \omega_1 = \frac{2.038}{L}\sqrt{\frac{E}{\rho}}, \quad \omega_2 = \frac{6.206}{L}\sqrt{\frac{E}{\rho}}.$$

The eigenvectors (or mode shapes) are given by

$$u_2^{(i)} = c_1^{(i)}\frac{x}{L} + c_2^{(i)}\frac{x^2}{L^2},$$

where $c_1^{(i)}$ and $c_2^{(i)}$ are calculated from the equations [see Eq. (1.11.100)]

$$\begin{bmatrix} 2 - \frac{\bar{\lambda}_i}{3} & 2 - \frac{\bar{\lambda}_i}{4} \\ 2 - \frac{\bar{\lambda}_i}{4} & \frac{7}{3} - \frac{\bar{\lambda}_i}{5} \end{bmatrix} \begin{Bmatrix} c_1^{(1)} \\ c_2^{(1)} \end{Bmatrix} = \begin{Bmatrix} 0 \\ 0 \end{Bmatrix}.$$

The above pair of equations is linearly dependent. Hence, one of the two equations can be used to determine c_2 in terms of c_1 (or vice versa) for each value of λ. We obtain

$$\bar{\lambda}_1 = 4.1545: \quad c_1^{(1)} = 1.0000, \quad c_2^{(1)} = -0.6399 \;\rightarrow\; u_2^{(1)}(x) = \frac{x}{L} - 0.6399\frac{x^2}{L^2}$$

$$\bar{\lambda}_2 = 38.512: \quad c_1^{(1)} = 1.0000, \quad c_2^{(1)} = -1.4207 \;\rightarrow\; u_2^{(2)}(x) = \frac{x}{L} - 1.4207\frac{x^2}{L^2}.$$

Plots of the two mode shapes are shown in Fig. 1.11.5.

The exact values of $\lambda = \omega^2$ are the roots of the transcendental equation (the reader may verify this by solving the problem analytically)

$$\lambda + \tan\lambda = 0,$$

whose first two roots are ($\omega^2 = \lambda$)

$$\omega_1 = \frac{2.02875}{L}\sqrt{\frac{E}{\rho}}, \quad \omega_2 = \frac{4.91318}{L}\sqrt{\frac{E}{\rho}}.$$

Note that the first approximate frequency is closer to the exact value than the second.

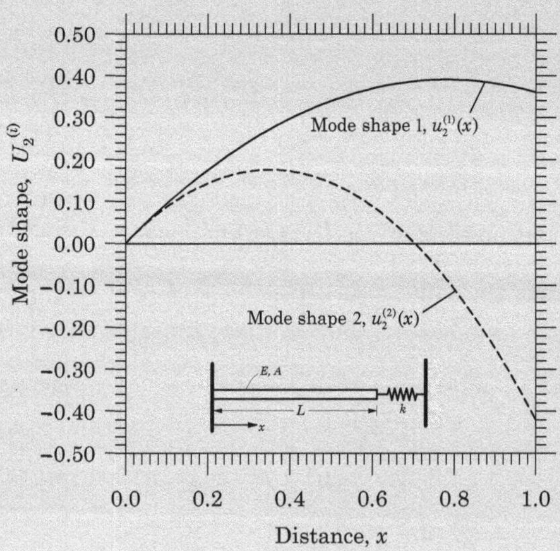

Fig. 1.11.5 First two mode shapes obtained by the Ritz method for the longitudinal natural frequencies of a spring-supported bar.

Example 1.11.6

Consider a straight beam of length $L = 12$ m and rectangular cross-section of width $b = 164$ mm and height $h = 400$ mm, and subjected to uniform distributed load of intensity $q_0 = 30$ kN/m, as shown in Fig. 1.11.6. Determine the transverse deflection using the Ritz method when the beam is simply supported and the beam modulus is $E = 200$ GPa.

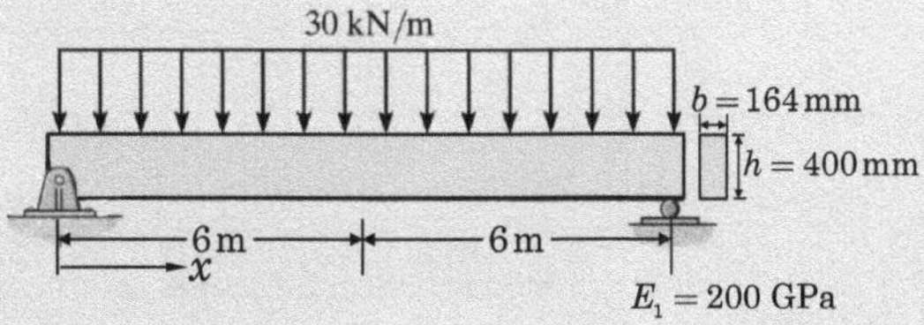

Fig. 1.11.6 A simply supported beam with uniform transverse load.

Solution: The governing differential equation according to the Euler–Bernoulli beam theory is given by Eq. (1.11.48) with $k = 0$ (because the beam is not on an elastic foundation) and $EI = 174.93$ MPa. The boundary conditions are

$$w(0) = M(0) \equiv -\frac{d^2w}{dx^2}\bigg|_{x=0} = 0, \quad w(L) = M(L) \equiv -\frac{d^2w}{dx^2}\bigg|_{x=L} = 0. \tag{1.11.101}$$

We note that $w(0) = w(L) = 0$ are essential boundary conditions, and $M(0) = M(L) = 0$ are natural boundary conditions. Both are homogeneous.

The weak form in this case is given by [cf. Eq. (1.11.53)]

$$0 = \int_0^L \left(EI \frac{d^2v}{dx^2}\frac{d^2w}{dx^2} - vq \right) dx, \tag{1.11.102}$$

where $v(x)$ is the weight function with the property $v(0) = v(L) = 0$.

The N-parameter Ritz approximation in this case is of the form (because $\phi_0(x) = 0$)

$$w_N(x) = \sum_{j=1}^{N} c_j\, \phi_j(x),$$

where $\phi_j(x)$ must be such that ϕ_j satisfies the conditions $\phi_j(0) = \phi_j(L) = 0$ for $j = 1, 2, \ldots, N$ and twice differentiable. For a choice of trigonometric functions, one may choose (because $\sin(j\pi x/L) = 0$ for $x = 0$ and $x = L$ for any integer j)

$$\phi_j(x) = \sin\frac{j\pi x}{L}. \tag{1.11.103}$$

For a choice of algebraic functions, it is clear that $\phi_1(x) = x(L - x)$ satisfies the conditions. Then $\phi_2(x) = x^2(L - x)$, $\phi_3(x) = x^3(L - x), \ldots,$

$$\phi_j(x) = x^j(L - x). \tag{1.11.104}$$

Trigonometric functions. The Ritz equations with the $\phi_j(x)$ given by Eq. (1.11.103) are

$$0 = \sum_{j=1}^{N} \left(\int_0^L EI \frac{d^2\phi_i}{dx^2} \frac{d^2\phi_j}{dx^2}\, dx \right) c_j - \int_0^L q_0\, \phi_i(x)\, dx \tag{1.11.105a}$$

$$= \sum_{j=1}^{N} \left[\int_0^L EI \left(\frac{i\pi}{L}\right)^2 \left(\frac{j\pi}{L}\right)^2 \sin\frac{i\pi x}{L} \sin\frac{j\pi x}{L} dx \right] c_j - \int_0^L q_0 \sin\frac{i\pi x}{L}\, dx$$

$$= \sum_{j=1}^{N} R_{ij} c_j - b_i, \tag{1.11.105b}$$

where

$$R_{ij} = \int_0^L EI \left(\frac{i\pi}{L}\right)^2 \left(\frac{j\pi}{L}\right)^2 \sin\frac{i\pi x}{L} \sin\frac{j\pi x}{L}\, dx, \quad b_i = \int_0^L q_0 \sin\frac{i\pi x}{L}\, dx. \tag{1.11.105c}$$

In view of the orthogonality of the sine functions

$$\int_0^L \sin\frac{i\pi x}{L} \sin\frac{j\pi x}{L}\, dx = \begin{cases} \frac{L}{2}, & i = j \\ 0, & i \neq j \end{cases} \tag{1.11.106}$$

and (for $i \geq 1$)

$$\int_0^L \sin\frac{i\pi x}{L}\, dx = \frac{L}{i\pi}\left[1 - (-1)^i\right] = \begin{cases} \frac{2L}{i\pi}, & i \text{ is odd} \\ 0, & i \text{ is even} \end{cases} \tag{1.11.107}$$

we have

$$R_{ij} = \begin{cases} EI\frac{i^2 j^2 \pi^4}{L^4}\frac{L}{2}, & i = j \\ 0, & i \neq j \end{cases}, \quad b_i = \begin{cases} \frac{2L}{i\pi}, & i \text{ is odd} \\ 0, & i \text{ is even} \end{cases}. \tag{1.11.108}$$

Thus, we have

$$c_i = \frac{4L^4}{i^5 \pi^5 EI} q_0, \quad \text{for odd values of} j \tag{1.11.109}$$

and the N-parameter Ritz solution is

$$w_N(x) = \sum_{j=1}^{N} \frac{q_0 L^4}{EI} \frac{4}{j^5 \pi^5} \sin\frac{j\pi x}{L}, \quad \text{for odd values of } j \text{ (i.e., } j = 1, 3, 5, \ldots). \tag{1.11.110}$$

This is a fast converging series, and for finitely few terms (i.e., $3 < N < 9$) the Ritz solution will be very close to the exact solution.

Algebraic functions. The Ritz equations with the $\phi_j(x)$ given by Eq. (1.11.104) are

$$0 = \sum_{j=1}^{N} \left(\int_0^L EI \frac{d^2\phi_i}{dx^2} \frac{d^2\phi_j}{dx^2}\, dx \right) c_j - \int_0^L q_0\, \phi_i(x)\, dx \tag{1.11.111a}$$

$$= \sum_{j=1}^{N} R_{ij} c_j - b_i, \tag{1.11.111b}$$

where

$$R_{ij} = \int_0^L EI \left[i(i-1)x^{i-2}L - i(i+1)x^{i-1} \right] \left[j(j-1)x^{j-2}L - j(j+1)x^{j-1} \right] dx$$

$$= EI(L)^{i+j-1} ij \left[\frac{(i-1)(j-1)}{i+j-3} - \frac{2(ij-1)}{i+j-2} + \frac{(i+1)(j+1)}{i+j-1} \right],$$

$$b_i = \int_0^L q_0\, x^i (L - x)\, dx = q_0 (L)^{i+2} \left(\frac{1}{i+1} - \frac{1}{i+2} \right) = q_0 L^{i+2} \frac{1}{(i+1)(i+2)}. \tag{1.11.111c}$$

For $N = 1$, we have $R_{11} = 4EIL$ and $b_1 = q_0L^3/6$, giving $c_1 = q_0L^2/24EI$; the one-parameter Ritz solution is

$$w_1(x) = \frac{q_0L^4}{24EI}\frac{x}{L}\left(1 - \frac{x}{L}\right).$$

For $N = 2$, we have

$$2EIL\begin{bmatrix} 2 & L \\ L & 2L^2 \end{bmatrix}\begin{Bmatrix} c_1 \\ c_2 \end{Bmatrix} = \frac{q_0L^3}{12}\begin{Bmatrix} 2 \\ L \end{Bmatrix}.$$

Solving for c_1 and c_2, we obtain

$$c_1 = \frac{q_0L^2}{24EI}, \quad c_2 = 0$$

and the solution remains the same as one-parameter Ritz solution.

For $N = 3$, we obtain the matrix equation

$$EIL\begin{bmatrix} 4 & 2L & 2L^2 \\ 2L & 4L^2 & 4L^3 \\ 2L^2 & 4L^3 & 4.8L^4 \end{bmatrix}\begin{Bmatrix} c_1 \\ c_2 \\ c_3 \end{Bmatrix} = \frac{q_0L^2}{12}\begin{Bmatrix} 2 \\ L \\ 0.6L^2 \end{Bmatrix}$$

The solution of these equations is

$$c_1 = \frac{q_0L^3}{12} \quad c_2 = \frac{q_0L^2}{12} \quad c_3 = -\frac{q_0L}{12}$$

and three-parameter solution becomes

$$w_3(x) = \frac{q_0L^4}{24EI}\left(\frac{x}{L} + \frac{x^2}{L^2} - \frac{x^3}{L^3}\right)\left(1 - \frac{x}{L}\right),$$

which coincides with the exact solution

$$w(x) = \frac{q_0L^4}{24EI}\left(\frac{x}{L} - 2\frac{x^3}{L^3} + \frac{x^4}{L^4}\right). \tag{1.11.112}$$

We note that the N-parameter solution obtained with the trigonometric functions is a series solution for any odd value of N because the exact solution can also be represented in sine series as

$$w(x) = \frac{4q_0L^4}{EI\pi^5}\sum_{n=1,3,5,\ldots}^{\infty}\frac{1}{n^5}\sin\frac{n\pi x}{L}. \tag{1.11.113}$$

The last example of this section deals with a BVP described by a PDE in two dimensions.

Example 1.11.7

Consider the following PDE governing two-dimensional heat transfer (and other problems like deflections of a membrane) in a square region:

$$-k\left(\frac{\partial^2 T}{\partial x^2} + \frac{\partial^2 T}{\partial y^2}\right) = g_0 \quad \text{in} \quad \Omega = \{(x,y) : 0 < (x,y) < 1\}, \tag{1.11.114}$$

where k is the conductivity and $T(x, y)$ is the temperature. The PDE is to be solved subject to the following boundary conditions:

$$T = 0 \quad \text{on sides } x = 1 \quad \text{and} \quad y = 1, \tag{1.11.115a}$$

$$\frac{\partial T}{\partial n} = 0 \quad \text{on sides } x = 0 \quad \text{and} \quad y = 0, \tag{1.11.115b}$$

where g_0 is the rate of uniform heat generation in the region. Equation (1.11.114) is often known as Poisson's equation. Determine an N-parameter Ritz solution of the form

$$T_N = \sum_{i,j=1}^{N} c_{ij} \cos \alpha_i x \, \cos \alpha_j y, \quad \alpha_i = \frac{1}{2}(2i - 1)\pi. \tag{1.11.116}$$

Note that Eq. (1.11.116) involves a double summation.

Solution: The PDE at hand is a special case of the problem considered in **Example 1.11.3**, with the change of notation $u = T$, $a_{xx} = a_{yy} = k$, and $f = g_0$. The boundary conditions in (1.11.115a) are essential (homogeneous) boundary conditions, and those in Eq. (1.11.115b) are the natural (homogeneous) boundary conditions. The variational problem associated with the PDE is [see Eqs. (1.11.68)–(1.11.70)]

$$B(w, T) = \ell(w), \tag{1.11.117a}$$

where w is the weight function and

$$\begin{aligned} B(w, T) &= \int_0^1 \int_0^1 k \left(\frac{\partial w}{\partial x} \frac{\partial T}{\partial x} + \frac{\partial w}{\partial y} \frac{\partial T}{\partial y} \right) dx dy \\ \ell(v) &= \int_0^1 \int_0^1 w g_0 \, dx dy. \end{aligned} \tag{1.11.117b}$$

Note that the series in Eq. (1.11.116) involves a double summation, and the approximation functions have double subscripts, $\phi_{ij}(x, y) = \cos \alpha_i x \, \cos \alpha_j y$. Since the boundary conditions are homogeneous, we take $\phi_0 = 0$. Incidentally, ϕ_{ij} also satisfies the natural boundary conditions of the problem, but that is not necessary to be admissible. While the choice $\hat{\phi}_{ij} = \sin i\pi x \sin j\pi y$ meets the essential boundary conditions, it is not complete, because it cannot be used to generate the solution that *does not* vanish on the sides $x = 0$ and $y = 0$. Hence, $\hat{\phi}_i$ are not admissible.

The coefficients R_{ij} and F_i can be computed by substituting Eq. (1.11.116) into Eqs. (1.11.117a) and (1.11.117b). Since the double finite Fourier series has two summations, we introduce the notation

$$\begin{aligned} R_{(ij)(k\ell)} &= k \int_0^1 \int_0^1 \big[(\alpha_i \sin \alpha_i x \cos \alpha_j y)(\alpha_k \sin \alpha_k x \cos \alpha_l y) \\ &\qquad + (\alpha_j \cos \alpha_i x \sin \alpha_j y)(\alpha_\ell \cos \alpha_k x \sin \alpha_\ell y) \big] dx dy \\ &= \begin{cases} 0, & \text{if } i \neq k \text{ or } j \neq \ell \\ \frac{1}{4} k (\alpha_i^2 + \alpha_j^2), & \text{if } i = k \text{ and } j = \ell \end{cases} \end{aligned} \tag{1.11.118a}$$

$$b_{ij} = g_0 \int_0^1 \int_0^1 \cos \alpha_i x \cos \alpha_j y \, dx dy = \frac{g_0}{\alpha_i \alpha_j} \sin \alpha_i \, \sin \alpha_j. \tag{1.11.118b}$$

In evaluating the integrals, the following orthogonality conditions were used

$$\int_0^1 \sin \alpha_i x \, \sin \alpha_j x \, dx = \begin{cases} 0 & \text{if } i \neq j \\ \frac{1}{2} & \text{if } i = j \end{cases} \tag{1.11.119}$$

$$\int_0^1 \cos \alpha_i x \cos \alpha_j x \, dx = \begin{cases} 0 & \text{if } i \neq j \\ \frac{1}{2} & \text{if } i = j \end{cases}. \tag{1.11.120}$$

Owing to the diagonal form of the coefficient matrix in Eq. (1.11.118a), we can readily solve for the coefficients c_{ij} (no sum on repeated indices):

$$c_{ij} = \frac{b_{ij}}{R_{(ij)(ij)}} = \frac{4g_0}{k} \frac{\sin \alpha_i \sin \alpha_j}{(\alpha_i^2 + \alpha_j^2)\alpha_i \alpha_j}. \tag{1.11.121}$$

The one- and two-parameter Ritz solutions are (the one-parameter solution has one term but the two-parameter solution has four terms)

$$T_1(x,y) = \frac{32 g_0}{k\pi^4} \cos \tfrac{1}{2}\pi x \cos \tfrac{1}{2}\pi y, \tag{1.11.122a}$$

$$T_2(x,y) = \frac{g_0}{k}\Big[0.3285 \cos \tfrac{1}{2}\pi x \cos \tfrac{1}{2}\pi y - 0.0219\big(\cos \tfrac{1}{2}\pi x \cos \frac{3}{2}\pi y + \cos \frac{3}{2}\pi x \cos \tfrac{1}{2}\pi y\big) + 0.0041 \cos \tfrac{3}{2}\pi x \cos \tfrac{3}{2}\pi y\Big]. \tag{1.11.122b}$$

If algebraic polynomials are to be used in the approximation of $T(x,y)$, one can choose $\phi_1(x,y) = (1-x)(1-y)$ or $\phi_1(x,y) = (1-x^2)(1-y^2)$, both of which satisfy the (homogeneous) essential boundary conditions but the latter also meets the natural boundary conditions of the problem. The one-parameter Ritz solution for the choice $\phi_1 = (1-x^2)(1-y^2)$ is

$$T_1(x,y) = \frac{5g_0}{16k}(1-x^2)(1-y^2). \tag{1.11.123}$$

The exact solution of Eqs. (1.11.114), (1.11.115a), and (1.11.115b) is

$$T(x,y) = \frac{g_0}{2k}\left[(1-y^2) + 4\sum_{n=1}^{\infty}(-1)^n \frac{\cos \alpha_n y \cosh \alpha_n x}{\alpha_n^3 \cosh \alpha_n}\right], \tag{1.11.124}$$

where $\alpha_n = (2n-1)\pi/2$. The Ritz solutions in Eqs. (1.11.122a), (1.11.122b), and (1.11.123) are compared with the exact solution in Eq. (1.11.124) in Fig. 1.11.7. The analytical solution is evaluated using 50 terms of the series in Eq. (1.11.124).

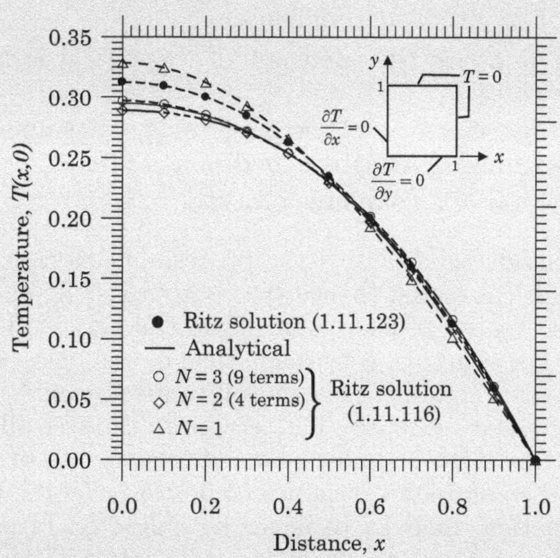

Fig. 1.11.7 Comparison of the Ritz solutions with the analytical solution of the Poisson equation in Eq. (1.11.114) with boundary conditions in Eqs. (1.11.115a) and (1.11.115b) in two dimensions.

1.12 Types of Errors

Solution of differential equations using numerical methods on computers will have errors of various kinds. Of course, the types of errors to be discussed here do not include mistakes introduced unintentionally in an algorithm written to implement a numerical approach. In general, the word "error" is used for the difference between the actual solution (which may not be known) of the differential equation and the numerical value obtained in a computer after executing the algorithm based on a numerical method.

We note that the difference between the numerical solution and experimentally observed/determined solution is a different type of error. This error is due to a number of factors, such as inadequate mathematical model to represent the physics observed in an experiment, incorrect experimental setup or measurement, and incorrect/inexact representation of the data (constitutive properties, boundary conditions, and loads).

Verification is the process of determining if the computational model is an accurate discrete analog of the mathematical model. Thus, if the round-off errors introduced due to finite arithmetic in a computer are negligible, the computational model with refinement (i.e., increasing the number of parameters of a convergent approximation) should give the exact solution of the mathematical model. During the verification, one uses a so-called benchmark problem. A *benchmark problem* is one that has a number of features to test a computer program for correctness. These features are as follows (additional discussion on verification and validation of simulation models can be found in [22, 24–26]). The benchmark problem

(1) has a set of standard reference results (analytical and numerical) that can be used to compare the accuracy and efficiency of computation;
(2) provides a list of pitfalls and difficulties;
(3) has a means to assess the adequacy of theoretical formulations and numerical algorithms;
(4) allows assessing robustness of the program (i.e. the ability of the program to handle ill-defined or sensitive conditions); and
(5) illustrates the use of a computer program.

On the other hand, *validation* is the process of determining the degree to which the mathematical model (hence the computer code that is verified) represents the physical reality of the system from the perspective of the intended uses of the model. For example, a mathematical model based on linear elasticity is adequate for determining linear elastic solutions of a solid but inadequate for determining its nonlinear response. The validation exercise allows one to modify the mathematical model to include the missing elements of the mathematical model that make the computed response come closer to the physical response. In fact, a mathematical model can never be validated because one does not know all the physics of the problem; it can only be invalidated. It is always a good idea, when developing a new computer program (also called a *code*), to undertake the verification exercise. Validation is a must when studying new problems.

An error is the difference between the actual value and its approximation. In addition to the error introduced in deriving the mathematical model using physical principles (because of the assumptions made), there are three sources of error in the numerical solution of differential equations that contribute to the error in the numerical solution when compared with the exact counterpart (see [27] and [28] for additional discussion):

(a) *Approximation errors.* Approximation errors can come from (i) the approximation of the solution in a numerical method (e.g., replacing a function and/or its derivative with a polynomial, including the truncated Taylor series); (ii) the approximation of the domain; and (iii) the approximation of the data input to the computer. Typically, the first two kinds of errors become smaller with an increase in the polynomial order (or when more terms in the Taylor series are included). The data errors are largely due to approximate representation of the coefficients in the differential equation. This in turn is related to approximate representation of the material and geometric properties of the system being analyzed, and they are partly responsible for the difference between the numerical and experimental solutions.

(b) *Round-off errors* are introduced due to inexact representation of certain quantities and/or the finite arithmetic in a computer. For example, quantities like π, $\sqrt{3}$, and $1/3$ represent specific numbers, but they cannot be expressed exactly by a finite number of digits (and computers can represent only a finite number of digits). They will be rounded off in a computer after a certain finite number of digits.

(c) *Truncation errors* are due to approximation used in computers to replace exact mathematical quantities. Transcendental functions like $\log x$, $\sin x$, and e^x are replaced in a computer or calculator with a truncated finite series [see series expansion in Eqs. (1.6.6)–(1.6.11)].

Approximation errors are introduced at the formulative stage and the round-off errors are introduced in actual computation. On the other hand, truncation errors may be introduced both at formulative level (by truncating the series representation of a function) as well as inside a computer (in replacing a transcendental function with finite series).

The truncation error decreases with a decrease in the step size (Δx or Δt). However, a small step size means more calculations, hence more round-off errors. Thus, the round-off errors and truncation errors bear an inverse relationship. The total error, which is a summation of round-off errors and truncation errors, decreases with an increase in step size up to some point and then increases with step size due to the increase in the round-off errors, as shown in Fig. 1.12.1. The challenge is to identify the point of diminishing returns where the round-off errors begin to negate the benefits of a decreased step size (see Chapra and Canale [27] for additional discussion).

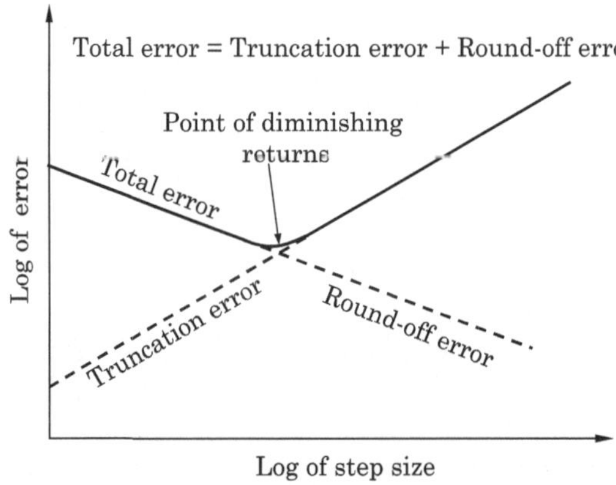

Fig. 1.12.1 Variation of the total error, the round-off error, and the truncation error as a function of the step size.

1.13 Summary

In this chapter, the following topics that are of interest in the later chapters are discussed:

(1) A brief discussion of most commonly used numerical methods (FDM, FVM, and FEM), along with the newly developed DMCDM is presented, giving the main features of each method.

(2) Types of differential equations (i.e., ODEs and PDEs) and the classification of problems described by differential equations into initial-value and boundary-value problems are discussed through a number of examples.

(3) The truncated Taylor series and its use in the approximation of the derivatives of functions are presented, and a brief review of matrices is included.

(4) Polynomial interpolation of functions is presented, and the Lagrange and Hermite type interpolations are discussed.

(5) Newton–Cotes and Gauss numerical integration methods are introduced and illustrated through an example.

(6) Solution of linear equations by direct methods and iterative methods is discussed. The Gaussian elimination and Gauss–Jordan direct methods and the Jacobi and Gauss–Seidel iteration methods are presented and illustrative examples are included.

(7) Solution of nonlinear equations using the Picard and Newton iterative methods are discussed. In the case of the Newton iteration method, the computation of the tangent matrix coefficients is also detailed.

(8) The method of manufactured solutions to differential equations in two dimensions is presented. The approach is valid even in three dimensions.

(9) Integral identities were presented and integrals formulations (weighted-integral and weak forms) are presented. A three-step procedure to develop weak forms of second- and higher-order differential equations is detailed. It is here that the concepts of duality and primary and secondary variables are introduced. The Ritz, Galerkin, subdomain, and least-squares methods are introduced to show how algebraic equations among the undetermined parameters of the approximation can be derived from the associated integral forms of the differential equations. Then the Ritz method, which is the foundation of most finite element models, is discussed in detail, and several numerical examples are presented.

(10) Various types of errors in solving the differential equations by a numerical method are discussed.

These mathematical preliminaries covered in this chapter will be useful in the subsequent chapters, and we will refer to them wherever we need them.

Problems

1.1 Consider the following set of three equations in three unknowns:

$$\begin{aligned} x_1 + x_2 + x_3 &= 2 \\ x_1 - x_2 - 3x_3 &= 3 \\ 3x_1 + x_2 - x_3 &= 1. \end{aligned}$$

Does this system of equations have a solution and why?

1.2 Consider the following set of three equations in three unknowns:

$$\begin{aligned} x_1 + 2x_2 + x_3 &= 1 \\ x_1 - x_2 - 2x_3 &= 3 \\ x_1 + x_2 &= 1. \end{aligned}$$

Does the system have a solution?

1.3 Consider the following set of three equations in three unknowns:

$$\begin{aligned} 2x_1 + 3x_2 - x_3 &= 4 \\ x_1 - x_2 + 2x_3 &= 2 \\ x_1 + 2x_2 - x_3 &= 1. \end{aligned}$$

Does the system have a solution and is it unique?

1.4 Consider the following set of three equations in three unknowns:

$$\begin{aligned} x_1 + x_2 + x_3 &= 1 \\ x_1 - x_2 - 3x_3 &= 2 \\ 3x_1 + x_2 - x_3 &= 3. \end{aligned}$$

Determine the solution using Cramer's rule.

1.5 Consider the following set of three equations in three unknowns:

$$\begin{aligned} 3x_1 - x_2 + 2x_3 &= 2 \\ 2x_1 + x_2 + x_3 &= -1 \\ x_1 + 3x_3 &= 2. \end{aligned}$$

Determine the solution using Cramer's rule.

1.6 Consider the following set of three equations in three unknowns:

$$\begin{aligned} -x_1 + 5x_2 + 4x_3 &= 0 \\ x_1 - x_2 + 2x_3 &= 2 \\ x_1 + 2x_2 - x_3 &= 1. \end{aligned}$$

Determine the solution using Cramer's rule.

1.7 Determine the solution of the system of equations in **Problem 1.5** using the Gauss elimination method.

1.8 Determine the solution of the system of equations in **Problem 1.5** using the Gauss elimination method.

1.9 Determine the solution of the system of equations in **Problem 1.5** using the Gauss–Jordan method.

1.10 Determine the solution of the system of equations in **Problem 1.6** using the Gauss–Jordan method.

1.11 Determine the solution of the system of equations in **Problem 1.5** using the Jacobi iteration method (you may write a MATLAB program to execute the steps).

1.12 Determine the solution of the system of equations in **Problem 1.6** using the Jacobi iteration method (you may write a MATLAB program to execute the steps).

1.13 Determine the solution of the system of equations in **Problem 1.5** using the Gauss–Seidel iteration method (you may write a MATLAB program to execute the steps).

1.14 Determine the solution of the system of equations in **Problem 1.5** using the Gauss–Seidel iteration method (you may write a MATLAB program to execute the steps).

1.15 If the matrix associated with Eq. (1.9.23) is identified as

$$[A] = \begin{bmatrix} 8\,x_3 & x_2 & 0 & 0 \\ x_1 & 4\,x_2^2 & x_3 & 1 \\ -x_1 & 2\,x_2\,x_4 & 6\,x_3 & -2 \\ 1 & -x_1\,x_2 & -x_2 & 4 \end{bmatrix}$$

determine the tangent matrix.

1.16 Develop the weak form of the following differential equation and boundary conditions:

$$-\frac{d}{dx}\left[(1+2x^2)\frac{du}{dx}\right] + u = x^2,$$

with the boundary conditions

$$u(0) = 1\,, \quad \left(\frac{du}{dx}\right)_{x=1} = 2.$$

1.17 Develop the weak forms of the following nonlinear coupled differential equations that arise in connection with the Euler–Bernoulli–von Kármán nonlinear theory of beams:

$$-\frac{d}{dx}\left\{EA\left[\frac{du}{dx} + \frac{1}{2}\left(\frac{dw}{dx}\right)^2\right]\right\} = f \quad \text{for} \quad 0 < x < L, \tag{1}$$

$$\frac{d^2}{dx^2}\left(EI\frac{d^2w}{dx^2}\right) - \frac{d}{dx}\left\{EA\frac{dw}{dx}\left[\frac{du}{dx} + \frac{1}{2}\left(\frac{dw}{dx}\right)^2\right]\right\} = q, \tag{2}$$

with the boundary conditions

$$u = w = 0 \quad \text{at} \quad x = 0,\ L; \quad \left(\frac{dw}{dx}\right)\Big|_{x=0} = 0; \quad \left(EI\frac{d^2w}{dx^2}\right)\Big|_{x=L} = M_0, \tag{3}$$

where EA, EI, f, and q are functions of x, and M_0 is a constant. Here u denotes the axial displacement and w the transverse deflection of the beam.

1.18 Develop the weak form of the following second-order partial differential equation:

$$-\frac{\partial}{\partial x}\left(a_{11}\frac{\partial u}{\partial x}+a_{12}\frac{\partial u}{\partial y}\right)-\frac{\partial}{\partial y}\left(a_{21}\frac{\partial u}{\partial x}+a_{22}\frac{\partial u}{\partial y}\right)+f=0 \quad \text{in} \quad \Omega$$

with the boundary conditions

$$u=u_0 \quad \text{on} \quad \Gamma_1, \quad \left(a_{11}\frac{\partial u}{\partial x}+a_{12}\frac{\partial u}{\partial y}\right)n_x+\left(a_{21}\frac{\partial u}{\partial x}+a_{22}\frac{\partial u}{\partial y}\right)n_y=t_0 \quad \text{on} \quad \Gamma_2,$$

where $a_{ij}=a_{ji}$ $(i,j=1,2)$ and f are given functions of position (x,y) in a two-dimensional domain Ω and u_0 and t_0 are known functions on portions Γ_1 and Γ_2 of the boundary Γ: $\Gamma_1+\Gamma_2=\Gamma$.

1.19 Develop the weak form of the following differential equation governing heat transfer in an axisymmetric geometry:

$$-\left[\frac{1}{r}\frac{\partial}{\partial r}\left(rk\frac{\partial T}{\partial r}\right)+\frac{\partial}{\partial z}\left(k\frac{\partial T}{\partial z}\right)\right]=g, \quad \text{in} \quad \Omega$$

with the boundary conditions

$$T=\hat{T} \quad \text{on} \quad \Gamma_T \quad \text{and} \quad q_n \equiv rk\left(\frac{\partial T_n}{\partial r}n_r+\frac{\partial T_n}{\partial z}n_z\right)=\hat{q}_n \quad \text{on} \quad \Gamma_q.$$

1.20 Governing equations for two-dimensional flow of viscous incompressible fluids:

$$-\mu\left[2\frac{\partial^2 v_x}{\partial x^2}+\frac{\partial}{\partial y}\left(\frac{\partial v_x}{\partial y}+\frac{\partial v_y}{\partial x}\right)\right]+\frac{\partial P}{\partial x}=f_x,$$

$$-\mu\left[2\frac{\partial^2 v_y}{\partial y^2}+\frac{\partial}{\partial x}\left(\frac{\partial v_x}{\partial y}+\frac{\partial v_y}{\partial x}\right)\right]+\frac{\partial P}{\partial y}=f_y, \quad \text{in } \Omega$$

$$\frac{\partial v_x}{\partial x}+\frac{\partial v_y}{\partial y}=0,$$

with the boundary conditions

$$v_x=\hat{v}_x, \quad v_y=\hat{v}_y \quad \text{on} \quad \Gamma_1,$$

and

$$\left.\begin{aligned}\left[\left(2\mu\frac{\partial v_x}{\partial x}-P\right)n_x+\mu\left(\frac{\partial v_x}{\partial y}+\frac{\partial v_y}{\partial x}\right)n_y\right]&=\hat{t}_x\\ \left[\left(2\mu\frac{\partial v_y}{\partial y}-P\right)n_y+\mu\left(\frac{\partial v_x}{\partial y}+\frac{\partial v_y}{\partial x}\right)n_x\right]&=\hat{t}_y.\end{aligned}\right\} \text{on } \Gamma_2,$$

1.21 Compute the coefficient matrix and the right-hand side of the n-parameter Ritz approximation of the equation

$$-\frac{d}{dx}\left[(1+x)\frac{du}{dx}\right]=0 \quad \text{for} \quad 0<x<1,$$

with the boundary conditions

$$u(0)=0, \quad u(1)=1.$$

Use algebraic polynomials for the approximation functions. Specialize your result for $n=2$ and compute the Ritz coefficients. *Answer:* $c_1=\frac{55}{131}$ and $c_2=-\frac{20}{131}$.

1.22 Use trigonometric functions for the two-parameter approximation of the equation in Problem 1.21 and obtain the Ritz coefficients. *Answer:* $c_1=-0.12407$ and $c_2=0.02919$ for a choice of functions.

1.23 A steel rod of diameter $d = 0.02$ m, length $L = 0.25$ m, and thermal conductivity $k = 50$ W/(m °C) is exposed to ambient air $T_\infty = 20\,°$C with a heat-transfer coefficient $\beta = 64$ W/(m^2 °C). Given that the left end of the rod is maintained at a temperature of $T_0 = 120\,°$C and the other end is exposed to the ambient temperature, determine the temperature distribution in the rod using a two-parameter Ritz approximation with polynomial approximation functions. The equation governing the problem is given by

$$-\frac{d^2\theta}{dx^2} + c\theta = 0 \quad \text{for} \quad 0 < x < 0.25\,\text{m},$$

where $\theta = T - T_\infty$, T is the temperature, and c is given by

$$c = \frac{\beta P}{Ak} = \frac{\beta \pi D}{\frac{1}{4}\pi D^2 k} = \frac{4\beta}{kD} = 256 \text{ m}^2$$

P being the perimeter and A the cross-sectional area of the rod. The boundary conditions are

$$\theta(0) = T(0) - T_\infty = 100\,°\text{C}, \qquad \left(k\frac{d\theta}{dx} + \beta\theta\right)\bigg|_{x=L} = 0.$$

Answer: For $L = 0.25$ m, $\phi_0 = 100$, $\phi_i = x^i$, the Ritz coefficients are $c_1 = -1,033.385$ and $c_2 = 2,667.261$.

1.24 Set up the equations for the n-parameter Ritz approximation of the following equations associated with a simply supported beam and subjected to a uniform transverse load $q = q_0$:

$$\frac{d^2}{dx^2}\left(EI\frac{d^2w}{dx^2}\right) = q \quad \text{for} \quad 0 < x < L,$$

with the boundary conditions

$$w = EI\frac{d^2w}{dx^2} = 0 \quad \text{at} \quad x = 0,\ L.$$

Identify (a) algebraic polynomials and (b) trigonometric functions for ϕ_0 and ϕ_i. Compute and compare the two-parameter Ritz solutions with the exact solution for uniform load of intensity q_0. *Answer:* (a) $c_1 = q_0L^2/(24EI)$ and $c_2 = 0$.

1.25 Repeat **Problem 1.24** for $q = q_0 \sin(\pi x/L)$, where the origin of the coordinate system is taken at the left end of the beam. *Answer:* $n = 2:\ \ c_1 = c_2L = 2q_0L^2/(3EI\pi^3)$.

1.26 Repeat **Problem 1.24** for $q = Q_0\delta(x - \frac{1}{2}L)$, where $\delta(x)$ is the Dirac delta function (i.e., a point load Q_0 is applied at the center of the beam).

1.27 Develop the n-parameter Ritz solution for a simply supported beam under uniform transverse load using the Timoshenko beam theory. The governing equations are given in Eqs. (1.11.55a) and (1.11.55b). Use Trigonometric functions to approximate w and ϕ_x.

1.28 Repeat **Problem 1.27** for a cantilever beam under uniform transverse load and an end moment M_0. Use algebraic polynomials to approximate w and ϕ_x.

1.29 Solve the Poisson equation governing heat conduction in a square region (see **Example 1.11.7**):

$$-k\nabla^2 T = g_0$$

with the boundary conditions

$$T = 0 \quad \text{on sides} \quad x = 1 \quad \text{and} \quad y = 1,$$

and

$$\frac{\partial T}{\partial n} = 0 \quad \text{(insulated) on sides} \quad x = 0 \quad \text{and} \quad y = 0.$$

Use a one-parameter Ritz approximation of the form

$$T_1(x, y) = c_1(1 - x^2)(1 - y^2).$$

Answer: $c_1 = \frac{5g_0}{16k}$.

1.30 Consider the (Neumann) boundary value problem

$$-\frac{d^2u}{dx^2} = f \quad \text{for} \quad 0 < x < L$$

with the boundary conditions

$$\left(\frac{du}{dx}\right)\Big|_{x=0} = \left(\frac{du}{dx}\right)\Big|_{x=L} = 0.$$

Find a two-parameter Galerkin approximation of the problem using trigonometric approximation functions, when (a) $f = f_0 \cos(\pi x/L)$ and (b) $f = f_0$. *Answer:* (a) $\phi_i = \cos(i\pi x/L)$, $c_1 = f_0 L^2/\pi^2$, $c_i = 0$ for $i \neq 1$.

1.31 Give a one-parameter Ritz solution of the equation

$$-\nabla^2 u = 1 \quad \text{in} \quad \Omega \quad (= \text{unit square})$$

with the boundary condition

$$u = 0 \quad \text{on} \quad \Gamma.$$

Use trigonometric approximation function.
Answer: $c_{ij} = \frac{16}{\pi^4} \frac{1}{ij(i^2+j^2)}$ $(i, j$ odd$)$, $\phi_{ij} = \sin i\pi x \sin j\pi y$.

1.32 Consider the differential equation

$$-\frac{d^2u}{dx^2} = \cos \pi x \quad \text{for} \quad 0 < x < 1$$

with the following three sets of boundary conditions:

(1) $u(0) = 0, \quad u(1) = 0.$
(2) $u(0) = 0, \quad \left(\frac{du}{dx}\right)_{x=1} = 0.$
(3) $\left(\frac{du}{dx}\right)_{x=0} = 0, \quad \left(\frac{du}{dx}\right)_{x=1} = 0.$

Determine a three-parameter solution, with trigonometric functions, using (a) the Ritz method, (b) the least-squares method, and (c) collocation at $x = \frac{1}{4}, \frac{1}{2}$, and $\frac{3}{4}$, and compare with the exact solutions:

(1) $u_0 = \pi^{-2}(\cos \pi x + 2x - 1).$
(2) $u_0 = \pi^{-2}(\cos \pi x - 1).$
(3) $u_0 = \pi^{-2} \cos \pi x.$

Answer: (1a) $c_i = \frac{4}{\pi^3 i(i^2-1)}$.

1.33 Consider a cantilever beam of variable flexural rigidity, $EI = a_0[2 - (x/L)^2]$ and carrying a distributed load, $q = q_0[1 - (x/L)]$. Find a three-parameter solution using the collocation method. *Answer:* $c_1 = -\frac{q_0 L^2}{4a_0}$ and $c_2 = \frac{q_0 L}{12a_0}$.

1.34 Find the first two eigenvalues associated with the differential equation

$$-\frac{d^2u}{dx^2} = \lambda u, \quad 0 < x < 1$$

with the boundary conditions

$$u(0) = 0, \quad u(1) + u'(1) = 0$$

using the Ritz method with algebraic polynomials. *Answer:* $\lambda_1 = 4.1545$ and $\lambda_2 = 38.512$.

1.35 Consider the Laplace equation

$$
\begin{gathered}
-\nabla^2 u = 0, \quad 0 < x < 1, \quad 0 < y < \infty, \\
u(0, y) = u(1, y) = 0 \quad \text{for} \quad y > 0, \\
u(x, 0) = x(1 - x), \quad u(x, \infty) = 0, \quad 0 \leq x \leq 1.
\end{gathered}
$$

Assuming one-parameter approximation of the form

$$
u(x, y) = c_1(y)x(1 - x)
$$

find the differential equation for $c_1(y)$ and solve it exactly.
Answer: $U_1(x, y) = (x - x^2)e^{-\sqrt{10}y}$.

2
Finite Difference Method

2.1 Finite Difference Formulas

2.1.1 Taylor's Series

This chapter is dedicated to the study of the finite difference method, which makes use of Taylor's series (see Section 1.6 for details). The finite difference method (FDM) is the oldest, perhaps the simplest, numerical method ever used to solve ODEs and PDEs. Until the arrival of the finite element method (FEM) and finite volume method (FVM) in the 70's and 80's, respectively, on the scene, the FDM ruled the numerical methods' domain. Most engineering and applied mathematics courses and their textbooks routinely introduced and used the FDM to solve differential equations. Even today, ideas from the FDM are used in the FEM and FVM.

There are different finite difference models of a differential equation, depending on the type of approximation (or the truncated Taylor series used). Recall from Section 1.6.2 that every analytic function (i.e., the function and its derivatives up to and including the nth derivative are continuous) in the interval containing x_0 and x can be expanded in a Taylor series (or use Taylor's formula) as [see Eq. (1.6.2)]

$$\begin{aligned} f(x) = f(x_0) + \frac{(x-x_0)}{1!} f'(x_0) + \frac{(x-x_0)^2}{2!} f''(x_0) \\ + \frac{(x-x_0)^3}{3!} f^{(3)}(x_i) + \cdots + \frac{(x-x_0)^n}{n!} f^{(n)}(x_0) + R_n(x), \end{aligned} \tag{2.1.1}$$

where $f^{(n)}(x_0)$ denotes the nth derivative of the function $f(x)$ evaluated at $x = x_0$ and R_n is remainder [see Eqs. (1.6.12) and (1.6.13)]

$$R_n(x) = \frac{(x-x_0)^{n+1}}{(n+1)!} f^{(n+1)}(\xi). \tag{2.1.2}$$

Let us define the "step size" h as the difference $h = x_{i+1} - x_i$ and write Eq. (2.1.1) at $x = x_{i+1} = x_i + h$ as

$$\begin{aligned} f(x_{i+1}) = f(x_i) + h\, f'(x_i) + \frac{h^2}{2!} f''(x_i) \\ + \frac{h^3}{3!} f^{(3)}(x_i) + \cdots + \frac{h^n}{n!} f^{(n)}(x_i) + R_n(x_{i+1}) \end{aligned} \tag{2.1.3}$$

with the remainder, which is the truncation error if the series is terminated after the nth derivative term, as

$$R_n(x) = \frac{h^{n+1}}{(n+1)!} f^{(n+1)}(\xi). \tag{2.1.4}$$

The error in the approximation of $f(x_{i+1})$ is said to be of the order h^{n+1} and written as "$O(h^{n+1})$."

For $x = x_{i-1} = x_i - h$ (i.e., $h = x_i - x_{i-1}$), the Taylor series takes the form

$$\begin{aligned} f(x_{i-1}) &= f(x_i) - h\, f'(x_i) + \frac{h^2}{2!} f''(x_i) \\ &\quad - \frac{h^3}{3!} f^{(3)}(x_i) + \cdots + (-1)^n \frac{h^n}{n!} f^{(n)}(x_i) + R_n(x_{i-1}). \end{aligned} \tag{2.1.5}$$

2.1.2 Difference Formulas for First and Second Derivatives

Equations (2.1.3) and (2.1.5) can be used to develop a number of finite difference formulas [see Eqs. (1.6.15)–(1.6.22)]. The first-order approximation of $f(x_{i+1})$ according to Eq. (2.1.3) is

$$f(x_{i+1}) \approx f(x_i) + h f'(x_i) \quad \text{or} \quad f_{i+1} = f_i + h\, f'_i, \tag{2.1.6}$$

where we have introduced the notation $f_i \equiv f(x_i)$ and $f'_i = f'(x_i)$. Equation (2.1.6) can be rewritten as

$$f'(x_i) = \left.\frac{df}{dx}\right|_{x_i} \approx \frac{f_{i+1} - f_i}{h} + O(h). \tag{2.1.7}$$

Thus, the right-hand side of the above equation provides an approximation of the slope of the function $f(x)$ at point x_i. The slope is determined using the function values at points $x = x_i$ and x_{i+1}, and it is known as the *forward difference formula* for the first derivative of a function.

Similarly, the first-order approximation of $f(x_{i+1})$ according to Eq. (2.1.5) is

$$f'(x_i) = \left.\frac{df}{dx}\right|_{x_i} \approx \frac{f_i - f_{i-1}}{h} + O(h). \tag{2.1.8}$$

Equation (2.1.8) is known as the *backward difference formula* for the first derivative of a function.

Adding Eqs. (2.1.7) and (2.1.8), we obtain the second-order accurate central difference formula for the first derivative

$$f'(x_i) = \left.\frac{df}{dx}\right|_{x_i} \approx \frac{f_{i+1} - f_{i-1}}{2h} + O(h^2). \tag{2.1.9}$$

By adding Eqs. (2.1.3) and (2.1.5) and solving for $f''(x_i)$, we obtain the following finite difference formula for the approximation of the second derivative of a function

$$f''(x_i) = \left.\frac{d^2 f}{dx^2}\right|_{x_i} \approx \frac{f_{i-1} - 2f_i + f_{i+1}}{h^2} + O(h^2). \tag{2.1.10}$$

Equation (2.1.9) is known as the *central difference formula* for the first derivative, and Eq. (2.1.10) is the central difference formula for the second derivative of a function.

All of the results developed can be written for nonuniform meshes by replacing h with h_{i+1} in the forward difference formula and with h_i in the backward difference formula. It is a bit complicated but possible to write finite difference formulas for other cases.

In general, it is possible to generate finite difference formulas for any order derivative and with any order accuracy with the help of a Taylor series expansion (see Hildebrand [29] and Chung [30] for additional finite difference schemes). As an example, consider the Taylor expansion of $f(x)$ with $x = x_{i-2} = x_i - 2h$,

$$\begin{aligned} f(x_{i-2}) = f(x_i) - 2h\, f'(x_i) + \frac{(2h)^2}{2!} f''(x_i) \\ - \frac{(2h)^3}{3!} f^{(3)}(x_i) + \cdots + (-1)^n \frac{(2h)^n}{n!} f^{(n)}(x_i) + R_n(x_{i-2}). \end{aligned} \quad (2.1.11)$$

Next, we express the linear combination of f_{i-2}, f_{i-1}, and f_i (one may also consider other linear combinations; e.g., f_{i-2}, f_i, and f_{i+1}) using Eqs. (2.1.5) and (2.1.11):

$$\begin{aligned} a\, f_i + b\, f_{i-1} + c\, f_{i-2} &= a\, f_i + b\Big[f(x_i) - h\, f'(x_i) + \frac{h^2}{2!} f''(x_i) \\ &\quad - \frac{h^3}{3!} f^{(3)}(x_i) + \cdots\Big] + c\Big[f(x_i) - 2h\, f'(x_i) + \frac{(2h)^2}{2!} f''(x_i) \\ &\quad - \frac{(2h)^3}{3!} f^{(3)}(x_i) + \cdots\Big] \\ &= (a+b+c) f(x_i) - (b+2c) h\, f'(x_i) + (b+4c)\frac{h^2}{2} f''(x_i) + O(h^3). \end{aligned} \quad (2.1.12)$$

Now we can choose the values of a, b, and c such that we obtain a finite difference formula for $f'(x_i)$ or $f''(x_i)$. To determine a formula for the first derivative at x_i, we set $a+b+c=0$, $b+4c=0$, and $b+2c=-1$ (the solution of these three algebraic equations is $a=1.5$, $b=-2$, and $c=0.5$) and obtain

$$f'(x_i) = \frac{df}{dx}\bigg|_{x_i} \approx \frac{3f_i - 4f_{i-1} + f_{i-2}}{2h} + O(h^2). \quad (2.1.13)$$

Similarly, to determine the finite difference formula for the second derivative at x_i, we set $a+b+c=0$, $b+2c=0$, and $b+4c=2$ (which gives $a=1$, $b=-2$, and $c=1$) and obtain the central difference formula [cf. Eq. (2.1.10)]

$$f''(x_i) = \frac{d^2 f}{dx^2}\bigg|_{x_i} \approx \frac{f_{i-2} - 2f_{i-1} + f_i}{h^2} + O(h). \quad (2.1.14)$$

We note that the formula in Eq. (2.1.10) is second-order accurate, while the one in Eq. (2.1.14) is only first-order accurate.

In addition to finite difference formulas derived using various Taylor expansions, one can also construct weighted averages of such formulas. For example, a weighted average of the first derivative of a function at two adjacent points, x_i and x_{i+1}, is approximated by the forward and backward difference formula

$$(1-\alpha)f'(x_i) + \alpha\, f'(x_{i+1}) \approx \frac{f_{i+1} - f_i}{h}$$

or

$$f_{i+1} = f_i + h\left[(1-\alpha)f'(x_i) + \alpha\, f'(x_{i+1})\right], \quad 0 \le \alpha \le 1. \tag{2.1.15}$$

Equation (2.1.15) is known as the α-family of approximation, which contains the following well-known formulas as special cases:

$$\begin{aligned} &\text{(a) } \alpha = 0: && \text{Forward difference formula.} \\ &\text{(b) } \alpha = 1: && \text{Backward difference formula.} \\ &\text{(c) } \alpha = 0.5: && \text{Crank–Nicolson formula.} \end{aligned} \tag{2.1.16}$$

We will discuss other such formulas in the coming sections.

Although the finite difference formulas derived here are valid for ODEs, they can be easily extended to multi-dimensions (i.e., to PDEs). For example, the forward difference formula in Eq. (2.1.7) can be written for the derivatives with respect to x and y for two-dimensional problems as

$$\left.\frac{\partial F}{\partial x}\right|_{x_i, y_i} \approx \frac{F_{i+1,j} - F_{i,j}}{h_x}, \tag{2.1.17a}$$

$$\left.\frac{\partial F}{\partial y}\right|_{x_i, y_i} \approx \frac{F_{i,j+1} - F_{i,j}}{h_y}, \tag{2.1.17b}$$

where $F_{i,j}$ denotes the value of $F(x, y)$ at point (x_i, y_i), $F_{i,j} = F(x_i, y_i)$, and $h_x = \Delta x$ and $h_y = \Delta y$ are the step sizes in the x and y coordinate directions, respectively (see Fig. 2.1.1 for a mesh or grid with fixed h_x and h_y step sizes).

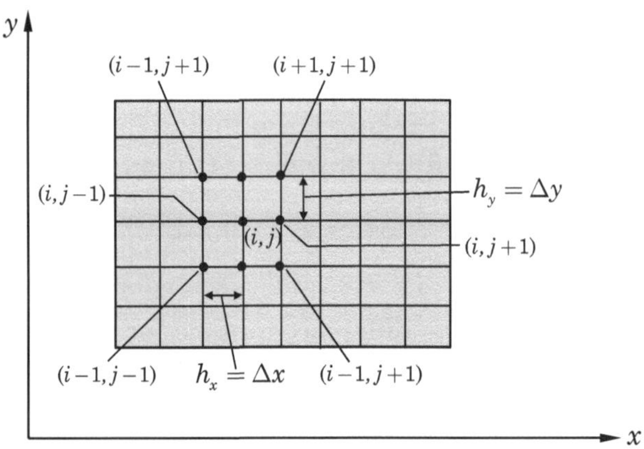

Fig. 2.1.1 Finite difference mesh or grid, indicating the step sizes along the x and y coordinate directions and the numbering of the mesh/grid points; (i, j) is the name given to the point (x_i, y_j) referred to as the (x, y)-coordinate system.

Similarly, the second-order derivatives can be approximated as

$$\left.\frac{\partial^2 F}{\partial x^2}\right|_{x_i,y_i} \approx \frac{F_{i-1,j} - 2F_{i,j} + F_{i+1,j}}{h_x^2} + O(h_x^2), \tag{2.1.18}$$

$$\left.\frac{\partial^2 F}{\partial y^2}\right|_{x_i,y_i} \approx \frac{F_{i,j-1} - 2F_{i,j} + F_{i,j+1}}{h_y^2} + O(h_y^2). \tag{2.1.19}$$

A simple, second-order accurate, approximation of the mixed derivative is given by

$$\left.\frac{\partial^2 F}{\partial x \partial y}\right|_{x_i,y_i} \approx \frac{F_{i+1,j+1} - F_{i+1,j-1} + F_{i-1,j-1} - F_{i-1,j+1}}{4h_x h_y} + O(h_x^2, h_y^2). \tag{2.1.20}$$

2.2 Solution of First-Order Ordinary Differential Equations

2.2.1 Euler's Method

Here, we consider a first-order ODE of the form

$$\frac{du}{dx} = f(x,u), \quad 0 \leq x \leq L, \quad u(0) = u_0, \tag{2.2.1}$$

where L is the length of the domain. Mathematically, f represents the slope of the function $u(x)$. If $f(x,u)$ is a nonlinear function of the dependent unknown u, then Eq. (2.2.1) is a nonlinear differential equation. As a part of the problem description, the specific form of f in terms of x and u will be known.

Integrating the above equation over a step size $h = x_{i+1} - x_i$, we obtain

$$u_{i+1} - u_i = \int_{x_i}^{x_{i+1}} f(x,u)\,dx, \tag{2.2.2}$$

where we presume that the solution $u_i = u(x_i)$ is known, and seek the solution $u_{i+1} = u(x_{i+1})$. The integration on the right-hand side can be carried out if one treats $f(x,u)$ as a constant in the interval $[x_i, x_{i+1}]$. The constant value of f (which is the slope of the function u) is determined by evaluating at a fixed point in the interval. The stability of the method and the complexity of the calculation depend on the choice of this point. For example, if $f(x,u)$ is evaluated at x_i, then Eq. (2.2.2) can be solved for u_{i+1} because everything on the right-hand side is known:

$$u_{i+1} = u_i + hf(x_i, u_i), \quad u_i = u(x_i). \tag{2.2.3}$$

This recursive formula begins with the known value u_0 at $x = x_0$:

$$u_1 = u_0 + h\, f(x_0, u_0). \tag{2.2.4}$$

Equation (2.2.4) is also known as the *Euler's method.*

Now suppose that $f(x, u)$ is evaluated at the point x_{i+1}, we obtain

$$u_{i+1} = u_i + h\, f(x_{i+1}, u_{i+1}). \tag{2.2.5}$$

In the above equation, we move terms involving u_{i+1} from the right side of the equality to the left side, and combining them will give a relation to solve for u_{i+1}. Thus, the coefficient of u_{i+1} in Eq. (2.2.5) will no longer be unity.

A generalized formulation can be expressed by a weighted-average of the slopes $f(x_i, u_i)$ and $f(x_{i+1}, u_{i+1})$:

$$u_{i+1} = u_i + h\left[(1-\alpha)f(x_i, u_i) + \alpha f(x_{i+1}, u_{i+1})\right], \quad 0 \le \alpha \le 1. \tag{2.2.6}$$

Equation (2.2.6) is termed as the *alpha-family of methods.* Depending on the value of α, the generalized formulation can reduce to Eq. (2.2.3) ($\alpha = 0$) or Eq. (2.2.5) ($\alpha = 1$).

2.2.2 Runge–Kutta Family of Methods

Runge–Kutta (RK) methods are a family of methods for numerically integrating ordinary differential equations [1, 27]. Runge–Kutta methods can be of nth order, where the value of the slope $f(x, y)$ is evaluated at n locations. Runge–Kutta methods achieve the accuracy of a Taylor series approach without requiring the calculations of higher derivatives. The nth order Runge–Kutta method (RKn) can be expressed as [27]

$$u_{i+1} = u_i + h\,\phi(x_i, u_i, h), \tag{2.2.7}$$

where $\phi(x_i, u_i, h)$ is termed an *increment function*, which represents a weighted average of slopes. In general, the increment function is of the form

$$\phi = a_1\, k_1 + a_2\, k_2 + \cdots + a_k\, k_n, \tag{2.2.8}$$

where a_i are constants and k_i are

$$\begin{aligned} k_1 &= f(x_i, u_i), \\ k_2 &= f(x_i + p_1 h, u_i + q_{11} k_1 h), \\ k_3 &= f(x_i + p_2 h, u_i + q_{21} k_1 h + q_{22} k_2 h), \\ &\vdots \\ k_n &= f(x_i + p_{n-1} h, u_i + q_{n-1,1}\, k_1 h + \cdots + q_{n-1,n-1} k_{n-1} k). \end{aligned} \tag{2.2.9}$$

The *first-order Runge–Kutta method* (RK1) is nothing but Euler's method. Here we consider the second-order and fourth-order Runge–Kutta methods because of their popularity in engineering applications.

The second-order Runge–Kutta method (RK2). In the second-order Runge–Kutta method, the slope is evaluated at two points ($a_1 = a_2 = 1/2$ and $p_1 = q_{11} = 1$) in the interval $[x_i, x_{i+1}]$:

$$\begin{aligned} u_{i+1} &= u_i + \frac{h}{2}(k_1 + k_2), \\ k_1 = f(x_i, u_i), &\quad k_2 = f(x_i + h, u_i + k_1\, h). \end{aligned} \tag{2.2.10}$$

The fourth-order Runge–Kutta method (RK4). The fourth-order Runge–Kutta method is given by:

$$\begin{aligned} u_{i+1} &= u_i + \tfrac{1}{6}h(k_1 + 2k_2 + 2k_3 + k_4), \\ k_1 &= f(x_i, u_i), \\ k_2 &= f(x_i + \tfrac{1}{2}h, u_i + \tfrac{1}{2}k_1\, h), \\ k_3 &= f(x_i + \tfrac{1}{2}h, u_i + \tfrac{1}{2}k_2\, h), \\ k_4 &= f(x_i + h, u_i + k_3\, h). \end{aligned} \tag{2.2.11}$$

Next, we consider an example of application of the first-order, second-order, and fourth-order Runge–Kutta methods to solve an IVP.

Example 2.2.1

Consider the problem of a falling parachutist. Newton's second law of motion states that the time rate of change of linear momentum (the product of mass times velocity) of a system is equal to the sum of the forces acting on the system:

$$\frac{d}{dt}(mv) = \sum_{i=1}^{n} F_i, \tag{2.2.12}$$

where t is time (s), m is the mass (kg), v is the velocity (m/s), and F_i are external forces (N or kg m/s^2) ($i = 1, 2, \ldots, n$). The external forces acting in the case of a parachutist are the gravitational force and the air resistance. The air resistance, acting upward (opposite to the fall), can be modeled as the force proportional to the velocity. Assume that the air resistance force is $F_a = -cv$, with c, called the drag coefficient (kg/s), to be a constant. The gravitational force is $F_g = mg$, m being the (fixed) mass of the parachutist and g is the gravitational constant ($g = 9.81$ m/s^2). Thus, we have ($n = 2$)

$$\frac{d}{dt}(mv) = F_g + F_a = mg - cv \quad \rightarrow \quad \frac{dv}{dt} = \frac{1}{m}(mg - cv). \tag{2.2.13}$$

Solve the first-order differential equation in Eq. (2.2.13) numerically using the first-, second-, and fourth-order Runge–Kutta methods and compare the solutions with the exact solution of the differential equation for initial velocity of $v(0) = 0$ (the reader is asked to verify the solution):

$$v(t) = \frac{mg}{c}\left[1 - e^{-(c/m)t}\right]. \tag{2.2.14}$$

Assume the following values of m, c, and time increment Δt: $m = 68$ kg, $c = 12.5$ kg/s, and $\Delta t = 2$ s.

Solution: In this problem, the dependent unknown is v, the independent variable (coordinate) is t, and the function $f(t, v)$ is given by

$$f(t, v) = \frac{1}{m}(mg - cv). \tag{2.2.15}$$

Programming Eqs. (2.2.5) for the RK1, (2.2.10) for the RK2, and (2.2.11) for the RK4 methods is straightforward (with a change of variables: $x = t$ and $y = v$). A Fortran source code is available from **http://mechanics.tamu.edu**. The numerical results are presented in Table 2.2.1 and Fig. 2.2.1. The finite difference solutions are close to the exact solution. The numerical solutions will improve with the reduction in the size of the time step.

Table 2.2.1 Comparison of the numerical solutions $v(t)$ with the exact solution (2.2.14) of Eq. (2.2.13); the numerical solutions are obtained with the first-, second-, and fourth-order Runge–Kutta methods with $\Delta t = 2$ s.

t	RK1	RK2	RK4	Exact
2.0	19.620	16.013	16.415	16.418
4.0	32.027	27.222	27.781	27.784
6.0	39.872	35.067	35.650	35.654
8.0	44.833	40.558	41.100	41.103
10.0	47.970	44.401	44.873	44.876
12.0	49.954	47.091	47.485	47.488
14.0	51.209	48.974	49.294	49.296
16.0	52.002	50.292	50.547	50.548
18.0	52.504	51.215	51.414	51.415
20.0	52.821	51.860	52.015	52.016
22.0	53.021	52.312	52.430	52.431
24.0	53.148	52.629	52.718	52.719
26.0	53.228	52.850	52.918	52.918
28.0	53.279	53.005	53.056	53.056
30.0	53.311	53.113	53.151	53.151
32.0	53.332	53.189	53.217	53.218
34.0	53.344	53.242	53.263	53.263
36.0	53.352	53.280	53.295	53.295
38.0	53.358	53.306	53.317	53.317
40.0	53.361	53.324	53.332	53.332
50.0	53.366	53.359	53.361	53.361
60.0	53.366	53.365	53.366	53.366

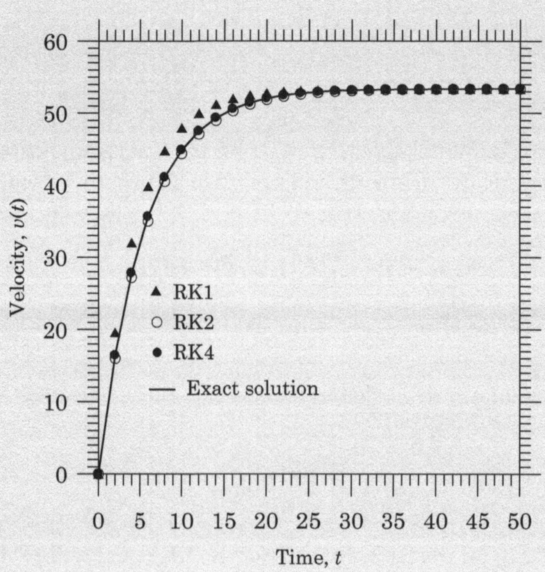

Fig. 2.2.1 Comparison of the numerical solutions $v(t)$ versus t, obtained with the RK1, RK2, and RK4 methods, with the exact solution.

2.2.3 Coupled System of First-Order Differential Equations

Here, consider a pair of coupled first-order ODEs of the form:

$$\frac{du_1}{dx} = f_1(x, u_1, u_2), \tag{2.2.16a}$$

$$\frac{du_2}{dx} = f_2(x, u_1, u_2), \tag{2.2.16b}$$

with the following conditions:

$$u_1(x_0) = u_1^0, \quad u_2(x_0) = u_2^0. \tag{2.2.16c}$$

The RK1 (Euler's) method applied to Eqs. (2.2.16a) and (2.2.16b) gives (here $k_{i,j}$ denotes the "k_i" of the jth equation)

$$\begin{aligned} u_1^{i+1} &= u_1^i + hk_{1,1}, \quad k_{1,1} = f_1(x_i, u_1^i, u_2^i), \\ u_2^{i+1} &= u_2^i + hk_{1,2}, \quad k_{1,2} = f_2(x_i, u_1^i, u_2^i), \end{aligned} \tag{2.2.17}$$

where $u_1^i = u_1(x_i)$, $u_2^i = u_2(x_i)$, $u_1^{i+1} = u_1(x_i + h)$, and $u_2^{i+1} = u_2(x_i + h)$.

The RK2 method applied to the pair of Eqs. (2.2.16a) and (2.2.16b) takes the form:

$$\begin{aligned} &u_1^{i+1} \equiv u_1(x_{i+1}) = u_1^i + \tfrac{1}{2}h(k_{1,1} + k_{2,1}), \\ &u_2^{i+1} \equiv u_2(x_{i+1}) = u_2^i + \tfrac{1}{2}h(k_{1,2} + k_{2,2}), \\ &k_{1,1} = f_1(x_i, u_1^i, u_2^i), \quad k_{2,1} = f_1(x_i + h, u_1^i + k_{1,1}\, h, u_2^i + k_{1,2}\, h), \\ &k_{1,2} = f_2(x_i, u_1^i, u_2^i), \quad k_{2,2} = f_2(x_i + h, u_1^i + k_{1,1}\, h, u_2^i + k_{1,2}\, h). \end{aligned} \tag{2.2.18}$$

Similarly, the RK4 method applied to Eqs. (2.2.16a) and (2.2.16b) has the following equations:

$$\begin{aligned} u_1^{i+1} &= u_1^i + \tfrac{1}{6}h(k_{1,1} + 2k_{2,1} + 2k_{3,1} + k_{4,1}), \\ u_2^{i+1} &= u_2^i + \tfrac{1}{6}h(k_{1,2} + 2k_{2,2} + 2k_{3,2} + k_{4,2}), \end{aligned} \tag{2.2.19}$$

where

$$\begin{aligned} k_{1,1} &= f_1(x_i, u_1^i, u_2^i), \quad k_{1,2} = f_2(x_i, u_1^i, u_2^i), \\ k_{2,1} &= f_1(x_i + \tfrac{1}{2}h, u_1^i + \tfrac{1}{2}k_{1,1}h, u_2^i + \tfrac{1}{2}k_{1,2}h), \\ k_{2,2} &= f_2(x_i + \tfrac{1}{2}h, u_1^i + \tfrac{1}{2}k_{1,1}h, u_2^i + \tfrac{1}{2}k_{1,2}h), \\ k_{3,1} &= f_1(x_i + \tfrac{1}{2}h, u_1^i + \tfrac{1}{2}k_{2,1}h, u_2^i + \tfrac{1}{2}k_{2,2}h), \\ k_{3,2} &= f_2(x_i + \tfrac{1}{2}h, u_1^i + \tfrac{1}{2}k_{2,1}h, u_2^i + \tfrac{1}{2}k_{2,2}h), \\ k_{4,1} &= f_1(x_i + h, u_1^i + k_{3,1}h, u_2^i + k_{3,2}h), \\ k_{2,4} &= f_2(x_i + h, u_1^i + k_{3,1}h, u_2^i + k_{3,2}h). \end{aligned} \tag{2.2.20}$$

Example 2.2.2

Consider a simple pendulum of string of length ℓ with mass m attached to the end of the string oscillating in a frictionless environment (see Fig. 1.5.2). Initially, the pendulum is displaced by a small angle θ_0 in the clockwise direction, and then the pendulum is allowed to oscillate under the influence of gravity. The governing differential equation and initial conditions for the simple pendulum are [see Eqs. (1.5.8) and (1.5.9)]:

$$m\frac{d^2\theta}{dt^2} + \frac{mg}{\ell}\sin\theta = 0, \tag{2.2.21a}$$

$$\theta(0) = \frac{\pi}{12}, \quad \left.\frac{d\theta}{dt}\right|_{t=0} = 0. \tag{2.2.21b}$$

Using the data $m = 1$ kg (does not enter the calculation), $\ell = 1$ m, and $g = 9.81$ m/s^2, and two different time step sizes of $\Delta t = 0.05$ s and $\Delta t = 0.01$ s, calculate angular displacement $\theta(t)$ and velocity $v(t) = \ell\frac{d\theta}{dt}$ as functions of time t using the RK1, RK2, and RK4 methods.

Solution: To use the Runge–Kutta methods, we first rewrite the second-order ODE as a pair of first-order ODEs by introducing velocity $v(t)$ as a dependent variable. Let

$$v = \ell\,\frac{d\theta}{dt} \tag{2.2.22}$$

so that we have

$$\frac{d\theta}{dt} = \frac{v}{\ell} \equiv f_1(t,\theta,v), \tag{2.2.23a}$$

$$\frac{dv}{dt} = -g\sin\theta \equiv f_2(t,\theta,v), \tag{2.2.23b}$$

with the initial conditions on θ and v as

$$\theta_0 = \theta(0) = \frac{\pi}{12}, \quad v_0 = v(0) = 0.0. \tag{2.2.24}$$

When the angular motion is assumed to be small, the governing ODE can be linearized for small angles, $\sin\theta \approx \theta$, giving

$$\frac{d\theta}{dt} = \frac{v}{\ell}, \quad \frac{dv}{dt} = -g\theta. \tag{2.2.25}$$

The exact solution of the linear problem for the initial conditions given in Eq. (2.2.24) is given by

$$\theta(t) = \theta_0\,\cos\alpha t, \quad v(t) = -\alpha\theta_0\,\ell\,\sin\alpha t, \quad \alpha = \sqrt{\frac{g}{\ell}}. \tag{2.2.26}$$

Here we calculate the solutions for the nonlinear case ($\theta_0 = 0.2618$ and $\sin\theta_0 = 0.2588$), but we expect the linear and nonlinear solutions to be close. We illustrate the calculations for the first time step using $\Delta t = 0.05$ s.

Euler's (RK1) Method: In this case, we have

$$\theta(0.05) = \theta_0 + \Delta t\, f_1(0,\theta_0,v_0) = \frac{\pi}{12} + 0.05 \times 0 = 0.2618 \text{ rad},$$
$$v(0.05) = v_0 + \Delta t\, f_2(0,\theta_0,v_0) = 0 - 0.05 \times 9.81 \times \sin(\tfrac{\pi}{12}) = -0.1269 \text{ rad/s}.$$

RK2 Method: We have

$$
\begin{aligned}
k_{11} &= f_1(t_0, \theta_0, v_0) = v_0 = 0, \\
k_{12} &= f_2(t_0, \theta_0, v_0) = -\frac{g}{\ell}\sin\theta_0 = -9.81\ \sin(\tfrac{\pi}{12}) = -2.5390, \\
k_{21} &= f_1(t_0 + h, \theta_0 + k_{11}\,h, v_0 + k_{12}\,h) = v_0 + k_{12}\,h = (0 + (-2.5390)0.05) = -0.1269, \\
k_{22} &= f_2(t_0 + h, \theta_0 + k_{11}\,h, v_0 + k_{12}\,h) = -9.81\ \sin(\tfrac{\pi}{12} + 0.05 \times 0) = -2.5390.
\end{aligned}
$$

Then $\theta(0.05)$ and $v(0.05)$ are given by

$$
\begin{aligned}
\theta(0.05) &= \theta_0 + \tfrac{1}{2}h(k_{11} + k_{21}) = \frac{\pi}{12} + \frac{0.05}{2}(0 - 0.1269) = 0.2586 \text{ rad}, \\
v(0.05) &= v_0 + \tfrac{1}{2}h(k_{12} + k_{22}) = 0 + \frac{0.05}{2}(-2.5364 - 2.5364) = -0.1269 \text{ rad/s}.
\end{aligned}
$$

RK4 Method: In this case, we have

$$
\begin{aligned}
k_{11} &= f_1(t_0, \theta_0, v_0) = v_0 = 0, \\
k_{12} &= f_2(t_0, \theta_0, v_0) = -\frac{g}{\ell}\sin\theta_0 = -9.81\ \sin(\tfrac{\pi}{12}) = -2.5390, \\
k_{21} &= f_1(t_0 + \tfrac{1}{2}\Delta t, \theta_0 + k_{11}\frac{h}{2} + v_0 + \tfrac{1}{2}k_{12}h) = v_0 + \tfrac{1}{2}k_{12}h = -0.0635, \\
k_{22} &= f_2(t_0 + \tfrac{1}{2}h, \theta_0 + \tfrac{1}{2}k_{11}h + v_0 + \tfrac{1}{2}k_{12}h) = -9.81\sin(\theta_0 + \tfrac{1}{2}k_{11}h) = -2.5390, \\
k_{31} &= f_1(t_0 + \tfrac{1}{2}h, \theta_0 + \tfrac{1}{2}k_{21}h + v_0 + \tfrac{1}{2}k_{22}h) = v_0 + \tfrac{1}{2}k_{22}h = -0.0635, \\
k_{32} &= f_2(t_0 + \tfrac{1}{2}h, \theta_0 + \tfrac{1}{2}k_{21}h + v_0 + \tfrac{1}{2}k_{22}h) = -9.81\sin(\theta_0 + \tfrac{1}{2}k_{21}h) = -2.5240, \\
k_{41} &= f_1(t_0 + h, \theta_0 + k_{31}\,h, v_0 + k_{32}\,h) = v_0 + k_{32}\,h = -0.1262, \\
k_{42} &= f_2(t_0 + h, \theta_0 + k_{31}\,h, v_0 + k_{32}\,h) = -9.81\sin[\tfrac{\pi}{12} + (-0.0635)0.05] = -2.5089.
\end{aligned}
$$

Therefore, we obtain

$$
\begin{aligned}
\theta(t = 0.05) &= \theta_0 + \frac{h}{6}(k_{11} + 2k_{21} + 2k_{31} + k_{41}) = 0.2586 \text{ rad}, \\
v(t = 0.05) &= v_0 + \frac{h}{6}(k_{12} + 2k_{22} + 2k_{32} + k_{42}) = -0.1264 \text{ rad/s}.
\end{aligned}
$$

The results for $\theta(t)$ and $v(t)$ at $t = 0.05$ s with two different time steps are summarized in Table 2.2.2. As one can see from the table, all three methods give positive values of the displacement. RK2 and RK4, unlike RK1, gave a value that is less than the initial displacement. This implies that the pendulum is still to the right of its neutral position ($\theta = 0$) at $t = 0.05$ s. Note that velocities predicted by all four methods at $t = 0.05$ s show negative values, indicating that the pendulum is swinging from right toward the neutral position. Plots of $\theta(t)$ versus t and $v(t)$ versus t are presented in Figs. 2.2.2 and 2.2.3, respectively.

Table 2.2.2 Results of the simple pendulum problem for $t = 0.05$ s.

	$\Delta t = 0.05$ s		$\Delta t = 0.01$ s	
Method	$\theta(0.05)$ [rad]	$v(0.05)$ [rad/s]	$\theta(0.05)$ [rad]	$v(0.05)$ [rad/s]
RK1	0.2618	−0.1269	0.2593	−0.1267
RK2	0.2586	−0.1269	0.2586	−0.1265
RK4	0.2586	−0.1264	0.2586	−0.1264
Exact	0.2586	−0.1279	0.2586	−0.1279

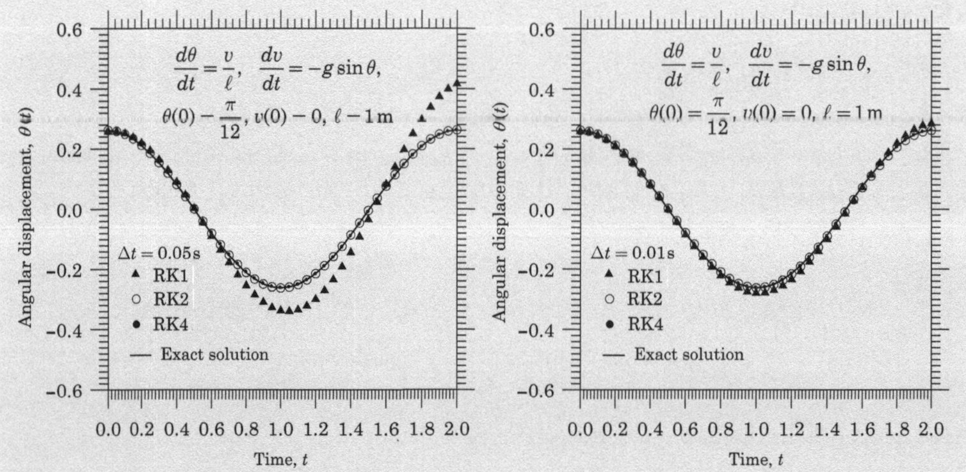

Fig. 2.2.2 Comparison of the numerical solutions $\theta(t)$ versus t, obtained with the RK1 (Euler's), RK2 (second-order Runge–Kutta), and RK4 (fourth-order Runge–Kutta) methods, with the exact solution of the simple pendulum problem ($g = 9.81$ m/s^2 and $\ell = 1$ m). The results are shown for two different time steps, $\Delta t = 0.05$ and $\Delta t = 0.01$. Smaller time step gives solutions closer to the analytical solution.

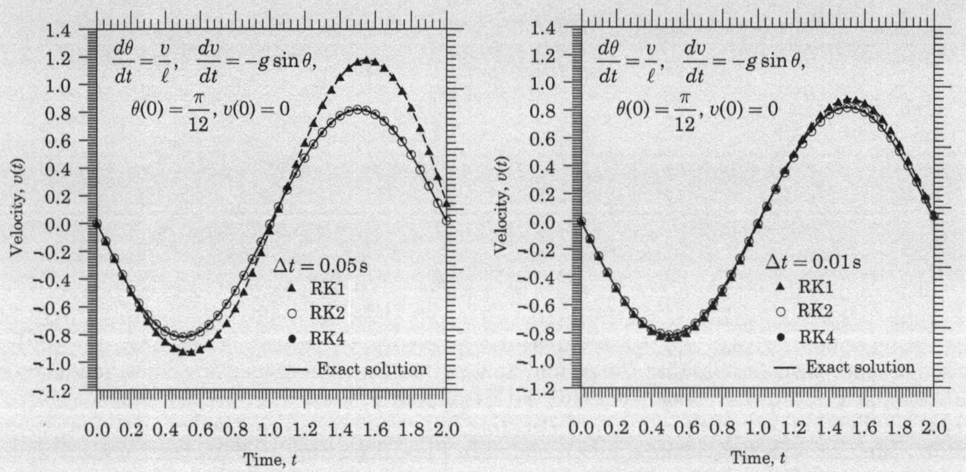

Fig. 2.2.3 Comparison of the numerical solutions $v(t)$ versus t, obtained with the RK1 (Euler's), RK2 (second-order Runge–Kutta), and RK4 (fourth-order Runge–Kutta) methods, with the exact solution of the simple pendulum problem ($g = 9.81$ m/s^2 and $\ell = 1$ m). The results are shown for two different time steps, $\Delta t = 0.05$ and $\Delta t = 0.01$. Smaller time step gives solutions closer to the analytical solution.

In this section, we have considered the numerical solution of a single first-order as well as a pair of coupled first-order ODEs. In the next section, we consider the numerical solution of second-order ODEs.

2.3 Solution of Second-Order Ordinary Differential Equations

Consider a second-order ODE of the form

$$-\frac{d^2u}{dx^2} = f\left(x, u, \frac{du}{dx}\right), \quad 0 < x < L. \tag{2.3.1}$$

When Eq. (2.3.1) defines a BVP, we must specify boundary conditions at $x = 0$ and $x = L$. The following boundary conditions provide an example:

$$u(0) = u_0, \quad \left[\frac{du}{dx}\right]_{x=L} = Q_L, \tag{2.3.2}$$

where u_0 and Q_L are specified boundary values.

If Eq. (2.3.1) defines an IVP, we have the following initial conditions:

$$u(0) = u_0, \quad \left[\frac{du}{dx}\right]_{x=0} = v_0, \tag{2.3.3}$$

where u_0 and v_0 are specified initial values. It is possible to rewrite a second-order equation as a pair of first-order equations, as illustrated with the simple pendulum problem (see **Example 2.2.1**).

Here we consider an example of a second-order differential equation governing a boundary-value problem of heat transfer.

Example 2.3.1

A steel rod of uniform diameter $d = 0.02$ m, length $L = 0.05$ m, and constant thermal conductivity $k = 50$ W/(m·°C) is exposed to ambient air at $T_\infty = 20$°C with a heat transfer coefficient $\beta = 100$ W/(m^2·°C). The left end of the rod is maintained at temperature $T_0 = 320$°C, and the other end is insulated, as shown in Fig. 2.3.1(a). Determine the temperature distribution and the heat input at the left end of the rod using uniform meshes of two and four intervals (i.e., three mesh points and five mesh points, respectively).

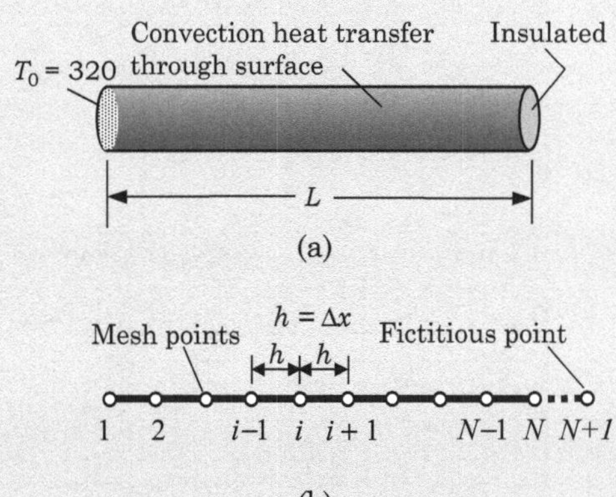

Fig. 2.3.1 (a) Heat transfer in a rod. (b) Finite difference mesh.

Solution: This boundary-value problem was discussed in Eqs. (1.5.14)–(1.5.21b). We shall use the dimensionless form of the governing equation in (1.5.16a):

$$-\frac{d^2u}{dx^2} + m^2 u = 0 \quad \text{for} \quad 0 < x < L, \tag{2.3.4}$$

where $u = T - T_\infty$, T being the temperature, and m^2 is given by

$$m^2 = \frac{\beta P}{Ak} = \frac{\beta \pi d}{\frac{1}{4}\pi d^2 k} = \frac{4\beta}{kd} = \frac{4 \times 100}{50 \times 0.02} = 400. \tag{2.3.5}$$

Assume the following boundary conditions of the problem [a special case of Set 2 boundary condition in Eq. (1.5.16c); i.e., $\beta = 0$]:

$$u(0) \equiv \hat{u}_0 = T(0) - T_\infty = 300°\text{C}, \quad \left.\frac{d\theta}{dx}\right|_{x=L} = 0. \tag{2.3.6}$$

The exact solution is given by [from Eq. (1.5.21b) with $\beta = 0.0$]

$$u(x) = \hat{u}_0 \frac{\cosh m(L-x)}{\cosh mL}, \quad q(0) = -\frac{du}{dx} = mu_0 \frac{\sinh mL}{\cosh mL}. \tag{2.3.7}$$

We identify a number of mesh points, including the boundary points $x = 0$ and $x = L$ (in the domain and on the boundary), as shown in Fig. 2.3.1(b) or Fig. 2.3.2(a). The second derivative of the function $u(x)$ at a mesh point x_I in the domain is approximated using the centered difference formula (2.1.10):

$$\left.\frac{d^2u}{dx^2}\right|_{x=x_i} \approx \frac{1}{h^2}\left(u_{i-1} - 2u_i + u_{i+1}\right), \tag{2.3.8}$$

where $h = \Delta x$ is the distance between two mesh points. We note that the coefficients in the finite difference equation (2.3.8) are real numbers (i.e., no evaluations of integrals is required). Substituting the above formula for the second derivative into Eq. (2.3.4), we arrive at

$$-u_{i-1} + (2 + m^2h^2)u_i - u_{i+1} = 0, \tag{2.3.9}$$

which is valid for any point where $u(x)$ is not specified. By applying the formula in Eq. (2.3.9) to nodes $2, 3, \ldots, N$, we obtain a set of linear algebraic relations among the values of $u(x)$ at the mesh points. Note that application of the formula to the last mesh point N brings the mesh point $N+1$, which is not a part of the domain, and it is called a "fictitious" mesh point. We shall discuss how to deal with this situation shortly.

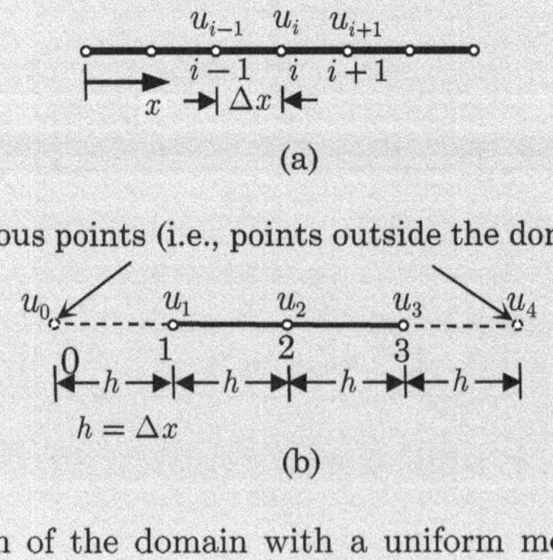

Fig. 2.3.2 (a) Discretization of the domain with a uniform mesh/grid of points. (b) Discretization of the domain with two subdivisions (i.e., two boundary points and one interior point).

First, we choose a grid of two intervals of equal length ($h = 0.025$), involving two boundary points and one point in the middle of the domain [see Fig. 2.3.2(b)]. Applying Eq. (2.3.9) at nodes 2 and 3, we obtain ($m^2 = 400$)

$$\begin{aligned} -u_1 + \left(2 + 400h^2\right) u_2 - u_3 = 0, \\ -u_2 + \left(2 + 400h^2\right) u_3 - u_4 = 0, \end{aligned} \tag{2.3.10}$$

where $u_1 = u(0) = 300°\text{C}$. Note that u_4 is the value of $u(x)$ at the fictitious node 4. To eliminate u_4, we can use one of the following formulas:

$$\left.\frac{du}{dx}\right|_{x=L} = \frac{u_4 - u_3}{h} = 0 \quad \text{(forward difference)} \tag{2.3.11a}$$

$$\left.\frac{du}{dx}\right|_{x=L} = \frac{u_4 - u_2}{2h} = 0 \quad \text{(central difference).} \tag{2.3.11b}$$

The latter is of order $O(h^2)$, consistent with the centered difference formula in Eq. (2.3.8). Using Eq. (2.3.11b), we set $u_4 = u_2$ in Eq. (2.3.10). Then Eq. (2.3.10) can be expressed in matrix form as

$$\begin{bmatrix} 2.25 & -1.00 \\ -2.00 & 2.25 \end{bmatrix} \begin{Bmatrix} u_2 \\ u_3 \end{Bmatrix} = \begin{Bmatrix} 300 \\ 0 \end{Bmatrix}. \tag{2.3.12}$$

The solution of these equations is

$$u_2 = 220.41°\ \text{C}, \quad u_3 = 195.92°\text{C}. \tag{2.3.13}$$

The exact values are $u(0.025) = 219.23°\text{C}$ and $u(0.05) = 194.42°\text{C}$. The heat at $x = 0$ is computed using the relation

$$\begin{aligned} q(0) &= -Ak\left.\frac{du}{dx}\right|_{x=0} \\ &= Ak\frac{u_1 - u_2}{h} = 3183.6\,Ak\ \text{W}. \end{aligned} \tag{2.3.14}$$

The actual value is $4569.56\,Ak$ W.

For a mesh/grid of four intervals (or five points; $h = 0.0125$), the application of Eq. (2.3.9) at mesh points 2, 3, 4, and 5, and using $u_6 = u_4$, gives

$$\begin{bmatrix} 2.0625 & -1 & 0 & 0 \\ -1 & 2.0625 & -1 & 0 \\ 0 & -1 & 2.0625 & -1 \\ 0 & 0 & -2 & 2.0625 \end{bmatrix} \begin{Bmatrix} u_2 \\ u_3 \\ u_4 \\ u_5 \end{Bmatrix} = \begin{Bmatrix} 300 \\ 0 \\ 0 \\ 0 \end{Bmatrix}. \tag{2.3.15}$$

The linear algebraic equations in Eq. (2.3.15) can be solved by any of the methods discussed in Section 1.9. Unless stated otherwise, we shall use the Gauss elimination method in the remainder of the book. The solution of these equations is

$$u_2 = 251.89°\text{C}, \quad u_3 = 219.53°\text{C}, \quad u_4 = 200.89°\text{C}, \quad u_5 = 194.80°\text{C}. \tag{2.3.16}$$

The heat at $x = 0$ is

$$\begin{aligned} q(0) &= -Ak\left.\frac{du}{dx}\right|_{x=0} \\ &= Ak\frac{u_1 - u_2}{h} = 3,848.8\,Ak\ \text{W}. \end{aligned} \tag{2.3.17}$$

A computer program can be written to implement the steps described in the preceding discussion. A comparison of the finite difference solutions with the solutions obtained with a various number of subdivisions, N ($N + 1$ mesh points), is presented in Table 2.3.1. The finite difference solution converges to the exact solution as the number of subdivisions is increased.

Table 2.3.1 Comparison of finite difference solutions with the exact solution of the steady-state heat transfer problem.

	Finite Difference Solution				Exact
$x \times 10^2$	$N = 2$	$N = 4$	$N = 8$	$N = 16$	Solution
0.31250	—	—	—	286.30	286.30
0.62500	—	—	273.74	273.72	273.71
0.93750	—	—	—	262.21	262.20
1.25000	—	251.89	251.75	251.72	251.71
1.56250	—	—	—	242.21	242.20
1.87500	—	—	233.70	233.66	233.64
2.18750	—	—	—	226.01	225.99
2.50000	220.41	219.53	219.30	219.25	219.23
2.81250	—	—	—	213.34	213.32
3.12500	—	—	208.33	208.25	208.25
3.43750	—	—	—	204.01	203.99
3.75000	—	200.89	200.61	200.55	200.52
4.06250	—	—	—	197.87	197.84
4.37500	—	—	196.03	195.96	195.94
4.68750	—	—	—	194.82	194.80
5.00000	195.92	194.80	194.51	194.44	194.42

Example 2.3.2

From **Example 1.11.1**, the bending of straight beams (without elastic foundation, i.e., $k = 0$) according to the Euler–Bernoulli beam theory (under the assumption of infinitesimal deformation) is governed by the following equations:

$$-\frac{d^2M}{dx^2} - q = 0, \quad \frac{d^2w}{dx^2} + \frac{M}{EI} = 0, \quad 0 < x < L, \tag{1}$$

where $w(x)$ is the transverse deflection, $M(x)$ is the bending moment, $q(x)$ is the distributed transverse load, and L and EI are the length and bending stiffness, respectively. For constant values of EI and $q = q_0$ (i.e., uniformly distributed load of intensity q_0), determine the deflection and bending moment of (a) simply supported [see Fig. 2.3.3(a)] and (b) clamped [see Fig. 2.3.3(b)] beams using the finite difference method.

Solution: Since both equations are second-order ODEs, we can use the procedure of **Example 2.3.1** to each of the equations. A discretized domain of N subdivisions is shown in Fig. 2.3.3(c). For each case of the two boundary conditions, we set up the necessary equations to determine the values of $w(x)$ and $M(x)$ at the mesh points.

The central difference approximation of the two equations in Eq. (1) are [see Eq. (2.1.10)]

$$-M_{i-1} + 2M_i - M_{i+1} = q_0(\Delta x)^2, \quad -w_{i-1} + 2w_i - w_{i+1} = M_i\frac{(\Delta x)^2}{EI}, \tag{2}$$

where Δx is the distance between mesh points. To begin with, we shall consider the case of four ($N = 4$) subdivisions in the domain (i.e., $\Delta x = 0.25L$).

(a) *Simply supported beam* For this case, the boundary conditions at the simply-supported ends $x = 0$ and $x = L$ are $w = 0$ and $M = 0$. The analytical (exact) solution for this case is given by ($\xi = x/L$)

$$w(x) = \frac{q_0L^4}{24EI}\left(\xi - 2\xi^3 + \xi^4\right), \quad M(x) = \frac{q_0L^2}{2}\xi(1 - \xi), \tag{3}$$

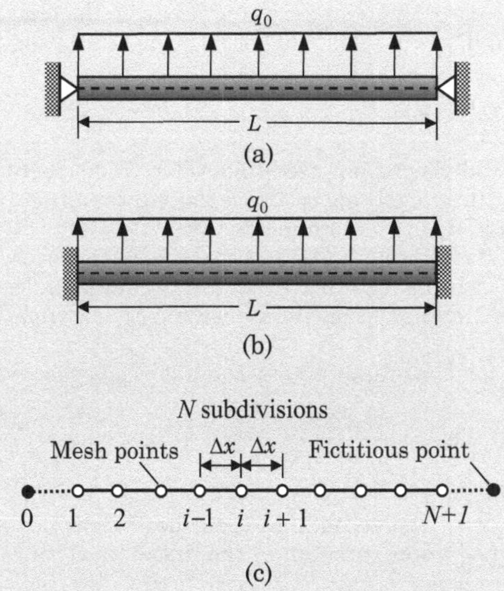

Fig. 2.3.3 (a) Simply supported beam, (b) clamped beam, and (c) discretization of the domain $(0, L)$ of the beam with a uniform mesh/grid of N subdivisions ($N + 1$ mesh points).

For the mesh of four subdivisions, we have $w_1 = w_5 = M_1 = M_5 = 0$. Therefore, we need to develop three algebraic equations among the mesh-point values of both w and M (i.e., three equations each). For the case of simply supported beams, the discretized equations for w and M are uncoupled, allowing us to solve first for M_1, M_2, and M_3, and then for w_1, w_2, and w_3.

Applying the first stencil in Eq. (2) to mesh points 2, 3, and 4, we obtain

$$\begin{bmatrix} 2 & -1 & 0 \\ -1 & 2 & -1 \\ 0 & -1 & 2 \end{bmatrix} \begin{Bmatrix} M_2 \\ M_3 \\ M_4 \end{Bmatrix} = q_0(\Delta x)^2 \begin{Bmatrix} 1 \\ 1 \\ 1 \end{Bmatrix}. \tag{4}$$

The solution of these equations (by Cramer's rule) gives

$$M_2 = \frac{6q_0(\Delta x)^2}{4} = \frac{3}{32}q_0L^2, \quad M_3 = \frac{8q_0(\Delta x)^2}{4} = \frac{1}{8}q_0L^2, \quad M_4 = \frac{6q_0(\Delta x)^2}{4} = \frac{3}{32}q_0L^2. \tag{5}$$

These values match with the exact values of the bending moment at the mesh points.

Applying the second stencil in Eq. (2) at mesh points 2, 3, and 4, and using the solution from Eq. (5), we obtain

$$\begin{bmatrix} 2 & -1 & 0 \\ -1 & 2 & -1 \\ 0 & -1 & 2 \end{bmatrix} \begin{Bmatrix} w_2 \\ w_3 \\ w_4 \end{Bmatrix} = \frac{(\Delta x)^2}{EI} \begin{Bmatrix} M_2 \\ M_3 \\ M_4 \end{Bmatrix} = \frac{q_0(\Delta x)^4}{2EI} \begin{Bmatrix} 3 \\ 4 \\ 3 \end{Bmatrix}. \tag{6}$$

The solution of these equations is

$$w_2 = \frac{5q_0(\Delta x)^4}{2} = \frac{5}{512}\frac{q_0L^4}{EI}, \quad w_3 = \frac{7q_0(\Delta x)^4}{2} = \frac{7}{512}\frac{q_0L^4}{EI} \quad w_4 = \frac{5q_0(\Delta x)^4}{2} = \frac{5}{512}\frac{q_0L^4}{EI}. \tag{7}$$

This solution for w does not match with the exact solution at the mesh points. With increased subdivisions of the domain, we hope to obtain more accurate solution.

(b) *Clamped beam* For this case, the boundary conditions at the clamped ends are that $w = 0$ and $dw/dx = 0$ at $x = 0$ and $x = L$. The analytical (exact) solution for this case is given by ($\xi = x/L$)

$$w(x) = \frac{q_0 L^4}{24EI}\xi^2 (1-\xi)^2, \quad M(x) = -\frac{q_0 L^2}{12}\left(1 - 6\xi + 6\xi^2\right). \tag{8}$$

For the mesh of four subdivisions, we need to develop three algebraic equations among the mesh-point values of w and five algebraic relations among the mesh-point values of M. First, we use the condition $dw/dx = 0$ to determine the values w_0 and w_6 [values at the fictitious nodes; see Fig. 2.3.3(c)] in terms of w_2 and w_4, respectively. However, we do not have any way to determine M_0 and M_6 in terms of other mesh point values (unless we have the condition $dM/dx = 0$ at mesh points 1 and 5) when the stencil for M is evaluated at mesh points 1 and 5.

The use of the stencil for w at mesh points 1 and 5 yields the equations

$$-w_0 + 2w_1 - w_2 = \frac{M_1(\Delta x)^2}{EI}, \quad -w_4 + 2w_5 - w_6 = \frac{M_5(\Delta x)^2}{EI}. \tag{9}$$

The boundary condition $dw/dx = 0$ at mesh points 1 and 5 yields, using the second-order accurate [to be consistent with the second-order accurate central difference formula used to obtain the stencils in Eq. (2)] approximation of the first derivative [see Eq. (2.1.9)]

$$\left.\frac{dw}{dx}\right|_{x=0} \approx \frac{w_2 - w_0}{2\Delta x} = 0, \quad \left.\frac{dw}{dx}\right|_{x=L} \approx \frac{w_6 - w_4}{2\Delta x} = 0, \tag{10}$$

which give $w_0 = w_2$ and $w_6 = w_4$, and Eq. (9) becomes (with the boundary conditions $w_1 = w_5 = 0$)

$$w_2 = -\frac{(\Delta x)^2}{2EI}M_1, \quad w_4 = -\frac{(\Delta x)^2}{2EI}M_5. \tag{11}$$

The use of the stencil for w from Eq. (2) at mesh points 2, 3, and 4 gives the matrix equations

$$\begin{bmatrix} 2 & -1 & 0 \\ -1 & 2 & -1 \\ 0 & -1 & 2 \end{bmatrix} \begin{Bmatrix} w_2 \\ w_3 \\ w_4 \end{Bmatrix} = \frac{(\Delta x)^2}{EI} \begin{Bmatrix} M_2 \\ M_3 \\ M_4 \end{Bmatrix}. \tag{12}$$

Thus, the use of the stencil for M at mesh points 2, 3, and 4 yields the equations

$$\begin{bmatrix} -1 & 2 & -1 & 0 & 0 \\ 0 & -1 & 2 & -1 & 0 \\ 0 & 0 & -1 & 2 & -1 \end{bmatrix} \begin{Bmatrix} M_1 \\ M_2 \\ M_3 \\ M_4 \\ M_5 \end{Bmatrix} = q_0(\Delta x)^2 \begin{Bmatrix} 1 \\ 1 \\ 1 \end{Bmatrix}. \tag{13}$$

Clearly, Eqs. (11)–(13) are coupled between the values of w and M at the mesh points. To solve these equations simultaneously, we write them into a single set of (unsymmetric) matrix equations as follows:

$$\begin{bmatrix} \alpha & 0 & 0 & 0 & 0 & 1 & 0 & 0 \\ -1 & 2 & -1 & 0 & 0 & 0 & 0 & 0 \\ 0 & -1 & 2 & -1 & 0 & 0 & 0 & 0 \\ 0 & 0 & -1 & 2 & -1 & 0 & 0 & 0 \\ 0 & 0 & 0 & 0 & \alpha & 0 & 0 & 1 \\ 0 & -2\alpha & 0 & 0 & 0 & 2 & -1 & 0 \\ 0 & 0 & -2\alpha & 0 & 0 & -1 & 2 & -1 \\ 0 & 0 & 0 & -2\alpha & 0 & 0 & -1 & 2 \end{bmatrix} \begin{Bmatrix} M_1 \\ M_2 \\ M_3 \\ M_4 \\ M_5 \\ w_2 \\ w_3 \\ w_4 \end{Bmatrix} = \begin{Bmatrix} 0 \\ q_0(\Delta x)^2 \\ q_0(\Delta x)^2 \\ q_0(\Delta x)^2 \\ q_0(\Delta x)^2 \\ 0 \\ 0 \\ 0 \end{Bmatrix}, \quad \alpha = \frac{(\Delta x)^2}{2EI}. \tag{14}$$

Table 2.3.2 contains a comparison of the dimensionless deflections and moments obtained using FDM with the exact solutions for simply supported and clamped beams at the mesh points of 4, 8, 16, 32, and 54 subdivisions. It is clear that the FDM is very slow in converging to the exact solution, especially for the clamped beams.

Table 2.3.2 Comparison of finite difference solutions for the dimensionless deflection $\bar{w}(\xi) = w(\xi)\times(EI/q_0L^4)$ and bending moment $\bar{M}(\xi) = M(\xi)/q_0L^2$ ($\xi = x/L$) with the exact solutions of simply supported and clamped beams bent by uniformly distributed transverse load of intensity q_0.

	Finite Difference Solution					
Variable	$N=4$	$N=8$	$N=16$	$N=32$	$N=64$	Exact
Simply Supported						
$w(0.125)\times 10^2$	——	0.5127	0.5074	0.5062	0.5057	0.5056
$w(0.250)\times 10^2$	0.9765	0.9399	0.9308	0.9285	0.9279	0.9277
$w(0.375)\times 10^2$	——	1.2207	1.2093	1.2064	1.2057	1.2054
$w(0.500)\times 10^2$	1.3672	1.3184	1.3062	1.3031	1.3023	1.3021
$M(0.125)\times 10$	——	0.5469	0.5469	0.5469	0.5469	0.5469
$M(0.250)\times 10$	0.9375	0.9375	0.9375	0.9375	0.9375	0.9375
$M(0.375)\times 10$	——	1.1719	1.1719	1.1719	1.1719	1.1719
$M(0.500)\times 10$	1.2500	1.2500	1.2500	1.2500	1.2500	1.2500
Clamped						
$w(0.125)\times 10^2$	——	0.0641	0.0534	0.0507	0.0501	0.0498
$w(0.250)\times 10^2$	0.2441	0.1709	0.1526	0.1480	0.1469	0.1465
$w(0.375)\times 10^2$	——	0.2594	0.2365	0.2308	0.2294	0.2289
$w(0.500)\times 10^2$	0.3906	0.2930	0.2604	0.2625	0.2609	0.2604
$M(0.000)\times 10$	−0.7813	−0.8203	−0.8301	−0.8325	−0.8331	−0.8333
$M(0.125)\times 10$	——	−0.2734	−0.2832	−0.2856	−0.2863	−0.2865
$M(0.250)\times 10$	0.1563	0.1172	0.1074	0.1050	0.1044	0.1042
$M(0.375)\times 10$	——	0.3516	0.3418	0.3394	0.3388	0.3385
$M(0.500)\times 10$	0.4688	0.4297	0.4199	0.4175	0.4169	0.4167

2.4 Solution of Partial Differential Equations

2.4.1 One-Dimensional Problems

Partial differential equations involving one dependent unknown and one spatial coordinate is necessarily an initial- and boundary-value problem. As an example, consider the following partial differential equation (PDE) governing unsteady-state heat transfer in a bar:

$$\rho c_p A\frac{\partial T}{\partial t} - \frac{\partial}{\partial x}\left(kA\frac{\partial T}{\partial x}\right) = f(x,t), \quad 0 < x < L, t > 0, \tag{2.4.1}$$

where $T(x,t)$ denotes the temperature (°C), f is the internal heat generation (W/m), A is the area of cross-section (m^2) of the bar, k is the conductivity of the material (W/(m·°C)), c_p is the specific heat at constant pressure (Joules/(kg °C)), and ρ is the density (kg/m^3). Assuming that $f = 0$ and kA and $\rho\, c_p A$ are constant, we can write

$$\frac{\partial T}{\partial t} - \alpha\,\frac{\partial^2 T}{\partial x^2} = 0, \quad 0 < x < L, \quad t > 0; \quad \alpha = \frac{kA}{\rho\, c_p A}. \tag{2.4.2}$$

The parameter α is known as the *thermal diffusivity*. Introducing the dimensionless variables

$$u = \frac{T - T_0}{T_1 - T_0}, \quad \tau = \frac{\alpha t}{L^2}, \quad \xi = \frac{x}{L}, \tag{2.4.3}$$

Eq. (2.4.2) can be expressed as

$$\frac{\partial u}{\partial \tau} - \frac{\partial^2 u}{\partial \xi^2} = 0. \tag{2.4.4}$$

2.4.1.1 Explicit scheme

The finite difference form of the PDE in Eq. (2.4.4) is [both time and spatial approximations are of second-order accurate; see Eqs. (2.1.9) and (2.1.10)]

$$\frac{1}{\Delta\tau}\left(u_{i,n+1} - u_{i,n}\right) - \frac{1}{(\Delta\xi)^2}\left(u_{i-1,n} - 2u_{i,n} + u_{i+1,n}\right) = 0, \tag{2.4.5}$$

or

$$u_{i,n+1} = \lambda\left(u_{i-1,n} + u_{i+1,n}\right) + (1 - 2\lambda)u_{i,n}, \quad \lambda = \frac{\Delta\tau}{(\Delta\xi)^2}, \tag{2.4.6}$$

where $u_{i,n} = u(\xi_i, \tau_n)$ and $\Delta\tau$ and $\Delta\xi$ are the time and spatial increments. This scheme is called an *explicit scheme* because $u_{i,n+1}$ can be solved for without inverting a matrix (because all terms on the right-hand side are known from the previous time step).

2.4.1.2 Implicit scheme

If we use the backward difference to approximate the time derivative (i.e., evaluating the derivatives at time $\tau = \tau_{n+1}$), we obtain

$$\frac{1}{\Delta\tau}\left(u_{i,n+1} - u_{i,n}\right) - \frac{1}{(\Delta\xi)^2}\left(u_{i-1,n+1} - 2u_{i,n+1} + u_{i+1,n+1}\right) = 0 \tag{2.4.7}$$

$$-\lambda\, u_{i-1,n+1} + (1 + 2\lambda)u_{i,n+1} - \lambda u_{i+1,n+1} = u_{i,n}, \quad \lambda = \frac{\Delta\tau}{(\Delta\xi)^2}. \tag{2.4.8}$$

This scheme is called an *implicit scheme* because Eq. (2.4.8) provides a relation among $u_{i-1,n+1}$, $u_{i,n+1}$, and $u_{i+1,n+1}$ at time τ_{n+1} in terms of the solution at τ_n, and thus requires inversion of a coefficient matrix to determine the solution.

2.4.2 Consistency, Stability, and Convergence

2.4.2.1 Consistency

The term *consistency* of a discretized equation means that the discretized equation may in fact approximate the actual PDE under study, and not the solution of some other PDE. To check for consistency, we must show that the difference between the discretized PDE and the actual PDE (i.e., the truncation error) goes to zero as Δt and Δx go to zero.

For example, consider the dimensionless heat conduction equation in Eq. (2.4.4). The truncation error in the explicit finite difference approximation is

$$\frac{1}{\Delta\tau}\left(u_{i,n+1}-u_{i,n}\right)-\frac{1}{(\Delta\xi)^2}\left(u_{i-1,n+1}-2u_{i,n+1}+u_{i+1,n+1}\right)-\left(\frac{\partial u}{\partial\tau}-\frac{\partial^2 u}{\partial\xi^2}\right)$$
$$=O\{(\Delta\tau)\}+O\{(\Delta\xi)^2\}. \tag{2.4.9}$$

Clearly, the truncation error goes to zero as $\Delta\tau$ and $\Delta\xi$ go to zero, and the explicit scheme in Eq. (2.4.5) is consistent with the PDE in Eq. (2.4.4).

2.4.2.2 Stability

In a time-marching scheme for a PDE, the solution of the discretized equation (i.e., the finite difference scheme) at a given time step depends on the solution from the previous time step. Any error introduced through the initial and boundary conditions or in the calculations can be amplified during the repeated use of the discretized equation. When such error grows without bounds, the solution of the discretized equation is said to be *unstable*. If the error remains bounded for any choice of the time and spatial grid sizes, Δt and Δx, the numerical solution is said to be *stable*. The numerical solution is said to be *conditionally stable* when the error in the numerical solution is bounded for some restriction on the ratio of Δt to Δx. It should be noted that stability does not mean that the error between the exact solution and the numerical solution is small; stability only implies the boundedness of the solution of the discretized equation.

2.4.2.3 Convergence

The term *convergence* means that the solution of a discretized PDE (in the absence of any round-off errors) tends to the exact solution of the PDE as the temporal and spatial grid sizes (i.e., Δt and Δx) tend to zero. *When a consistency criterion is satisfied for any discretized equation, stability is both a necessary and sufficient condition for convergence* (see Lax and Richtmyer [31] and Richtmyer [32]). Thus, the accuracy of a stable numerical solution is dependent on the temporal and spatial grid sizes (the smaller the grid sizes, the more accurate the solution would be).

The von Neumann stability analysis of finite difference approximations was discussed by O'Brien, Hyman, and Kaplan [33], and it is outlined briefly here. When a separation of time and space variables can be made, the solution $u_h(x,t)$ to the finite difference equation is assumed to be of the form

$$u_h(x,t)=u_0(t)\,e^{i\beta x},\quad i=\sqrt{-1}, \tag{2.4.10}$$

where β is a positive constant. By substituting Eq. (2.4.10) into the finite difference equation, the form of $u_0(t)$ can be found, and a stability criterion is established by requiring that $u_0(t)$ is bounded as t becomes large. We shall illustrate the procedure in connection with the solution of a PDE in one dimension.

Example 2.4.1

An infinite (into the plane of the paper) slab of constant thickness and width L and thermal diffusivity α is initially (i.e., at $t = 0$) at a uniform temperature T_0. The left and right faces of the slab are maintained at a constant temperature T_1 (see Fig. 2.4.1). Assuming one-dimensional heat flow along the width, determine how the temperature in the slab varies with time and position.

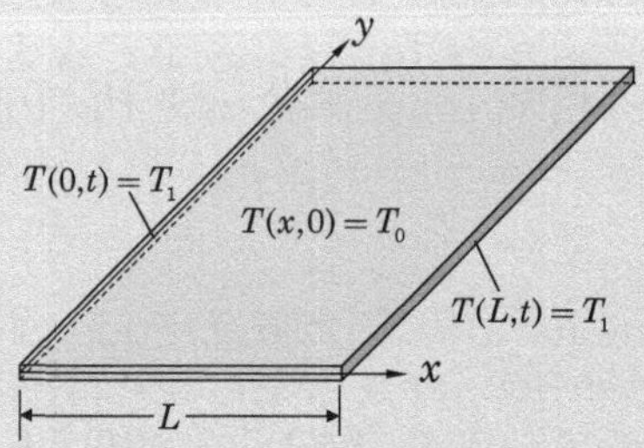

Fig. 2.4.1 Transient heat transfer in an infinite slab of finite width.

Solution: The assumption of no heat flow along the direction (y) into the plane of the paper is justified because of the negligible thermal gradient with respect to y. The governing equation in dimensionless form is given by Eq. (2.4.4). The boundary and initial conditions are

$$u(0,\tau) = u(1,\tau) = 1, \quad \tau > 0; \quad u(\xi,0) = 0, \quad 0 \leq \xi \leq 1. \tag{2.4.11}$$

The exact solution to the problem described by Eqs. (2.4.4) and (2.4.11) is given by

$$u(\xi,\tau) = 1 - \frac{4}{\pi}\sum_{n=1}^{\infty} \frac{\sin \mu_n \xi}{(2n-1)} e^{-\mu_n^2 \tau}; \quad \mu_n = (2n-1)\pi. \tag{2.4.12}$$

Clearly, there is a conflict between the initial and boundary conditions at mesh points $\xi = 0$ and $\xi = 1$. To circumvent this, we shall use the initial condition at all mesh points to be $u_{i,0} = 0.0$; at subsequent times, τ_n, we impose the boundary conditions $u_{1,n} = u_{M+1,n} = 1.0$, $n > 0$, where M is the number of subdivisions of the spatial interval $(0,1)$ (i.e., there are $M+1$ mesh points, as shown in Fig. 2.4.2).

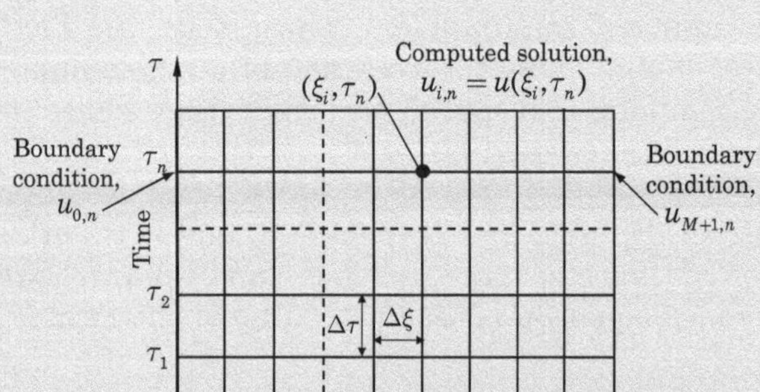

Fig. 2.4.2 The finite difference time-marching scheme for the transient response of a one-dimensional heat transfer problem. At a given time τ_n, the temperature values $u_{i,n}$ for all spatial points ξ_i ($i = 1, 2, \ldots, M+1$) are obtained by solving Eq. (2.4.5).

Applying Eq. (2.4.5) at mesh points $i = 2, 3, \ldots M$, we obtain the necessary algebraic equations to solve for the dimensionless temperatures at nodes $2, 3, \ldots, M$ for all times $\tau_n > 0$. The time-marching scheme begins with $n = 0$, with $u_{i,0} = 0$ at all mesh points of the domain. The solution after the first time step is $u_{1,1} = u_{M+1,1} = 1.0$, and the temperatures at all other mesh points will be zero. The solution develops at the interior mesh points as we march in time.

The von Neumann stability analysis of Eq. (2.4.5) is as follows. Substituting Eq. (2.4.10) into Eq. (2.4.5), we obtain

$$\frac{e^{i\beta\xi}}{\Delta\tau}\left[u_0(\tau+\Delta\tau)-u_0(\tau)\right]=\frac{u_0(\tau)}{(\Delta\xi)^2}\left[e^{i\beta(\xi-\Delta\xi)}-2e^{i\beta\xi}+e^{i\beta(\xi+\Delta\xi)}\right] \tag{2.4.13}$$

from which we obtain

$$u_0(\tau+\Delta\tau)=u_0(\tau)\left[1-4\lambda\sin^2\frac{\beta\,\Delta\xi}{2}\right]. \tag{2.4.14}$$

The solution to Eq. (2.4.14) is obtained as (see **Problem 2.7**)

$$u_0(\tau)=\left[1-4\lambda\sin^2\frac{\beta\,\Delta\xi}{2}\right]^{\frac{\tau}{\Delta\tau}}. \tag{2.4.15}$$

For stability, the function $u_0(\tau)$ should be bounded for all values of β. This condition leads to (because the largest value $\sin(\beta\,\Delta\xi/2)$ can take is unity) the requirement that

$$\left|1-4\lambda\sin^2\frac{\beta\,\Delta\xi}{2}\right|\le 1 \quad\text{or}\quad \lambda=\frac{\Delta\tau}{(\Delta\xi)^2}\le\frac{1}{2}. \tag{2.4.16}$$

For the choice of $\lambda = 0.4$, $\Delta\tau = 0.001$, and $\Delta\xi = 0.05$, Fig. 2.4.3 shows the evolution of the dimensionless temperature u with dimensionless time τ, while Table 2.4.1 contains the dimensionless temperature values at selective times and mesh points. Note that the solution is symmetric about $\xi = 0.5$. The steady-state solution of $u_{i,n} = 1.0$ ($i = 1, 2, \ldots, M+1$) will be reached for $\tau_n \approx 0.9$ (rounding to the fourth decimal point).

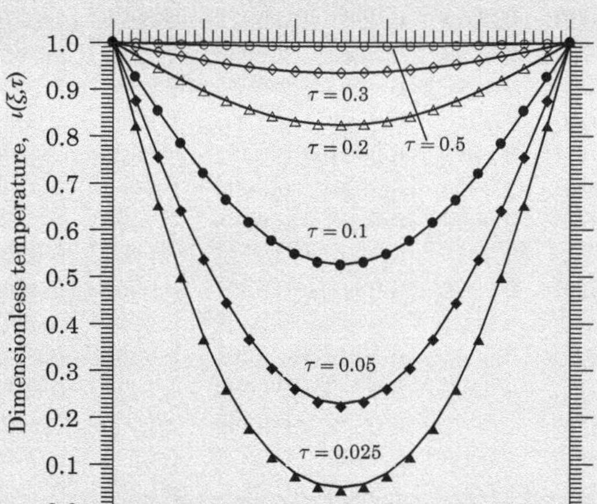

Fig. 2.4.3 Dimensionless temperature $u(\xi, \tau)$ as a function of ξ for different values of the dimensionless time τ.

Table 2.4.1 The dimensionless temperatures obtained with the explicit finite difference scheme applied to the transient heat transfer problem.

	Finite Difference Solution, $u_{i,n}$						Exact
τ_n	$u_{1,n}$	$u_{3,n}$	$u_{5,n}$	$u_{7,n}$	$u_{9,n}$	$u_{11,n}$	$u_{11,n}$
0.000	0.0000	0.0000	0.0000	0.0000	0.0000	0.0000	0.0000
0.025	1.0000	0.6523	0.3667	0.1754	0.0745	0.0443	0.0507
0.050	1.0000	0.7544	0.5356	0.3656	0.2588	0.2225	0.2277
0.075	1.0000	0.8109	0.6405	0.5057	0.4194	0.3897	0.3932
0.100	1.0000	0.8526	0.7196	0.6142	0.5465	0.5231	0.5255
0.125	1.0000	0.8849	0.7811	0.6987	0.6459	0.6276	0.6292
0.150	1.0000	0.9102	0.8291	0.7648	0.7235	0.7093	0.7103
0.175	1.0000	0.9299	0.8666	0.8163	0.7841	0.7730	0.7736
0.200	1.0000	0.9452	0.8958	0.8566	0.8314	0.8228	0.8231
0.225	1.0000	0.9572	0.9187	0.8880	0.8684	0.8616	0.8618
0.250	1.0000	0.9666	0.9365	0.9126	0.8972	0.8919	0.8920
0.275	1.0000	0.9739	0.9504	0.9317	0.9198	0.9156	0.9156
0.300	1.0000	0.9796	0.9613	0.9467	0.9374	0.9341	0.9341
0.325	1.0000	0.9841	0.9698	0.9584	0.9511	0.9486	0.9485
0.350	1.0000	0.9876	0.9764	0.9675	0.9618	0.9598	0.9598
0.375	1.0000	0.9903	0.9816	0.9746	0.9702	0.9686	0.9686
0.400	1.0000	0.9924	0.9856	0.9802	0.9767	0.9755	0.9754
0.425	1.0000	0.9941	0.9888	0.9845	0.9818	0.9809	0.9808
0.450	1.0000	0.9954	0.9912	0.9879	0.9858	0.9851	0.9850
0.475	1.0000	0.9964	0.9932	0.9906	0.9889	0.9883	0.9883
0.500	1.0000	0.9972	0.9947	0.9926	0.9913	0.9909	0.9908
0.525	1.0000	0.9978	0.9958	0.9943	0.9932	0.9929	0.9929
0.550	1.0000	0.9983	0.9967	0.9955	0.9947	0.9945	0.9944
0.575	1.0000	0.9987	0.9975	0.9965	0.9959	0.9957	0.9956
0.600	1.0000	0.9990	0.9980	0.9973	0.9968	0.9966	0.9966
0.625	1.0000	0.9992	0.9984	0.9979	0.9975	0.9974	0.9973
0.650	1.0000	0.9994	0.9988	0.9983	0.9980	0.9979	0.9979
0.675	1.0000	0.9995	0.9991	0.9987	0.9985	0.9984	0.9984
0.700	1.0000	0.9996	0.9993	0.9990	0.9988	0.9987	0.9987
0.725	1.0000	0.9997	0.9994	0.9992	0.9991	0.9990	0.9990
0.750	1.0000	0.9998	0.9995	0.9994	0.9993	0.9992	0.9992
0.775	1.0000	0.9998	0.9996	0.9995	0.9994	0.9994	0.9994
0.800	1.0000	0.9999	0.9997	0.9996	0.9996	0.9995	0.9995
0.825	1.0000	0.9999	0.9998	0.9997	0.9997	0.9996	0.9996
0.850	1.0000	0.9999	0.9998	0.9998	0.9997	0.9997	0.9997
0.875	1.0000	0.9999	0.9999	0.9998	0.9998	0.9998	0.9998
0.900	1.0000	0.9999	0.9999	0.9999	0.9998	0.9998	0.9998
0.925	1.0000	1.0000	0.9999	0.9999	0.9999	0.9999	0.9999
0.950	1.0000	1.0000	0.9999	0.9999	0.9999	0.9999	0.9999
0.975	1.0000	1.0000	1.0000	0.9999	0.9999	0.9999	0.9999
1.000	1.0000	1.0000	1.0000	0.9999	0.9999	0.9999	0.9999

It can be established that the implicit finite difference equation (2.4.8) is stable, independent of $\lambda = \Delta\tau/(\Delta\xi)^2$. To see this, substitute $u(\xi,\tau) = u_0(\tau)\,e^{i\beta\xi}$ into Eq. (2.4.8) and obtain

$$\left[1 + 4\lambda\sin^2\frac{\beta\,\Delta\xi}{2}\right] u_0(\tau+\Delta\tau) = u_0(\tau). \tag{2.4.17}$$

The solution to Eq. (2.4.17) is obtained as

$$u_0(\tau) = \left[\frac{1}{1 + 4\lambda \sin^2 \frac{\beta \Delta \xi}{2}}\right]^{\frac{\tau}{\Delta \tau}}. \tag{2.4.18}$$

For stability, the function $u_0(\tau)$ should be bounded for all values of β. This condition is clearly satisfied without any restriction on λ.

Solutions of Eqs. (2.4.4) and (2.4.11) using the implicit finite difference equation (2.4.8) with various sets of values of $\Delta\tau$ and $\Delta\xi$ are presented in Table 2.4.2. We note that the implicit finite difference equation is not subjected to any stability criterion (i.e., $\Delta\tau$ and $\Delta\xi$ can be arbitrary).

Table 2.4.2 The dimensionless temperatures obtained with the implicit finite difference scheme applied to the transient heat transfer problem.

	Finite Difference Solution, $u_{i,n}$			
ξ	$\Delta\tau = 0.0125$, $\Delta\xi = 0.01$, $\lambda = 1.25$	$\Delta\tau = 0.01$, $\Delta\xi = 0.05$, $\lambda = 4.0$	$\Delta\tau = 0.001$, $\Delta\xi = 0.05$, $\lambda = 0.4$	Exact
		$\tau = 0.05$		
0.0000	1.0000	1.0000	1.0000	1.0000
0.1000	0.7339	0.7393	0.7543	0.7558
0.2000	0.5105	0.5168	0.5364	0.5384
0.3000	0.3503	0.3523	0.3683	0.3696
0.4000	0.2567	0.2551	0.2635	0.2637
0.5000	0.2261	0.2232	0.2280	0.2277
		$\tau = 0.10$		
0.0000	1.0000	1.0000	1.0000	1.0000
0.1000	0.8439	0.8459	0.8526	0.8533
0.2000	0.7039	0.7073	0.7196	0.7210
0.3000	0.5939	0.5980	0.6142	0.6161
0.4000	0.5239	0.5281	0.5466	0.5487
0.5000	0.4999	0.5042	0.5233	0.5255

2.4.3 Two-Dimensional Problems

Partial differential equations in two dimensions can be either a boundary-value problem or an initial- and boundary-value problem. For example, consider the problem of finding steady-state heat transfer in a two-dimensional body. The governing equation is of the form

$$-\frac{\partial}{\partial x}\left(k_{xx}\frac{\partial T}{\partial x}\right) - \frac{\partial}{\partial y}\left(k_{yy}\frac{\partial T}{\partial y}\right) = g(x,y) \text{ in } \Omega, \tag{2.4.19}$$

where Ω is a two-dimensional bounded domain with closed boundary Γ, k_{xx} and k_{yy} are conductivities (W/m·°C) in the x and y coordinate directions (i.e., the body is made of orthotropic material, whose material coordinates coincide with the x and y coordinates) and $g(x, y)$ is the internal heat generation (W/m^3).

Equation (2.4.19) is to be solved with a set of boundary conditions, which involve specifying either the temperature or the flux normal to the boundary:

$$T = \hat{T} \quad \text{or} \quad q_n \equiv \left(k_{xx}\frac{\partial T}{\partial x}\right) n_x + \left(k_{yy}\frac{\partial T}{\partial y}\right) n_y = \dot{q} \quad \text{on } \Gamma, \tag{2.4.20}$$

where (n_x, n_y) are the direction cosines of the unit normal vector $\hat{\mathbf{n}}$ on the boundary.

The finite difference approximation of Eq. (2.4.19) with $k_{xx} = k_{yy} = k$, a constant, is [see Eqs. (2.1.18) and (2.1.19)]

$$\frac{k}{(\Delta x)^2}\left(u_{i-1,j} - 2u_{i,j} + u_{i+1,j}\right) + \frac{1}{(\Delta y)^2}\left(u_{i,j-1} - 2u_{i,j} + u_{i,j+1}\right) = g_{i,j}, \tag{2.4.21}$$

where $u_{i,j} = u(x_i, y_j)$ and Δx and Δy are the spatial increments in the x and y directions. Equation (2.4.21) must be replaced with a suitable forward or backward approximation of the gradient boundary conditions on the boundary. An example of application of this equation is presented in **Example 2.4.2**.

Additional examples of a PDE in two dimensions are provided by the transient heat transfer in a two-dimensional domain

$$\rho c_p \frac{\partial T}{\partial t} - \frac{\partial}{\partial x}\left(k_{xx}\frac{\partial T}{\partial x}\right) - \frac{\partial}{\partial y}\left(k_{yy}\frac{\partial T}{\partial y}\right) = g(x,y,t) \text{ in } \Omega \quad \text{for } t > 0, \tag{2.4.22}$$

and the transient deflections of a membrane

$$\rho\frac{\partial^2 u}{\partial t^2} - a_0\left(\frac{\partial^2 u}{\partial x^2} + \frac{\partial^2 u}{\partial y^2}\right) = f(x,y,t) \text{ in } \Omega \quad \text{for } t > 0, \tag{2.4.23}$$

where $u(x, y, t)$ is the transverse deflection, a_0 is the initial tension in the membrane, and $f(x, y, t)$ is the externally applied force. The finite difference approximations of the equations can be developed along the same lines as given in Eqs. (2.4.5) and (2.4.21).

Example 2.4.2

Consider an infinite (into the plane of the paper) slab of rectangular cross-section of sides $6a$ by $4a$, as shown in Fig. 2.4.4(a). The slab is assumed to be made of isotropic material with conductivity k (W/m °C). The bottom and top surfaces are maintained at a temperature $T = T_1(x) = T_0 \cos\frac{\pi x}{6a}$, while the left and right surfaces are maintained at temperature $T = T_2 = 0$, as shown in Fig. 2.4.4(b). Assuming that heat flow along the length (into the plane of the paper) is negligible, determine how the temperature in the slab varies with position (x, y).

Solution: The governing equation for the problem at hand is

$$k\left(\frac{\partial^2 T}{\partial x^2} + \frac{\partial^2 T}{\partial y^2}\right) = 0. \tag{2.4.24}$$

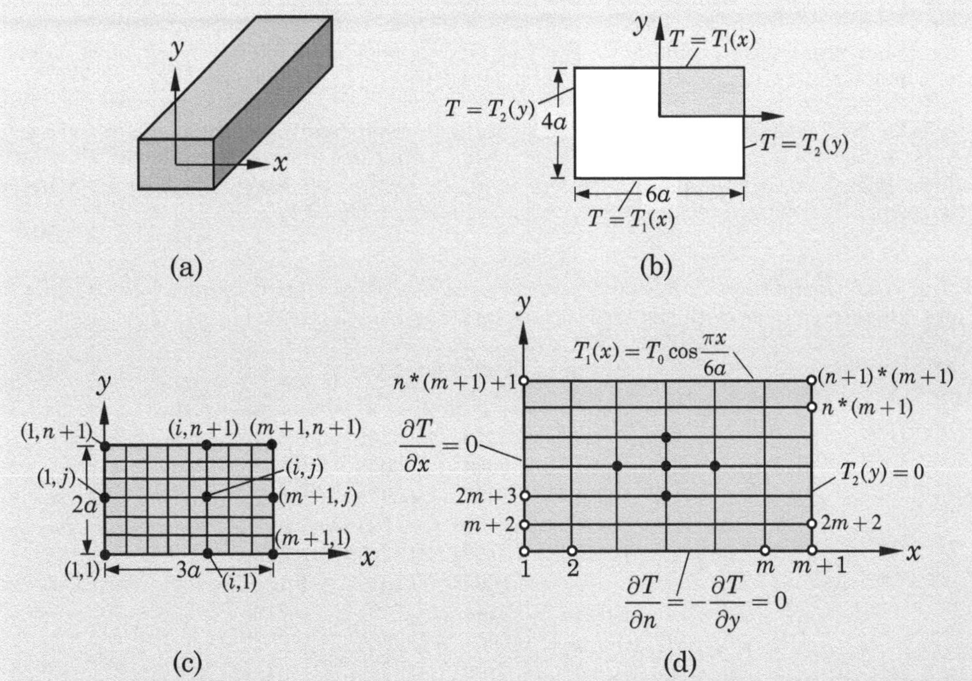

Fig. 2.4.4 (a) An infinite slab of rectangular cross-section with (b) specified surface temperatures. (c) Mesh points in a typical finite difference grid. (d) Actual numbering of the mesh points of the finite difference grid.

Because of the biaxial symmetry of the solution, one can identify one quadrant of the domain, such as the one shown in Fig. 2.4.4(b). For a uniform mesh/grid of m subdivisions in the x direction and n subdivisions in the y direction, the mesh points of the central difference scheme will have two indices, one for the x direction and another for the y direction (i.e., $T_{i,j} = T(x_i, y_j)$, as shown in Fig. 2.4.4(c). However, for a mesh of $m \times n$ subdivisions in the quadrant, there will be $(m+1)$ times $(n+1)$ mesh points in the computational domain, and the mesh points are numbered sequentially, from the left to right and from the bottom to the top, as indicated in Fig. 2.4.4(d).

The boundary conditions for the computational domain are

$$\left.\frac{\partial T}{\partial x}\right|_{(0,y)} = \left.\frac{\partial T}{\partial y}\right|_{(x,0)} = 0; \quad T(3a, y) = 0, \quad T(x, 2a) = T_0 \cos\frac{\pi x}{6a} \tag{2.4.25}$$

The exact solution of Eqs. (2.4.24) and (2.4.25) is

$$T(x,y) = T_0\,\frac{\cosh(\pi y/6a)\cos(\pi x/6a)}{\cosh(\pi/3)}. \tag{2.4.26}$$

We note that the solution does not depend on k because the source term, $g(x,y)$, in the differential equation is zero for the problem at hand.

The central finite difference scheme of the PDE is

$$\frac{k}{(\Delta x)^2}\left(T_{i-1,j} - 2T_{i,j} + T_{i+1,j}\right) + \frac{k}{(\Delta y)^2}\left(T_{i,j-1} - 2T_{i,j} + T_{i,j+1}\right) = 0. \tag{2.4.27}$$

Since the temperature values at the mesh points along lines $x = 3a$ and $y = 2a$ are known, we only need to write the finite difference equations for mesh points along the symmetry lines and mesh points inside the domain. Thus, Eq. (2.4.27) is applied at all interior mesh points. Along the symmetry lines (i.e., $x = 0$ and $y = 0$ lines), we use $T_{i-1,j} = T_{i+1,j}$ and $T_{i,j-1} = T_{i,j+1}$ in Eq. (2.4.27).

Table 2.4.3 contains a comparison of the normalized temperatures $T(x, y)/T_0$ at the mesh points as predicted by the central difference scheme with three different meshes with the exact values. It is clear that the finite difference solutions tend to the exact values with the mesh refinements. A contour plot of the solution is included in Fig. 2.4.5.

Table 2.4.3 Comparison of the nodal temperatures $T(x, y)/T_0$, obtained using various uniform finite difference meshes with the analytical solution in Eq. (2.4.26).

x	y	Finite Difference Solutions 3×2	6×4	12×8	Exact Solution
0.0	0.0	0.6362	0.6278	0.6256	0.6249
0.5	0.0	——	0.6064	0.6043	0.6036
1.0	0.0	0.5510	0.5437	0.5418	0.5412
1.5	0.0	——	0.4439	0.4424	0.4419
2.0	0.0	0.3181	0.3139	0.3128	0.3124
2.5	0.0	——	0.1625	0.1619	0.1617
0.0	1.0	0.7214	0.7148	0.7131	0.7125
0.5	1.0	——	0.6904	0.6888	0.6882
1.0	1.0	0.6248	0.6190	0.6176	0.6171
1.5	1.0	——	0.5054	0.5042	0.5038
2.0	1.0	0.3607	0.3574	0.3565	0.3563
2.5	1.0	——	0.1850	0.1846	0.1844

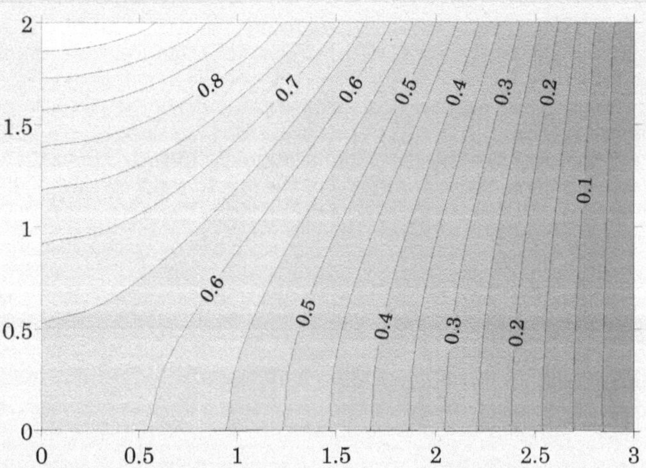

Fig. 2.4.5 A contour plot (i.e., isotherms) of the temperature field.

2.5 Summary

In this chapter, the finite difference solutions of both boundary-value and initial-value problems in one and two dimensions are discussed. Numerical examples using various finite difference methods such as Euler's, forward, backward, central, and Runge–Kutta finite difference schemes are presented.

The finite difference grids used here are uniform (i.e., the temporal and spatial grid sizes in each coordinate direction are constant). It is not common to find variable grid sizes in the same finite difference formula (e.g., the grid sizes on the left and right sides of a mesh point being different). Also, imposition of gradient boundary conditions along an arbitrary boundary line (i.e., not a line of symmetry) is approximate and involves values of the function from the interior (and possibly exterior) points. Such difficulties are not faced in the finite element method or the dual mesh control domain method.

Problems

2.1 (a) Write Eqs. (1.5.14) and (1.5.15b) in the following alternative form using $u = T - T_\infty$ (see Eqs. (1.5.16a) and (1.5.16c)):

$$-\frac{d^2u}{dx^2} + m^2\, u = 0, \quad m = \sqrt{\frac{\beta P}{kA}}, \quad 0 < x < L, \tag{1}$$

with the boundary conditions

$$u(0) = u_0, \quad \left[\frac{du}{dx} + \frac{\beta}{k}u\right]_{x=L} = 0. \tag{2}$$

(b) Use the centered finite difference approximation to develop the discrete equation. Use the diameter of the rod to be $d = 0.02$ m, length $L = 0.05$ m, and thermal conductivity $k = 50$ W/(m °C); take $T_0 = 320$°C, $T_\infty = 20$°C, and $\beta = 100$ W/(m^2 · °C), and determine the solution for four and eight intervals.

2.2 Solve **Problem 2.1** for 16 subdivisions and compare the finite difference solution with the analytical solution. *Answer:* The finite difference solutions are (in °C)

$$\{\theta\} = \{285.36,\ 271.84,\ 259.37,\ 247.92,\ 237.44,\ 227.89,\ 219.22, 211.42,\ \cdots, 176.78\}^{\mathrm{T}}. \tag{1}$$

The analytical solution at the same points is (in °C)

$$\{\theta\} = \{285.56,\ 272.25,\ 259.99,\ 248.75,\ 238.48,\ 229.15,\ 220.71, 213.13,\ \cdots, 180.66\}^{\mathrm{T}}. \tag{2}$$

2.3 An improvement of Euler's scheme is provided by Heun's scheme, which uses the average of the derivatives at the two ends of the interval to estimate the slope. Applied to the equation

$$\frac{du}{dt} = f(t, u). \tag{1}$$

Heun's scheme has the form

$$u_{i+1} = u_i + \frac{\Delta t}{2}\left[f(t_i, u_i) + f(t_{i+1}, \overset{0}{u}_{i+1})\right], \quad \overset{0}{u}_{i+1} = u_i + \Delta t\ f(t_i, u_i). \tag{2}$$

The second equation is known as the *predictor* equation and the first equation is called the *corrector* equation. Apply Heun's scheme to Eqs. (2.2.21a) and (2.2.21b) with the initial conditions $\theta(0) = \theta_0 = \pi/4$ and $v(0) = v_0 = 0$, and obtain the numerical solution for $\Delta t = 0.05$.

2.4 Derive a second-order accurate approximation of the first derivative of a function $f(x)$ at x_i in terms of the function values at x_{i-2}, x_i and x_{i+1}.

2.5 Show that a second-order accurate approximation of the derivative du/dx at the left end is given by

$$\left.\frac{du}{dx}\right|_w \approx \frac{9U_2 - U_3 - 8U_1}{3\Delta x}. \tag{1}$$

2.6 Use the second-order approximation derived in **Problem 2.5** to solve the problem described by Eq. (2.3.4). Use two control volumes of equal length, $h = 0.5$.

2.7 Solve the beam problems of **Example 2.3.2** exploiting the symmetry about the centerline of the beam. Use four subdivisions in the half-beam. *Hint:* The boundary conditions at the line of symmetry (i.e., at $x = L/2$) are: $dw/dx = 0$ and $V = dM/dx = 0$.

2.8 Establish the result in Eq. (2.4.15).

2.9 Establish the result in Eq. (2.4.17).

2.10 Show the approximation used for the first-order derivative at the boundary faces is only first-order accurate.

3
Finite Volume Method

3.1 General Idea

The present chapter deals with the finite volume method (FVM), which is currently the most popular numerical method used in fluid dynamics. Most literature on the FVM cites the book by Patankar [2] for the basic idea behind the method and uses the terminology introduced there. As per the book, the FVM has some similarity to the subdomain method. The subdomain method is a weighted–residual method in which the integral of the differential equation, more precisely, the residual due to the approximation of the differential equation, is set to zero (with the weight function being unity). However, in the subdomain method, the dependent variable(s) of the problem are approximated as linear combinations of arbitrary parameters and approximation functions that meet certain requirements (see Chapter 1 for details). The similarity between the subdomain method and the FVM is only in setting up the integral statement of the differential equation to be solved. Beyond that, there is no connection between the two methods.

In the finite volume method, the domain Ω is divided into a collection of nonoverlapping subdomains, called *control volumes* (CVs); the interfaces of the CVs are points in one dimension, lines in two dimensions, and surfaces in three dimensions. The CVs need not be of the same size (a generalization from the traditional finite difference methods). Then a mesh point (or node) is identified at the center of each CV. Figure 3.1.1 shows an arbitrary two-dimensional domain Ω and its discretization into CVs of quadrilateral shape, with each CV centered around a nodal point.

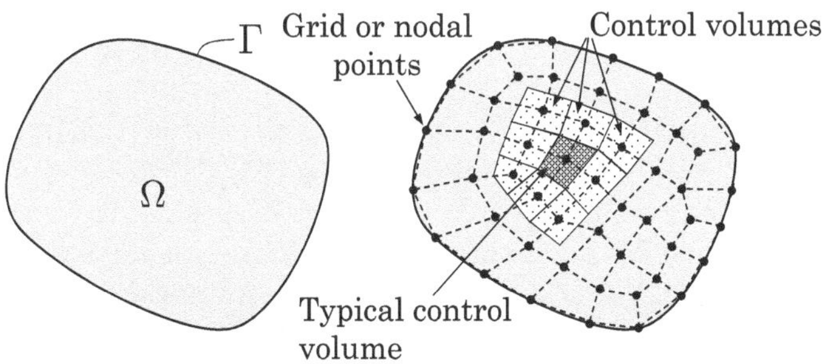

Fig. 3.1.1 Representation of a two-dimensional domain with meshes of node-centered quadrilateral CVs.

The discretized equations of a given differential equation over a typical CV are derived by setting the integral form of the equation over the CV to zero and replacing the flux across the boundaries of the CV with finite Taylor series expansions. The discretized equations thus obtained are in terms of the dependent unknown at the node of the CV as well as at the nodes of the surrounding CVs. In general, the FVM discretization of a domain consists of two different meshes, namely, a mesh of CVs on which the governing differential equations are satisfied (or balanced) and a mesh of nodal points at which the dependent variable is evaluated. The details of this process will be explained for differential equations in one and two dimensions.

3.2 One-Dimensional Problems

3.2.1 Model Differential Equation and Domain Discretization

Consider the following model equation which arises, among many others (see Reddy [8]), in a one-dimensional heat flow problem [see Eq. (1.5.14)]:

$$-\frac{d}{dx}\left(a\frac{du}{dx}\right) + cu = f, \quad 0 < x < L \tag{3.2.1}$$

where u is the dependent variable, and a, c, and f are known functions of x, called the data of the problem. Equation (3.2.1) is to be solved in conjunction with suitable boundary conditions at $x = 0$ and $x = L$. For one-dimensional heat flow, where u denotes temperature above a reference temperature; $a = kA$, k being the conductivity, and A is the cross-sectional area of the bar (or fin); b is typically zero except when we consider advection–diffusion phenomenon (e.g., heat transfer in the presence of flow), $c = P\beta$, P being the perimeter of the bar and β is the convective heat transfer coefficient; and $f(x)$ denotes the internal heat generation. In the remainder of this section, we assume that $b = 0$, and consider an example in Section 3.3 where $b \neq 0$.

The notation used to number the nodes and CVs in most FVM books is that found in the book by Patankar [2]; for example, the terminology of *W*est side for the left side, *E*ast side for the right side, and *N*orth and *S*outh sides for nodes that are above and below the node of interest, respectively, was used. In this book, a different notation that is believed to be convenient and common to other methods is adopted.

There are two different ways to generate the grid/mesh in the FVM: (1) half-control volume approach (HFVM) and (2) zero-thickness control volume (ZFVM) approach. In the HFVM approach, we divide the one-dimensional domain $\Omega = (0, L)$ into a set of N subdivisions, $\Omega^{(1)}$, $\Omega^{(2)}$, $\cdots$, $\Omega^{(N)}$, with each interval having two end nodes. Thus, there are $N + 1$ nodes in the mesh. We identify $N - 1$ interior CVs such that the $N - 1$ interior nodes are inside (e.g., at the center of) these CVs, and two half-control volumes, each containing one boundary node at the "free" end of the CV, as shown in Fig. 3.2.1. The CV sizes can be varied by varying the distance between the nodes. We denote the size of the Ith CV with Δx_I and the distance between nodes $I - 1$ and I with δx_{I-1}. For a uniform mesh, all intervals are of the same size $\Delta x = L/N$, each interior CV is centered around a node, the distance between any two nodes is

δx, and $\delta x = \Delta x$. For a nonuniform mesh, $\Delta x_I = 0.5(\delta x_{I-1} + \delta x_I)$. The CVs and the nodes are numbered sequentially from the left to the right. In this approach, a refined mesh includes the nodes of the previous mesh, as can be seen from Fig. 3.2.2.

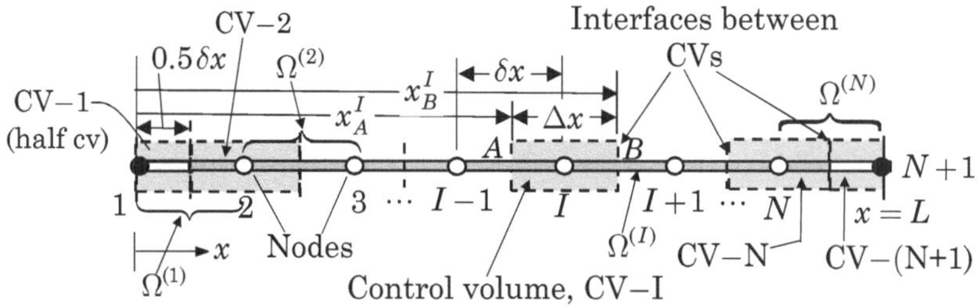

Fig. 3.2.1 The half-control finite volume discretization of a one-dimensional domain; in this figure, there are $N = 7$ subdivisions, $N+1 = 8$ nodes, and $NCV = N+1$ CVs with $N-1 = 6$ internal CVs, and 2 boundary half-CVs. The Ith node is inside the Ith CV.

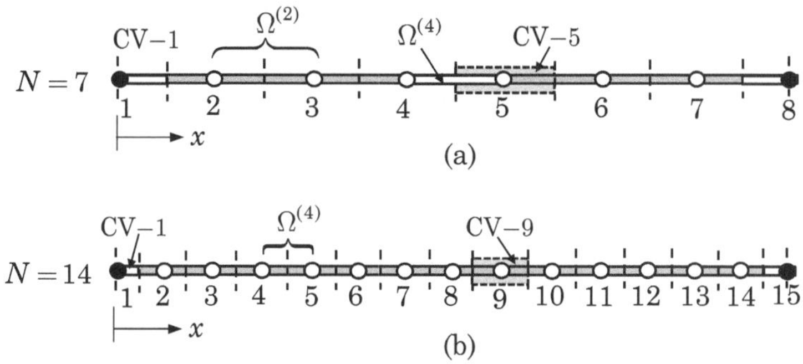

Fig. 3.2.2 (a) Mesh 1 from Fig. 3.2.1, which has $N = 7$ subdivisions and 8 nodes. (b) Mesh 2, a refined mesh that doubles the number of subdivisions in (a), creating 14 subdivisions, 15 nodes, and 15 CVs.

In the second approach (see Reddy, Anand, and Roy [7]), called the "zero-thickness control volume" (ZFVM) approach, the domain is divided into N CVs, with a node at the center of each CV, as shown in Fig. 3.2.3. In addition, the boundary points are also treated as nodes but with a CV that has "zero thickness." Thus, there are $N + 2$ nodal points, as illustrated in Fig. 3.2.3 for the case of $N = 6$.

3.2.2 Integral Representation of the Governing Equation

We consider a typical control volume between points A and B, with its node I and control volumes of different sizes on either side of the CV [it is CV-I in the half-control volume grid and CV-(I-1) in the ZFVM grid], as shown in Figs. 3.2.4(a) and 3.2.4(b), respectively.

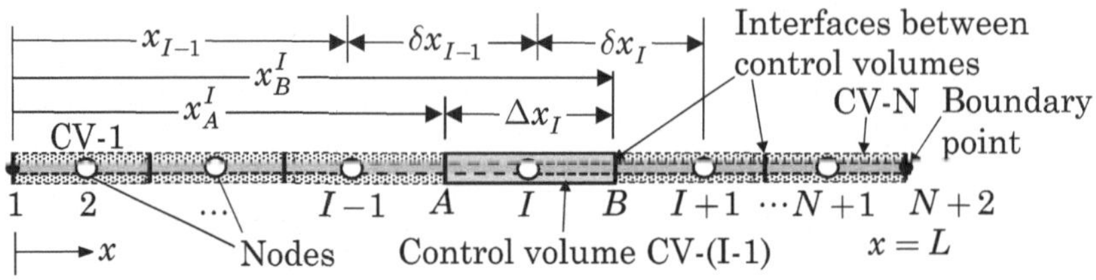

Fig. 3.2.3 ZFVM discretization of a one-dimensional domain; in this figure, there are six subdivisions, $NCV = 6$ CVs, $N = NCV + 2$ nodes (including two boundary nodes). The Ith node is inside the $I - 1$st CV.

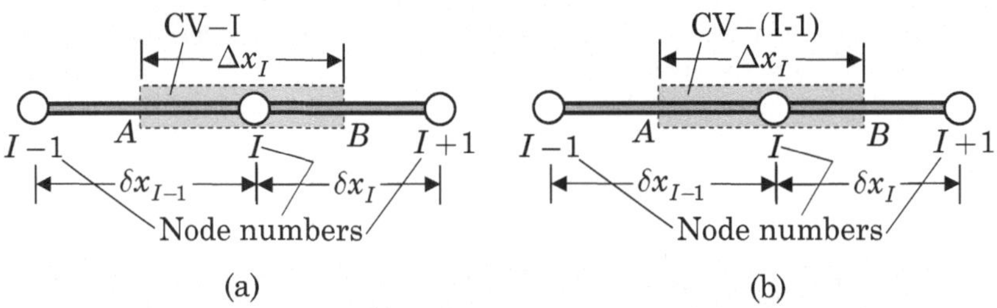

Fig. 3.2.4 Typical one-dimensional control volume with its node I and control volumes on the left (west) and right (east) sides. (a) Control volume CV-I in the HFVM formulation. (b) Control volume CV-(I-1) in the ZFVM formulation. The two internal control volumes differ only in the numbering.

In order to derive the discretized equations associated with the governing differential equation, we begin with the integral statement of Eq. (3.2.1) over the control volume occupying the location between points A and B (once again, except for the numbering of the CVs, both ZFVM and HFVM approaches will have the same discretized equations for an interior control volume):

$$0 = \int_A^B \left[-\frac{d}{dx}\left(a\frac{du}{dx} \right) + cu - f \right] dx. \tag{3.2.2}$$

Points A and B refer to the left and right interfaces, respectively, of the control volume, and x_A^I and x_B^I are coordinates of points A and B, respectively. We weaken the differentiability on u by carrying out the indicated integration and obtain

$$\begin{aligned} 0 &= \int_{x_A^I}^{x_B^I} \left[-\frac{d}{dx}\left(a\frac{du}{dx} \right) + cu - f \right] dx \\ &= -\left[-a\frac{du}{dx} \right]_{x_A^I} - \left[a\frac{du}{dx} \right]_{x_B^I} + \int_{x_A^I}^{x_B^I} (cu - f)\, dx \end{aligned} \tag{3.2.3}$$

or

$$0 = -Q_A^I - Q_B^I + G_I(C_I, U_I) - F_I, \quad Q_A^I \equiv \left[-a\frac{du}{dx}\right]_{x_A^I}, \quad Q_B^I = \left[a\frac{du}{dx}\right]_{x_B^I}, \tag{3.2.4}$$

where the integral of cu and f are approximated as follows:

$$\int_{x_A^I}^{x_B^I} c(x)\, u(x)\, dx \approx G_I(C_I, U_I); \quad \int_{x_A^I}^{x_B^I} f\, dx \approx F_I, \tag{3.2.5}$$

where U_I denotes the value of $u(x)$ at $x = x_I$ and C_I are coefficients whose value depends on the nature of $c(x)$, assumed variation of $u(x)$, and the domain length. For example, the integral of $f(x)$ can be evaluated using the trapezoidal rule or Simpson's rule. Physically, Q_A^I denotes the heat at the left interface and Q_B^I is the heat at the right interface; for axial deformation, Q_A^I denotes the force at the left interface and Q_B^I is the force at the right interface.

3.2.3 Evaluation of Domain Integrals

In some books on the FVM (see, e.g., [7]), the domain integrals of the source as well as dependent variables are evaluated as if they are uniform over the control volume:

$$\int_{x_A^I}^{x_B^I} u(x)\, dx \approx u(x_I)\left(x_B^I - x_A^I\right) = U_I\, \Delta x. \tag{3.2.6}$$

This kind of approximation can introduce additional error into the numerical solution, especially when coarse meshes are used.

Alternatively, one may treat all functions as linear between nodes (consistent with the Taylor series approximation of the first derivatives being used). For example, to evaluate the integral of $u(x)$, we assume that $u(x)$ is linear between the nodes I and $I-1$, as illustrated in Fig. 3.2.5(a):

$$u(\bar{x}) \approx U_{I-1} + \frac{U_I - U_{I-1}}{h}\bar{x}, \quad 0 \le \bar{x} \le h. \tag{3.2.7}$$

This is equivalent to using the trapezoidal rule to evaluate the integral of a function.

The following integral identities hold when $u(x)$ is treated as a linear function over each control volume [see Fig. 3.2.5(b)]:

$$\begin{aligned} \int_{0.5h}^{h} u(\bar{x})\, d\bar{x} &= \frac{h}{8}\left(U_{I-1} + 3U_I\right), \\ \int_{0}^{0.5h} u(\bar{x})\, d\bar{x} &= \frac{h}{8}\left(3U_I + U_{I+1}\right), \end{aligned} \tag{3.2.8}$$

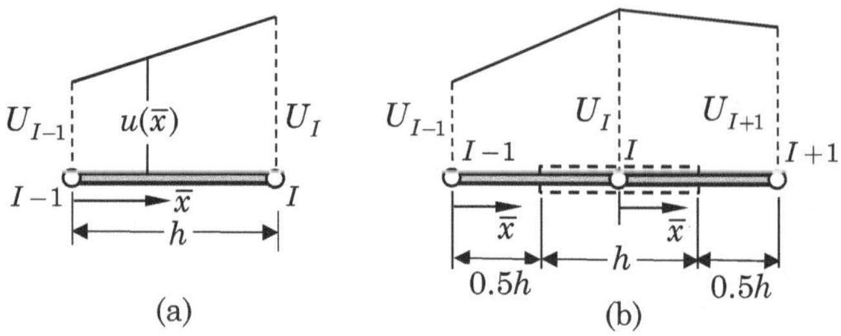

Fig. 3.2.5 (a) Assumed linear variation of the function $u(x)$ between nodes. (b) A typical control volume around node I, which spans two neighboring subdivisions.

and

$$\int_{x_A^I}^{x_I} u(x)dx + \int_{x_I}^{x_B^I} u(x)dx = \frac{h}{8}\left(U_{I-1} + 6U_I + U_{I+1}\right)$$
$$\equiv C_{I-1}\,U_{I-1} + C_I\,U_I + C_{I+1}\,U_{I+1}, \tag{3.2.9a}$$

where

$$C_{I-1} = \frac{h}{8},\quad C_I = \frac{6h}{8},\quad C_{I+1} = \frac{h}{8}. \tag{3.2.9b}$$

3.2.4 Approximation of the First Derivatives

3.2.4.1 Internal to the domain

Toward deriving the discretized equations among the nodal values of the dependent variable u, we must first approximate du/dx in terms of the nodal values of u. The Taylor series expansion of $u(x_I) \equiv U_I$ about $u(x_B^{(I)})$ is

$$u(x_I) = u(x_B^{(I)}) - \frac{\delta x_I}{2}\frac{du}{dx}\bigg|_{x=x_B^{(I)}} + \frac{1}{2!}\left(\frac{\delta x_I}{2}\right)^2 \frac{d^2u}{dx^2}\bigg|_{x=x_B^{(I)}}$$
$$- \frac{1}{3!}\left(\frac{\delta x_I}{2}\right)^3 \frac{d^3u}{dx^3}\bigg|_{x=x_B^{(I)}} + \cdots . \tag{3.2.10}$$

A Taylor series expansion of $u(x_{I+1}) \equiv U_{I+1}$ about $u(x_B^I)$ is

$$u(x_{I+1}) = u(x_B^{(I)}) + \frac{\delta x_I}{2}\frac{du}{dx}\bigg|_{x=x_B^{(I)}} + \frac{1}{2!}\left(\frac{\delta x_I}{2}\right)^2 \frac{d^2u}{dx^2}\bigg|_{x=x_B^{(I)}}$$
$$+ \frac{1}{3!}\left(\frac{\delta x_I}{2}\right)^3 \frac{d^3u}{dx^3}\bigg|_{x=x_B^{(I)}} + \cdots . \tag{3.2.11}$$

By subtracting Eq. (3.2.10) from Eq. (3.2.11) and solving for du/dx at $x = x_B^{(I)}$, we obtain

$$\frac{du}{dx}\bigg|_{x=x_B^{(I)}} = \frac{U_{I+1} - U_I}{\delta x_I} + \frac{(\delta x_I)^2}{24}\frac{d^3U}{dx^3}\bigg|_{x=x_B^{(I)}} + \cdots = \frac{U_{I+1} - U_I}{\delta x_I} + O(\delta x_I)^2. \tag{3.2.12}$$

Similarly, using the Taylor series expansions of U_I and U_{I-1} about $u(x_A^{(I)})$, subtracting the second expansion from the first one, and solving for du/dx at $x = x_A^{(I)}$, we obtain

$$\left.\frac{du}{dx}\right|_{x=x_A^{(I)}} = \frac{U_I - U_{I-1}}{\delta x_{I-1}} + O(\delta x_{I-1})^2. \tag{3.2.13}$$

Since we have used two consecutive nodes to derive the expressions in Eqs. (3.2.11) and (3.2.13), the formulation is termed a *two-node formulation.* We note that the first-derivative approximations are second-order accurate.

3.2.4.2 At the boundaries of the domain

The first derivatives of a function on the boundary are based on either forward difference (for the left boundary) or backward difference (for the right boundary) so that no fictitious points are introduced. The forward and backward difference formulas are only first-order accurate [see Eqs. (2.1.7) and (2.1.8)]. Second-order accurate formulas for the first derivative at the boundary point can be derived (see Eq. (2.1.9) and **Problem 3.5**; $x_A^1 = 0$ and $x_B^N = x_{N+2} = L$, L being the length of the domain):

$$\left.\frac{du}{dx}\right|_{x=x_A^{(1)}} \approx \frac{9U_2 - U_3 - 8U_1}{3\Delta x}, \quad \left.\frac{du}{dx}\right|_{x=x_B^{(N)}} \approx -\frac{9U_{N+1} - U_N - 8U_{N+2}}{3\Delta x}. \tag{3.2.14}$$

3.2.5 Discretized Equations for Interior Nodes

First, we presume that we can express the integral of $c(x)u(x)$ as

$$\int_{x_A^{(I)}}^{x_I} c(x)\, u(x)\, dx = C_{I-1}U_{I-1} + C_I U_I + C_{I+1}U_{I+1}. \tag{3.2.15}$$

When c is a constant, Eq. (3.2.6) gives $C_{I-1} = C_{I+1} = 0$ and $C_I = ch$. On the other hand, if we use Eqs. (3.2.9a) and (3.2.9b), we have $C_{I-1} = ch/8$, $C_I = 6ch/8$, and $C_{I+1} = ch/8$.

Next, we replace quantities $Q_A^{(I)}$ and $Q_B^{(I)}$ in Eq. (3.2.4) using the results in Eqs. (3.2.12) and (3.2.13):

$$\begin{aligned} Q_A^{(I)} &\equiv -\left[a\frac{du}{dx}\right]_A = -\left[a\frac{du}{dx}\right]_{x_A^{(I)}} = -a_A^{(I)}\left(\frac{U_I - U_{I-1}}{\Delta x_{I-1}}\right) \\ Q_B^{(I)} &\equiv \left[a\frac{du}{dx}\right]_B = \left[a\frac{du}{dx}\right]_{x_B^{(I)}} = a_B^{(I)}\left(\frac{U_{I+1} - U_I}{\Delta x_I}\right). \end{aligned} \tag{3.2.16}$$

Then, substituting the approximations in Eq. (3.2.15) and F_I from Eq. (3.2.5) into Eq. (3.2.4), we obtain (for $I = 2, 3, \ldots, N$):

$$A_{I-1}U_{I-1} + A_I U_I + A_{I+1}U_{I+1} = F_I, \tag{3.2.17}$$

where, for $I = 2, 3, \ldots, N$, we have:

$$A_{I-1} = C_{I-1} - \frac{a_A^{(I)}}{\Delta x_{I-1}}, \quad A_I = C_I + \frac{a_A^{(I)}}{\Delta x_{I-1}} + \frac{a_B^{(I)}}{\Delta x_I}, \quad A_{I+1} = C_{I+1} - \frac{a_B^{(I)}}{\Delta x_I} \tag{3.2.18a}$$

and

$$a_A^{(I)} = a(x_A^{(I)}), \quad a_B^{(I)} = a(x_B^{(I)}), \quad F_I = \int_{x_A^{(I)}}^{x_B^{(I)}} f(x)\,dx$$

$$\int_{x_A^{(I)}}^{x_I} c(x)u(x)dx + \int_{x_I}^{x_B^{(I)}} c(x)u(x)dx = C_{I-1}U_{I-1} + C_I U_I + C_{I+1}U_{I+1}. \tag{3.2.18b}$$

The C_{I-1}, C_I, and C_{I+1} are defined in Eq. (3.2.9b). When Eq. (3.2.6) is used (i.e., treating cu as uniform over a CV), $C_{I-1} = C_{I+1} = 0$ and $C_I = c(x_I)h$. Equation (3.2.17) is valid for any interior CV [i.e., CV-2 through CV-(N-1)] and nodes $I = 3, 4, \ldots, N$.

3.2.6 Discretized Equations for Boundary Nodes

3.2.6.1 Half-control volumes at the boundary

For a boundary node, the discrete equations will be different and they have to be formulated. For node 1 [see Fig 3.2.6(a)], we have $x_A^{(1)} = x_1 = 0$, and Eq. (3.2.4) takes the form ($Q_A^{(1)} = Q_1^{(1)}$)

$$0 = -Q_1^{(1)} - a_B^{(1)}\left(\frac{U_2 - U_1}{\Delta x_1}\right) + \int_0^{x_B^{(1)}} (cu - f)dx$$

or

$$-Q_1^{(1)} + A_1 U_1 + A_2 U_2 - F_1 = 0, \tag{3.2.19}$$

where (A_I defined here are different from those for the interior nodes)

$$A_1 = C_1 + \frac{a_B^{(1)}}{\Delta x_1}, \quad A_2 = C_2 - \frac{a_B^{(1)}}{\Delta x_1}, \quad F_1 = \int_0^{x_B^{(1)}} f(x)\,dx,$$

$$\int_0^{x_B^{(1)}} c(x)u(x)dx = C_1 U_1 + C_2 U_2 \tag{3.2.20a}$$

and, assuming linear variation of $u(x)$ and constant c, we have

$$C_1 = \frac{3ch}{8}, \quad C_2 = \frac{ch}{8}. \tag{3.2.20b}$$

Again, if Eq. (3.2.6) is used, $C_1 = ch_1$ and $C_2 = 0$ (h_1 is the size of CV-1).

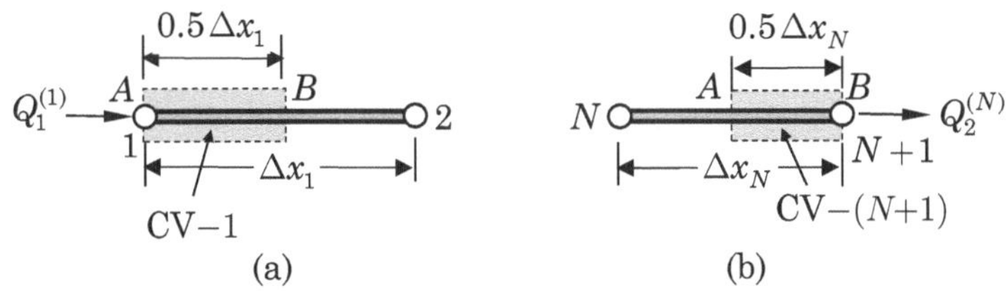

Fig. 3.2.6 Half-control volumes at the boundary points. (a) Boundary at node 1. (b) Boundary at node $N+1$.

When $Q_1^{(1)}$ is specified, say, as $Q_1^{(1)} = q_L$ (the Neumann boundary condition) or as $Q_1^{(1)} + \beta(u(0) - u_0) = q_L$ (the mixed or Newton boundary condition), where β, u_0, and q_L are specified values, we replace it in terms of q_L in the case of the Neumann boundary condition or as $Q_1^{(1)} = -\beta(U_1 - u_0) + q_L$ in the case of the mixed boundary condition. We note that the mixed boundary condition includes the Neumann boundary condition as a special case (by setting $\beta = 0$).

Similarly, for node $N+1$ at the right end of the domain [see Fig 3.2.6(b)], we have

$$0 = a_1^{(N)} \left(\frac{U_{N+1} - U_N}{\delta x_N} \right) - Q_2^{(N)} + C_N U_N + C_{N+1} U_{N+1} - F_{N+1}$$

or

$$-Q_2^{(N)} + A_N U_N + A_{N+1} U_{N+1} - F_{N+1} = 0, \tag{3.2.21}$$

where

$$A_N = C_N - \frac{a_A^{(N)}}{\delta x_N}, \quad A_{N+1} = C_{N+1} + \frac{a_A^{(N)}}{\delta x_N}, \quad F_{N+1} = \int_{x^{(N)}}^{x_{N+1}} f(x)\, dx, \tag{3.2.22a}$$

$$\int_{x_N}^{x_{N+1}} c(x)u(x)dx = C_N U_N + C_{N+1} U_{N+1},$$

and, when $u(x)$ is assumed to be linear over CV-N,

$$C_N = \frac{ch}{8}, \quad C_{N+1} = \frac{3ch}{8}. \tag{3.2.22b}$$

Otherwise, i.e., if Eq. (3.2.6) is used, $C_N = 0$ and $C_{N+1} = ch_N$.

The Neumann or mixed type boundary condition can be handled as explained previously.

3.2.6.2 Zero-thickness control volumes at the boundary

For the case of of zero-thickness control volumes at the boundary, the numbering of the nodes is different from that of the half-control volumes at the boundary. In the ZFVM approach, the boundary nodes do not have their own control volumes on which the governing equation is satisfied in the integral sense. The first control volume is that surrounding node 2 on the left and node $N+1$ on the right side of the domain. Thus, it amounts to having "zero" dimension control volumes for the boundary nodes, as can be seen from Fig. 3.2.7.

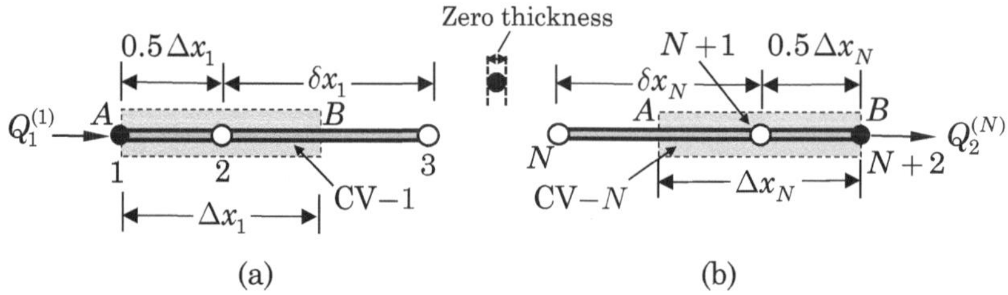

Fig. 3.2.7 Zero-thickness control volumes at the boundary points. (a) Boundary at node 1. (b) Boundary at node $N+2$. Each of the boundary nodes have a "zero-thickness" control volume, which is not visible.

For CV-1, Eq. (3.2.3) takes the form (point A is the same as node 1 with $x = x_1 = 0$, and point B is in the middle of nodes 2 and 3 with $\delta x_B = \Delta x$)

$$0 = -\left[-a\frac{du}{dx}\right]_A - a_B^{(1)}\frac{U_3 - U_2}{\Delta x} + C_1U_1 + C_2U_2 + C_3U_3 - F_2\Delta x, \qquad (3.2.23a)$$

$$C_1 = \frac{ch}{4} \quad C_2 = \frac{5ch}{8} \quad C_3 = \frac{ch}{8}, \qquad (3.2.23b)$$

where $(du/dx)_B$ (at an interior point) is replaced with the second-order accurate formula (3.2.12) in terms of U_2 and U_3. On the left end of the domain, either $(a\,du/dx)_A$ is specified or $u_A = U_1$ is specified. When $(a\,du/dx)_A$ is specified (either as Neumann or Newton type boundary condition), it is replaced in terms of q_L and U_1, as explained before. When $u(0) = U_1$ is specified, $(a\,du/dx)_A$ is replaced with a suitable approximation that brings U_1 into the equation.

To further discuss the case in which u is specified at the left end of the domain, say as $u(0) = u_0$, we can use either the first-order or the second-order [see Eq. (3.2.14)] accurate representation of $(a\,du/dx)_A$ at the left interface (i.e., at $x = 0$), although this is inconsistent with the second-order accurate representation used on the right face (which is an interior point) of the CV-1.

First-order accurate approximation of the derivatives at the boundary In this case, the forward difference formula in Eq. (2.1.7) is used (with $i = 1$) and $U_1 = u_0$, $\delta x_A = 0.5\Delta x$, and $\delta x_B = \Delta x$ [see Fig. 3.2.7(a)]. This approach is known as the "zero-thickness" control volume at the boundaries. Thus, we have (for CV-1)

$$\left(\frac{a_A^{(1)}}{\delta x_A}+\frac{a_B^{(1)}}{\delta x_B}\right)U_2-\left(\frac{a_B^{(1)}}{\delta x_B}\right)U_3+C_2U_2+C_3U_3 = F_2\,\Delta x+\left(\frac{a_A^{(1)}}{\delta x_A}-C_1\right)u_0 \quad (3.2.24)$$

or

$$A_2\,U_2 + A_3\,U_3 = \hat{F}_2, \qquad (3.2.25a)$$

where ($\delta x_A = \delta x_1 = 0.5\Delta x$, $\delta x_B = \delta x_2 = \Delta x$, $a_A^{(1)} = a_1$, and $a_B^{(1)} = a_2$)

$$A_2 = C_2 + \frac{2a_1}{\Delta x} + \frac{a_2}{\Delta x}, \quad A_3 = C_3 - \frac{a_2}{\Delta x}, \quad \hat{F}_2 = F_2\,\Delta x + \left(\frac{2a_1}{\Delta x} - C_1\right)u_0. \quad (3.2.25b)$$

For CV-N, Eq. (3.2.3) can be expressed as [see Fig. 3.2.7(b)]

$$0 = a_A^{(N)}\frac{U_{N+1} - U_N}{\delta x_A} - \left(a\frac{du}{dx}\right)_B + C_{N-1}U_{N-1} + C_N U_N + C_{N+1}U_{N+1} + F_{N+1}\Delta x, \quad (3.2.26a)$$

$$C_N = \frac{ch}{8} \quad C_{N+1} = \frac{5ch}{8} \quad C_{N+2} = \frac{ch}{4}, \quad (3.2.26b)$$

where $(du/dx)_A$ (which is at an interior point) is replaced with the second-order accurate formula (3.2.13). Again, either the gradient $(a\,du/dx)_B$ is specified or the function u_B is specified at node $N+1$. When $(a\,du/dx)_B$ is specified (either through the Neumann or Newton type boundary condition), we replace it in terms of the specified values.

In the case where u is specified at the right end of the domain, say as $u(L) = u_L$, we use the first-order accurate approximation at interface B [i.e., the backward difference formula in Eq. (2.1.8)], with $U_{N+2} = u_L$, $\delta x_B = 0.5\Delta x$ and $\delta x_A = \Delta x$. Thus, we have

$$\left(\frac{a_A^{(N)}}{\delta x_A} + \frac{a_B^{(N)}}{\delta x_B}\right)U_{N+1} - \left(\frac{a_A^{(N)}}{\delta x_A}\right)U_N = F_{N+1}\,\Delta x + \left(\frac{a_B^{(N)}}{\delta x_B}\right)u_L + C_N U_N + C_{N+1}U_{N+1} + C_{N+2}U_{N+2} \quad (3.2.27)$$

or

$$A_{N+1}\,U_{N+1} + A_N\,U_N = \hat{F}_{N+1}, \quad (3.2.28a)$$

where

$$A_N = -\frac{a_A^{(N)}}{\Delta x} + C_N\,, \quad A_{N+1} = \frac{a_A^{(N)}}{\Delta x} + C_N + \frac{2a_B^{(N)}}{\Delta x} + C_{N+1} \quad (3.2.28b)$$

$$\hat{F}_{N+1} = F_{N+1} + \left(\frac{2a_B^{(N)}}{\Delta x} + C_{N+2}\right)u_L. \quad (3.2.28c)$$

Once again, if Eq. (3.2.6) is used (i.e., assume that $cu(x)$ is uniform over each CV), the values of C_1, C_2, C_N, and C_{N+1} will be different as explained in the case of half-control volume formulation.

Second-order accurate approximation of the derivatives at the boundary Let us consider the cases in which the boundary points have the Dirichlet boundary condition specified (i.e., u at $x = 0$ is specified as u_0). For boundary point at node 1, we replace the derivative at point A with the formula in Eq. $(3.2.14)_1$ and obtain

$$A_2 U_2 + A_3 U_3 = \hat{F}_2, \tag{3.2.29a}$$

where

$$\begin{aligned} A_2 &= C_2 + \frac{9a_A^{(1)}}{3\Delta x} + \frac{a_B^{(1)}}{\Delta x}, \quad A_3 = C_3 - \frac{a_A^{(1)}}{3\Delta x} - \frac{a_B^{(1)}}{\Delta x}, \\ \hat{F}_2 &= F_2 \, \Delta x + \left(\frac{8a_A^{(1)}}{3\Delta x} - C_1 \right) u_0. \end{aligned} \tag{3.2.29b}$$

Similarly, when u at $x = 0$ is specified as u_L at node $N + 2$, we replace the derivative at point B (which is the right-hand boundary point) with the formula in Eq. $(3.2.14)_2$ and obtain

$$A_N U_N + A_{N+1} U_{N+1} = \hat{F}_{N+1}, \tag{3.2.30a}$$

where

$$\begin{aligned} A_N &= C_N - \frac{a_A^{(N)}}{3\Delta x} - \frac{a_B^{(N)}}{\Delta x}, \quad A_{N+1} = C_{N+1} + \frac{9a_A^{(N)}}{3\Delta x} + \frac{a_B^{(N)}}{\Delta x}, \\ \hat{F}_{N+1} &= F_{N+1} \, \Delta x + \left(\frac{8a_A^{(N)}}{3\Delta x} - C_{N+2} \right) u_L. \end{aligned} \tag{3.2.30b}$$

3.3 Numerical Examples

In this section, we consider several numerical examples to illustrate the application of the FVM to the solution of boundary value problems with different types of boundary conditions. In the examples considered here, we provide the finite volume discretized equations necessary to solve (often, in the present book, using the Gauss elimination method) for the nodal values of the field variable. Both the HFVM and ZFVM approaches will be discussed, an aspect that is not found in other books. As already pointed out, mesh refinements in the ZFVM do not contain a less refined mesh as a subset (i.e., the nodal locations in a refined mesh do not match with those of a crude mesh). As a result, the numerical solutions obtained with various meshes in the ZFVM can only be compared in graphical form. The discretized equations developed for the model equation are implemented into a computer program. The Fortran and MATLAB source codes are available from the author's website, **http://mechanics.tamu.edu**.

Example 3.3.1

Consider the boundary value problem described by the following second-order linear differential equation

$$-\frac{d^2u}{dx^2} = f_0 \cos x, \quad 0 < x < 1 \tag{3.3.1}$$

and the following two cases of boundary conditions:

$$\textbf{Case 1:} \quad u(0) = u(1) = 0; \quad \textbf{Case 2:} \quad u(0) = \left.\frac{du}{dx}\right|_{x=1} = 0. \tag{3.3.2}$$

Determine the finite volume solution of the problem using uniform meshes of 4 and 8 subdivisions of the domain and the formulations discussed in this chapter. Compare the results with the exact solutions for $f_0 = 10$. Evaluate the source term using the trapezoidal rule.

The exact solutions of the two boundary-value problems described above are given by

$$\textbf{Case 1:}\quad u(x) = f_0\left[-(1-\cos x) + x(1-\cos 1)\right];$$
$$q(x) \equiv -\frac{du}{dx} = f_0\left[\sin x - (1-\cos 1)\right]. \tag{3.3.3a}$$
$$\textbf{Case 2:}\quad u(x) = f_0\left[-(1-\cos x) + x\sin 1\right];$$
$$q(x) \equiv -\frac{du}{dx} = f_0\left(\sin x - \sin 1\right)). \tag{3.3.3b}$$

Solution: Equation (3.3.1) is a special case of the model equation in Eq. (3.2.1), with $a = 1$, $c = 0$, and $f(x) = 10\cos x$. A BVP with Case 1 boundary conditions is often known as the *Dirichlet boundary value problem*, a BVP with Case 2 boundary conditions is known as a *mixed boundary-value problem*.

The source term $f(x) = f_0 \cos x$ is evaluated using the trapezoidal rule within each control volume (i.e., $F_I = 0.5\Delta x[f(x_A^I) + f(x_B^I)]$) in both formulations. Here x_A^I and x_B^I denote the coordinates of points A and B (i.e., left and right interfaces; see Fig. 3.2.4).

For the uniform mesh of four subdivisions ($\Delta x = 0.25$), the half-control volume formulation will have five nodes ($N = 5$), with a control volume around each interior node and a half-control volume at each of the boundary nodes (i.e., five control volumes), as shown in Fig. 3.3.1(a). In the ZFVM, each of the four subdivisions forms a control volume (i.e., $N = 4$) with a node at the center of each subdivision. Thus, the interior four nodes plus two boundary nodes constitute six nodes ($N + 2$) in the ZFVM, as shown in Fig. 3.3.1(b).

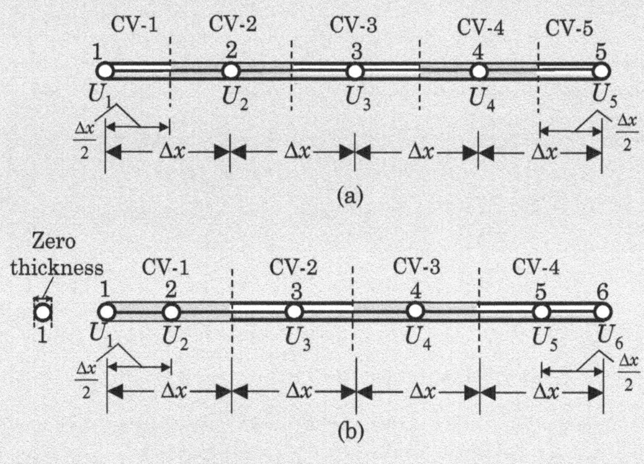

Fig. 3.3.1 Uniform mesh of four subdivisions in the (a) HFVM and (b) ZFVM formulations.

Case 1 $[u(0) = u(1) = 0]$

ZFVM formulation For this case, we need four equations for the four unknowns (U_2, U_3, U_4, U_5) at the four interior nodes. The four algebraic equations for the first-order accurate ZFVM formulation are obtained as follows. The first equation associated with the second node [equation to post-determine the unknown $Q_1^{(1)}$ at node 1 can be obtained from Eq. (3.2.19)] is obtained using Eqs. (3.2.25a) and (3.2.25b) with $\delta x_1 = \Delta x/2$ and $2/\Delta x = 8$:

$$A_2 = \frac{2}{\Delta x} + \frac{1}{\Delta x} = 12, \quad A_3 = -\frac{1}{\Delta x} = -4, \tag{1}$$

$$F_2\Delta x = 0.5 f_0 \Delta x\ [\cos(0) + \cos(\Delta x)] = 5 \times 0.25 \times (1.0 + 0.9689) = 2.4611. \tag{2}$$

Hence, the first equation is $12U_2 - 4U_3 = 2.4611 + 8U_1 = 2.4611$ (because $U_1 = 0$). The fourth equation (associated with the fifth node) is obtained using Eqs. (3.2.28a) and (3.2.28b) for the right boundary node. We have ($N = 4$)

$$A_N = A_4 = -\frac{1}{\Delta x} = -4, \quad A_{N+1} = A_5 = \frac{1}{\Delta x} + \frac{2}{\Delta x} = 12, \tag{3}$$

$$F_{N+1} = F_5 = 5 \times 0.25 \times [\cos(0.75) + \cos(1.0)] = 1.25(0.7317 + 0.5403) = 1.5900 \tag{4}$$

giving the fourth equation $-4U_4 + 12U_5 = 1.5900 + 8U_6 = 1.5900$ (because $U_6 = 0$).

The second and third equations are obtained from Eqs. (3.2.17), (3.2.18a), and (3.2.18b). We have

$$A_{I-1} = -\frac{1}{\Delta x} = -4, \quad A_I = \frac{1}{\Delta x} + \frac{1}{\Delta x} = 8, \quad A_{I+1} = -\frac{1}{\Delta x} = -4, \tag{5}$$

$$F_I = 0.5 f_0 \Delta x[\cos(x_A^I) + \cos(x_B^I)] \tag{6}$$

for $I = 3$ and $I = 4$. Note that $x_A^3 = 0.25$, $x_B^3 = 0.5$, $x_A^4 = 0.5$, and $x_B^4 = 0.75$. Hence, equations 2 and 3 are given by $-4U_2 + 8U_3 - 4U_4 = 2.3081$ and $-4U_3 + 8U_4 - 4U_5 = 2.0116$.

In summary, the four equations for the ZFVM with first-order accurate formulation and four subdivisions are

$$\begin{bmatrix} 12 & -4 & 0 & 0 \\ -4 & 8 & -4 & 0 \\ 0 & -4 & 8 & -4 \\ 0 & 0 & -4 & 12 \end{bmatrix} \begin{Bmatrix} U_2 \\ U_3 \\ U_4 \\ U_5 \end{Bmatrix} = \begin{Bmatrix} 8U_1 \\ 0 \\ 0 \\ 8U_6 \end{Bmatrix} + \begin{Bmatrix} 2.4611 \\ 2.3081 \\ 2.0116 \\ 1.5900 \end{Bmatrix}, \tag{7}$$

whose solution is ($U_1 = U_6 = 0$): $U_2 = 0.5686, U_3 = 1.0906, U_4 = 1.0356$, and $U_5 = 0.4777$.

The first algebraic equation for the second-order accurate ZFVM formulation is obtained using Eq. $(3.2.14)_1$ in Eq. (3.2.23a). We have ($C_I = 0$ and $a_A = a_B = 1$)

$$\begin{aligned} 0 &= -\left[-a\frac{du}{dx}\right]_A - a_B\frac{U_3 - U_2}{\delta x} - F_2\Delta x \\ &= \frac{9U_2 - U_3 - 8U_1}{3\Delta x} - \frac{U_3 - U_2}{\Delta x} - F_2\Delta x \\ &= -\frac{4}{3\Delta x}U_3 + \frac{4}{\Delta x}U_2 - \frac{8}{3\Delta x}U_1 - F_2\Delta x \end{aligned} \tag{3.3.4}$$

With $\Delta x = 0.25$, the first equation becomes $-\frac{32}{3}U_1 + 16U_2 - \frac{16}{3}U_3 = 2.4611$. Similarly, the fourth equation is obtained using Eq. $(3.2.14)_2$ in Eq. (3.2.26a). We have

$$\begin{aligned} 0 &= a_A\frac{U_{N+1} - U_N}{\delta x_A} - \left[a\frac{du}{dx}\right]_B - F_{N+1}\Delta x \\ &= \frac{U_{N+1} - U_N}{\Delta x} - \frac{-9U_{N+1} + U_N + 8U_{N+2}}{3\Delta x} - F_{N+1}\Delta x \\ &= \frac{4}{\Delta x}U_{N+1} - \frac{4}{3\Delta x}U_N - \frac{8}{3\Delta x}U_{N+2} - F_{N+1}\Delta x. \end{aligned} \tag{3.3.5}$$

With $\Delta x = \delta x = 0.25$, the fourth equation becomes $-\frac{32}{3}U_6 + 16U_5 - \frac{16}{3}U_4 = 1.5900$. The second and third equations remain the same as those in the first-order approximation. Thus, we have ($U_1 = U_6 = 0$)

$$\begin{bmatrix} 16 & -\frac{16}{3} & 0 & 0 \\ -4 & 8 & -4 & 0 \\ 0 & -4 & 8 & -4 \\ 0 & 0 & -\frac{16}{3} & 16 \end{bmatrix} \begin{Bmatrix} U_2 \\ U_3 \\ U_4 \\ U_5 \end{Bmatrix} = \begin{Bmatrix} 2.4611 \\ 2.3081 \\ 2.0116 \\ 1.5900 \end{Bmatrix}, \tag{8}$$

whose solution is $U_2 = 0.4951$, $U_3 = 1.0240$, $U_4 = 0.9757$, and $U_5 = 0.4246$.

Half-control volume formulation For the half-control volume formulation, Case 1 ($U_1 = U_5 = 0$) requires three equations among the three unknowns (U_2, U_3, U_4) at the three internal nodes when four subdivisions are used [see Fig. 3.3.1(a)]; all three equations are obtained from Eqs. (3.2.18a) and (3.2.18b):

$$\begin{bmatrix} 8 & -4 & 0 \\ -4 & 8 & -4 \\ 0 & -4 & 8 \end{bmatrix} \begin{Bmatrix} U_2 \\ U_3 \\ U_4 \end{Bmatrix} = \begin{Bmatrix} 4U_1 \\ 0 \\ 4U_5 \end{Bmatrix} + \begin{Bmatrix} 2.4034 \\ 2.1768 \\ 1.8149 \end{Bmatrix}. \tag{9}$$

The solution of these equations is ($U_1 = U_5 = 0$): $U_2 = 0.8362, U_3 = 1.0715$, and $U_4 = 0.7626$.

The numerical values obtained with various formulations are compared in Table 3.3.1 for $u(x)$; it also contains values of $q(x) = -du/dx$ at $x = 0$ and $x = 1$; the underlined numbers are *linearly* interpolated values at the CV interfaces. Clearly, the HFVM formulation gives more accurate results than the first-order (FVM1) and second-order (FVM2) accurate ZFVM formulations.

Table 3.3.1 Comparison of various FVM solutions with the exact solution (u) of: $-\frac{d^2u}{dx^2} = 10\cos x;\quad 0 < x < 1;\quad u(0) = u(1) = 0.$

		ZFVM				HFVM	
x	Exact	FVM1(4)	FVM2(4)	FVM1(8)	FVM2(8)	HFVM(4)	HFVM(8)
0.0625	0.2678	——	——	0.2866	0.2676	——	0.2481
0.1250	0.4966	0.5686	0.4951	0.4953	0.4768	0.4181	0.4963
0.1875	0.6867	——	——	0.7040	0.6861	——	0.6670
0.2500	0.8384	0.8296	0.7595	0.8362	0.8187	0.8362	0.8378
0.3125	0.9522	——	——	0.9683	0.9513	——	0.9330
0.3750	1.0289	1.0907	1.0240	1.0263	1.0098	0.9538	1.0283
0.4375	1.0693	——	——	1.0842	1.0682	——	1.0509
0.5000	1.0743	1.0631	0.9998	1.0715	1.0560	1.0715	1.0736
0.5625	1.0450	——	——	1.0588	1.0438	——	1.0279
0.6250	0.9827	1.0356	0.9757	0.9802	0.9657	0.9171	0.9821
0.6875	0.8888	——	——	0.9015	0.8875	——	0.8731
0.7500	0.7646	0.7567	0.7002	0.7626	0.7491	0.7626	0.7641
0.8125	0.6119	——	——	0.6237	0.6107	——	0.5981
0.8750	0.4323	0.4777	0.4246	0.4312	0.4187	0.3813	0.4320
0.9375	0.2277	——	——	0.2387	0.2266	——	0.2160
$-q(0)$	4.5970	4.5492	4.5764	4.5850	4.5929	4.5898	4.5946
$q(1)$	3.8177	3.8217	3.7944	3.8187	3.8108	3.7888	3.8101

Figures 3.3.2 and 3.3.3 contain a comparison of the numerical solutions obtained from the ZFVM formulation using first-order (FVM1) and second-order (FVM2) accurate schemes and the HFVM formulation with the exact solutions for $u(x)$ and $q(x)$. Solutions obtained using meshes of four and eight subdivisions are presented. As expected, the second-order accurate scheme performs better than the first-order accurate scheme, although with mesh refinements they all converge to the exact solutions. The first-order accurate solution for coarse meshes has more error in u as well as q (especially at the boundary). The HFVM results shown in Figs. 3.3.2 and 3.3.3 match the exact solutions.

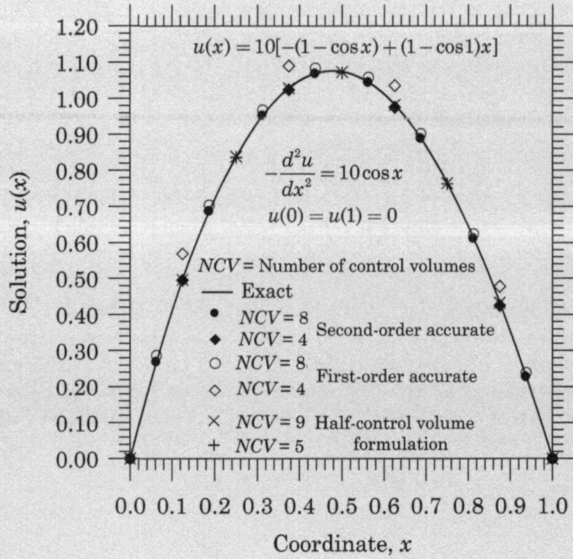

Fig. 3.3.2 Comparison of the HFVM and ZFVM (first- and second-order accurate) solutions for $u(x)$ with the exact solution of Case 1 boundary conditions.

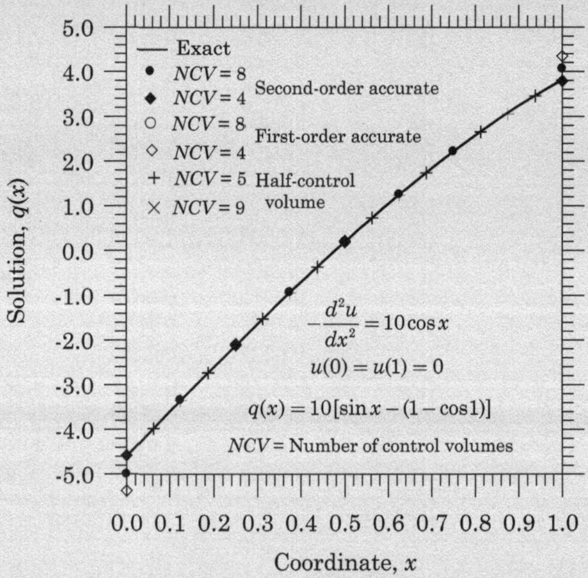

Fig. 3.3.3 Comparison of the ZFVM (first- and second-order accurate) solutions for $q(x) = -du/dx$ with the exact solution of Case 1.

Case 2 $\left[u(0) = 0, \ \left.\frac{du}{dx}\right|_{x=1} = 0\right]$

For Case 2, the boundary condition at the right end requires that the gradient of u be zero, making u at $x = L$ an unknown.

ZFVM formulation For the first-order accurate ZFVM formulation, there are five unknowns $(U_2, U_3, U_4, U_5, U_6)$ for the mesh of four subdivisions. The first four equations remain the same as in Case 1. The fifth equation is obtained by expressing the gradient boundary condition with the first-order (backward difference) approximation

$$0 = \left.\frac{du}{dx}\right|_{x=1} \approx \frac{U_6 - U_5}{\Delta x} \quad \rightarrow \quad U_6 = U_5 \tag{10}$$

Thus, the values of u at nodes 6 and 5 are the same in the ZFVM formulation in the case of zero gradient boundary condition. Thus, we have (with $U_1 = 0$)

$$\begin{bmatrix} 12 & -4 & 0 & 0 \\ -4 & 8 & -4 & 0 \\ 0 & -4 & 8 & -4 \\ 0 & 0 & -4 & 4 \end{bmatrix} \begin{Bmatrix} U_2 \\ U_3 \\ U_4 \\ U_5 \end{Bmatrix} = \begin{Bmatrix} 2.4611 \\ 2.3081 \\ 2.0116 \\ 1.5900 \end{Bmatrix}, \tag{11}$$

whose solution is $(U_1 = 0)$:

$$U_2 = 1.0464, \quad U_3 = 2.5238, \quad U_4 = 3.4242, \quad U_5 = U_6 = 3.8217. \tag{12}$$

Use of the second-order accurate approximation at boundary node 1 and the vanishing gradient boundary condition at node $N + 2$ in the ZFVM yields the equations $(U_1 = 0)$

$$\begin{bmatrix} 16 & -\frac{16}{3} & 0 & 0 \\ -4 & 8 & -4 & 0 \\ 0 & -4 & 8 & -4 \\ 0 & 0 & -4 & 4 \end{bmatrix} \begin{Bmatrix} U_2 \\ U_3 \\ U_4 \\ U_5 \end{Bmatrix} = \begin{Bmatrix} 2.4611 \\ 2.3081 \\ 2.0116 \\ 1.5900 \end{Bmatrix}, \tag{13}$$

whose solution is $U_2 = 0.9694$, $U_3 = 2.4469$, $U_4 = 3.3473$, and $U_5 = 3.7448$. The value U_6 is calculated using the second-order accurate formula

$$\frac{-9U_5 + U_4 + 8U_6}{3\Delta x} = 0 \rightarrow U_6 = \frac{9U_5 - U_4}{8} = 3.7945. \tag{14}$$

The exact solution at the same points is $u(0.125) = 0.9738$, $u(0.375) = 2.4606$, $u(0.625) = 3.3688$, $u(0.875) = 3.7728$, and $u(1.0) = 3.8177$.

Half-control volume formulation The half-control volume formulation with four subdivisions gives the following equations among the four unknowns (U_2, U_3, U_4, U_5):

$$\begin{bmatrix} 8 & -4 & 0 & 0 \\ -4 & 8 & -4 & 0 \\ 0 & -4 & 8 & -4 \\ 0 & 0 & -4 & 4 \end{bmatrix} \begin{Bmatrix} U_2 \\ U_3 \\ U_4 \\ U_5 \end{Bmatrix} = \begin{Bmatrix} 2.4034 \\ 2.1768 \\ 1.8149 \\ 0.7383 \end{Bmatrix}. \tag{15}$$

The solution of these equations is: $U_2 = 1.7834$, $U_3 = 2.9659$, $U_4 = 3.6042$, and $U_5 = 3.7888$ and the exact solution at the same points is $u(0.25) = 1.7928$, $u(0.5) = 2.9832$, $u(0.75) = 3.6279$, and $u(1.0) = 3.8177$.

Figures 3.3.4 and 3.3.5 contains a comparison of the numerical solutions obtained from the ZFVM (first- and second-order accurate schemes) and HFVM formulations with the exact solutions $u(x)$ and $q(x) = -du/dx$ for meshes of four and eight subdivisions (i.e., $NCV = 6$ and $NCV = 8$ for the ZFVM formulation and $NCV = 5$ and $NCV = 9$ for the HFVM formulation). All formulations yield numerical solutions that converge to the exact solution with mesh refinement. Once again, we note that the HFVM formulation yields more accurate solutions.

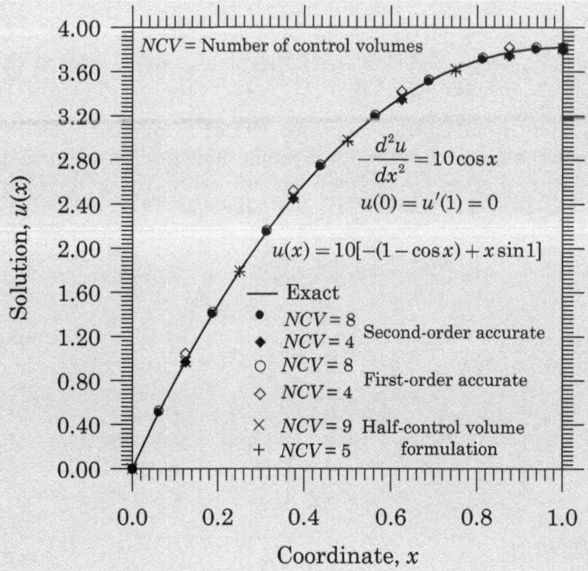

Fig. 3.3.4 Comparison of the HFVM and ZFVM (first- and second-order accurate) solutions for $u(x)$ with the exact solution of Case 2 boundary conditions.

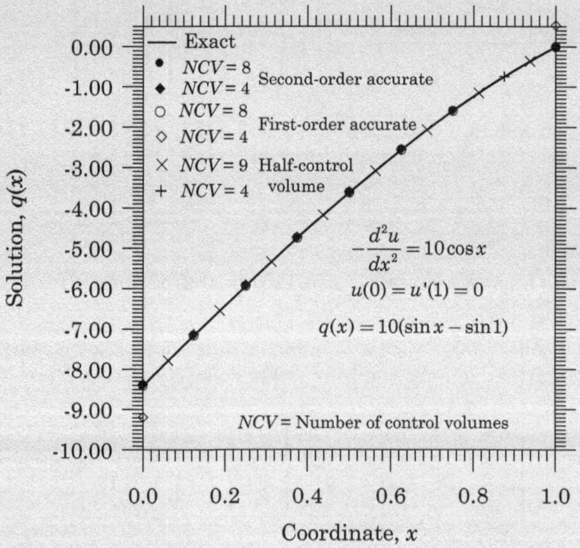

Fig. 3.3.5 Comparison of the HFVM and ZFVM (first- and second-order accurate) solutions for $q(x) = -du/dx$ with the exact solution of Case 2 boundary conditions.

The next example deals with heat transfer in a steel rod. Here, we consider ZFVM and HFVM formulations and compare the numerical results with those obtained using the FDM (see **Example 2.3.1**).

Example 3.3.2

A steel rod of uniform diameter $D = 0.02$ m, length $L = 0.05$ m, and constant thermal conductivity $k = 50$ W/(m· °C) is exposed to ambient air at $T_\infty = 20°$C with a heat transfer coefficient $\beta = 100$ W/(m^2 · °C). The left end of the rod is maintained at temperature $T_0 = 320°$C and the other end is either (1) insulated or (2) exposed to the ambient air. Illustrate the following six FVM formulations using a uniform mesh of five subdivisions:

- FVM11 = ZFVM with the first-order approximation and uniform representation of u.
- FVM12 = ZFVM with the first-order approximation and linear representation of u.
- FVM21 = ZFVM with the second-order approximation and uniform representation of u.
- FVM22 = ZFVM with the second-order approximation and linear representation of u.
- FVM31 = HFVM with the second-order approximation and uniform representation of u.
- FVM32 = HFVM with the second-order approximation and linear representation of u.

Plot the FVM solutions along with the exact solutions for u and $q = -du/dx$.

Solution: The governing equation is given by (3.3.1) [see also, Eqs. (2.3.4)–(2.3.7)]

$$-\frac{d^2u}{dx^2} + m^2 u = 0 \quad \text{for} \quad 0 < x < L \tag{3.3.6}$$

with the following two cases of boundary conditions (see **Example 2.3.1** and Fig. 2.3.1 for Case 1 boundary conditions):

$$\textbf{Case 1:}\ u(0) = u_0 = 300°\text{C}; \quad \left[-\frac{du}{dx}\right]_{x=L} = 0. \tag{3.3.7}$$

$$\textbf{Case 2:}\ u(0) = u_0 = 300°\text{C}; \quad \left[\frac{du}{dx} + \frac{\beta}{k}u\right] = 0. \tag{3.3.8}$$

Here u denotes $u(x) = T(x) - T_\infty$ and m^2 is given by

$$m^2 = \frac{\beta P}{Ak} = \frac{\beta \pi D}{\frac{1}{4}\pi D^2 k} = \frac{4\beta}{kD} = \frac{4 \times 100}{50 \times 0.02} = 400.$$

Comparing Eq. (3.3.6) with Eq. (3.2.1), we have $a = 1$, $b = 0$, $c = m^2 = 400$, and $f = 0$. The main difference between **Example 3.3.1** and the present example is that $c \neq 0$ (a constant), requiring the evaluation of integral of $u(x)$ over each control volume. We will consider two options to evaluate the integral of $u(x)$ over a control volume: (1) treat u as uniform, resulting in $U_I \Delta x$ or (2) linear variation, resulting in expressions presented in Eqs. (3.2.18b)–(3.2.28b) (depending on whether the node is interior or on the boundary and also on the formulation).

The exact solution for Case 2 is given by [see Eq. (1.5.21b)]

$$\begin{aligned} u(x) &= u_0 \left[\frac{\cosh m(L-x) + (\beta/mk)\sinh m(L-x)}{\cosh mL + (\beta/mk)\sinh mL}\right], \\ q(x) &\equiv -Ak\frac{du}{dx} = Akmu_0 \left[\frac{\sinh m(L-x) + (\beta/mk)\cosh m(L-x)}{\cosh mL + (\beta/mk)\sinh mL}\right] \end{aligned} \tag{3.3.9}$$

and the exact solution for Case 1 is obtained by setting $\beta = 0$ in Eq. (3.3.9) [see Eq. (1.5.20b)]

$$u(x) = u_0 \frac{\cosh m(L-x)}{\cosh mL}, \quad q(x) \equiv -Ak\frac{du}{dx} = Akmu_0 \frac{\sinh m(L-x)}{\cosh mL}. \tag{3.3.10}$$

For a uniform mesh of five ($N = 5$) subdivisions, we have $\Delta x = h = 0.01$, $a = 1$, and $c = 400$ ($ch = 4$ and $a/h = 100$). Depending on the formulation, the number of control volumes (NCV) and number of nodes (ND) is different. In the HFVM mesh, NCV is equal to the number of nodes $ND = N + 1$ [see Fig. 3.3.6(a)], while in the ZFVM mesh, NCV is equal to the number of subdivisions N and the number of nodes is equal to $ND = NCV + 2 = N + 2$ [see Fig. 3.3.6(b)]. The number of equations required is always equal to the number of nodes (for a single degree-of-freedom problem) and there will be the same number of unknowns.

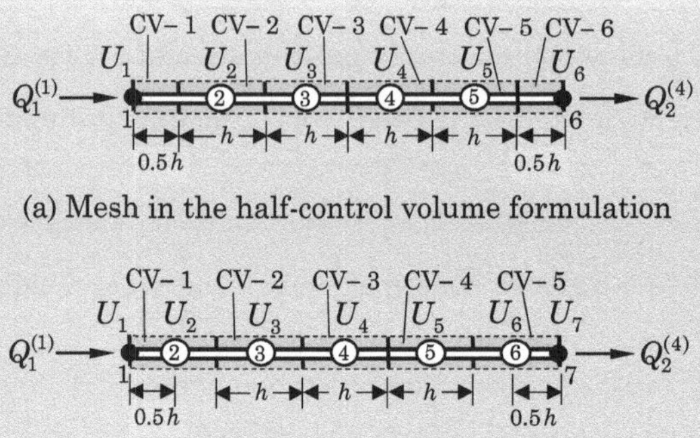

(a) Mesh in the half-control volume formulation

(b) Mesh in the zero-thickness control volume formulation

Fig. 3.3.6 Uniform meshes of five ($N = 5$) subdivisions in (a) HFVM formulation ($NCV = 6$) and (b) ZFVM formulation ($NCV = 7$).

Case 1: The half-control volume formulation-Models FVM31 and FVM32

For the mesh of five subdivisions, we need six equations to determine the six unknowns: $Q_1^{(1)}$, U_2, U_3, U_4, U_5, and U_6. The first equation to determine $Q_1^{(1)}$ is obtained from Eqs. (3.2.19); and (3.2.17) with Eqs. (3.2.8)–(3.2.9b) ($h = 0.01$, $ch = 4$, $a/h = 100$, and $f = 0$) can be used to determine the remaining equations.

Model FVM31: Uniform representation of $u(x)$ For this case, the first equation is given by (to determine the unknown heat at node 1)

$$0 = -Q_1^{(1)} + A_1 U_1 + A_2 U_2, \quad A_1 = c(0.5h) + \frac{a}{h} = 102, \quad A_2 = -\frac{a}{h} == -100. \tag{1}$$

The next four equations come from interior nodes $I = 2, 3, 4, 5$; using Eq. (3.2.17) with $C_I = ch$ and $C_{I-1} = C_{I+1} = 0$. We have

$$A_{I-1} = -\frac{a}{h} = -100, \quad A_I = ch + \frac{2a}{h} = 204, \quad A_{I+1} = -\frac{a}{h} = -100. \tag{2}$$

Thus, Eq. (3.2.17) gives the following four equations in the present case:

$$\begin{aligned}
-100\,U_1 + 204\,U_2 - 100\,U_3 &= 0,\\
-100\,U_2 + 204\,U_3 - 100\,U_4 &= 0,\\
-100\,U_3 + 204\,U_4 - 100\,U_5 &= 0,\\
-100\,U_4 + 204\,U_5 - 100\,U_6 &= 0.
\end{aligned} \tag{3}$$

The last equation is obtained using Eq. (3.2.21) with $Q_2^{(5)} = 0$. We have

$$0 = -Q_2^{(5)} + A_5 U_5 + A_6 U_6, \quad A_5 = -\frac{1}{h} = -100, \quad A_6 = C(0.5h) + \frac{1}{h} = 102. \tag{4}$$

Writing the six equations derived in matrix form, we obtain

$$\begin{bmatrix} 102 & -100 & 0.0 & 0.0 & 0.0 & 0.0 \\ -100 & 204 & -100 & 0.0 & 0.0 & 0.0 \\ 0.0 & -100 & 204 & -100 & 0.0 & 0.0 \\ 0.0 & 0.0 & -100 & 204 & -100 & 0.0 \\ 0.0 & 0.0 & 0.0 & -100 & 204 & -100 \\ 0.0 & 0.0 & 0.0 & 0.0 & -100 & 102 \end{bmatrix} \begin{Bmatrix} U_1 \\ U_2 \\ U_3 \\ U_4 \\ U_5 \\ U_6 \end{Bmatrix} = \begin{Bmatrix} Q_1^{(1)} \\ 0 \\ 0 \\ 0 \\ 0 \\ Q_2^{(5)} \end{Bmatrix} \tag{5}$$

Using the boundary conditions $U_1 = 300$ and $Q_2^{(5)} = 0$, we can solve Eq. (5) for the unknowns. Typically, we solve the last five equations for the unknown temperature and then use the first equation to solve for the unknown heat. The solution for the nodal temperatures is

$$U_2 = 260.12°\text{C}, \quad U_3 = 230.64°\text{C}, \quad U_4 = 210.39°\text{C}, \quad U_5 = 198.56°\text{C}, \quad U_6 = 194.66°\text{C}. \tag{6}$$

and the heat at the left end is ($kA = 157.0796 \times 10^{-4}$)

$$Q(0) = kA\, Q_1^{(1)} = kA\, 4588.15 = 72.07\,\text{W} \tag{7}$$

The exact solution for the temperatures at the same locations and the heat at $x = 0$ are

$$\begin{aligned} &u(0.01) = 260.02°\text{C}, \quad u(0.02) = 230.47°\text{C}, \quad u(0.03) = 210.18°\text{C} \\ &u(0.04) = 198.32°\text{C}, \quad u(0.05) = 194.42°\text{C}, \quad Q(0) = 71.78\,\text{W}. \end{aligned} \tag{8}$$

Model FVM32: Linear representation of $u(x)$ The first equation is

$$\begin{aligned} 0 &= -Q_1^{(1)} + A_1U_1 + A_2U_2 \\ A_1 &= C_1 + \frac{a}{h} = \frac{3ch}{8} + \frac{a}{h} = 101.5, \\ A_2 &= C_2 - \frac{a}{h} = \frac{ch}{8} - \frac{a}{h} = -99.5. \end{aligned} \tag{9}$$

The next four equations come from interior nodes $I = 2, 3, 4, 5$; using Eqs. (3.2.17), (3.2.18a), and (3.2.18b). We obtain (with $a = 1$, $c = 400$, and $h = 0.01$)

$$\begin{aligned} A_{I-1} &= C_{I-1} - \frac{a}{h} = \frac{ch}{8} - \frac{a}{h} = -99.5, \\ A_I &= C_I + \frac{2a}{h} = \frac{6ch}{8} + \frac{2}{h} = 203.0, \\ A_{I+1} &= C_{I+1} - \frac{a}{h} = \frac{ch}{8} - \frac{a}{h} = -99.5. \end{aligned} \tag{10}$$

Thus, Eq. (3.2.17) gives the following four equations in the present case:

$$\begin{aligned} -99.5\,U_1 + 203.0\,U_2 - 99.5\,U_3 &= 0, \\ -99.5\,U_2 + 203.0\,U_3 - 99.5\,U_4 &= 0, \\ -99.5\,U_3 + 203.0\,U_4 - 99.5\,U_5 &= 0, \\ -99.5\,U_4 + 203.0\,U_5 - 99.5\,U_6 &= 0. \end{aligned} \tag{11}$$

The last equation is obtained using Eq. (3.2.21) with $Q_2^{(5)} = 0$. We have

$$\begin{aligned} 0 &= -Q_2^{(5)} + A_5U_5 + A_6U_6, \\ A_5 &= C_5 - \frac{1}{h} = \frac{ch}{8} - \frac{1}{h} = -99.5, \\ A_6 &= C_6 + \frac{1}{h} = \frac{3ch}{8} + \frac{1}{h} = 101.5. \end{aligned} \tag{12}$$

Writing the six equations derived in matrix form, we obtain

$$\begin{bmatrix} 101.5 & -99.5 & 0.0 & 0.0 & 0.0 & 0.0 \\ -99.5 & 203.0 & -99.5 & 0.0 & 0.0 & 0.0 \\ 0.0 & -99.5 & 203.0 & -99.5 & 0.0 & 0.0 \\ 0.0 & 0.0 & -99.5 & 203.0 & -99.5 & 0.0 \\ 0.0 & 0.0 & 0.0 & -99.5 & 203.0 & -99.5 \\ 0.0 & 0.0 & 0.0 & 0.0 & -99.5 & 101.5 \end{bmatrix} \begin{Bmatrix} U_1 \\ U_2 \\ U_3 \\ U_4 \\ U_5 \\ U_6 \end{Bmatrix} = \begin{Bmatrix} Q_1^{(1)} \\ 0 \\ 0 \\ 0 \\ 0 \\ Q_2^{(5)} \end{Bmatrix} \tag{13}$$

Using the boundary conditions $U_1 = u(0) = 300$ and $Q_2^{(5)} = 0$, we write the last five equations (i.e., omit the first equation) of the system in Eq. (13) for the unknown nodal temperatures U_2 through U_6 as

$$\begin{bmatrix} 203.0 & -99.5 & 0.0 & 0.0 & 0.0 \\ -99.5 & 203.0 & -99.5 & 0.0 & 0.0 \\ 0.0 & -99.5 & 203.0 & -99.5 & 0.0 \\ 0.0 & 0.0 & -99.5 & 203.0 & -99.5 \\ 0.0 & 0.0 & 0.0 & -99.5 & 101.5 \end{bmatrix} \begin{Bmatrix} U_2 \\ U_3 \\ U_4 \\ U_5 \\ U_6 \end{Bmatrix} = \begin{Bmatrix} 99.5\, U_1 \\ 0 \\ 0 \\ 0 \\ 0 \end{Bmatrix}. \tag{14}$$

The system of equations for the primary nodal unknowns (i.e., temperatures) is termed *condensed* equations. Clearly, the condensed equations are obtained from the total system by omitting the equations associated with known primary variables (in the present case U_1) and accounting for the known values of Us in the other equations.

The solution of Eq. (14) is given by

$$U_2 = 259.97^\circ\text{C},\quad U_3 = 230.39^\circ\text{C},\quad U_4 = 210.07^\circ\text{C},\quad U_5 = 198.20^\circ\text{C},\quad U_6 = 194.29^\circ\text{C} \tag{15}$$

and

$$Q_1^{(1)} = 101.5 \times 300 - 99.5 \times 259.97 = 4583 \tag{16}$$

The heat at the left end is

$$Q(0) = kA\, Q_1^{(1)} = kA\, 4583.7 = 71.99\,\text{W}. \tag{17}$$

The solution obtained with a linear representation of $u(x)$ is only slightly better than that obtained with a uniform representation of $u(x)$.

Case 1: ZFVM formulation

For a uniform mesh of $N = 5$ subdivisions in the ZFVM formulation [see Fig. 3.3.6(b)], there are $NCV = N = 5$ control volumes and $ND = NCV + 2 = 7$ unknowns, namely, $Q_1^{(1)}$, $U_2, U_3, \ldots U_7$. The value of $Q_1^{(1)}$ in this formulation is determined by suitable representation of the derivative du/dx at $x = 0$ (often, the first-order accurate representation). The remaining six equations are determined as discussed next.

Model FVM11: First-order approximation at the boundary nodes with uniform representation of $u(x)$ Since u is specified, we do not write an equation for node 1. The equation associated with node 2, with the first-order accurate approximation of du/dx at node 1, is given by Eqs. (3.2.25a) and (3.2.25b) with ($a = 1$, $c = 400$, and $h = 0.01$).

$$\begin{aligned} 0 &= A_1U_1 + A_2U_2 + A_3U_3, \\ A_1 &= -\frac{2a}{h} = -\frac{2a}{h} = -200, \\ A_2 &= ch + \frac{2a}{h} + \frac{a}{h} = 304, \\ A_3 &= -\frac{a}{h} = -100. \end{aligned} \tag{18}$$

Equations associated with nodes $I = 3, 4$, and 5 (or CV-2, CV-3, and CV-4) are obtained from Eqs. (3.2.17), (3.2.18a), and (3.2.18b) (the same as those in the HFVM):

$$\begin{aligned} 0 &= A_{I-1}U_{I-1} + A_I U_I + A_{I+1}U_{I+1}, \\ A_{I-1} &= -\frac{a}{h} = -\frac{a}{h} = -100, \\ A_I &= ch + \frac{2a}{h} = 204, \\ A_{I+1} &= -\frac{a}{h} = \frac{ch}{8} - \frac{a}{h} = -100. \end{aligned} \tag{19}$$

Thus, we have the following three equations:

$$\begin{aligned} -100\,U_2 + 204\,U_3 - 100\,U_4 &= 0, \\ -100\,U_3 + 204\,U_4 - 100\,U_5 &= 0, \\ -100\,U_4 + 204\,U_5 - 100\,U_6 &= 0. \end{aligned} \tag{20}$$

The equation associated with the CV-5 (i.e., node 6) is (noting that $du/dx = 0$ at $x = L$)

$$\begin{aligned} 0 &= A_5 U_5 + A_6 U_6 + A_7 U_7, \\ A_5 &= -\frac{a}{h} = -100, \\ A_6 &= ch + \frac{a}{h} + \frac{2a}{h} = 304, \\ A_7 &= -- \frac{2a}{h} = -200. \end{aligned} \tag{21}$$

The sixth and final equation is determined using the first-order accurate representation of du/dx at boundary node 7:

$$0 = \left.\frac{du}{dx}\right|_{x=L} = \frac{U_7 - U_6}{0.5h} = -200U_6 + 200U_7. \tag{22}$$

In summary, the set of six algebraic equations for the ZFVM with first-order accurate formulation and uniform representation of $u(x)$ are

$$\begin{bmatrix} 304 & -100 & 0.0 & 0.0 & 0.0 & 0.0 \\ -100 & 204 & -100 & 0.0 & 0.0 & 0.0 \\ 0.0 & -100 & 204 & -100 & 0.0 & 0.0 \\ 0.0 & 0.0 & -100 & 204 & -100 & 0.0 \\ 0.0 & 0.0 & 0.0 & -100 & 304 & -200 \\ 0.0 & 0.0 & 0.0 & 0.0 & -200 & 200 \end{bmatrix} \begin{Bmatrix} U_2 \\ U_3 \\ U_4 \\ U_5 \\ U_6 \\ U_7 \end{Bmatrix} = \begin{Bmatrix} 200\,U_1 \\ 0 \\ 0 \\ 0 \\ 0 \\ Q_2^{(5)} \end{Bmatrix}. \tag{23}$$

The solution of the equations in Eq. (23) is

$$U_2 = 277.29°\text{C},\ U_3 = 242.95°\text{C},\ U_4 = 218.33°\text{C},\ U_5 = 202.45°\text{C},\ U_6 = U_7 = 194.66°\text{C}. \tag{24}$$

The heat at node 1 is calculated using

$$Q(0) = -kA\left.\frac{du}{dx}\right|_{x=0} = kA\frac{U_1 - U_2}{0.5h} = 71.34 \text{ W}. \tag{25}$$

The exact solution at the same locations as the ZFVM mesh points is

$$\begin{aligned} &u(0.005) = 278.02°\text{C}, \quad u(0.015) = 244.03°\text{C}, \quad u(0.025) = 219.23°\text{C} \\ &u(0.035) = 203.23°\text{C}, \quad u(0.045) = 195.39°\text{C}, \quad u(0.05) = 194.42°\text{C} \end{aligned} \tag{26}$$

Model FVM12: First-order approximation at the boundary with linear representation of $u(x)$ In this case, the integral of $c\,u(x)$ (with constant c) over the control volume is carried out as in Eqs. (3.2.8)–(3.2.9a). For node 2 equation (over CV-1), we have:

$$
\begin{aligned}
c\int_0^{0.5h} u(\bar{x})\,d\bar{x} &+ c\int_0^{0.5h} u(\bar{x})\,d\bar{x}\\
&= c\int_0^{0.5h}\left[U_1+\frac{U_2-U_1}{0.5h}\bar{x}\right]d\bar{x} + c\int_0^{0.5h}\left[U_2+\frac{U_3-U_2}{h}\bar{x}\right]d\bar{x}\\
&= \frac{ch}{4}(U_1+U_2)+\frac{ch}{8}(3U_2+U_3) = \frac{ch}{4}U_1+\frac{5ch}{8}U_2+\frac{ch}{8}U_3, \qquad (3.3.11)
\end{aligned}
$$

where $\bar{x}$ is a local coordinate with origin at each node. Thus, we have ($F_I = 0$ because $f = 0$)

$$
\begin{aligned}
0 &= A_1U_1 + A_2U_2 + A_3U_3,\\
A_1 &= C_1 - \frac{2a}{h} = \frac{ch}{4} - \frac{2a}{h} = 1 - 200 = -199,\\
A_2 &= C_2 + \frac{2a}{h} + \frac{a}{h} = \frac{5ch}{8} + \frac{3a}{h} = 2.5 + 300 = 302.5,\\
A_3 &= C_3 - \frac{a}{h} = \frac{ch}{8} + \frac{a}{h} = 0.5 - 100 = -99.5.
\end{aligned} \qquad (27)
$$

Equations associated with nodes $I = 3, 4$, and 5 (or CV-2, CV-3, and CV-4) are obtained from Eqs. (3.2.17), (3.2.18a), and (3.2.18b) (the same as those in the HFVM):

$$
\begin{aligned}
0 &= A_{I-1}U_{I-1} + A_IU_I + A_{I+1}U_{I+1},\\
A_{I-1} &= C_{I-1} - \frac{a}{h} = \frac{ch}{8} - \frac{a}{h} = -99.5,\\
A_I &= C_I + \frac{a}{h} + \frac{a}{h} = \frac{6ch}{8} + \frac{2}{h} = 203.0,\\
A_{I+1} &= C_{I+1} - \frac{a}{h} = \frac{ch}{8} - \frac{a}{h} = -99.5.
\end{aligned} \qquad (28)
$$

Thus, we have the following three equations:

$$
\begin{aligned}
-99.5\,U_2 + 203.0\,U_3 - 99.5\,U_4 &= 0,\\
-99.5\,U_3 + 203.0\,U_4 - 99.5\,U_5 &= 0,\\
-99.5\,U_4 + 203.0\,U_5 - 99.5\,U_6 &= 0.
\end{aligned} \qquad (29)
$$

The equation associated with the CV-5 is

$$
\begin{aligned}
0 &= A_5U_5 + A_6U_6 + A_7U_7,\\
A_5 &= C_5 - \frac{a}{h} = \frac{ch}{8} - \frac{a}{h} = 0.5 - 100 = -99.5,\\
A_6 &= C_6 + \frac{a}{h} + \frac{2a}{h} = \frac{5ch}{8} + \frac{3a}{h} = 2.5 + 300 = 302.5,\\
A_7 &= C_7 - \frac{2a}{h} = \frac{ch}{4} - \frac{2a}{h} = -199.0.
\end{aligned} \qquad (30)
$$

The sixth and final equation is determined using the first-order accurate representation of du/dx:

$$
0 = \left.\frac{du}{dx}\right|_{x=L} = \frac{U_7 - U_6}{0.5h} = -200U_6 + 200U_7. \qquad (31)
$$

In summary, the set of algebraic equations for the ZFVM with first-order accurate formulation are

$$\begin{bmatrix} 302.5 & -99.5 & 0.0 & 0.0 & 0.0 & 0.0 \\ -99.5 & 203.0 & -99.5 & 0.0 & 0.0 & 0.0 \\ 0.0 & -99.5 & 203.0 & -99.5 & 0.0 & 0.0 \\ 0.0 & 0.0 & -99.5 & 203.0 & -99.5 & 0.0 \\ 0.0 & 0.0 & 0.0 & -99.5 & 302.5 & -199.0 \\ 0.0 & 0.0 & 0.0 & 0.0 & -200.0 & 200.0 \end{bmatrix} \begin{Bmatrix} U_2 \\ U_3 \\ U_4 \\ U_5 \\ U_6 \\ U_7 \end{Bmatrix} = \begin{Bmatrix} 199\,U_1 \\ 0 \\ 0 \\ 0 \\ 0 \\ Q_2^{(5)} \end{Bmatrix} \tag{32}$$

The solution is given by

$$U_2 = 277.20°\text{C}, \quad U_3 = 242.74°\text{C}, \quad U_4 = 218.04°\text{C}, \quad U_5 = 202.10°\text{C}, \quad U_6 = U_7 = 194.29°\text{C}. \tag{33}$$

Model FVM21: Second-order approximation at the boundary nodes with uniform representation of $u(x)$ In this case, only the discretized equations associated with the boundary change. Thus, Eq. (23) becomes

$$\begin{bmatrix} 404 & -133.33 & 0.0 & 0.00 & 0.0 & 0.00 \\ -100 & 204.00 & -100 & 0.00 & 0.0 & 0.00 \\ 0.0 & -100.00 & 204 & -100.00 & 0.0 & 0.00 \\ 0.0 & 0.00 & -100 & 204.00 & -100 & 0.00 \\ 0.0 & 0.00 & 0.0 & -133.33 & 404 & -266.67 \\ 0.0 & 0.00 & 0.0 & 33.33 & -300 & 266.67 \end{bmatrix} \begin{Bmatrix} U_2 \\ U_3 \\ U_4 \\ U_5 \\ U_6 \\ U_7 \end{Bmatrix} = \begin{Bmatrix} 266.67\,U_1 \\ 0 \\ 0 \\ 0 \\ 0 \\ Q_2^{(5)} \end{Bmatrix}. \tag{34}$$

The solution of the equations in Eq. (34) is

$$\begin{aligned} &U_2 = 278.57°\text{C}, \quad U_3 = 244.08°\text{C}, \quad U_4 = 219.35°\text{C}, \\ &U_5 = 203.39°\text{C}, \quad U_6 = 195.57°\text{C}, \quad U_7 = 194.59°\text{C}. \end{aligned} \tag{35}$$

The heat at node 1 is calculated to be $Q(0) = 71.69$ W.

Model FVM22: Second-order approximation at the boundary nodes with linear representation of $u(x)$ In this case, Eq. (32) becomes

$$\begin{bmatrix} 402.5 & -132.83 & 0.0 & 0.0 & 0.0 & 0.0 \\ -99.5 & 203.0 & -99.5 & 0.0 & 0.0 & 0.0 \\ 0.0 & -99.5 & 203.0 & -99.5 & 0.0 & 0.0 \\ 0.0 & 0.0 & -99.5 & 203.0 & -99.5 & 0.0 \\ 0.0 & 0.0 & 0.0 & -132.83 & 402.5 & -265.67 \\ 0.0 & 0.0 & 0.0 & 33.33 & -300.0 & 266.67 \end{bmatrix} \begin{Bmatrix} U_2 \\ U_3 \\ U_4 \\ U_5 \\ U_6 \\ U_7 \end{Bmatrix} = \begin{Bmatrix} 265.67\,U_1 \\ 0 \\ 0 \\ 0 \\ 0 \\ Q_2^{(5)} \end{Bmatrix} \tag{36}$$

The solution is given by

$$\begin{aligned} &U_2 = 278.50°\text{C}, \quad U_3 = 243.89°\text{C}, \quad U_4 = 219.08°\text{C}, \\ &U_5 = 203.07°\text{C}, \quad U_6 = 195.23°\text{C}, \quad U_7 = 194.25°\text{C}. \end{aligned} \tag{37}$$

The heat at node 1 is calculated as $Q(0) = 71.93$ W.

A comparison of the nodal values of u obtained by using the six FVM formulations with the exact values for Case 1 boundary conditions is presented in Table 3.3.2. The numerical solutions are obtained with uniform mesh of five ($N = 5$) subdivisions. All solutions are in good agreement with the exact solution.

A comparison of the nodal values of u obtained by the FDM, FVM22 (ZFVM), and FVM32 (HFVM) methods, obtained with uniform meshes of five ($N = 5$) and ten ($N = 10$) subdivisions, with the exact values for Case 1 boundary conditions is presented in Table 3.3.3. All solutions are in good agreement with the exact solution, with the HFVM being the most accurate.

Table 3.3.2 Comparison of the solutions obtained with various FVM formulations (using uniform mesh of five subdivisions) with the exact solution of Eqs. (3.3.6) and (3.3.7) for Case 1.

$$-\frac{d^2u}{dx^2} + 400\,u = 0,\ 0 < x < 0.05;\ \ u(0) = 300,\ \ \frac{du}{dx}\bigg|_{x=0.05} = 0.$$

x	Exact	FVM11	FVM12	FVM21	FVM22	FVM31	FVM32
0.005	278.62	277.29	277.20	278.57	278.50	280.06	279.98
0.010	260.02	260.12	259.97	261.33	261.19	260.12	259.97
0.015	244.03	242.95	242.74	244.08	243.89	245.38	245.18
0.020	230.47	279.98	230.39	231.71	231.48	230.64	230.39
0.025	219.23	218.33	218.04	219.35	219.08	220.52	220.23
0.030	210.18	210.39	210.07	211.37	211.07	210.39	210.07
0.035	203.23	202.45	202.10	203.39	203.07	204.47	204.13
0.040	198.32	198.56	198.20	199.48	199.15	198.56	198.20
0.045	195.39	194.66	194.29	195.57	195.23	196.61	196.24
0.050	194.42	194.66	194.29	194.59	194.25	194.66	194.29

Table 3.3.3 Comparison of the FDM, ZFVM (FVM22), and HFVM (FVM32) solutions with the exact solution of Eqs. (3.3.6) and (3.3.7) for Case 1.

$$-\frac{d^2u}{dx^2} + 400\,u = 0,\ 0 < x < 0.05;\ \ u(0) = 300,\ \ \frac{du}{dx}\bigg|_{x=0.05} = 0.$$

x	Exact Solution	FDM Solution		FVM22 Solution		FVM32 Solution	
		$N=5$	$N=10$	$N=5$	$N=10$	$N=5$	$N=10$
0.005	278.62	——	278.63	278.50	278.95	279.98	278.61
0.010	260.02	260.12	260.04	261.19	260.32	259.97	260.01
0.015	244.03	——	244.06	243.89	244.30	245.18	244.01
0.020	230.47	230.64	230.52	231.48	230.73	230.39	230.45
0.025	219.23	——	219.28	219.08	219.47	220.23	219.20
0.030	210.18	210.39	210.23	211.07	210.41	210.07	210.15
0.035	203.23	——	203.29	203.07	203.45	204.13	203.20
0.040	198.32	198.56	198.38	199.15	198.53	198.20	198.29
0.045	195.39	——	195.45	195.23	195.60	196.24	195.36
0.050	194.42	194.66	194.48	194.25	194.38	194.29	194.39

Case 2: HFVM

The main difference between Case 1 and Case 2 is that the boundary condition at $x = L$, which is of the mixed kind

$$\left[\frac{du}{dx} + \frac{\beta}{k}u\right]_{x=L} = 0 \rightarrow \frac{du}{dx}\bigg|_{x=L} = -\frac{\beta}{k}u(L). \tag{38}$$

Thus, the diagonal element of the row corresponding to the boundary condition of the system is increased by 2 in all six types discussed for Case 1 boundary conditions. As an example, consider Eq. (13) of model FVM32 for the case of five subdivisions. We replace $Q_2^{(5)} = -(\beta/k)U_6 = -2U_6$ [i.e., the value $\beta/k = 2$ gets added to the coefficient in location (5,5)] and obtain the condensed equations [cf. Eq. (14)]

$$\begin{bmatrix} 203.0 & -99.5 & 0.0 & 0.0 & 0.0 \\ -99.5 & 203.0 & -99.5 & 0.0 & 0.0 \\ 0.0 & -99.5 & 203.0 & -99.5 & 0.0 \\ 0.0 & 0.0 & -99.5 & 203.0 & -99.5 \\ 0.0 & 0.0 & 0.0 & -99.5 & \mathbf{103.5} \end{bmatrix} \begin{Bmatrix} U_2 \\ U_3 \\ U_4 \\ U_5 \\ U_6 \end{Bmatrix} = \begin{Bmatrix} 99.5\,U_1 \\ 0 \\ 0 \\ 0 \\ 0 \end{Bmatrix}, \tag{39}$$

whose solution is

$$\begin{aligned} &U_2 = 257.62^\circ\mathrm{C}, U_3 = 225.59^\circ\mathrm{C}, U_4 = 202.64^\circ\mathrm{C}, \\ &U_5 = 187.83^\circ\mathrm{C}, U_6 = 180.57^\circ\mathrm{C} \end{aligned} \tag{40}$$

and the exact solution at the same points is

$$\begin{aligned} &u(0.01) = 257.66^\circ\mathrm{C}, \quad u(0.02) = 225.66^\circ\mathrm{C}, \quad u(0.03) = 202.72^\circ\mathrm{C}, \\ &u(0.04) = 187.92^\circ\mathrm{C}, \quad u(0.05) = 180.66^\circ\mathrm{C}. \end{aligned} \tag{41}$$

Similarly, equations of Model FVM31 can be obtained from Eq. (5).

Case 2: ZFVM formulation

Here we discuss the modification to model FVM12, namely, modification of Eq. (32):

$$\begin{bmatrix} 302.5 & -99.5 & 0.0 & 0.0 & 0.0 & 0.0 \\ -99.5 & 203.0 & -99.5 & 0.0 & 0.0 & 0.0 \\ 0.0 & -99.5 & 203.0 & -99.5 & 0.0 & 0.0 \\ 0.0 & 0.0 & -99.5 & 203.0 & -99.5 & 0.0 \\ 0.0 & 0.0 & 0.0 & -99.5 & 102.5 & \mathbf{3.0} \\ 0.0 & 0.0 & 0.0 & 0.0 & -200.0 & \mathbf{202.0} \end{bmatrix} \begin{Bmatrix} U_2 \\ U_3 \\ U_4 \\ U_5 \\ U_6 \\ U_7 \end{Bmatrix} = \begin{Bmatrix} 199\,U_1 \\ 0 \\ 0 \\ 0 \\ 0 \\ Q_2^{(5)} \end{Bmatrix}. \tag{42}$$

The solution is given by

$$\begin{aligned} &U_2 = 276.03^\circ\mathrm{C}, \quad U_3 = 239.19^\circ\mathrm{C}, \quad U_4 = 211.96^\circ\mathrm{C} \\ &U_5 = 193.25^\circ\mathrm{C}, \quad U_6 = 182.31^\circ\mathrm{C}, \quad U_7 = 180.05^\circ\mathrm{C}, \end{aligned} \tag{43}$$

and the exact solution at the same points is

$$\begin{aligned} u(0.005) = 277.44^\circ\mathrm{C}, \quad u(0.015) = 240.46^\circ\mathrm{C}, \quad u(0.025) = 213.13^\circ\mathrm{C}, \\ u(0.045) = 183.37^\circ\mathrm{C}, \quad u(0.05) = 180.66^\circ\mathrm{C}. \end{aligned} \tag{44}$$

A comparison of the nodal values of u obtained by using the six FVM formulations with the exact values for Case 2 boundary conditions is presented in Table 3.3.4. The numerical solutions are obtained with uniform mesh of five ($N = 5$) subdivisions. All solutions are in good agreement with the exact solution.

Table 3.3.4 Comparison of the solutions obtained with various FVM formulations (using the uniform mesh of five subdivisions) with the exact solution of Eqs. (3.3.6) and (3.3.8) for Case 2.

$$-\frac{d^2u}{dx^2} + 400\,u = 0,\ 0 < x < 0.05;\ \ u(0) = 300,\ \ \left.\frac{du}{dx}\right|_{x=0.05} + 2u(0.05) = 0.$$

x	Exact	FVM11	FVM12	FVM21	FVM22	FVM31	FVM32
0.005	277.44	276.11	276.03	277.40	277.32	278.89	278.81
0.010	257.66	257.75	257.61	258.95	258.83	257.77	257.62
0.015	240.46	239.38	239.19	240.51	240.33	241.81	241.61
0.020	225.66	225.81	225.57	226.87	226.66	225.85	225.59
0.025	213.13	212.23	211.96	213.24	212.99	214.41	214.12
0.030	202.72	202.90	202.60	203.87	203.60	202.97	202.64
0.035	194.35	193.56	193.25	194.50	194.21	195.58	195.23
0.040	187.02	188.10	187.78	189.02	188.73	188.20	187.83
0.045	183.37	182.64	182.31	183.54	183.25	184.58	184.20
0.050	180.66	180.83	180.50	180.82	180.52	180.96	180.57

A comparison of the nodal values of u obtained by the FDM and the finite volume formulations FVM12 and FVM32 with the exact values for Case 2 boundary conditions is presented in Table 3.3.5. The numerical solutions are obtained with uniform meshes of five ($N = 5$) and ten ($N = 10$) subdivisions. All three solutions are in good agreement with the exact solution.

Table 3.3.5 Comparison of the FDM and FVM (FVM12 and FVM32) solutions (obtained with the uniform meshes of five and ten subdivisions) with the exact solution of Eqs. (3.3.6) and (3.3.8) for Case 2.

x	Exact Solution	FDM Solution		FVM12 Solution		FVM32 Solution	
		$N=5$	$N=10$	$N=5$	$N=10$	$N=5$	$N=10$
0.005	277.44	——	277.46	276.03	277.44	278.81	277.44
0.010	257.66	257.77	257.69	257.61	257.65	257.62	257.65
0.015	240.46	——	240.50	239.19	240.44	241.61	240.45
0.020	225.66	225.85	225.71	225.57	225.64	225.59	225.65
0.025	213.13	——	213.18	211.96	213.10	214.12	213.11
0.030	202.72	202.97	202.78	202.60	202.69	202.64	202.70
0.035	194.35	——	194.42	193.25	194.32	195.23	194.33
0.040	187.92	188.20	187.99	187.78	187.88	187.83	187.90
0.045	183.37	——	183.44	182.31	183.33	184.20	183.35
0.050	180.66	180.96	180.73	180.50	180.62	180.57	180.64

Figure 3.3.7 shows a comparison of numerical solutions obtained with various FVM formulations and the exact solution for $u(x)$. The HFVM and second-order accurate ZFVM give the closest solutions to the exact solution.

Table 3.3.6 contains a comparison of the heat flux, $q(x) = -(du/dx)$, computed at the control volume interfaces (including the boundary points) with a mesh of ten subdivisions and FVM12, FVM22, and FVM32 formulations. We note that the locations of the control volume interfaces for the HFVM and ZFVM are different. The flux at the interfaces is computed using Eq. (3.2.14) for the FVM22 and discretized equations for the first and last nodes of the FVM32.

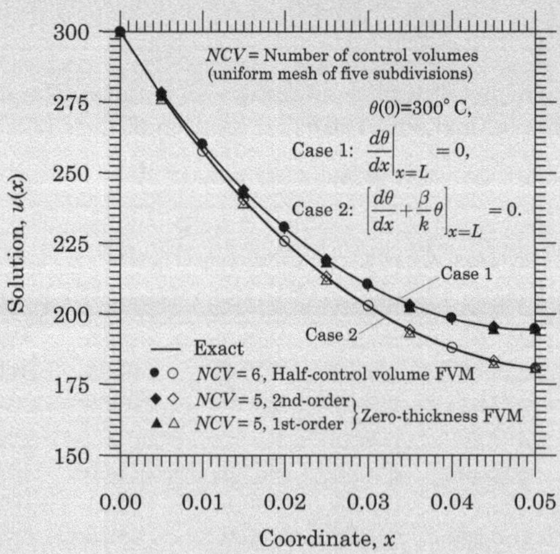

Fig. 3.3.7 Comparison of the HFVM and ZFVM (with first- and second-order accurate) solutions with the exact solution for $u(x)$. Results (obtained with the uniform mesh of five subdivisions) for the two cases of boundary conditions are included.

Table 3.3.6 Comparison of the FVM (FVH and FVZ) solutions (obtained with the uniform mesh of ten subdivisions) with the exact solution $\bar{q} = q(x) \times 10^{-4}$ ($q(x) = -du/dx$) of Eqs. (3.3.6), (3.3.7), and (3.3.8) for Cases 1 and 2.

	Case 1				Case 2			
x	Exact	FVM12	FVM22	FVM32	Exact	FVM12	FVM22	FVM32
0.0000	0.4570	0.4567	0.4572	0.4578	0.4804	0.4801	0.4807	0.4813
0.0025	0.4275	——	——	0.4278	0.4510	——	——	0.4513
0.0050	0.3991	0.3989	0.3994	——	0.4227	0.4225	0.4229	——
0.0075	0.3718	——	——	0.3720	0.3954	——	——	0.3957
0.0100	0.3453	0.3451	0.3455	——	0.3692	0.3690	0.3694	——
0.0125	0.3197	——	——	0.3200	0.3439	——	——	0.3441
0.0150	0.2950	0.2948	0.2951	——	0.3194	0.3193	0.3196	——
0.0175	0.2709	——	——	0.2711	0.2958	——	——	0.2960
0.0200	0.2476	0.2474	0.2477	——	0.2729	0.2727	0.2730	——
0.0225	0.2248	——	——	0.2250	0.2506	——	——	0.2508
0.0250	0.2026	0.2025	0.2027	——	0.2290	0.2289	0.2291	——
0.0275	0.1809	——	——	0.1810	0.2080	——	——	0.2081
0.0300	0.1597	0.1596	0.1598	——	0.1875	0.1874	0.1876	——
0.0325	0.1389	——	——	0.1390	0.1674	——	——	0.1675
0.0350	0.1184	0.1183	0.1185	——	0.1478	0.1477	0.1479	——
0.0375	0.0982	——	——	0.0983	0.1285	——	——	0.1286
0.0400	0.0783	0.0782	0.0783	——	0.1096	0.1096	0.1096	——
0.0425	0.0585	——	——	0.0586	0.0909	——	——	0.0910
0.0450	0.0389	0.0389	0.0389	——	0.0725	0.0725	0.0725	——
0.0475	0.0194	——	——	0.0195	0.0542	——	——	0.0542
0.0500	0.0000	0.0000	0.0000	0.0000	0.0361	0.0361	0.0722	0.0362

The next example deals with axisymmetric heat transfer in a long cylinder; the assumptions of axisymmetry and *long* cylinder reduce the three-dimensional heat transfer to one-dimensional heat flow in the radial direction (see Fig. 3.3.8).

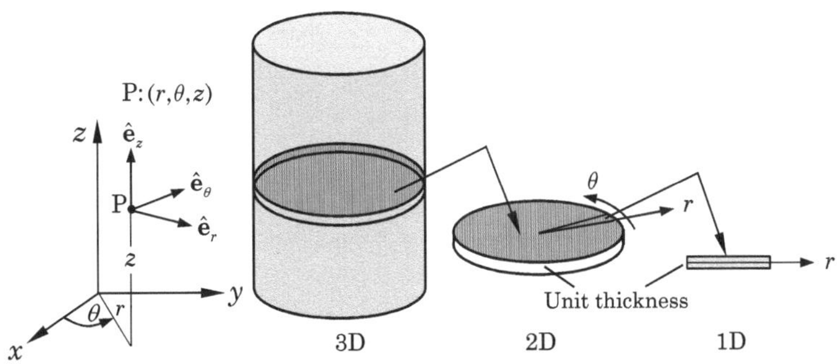

Fig. 3.3.8 The reduction of heat flow in a long cylinder with axisymmetry (i.e., material properties, boundary conditions, and source are independent of z and the radial coordinate θ) from a three-dimensional problem to one-dimensional problem.

Example 3.3.3

Consider a long, homogeneous, isotropic solid circular cylinder of outside radius $R_0 = 0.01$ m, conductivity $k = 20$ W/(m·°C), and a constant rate of internal heat generation $g_0 = 2 \times 10^8$ W/m^3. Suppose that the boundary surface at $r = a$ is maintained at $T_0 = 100$°C. (a) Develop the discretized equations using the HFVM and ZFVM using N subdivisions. (b) Specialize the results for the case of $N = 4$ and calculate the temperatures at the nodes and heat at $r = R_0$ in the two formulations.

Solution: When material properties (e.g., conductivity k) and source g are independent of the coordinates z and θ (z is taken along the length of the cylinder, which is assumed to be very large compared to the radius of the cylinder), the heat flows only radially along the radius (i.e., no flow along z or θ), reducing the three-dimensional problem to a one-dimensional problem, as illustrated in Fig. 3.3.8.

The governing equation for this axisymmetric heat flow (i.e., the temperature and heat flow are only functions of the radial coordinate r, and all radial material lines of the cylinder experience the same state of temperature and heat flow) is given by

$$-\frac{1}{r}\frac{d}{dr}\left(kr\frac{dT}{dr}\right) = g(r), \quad 0 < r < R_0, \tag{3.3.12}$$

where T is the temperature, k is the conductivity [W/(m°C)], and g is the internal heat generation (W/m^3). The volume element of a cylindrical object is $dv = r\,dr\,d\theta\,dz$. Due to the fact that the T is only a function of r, integration with respect to θ and z results in a constant multiple of $2\pi R_0$. Therefore, Eq. (3.3.12) becomes a special case of Eq. (3.2.1) with $x = r$, $a = r\,k$, $c = 0$, and $f = r\,g$.

For a solid cylinder, due to symmetry, the temperature is finite and heat flow is zero at $r = 0$. At the outer surface of the cylinder, one may specify the temperature (as is the case in the present problem), heat flow, or a convection boundary condition. Thus, for the present problem, the boundary conditions are

$$\left.\left(2\pi kr\frac{dT}{dr}\right)\right|_{r=0} = 0, \quad T(R_0) = T_0. \tag{3.3.13}$$

The exact solution of Eq. (3.3.12) with the boundary conditions in Eq. (3.3.13) is

$$T(r) = \frac{g_0 R_0^2}{4k}\left[1 - \left(\frac{r}{R_0}\right)^2\right] + T_0 \text{ (°C)}, \tag{3.3.14a}$$

$$q(r) = -k\frac{dT}{dr} = \frac{1}{2}g_0 r \text{ (W/m}^2\text{)}, \tag{3.3.14b}$$

$$Q(R_0) = -\left.\left(2\pi kr\frac{dT}{dr}\right)\right|_{R_0} = \pi g_0 R_0^2 \text{ (W)}. \tag{3.3.14c}$$

(a) The integral statement of Eq. (3.3.12) is similar to (3.2.4). Let $r_A^{(I)}$ and $r_B^{(I)}$ denote the coordinates of the left and right interfaces of a typical control volume (i.e., a line segment between coordinates $r_A^{(I)}$ and $r_B^{(I)}$) with r_I being the coordinate of the node I at the center of the control volume. We obtain the statement

$$0 = \int_0^{2\pi}\int_{r_A^{(I)}}^{r_B^{(I)}}\left[-\frac{1}{r}\frac{d}{dr}\left(kr\frac{dT}{dr}\right) - g\right] rdrd\theta$$

$$= 2\pi\left[kr\frac{dT}{dr}\right]_{r_A^{(I)}} - 2\pi\left[kr\frac{dT}{dr}\right]_{r_B^{(I)}} - 2\pi\int_{r_A^{(I)}}^{r_B^{(I)}} rg(r)\,dr, \tag{3.3.15a}$$

or (canceling 2π out)

$$-Q_A^{(I)} - Q_B^{(I)} - \int_{r_A^{(I)}}^{r_B^{(I)}} rg(r)\,dr = 0, \tag{3.3.15b}$$

where [see Eqs. (3.2.11) and (3.2.12); variable length subdivisions are assumed]

$$Q_A^{(I)} \equiv \left[-kr\frac{dT}{dr}\right]_{r_A^{(I)}} = a_A^{(I)}\left(\frac{T_{I-1}-T_I}{h_{I-1}}\right), \tag{3.3.16a}$$

$$Q_B^{(I)} \equiv \left[kr\frac{dT}{dr}\right]_{r_B^{(I)}} = a_B^{(I)}\left(\frac{T_I-T_{I+1}}{h_I}\right), \tag{3.3.16b}$$

and

$$a_A^{(I)} = r_A^{(I)}\,k,\;\; a_B^{(I)} = r_B^{(I)}\,k;\;\; r_A^{(I)} = r_I - 0.5h_{I-1},\;\; r_B^{(I)} = r_I + 0.5h_I. \tag{3.3.16c}$$

Substituting the approximations in Eqs. (3.3.16a) and (3.3.16b) into Eq. (3.3.15b), we obtain (for $I = 2, 3, \ldots, N$)

$$a_A^{(I)}\left(\frac{T_I-T_{I-1}}{h_{I-1}}\right) - a_B^{(I)}\left(\frac{T_{I+1}-T_I}{h_I}\right) = G_I, \tag{3.3.17a}$$

where

$$G_I = \int_{r_A^{(I)}}^{r_B^{(I)}} rg(r)\,dr. \tag{3.3.17b}$$

Equation (3.3.17a), for any interior node $I = 2, 3, \ldots, N$, can be expressed as

$$A_{I-1}T_{I-1} + A_IT_I + A_{I+1}T_{I+1} = G_I, \tag{3.3.18a}$$

where

$$A_{I-1} = -\frac{a_A^{(I)}}{h_{I-1}},\quad A_I = \frac{a_A^{(I)}}{h_{I-1}} + \frac{a_B^{(I)}}{h_I},\quad A_{I+1} = -\frac{a_B^{(I)}}{h_I}. \tag{3.3.18b}$$

In this example, because of the simple nature of the integral, we evaluate G_I exactly. For a uniform mesh of N subdivisions (i.e., $h_1 = h_2 = \cdots = h_N = h$), $r_A^{(I)} = r_I - 0.5h$, $r_B^{(I)} = r_I + 0.5h$, and $g = g_0$, we have

$$G_I = \int_{r_A^{(I)}}^{r_B^{(I)}} rg_0\,dr = \frac{g_0}{2}\left[\left(r_B^{(I)}\right)^2 - \left(r_A^{(I)}\right)^2\right] = 0.5g_0\left(r_A^{(I)} + r_B^{(I)}\right)\left(r_B^{(I)} - r_A^{(I)}\right). \tag{3.3.19}$$

We note that this result is the same as if we replaced the integral with $G_I = r_I g_0 \Delta r$ because $r_I = 0.5(r_A^{(I)} + r_B^{(I)})$ and $\Delta r = r_B^{(I)} - r_A^{(I)}$.

Equations (3.3.18a), (3.3.18b), and (3.3.19) are valid for both the HFVM and ZFVM formulations. The two formulations only differ in terms of the number of nodes (hence their coordinates) and the number of internal control volumes. Thus, $I = 2, 3, \ldots, N$ for the HFVM formulation and $I = 3, 4, \ldots, N$ for the ZFVM formulation. The equations for the first and the last nodes are different in the two formulations, as explained next.

Half-control volume formulation. For a mesh of N subdivisions, there are $N+1$ nodes, requiring $N+1$ equations. Eqs. (3.3.18a) and (3.3.18b) provide $N-1$ equations. The remaining two are provided by the node 1 and node $N+1$ equations, as discussed next.

For node 1, the integral statement over the half-control volume yields

$$-Q_A^{(1)} + \bar{A}_1T_1 + \bar{A}_2T_2 = G_1, \tag{1}$$

where

$$\bar{A}_1 = \frac{a_B^{(1)}}{h_1}, \quad \bar{A}_2 = -\frac{a_B^{(1)}}{h_1}, \quad G_1 = 0.5g_0\left(\frac{\Delta r}{2}\right)^2. \tag{2}$$

For the $N+1$st node, we have

$$Q_B^{(N)} + \hat{A}_N T_N + \hat{A}_{N+1} T_{N+1} = G_{N+1}, \tag{3a}$$

where

$$\hat{A}_N = -\frac{a_A^{(N)}}{h_N}, \quad \hat{A}_{N+1} = \frac{a_A^{(N)}}{h_N}, \quad G_{N+1} = 0.5g_0\left(r_{N+1}^2 - r_N^2\right). \tag{3b}$$

When T_{N+1} is specified, Eq. (3a) is used to determine $Q_B^{(N)}$; if $Q_B^{(N)}$ is specified as a mixed boundary condition, which includes the Neumann boundary condition as a special case, we simply replace it in Eq. (3a) with

$$Q_B^{(N)} + \beta_L(T_{N+1} - T_\infty) = Q_L \;\rightarrow\; Q_B^{(N)} = -\beta_L(T_{N+1} - T_\infty) + Q_L. \tag{4}$$

ZFVM formulation. For a mesh of N subdivisions, there are N control volumes and $N+2$ nodes, requiring $N+2$ equations. Equations (3.3.18a) and (3.3.18b) yield $N-2$ equations (for $I = 3, 4, \ldots, N$), requiring additional four equations, two equations each from the first and last control volumes. The condition $Q_A^{(1)} = 0$ at node 1 is used to write the first equation (with first-order accurate scheme)

$$Q_A^{(1)} = -\frac{T_2 - T_1}{0.5\Delta r} = 0 \;\rightarrow\; T_1 = T_2, \quad Q_B^{(N)} = -\beta_L(T_{N+1} - T_\infty) + Q_L. \tag{5}$$

The second equation is obtained from Eqs. (3.3.18a) and (3.3.18b) by setting $I = 2$ and replacing $Q_A^{(1)}$ by a suitable approximation (first-order or second-order accurate). We shall use the first-order accurate scheme and write

$$A_1 T_1 + A_2 T_2 + A_3 T_3 = G_2, \tag{6a}$$

where ($r_A^{(2)} = 0$ and $r_B^{(2)} = \Delta r$)

$$A_1 = -\frac{kr_A^{(2)}}{0.5\Delta r} = 0, \quad A_2 = \frac{kr_A^{(2)}}{0.5\Delta r} + \frac{kr_B^{(2)}}{\Delta r} = \frac{kr_B^{(2)}}{\Delta r}, \quad A_3 = -\frac{kr_B^{(2)}}{\Delta r}. \tag{6b}$$

For node $N+2$, the first equation is provided by the specified temperature $T_{N+2} = T_0$. The second equation is obtained from Eqs. (3.3.18a) and (3.3.18b) by setting $I = N+1$:

$$A_N T_N + A_2 T_{N+1} + A_{N+2} T_{N+2} = G_2, \tag{7a}$$

where ($r_A^{N+1} = R_0 - \Delta r$ and $r_B^{N+1} = R_0$)

$$A_N = -\frac{kr_A^{N+1}}{\Delta r} = 0, \quad A_{N+1} = \frac{kr_A^{N+1}}{\Delta r} + \frac{kr_B^{N+1}}{\Delta r}, \quad A_{N+2} = -\frac{kr_B^{N+1}}{0.5\Delta r}. \tag{7b}$$

(b) We now illustrate the discrete equations developed for the uniform mesh of four subdivisions, $N = 4$ (i.e., $\Delta r = h = R_0/4 = 0.0025$). There are five equations for the HFVM formulation and six equations in the ZFVM formulation before the boundary conditions are imposed.

Half-control volume formulation. The five equations are

$$
\begin{aligned}
-Q_A^{(1)} + \frac{kh}{2}\frac{(T_2 - T_1)}{h} &= \frac{g_0}{2}\frac{h}{2}\frac{h}{2}, \\
\frac{kh}{2}\frac{(T_2 - T_1)}{h} - \frac{3kh}{2}\frac{(T_3 - T_2)}{h} &= \frac{g_0}{2}h\left(\frac{h}{2} + \frac{3h}{2}\right), \\
\frac{3kh}{2}\frac{(T_3 - T_2)}{h} - \frac{5kh}{2}\frac{(T_4 - T_3)}{h} &= \frac{g_0}{2}h\left(\frac{3h}{2} + \frac{5h}{2}\right), \\
\frac{5kh}{2}\frac{(T_4 - T_3)}{h} - \frac{7kh}{2}\frac{(T_5 - T_4)}{h} &= \frac{g_0}{2}h\left(\frac{5h}{2} + \frac{7h}{2}\right), \\
\frac{7kh}{2}\frac{(T_5 - T_4)}{h} - Q_B^{(4)} &= \frac{g_0}{2}\frac{h}{2}\left(\frac{7h}{2} + 4h\right).
\end{aligned}
\tag{8}
$$

or

$$
\begin{bmatrix} 10 & -10 & 0 & 0 & 0 \\ -10 & 40 & -30 & 0 & 0 \\ 0 & -30 & 80 & -50 & 0 \\ 0 & 0 & -50 & 120 & -70 \\ 0 & 0 & 0 & -70 & 70 \end{bmatrix}
\begin{Bmatrix} T_1 \\ T_2 \\ T_3 \\ T_4 \\ T_5 \end{Bmatrix}
=
\begin{Bmatrix} Q_A^{(1)} + 156.25 \\ 1250.0 \\ 2500.0 \\ 3750.0 \\ Q_B^{(4)} + 2343.75 \end{Bmatrix}
\tag{9}
$$

Using the boundary conditions, $T_5 = 100$ and $Q_1^{(1)} = 0$, and omitting the last equation (because T_5 is known), the condensed equations are ($h^2 g_0 = 1250$):

$$
\begin{bmatrix} 10 & -10 & 0 & 0 \\ -10 & 40 & -30 & 0 \\ 0 & -30 & 80 & -50 \\ 0 & 0 & -50 & 120 \end{bmatrix}
\begin{Bmatrix} T_1 \\ T_2 \\ T_3 \\ T_4 \end{Bmatrix}
=
\begin{Bmatrix} 156.25 \\ 1250.00 \\ 2500.00 \\ 3750 + 7000 \end{Bmatrix}.
\tag{10}
$$

The solution of these equations is given by

$$
T_1 = 350.0^\circ\text{C}, \quad T_2 = 334.375^\circ\text{C}, \quad T_3 = 287.5^\circ\text{C}, \quad T_4 = 209.375^\circ\text{C}. \tag{11}
$$

The heat at $r = R_0$ is computed using the last equation of Eq. (9),

$$
Q(R_0) = Q_B^{(4)} = 70\,T_4 - 70\,T_5 + 2343.75 = 10^4. \tag{12}
$$

The numerical solution matches with the exact solution at the nodes.

ZFVM formulation. The six equations (note the first equation is based on $dT/dr = 0$ and not on the true boundary condition, $rk(dT/dr) = 0$ at $r = 0$; this is technically incorrect for this formulation, but $dT/dr = 0$ also satisfies the boundary condition), with the first-order accurate representation of the derivative at $x = L$, are

$$
\begin{aligned}
k\frac{(T_2 - T_1)}{0.5h} &= 0, \\
-kh\frac{(T_3 - T_2)}{h} &= \frac{g_0 h^2}{2}, \\
kh\frac{(T_3 - T_2)}{h} - 2kh\frac{(T_4 - T_3)}{h} &= \frac{g_0}{2}\left(4h^2 - h^2\right), \\
2kh\frac{(T_4 - T_3)}{h} - 3kh\frac{(T_5 - T_4)}{h} &= \frac{g_0}{2}\left(9h^2 - 4h^2\right), \\
3kh\frac{(T_5 - T_4)}{h} - 4kh\frac{(T_6 - T_5)}{0.5h} &= \frac{g_0}{2}\left(16h^2 - 9h^2\right), \\
T_6 &= T_0,
\end{aligned}
\tag{13}
$$

or

$$\begin{bmatrix} 16\times10^3 & -16\times10^3 & 0 & 0 & 0 & 0 \\ 0 & 20 & -20 & 0 & 0 & 0 \\ 0 & -20 & 60 & -40 & 0 & 0 \\ 0 & 0 & -40 & 100 & -60 & 0 \\ 0 & 0 & 0 & -60 & 220 & -160 \\ 0 & 0 & 0 & 0 & 0 & 1 \end{bmatrix} \begin{Bmatrix} T_1 \\ T_2 \\ T_3 \\ T_4 \\ T_5 \\ T_6 \end{Bmatrix} = \begin{Bmatrix} 0 \\ 625 \\ 1875 \\ 3125 \\ 4375 \\ 100 \end{Bmatrix}. \tag{14}$$

The solution of these equations is (at points $r = 0, 0.00125$, 0.00375, 0.00625, and 0.00875, respectively):

$$T_1 = 350.0^\circ\text{C},\quad T_2 = 350.0^\circ\text{C},\quad T_3 = 318.75^\circ\text{C},\quad T_4 = 256.25^\circ\text{C}.\quad T_5 = 162.5^\circ\text{C}. \tag{15}$$

The heat at $r = R_0$ is

$$Q(R_0) = -kR_0\frac{T_6 - T_5}{0.5h} = 10^4. \tag{16}$$

The exact solution in Eq. (3.3.14a) at the same locations yields

$$T(0) = 350.0^\circ\text{C},\quad T(0.00125) = 346.09^\circ\text{C},\quad T(0.00375) = 314.84^\circ\text{C},$$
$$T(0.00625) = 252.34^\circ\text{C},\quad T(0.00875) = 158.59^\circ\text{C}. \tag{17}$$

The finite volume solutions obtained with the two formulations and uniform meshes of four, eight, and sixteen subdivisions are compared with the exact solution in Table 3.3.7. It is clear that the nodal values for the temperatures and heats predicted by the HFVM coincide with the exact solution at the nodes for any number of subdivisions. The ZFVM formulation is less accurate at the nodes, while it is very accurate at the control volume interfaces. In both formulations, the fluxes computed at the interfaces and the boundary nodes match with the exact solutions.

Plots of the numerical solutions obtained using 5, 10, and 20 subdivisions in the FVM22 (ZFVM with second-order accurate formulation) for the dimensionless temperature $\bar{T} = (T - T_0)k/g_0R_0^2$ and heat flux $\bar{Q}(r) = -krdT/dr/(R_0g_0)$ (plotted versus $\bar{r} = r/R_0$) along with the respective exact solutions are shown in Figs. 3.3.9 and 3.3.10. As noted before, the FVM32

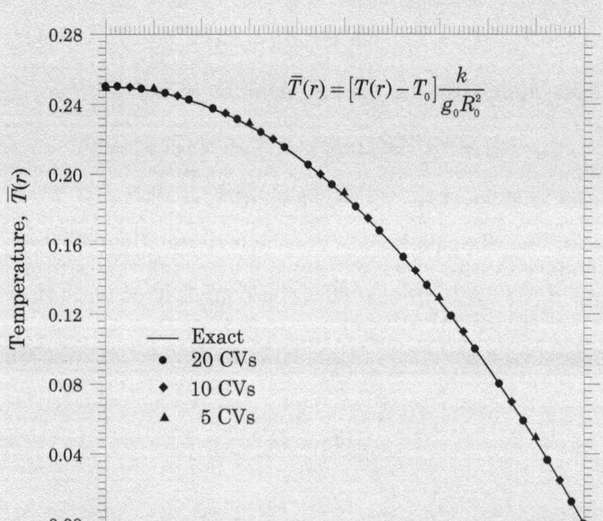

Fig. 3.3.9 Comparison of the FVM solutions with the exact solution for the temperature $\bar{T}(r)$ in a radially symmetric heat transfer in a long cylinder.

solutions for all meshes match the exact solution at the nodes and, therefore, not included in the plots. The convergence of the FVM solutions with mesh refinement is obvious from the plots.

Table 3.3.7 Comparison of the FVM solutions* with the exact solutions for temperature distribution in an axisymmetric circular cylinder [$R_0 = 0.01$ m, $k = 20$ W/(m·°C), $g_0 = 2\times10^8$ W/m^3, and $T_0 = 100$°C].

$$-\frac{1}{r}\frac{d}{dr}\left(kr\frac{dT}{dr}\right) = g(r), \quad 0 < r < R_0; \quad \left(kr\frac{dT}{dr}\right)\Big|_{r=0} = 0, \quad T(R_0) = T_0.$$

		FVM32 (HFVM)			FVM12 (ZFVM)		
r/R_0	Exact	$N=4$	$N=8$	$N=16$	$N=4$	$N=8$	$N=16$
0.00000	350.00	350.00	350.00	350.00	350.00	350.00	350.00
0.03125	349.76		349.02	349.51	350.00	350.00	350.00
0.06250	349.02	346.09	348.05	349.02	350.00	350.00	349.02
0.09375	347.80		347.07	347.56	350.00	348.05	348.05
0.12500	346.09	342.19	346.09	346.09	350.00	346.09	346.09
0.15625	343.90		343.16	343.65		344.14	344.14
0.18750	341.21	338.28	340.23	314.21	342.19	342.19	341.21
0.21875	338.04		337.30	337.79		338.28	338.28
0.25000	334.38	334.38	334.38	334.38	334.37	334.37	334.37
0.28125	330.22		329.49	329.98		330.47	330.47
0.31250	325.59	322.66	324.61	325.59	326.56	326.56	325.59
0.34375	320.46		319.73	320.21		320.70	320.70
0.37500	314.84	310.94	314.84	314.84	318.75	314.84	314.84
0.40625	308.74		308.01	308.50		308.98	308.98
0.43750	302.15	299.22	301.17	302.15	303.12	303.12	302.15
0.46875	295.07		294.34	294.82		295.31	295.31
0.50000	287.50	287.50	287.50	287.50	287.50	287.50	287.50
0.53125	279.44		278.71	279.20		279.69	279.69
0.56250	270.90	267.97	269.92	270.90	271.87	271.87	270.90
0.59375	261.87		261.13	261.62		262.11	262.11
0.62500	252.34	248.44	248.44	248.44	256.25	252.34	252.34
0.65625	242.33		241.60	242.09		242.58	242.58
0.68750	231.84	228.91	230.86	231.84	232.81	232.81	231.84
0.71875	220.85		220.12	220.61		221.09	221.09
0.75000	209.37	209.37	209.37	209.37	209.37	209.37	209.37
0.78125	197.41		196.68	197.17		197.66	197.66
0.81250	184.96	182.03	183.98	184.96	185.94	185.94	184.96
0.84375	172.02		171.29	171.78		172.27	172.27
0.87500	158.59	154.69	154.69	154.69	162.50	158.59	158.59
0.90625	144.68		143.94	144.43	146.87	144.92	144.92
0.93750	130.27	127.34	129.30	130.27	131.25	131.25	130.27
0.96875	115.38		114.65	115.14	115.62	115.62	115.62
1.00000	100.00	100.00	100.00	100.00	100.00	100.00	100.00

*The underlined terms are the linearly interpolated values between the consecutive nodal values.

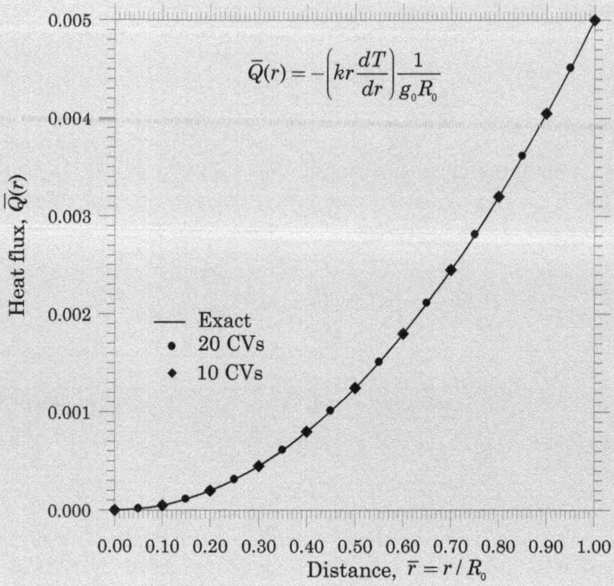

Fig. 3.3.10 Comparison of the FVM solutions with the exact solution for the heat flux $\bar{Q}(r)$ in axisymmetric heat transfer in a long cylinder.

Example 3.3.4

The advection–diffusion equation in one dimension can be expressed in general form as

$$b\frac{d\phi}{dx} - \frac{d}{dx}\left(a\frac{d\phi}{dx}\right) - g = 0, \quad 0 < x < L, \tag{3.3.20}$$

where a and b are constants (in this example), and g is the source term. For example, heat transfer in the presence of fluid flow, we have $b = \rho c_p u$ and $a = k$, where ρ is the mass density of the fluid, c_p is the specific heat at constant pressure, u is the velocity along the length of the channel, k is the conductivity of the medium, and ϕ is the temperature.

Equation (3.3.20) can be expressed in dimensionless form when k is a constant. We introduce the following quantities:

$$\bar{x} = \frac{x}{L}, \quad Pe = \frac{\rho u L}{k}, \quad \bar{g} = \frac{L}{\rho u} g, \tag{3.3.21}$$

where Pe is known as the *Péclet number*. Then Eq. (3.3.20) becomes

$$\frac{d\phi}{d\bar{x}} - \frac{1}{Pe}\frac{d^2\phi}{d\bar{x}^2} - \bar{g} = 0, \quad 0 < \bar{x} < 1, \tag{3.3.22}$$

and the dimensionless diffusion flux is

$$\bar{q}_n = -\frac{1}{Pe}\frac{d\phi}{d\bar{x}}, \tag{3.3.23}$$

We note that the dimensionless form of the governing equation is a special case of the original equation with $b = 1$ and $a = 1/Pe$. In the remainder of this discussion, omit the bar over the quantities for brevity.

Develop the (a) finite difference and (b) half-control volume FVM formulation of Eq. (3.3.20) and investigate the behavior of the numerical solutions for various values of the Péclet number and number of subdivisions of the domain when $g = 0$ and the boundary conditions are

$$\phi(0) = 1, \quad \phi(1) = 0. \tag{3.3.24}$$

The exact solution of Eq. (3.3.20) with the boundary conditions in Eq. (3.3.24) can be shown to be

$$\phi(x) = \frac{e^{Pe\,x} - e^{Pe}}{1 - e^{Pe}}, \quad \frac{d\phi}{dx} = \frac{Pe\,e^{Pe\,x}}{1 - e^{Pe}}, \quad q_n(x) = -\frac{e^{Pe\,x}}{1 - e^{Pe}}. \tag{3.3.25}$$

The solution rapidly decreases from unity to zero in a very narrow region around $x = 1$. Typically, numerical methods have difficulty in producing numerical solutions that are close to the exact solution in this narrow region.

Solution: First, we develop the discretized equations using the FDM and FVM for uniform mesh of N subdivisions (both methods will have $N + 1$ mesh points). Then we study the numerical character of the discretized equations by carrying out stability analysis.

(a) **Finite Difference Equations** When the central difference approximations are used to approximate the first- and second-order derivatives of Eq. (3.3.21), we obtain

$$b\left(\frac{\Phi_{I+1} - \Phi_{I-1}}{2h}\right) - a\left(\frac{\Phi_{I-1} - 2\Phi_I + \Phi_{I+1}}{h^2}\right) = 0, \tag{3.3.26}$$

where $b = 1$, $a = 1/Pe$ and $\Phi_I = \phi(x_I)$, and h is spatial step size. Equation (3.3.26) can be rewritten in the form

$$\left(-\frac{b}{2} - \frac{a}{h}\right)\Phi_{I-1} + \frac{2a}{h}\phi_I + \left(\frac{b}{2} - \frac{a}{h}\right)\Phi_{I+1} = 0. \tag{3.3.27}$$

(b) **Finite Volume Equations** We begin with the integral statement of Eq. (3.3.20) over the Ith control volume centered around node I, occupying the domain between points A and B (see Fig. 3.3.11):

$$0 = \int_{x_A^{(I)}}^{x_B^{(I)}} \left[\frac{d}{dx}\left(b\phi - a\frac{d\phi}{dx}\right) - g\right] dx = \left[b\phi - a\frac{d\phi}{dx}\right]_A^B - \int_{x_A^{(I)}}^{x_B^{(I)}} g\,dx, \tag{3.3.28}$$

where $x_A^{(I)}$ and $x_B^{(I)}$ refer to the left and right end locations (control volume interfaces), respectively, of the Ith control volume. We have weakened the differentiability on ϕ by carrying out the indicated integration. Equation (3.3.28) can be expressed as

$$0 = -Q_A^{(I)} + Q_B^{(I)} - b\phi\Big|_{x_A^{(I)}} + b\phi\Big|_{x_B^{(I)}} - \int_{x_A^{(I)}}^{x_B^{(I)}} g\,dx, \tag{3.3.29a}$$

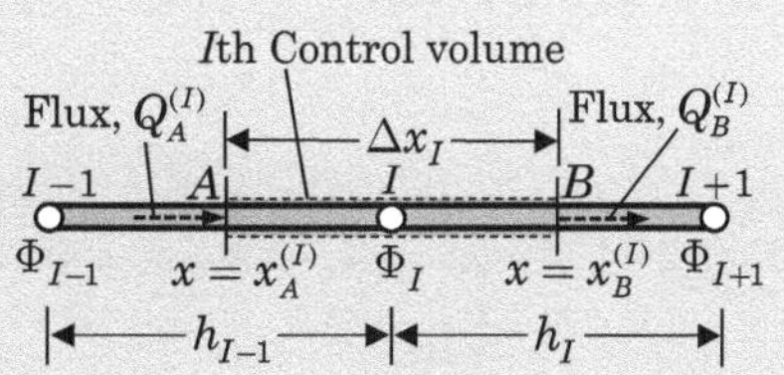

Fig. 3.3.11 A typical control volume around the Ith node.

where

$$Q_A^{(I)} \equiv \left[-a\frac{d\phi}{dx}\right]_{x_A^{(I)}}, \qquad Q_B^{(I)} \equiv \left[-a\frac{d\phi}{dx}\right]_{x_B^{(I)}}. \tag{3.3.29b}$$

Here $Q_A^{(I)}$ and $Q_B^{(I)}$ denote the secondary variables (e.g., diffusion fluxes) at the left and right interfaces (A and B, respectively) of the control volume centered at node I.

We can express the heat fluxes $Q_A^{(I)}$ and $Q_B^{(I)}$ defined in Eq. (3.3.29b) as

$$Q_A^{(I)} \equiv \left[-a\frac{d\phi}{dx}\right]_{x_A^{(I)}} = a\left(\frac{\Phi_{I-1} - \Phi_I}{h_{I-1}}\right), \tag{3.3.30a}$$

$$Q_B^{(I)} \equiv \left[-a\frac{d\phi}{dx}\right]_{x_B^{(I)}} = a\left(\frac{\Phi_I - \Phi_{I+1}}{h_I}\right). \tag{3.3.30b}$$

In addition, we use the following identities:

$$\phi\Big|_{x_A^{(I)}} = \frac{1}{2}\left(\Phi_{I-1} + \Phi_I\right), \quad \phi\Big|_{x_B^{(I)}} = \frac{1}{2}\left(\Phi_I + \Phi_{I+1}\right). \tag{3.3.31}$$

Substituting the approximations in Eqs. (3.3.30a), (3.3.30b), and (3.3.31) into Eq. (3.3.29a), we obtain (for $I = 2, 3, \ldots, N$)

$$a\left[\left(\frac{\Phi_I - \Phi_{I-1}}{h_{I-1}}\right) + a\left(\frac{\Phi_I - \Phi_{I+1}}{h_I}\right)\right] - \frac{b}{2}\Phi_{I-1} + \frac{b}{2}\Phi_{I+1} = 0$$

or

$$A_{I-1}\Phi_{I-1} + A_I\Phi_I + A_{I+1}\Phi_{I+1} = 0, \tag{3.3.32a}$$

where

$$A_{I-1} = -\frac{b}{2} - \frac{a}{h_{I-1}}, \quad A_I = \frac{a}{h_{I-1}} + \frac{a}{h_I}, \quad A_{I+1} = \frac{b}{2} - \frac{a}{h_I}. \tag{3.3.32b}$$

Equations (3.3.32a) and (3.3.32b) are valid for any interior node $I = 2, 3, \ldots, N$. We note that this equation is precisely the same as Eq. (3.3.27) (with $b = 1$ and $a = 1/Pe$) derived using the FDM.

For a boundary node, the discrete equations will be different when ϕ is not specified there, and these discrete equations are used to determine $Q_A^{(1)}$ and $Q_B^{(N)}$ at the boundary nodes. For node 1, we have $x_A^{(1)} = 0$, and Eq. (3.3.21) takes the form

$$0 = -Q_A^{(1)} + \left(a\frac{\Phi_1 - \Phi_2}{h_1}\right) - \frac{b}{2}\left(\Phi_1 + \Phi_2\right) \tag{3.3.33}$$

or

$$-Q_A^{(1)} + \bar{A}_1\Phi_1 + \bar{A}_2\Phi_2 = 0 \tag{3.3.34a}$$

where

$$\bar{A}_1 = -\frac{b}{2} + \frac{a}{h_1}, \quad \bar{A}_2 = \frac{b}{2} - \frac{a}{h_1} \tag{3.3.34b}$$

Similarly, for node $N+1$, we have

$$Q_B^{(N)} + \hat{A}_N\Phi_N + \hat{A}_{N+1}\Phi_{N+1} = 0 \tag{3.3.35a}$$

where

$$\hat{A}_N = -\frac{b}{2} - \frac{a}{h_N}, \quad \hat{A}_{N+1} = \frac{b}{2} + \frac{a}{h_N}, \tag{3.3.35b}$$

Stability analysis Using $b = 1$ and $a = 1/Pe$ and noting that the mesh is uniform, i.e., $h_1 = h_2 = \cdots = h_{I-1} = h_I = h_{I+1}$, Eqs. (3.3.32a) and (3.3.32b) together give

$$-\left(\frac{1}{2} + \frac{1}{h\,Pe}\right)\Phi_{I-1} + \frac{2}{h\,Pe}\Phi_I + \left(\frac{1}{2} - \frac{1}{h\,Pe}\right)\Phi_{I+1} = 0. \tag{3.3.36}$$

Equation (3.3.36) can be written as

$$\left(1 + \frac{h\,Pe}{2}\right)\Phi_{I-1} - 2\Phi_I + \left(1 - \frac{h\,Pe}{2}\right)\Phi_{I+1} = 0. \tag{3.3.37}$$

The second-order difference equation in Eq. (3.3.37) can be solved analytically by seeking solution in the form (for details, see Surana and Reddy [14], pages 371–378),

$$\Phi_I = \xi^{I-1} \tag{3.3.38}$$

where ξ is a discrete coordinate. Substituting Eq. (3.3.38) in Eq. (3.3.37), we obtain

$$\left(1 + \frac{h\,Pe}{2}\right)\xi^{I-2} - 2\,\xi^{I-1} + \left(1 - \frac{h\,Pe}{2}\right)\xi^{I} = 0$$

and dividing throughout by ξ^{I-2} yields ($a\xi^2 - 2b\xi + c = 0$)

$$\left(\frac{h\,Pe}{2} + 1\right) - 2\,\xi + \left(1 - \frac{h\,Pe}{2}\right)\xi^2 = 0. \tag{3.3.39}$$

Equation (3.3.39) is quadratic in ξ with the following two roots ξ_1 and ξ_2:

$$\xi_1 = 1, \quad \xi_2 = \frac{1 + \frac{h\,Pe}{2}}{1 - \frac{h\,Pe}{2}} \tag{3.3.40}$$

Hence, the solution Φ_I consists of a linear combination of the roots ξ_1^{I-1} and ξ_2^{I-1}, and the solution can be expressed as

$$\Phi_I(\xi) = K_1, \xi_1^{I-1} + K_2\,\xi_2^{I-1} = K_1 + K_2\xi_2^{I-1} \tag{3.3.41}$$

where K_1 and K_2 are the constants, determined by imposing the boundary conditions $U_1 = 1$ and $\Phi_{N+1} = 0$ ($K_1 + K_2 = 1$ and $K_1 + K_2\xi_2^N = 0$):

$$K_1 = -\frac{\xi_2^N}{1 - \xi_2^N}, \quad K_2 = \frac{1}{1 - \xi_2^N} \tag{3.3.42}$$

An examination of Eq. (3.3.41) [with the constants K_1 and K_2 defined in Eq. (3.3.42)], we note that ($h = 1/N$ for a domain of unit length)

(1) If $\frac{h\,Pe}{2} < 1$ or $N > \frac{Pe}{2}$, then ξ_2 is positive and thus ξ_2^{I-1} is positive for all values of I.

(2) If $\frac{h\,Pe}{2} \geq 1$ or $N \leq \frac{Pe}{2}$, then ξ_2 is negative and hence ξ_2^{I-1} changes sign with each successive values of I. This implies that the solution given by Eq. (3.3.33) or Eq. (3.3.38) exhibits oscillations in going from one node to the next.

Numerical results First we consider the case of $Pe = 10$. For a uniform mesh of four subdivisions ($N = 4$ and $h = \frac{1}{N} = 0.25$) in the half-control volume formulation, we have five control volumes and five nodes. Since ϕ is specified at the boundary points (i.e., $\Phi_1 = 1.0$ and $\Phi_5 = 0$), we obtain three equations among Φ_2, Φ_3, and Φ_4 associated with the three interior nodes. The three equations are obtained from Eqs. (3.3.32a) and (3.3.32b) as

$$-0.9\,\Phi_{I-1} + 0.8\,\Phi_I + 0.1\,\Phi_{I+1} = 0. \tag{1}$$

For $I = 2, 3$, and 4 we obtain three equations, which are expressed in matrix form as

$$\begin{bmatrix} 0.8 & 0.1 & 0.0 \\ -0.9 & 0.8 & 0.1 \\ 0.0 & -0.9 & 0.8 \end{bmatrix} \begin{Bmatrix} \Phi_2 \\ \Phi_3 \\ \Phi_4 \end{Bmatrix} = \begin{Bmatrix} 0.9\,\Phi_1 \\ 0.0 \\ -0.1\Phi_5 \end{Bmatrix} = \begin{Bmatrix} 0.9 \\ 0.0 \\ 0.0 \end{Bmatrix}. \tag{2}$$

The solution of these equations is

$$\Phi_1 = 1.0000, \quad \Phi_2 = 1.0015, \quad \Phi_3 = 0.9878, \quad \Phi_4 = 1.1113, \quad \Phi_5 = 0.0000. \tag{3}$$

Thus, the solution in this case ($N < 5$) exhibits oscillatory behavior, which is in conformity with the stability criterion.

For a uniform mesh of five subdivisions ($N = 5$ and $h = 0.2$), we obtain the simple relation

$$-10\,\Phi_{I-1} + 10\,\Phi_I = 0. \tag{4}$$

We obtain the following four equations, expressed in matrix form, from Eq. (4) for $I = 2, 3, 4$, and 5:

$$\begin{bmatrix} 10.0 & 0.0 & 0.0 & 0.0 \\ -10.0 & 10.0 & 0.0 & 0.0 \\ 0.0 & -10.0 & 10.0 & 0.0 \\ 0.0 & 0.0 & -10.0 & 10.0 \end{bmatrix} \begin{Bmatrix} \Phi_2 \\ \Phi_3 \\ \Phi_4 \\ \Phi_5 \end{Bmatrix} = \begin{Bmatrix} 10\,\Phi_1 \\ 0.0 \\ 0.0 \\ 0.0 \end{Bmatrix} = \begin{Bmatrix} 10.0 \\ 0.0 \\ 0.0 \\ 0.0 \end{Bmatrix}, \tag{5}$$

whose solution is $\Phi_1 = 1.0$, $\Phi_2 = 1.0$, $\Phi_3 = 1.0$, $\Phi_4 = 1.0$, $\Phi_5 = 1.0$, and $\Phi_6 = 0.0$. Clearly, the solution is not oscillatory because the $N = 5$ meets the stability criterion. Obviously, with an increased number of subdivisions, the solution will be increasingly close (without oscillations) to the exact solution.

Table 3.3.8 contains a summary of numerical solutions obtained with various meshes for different values of the Péclet number. Figure 3.3.12 shows a comparison of the exact solution (solid line) with the FVM solutions (the same as the FDM) for a mesh of $N = 4$ ($< Pe/2$) and $N = 10$ ($> Pe/2$) elements. Similar results are also presented for $Pe = 30$ in the same figure. Figure 3.3.13 shows a comparison of the exact solution (solid line) with the FVM (the same as the FDM) solutions for a mesh of $N = 100$ subdivisions for $Pe = 10, 100$, and 250. Clearly, for $N = 100$ and $Pe = 250$, the numerical solution (dark circles) exhibits oscillations because $N < Pe/2$. When 250 elements are used, the solution (open circles) does not exhibit oscillations. In other words, one must use $N > Pe/2$ subdivisions to obtain accurate solutions. Thus, for very large Péclet number problems, the number of subdivisions to be used goes up because the number of subdivisions (N) should be at least $N = Pe/2$ to satisfy the stability requirement. Also, a nonuniform mesh is desirable with very small elements close to $x = 1.0$ because of the large gradient of the solution there.

In closing this example, we recall that the discretized equation of the advection–diffusion equation, Eq. (3.3.32a), is the same as that in Eq. (3.3.27) for the Ith mesh point, irrespective of the method. In particular, as will be seen in Chapter 4, the FEM with a uniform mesh of linear elements also gives the same equation for a global node I [in the FEM, one needs to identify the structure of the finite element equation associated with the global node I, after the assembly of element equations, to be the same as that given in Eq. (3.3.22)]. Therefore, the stability analysis performed in this example is also valid for the finite element equations obtained using meshes of linear finite elements. For quadratic elements, one may carry out the stability analysis by identifying the finite element equation associated with a typical global node I.

Table 3.3.8 Comparison of the FVM32 solutions* with the exact solutions for the field variable ϕ of the advection–diffusion equation.

$$\frac{d\phi}{d\xi} - \frac{1}{Pe}\frac{d^2\phi}{d^2\xi} = 0, \quad 0 < \xi < 1; \qquad \phi(0) = 1, \quad \phi(1) = 0.$$

	$Pe = 10$				$N = 100$			
ξ	Exact	$N = 5$	$N = 10$	ξ	Exact	$Pe = 100$	Exact	$Pe = 250$
0.200	1.0000	**1.0000**	**0.9999**	0.20	**1.0000**	**1.0000**	1.0000	**1.0000**
0.300	1.0000	1.0000	0.9996	0.50	**1.0000**	**1.0000**	1.0000	**1.0000**
0.350	1.0000	1.0000	0.9996	0.80	1.0000	**1.0000**	1.0000	**1.0000**
0.400	1.0000	**1.0000**	**0.9986**	0.85	1.0000	**1.0000**	1.0000	**1.0000**
0.450	1.0000	1.0000	0.9973	0.86	1.0000	**1.0000**	1.0000	**1.0000**
0.500	1.0000	1.0000	**0.9959**	0.87	1.0000	**1.0000**	1.0000	**1.0000**
0.550	1.0000	1.0000	0.9918	0.88	1.0000	**1.0000**	1.0000	**1.0000**
0.600	1.0000	**1.0000**	**0.9877**	0.89	1.0000	**1.0000**	1.0000	**1.0000**
0.650	1.0000	1.0000	0.9753	0.90	1.0000	**1.0000**	1.0000	**1.0000**
0.700	1.0000	1.0000	**0.9630**	0.91	0.9999	**0.9999**	1.0000	**1.0000**
0.750	1.0000	1.0000	0.9259	0.92	0.9997	**0.9998**	1.0000	**1.0000**
0.800	0.8647	**1.0000**	**0.8889**	0.93	0.9991	**0.9995**	1.0000	**1.0000**
0.850	0.7769	0.7500	0.7778	0.94	0.9975	**0.9986**	1.0000	**1.0000**
0.900	0.6321	0.5000	**0.6667**	0.95	0.9933	**0.9959**	1.0000	**1.0000**
0.925	0.5277	0.3750	0.5000	0.96	0.9817	**0.9876**	1.0000	0.9998
0.950	0.3935	0.2500	0.3333	0.97	0.9502	**0.9630**	1.0000	1.0014
0.975	0.2212	0.1250	0.1667	0.98	0.9179	**0.8889**	0.9981	0.9876
0.995	0.0488	0.0250	0.0333	0.99	0.3935	**0.6667**	0.7135	1.1111

*The numbers in bold are the nodal values; other numbers are the interpolated values.

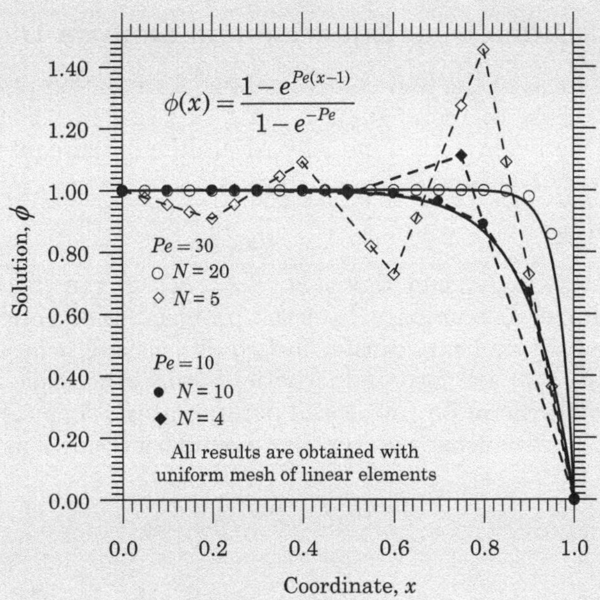

Fig. 3.3.12 A comparison of the numerical solutions against the exact solution for $N = 4$ and $N = 10$ for $Pe = 10$ and $Pe = 30$.

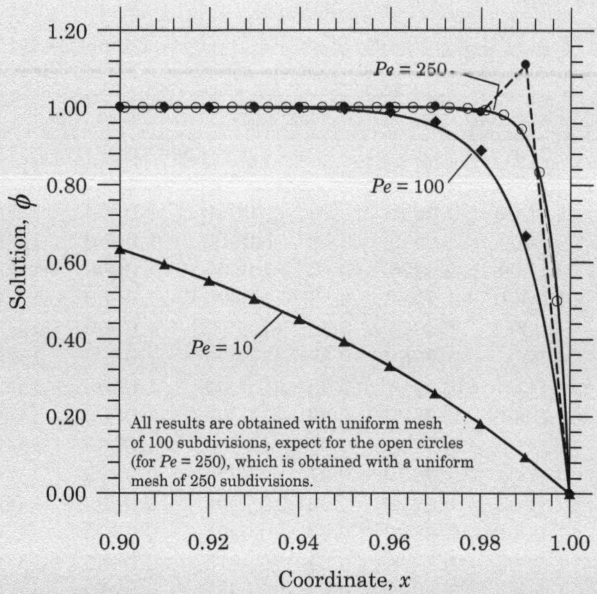

Fig. 3.3.13 A comparison of the numerical solutions against the exact solutions for $Pe = 10, 100$, and 250.

3.4 Two-Dimensional Problems

3.4.1 Model Differential Equation and Domain Discretization

Extension of the ideas of the half-control volume formulation presented for one-dimensional problems to rectangular meshes of two-dimensional problems is presented here. We begin with a model partial differential equation

$$-\left[\frac{\partial}{\partial x}\left(a_{xx}\frac{\partial u}{\partial x}\right)+\frac{\partial}{\partial y}\left(a_{yy}\frac{\partial u}{\partial y}\right)\right]=f(x,y) \quad \text{in } \Omega, \tag{3.4.1}$$

where $u(x,y)$ is the dependent unknown, (a_{xx}, a_{yy}, g) are the data, and Ω is the domain with closed boundary Γ of the problem. The equation arises, for example, in steady-state heat transfer in two dimensions, where u denotes the temperature, (a_{xx}, a_{yy}) are the conductivities of an orthotropic medium, and g is the internal heat generation (measured per unit area). Typically u is required to satisfy one of the following two types of boundary conditions at a boundary point

$$u = \hat{u}(s) \quad \text{on} \quad \Gamma_u \tag{3.4.2a}$$

and

$$\left(a_{xx}\frac{\partial u}{\partial x}n_x + a_{yy}\frac{\partial u}{\partial y}n_y\right) + \beta(u - u_\infty) = \hat{q}_n(s) \quad \text{on} \quad \Gamma_q \tag{3.4.2b}$$

where Γ_u and Γ_q are disjoint portions of the total boundary Γ such that $\Gamma = \Gamma_u \cup \Gamma_q$ and (n_x, n_y) are the direction cosines of the unit normal vector $\hat{\mathbf{n}}$ on the boundary Γ; in the case of heat transfer, β is the heat transfer coefficient and u_∞ the temperature of the medium surrounding the body. Of course, there are many problems of engineering science that are governed by the model equation (see Table 9.1.1 of the book by Reddy [8] for additional examples). In the following discussion we shall assume that the domain is homogeneous and orthotropic with respect to the global coordinate axes (x, y) (i.e., a_{xx} and a_{yy} are constant throughout the domain).

In the following discussion, we shall assume that the domain Ω of the model problem is of rectangular geometry. The rectangular domain is subdivided into $N \times M$ mesh of cells (i.e., N subdivisions along the x-axis and M subdivisions along the y-axis). As was the case with a one-dimensional domain, in the HFVM, the mesh of control volumes is created in such a way that each of the internal nodes is at the center of a control volume, and the nodes on the boundary of the domain are inside (not necessarily at the center of) the quarter or half-control volumes, as shown in Fig. 3.4.1. In the ZFVM, the $N \times M$ cells are taken as the control volumes and nodes are placed at the center of each control volume, and the boundary nodes are placed with "zero-thickness" control volumes on the boundary, as shown in Fig. 3.4.2. We can follow the same procedure as in one-dimensional model equation, Eq. (3.2.1), to derive the algebraic equations for all of the nodes in the mesh of either formulation. In the following discussion we shall assume, for the sake of simplicity, that the mesh is uniform with each rectangular subdivision being of the size $a \times b$.

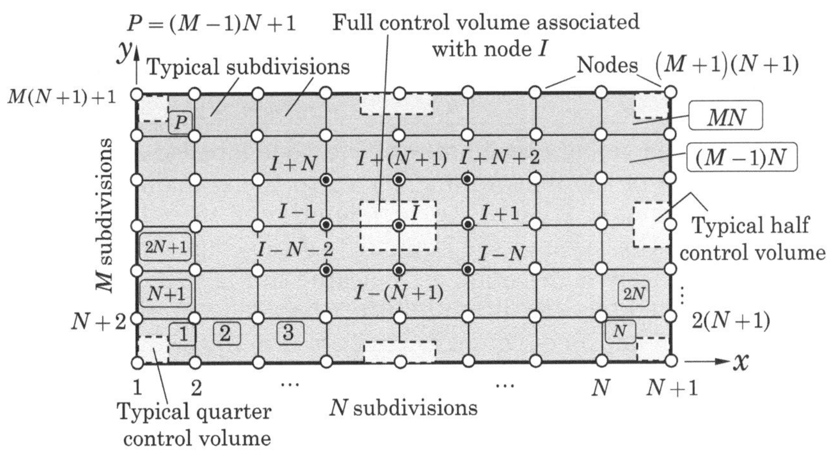

Fig. 3.4.1 An $N \times M$ mesh of subdivisions in the HFVM over a rectangular domain.

3.4.2 Integral Statement over a Typical Control Volume

We consider an arbitrary control domain from the mesh shown in Fig. 3.4.1 to set up the integral statement of the governing equation. A domain whose center node is labelled as I is considered (see Fig. 3.4.3). The integral statement

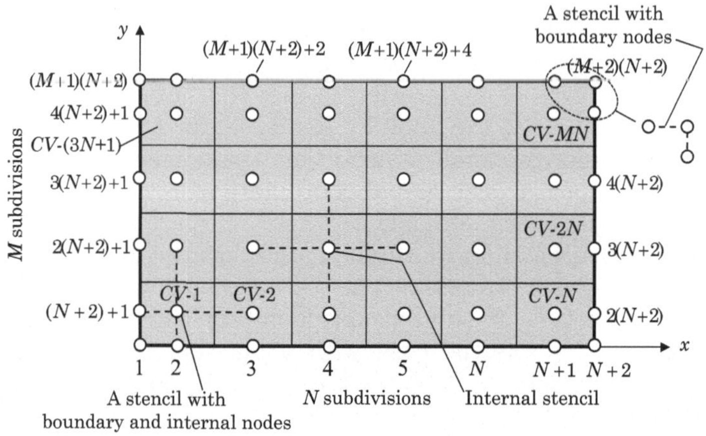

Fig. 3.4.2 A $N \times M$ mesh of subdivisions in the ZFVM over a two-dimensional rectangular domain.

of Eq. (3.4.1) over a typical rectangular control volume is [we note that the local coordinate system $(\bar{x}, \bar{y})$ used for each subdivision is only a translation of (x, y); the bars on x and y are omitted]

$$
\begin{aligned}
0 &= -\int_{x_I-0.5a}^{x_I+0.5a} \int_{y_I-0.5b}^{y_I+0.5b} \left\{ \left[\frac{\partial}{\partial x}\left(a_{xx}\frac{\partial u}{\partial x}\right) + \frac{\partial}{\partial y}\left(a_{yy}\frac{\partial u}{\partial y}\right) \right] + f \right\} dxdy \\
&= -\oint_\Gamma \left[n_x \left(a_{xx}\frac{\partial u}{\partial x}\right) + n_y \left(a_{yy}\frac{\partial u}{\partial y}\right) \right] ds - \int_{x_I-0.5a}^{x_I+0.5a} \int_{y_I-0.5b}^{y_I+0.5b} f\, dxdy \quad (3.4.3)
\end{aligned}
$$

where (x_I, y_I) are the global coordinates of the node labelled as I, (n_x, n_y) are the direction cosines of the unit normal vector to the boundary of the control volume with node number I, and Γ is the boundary of the rectangular control volume. The integration around the boundary of the control volume is carried out in the counterclockwise direction as indicated in Fig. 3.4.3.

The boundary integrals can be simplified using the values of the direction cosines on each boundary line segment. We have $\hat{\mathbf{n}} = (n_x, n_y)$; $\hat{\mathbf{n}} = (0, -1)$ and $ds = dx$ on AB; $\hat{\mathbf{n}} = (1, 0)$ and $ds = dy$ on BC; $\hat{\mathbf{n}} = (0, 1)$ and $ds = -dx$ on CD; and $\hat{\mathbf{n}} = (-1, 0)$ and $ds = -dy$ on DA. Then the boundary integral in Eq. (3.4.3) simplifies to

$$
\begin{aligned}
&\oint_\Gamma \left[n_x \left(a_{xx}\frac{\partial u}{\partial x}\right) + n_y \left(a_{yy}\frac{\partial u}{\partial y}\right) \right] ds \\
&= -\int_{x_I-0.5a}^{x_I+0.5a} \left[a_{yy}\frac{\partial u}{\partial y} \right]_{y=y_I-0.5b} dx + \int_{y_I-0.5b}^{y_I+0.5b} \left[a_{xx}\frac{\partial u}{\partial x} \right]_{x=x_I+0.5a} dy \\
&\quad + \int_{x_I-0.5a}^{x_I+0.5a} \left[a_{yy}\frac{\partial u}{\partial y} \right]_{y=y_I+0.5b} dx - \int_{y_I-0.5b}^{y_I+0.5b} \left[a_{xx}\frac{\partial u}{\partial x} \right]_{x=x_I-0.5a} dy \qquad (3.4.4)
\end{aligned}
$$

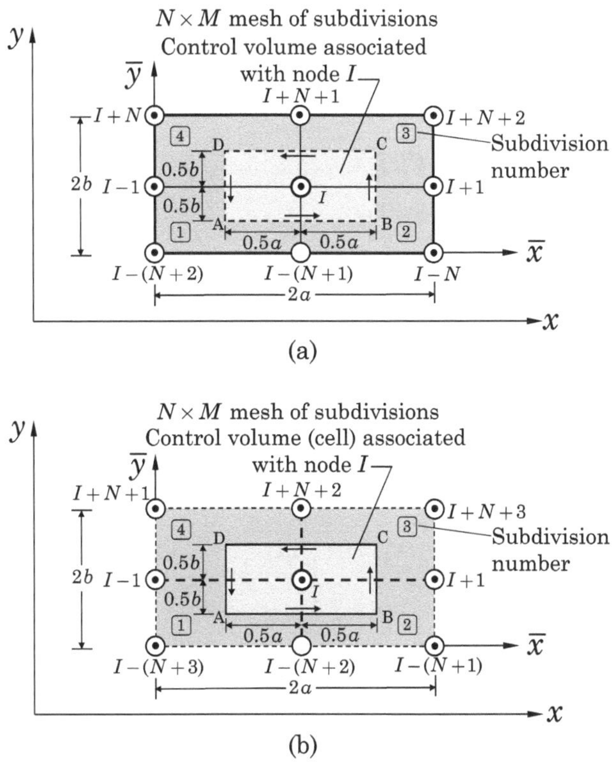

Fig. 3.4.3 A typical control volume associated with an interior node I in the (a) HFVM and (b) ZFVM. The control volume domain is the same in both formulations. Only the node numbers are different.

Then the integral form in Eq. (3.4.3) becomes

$$
\begin{aligned}
0 = & \int_{x_I-0.5a}^{x_I+0.5a} \left[a_{yy} \frac{\partial u}{\partial y} \right]_{y=y_I-0.5b} dx - \int_{y_I-0.5b}^{y_I+0.5b} \left[a_{xx} \frac{\partial u}{\partial x} \right]_{x=x_I+0.5a} dy \\
& - \int_{x_I-0.5a}^{x_I+0.5a} \left[a_{yy} \frac{\partial u}{\partial y} \right]_{y=y_I+0.5b} dx + \int_{y_I-0.5b}^{y_I+0.5b} \left[a_{xx} \frac{\partial u}{\partial x} \right]_{x=x_I-0.5a} dy \\
& - \int_{x_I-0.5a}^{x_I+0.5a} \int_{y_I-0.5b}^{y_I+0.5b} f(x,y)\, dxdy \qquad (3.4.5)
\end{aligned}
$$

3.4.3 Discretized Equations for Half-Control Volume Formulation

We discretize Eq. (3.4.5) for interior and boundary nodes by replacing various derivatives of u with Taylor series representations ($\Delta x = a$ and $\Delta y = b$).

3.4.3.1 Equations for an interior node

We now make the following approximations [see Figs. 3.4.3(a) and 3.4.4]:

$$\int_{x_I-0.5a}^{x_I+0.5a}\left[a_{yy}\frac{\partial u}{\partial y}\right]_{y=y_I-0.5b} dx = \Delta x\left[a_{yy}\frac{\partial u}{\partial y}\right]^{(B)}$$

$$= \frac{a}{b}a_{yy}^{(B)}\left(U_I - U_{I-N-1}\right) \tag{3.4.6a}$$

$$\int_{x_I-0.5a}^{x_I+0.5a}\left[a_{yy}\frac{\partial u}{\partial y}\right]_{y=y_I+0.5b} dx = \Delta x\left[a_{yy}\frac{\partial u}{\partial y}\right]^{(T)}$$

$$= \frac{a}{b}a_{yy}^{(T)}\left(U_{I+N+1} - U_I\right) \tag{3.4.6b}$$

$$\int_{y_I-0.5b}^{y_I+0.5b}\left[a_{xx}\frac{\partial u}{\partial x}\right]_{x=x_I-0.5a} dy = \Delta y\left[a_{xx}\frac{\partial u}{\partial x}\right]^{(L)}$$

$$= \frac{b}{a}a_{xx}^{(L)}\left(U_I - U_{I-1}\right) \tag{3.4.6c}$$

$$\int_{y_I-0.5b}^{y_I+0.5b}\left[a_{xx}\frac{\partial u}{\partial x}\right]_{x=x_I+0.5a} dy = \Delta y\left[a_{xx}\frac{\partial u}{\partial x}\right]^{(R)}$$

$$= \frac{b}{a}a_{xx}^{(R)}\left(U_{I+1} - U_I\right), \tag{3.4.6d}$$

where superscripts on a_{xx} and a_{yy} denote the face or edge [B for the bottom (or south) face AB, T for the top (north) face CD, R for the right (east) face BC, and L for the left (west) face DA] of the control volume where they are evaluated. In addition, we approximate the source term as

$$\int_{x_I-0.5a}^{x_I+0.5a}\int_{y_I-0.5b}^{y_I+0.5b} f(x,y)\,dxdy = f_0\,\Delta x\,\Delta y = f_0\,ab \tag{3.4.7}$$

where f_0 denotes the average value of $f(x,y)$ over the control volume.

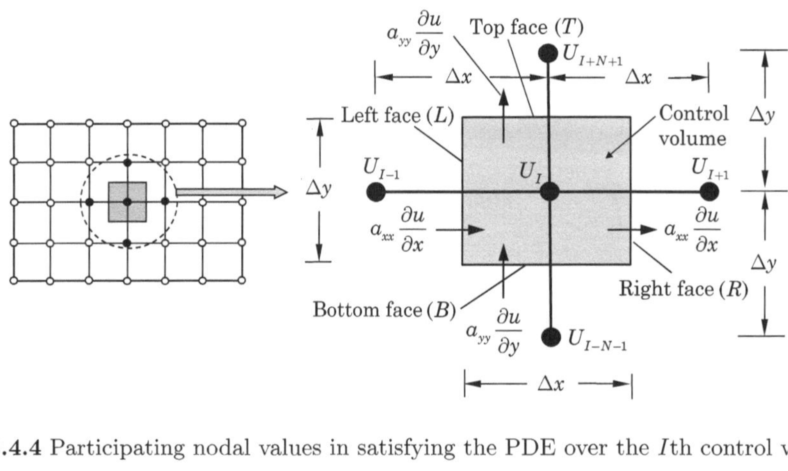

Fig. 3.4.4 Participating nodal values in satisfying the PDE over the Ith control volume.

In view of the approximations in Eqs. (3.4.6a)–(3.4.6d) and Eq. (3.4.7), Eq. (3.4.5) can be expressed symbolically as

$$\begin{aligned} &A_{I-N-1}\,U_{I-N-1} + A_{I-1}\,U_{I-1} + A_I\,U_I + A_{I+1}\,U_{I+1} \\ &\qquad + A_{I+N+1}\,U_{I+N+1} = F_I \end{aligned} \tag{3.4.8a}$$

where the coefficients A_K and F_I are defined as follows:

$$\begin{aligned} &A_{I-N-1} = -\frac{a}{b}\,a_{yy}^{(B)},\quad A_{I-1} = -\frac{b}{a}\,a_{xx}^{(L)},\quad A_{I+1} = -\frac{b}{a}\,a_{xx}^{(R)},\quad A_{I+N+1} = -\frac{a}{b}\,a_{yy}^{(T)}, \\ &A_I = \frac{a}{b}\left(a_{yy}^{(B)} + a_{yy}^{(T)}\right) + \frac{b}{a}\left(a_{xx}^{(R)} + a_{xx}^{(L)}\right),\quad F_I = f_0\,ab. \end{aligned} \tag{3.4.8b}$$

For uniform mesh and constant data (i.e., a_{xx} and a_{yy} are not functions of x and y), the coefficients A_K and F_I are given by

$$\begin{aligned} &A_{I-N-1} = -\frac{a}{b}\,a_{yy},\quad A_{I-1} = -\frac{b}{a}\,a_{xx},\quad A_{I+N-1} = -\frac{a}{b}\,a_{yy}, \\ &A_{I+1} = -\frac{b}{a}\,a_{xx},\quad A_I = \frac{2a}{b}\,a_{yy} + \frac{2b}{a}\,a_{xx},\quad F_I = f_0\,ab. \end{aligned} \tag{3.4.9}$$

3.4.3.2 Equations for nodes on the boundary

Next, we consider nodes that are on the boundary of a rectangular domain, as shown in Fig. 3.4.5. Each is considered separately.

Nodes on the bottom boundary [see Fig. 3.4.5(a)] For this case, the integral form in Eq. (3.4.5) takes the form ($y_I = 0$)

$$\begin{aligned} 0 = &\int_{x_I-0.5a}^{x_I+0.5a} \left[a_{yy}\frac{\partial u}{\partial y}\right]^{(B)} dx - \int_{y_I}^{y_I+0.5b} \left[a_{xx}\frac{\partial u}{\partial x}\right]^{(R)} dy \\ &- \int_{x_I-0.5a}^{x_I+0.5a} \left[a_{yy}\frac{\partial u}{\partial y}\right]^{(T)} dx + \int_{y_I}^{y_I+0.5b} \left[a_{xx}\frac{\partial u}{\partial x}\right]^{(L)} dy \\ &- \int_{x_I-0.5a}^{x_I+0.5a}\int_{y_I}^{y_I+0.5b} f(x,y)\,dxdy \end{aligned} \tag{3.4.10}$$

or

$$A_{I-1}U_{I-1} + A_I U_I + A_{I+1}U_{I+1} + A_{I+N+1}U_{I+N+1} = F_I + Q_I, \tag{3.4.11a}$$

where

$$\begin{aligned} &A_{I-1} = -\frac{b}{2a}\,a_{xx}^{(L)},\quad A_I = \frac{b}{2a}\left(a_{xx}^{(L)} + a_{xx}^{(R)}\right) + \frac{a}{b}\,a_{yy}^{(T)},\quad A_{I+1} = -\frac{b}{2a}\,a_{xx}^{(R)}, \\ &A_{I+N+1} = -\frac{a}{b}\,a_{yy}^{(T)},\quad Q_I = -\int_{x_I-0.5a}^{x_I+0.5a} \left[a_{yy}\frac{\partial u}{\partial y}\right]^{(B)} dx,\quad F_I = f_0\frac{ab}{2}. \end{aligned} \tag{3.4.11b}$$

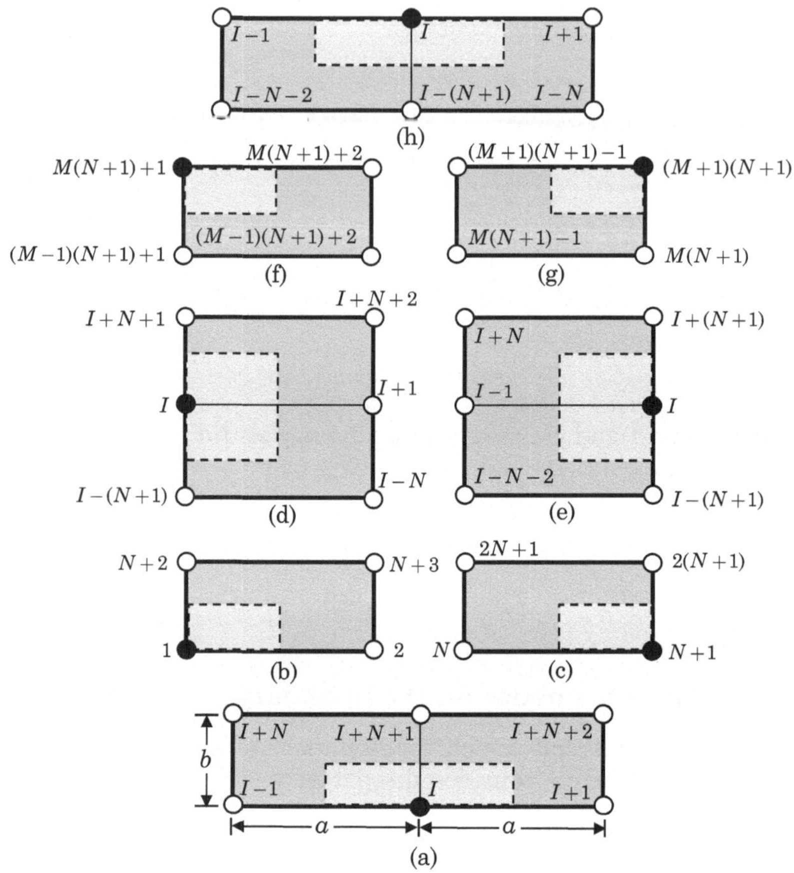

Fig. 3.4.5 Various possible boundary control volumes in a rectangular domain. A dark circle in each control volume denotes the primary node associated with that control volume.

Nodes on the top boundary [see Fig. 3.4.5(h)] The integral statement in Eq. (3.4.5) takes the form

$$
\begin{aligned}
0 = &\int_{x_I-0.5a}^{x_I+0.5a} \left[a_{yy}\frac{\partial u}{\partial y}\right]^{(B)} dx - \int_{x_I-0.5a}^{x_I+0.5a} \left[a_{yy}\frac{\partial u}{\partial y}\right]^{(T)} dx \\
&- \int_{y_I-0.5b}^{y_I} \left[a_{xx}\frac{\partial u}{\partial x}\right]^{(R)} dy + \int_{y_I-0.5b}^{y_I} \left[a_{xx}\frac{\partial u}{\partial x}\right]^{(L)} dy \\
&- \int_{x_I-0.5a}^{x_I+0.5a} \int_{y_I-0.5b}^{y_I+0.5b} f(x,y)\, dxdy, \qquad (3.4.12)
\end{aligned}
$$

which can be expressed as

$$
A_{I-1}U_{I-1} + A_I U_I + A_{I+1}U_{I+1} + A_{I-N-1}U_{I-N-1} = F_I + Q_I, \qquad (3.4.13a)
$$

where

$$A_{I-1} = -\frac{b}{2a}\,a_{xx}^{(L)}, \quad A_I = \frac{b}{2a}\left(a_{xx}^{(L)} + a_{xx}^{(R)}\right) + \frac{a}{b}\,a_{yy}^{(B)}, \quad A_{I+1} = -\frac{b}{2a}\,a_{xx}^{(R)},$$
$$A_{I-N-1} = -\frac{a}{b}\,a_{yy}^{(B)}, \quad F_I = f_0\frac{ab}{2}, \quad Q_I = \int_{x_I-0.5a}^{x_I+0.5a}\left[a_{yy}\,\frac{\partial u}{\partial y}\right]^{(T)} dx. \quad (3.4.13b)$$

Nodes on the left boundary [see Fig. 3.4.5(d)] The integral statement for this case is given by ($x_I = 0.0$)

$$\begin{aligned}0 = &\int_{x_I}^{x_I+0.5a}\left[a_{yy}\frac{\partial u}{\partial y}\right]^{(B)} dx - \int_{y_I-0.5b}^{y_I+0.5b}\left[a_{xx}\frac{\partial u}{\partial x}\right]^{(R)} dy\\ &- \int_{x_I}^{x_I+0.5a}\left[a_{yy}\frac{\partial u}{\partial y}\right]^{(T)} dx + \int_{y_I-0.5b}^{y_I+0.5b}\left[a_{xx}\frac{\partial u}{\partial x}\right]^{(L)} dy\\ &- \int_{x_I}^{x_I+0.5a}\int_{y_I-0.5b}^{y_I+0.5b} f(x,y)\,dxdy, \end{aligned} \quad (3.4.14)$$

which can be expressed as

$$A_{I-N-1}U_{I-N-1} + A_I U_I + A_{I+1}U_{I+1} + A_{I+N+1}U_{I+N+1} = F_I + Q_I, \quad (3.4.15a)$$

where

$$A_{I-N-1} = -\frac{a}{2b}\,a_{yy}^{(L)}, \quad A_I = \frac{a}{2b}\left(a_{yy}^{(B)} + a_{yy}^{(T)}\right) + \frac{b}{a}\,a_{xx}^{(R)}, \quad A_{I+1} = -\frac{b}{a}\,a_{xx}^{(R)},$$
$$A_{I+N+1} = -\frac{a}{2b}\,a_{yy}^{(T)}, \quad F_I = f_0\frac{ab}{2}, \quad Q_I = \int_{y_I-0.5b}^{y_I+0.5b}\left[a_{xx}\,\frac{\partial u}{\partial x}\right]^{(L)} dy. \quad (3.4.15b)$$

Nodes on the right boundary [see Fig. 3.4.5(e)] The integral statement for this case is

$$\begin{aligned}0 = &\int_{x_I-0.5a}^{x_I}\left[a_{yy}\frac{\partial u}{\partial y}\right]^{(B)} dx - \int_{y_I-0.5b}^{y_I+0.5b}\left[a_{xx}\frac{\partial u}{\partial x}\right]^{(R)} dy\\ &+ \int_{y_I-0.5b}^{y_I+0.5b}\left[a_{xx}\frac{\partial u}{\partial x}\right]^{(T)} dy - \int_{x_I-0.5a}^{x_I}\left[a_{yy}\frac{\partial u}{\partial y}\right]^{(L)} dx\\ &- \int_{x_I-0.5a}^{x_I}\int_{y_I-0.5b}^{y_I+0.5b} f(x,y)\,dxdy, \end{aligned} \quad (3.4.16)$$

which can be expressed as

$$A_{I-N-1}U_{I-N-1} + A_{I-1}U_{I-1} + A_I U_I + A_{I+N+1}U_{I+N+1} = F_I - Q_I, \quad (3.4.17a)$$

where

$$A_{I-N-1} = -\frac{a}{2b}\, a_{yy}^{(B)}, \quad A_{I-1} = -\frac{b}{a}\, a_{xx}^{(L)}, \quad A_I = \frac{a}{2b}\left[a_{yy}^{(B)} + a_{yy}^{(T)}\right] + \frac{b}{a}\, a_{xx}^{(L)},$$

$$A_{I+N+1} = -\frac{a}{2b}\, a_{yy}^{(T)}, \quad F_I = f_0\,\frac{ab}{2}, \quad Q_I = -\int_{y_I-0.5a}^{y_I+0.5a} q_x(x_I, y)\, dy. \quad (3.4.17b)$$

Node on the bottom left corner [see Fig. 3.4.5(b)] The integral statement for this case can be expressed as

$$\begin{aligned} 0 = &\int_0^{0.5a}\left[a_{yy}\frac{\partial u}{\partial y}\right]^{(B)} dx + \int_0^{0.5b}\left[a_{xx}\frac{\partial u}{\partial x}\right]^{(R)} dy \\ &- \int_0^{0.5a}\left[a_{yy}\frac{\partial u}{\partial y}\right]^{(T)} dx - \int_0^{0.5b}\left[a_{xx}\frac{\partial u}{\partial x}\right]^{(L)} dy \\ &- \int_0^{0.5a}\int_0^{0.5b} f(x,y)\, dxdy, \end{aligned} \quad (3.4.18)$$

which can be expressed as

$$A_1 U_1 + A_2 U_2 + A_{N+2} U_{N+2} = F_1 + Q_1, \quad (3.4.19a)$$

where

$$A_1 = \frac{b}{2a}\, a_{xx}^{(R)} + \frac{a}{2b}\, a_{yy}^{(T)}, \quad A_2 = -\frac{b}{2a}\, a_{xx}^{(R)}, \quad A_{N+2} = -\frac{a}{2b}\, a_{yy}^{(T)}$$

$$F_1 = f_0\frac{ab}{4}, \quad Q_1 = \int_0^{0.5a}\left[a_{yy}\frac{\partial u}{\partial y}\right]^{(B)} dx + \int_0^{0.5b}\left[a_{xx}\frac{\partial u}{\partial x}\right]^{(L)} dy. \quad (3.4.19b)$$

Node on the bottom right corner [see Fig. 3.4.5(c)] The integral statement is given by

$$\begin{aligned} 0 = &\int_{x_{N+1}-0.5a}^{x_{N+1}}\left[a_{yy}\frac{\partial u}{\partial y}\right]^{(B)} dx - \int_0^{0.5b}\left[a_{xx}\frac{\partial u}{\partial x}\right]^{(R)} dy \\ &- \int_{x_{N+1}-0.5a}^{x_{N+1}}\left[a_{yy}\frac{\partial u}{\partial y}\right]^{(T)} dx + \int_0^{0.5b}\left[a_{xx}\frac{\partial u}{\partial x}\right]^{(L)} dy, \end{aligned} \quad (3.4.20)$$

which can be expressed as

$$A_N U_N + A_{N+1} U_{N+1} + A_{2N+2} U_{2N+2} = F_{N+1} + Q_{N+1}, \quad (3.4.21a)$$

where

$$A_N = -\frac{b}{2a}\, a_{xx}^{(L)}, \quad A_{N+1} = \frac{b}{2a}\, a_{xx}^{(L)} + \frac{a}{2b}\, a_{yy}^{(T)}, \quad A_{2N+2} = -\frac{a}{2b}\, a_{yy}^{(T)}$$

$$F_{N+1} = f_0\frac{ab}{4}, \quad Q_{N+1} = -\int_{x_{N+1}-0.5a}^{x_{N+1}}\left[a_{yy}\frac{\partial u}{\partial y}\right]^{(B)} dx + \left[a_{xx}\frac{\partial u}{\partial x}\right]^{(R)} dy.$$

(3.4.21b)

Node on the top left corner [see Fig. 3.4.5(f)] The integral statement for this case can be expressed as ($x_L = 0$)

$$
\begin{aligned}
0 = & \int_0^{0.5a} \left[a_{yy} \frac{\partial u}{\partial y} \right]^{(B)} dx - \int_{y_L - 0.5b}^{y_L} \left[a_{xx} \frac{\partial u}{\partial x} \right]^{(R)} dy \\
& - \int_0^{0.5a} \left[a_{yy} \frac{\partial u}{\partial y} \right]^{(T)} dx + \int_{y_L - 0.5b}^{y_L} \left[a_{xx} \frac{\partial u}{\partial x} \right]^{(L)} dy \\
& - \int_0^{0.5a} \int_{y_L - 0.5b}^{y_L} f(x,y)\, dx, dy, \qquad (3.4.22)
\end{aligned}
$$

which can be expressed as

$$
A_K U_K + A_L U_L + A_{L+1} U_{L+1} = F_L + Q_L, \qquad (3.4.23a)
$$

where $K = (M-1)(N+1)+1$ and $L = M(N+1)+1$, and

$$
\begin{aligned}
& A_K = -\frac{a}{2b}\, a_{yy}^{(B)}, \quad A_L = \frac{a}{2b}\, a_{yy}^{(B)} + \frac{b}{2a}\, a_{xx}^{(R)}, \quad A_{L+1} = -\frac{b}{2a}\, a_{xx}^{(R)} \\
& F_L = f_0 \frac{ab}{4}, \quad Q_L = -\int_{y_L - 0.5b}^{y_L} \left[a_{xx} \frac{\partial u}{\partial x} \right]^{(L)} dy + \int_{x_L}^{x_L + 0.5a} \left[a_{yy} \frac{\partial u}{\partial y} \right]^{(T)} dx. \\
& \qquad (3.4.23b)
\end{aligned}
$$

Node on the top right corner [see Fig. 3.4.5(g)] The integral statement for this case can be expressed as [$P = (M+1)(N+1)$],

$$
\begin{aligned}
0 = & \int_{x_P - 0.5a}^{x_P} \left[a_{yy} \frac{\partial u}{\partial y} \right]^{(B)} dx - \int_{y_P - 0.5b}^{y_P} \left[a_{xx} \frac{\partial u}{\partial x} \right]^{(R)} dy \\
& - \int_{x_P - 0.5a}^{x_P} \left[a_{yy} \frac{\partial u}{\partial y} \right]^{(T)} dx + \int_{y_P - 0.5b}^{y_P} \left[a_{xx} \frac{\partial u}{\partial x} \right]^{(L)} dy \\
& - \int_{x_P - 0.5a}^{x_P} \int_{y_P - 0.5b}^{y_P} f(x,y)\, dxdy, \qquad (3.4.24)
\end{aligned}
$$

which can be expressed as

$$
A_K U_K + A_{K+1} U_{K+1} + A_L U_L + A_{L+1} U_{L+1} = F_{L+1} + Q_{L+1}, \qquad (3.4.25a)
$$

where $K = M(N+1)$ and $L = (M+1)(N+1)$, and

$$
\begin{aligned}
& A_K = -\frac{a}{2b}\, a_{yy}^{(B)}, \quad A_{L-1} = -\frac{b}{2a}\, a_{xx}^{(L)}, \quad A_L = \frac{a}{2b}\, a_{yy}^{(B)} + \frac{b}{2a}\, a_{xx}^{(L)} \\
& F_L = f_0 \frac{ab}{4}, \quad Q_L = \int_{y_P - 0.5b}^{y_P} \left[a_{xx} \frac{\partial u}{\partial x} \right]^{(R)} dy + \int_{x_P - 0.5a}^{x_P} \left[a_{xx} \frac{\partial u}{\partial x} \right]^{(T)} dx. \\
& \qquad (3.4.25b)
\end{aligned}
$$

This completes the derivation of the HFVM equations for the Poisson equation with constant properties on rectangular domains with uniform meshes. The derivations can be generalized, for example, for nonuniform meshes and element-wise constant properties (i.e., a_{xx}, a_{yy}, and f are constant but different for different cells).

3.4.4 Discretized Equations for ZFVM

The integral statement for the ZFVM remains the same as that for the HFVM [i.e., Eq. (3.4.5) is valid]. Equations (3.4.6a)–(3.4.6d) are also valid with a change of the labels on U. We have

$$\begin{aligned}\int_{x_I-0.5a}^{x_I+0.5a}\left[a_{yy}\frac{\partial u}{\partial y}\right]_{y=y_I-0.5b} dx &= \Delta x\left[a_{yy}\frac{\partial u}{\partial y}\right]^{(B)}\\ &= \frac{a}{b}a_{yy}^{(B)}\left(U_I - U_{I-N-2}\right)\end{aligned} \tag{3.4.26a}$$

$$\begin{aligned}\int_{x_I-0.5a}^{x_I+0.5a}\left[a_{yy}\frac{\partial u}{\partial y}\right]_{y=y_I+0.5b} dx &= \Delta x\left[a_{yy}\frac{\partial u}{\partial y}\right]^{(T)}\\ &= \frac{a}{b}a_{yy}^{(T)}\left(U_{I+N+2} - U_I\right)\end{aligned} \tag{3.4.26b}$$

$$\begin{aligned}\int_{y_I-0.5b}^{y_I+0.5b}\left[a_{xx}\frac{\partial u}{\partial x}\right]_{x=x_I-0.5a} dy &= \Delta y\left[a_{xx}\frac{\partial u}{\partial x}\right]^{(L)}\\ &= \frac{b}{a}a_{xx}^{(L)}\left(U_I - U_{I-1}\right)\end{aligned} \tag{3.4.26c}$$

$$\begin{aligned}\int_{y_I-0.5b}^{y_I+0.5b}\left[a_{xx}\frac{\partial u}{\partial x}\right]_{x=x_I+0.5a} dy &= \Delta y\left[a_{xx}\frac{\partial u}{\partial x}\right]^{(R)}\\ &= \frac{b}{a}a_{xx}^{(R)}\left(U_{I+1} - U_I\right),\end{aligned} \tag{3.4.26d}$$

Then Eq. (3.4.5) can be expressed symbolically as

$$\begin{aligned}A_{I-N-2}\,U_{I-N-2} + A_{I-1}\,U_{I-1} + A_I\,U_I + A_{I+1}\,U_{I+1}\\ +A_{I+N+2}\,U_{I+N+2} = F_I\end{aligned} \tag{3.4.27a}$$

where the coefficients A_K and F_I are defined as follows:

$$\begin{aligned}A_{I-N-2} = -\frac{a}{b}\,a_{yy}^{(B)},\quad A_{I-1} = -\frac{b}{a}\,a_{xx}^{(L)},\quad A_{I+1} = -\frac{b}{a}\,a_{xx}^{(R)},\quad A_{I+N+2} = -\frac{a}{b}\,a_{yy}^{(T)},\\ A_I = \frac{a}{b}\left(a_{yy}^{(B)} + a_{yy}^{(T)}\right) + \frac{b}{a}\left(a_{xx}^{(R)} + a_{xx}^{(L)}\right),\quad F_I = f_0\,ab.\end{aligned} \tag{3.4.27b}$$

The main difference between the ZFVM and HFVM lies in the discretized equations of the boundary nodes. Suppose that u is specified to be $u_0(y)$ on the right face of the domain. Then the values of $U_{N+2}, U_{2(N+2)}, \ldots, U_{(M+2)(N+2)}$ will be known by evaluating $u_0(y)$ at the respective nodes (i.e., $U_{N+2} = u_0(0)$, $U_{2(N+2)} = u_0(\Delta y)$, etc.; see Fig. 3.4.2). No discretized equations are required for these nodes.

If the Neumann or mixed boundary condition is specified, then we need to write the discretized equations for all nodes that bring boundary nodes into play. For illustrative purposes, let us consider writing the discretized equations for nodes $N+1$, $2N+3$, and $3N+5$.

Node $N+1$ On the bottom face (where node $N+1$ is), we have $n_x = 0$ and $n_y = -1$. Hence, we have (note that the source term does not enter the calculation because of the zero-thickness assumption)

$$\begin{aligned}\oint_\Gamma q_n\, ds &= \oint_\Gamma \left[n_x\left(a_{xx}\frac{\partial u}{\partial x}\right) + n_y\left(a_{yy}\frac{\partial u}{\partial y}\right)\right] ds\\ &= -\int_0^{\Delta x}\left(a_{yy}\frac{\partial u}{\partial y}\right) dx = \frac{2a}{b}a_{yy}^{(B)}\left(U_{2N+3} - U_{N+1}\right) \qquad (3.4.28a)\end{aligned}$$

If q_n is specified to be zero, we will have $U_{2N+3} = U_{N+1}$. If $q_n + \beta(u - u_\infty) = 0$, then we will have

$$\frac{2a}{b}a_{yy}^{(B)}\left(U_{2N+3} - U_{N+1}\right) - \beta\, a\left(U_{N+1} - u_\infty\right) = 0. \qquad (3.4.28b)$$

Similar expressions can be written for nodes $N+2$ (on the bottom face as well as on the right face) and $2(N+2)$.

<u>**Node** $2N+3$</u> For this case, the balance equation in Eq. (3.4.5) is used with approximations in Eqs. (3.4.26a)–(3.4.26d). We have

$$\begin{aligned}0 &= \oint_\Gamma \left[n_x\left(a_{xx}\frac{\partial u}{\partial x}\right) + n_y\left(a_{yy}\frac{\partial u}{\partial y}\right)\right] ds + \int_\Omega f\, dxdy\\ &= -a_{yy}^{(B)}\tfrac{2a}{b}\left(U_{2N+3} - U_{N+1}\right) + a_{xx}^{(R)}\tfrac{2b}{a}\left(U_{2N+4} - U_{2N+3}\right)\\ &\quad + a_{yy}^{(T)}\tfrac{a}{b}\left(U_{2N+5} - U_{2N+3}\right) - a_{xx}^{(L)}\tfrac{b}{a}\left(U_{2N+3} - U_{2N+2}\right) +\\ &= \tfrac{2a}{b}a_{yy}^{(B)}U_{N+1} + \tfrac{2b}{a}a_{xx}^{(L)}U_{2N+2} - \left(\tfrac{2a}{b}a_{yy}^{(B)} + \tfrac{2b}{a}a_{xx}^{(R)} + \tfrac{a}{b}a_{yy}^{(T)} + \tfrac{b}{a}a_{yy}^{(L)}\right)U_{2N+3}\\ &\quad + \tfrac{b}{a}a_{xx}^{(R)}U_{2N+4} + \tfrac{a}{b}a_{yy}^{(T)}U_{3N+5} + f_0 ab. \qquad (3.4.29)\end{aligned}$$

<u>**Node** $3N+5$</u> Similarly, we have

$$\begin{aligned}0 &= \oint_\Gamma \left[n_x\left(a_{xx}\frac{\partial u}{\partial x}\right) + n_y\left(a_{yy}\frac{\partial u}{\partial y}\right)\right] ds + \int_\Omega f\, dxdy\\ &= -a_{yy}^{(B)}\tfrac{a}{b}\left(U_{2N+5} - U_{2N+3}\right) + a_{xx}^{(R)}\tfrac{2b}{a}\left(U_{3N+6} - U_{3N+5}\right)\\ &\quad + a_{yy}^{(T)}\tfrac{2a}{b}\left(U_{4N+7} - U_{3N+5}\right) - a_{xx}^{(L)}\tfrac{2b}{a}\left(U_{3N+5} - U_{3N+4}\right) + f_0 ab\\ &= \tfrac{a}{b}a_{yy}^{(B)}U_{2N+3} + \tfrac{b}{a}a_{xx}^{(L)}U_{3N+4} - \left(\tfrac{a}{b}a_{yy}^{(B)} + \tfrac{2b}{a}a_{xx}^{(R)} + \tfrac{a}{b}a_{yy}^{(T)} + \tfrac{b}{a}a_{yy}^{(L)}\right)U_{3N+5}\\ &\quad + \tfrac{2b}{a}a_{xx}^{(R)}U_{3N+6} + \tfrac{a}{b}a_{yy}^{(T)}U_{4N+7} + f_0 ab. \qquad (3.4.30)\end{aligned}$$

3.4.5 Numerical Examples

Here we present a couple of illustrative example problems described by the model differential equation. We make use of the discretized equations developed in the previous subsection. Of course, one needs to write a computer program to generate the mesh, define all coefficients for every node in the mesh, impose boundary conditions, and solve the linear equations.

Example 3.4.1

Consider steady-state heat conduction in an isotropic rectangular region of dimensions $3a \times 2a$, as shown in Fig. 3.4.6(a). Boundaries $x = 0$ and $y = 0$ are insulated (i.e., heat flux normal to the boundary is zero, $q_n = 0$), boundary $x = 3a$ is maintained at zero temperature, and boundary $y = 2a$ is maintained at temperature $T = T_0 \cos(\pi x/6a)$. Determine the temperature distribution using the (a) HFVM and (b) ZFVM with the mesh shown in Figs. 3.4.6(b) and 3.4.6(c), respectively.

Solution: The governing equation is a special case of the model equation (3.4.1) with zero internal heat generation $f = 0$ and coefficients $a_{xx} = a_{yy} = k$. Thus, Eq. (3.4.1) takes the form

$$-k\left(\frac{\partial^2 T}{\partial x^2} + \frac{\partial^2 T}{\partial y^2}\right) = 0. \tag{1a}$$

The boundary conditions for the computational domain are

$$\left.\frac{\partial T}{\partial x}\right|_{(0,y)} = \left.\frac{\partial T}{\partial y}\right|_{(x,0)} = 0; \quad T(3a, y) = 0, \quad T(x, 2a) = T_0 \cos\frac{\pi x}{6a}. \tag{1b}$$

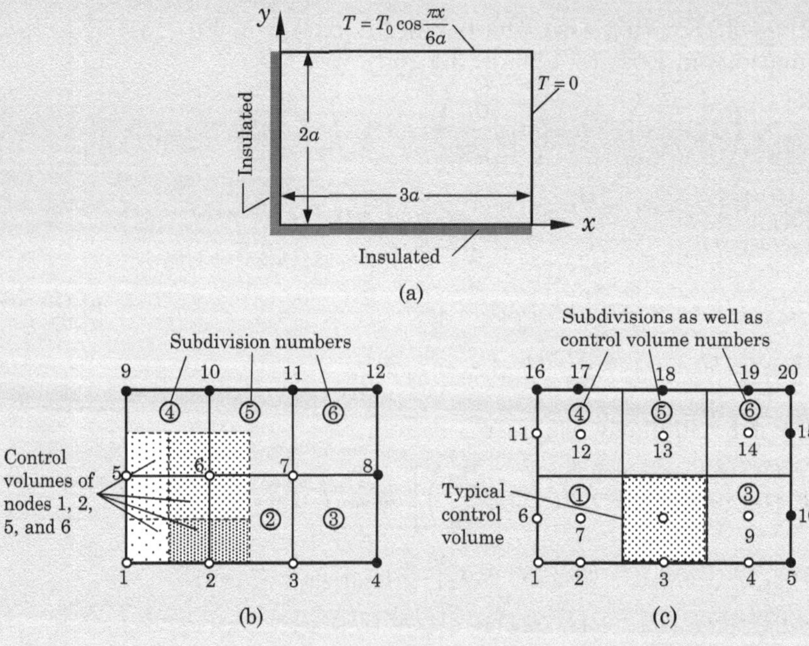

Fig. 3.4.6 Analysis of a heat conduction problem over a rectangular domain using the FVM: (a) domain, (b) 3×2 uniform mesh for the HFVM, and (c) 3×2 uniform mesh for the ZFVM. Nodes with dark circles indicate that they have known values of the temperature.

The exact solution of Eq. (1) for the boundary conditions shown in Fig. 3.4.6(a) is

$$T(x,y) = T_0 \frac{\cosh(\pi y/6a)\cos(\pi x/6a)}{\cosh(\pi/3)}. \tag{2}$$

(a) **HFVM formulation** Comparing the mesh shown in Fig. 3.4.6(b) with that shown in Fig. 3.4.1, we note that $N = 3$ and $M = 2$. Because of the fact that $f = 0$ and the specified boundary conditions, we have $F_I = 0$ at all nodes; $Q_I = 0$ at nodes 1, 2, 3, and 5; $T_4 = 0$, $T_8 = 0$, and $T_{12} = 0$; and $T_9 = T_0$, $T_{10} = \sqrt{3}/2T_0$, and $T_{11} = 0.5T_0$. The algebraic equations associated with nodes with known temperatures can be used to determine the unknown heats Q_I at those nodes. These equations are not needed to determine the unknown temperatures. Thus, we need to determine equations associated with nodes (at which the temperatures are unknown) 1, 2, 3, 5, 6, and 7 of the mesh shown in Fig. 3.4.6(b).

Equation (3.4.8a) is valid for the interior nodes 6 and 7. For node $I = 6$, Eq. (3.4.8a) takes the form

$$A_2T_2 + A_5T_5 + A_6T_6 + A_7T_7 + A_{10}T_{10} = 0 \tag{3a}$$

with

$$A_2 = -k, \quad A_5 = -k, \quad A_6 = 4k, \quad A_7 = -k, \quad A_{10} = -k. \tag{3b}$$

Similarly, for node $I = 7$, we have

$$A_3T_3 + A_6T_6 + A_7T_7 + A_8T_8 + A_{11}T_{11} = 0 \tag{4a}$$

with

$$A_3 = -k, \quad A_6 = -k, \quad A_7 = 4k, \quad A_8 = -k, \quad A_{11} = -k, \tag{4b}$$

For boundary nodes 2 and 3 at the bottom, we use Eq. (3.4.11a):

$$A_1T_1 + A_2T_2 + A_3T_3 + A_6T_6 = Q_2 = 0 \tag{5a}$$

with

$$A_1 = -\tfrac{1}{2}k, \quad A_2 = 2k, \quad A_3 = -\tfrac{1}{2}k, \quad A_6 = -k, \tag{5b}$$

and

$$A_2T_2 + A_3T_3 + A_4T_4 + A_7T_7 = Q_3 = 0 \tag{6a}$$

with

$$A_2 = -\tfrac{1}{2}k, \quad A_3 = 2k, \quad A_4 = -\tfrac{1}{2}k, A_7 = -k. \tag{6b}$$

The remaining two algebraic equations are provided by considering equations of nodes 1 and 5, which are obtained from Eqs. (3.4.19a) and (3.4.15a), respectively:

$$\tfrac{1}{2}k\,(2T_1 - T_2 - T_5) = 0, \quad \tfrac{1}{2}k\,(-T_1 + 4T_5 - 2T_6 - T_9) = 0 \tag{7}$$

Stencils of the equations derived are shown in Fig. 3.4.7. Stencils in Figs. 3.4.7(b) and (d) are used twice (giving four equations), and the stencils in Figs. 3.4.7(a) and (c) are used once each (giving two equations).

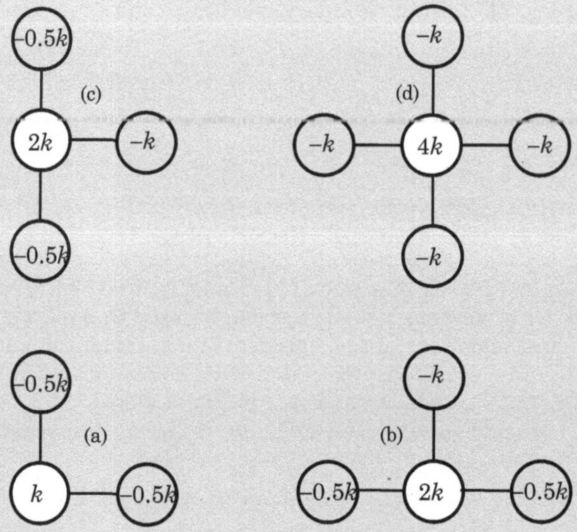

Fig. 3.4.7 Stencils of the half-control volume formulation (for $a = b$). (a) Stencil for a left corner node (the same for the right corner node when flipped about the vertical). (b) Stencil for a boundary node at the bottom (the same for the top nodes when flipped about the horizontal). (c) Stencil for a node on the left boundary (the same for the right boundary nodes when flipped about the vertical). (d) Stencil for an interior node.

Writing all of the relations (in the order of the sequential node numbers) in matrix form, we obtain (known temperatures are moved to the right-hand side)

$$\frac{k}{2}\begin{bmatrix} 2 & -1 & 0 & -1 & 0 & 0 \\ -1 & 4 & -1 & 0 & -2 & 0 \\ 0 & -1 & 4 & 0 & 0 & -2 \\ -1 & 0 & 0 & 4 & -2 & 0 \\ 0 & -2 & 0 & -2 & 8 & -2 \\ 0 & 0 & -2 & 0 & -2 & 8 \end{bmatrix} \begin{Bmatrix} T_1 \\ T_2 \\ T_3 \\ T_5 \\ T_6 \\ T_7 \end{Bmatrix} = \frac{k}{2}\begin{Bmatrix} 0 \\ 0 \\ 0 \\ T_0 \\ \sqrt{3}T_0 \\ T_0 \end{Bmatrix} \tag{8}$$

The solution of these equations is (in °C)

$$\begin{aligned} &T_1 = 0.6362\,T_0, \quad T_2 = 0.5510\,T_0, \quad T_3 = 0.3181\,T_0, \\ &T_5 = 0.7214\,T_0, \quad T_6 = 0.6248\,T_0, \quad T_7 = 0.3607\,T_0. \end{aligned} \tag{9}$$

The exact solution at the same locations is

$$\begin{aligned} &T(0,0) = 0.6249\,T_0, \quad T(1,0) = 0.5412\,T_0, \quad T(2,0) = 0.3124\,T_0, \\ &T(0,1) = 0.7125\,T_0, \quad T(1,1) = 0.6171\,T_0, \quad T(2,1) = 0.3563\,T_0. \end{aligned} \tag{10}$$

(b) **ZFVM formulation** For this case, the number of unknown temperatures is 12 [see Fig. 3.4.6(c)]. The zero flux conditions, $\partial T/\partial x = 0$ at $x = 0$ and $\partial T/\partial y = 0$ at $y = 0$, give the following relations for the 3×2 mesh shown in Fig. 3.4.6(c):

$$T_1 = T_2 = T_6 = T_7, \quad T_3 = T_8, \quad T_4 = T_9, \quad T_{11} = T_{12}. \tag{11}$$

Thus, the only unknowns are T_7, T_8, T_9, T_{12}, T_{13}, and T_{14}. The six equations needed to solve for the six unknowns are provided by equations of the form in Eqs. (3.4.29) and (3.4.30).

Figure 3.4.8 shows stencils of representative interior nodes. Thus, we obtain (by using the stencils at the six interior nodes and noting that $\Delta x = a = \Delta y = b$, $a/b = 1$, and $b/a = 1$)

$$\begin{aligned} 6kT_7 - 2kT_6 - kT_8 - 2kT_2 - kT_{12} &= 0, \\ 5kT_8 - kT_7 - kT_9 - 2kT_3 - kT_{13} &= 0, \\ 6kT_9 - kT_8 - 2kT_{10} - 2kT_4 + kT_{14} &= 0, \\ 6kT_{12} - 2kT_{11} - kT_{13} - 2kT_{17} - kT_7 &= 0, \\ 5kT_{13} - kT_{12} - kT_{14} - 2kT_{18} - kT_8 &= 0, \\ 6kT_{14} - kT_{13} - 2kT_{15} - 2kT_{19} - kT_9 &= 0, \end{aligned} \tag{12}$$

where $T_5 = T_{10} = T_{15} = T_{20} = 0$, $T_{16} = T_0$, $T_{17} = 0.965926T_0$, $T_{18} = 0.707107T_0$, and $T_{19} = 0.258819T_0$ (from the specified temperatures). In view of the relations in Eq. (11) and known values of the temperature, the six equations in Eq. (12) can be expressed in matrix form as (k is cancelled on both sides of the equation)

$$\begin{bmatrix} 2.0 & -1.0 & 0.0 & -1.0 & 0.0 & 0.0 \\ -1.0 & 3.0 & -1.0 & 0.0 & -1.0 & 0.0 \\ 0.0 & -1.0 & 4.0 & 0.0 & 0.0 & -1.0 \\ -1.0 & 0.0 & 0.0 & 4.0 & -1.0 & 0.0 \\ 0.0 & -1.0 & 0.0 & -1.0 & 5.0 & -1.0 \\ 0.0 & 0.0 & -1.0 & 0.0 & -1.0 & 6.0 \end{bmatrix} \begin{Bmatrix} T_7 \\ T_8 \\ T_9 \\ T_{12} \\ T_{13} \\ T_{14} \end{Bmatrix} = \begin{Bmatrix} 0.00000 \\ 0.00000 \\ 0.00000 \\ 1.93185 \\ 1.41421 \\ 0.51764 \end{Bmatrix} T_0. \tag{13}$$

The solution of Eq. (12) is

$$\begin{aligned} &T_7 = 0.6145\,T_0, \quad T_8 = 0.4499\,T_0, \quad T_9 = 0.1647\,T_0, \\ &T_{12} = 0.7792\,T_0, \quad T_{13} = 0.5704\,T_0, \quad T_{14} = 0.2088\,T_0, \end{aligned} \tag{14}$$

whereas the exact solution at the same locations is

$$\begin{aligned} &T(0.5a, 0.5a) = 0.6244\,T_0, \quad T(1.5a, 0.5a) = 0.4571\,T_0, \quad T(2.5a, 0.5a) = 0.1673\,T_0, \\ &T(0.5a, 1.5a) = 0.7995\,T_0, \quad T(1.5a, 1.5a) = 0.5853\,T_0, \quad T(2.5a, 1.5a) = 0.2142\,T_0. \end{aligned} \tag{15}$$

If we refine the mesh by doubling the subdivisions in the x and y directions (i.e., $N = 6$ and $M = 4$), we obtain 24 equations for the 24 unknown nodal temperatures for the HFVM formulation [see Fig. 3.4.9(a)]. These equations can be readily obtained from the general relations given for a $N \times M$ mesh. Typical CVs for the 6×4 mesh are shown in Fig. 3.4.9(a). Similarly, for a uniform mesh of 12×8 subdivisions (i.e., $N = 12$ and $M = 8$), we will have

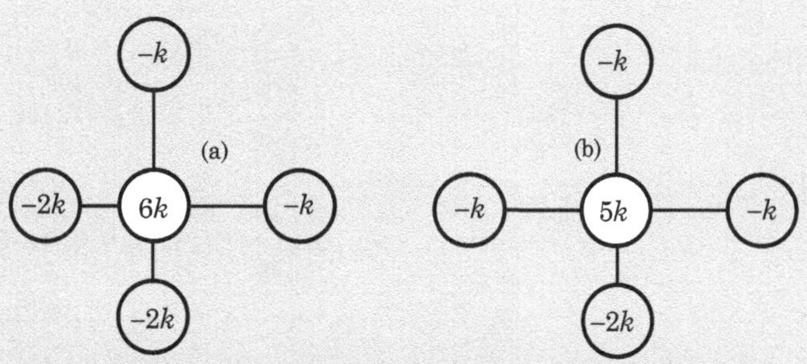

Fig. 3.4.8 Stencils of nodes adjacent to the boundaries in the ZFVM formulation (for $a = b$). (a) Stencil for a node near the left corner node (the same for the right corner node when flipped about the vertical). (b) Stencil for a node above the bottom boundary (the same for the top nodes when flipped about the horizontal). Stencil for an interior node remains the same as that in Fig. 3.4.7(d).

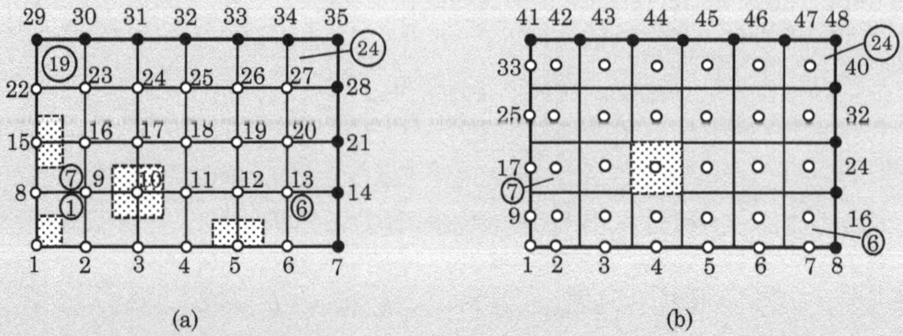

Fig. 3.4.9 A 6×4 uniform mesh for the heat transfer problem of **Example 3.4.1**. (a) HFVM formulation. (b) ZFVM formulation. Nodes with dark circles have specified temperatures (hence, boundary equations at these nodes are not used). Note that the 6×4 mesh contains the 3×2 mesh as a subset only for the HFVM formulation.

117 nodes for the 96 unknown nodal temperatures. In the case of the ZFVM formulation, a 6×4 mesh will have 35 nodal unknowns [see Fig. 3.4.9(b)], and it will not contain the nodes of the 3×2 mesh as a subset.

Table 3.4.1 contains a comparison of the ZFVM and HFVM solutions with the FDM solution (see Table 2.4.3) and the analytical solution in Eq. (2) for three different meshes,

Table 3.4.1 Comparison of the nodal temperatures $T(x, 0)/T_0$, obtained using various meshes in the ZFVM and HFVM formulations with the FDM solution.

	FVM Solutions							
	3×2		6×4		12×8		12×8 FDM	Analyt. solution
x	ZFVM	HFVM	ZFVM	HFVM	ZFVM	HFVM		
0.000	0.6145	0.6362	0.6224	0.6278	0.6243	0.6256	0.6256	0.6249
0.125	——	——	——	——	0.6243	——	——	0.6249
0.250	——	——	0.6224	——	——	0.6202	0.6202	0.6195
0.375	——	——	——	——	0.6136	——	——	0.6129
0.500	0.6145	——	——	0.6064	——	0.6043	0.6043	0.6036
0.625	——	——	——	——	0.5924	——	——	0.5917
0.750	——	——	0.5800	——	——	0.5780	0.5780	0.5773
0.875	——	——	——	——	0.5611	——	——	0.5604
1.000	——	0.5510	——	0.5437	——	0.5418	0.5418	0.5412
1.125	——	——	——	——	0.5202	——	——	0.5196
1.250	——	——	0.4981	——	——	0.4963	0.4963	0.4958
1.375	——	——	——	——	0.4704	——	——	0.4698
1.500	0.4499	——	——	0.4439	——	0.4424	0.4424	0.4419
1.625	——	——	——	——	0.4125	——	——	0.4120
1.750	——	——	0.3822	——	——	0.3808	0.3808	0.3804
1.875	——	——	——	——	0.3476	——	——	0.3472
2.000	——	0.3181	——	0.3139	——	0.3128	0.3128	0.3124
2.125	——	——	——	——	0.2767	——	——	0.2764
2.250	——	——	0.2402	——	——	0.2394	0.2394	0.2391
2.375	——	——	——	——	0.2011	——	——	0.2009
2.500	0.1647	——	——	0.1625	——	0.1619	0.1619	0.1617
2.625	——	——	——	——	0.1220	——	——	0.1219
2.750	——	——	0.0819	——	——	0.0817	0.0817	0.0816
2.875	——	——	——	——	0.0409	——	——	0.0409

3×2, 6×4, and 12×8 (results are independent of a and k). Clearly, with a mesh refinement, the FVM solutions get closer to the analytical solution, indicating the convergence of the FVM solutions to the exact solution. We also note that the HFVM and FDM solutions for the 12×8 mesh are the same.

Figure 3.4.10 contains a plots of the temperature $T(x,0)/T_0$ vs. x obtained with the ZFVM and the analytical solution. The numerical convergence of the ZFVM solutions to the exact solution can be seen as the mesh is refined.

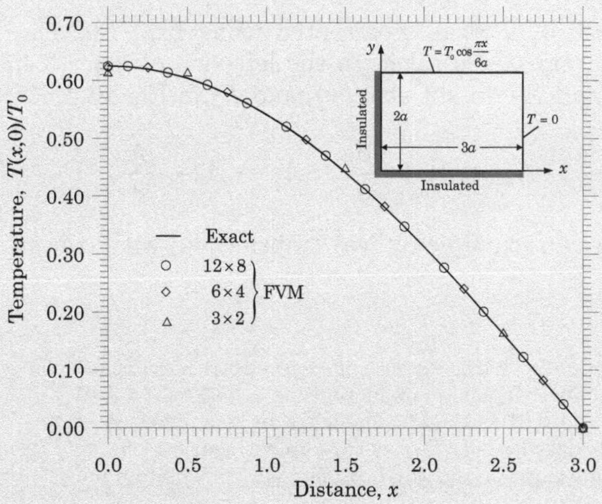

Fig. 3.4.10 Plots of numerical (ZFVM) and analytical solutions $T(x,0)/T_0$ as a function of x for the heat transfer problem of **Example 3.4.1**.

When a boundary condition of the mixed type [see Eq. (3.4.2b)] is specified in a problem, the coefficients A_K and Q_K in the discretized equations associated with the boundary nodes of the FVM get modified. The mixed boundary condition modifies the discrete equations of only the control volumes around the nodes that are on the boundary with convection. For example, suppose that the edge $y = 0$ is exposed to an ambient temperature of u_∞ [see Fig. 3.4.5(a) with $I = 2$ and $\hat{q}_n = 0$]. Then, in the half-control volume formulation, Eq. (3.4.2b) can be used to write

$$\begin{aligned} 0 = &\int_{0.5a}^{1.5a} \left[a_{yy} \frac{\partial u}{\partial y} \right]^{(B)} dx - \int_0^{0.5b} \left[a_{xx} \frac{\partial u}{\partial x} \right]^{(R)} dy \\ &- \int_{0.5a}^{1.5a} \left[a_{yy} \frac{\partial u}{\partial y} \right]^{(T)} dx + \int_0^{0.5b} \left[a_{xx} \frac{\partial u}{\partial x} \right]^{(L)} dy \\ &- \int_{0.5a}^{1.5a} \int_0^{0.5b} f(x,y)\, dxdy \end{aligned} \tag{3.4.31}$$

or

$$A_1 U_1 + A_2 U_2 + A_3 U_3 + A_6 U_6 = F_2 + Q_2, \tag{3.4.32a}$$

where A_i $(i = 1, 2, 3, 6)$ and F_2 are defined in Eq. (3.4.11b) with $I = 2$, and

$$Q_2 = -\int_{0.5u}^{1.5a} \left[a_{yy}\frac{\partial u}{\partial y}\right]^{(B)} dx = -\int_{0.5a}^{1.5a} \left[\beta\left(u - u_\infty\right)\right]^{(B)} dx$$
$$\approx -\frac{\beta a}{8}\left(U_1 + 6U_2 + U_3\right) + a\beta\, u_\infty, \tag{3.4.32b}$$

where the integral of $u(x)$ is represented using Eq. (3.2.7). The expression involving U_1, U_2, and U_3 is taken to the left side of Eq. (3.4.32a), modifying A_1, A_2, A_3 (A_6 and A_7 do not change), and F_2 in Eq. (3.4.32a):

$$A_1 \to A_1 + \frac{\beta a}{8}, \quad A_2 \to A_2 + \frac{6\beta a}{8}, \quad A_3 \to A_3 + \frac{\beta a}{8}, \quad F_2 \to F_2 + a\beta\, u_\infty.$$

Next, we consider an example of heat transfer [34] with convection.

Example 3.4.2

The bus bar shown in Fig. 3.4.11 carries sufficient electrical current to have a heat generation of $g = 10^6$ W/m^3. The bar has a conductivity of $a_{xx} = a_{yy} = k$ W/(m K) and dimensions 0.10 m by 0.05 m (and 0.01 m thick). The left side is maintained at 40°C and the right side at 10°C. Assuming that the heat flow is two-dimensional (or one may assume that the front and back faces are insulated), and the bottom edge is insulated and the top edge is exposed to ambient air temperature of $T_\infty = 0$°C with a heat transfer coefficient of 75 W/(m^2 K), and a uniform conductivity of $k = k_0 = 20$ W/(m· °C), determine the steady-state temperature distribution with 10 × 5 and 20 × 10 uniform meshes using the half-control finite volume formulation (HFVM).

Solution: The discretized equations for the problem at hand can be obtained as described previously and illustrated in Example 3.4.1. For example, for the 10 × 5 mesh, there 66 nodes and 66 control volumes (including the half- and quarter-control volumes). Nodes 1, 12, 23, 34, 45, 56 have a specified temperature of 40°C and nodes 11, 22, 33, 44, 55, and 66 have a specified temperature of 40°C. Equations associated with nodes 57–65 will have contribution due to the convective boundary condition.

The numerical solutions obtained with the HFVM for $T(x, 0)$ and $T(x, 0.05)$ versus x are presented in Table 3.4.2. Figure 3.4.12 contains plots of isotherms obtained with the 20 × 10 mesh, while the plots of the temperature at the bottom and top surfaces are presented in Fig. 3.4.13. The solutions obtained with 10 × 5 and 20 × 10 are indistinguishable in the plots.

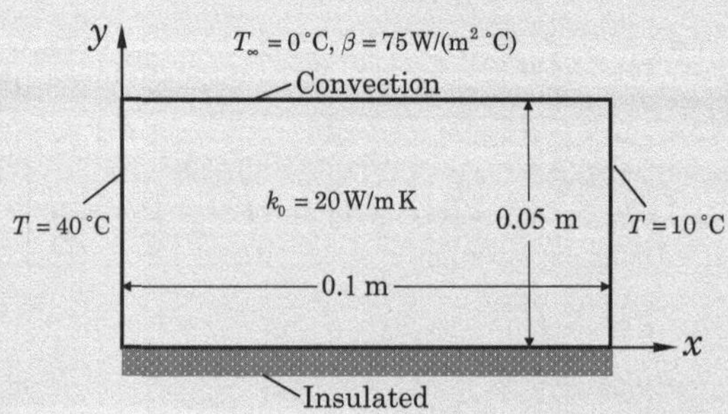

Fig. 3.4.11 Domain and boundary conditions for convective heat transfer in a bus bar. The bottom is insulated, while the top is exposed to ambient temperature of 0°C; the left side is kept at 40°C and the right face is maintained at 10°C.

Table 3.4.2 The HFVM solutions (temperature $T(x, y)$ in °C) of a two-dimensional heat conduction problem with convection boundary condition.

x	$y = 0.0$		$y = 0.05$	
	10×5	20×10	10×5	20×10
0.01	58.099	58.099	55.169	55.025
0.02	71.352	71.349	66.692	66.578
0.03	79.885	79.879	74.205	74.103
0.04	83.786	83.777	77.622	77.525
0.05	83.102	83.091	76.918	76.825
0.06	77.842	77.831	72.067	71.978
0.07	67.976	67.967	63.024	62.940
0.08	53.445	53.439	49.720	49.644
0.09	34.158	34.155	32.069	32.003

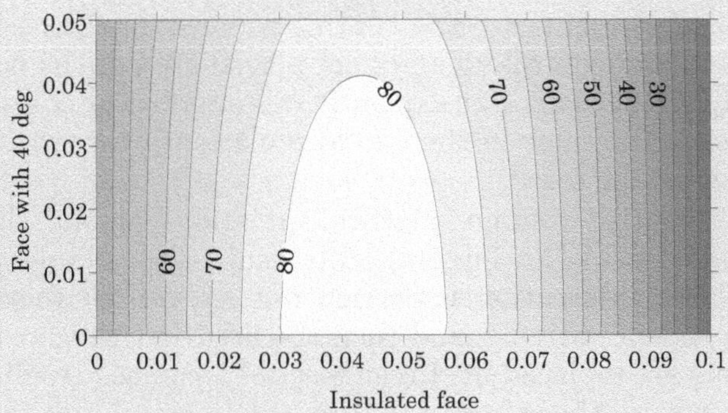

Fig. 3.4.12 Isotherms for the temperature field in the bar of **Example 3.4.2**.

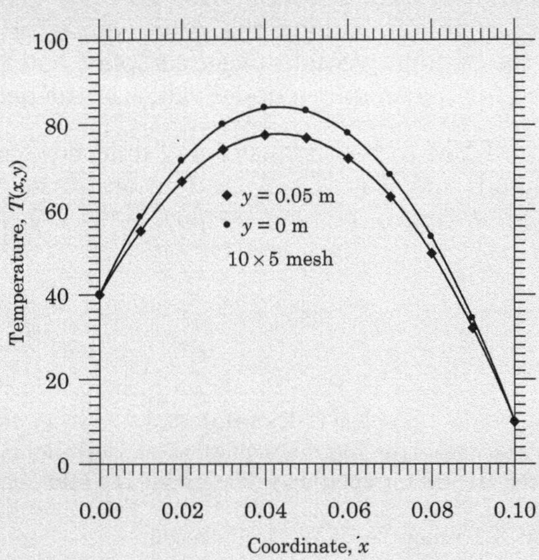

Fig. 3.4.13 Plots of temperature distributions at the bottom and top surfaces of the bar.

3.5 Summary

In this chapter, zero-thickness (ZFVM) and half-control finite volume (HFVM) formulations of model differential equations in one- and two-dimensions involving a single unknown are developed. Detailed discussions of the integral formulation, discretized equations for interior and boundary nodes are presented. Several numerical examples, mostly taken from heat transfer, are presented to illustrate the development of the discretized equations and their solutions.

In summary, the FVM involves the following basic steps:

(1) The whole domain Ω is represented as a collection of N interconnected and nonoverlapping subdomains, Ω^e $(e = 1, 2, \ldots, N)$. The intersection points of the subdomains as well as the end points are identified as the *nodes* in the HFVM formulation, whereas in the ZFVM formulation nodes are placed at the center of the subdomains as well as on the boundary. Thus, the subdomains are the control volumes in the case of the ZFVM formulation; and control volumes are placed around the nodes in the case of HFVM formulation. Consequently, the number of control volumes and nodes are not the same in the two formulations even when the number of subdomains is the same.

(2) Over each control volume, algebraic relations among the nodal values are developed using an integral statement of a second-order differential equation. An integration is carried out to weaken the differentiability of the dependent variable, and then the first derivatives appearing at the mesh points are replaced by Taylor's series formulas, resulting in algebraic relations between the values of the dependent variable at the enclosed node as well as nodes (neighbors) in the inter-connected neighboring control volumes. Similar relations are also derived for nodes on the boundary of the control volumes.

(3) Using the algebraic relations of all nodes (inside as well as on the boundary), necessary equations among the nodal values of the mesh are obtained. Then the boundary conditions are applied and the algebraic equations are solved for the unknown nodal values of the dependent unknown.

Application of the FVM to heat transfer and fluid flow can be found in the book by Reddy, Anand, and Roy [7]. An extension of the half-control FVM, termed as the *dual mesh control domain method*, is discussed in Chapter 5.

Problems

ONE-DIMENSIONAL PROBLEMS

3.1 Consider the heat transfer problem of **Example 3.3.2**. Use a uniform mesh of two and four subdivisions to derive the discrete equations and determine the temperatures at the nodes. Use the HFVM formulation to determine the solutions.

3.2 Repeat **Problem 3.1** using the ZFVM formulation.

3.3 Consider axisymmetric heat transfer in a long cylinder (i.e., **Example 3.3.3**). Using the HFVM formulation, (a) write the discretized equations for a uniform mesh of two subdivisions, (b) solve the condensed equations (i.e., after imposing the boundary conditions), and (c) compare the results with the exact solution at the nodes.

3.4 Give the explicit forms of the coefficients A_{I-1}, A_I, A_{I+1}, and F_I [see Eq. (3.2.17)] for the model equation (3.2.1) when the coefficients a, c , and f are linear functions of x:

$$a(x) = a_0 + a_1\, x; \quad c(x) = c_0 + c_1\, x; \quad f(x) = f_0 + f_1\, x. \tag{1}$$

3.5 An insulating wall is constructed of three homogeneous layers with conductivities k_1, k_2, and k_3 and in intimate contact (see Fig. P3.5). Under steady-state conditions, the temperatures of the surroundings on the left and right, T_L and T_R, respectively, enter the convection boundary conditions. Formulate the problem using the half-control volume FVM to determine the temperatures T_i $(i = 1, 2, 3, 4)$ when the ambient temperatures T_L and T_R (at the left and right, respectively) and the (surface) film coefficients β_L and β_R are known for the following cases. Case 1: $T_L = T_R = 20°$C and $\beta_L = \beta_R = 500$ W/m^2·C. Case 2: $T_L = 120°$C, $T_R = 20°$C, $\beta_L = 500$ W/m^2·C, and $\beta_R = 560$ W/m^2·C. Case 3: $T_1 = 100°$C, $T_R = 20°$C, and $\beta_R = 500$ W/m^2·C.

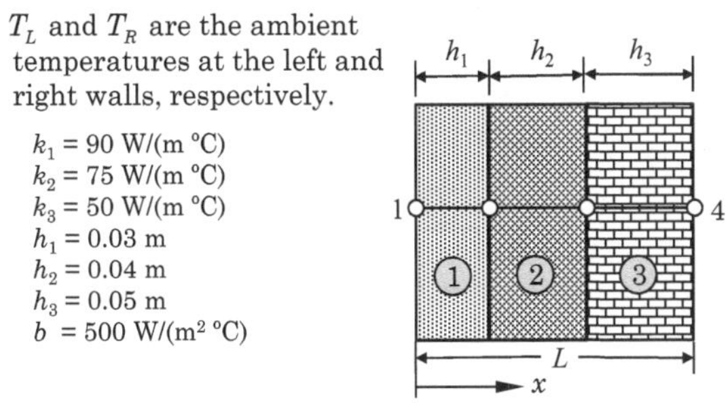

Fig. P3.5

TWO-DIMENSIONAL PROBLEMS

3.6 Consider the partial differential equation governing heat transfer in an axisymmetric geometry

$$-\frac{1}{r}\frac{\partial}{\partial r}\left(r k_{rr}\frac{\partial T}{\partial r}\right) - \frac{\partial}{\partial z}\left(k_{zz}\frac{\partial T}{\partial z}\right) = f(r, z) \tag{1}$$

where $(k_{rr},\ k_{zz})$ and f are the conductivities and internal heat generation per unit volume, respectively. In developing the integral form, we integrate over the elemental volume of the axisymmetric geometry: $r\, dr\, d\theta\, dz$. Develop the integral form and associated discretized equations using the half-control volume formulation.

3.7 Consider the steady-state heat transfer in a square region shown in Fig. P3.7. The governing equation is given by

$$-\frac{\partial}{\partial x}\left(k\frac{\partial u}{\partial x}\right) - \frac{\partial}{\partial y}\left(k\frac{\partial u}{\partial y}\right) = f_0 \tag{1}$$

The boundary conditions for the problem are:

$$u(0, y) = y^2\ ,\ \ u(x, 0) = x^2\ ,\ \ u(1, y) = 1 - y\ ,\ \ u(x, 1) = 1 - x \tag{2}$$

Assuming $k = 1$ and $f_0 = 2$, determine the unknown nodal value of u using the mesh shown in Fig. P3.7 and the half-control volume formulation.

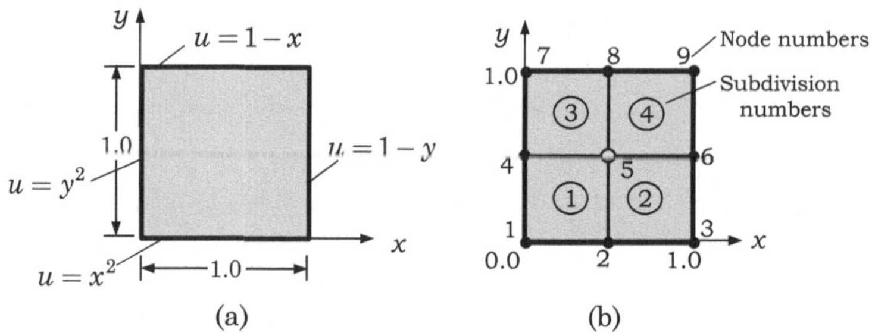

Fig. P3.7

3.8 Consider **Problem 3.7** with the mesh of triangles shown in Fig. P3.8. Write the control volume equations for the center node using the square control volume.

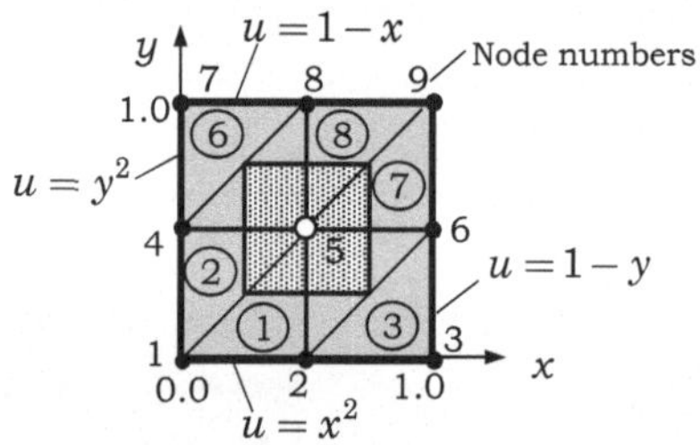

Fig. P3.8

4 Finite Element Method

4.1 Introduction

4.1.1 Analysis Steps

This chapter introduces the finite element method (FEM) as a technique of solving differential equations governing a single dependent variable. A detailed account of the method can be found in the books by Reddy [8, 13], Reddy and Gartling [10], Surana and Reddy [14, 15], and Reddy, Anand, and Roy [7], among many other books on the subject.

The FEM has the following five major analysis steps:

(1) *Discretization of geometry and solution.* In general, as in the FVM, there are two different discretizations in the FEM. First, the geometry of the domain Ω is discretized as a collection of interconnected and nonoverlapping subdomains Ω^e, called *finite elements*[1]. Second, the solution of the governing equation is also approximated on each of another set of nonoverlapping finite elements that cover the domain. Typically, the two meshes overlap each other; in fact, no works can be found in the literature where the two meshes do not overlap, although one mesh may contain the other. When the meshes are identical and overlap, the formulation is known as *isoparametric*, and no distinction is made between the two meshes. The first mesh is used to satisfy the governing equation in an integral sense (to be discussed in detail in the coming sections) and the geometry, while the second mesh is used to approximate the solution.

(2) *Derivation of discretized equations.* Over each representative finite element Ω^e of the first mesh, we develop algebraic relations among the duality pairs (e.g., relations between pairs of dual variables such as "temperatures" and "fluxes" or "heats," and "displacements" and "forces," and so on) using a *weighted*-integral statement equivalent to the governing equation. The set of algebraic equations among the nodal values of the typical element Ω^e is called a *finite element model.* There are several possible ways, including the subdomain method, least-squares method, and so on, to satisfy the governing equation in an integral sense. Each method results in a different finite element model (e.g., subdomain finite element model, least-squares finite element model, and so on) of the same equation. The approximation functions of the two meshes play a role in the evaluation of the integral statements.

[1]Since a domain can only be divided into a finite number of subdomains, the size of the element is finite, hence the phrase *finite element.*

(3) *Assembly of element equations.* The algebraic relations of all elements are put together, which is called an *assembly*, using certain physical and mathematical requirements. The assembled equations represent relations between the duality pairs of the nodes in the whole domain.
(4) *Solution of discretized equations.* Apply the known (boundary) conditions, which have to be converted to known values of the duality pairs at the nodes, and solve the algebraic equations for the nodal values.
(5) *Post-computation.* The solution and any auxiliary quantities that depend on the solution are determined through the post-computation.

4.1.2 Remarks on the Analysis Steps

While more details will follow in the coming sections, it is informative to understand a few basic ideas related to the method.

(1) A finite element is a geometric shape that allows the unique derivation of polynomial approximation functions. For the purpose of defining the geometry of an element, we identify points in the element, called *nodes*. For example, three points in a plane define a triangle uniquely, and the three points are necessarily the vertexes of the triangle. The collection of elements is termed a *finite element mesh.*

(2) In this book, we will develop the finite element equations (i.e., the discretized equations over a typical finite element) using the so-called weak form (equivalent to the Ritz method [17]; see Section 1.11.3.6), which will be discussed in detail in Section 4.2.4. Central to this development is the introduction of the concept of *duality pairs* (i.e., cause and effect), which exists in every physical process. The FVM, in a broad sense, can be viewed as a different finite element model in which an integral statement, not a weighted-integral statement, of the governing equation is used.

(3) When the meshes are different (i.e., meshes with different kinds of elements), the approximations used for the geometry and the dependent variable are different. In the second step, the conversion of the integral statement to algebraic equations among the nodal values requires an interpolation of the dependent variables of the governing differential equation(s). The approximation functions of the second mesh actually interpolate the function values at the nodes of the mesh (and thus are called *interpolation functions*). The approximation functions of the first mesh are used to evaluate the integrals appearing in the weak form. The set of algebraic equations, that is, the finite element model, represents a discrete model of the original differential equation.

(4) The discretized equations over an element contain the nodal values of the element only (a major difference between FEM and FVM). The third step guarantees that the elements are interconnected (i.e., the discrete equations from different elements are superposed or assembled) in such a way that the interelement continuity and balance conditions are satisfied.

(5) The fourth step is common to all numerical methods: the end result of the application of a numerical method is a set of algebraic equations. The physical boundary conditions are expressed in terms of the values of the discretized variables and imposed on the system before solving with any of the methods discussed in Section 1.10. Thus, the variables of the formulation in any numerical method must necessarily be those which allow the imposition of boundary conditions.

In this chapter, we introduce the finite element method as applied to model (i.e., typical) differential equations in one and two dimensions, involving a single unknown. A detailed discussion of the steps that convert the weighted-integral statement of the model differential equation to an associated weak-form, derivation of the finite element equations (i.e., set of algebraic equations among the nodal values), assembly of element equations, imposition of boundary conditions, and the post-computation of the secondary variables is presented. The solution of algebraic equations by any of the direct or iterative methods discussed in Chapter 1 is possible. For additional discussion and applications of the method, the reader is asked to consult the books by Reddy [8, 13] and Reddy and Gartling [10].

4.2 One-Dimensional Problems

4.2.1 Model Differential Equation

In this chapter, the finite element analysis steps are illustrated using the following differential equation [same as that considered in Eq. (3.2.1)]

$$-\frac{d}{dx}\left(a\frac{du}{dx}\right)+b\frac{du}{dx}+cu=f \qquad \text{for} \qquad 0<x<L, \tag{4.2.1}$$

where $a = a(x)$, $b = b(x)$, $c = c(x)$, and $f = f(x)$ are the *data* (i.e., known quantities) of the problem and $u(x)$ is the dependent variable. As in Chapter 3, Eq. (4.2.1) is solved numerically with physically admissible end or boundary conditions, such as specifying u or its dual, namely, $Q = a(du/dx)$ at a boundary point.

4.2.2 Finite Element Mesh of the Geometry

The domain $\Omega = (0, L)$ of the uniaxial member (assumed here to be a line interval of length L) consists of all points between $x = 0$ and $x = L$, and not including the end points. The points $x = 0$ and $x = L$ are the boundary points of the domain. In the FEM, the domain $(0, L)$ is divided into a set of line intervals, $\Omega^e = (x_a^e, x_b^e)$, where x_a^e and x_b^e denote the coordinates of the end points of the line segment Ω^e with respect to the coordinate x, as shown in Fig. 4.2.1. A typical line segment occupying the interval (x_a^e, x_b^e) is called a *finite element* with $h_e = x_b^e - x_a^e$ being its length. An arbitrary point x in the interval (x_a^e, x_b^e) can be expressed as

$$x = x_a^e + \bar{x} = x_a^e\left(1-\frac{\bar{x}}{h_e}\right)+x_b^e\frac{\bar{x}}{h_e} \tag{4.2.2}$$

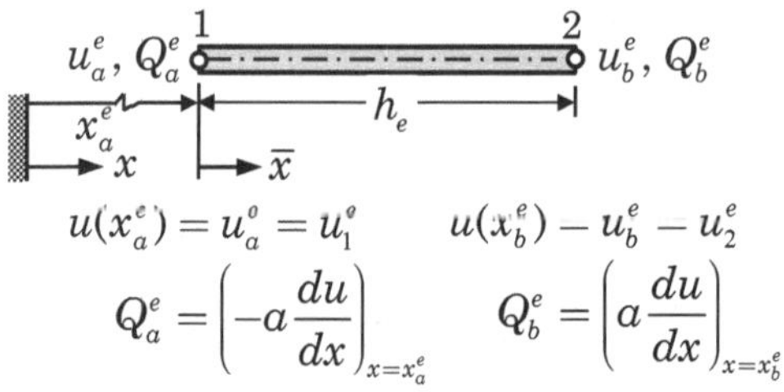

Fig. 4.2.1 A typical (line) finite element in one dimension. The duality pairs (u_a^e, Q_a^e) and (u_b^e, Q_b^e) will be identified during the weak-form development.

where $\bar{x}$ is the distance measured from the left end of the element Ω^e. Clearly, the representation in Eq. (4.2.2) is linear.

4.2.3 Approximation of the Solution over the Element

In the FEM, we seek an approximate solution $u_h^e(x)$ to the actual solution $u(x)$ over element $\Omega^e = (x_a^e, x_b^e)$. The finite element approximation $u_h^e(x)$ is sought in the form of a *complete* algebraic polynomial. The word "complete" is used here to mean that all order terms, including the constant term, are included in the polynomial so that a solution that is a constant, linear, quadratic, and higher-order can be represented. For example, $u_h^e(x) = c_1 x + c_2 x^2$ and $u_h^e(x) = c_0 + c_2 x^2$ are incomplete because the constant term in the first and linear term in the second are missing. Examples of complete polynomials are provided by $u_h^e(x) = c_0 + c_1 x$, $u_h(x) = c_0 + c_1 x + c_2 x^2$, $u_h(x) = c_0 + c_1 x + c_2 x^2 + c_3 x^3$, and so on. Thus, the $(n-1)$th-degree complete polynomial (contains n terms) is given by

$$u(x) \approx u_h^e(x) = c_0^e + c_2^e x + \cdots + c_{n-1}^e x^{n-1}, \tag{4.2.3}$$

where c_j^e are constants to be determined such that $u_h^e(x)$ satisfies the differential equation (4.2.1) and appropriate *end conditions* of the element. Since there are n unknown parameters, n linearly independent algebraic relations are needed to determine them. As we will see shortly, it is both necessary and useful to express the parameters c_i^e in terms of the values of u at n selected points so that the end conditions of the element, $u(x_a^e) = u_1^e$ and $u(x_b^e) = u_n^e$, are also met. That is, Eq. (4.2.3) is of the form

$$u_h^e(x) = \sum_{j=1}^{n} u_j^e \psi_j^e(x), \tag{4.2.4}$$

where $\{\psi_j^e(x)\}_{j=1}^n$ are a set of interpolation functions systematically derived so that $u_h^e(x_j^e) = u_j^e$ for $j = 1, 2, \ldots, n$ [i.e., $\psi_i^e(x)$ satisfy the interpolation property, $\psi_i^e(x_j^e) = \delta_{ij}$]. This requires the identification of n nodes in the element (including the end points). We will return to the derivation of $\psi_j^e(x)$ in Section 4.2.6.

Substituting the approximate solution in Eq. (4.2.3) into the Eq. (4.2.1), we obtain (recall the discussion from Section 1.11.3)

$$-\frac{d}{dx}\left(a\frac{du_h^e}{dx}\right)+b\frac{du_h^e}{dx}+cu_h^e(x)-f(x)\equiv \mathbb{R}^e(x,u_1^e,u_2^e,\ldots,u_n^e)\neq 0.$$

We wish to determine u_j^e, $j=1,2,\ldots,n$ (appearing in $u_h^e(x)$) such that the residual $\mathbb{R}^e$ is zero over the element Ω^e in some acceptable sense (as already discussed in Section 1.11.3).

We note that the residual $\mathbb{R}^e$ contains the same order derivatives of the dependent unknown $u(x)$ as in the differential equation, Eq. (4.2.1), and hence requires at least a quadratic representation of $u_h^e(x)$. To reduce or weaken the differentiability of ψ_i^e, we mathematically distribute differentiation equally (when the given differential equation is of even order) between w_i^e and u_h^e, and the resulting integral form is known as the *weak form*, because the differentiability on u_h^e (hence on ψ_i^e) is reduced or weakened. The three-step weak-form development presented in Section 1.11.3.6 is revisited in Section 4.2.4. Most finite element models in the literature as well as in the commercial codes are based on weak formulations of their governing differential equations.

4.2.4 Derivation of the Weak Form: The Three-Step Procedure

The three-step procedure of constructing the weak form beginning with the governing differential equation, Eq. (4.2.1), is presented here.

Step 1. *Weighted-integral statement*

$$0=\int_{x_a^e}^{x_b^e} w_i^e\left[-\frac{d}{dx}\left(a\frac{du_h^e}{dx}\right)+b\frac{du_h^e}{dx}+cu_h^e-f\right]dx. \tag{4.2.5}$$

The *primary variable* is the variable (u_h^e) appearing in the governing differential equation (for higher-order equations, additional primary variables are introduced through the weak-form development, as will be seen in Chapter 7 on beams). The specification of a primary variable on the boundary constitutes the *essential* or *Dirichlet boundary condition.*

Step 2. *Trading the differentiation between the dependent unknown u_h^e and the weight function w_i^e and the identification of the secondary variables*

$$0=\int_{x_a^e}^{x_b^e}\left(a\frac{dw_i^e}{dx}\frac{du_h^e}{dx}+bw_i^e\frac{du_h^e}{dx}+cw_i^eu_h^e-w_i^ef\right)dx-\left[w_i^e\cdot a\frac{du_h^e}{dx}\right]_{x_a^e}^{x_b^e}. \tag{4.2.6}$$

To identify the dual variable to the primary variable u_h^e, consider the boundary term appearing in Eq. (4.2.6), namely, the expression

$$\left[w_i^e\cdot a\frac{du_h^e}{dx}\right]_{x_a^e}^{x_b^e}.$$

The coefficient of the weight function w_i^e in the boundary expression, $a(du_h^e/dx)$, is the dual variable, and it is termed the *secondary variable.* The specification of a secondary variable constitutes the *natural* or *Neumann* boundary condition. Thus, for the model equation at hand, the primary and secondary variables are

$$\text{Primary variable: } u_h^e \;; \quad \text{Secondary variable: } n_x \left(a \frac{du_h^e}{dx} \right) \equiv Q_h^e(x), \quad (4.2.7)$$

where $n_x = -1$ at the left end of the element and $n_x = 1$ at the right. In solid and structural mechanics problems, the product of the primary and dual variables (e.g., force multiplied by the displacement) represents the work done.

Step 3. *Final weak form*

$$0 = \int_{x_a^e}^{x_b^e} \left(a \frac{dw_i^e}{dx} \frac{du_h^e}{dx} + bw_i^e \frac{du_h^e}{dx} + cw_i^e u_h^e - w_i^e f \right) dx - w_i^e(x_a^e) Q_a^e - w_i^e(x_b^e) Q_b^e, \quad (4.2.8)$$

where (see Fig. 4.2.1)

$$Q_a^e = Q(x_a^e) = - \left(a \frac{du_h^e}{dx} \right)_{x_a^e}, \qquad Q_b^e = Q(x_b^e) = \left(a \frac{du_h^e}{dx} \right)_{x_b^e}. \quad (4.2.9)$$

The primary and secondary variables at the nodes are shown on the typical element in Fig. 4.2.1. It can be viewed as the *free-body diagram* of a typical but arbitrary element of the mesh, with Q_a^e and Q_b^e denoting the nodal secondary variables. For heat conduction problems, Q_a^e denotes the heat at the left end of the element, and Q_b^e denotes the heat at the right end. For axial deformation of a bar, Q_a^e denotes the compressive force at the left end of the element and Q_b^e denotes the tensile force at the right end. Although Q_h^e replaced $a(du_h^e/dx)$, Q_h^e is *not* considered to be a function of u_h^e; it is a variable that is dual to u_h^e and can be specified at a point whenever u_h^e is not specified at the same point. This completes the three-step procedure of constructing a weak form of a second-order differential equation.

4.2.5 Remarks on the Weak Form

(1) As noted in Section 1.11.3.6, the integration-by-parts used in Step 2 has two computational benefits: first, the weak form has weakened (i.e., reduced) differentiability requirement on u_h^e (and admits lower-order approximation than otherwise) and, second, contains secondary variables Q_a^e and Q_b^e that are physically meaningful in the sense that they can be specified at a point whenever the primary variable u_h^e at the same point is not specified there. *If a secondary variable Q_h^e resulting from the integration by parts is not a physical quantity, then integration by parts should not be carried out even if it results in the weakening of the differentiability of u_i^e.*

(2) Although the weak form is derived starting with a weighted-residual statement of the governing equation, the resulting weak form may have

additional meaning. In the case of solid mechanics, the weak form is indeed the same as the principle of virtual displacements with w_i^e interpreted as the virtual displacement δu_h^e (see Reddy [17]). Thus, the weak form for the axial deformation of bars, for example, also represents a physical principle, whereas for problems for which such an interpretation is not available, the weak form is just a mathematical statement equivalent to the original differential equation.

(3) Weak-form formulation allows the identification of the duality pairs. One element of a duality pair (primary variable) is often a single-valued observable or measurable quantity at every point of the domain, while the other (secondary variable) is a derivable quantity. When the primary variable is single-valued, it is taken as the nodal variable and made continuous across the elements during the assembly.

(4) Although a weak form admits a variety of approximation functions to represent a primary variable, for simplicity and practicality, most finite element approximations used in the literature have been algebraic polynomials. It is easier to numerically evaluate the integrals of finite-degree polynomial expressions than those of trigonometric or hyperbolic functions (knowing that such functions are replaced by a finite-term series during the evaluation). Approximation functions used in so-called "meshless methods" require a large number of integration points to numerically evaluate their integrals.

4.2.6 Interpolation Functions

The approximation of a primary variable $u_h^e(x)$ should be selected such that (a) the differentiability implied by the weak form is met (i.e., the derivatives of $u_h^e(x)$ appearing in the weak form should not be zero), (b) $u_h^e(x)$ is continuous at points common to elements, and (c) $u_h^e(x)$ be a complete polynomial (when polynomials are used). Since the weak form in the present case contains the first-order derivative of u_h^e, any function with a nonzero first derivative would be a candidate for u_h^e. The continuity of primary variables between elements can be satisfied by equating their nodal values at the nodes common to elements. The easiest way to impose the interelement continuity of a primary variable is to equate their nodal values at the nodes common to elements. This in turn requires the approximation functions to be interpolants of u_h^e through the nodal points (i.e., u_h^e must be equal to u_a^e at x_a^e and u_b^e at x_b^e). The end points that define the geometry and satisfy the continuity requirement must be included as the interpolation points.

Linear approximation. A complete linear polynomial is

$$u_h^e(x) = c_1^e + c_2^e\, x, \tag{4.2.10}$$

which is admissible if one can select c_1^e and c_2^e such that

$$u_h^e(x_a^e) \equiv u_1^e = c_1^e + c_2^e\, x_a^e, \qquad u_h^e(x_b^e) \equiv u_2^e = c_1^e + c_2^e\, x_b^e, \tag{4.2.11}$$

which can be expressed in matrix form as

$$\begin{bmatrix} 1 & x_a^e \\ 1 & x_b^e \end{bmatrix} \begin{Bmatrix} c_1^e \\ c_2^e \end{Bmatrix} = \begin{Bmatrix} u_1^e \\ u_2^e \end{Bmatrix}. \tag{4.2.12a}$$

The solution of Eq. (4.2.12a) using Cramer's rule yields

$$c_1^e = \frac{u_1^e\, x_b^e - u_2^e\, x_a^e}{x_b^e - x_a^e}, \quad c_2^e = \frac{u_2^e - u_1^e}{x_b^e - x_a^e}. \tag{4.2.12b}$$

Substitution of Eq. (4.2.12b) for c_i^e into Eq. (4.2.10) gives

$$\begin{aligned} u_h^e(x) &= c_1^e + c_2^e\, x = \frac{u_a^e\, x_b^e - u_b^e\, x_a^e}{x_b^e - x_a^e} + \frac{u_b^e - u_a^e}{x_b^e - x_a^e}\, x \\ &= \psi_1^e(x)\, u_1^e + \psi_2^e(x)\, u_2^e = \sum_{j=1}^{2} \psi_j^e(x)\, u_j^e, \end{aligned} \tag{4.2.13}$$

where (noting that $h_e = x_b^e - x_a^e$) the functions ψ_j^e are known as the *linear Lagrange interpolation functions* [see Fig. 4.2.2(a)]

$$\psi_1^e(x) = \frac{x_b^e - x}{h_e} = 1 - \frac{\bar{x}}{h_e}, \qquad \psi_2^e(x) = \frac{x - x_a^e}{h_e} = \frac{\bar{x}}{h_e}. \tag{4.2.14}$$

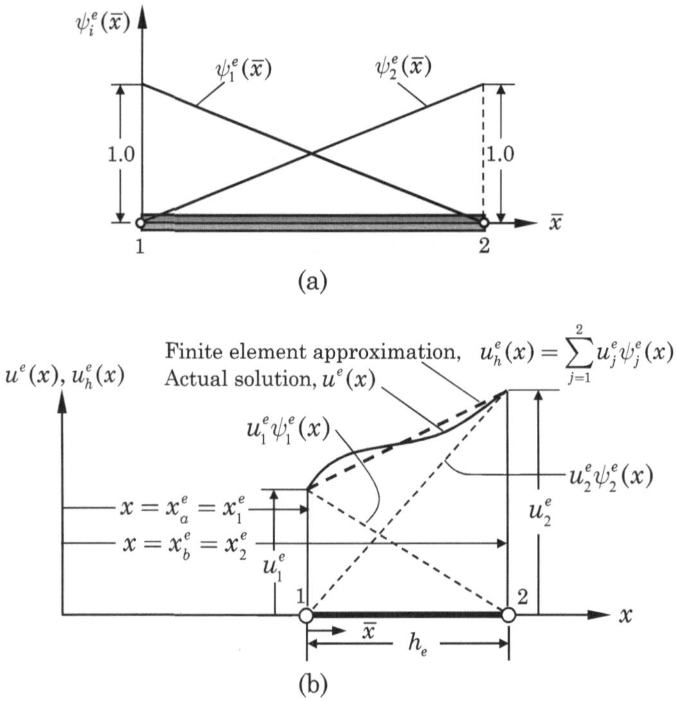

Fig. 4.2.2 (a) Linear interpolation functions. (b) Linear approximation of a function $u(x)$ over a finite element.

Here $\bar{x}$ is the element (or local) coordinate, $\bar{x} = x - x_a^e$, and

$$u_h^e(x_a^e) = u_1^e = u_a^e, \qquad u_h^e(x_b^e) = u_2^e = u_b^e \tag{4.2.15}$$

An element with linear approximation [see Fig. (4.2.2)(b)] is called a *linear element.*

We note that the function $\psi_j^e(x)$ satisfies the *interpolation property*:

$$\psi_j^e(x_i^e) = \delta_{ij} = \begin{cases} 1, & \text{if } i = j \\ 0, & \text{if } i \neq j, \end{cases} \tag{4.2.16}$$

where $x_1^e = x_a^e$ and $x_2^e = x_b^e$. In addition, the Lagrange interpolation functions satisfy the property known as the *partition of unity* :

$$1 = \psi_1^e(x) + \psi_2^e(x) + \cdots + \psi_n^e(x) = \sum_{j=1}^{n} \psi_j^e(x). \tag{4.2.17}$$

Quadratic approximation. If we wish to approximate $u(x)$ with a quadratic polynomial, we write

$$u_h^e(x) = c_1^e + c_2^e x + c_3^e x^2 \tag{4.2.18}$$

and express the three parameters (c_1^e, c_2^e, c_3^e) in terms of three values of u_h^e at three nodes, including the two end nodes of the element. Choosing the third node at the center of the element, as indicated in Fig. 4.2.3(a), we obtain

$$\begin{aligned} u_1^e &\equiv u_h^e(x_1^e) = c_1^e + c_2^e x_1^e + c_3^e (x_1^e)^2 \\ u_2^e &\equiv u_h^e(x_2^e) = c_1^e + c_2^e x_2^e + c_3^e (x_2^e)^2 \\ u_3^e &\equiv u_h^e(x_3^e) = c_1^e + c_2^e x_3^e + c_3^e (x_3^e)^2, \end{aligned} \tag{4.2.19}$$

where x_i^e ($i = 1, 2, 3$) are the global coordinates of the nodes of the element (with $x_2^e = x_1^e + 0.5h_e$). Equation (4.2.19) can be solved for (c_1^e, c_2^e, c_3^e) in terms of (u_1^e, u_2^e, u_3^e) and substituted into Eq. (4.2.18) to obtain

$$u_h^e(x) = \psi_1^e(x)u_1^e + \psi_2^e(x)u_2^e + \psi_3^e(x)u_3^e = \sum_{j=1}^{3} \psi_j^e(x)u_j^e, \tag{4.2.20}$$

where $\psi_i^e(x)$ are the quadratic Lagrange interpolation functions shown in Fig. 4.2.3(b), which can be expressed in terms of the element (or local) coordinate $\bar{x} = x - x_a^e$ as

$$\begin{aligned} \psi_1^e(\bar{x}) &= \left(1 - \frac{\bar{x}}{h_e}\right)\left(1 - \frac{2\bar{x}}{h_e}\right) \\ \psi_2^e(\bar{x}) &= 4\frac{\bar{x}}{h_e}\left(1 - \frac{\bar{x}}{h_e}\right) \\ \psi_3^e(\bar{x}) &= -\frac{\bar{x}}{h_e}\left(1 - \frac{2\bar{x}}{h_e}\right), \end{aligned} \tag{4.2.21}$$

which also satisfy the properties in Eqs. (4.2.16) and (4.2.17). An element with quadratic approximation of the field variable $u_h^e(x)$ is called a *quadratic element.*

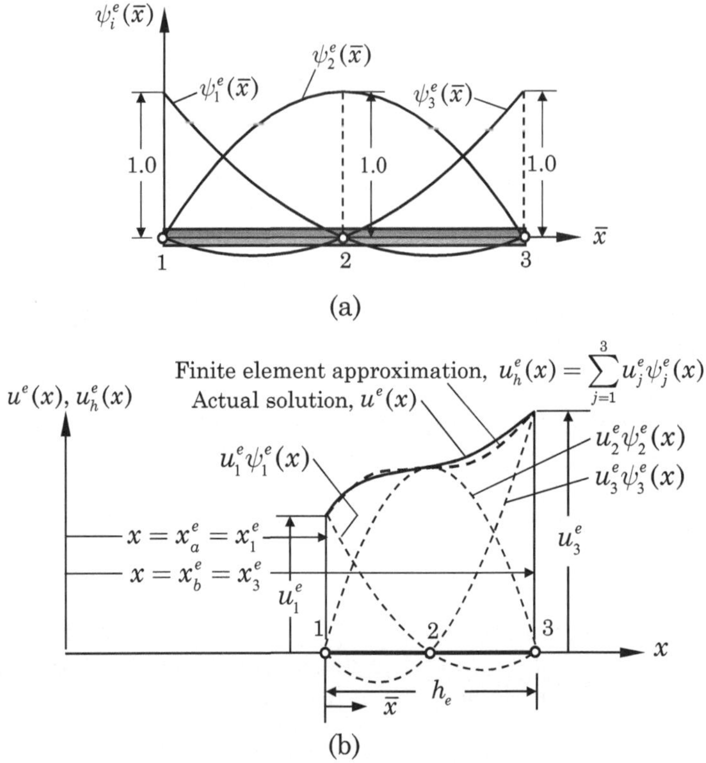

Fig. 4.2.3 (a) Quadratic finite element and (b) associated interpolation functions.

Higher-order Lagrange interpolations of $u(x)$ can be developed along similar lines. Thus, an $(n-1)$-degree Lagrange interpolation of $u(x)$ is given by

$$u_h^e(x) = \psi_1^e(x)u_1^e + \psi_2^e(x)u_2^e + \cdots + \psi_n^e(x)u_n^e = \sum_{j=1}^{n} \psi_j^e(x)u_j^e, \tag{4.2.22}$$

where the Lagrange interpolation functions of degree $(n-1)$ are given by

$$\psi_j^e(x) = \prod_{i=1, i\neq j}^{n} \left(\frac{x - x_i^e}{x_j^e - x_i^e} \right). \tag{4.2.23}$$

The element contains n nodes. In general, the nodes are not equally spaced.

4.2.7 Remarks on the Interpolation Functions

The finite element solution $u_h^e(x)$ must fulfill certain requirements for it to be convergent to the actual solution $u(x)$ as the number of elements (h refinement) or the degree of the polynomials (p refinement) is increased. These requirements are listed next.

(1) The approximate solution should be continuous and differentiable as required by the weak form in Eq. (4.2.6).
(2) It should be a complete polynomial, that is, the polynomial includes all lower-order terms, including the constant term, up to the highest-order term used.
(3) It should be an interpolant of the primary variable u_h^e at the nodes of the finite element (i.e., it should at least interpolate the solution at the boundary nodes of the element).

The first requirement ensures that every term of the weak form has a nonzero contribution. The second requirement is necessary in order to capture all possible states, say, constant, linear, and so on, of the actual solution. For example, if a linear polynomial without the constant term is used to represent the temperature distribution in a one-dimensional system, the approximate solution can never be able to represent a uniform state of temperature field in the element. The third requirement is necessary in order to enforce continuity of the primary variable u_h^e at nodes where the element is connected to its neighboring elements.

4.2.8 Finite Element Model

First, recall that the phrase "finite element model" means that it is a set of algebraic equations obtained, after the substitution of the interpolation functions, from an integral statement, such as the weak form, of the governing differential equation. Algebraic equations obtained by substituting admissible approximation of u_h^e into the weighted-integral form in Eq. (4.2.5) is termed a *weighted-integral finite element model* (or Galerkin model if w_i^e is replaced with ψ_i^e). The finite element model obtained using a weak form is termed the *weak-form Galerkin finite element model.*

Substitution of the approximation from Eq. (4.2.22) into the weak form in Eq. (4.2.8) gives n *algebraic* equations among the n nodal values u_i^e ($i = 1, 2, \ldots, n$) and Q_i^e of the element. In order to formulate the finite element model (i.e., derive a set of algebraic relations among u_i^e and Q_i^e) based on the weak form in Eq. (4.2.8), it is *not* necessary to decide the degree of approximation of $u_h^e(x)$ a priori. The equations can be written in generic form in terms of ψ_i^e.

For $n > 2$, the weak form in Eq. (4.2.8) must be modified to include nonzero secondary variables, if any, at the nodes interior to the element (e.g., node 2 in the case of the quadratic element):

$$0 = \int_{x_a}^{x_b} \left(a\frac{dw_i^e}{dx}\frac{du_h^e}{dx} + bw_i^e\frac{du_h^e}{dx} + cw_i^e u_h^e \right) dx - \int_{x_a}^{x_b} w_i^e f\, dx - \sum_{j=1}^{n} w_i^e(x_j^e)\, Q_j^e, \tag{4.2.24}$$

where x_i^e is the global coordinate of the ith node of element $\Omega^e = (x_a^e, x_b^e)$. If nodes 1 and n denote the end points of the element, then $Q_1^e = Q_a^e$ and $Q_n^e = Q_b^e$ represent the *unknown* point sources, and all other Q_i^e ($i = 2, 3, \ldots, n-1$) are the externally applied (hence, known) point sources at nodes $2, 3, \ldots, n-1$, respectively.

Substituting Eq. (4.2.22) for u_h^e and $w_1^e = \psi_1^e$, $w_2^e = \psi_2^e, \ldots,$ $w_i^e = \psi_i^e, \ldots,$ $w_n^e = \psi_n^e$ into the weak form Eq. (4.2.24), we obtain n algebraic equations. The equations are numbered in such a way that the algebraic equation obtained from the weak form with $w_i^e = \psi_i^e$ is numbered as the ith equation of the set of n equations. Thus, the ith algebraic equation is

$$0 = \sum_{j=1}^{n} \left[\int_{x_a^e}^{x_b^e} \left(a\frac{d\psi_i^e}{dx}\frac{d\psi_j^e}{dx} + b\psi_i^e \frac{d\psi_j^e}{dx} + c\psi_i^e\psi_j^e \right) dx \right] u_j^e - \int_{x_a^e}^{x_b^e} \psi_i^e f\, dx - Q_i^e$$

$$0 = \sum_{j=1}^{n} K_{ij}^e u_j^e - f_i^e - Q_i^e \tag{4.2.25}$$

for $i = 1, 2, \ldots, n$, where

$$K_{ij}^e = \int_{x_a^e}^{x_b^e} \left(a\frac{d\psi_i^e}{dx}\frac{d\psi_j^e}{dx} + b\psi_i^e \frac{d\psi_j^e}{dx} + c\psi_i^e\psi_j^e \right) dx, \quad f_i^e = \int_{x_a^e}^{x_b^e} f\psi_i^e\, dx. \tag{4.2.26}$$

Note that the interpolation property in Eq. (4.2.16) is used to write

$$\sum_{j=1}^{n} \psi_i^e(x_j^e) Q_j^e = \sum_{j=1}^{n} \delta_{ij}\, Q_j^e = Q_i^e, \tag{4.2.27}$$

where δ_{ij} is the Kronecker delta symbol . In matrix notation, these algebraic equations can be expressed as

$$\mathbf{K}^e\mathbf{u}^e = \mathbf{f}^e + \mathbf{Q}^e \equiv \mathbf{F}^e. \tag{4.2.28}$$

The matrix $\mathbf{K}^e$, which is unsymmetric in the present case (it is symmetric only when $b^e = 0$), is called the *coefficient matrix*, and the column vector $\mathbf{f}^e$ is the *source vector*, and it is a nodal representation of the distributed source $f(x)$.

As already stated, the choice of replacing w_i^e with ψ_i^e amounts to using the Galerkin method of approximation (see Section 1.11.3.5). It is important to note that the classical Galerkin method was based on the weighted-integral statement of the residual – not on the weak form. In fact, utilizing the weak form amounts to using the Ritz method of approximation (see Section 1.11.3.5; Reddy [8, 17]). Therefore, the finite element model should be called the *Ritz* or the *weak-form Galerkin finite element model*, but not as the Galerkin finite element model.

Equation (4.2.28) consists of n equations among $n+2$ unknowns, namely, n primary nodal values $(u_1^e, u_2^e, \ldots, u_n^e)$ and two secondary nodal values (Q_1^e, Q_n^e); the values $(u_1^e, u_2^e, \ldots, u_n^e)$ are also called the element *primary nodal degrees of freedom*. Because there are more unknowns than the number of equations, the element equations cannot be solved without assembling all elements of the total domain. The assembly of elements (i.e., putting the elements together) is carried out by imposing the *continuity of the primary variables* and *equilibrium of secondary variables* at nodes common to different elements. Upon assembly and imposition of boundary conditions, we shall obtain exactly the same number of algebraic equations as the total number of unknown primary and secondary nodal degrees of freedom. This is explained next.

The coefficient matrix $\mathbf{K}^e$ and source vector $\mathbf{f}^e$ can be evaluated for a given interpolation of the variable u_h^e and data (a, c, and f). When a, c, and f

are functions of x, it may be necessary to evaluate $\mathbf{K}^e$ and $\mathbf{f}^e$ using numerical integration. When the data (a, c, and f) is element-wise constant, we can evaluate the integrals exactly. Suppose that a_e, c_e, and f_e denote the element-wise constant values of $a(x)$, $c(x)$, and $f(x)$, respectively, over a typical element $\Omega^e = (x_a^e, x_b^e)$. Then the element (matrix) equations are obtained by analytically evaluating the integrals appearing in K_{ij}^e and f_i^e of Eq. (4.2.26). The results are summarized here for a typical line element with linear and quadratic approximations.

Linear element (i.e., element with linear approximation)

$$\psi_1^e(\bar{x}) = 1 - \frac{\bar{x}}{h_e}, \quad \psi_2^e(\bar{x}) = \frac{\bar{x}}{h_e}, \quad \frac{d\psi_1^e}{d\bar{x}} = -\frac{1}{h_e}, \quad \frac{d\psi_2^e}{d\bar{x}} = \frac{1}{h_e}, \tag{4.2.29a}$$

$$\begin{aligned}
\int_0^{h_e} \psi_i^e(\bar{x})\psi_j^e(\bar{x})\, d\bar{x} &= \begin{cases} \frac{h_e}{3}, & \text{if } i = j \ (= 1 \text{ or } 2), \\ \frac{h_e}{6}, & \text{if } i \neq j \ (i, j = 1, 2), \end{cases} \\
\int_0^{h_e} \psi_i^e\, d\bar{x} &= \frac{h_e}{2} \ (i = 1, 2), \\
\int_0^{h_e} \frac{d\psi_i^e}{d\bar{x}} \frac{d\psi_j^e}{d\bar{x}}\, d\bar{x} &= (-1)^{i+j} \frac{1}{h_e}, \\
\int_0^{h_e} \psi_i^e \frac{d\psi_j^e}{d\bar{x}}\, d\bar{x} &= \frac{(-1)^j}{2} \ (i, j = 1, 2),
\end{aligned} \tag{4.2.29b}$$

$$\begin{aligned}
&\left(\frac{a_e}{h_e} \begin{bmatrix} 1 & -1 \\ -1 & 1 \end{bmatrix} + \frac{b_e}{2} \begin{bmatrix} -1 & 1 \\ -1 & 1 \end{bmatrix} + \frac{c_e h_e}{6} \begin{bmatrix} 2 & 1 \\ 1 & 2 \end{bmatrix} \right) \begin{Bmatrix} u_1^e \\ u_2^e \end{Bmatrix} \\
&\qquad = \frac{f_e h_e}{2} \begin{Bmatrix} 1 \\ 1 \end{Bmatrix} + \begin{Bmatrix} Q_1^e \\ Q_2^e \end{Bmatrix}.
\end{aligned} \tag{4.2.29c}$$

Quadratic element (i.e., element with quadratic approximation)

$$\begin{aligned}
\psi_1^e(\bar{x}) &= \left(1 - \frac{2\bar{x}}{h_e}\right)\left(1 - \frac{\bar{x}}{h_e}\right), \\
\psi_2^e(\bar{x}) &= \frac{4\bar{x}}{h_e}\left(1 - \frac{\bar{x}}{h_e}\right), \\
\psi_3^e(\bar{x}) &= -\frac{\bar{x}}{h_e}\left(1 - \frac{2\bar{x}}{h_e}\right),
\end{aligned} \tag{4.2.30a}$$

$$\frac{d\psi_1^e}{d\bar{x}} = \frac{1}{h_e}\left(3 - 4\frac{\bar{x}}{h_e}\right), \quad \frac{d\psi_2^e}{d\bar{x}} = \frac{4}{h_e}\left(1 - 2\frac{\bar{x}}{h_e}\right), \quad \frac{d\psi_3^e}{d\bar{x}} = -\frac{1}{h_e}\left(1 - 4\frac{\bar{x}}{h_e}\right), \tag{4.2.30b}$$

$$\int_0^{h_e} \frac{d\psi_1^e}{d\bar{x}}\frac{d\psi_1^e}{d\bar{x}}\,d\bar{x} = \frac{7}{3h_e}, \quad \int_0^{h_e} \frac{d\psi_1^e}{d\bar{x}}\frac{d\psi_2^e}{d\bar{x}}\,d\bar{x} = -\frac{8}{3h_e}, \quad \int_0^{h_e} \frac{d\psi_1^e}{d\bar{x}}\frac{d\psi_3^e}{d\bar{x}}\,d\bar{x} = \frac{1}{3h_e}, \tag{4.2.30c}$$

$$\int_0^{h_e} \frac{d\psi_2^e}{d\bar{x}}\frac{d\psi_2^e}{d\bar{x}}\,d\bar{x} = \frac{16}{3h_e}, \quad \int_0^{h_e} \frac{d\psi_2^e}{d\bar{x}}\frac{d\psi_3^e}{d\bar{x}}\,d\bar{x} = -\frac{8}{3h_e}, \quad \int_0^{h_e} \frac{d\psi_3^e}{d\bar{x}}\frac{d\psi_3^e}{d\bar{x}}\,d\bar{x} = \frac{7}{3h_e},$$

$$\int_0^{h_e} \psi_1^e\frac{d\psi_1^e}{d\bar{x}}\,d\bar{x} = -\frac{3}{6}, \quad \int_0^{h_e} \psi_1^e\frac{d\psi_2^e}{d\bar{x}}\,d\bar{x} = \frac{4}{6}, \quad \int_0^{h_e} \psi_1^e\frac{d\psi_3^e}{d\bar{x}}\,d\bar{x} = -\frac{1}{6},$$

$$\int_0^{h_e} \psi_2^e\frac{d\psi_1^e}{d\bar{x}}\,d\bar{x} = -\frac{4}{6}, \quad \int_0^{h_e} \psi_2^e\frac{d\psi_2^e}{d\bar{x}}\,d\bar{x} = 0, \quad \int_0^{h_e} \psi_2^e\frac{d\psi_3^e}{d\bar{x}}\,d\bar{x} = \frac{4}{6}, \tag{4.2.30d}$$

$$\int_0^{h_e} \psi_3^e\frac{d\psi_1^e}{d\bar{x}}\,d\bar{x} = \frac{1}{6}, \quad \int_0^{h_e} \psi_3^e\frac{d\psi_2^e}{d\bar{x}}\,d\bar{x} = -\frac{4}{6}, \int_0^{h_e} \psi_3^e\frac{d\psi_3^e}{d\bar{x}}\,d\bar{x} = \frac{3}{6},$$

$$\int_0^{h_e} \psi_2^e(\bar{x})\,d\bar{x} = \frac{4h_e}{6}, \quad \int_0^{h_e} \psi_1^e(\bar{x})\,d\bar{x} = \int_0^{h_e} \psi_3^e(\bar{x})\,d\bar{x} = \frac{h_e}{6}, \tag{4.2.30e}$$

$$\left(\frac{a_e}{3h_e}\begin{bmatrix} 7 & -8 & 1 \\ -8 & 16 & -8 \\ 1 & -8 & 7 \end{bmatrix} + \frac{b_e}{6}\begin{bmatrix} -3 & 4 & -1 \\ -4 & 0 & 4 \\ 1 & -4 & 3 \end{bmatrix} + \frac{c_e h_e}{30}\begin{bmatrix} 4 & 2 & -1 \\ 2 & 16 & 2 \\ -1 & 2 & 4 \end{bmatrix}\right)\begin{Bmatrix} u_1^e \\ u_2^e \\ u_3^e \end{Bmatrix}$$
$$= \frac{f_e h_e}{6}\begin{Bmatrix} 1 \\ 4 \\ 1 \end{Bmatrix} + \begin{Bmatrix} Q_1^e \\ Q_2^e \\ Q_3^e \end{Bmatrix}. \tag{4.2.30f}$$

We note that the contribution of a uniform source to the nodes in a quadratic element is nonuniform, that is, $f_i^e \neq f_e h_e/3$; also, the source vector of a quadratic element of length h_e is *not* equivalent to that of two linear elements of length $h_e/2$ each. We also note that there are more unknowns (because of the secondary variables Q_i^e at the end nodes) than the number of equations, independent of whether the element is linear or quadratic. Therefore, the element equations cannot be solved without properly assembling finite element equations associated with all elements in the mesh. When only one element is used, there will be only n unknowns because two known values are provided by the boundary conditions.

The one-dimensional finite element formulations presented in this section can be implemented into a computer program. Interested readers can consult the textbook by Reddy [8] for the details; the book also contains a description of the finite element program **FEM1D** and its use in the solution of a variety of one-dimensional problems. The Fortran as well as MATLAB versions of FEM1D and FEM2D can be downloaded from the website, **http://mechanics.tamu.edu**.

Next, we consider the same example problems solved in Chapter 3 to illustrate the ideas presented in the preceding sections to solve one-dimensional heat transfer problems using the FEM and compare the results obtained with the FVM and FDM.

Example 4.2.1

Consider the boundary value problem described by the following second-order linear differential equation (see **Example 3.3.1**)

$$-\frac{d^2u}{dx^2} = f_0 \cos x, \quad 0 < x < 1 \tag{1}$$

and the following two cases of boundary conditions:

$$\textbf{Case 1}: \quad u(0) = u(1) = 0; \quad \textbf{Case 2}: \quad u(0) = \left.\frac{du}{dx}\right|_{x=1} = 0. \tag{2}$$

Determine the finite element solution of the problem using uniform meshes of 4 and 8 linear elements and 2 and 4 quadratic elements of the domain. Compare the FEM solutions with the exact and HFVM solutions for $f_0 = 10$. The exact solutions for the two cases of boundary conditions are given in Eqs. (3.3.3a) and (3.3.3b).

Solution: Equation (1) is a special case of the model equation in Eq. (4.2.1), with $a = 1$, $b = 0$, $c = 0$, and $f(x) = 10 \cos x$. We illustrate the finite element analysis steps for meshes of four linear elements and two quadratic elements only but present numerical results for four and eight linear elements and two and four quadratic elements.

Mesh of four linear elements. The finite element equations for a typical element of the mesh are given by Eq. (4.2.29c) (except for f_i^e) with $a = 1$, $b = 0$, and $c = 0$ as

$$\frac{1}{h_e}\begin{bmatrix} 1 & -1 \\ -1 & 1 \end{bmatrix}\begin{Bmatrix} u_1^e \\ u_2^e \end{Bmatrix} = \begin{Bmatrix} f_1^e \\ f_2^e \end{Bmatrix} + \begin{Bmatrix} Q_1^e \\ Q_2^e \end{Bmatrix}, \tag{3}$$

where

$$f_i^e = \int_{x_a^e}^{x_b^e} f_0 \cos x \, \psi_i^e(x)\, dx \tag{4}$$

and $\psi_i^e(x)$ are the linear interpolation functions listed in Eq. (4.2.29a). The integral in Eq. (4) is evaluated using the two-point Gauss quadrature, which amounts to treating the integrand as a third-degree polynomial (see Section 1.9). Of course, one may use more Gauss points, but the final solution will not change significantly (something the user may investigate).

The equations of each finite element are ($h = 0.25$ or $1/h = 4$)

$$\begin{bmatrix} 4 & -4 \\ -4 & 4 \end{bmatrix}\begin{Bmatrix} u_1^{(1)} \\ u_2^{(1)} \end{Bmatrix} = \begin{Bmatrix} 1.2435 \\ 1.2305 \end{Bmatrix} + \begin{Bmatrix} Q_1^{(1)} \\ Q_2^{(1)} \end{Bmatrix}, \tag{5a}$$

$$\begin{bmatrix} 4 & -4 \\ -4 & 4 \end{bmatrix}\begin{Bmatrix} u_1^{(2)} \\ u_2^{(2)} \end{Bmatrix} = \begin{Bmatrix} 1.1792 \\ 1.1410 \end{Bmatrix} + \begin{Bmatrix} Q_1^{(2)} \\ Q_2^{(2)} \end{Bmatrix}, \tag{5b}$$

$$\begin{bmatrix} 4 & -4 \\ -4 & 4 \end{bmatrix}\begin{Bmatrix} u_1^{(3)} \\ u_2^{(3)} \end{Bmatrix} = \begin{Bmatrix} 1.0415 \\ 0.9806 \end{Bmatrix} + \begin{Bmatrix} Q_1^{(3)} \\ Q_2^{(3)} \end{Bmatrix}, \tag{5c}$$

$$\begin{bmatrix} 4 & -4 \\ -4 & 4 \end{bmatrix}\begin{Bmatrix} u_1^{(4)} \\ u_2^{(4)} \end{Bmatrix} = \begin{Bmatrix} 0.8391 \\ 0.7592 \end{Bmatrix} + \begin{Bmatrix} Q_1^{(4)} \\ Q_2^{(4)} \end{Bmatrix}, \tag{5d}$$

Assembly of elements equations. In a mesh of four linear elements there are five nodes, and the global (i.e., the whole mesh) system of equations consists of five equations. Noting that (amounts to imposing the continuity of the primary variables at nodes common to the elements)

$$u_2^{(1)} = u_1^{(2)}, \quad u_2^{(2)} = u_1^{(3)}, \quad u_2^{(3)} = u_1^{(4)}, \tag{6}$$

and denoting the value of $u(x)$ at the Ith global node as U_I:

$$u_1^{(1)} = U_1, \quad u_2^{(1)} = u_1^{(2)} = U_2, \quad u_2^{(2)} = u_1^{(3)} = U_3, \quad u_2^{(3)} = u_1^{(4)} = U_4, \quad u_2^{(4)} = U_5. \tag{7}$$

We can express element 1 equations (which contain only U_1 and U_2) as

$$\begin{bmatrix} 4 & -4 & 0 & 0 & 0 \\ -4 & 4 & 0 & 0 & 0 \\ 0 & 0 & 0 & 0 & 0 \\ 0 & 0 & 0 & 0 & 0 \\ 0 & 0 & 0 & 0 & 0 \end{bmatrix} \begin{Bmatrix} U_1 \\ U_2 \\ U_3 \\ U_4 \\ U_5 \end{Bmatrix} = \begin{Bmatrix} 1.2435 \\ 1.2305 \\ 0.0000 \\ 0.0000 \\ 0.0000 \end{Bmatrix} + \begin{Bmatrix} Q_1^{(1)} \\ Q_2^{(1)} \\ 0.0 \\ 0.0 \\ 0.0 \end{Bmatrix}. \tag{8a}$$

Similarly, the equations of elements 2, 3, and 4 can be placed into their proper locations in the global system as

$$\begin{bmatrix} 0 & 0 & 0 & 0 & 0 \\ 0 & 4 & -4 & 0 & 0 \\ 0 & -4 & 4 & 0 & 0 \\ 0 & 0 & 0 & 0 & 0 \\ 0 & 0 & 0 & 0 & 0 \end{bmatrix} \begin{Bmatrix} U_1 \\ U_2 \\ U_3 \\ U_4 \\ U_5 \end{Bmatrix} = \begin{Bmatrix} 0.0000 \\ 1.1792 \\ 1.1410 \\ 0.0000 \\ 0.0000 \end{Bmatrix} + \begin{Bmatrix} 0.0 \\ Q_1^{(2)} \\ Q_2^{(2)} \\ 0.0 \\ 0.0 \end{Bmatrix}, \tag{8b}$$

$$\begin{bmatrix} 0 & 0 & 0 & 0 & 0 \\ 0 & 0 & 0 & 0 & 0 \\ 0 & 0 & 4 & -4 & 0 \\ 0 & 0 & -4 & 4 & 0 \\ 0 & 0 & 0 & 0 & 0 \end{bmatrix} \begin{Bmatrix} U_1 \\ U_2 \\ U_3 \\ U_4 \\ U_5 \end{Bmatrix} = \begin{Bmatrix} 0.0000 \\ 0.0000 \\ 1.0415 \\ 0.9806 \\ 0.0000 \end{Bmatrix} + \begin{Bmatrix} 0.0 \\ 0.0 \\ Q_1^{(3)} \\ Q_2^{(3)} \\ 0.0 \end{Bmatrix}, \tag{8c}$$

$$\begin{bmatrix} 0 & 0 & 0 & 0 & 0 \\ 0 & 0 & 0 & 0 & 0 \\ 0 & 0 & 0 & 0 & 0 \\ 0 & 0 & 0 & 4 & -4 \\ 0 & 0 & 0 & -4 & 4 \end{bmatrix} \begin{Bmatrix} U_1 \\ U_2 \\ U_3 \\ U_4 \\ U_5 \end{Bmatrix} = \begin{Bmatrix} 0.0000 \\ 0.0000 \\ 0.0000 \\ 0.8391 \\ 0.7592 \end{Bmatrix} + \begin{Bmatrix} 0.0 \\ 0.0 \\ 0.0 \\ Q_1^{(4)} \\ Q_2^{(4)} \end{Bmatrix}. \tag{8d}$$

Now superposing (i.e., adding or assembling) the system of equations in Eqs. (8a)–(8d), we obtain the following system of global (assembled) equations:

$$\begin{bmatrix} 4 & -4 & 0 & 0 & 0 \\ -4 & 4+4 & -4 & 0 & 0 \\ 0 & -4 & 4+4 & -4 & 0 \\ 0 & 0 & -4 & 4+4 & -4 \\ 0 & 0 & 0 & -4 & 4 \end{bmatrix} \begin{Bmatrix} U_1 \\ U_2 \\ U_3 \\ U_4 \\ U_5 \end{Bmatrix} = \begin{Bmatrix} 1.2435 \\ 1.2305 + 1.1792 \\ 1.1410 + 1.0415 \\ 0.9806 + 0.8391 \\ 0.7592 \end{Bmatrix} + \begin{Bmatrix} Q_1^{(1)} \\ Q_2^{(1)} + Q_1^{(2)} \\ Q_2^{(2)} + Q_1^{(3)} \\ Q_2^{(3)} + Q_1^{(4)} \\ Q_2^{(4)} \end{Bmatrix}. \tag{9}$$

We note that the coefficients from elements that share the nodes are being added.

Mesh of two quadratic elements. The uniform mesh of two quadratic elements will also have five nodes with five global equations. The finite element equations of a typical element are given by Eq. (4.2.30f) (except for f_i^e):

$$\frac{1}{3h}\begin{bmatrix} 7 & -8 & 1 \\ -8 & 16 & -8 \\ 1 & -8 & 7 \end{bmatrix} \begin{Bmatrix} u_1^e \\ u_2^e \\ u_3^e \end{Bmatrix} = \begin{Bmatrix} f_1^e \\ f_2^e \\ f_3^e \end{Bmatrix} + \begin{Bmatrix} Q_1^e \\ Q_2^e \\ Q_3^e \end{Bmatrix} \tag{10}$$

and f_i^e are determined using Eq. (4) with ψ_i^e being the quadratic interpolation functions listed in Eq. (4.2.30a). The integral in Eq. (10) is evaluated using the three-point Gauss quadrature, which amounts to treating the integrand as a fifth-degree polynomial. The element equations for the mesh of two quadratic finite elements are ($h = 0.5$)

$$\frac{2}{3}\begin{bmatrix} 7 & -8 & 1 \\ -8 & 16 & -8 \\ 1 & -8 & 7 \end{bmatrix}\begin{Bmatrix} u_1^{(1)} \\ u_2^{(1)} \\ u_3^{(1)} \end{Bmatrix} = \begin{Bmatrix} 0.8436 \\ 3.2096 \\ 0.7411 \end{Bmatrix} + \begin{Bmatrix} Q_1^{(1)} \\ Q_2^{(1)} \\ Q_3^{(1)} \end{Bmatrix} \tag{11a}$$

$$\frac{2}{3}\begin{bmatrix} 7 & -8 & 1 \\ -8 & 16 & -8 \\ 1 & -8 & 7 \end{bmatrix}\begin{Bmatrix} u_1^{(2)} \\ u_2^{(2)} \\ u_3^{(2)} \end{Bmatrix} = \begin{Bmatrix} 0.7395 \\ 2.4238 \\ 0.4572 \end{Bmatrix} + \begin{Bmatrix} Q_1^{(2)} \\ Q_2^{(2)} \\ Q_3^{(2)} \end{Bmatrix} \tag{11b}$$

Assembled equations. The assembled system of equations is given by [$u_1^{(1)} = U_1$, $u_2^{(1)} = U_2$, $u_3^{(1)} = u_1^{(2)} = U_3$, $u_2^{(2)} = U_4$, and $u_3^{(2)} = U_5$]

$$\frac{2}{3}\begin{bmatrix} 7 & -8 & 1 & 0 & 0 \\ -8 & 16 & -8 & 0 & 0 \\ 1 & -8 & \mathbf{14} & -8 & -1 \\ 0 & 0 & -8 & 16 & -8 \\ 0 & 0 & 1 & -8 & 7 \end{bmatrix}\begin{Bmatrix} U_1 \\ U_2 \\ U_3 \\ U_4 \\ U_5 \end{Bmatrix} = \begin{Bmatrix} 0.8436 \\ 3.2096 \\ \mathbf{1.4806} \\ 2.4238 \\ 0.4572 \end{Bmatrix} + \begin{Bmatrix} Q_1^{(1)} \\ Q_2^{(1)} \\ Q_3^{(1)} + Q_1^{(2)} \\ Q_2^{(2)} \\ Q_3^{(2)} \end{Bmatrix} \tag{12}$$

Imposition of boundary conditions and the determination of the nodal unknowns. At this point, it is important to note that imposition of boundary conditions on the primary or secondary variables did not enter the assembly process. In fact, Eq. (9) is valid for any set of boundary conditions for the four-element mesh of the problem at hand.

All physical processes allow the specification of only one element of the duality pair at a point of the domain, and the other element is not known (and determined in the post-computation). In numerical methods in which the discretized equations are among the duality pairs at all mesh points (i.e., nodes), one is required to know which element of the duality pair is specified at each node.

In the present problem, the secondary variable at all interior points is specified to be zero in both Case 1 and Case 2 of boundary conditions. The "balance law" requires that the secondary variables at a node shared by different elements must add up to the specified secondary variable at that node. If nothing is specified externally at a node, the sum at that node is zero. Thus, in the present case, we have no point sources external to the domain, requiring

$$Q_2^{(1)} + Q_1^{(2)} = 0, \quad Q_2^{(2)} + Q_1^{(3)} = 0, \quad Q_2^{(3)} + Q_1^{(4)} = 0. \tag{13}$$

Next we consider the two cases of boundary conditions separately.

Case 1 [$u(0) = u(1) = 0$]. The boundary conditions in this case translate to

$$U_1 = 0, \quad U_5 = 0. \tag{14}$$

Hence, $Q_1^{(1)}$ and $Q_2^{(4)}$ are unknown and will be determined in the post-computation (one way to compute $Q_1^{(1)}$ and $Q_2^{(4)}$ is to use the first and last equations respectively, of the assembled system).

For the case of mesh with four linear elements, use of Eqs. (13) and (14) in Eq. (9) will yield five equations in five unknowns (in matrix form): $Q_1^{(1)}$, U_2, U_3, U_4, and $Q_2^{(4)}$. The first row (i.e., the first equation) and last row (i.e., the last equation) of the matrix equations in Eq. (9) can be solved for $Q_1^{(1)}$ and $Q_2^{(4)}$ once U_2, U_3, and U_4 are known (so that all nodal values of u are known). Therefore, we use the fact that $U_1 = 0$ and $U_5 = 0$ to obtain [by selecting rows 2, 3, and 4 of Eq. (9) and discarding rows 1 and 5] the following three coupled equations among U_2, U_3, and U_4:

$$\begin{bmatrix} 8 & 4 & 0 \\ -4 & 8 & -4 \\ 0 & -4 & 8 \end{bmatrix} \begin{Bmatrix} U_2 \\ U_3 \\ U_4 \end{Bmatrix} = \begin{Bmatrix} 2.4097 \\ 2.1826 \\ 1.8197 \end{Bmatrix} + \begin{Bmatrix} 4U_1 \\ 0 \\ 4U_5 \end{Bmatrix} = \begin{Bmatrix} 2.4097 \\ 2.1826 \\ 1.8197 \end{Bmatrix}. \tag{15}$$

This system is called *condensed* equations for the primary nodal variables. The solution of Eq. (15) is ($U_1 = U_5 = 0$): $U_2 = 0.8384$, $U_3 = 1.0743$, and $U_4 = 0.7646$.

Using the first and last equations of the assembled system in Eq. (9), we can determine the secondary variables at nodes 1 and 5 as (the actual, not truncated to four decimal points, values of U_I are used here)

$$Q_1^{(1)} = 4U_1 - 4U_2 - 1.2435 = -4.5970, \quad Q_2^{(4)} = -4U_4 + 4U_5 - 0.7592 = -3.8177. \tag{16}$$

We can also determine (which is most common practice with commercial finite element programs) using the definition of $Q_1^{(1)}$ and $Q_2^{(4)}$ [see Eq. (4.2.9)] (noting that $q = du/dx$ and $h = L/4 = 0.25$):

$$\begin{aligned} Q_1^{(1)}\Big|_{\text{def}} &= -\frac{du_h}{dx}\Big|_{x=0} = \frac{1}{h}(U_1 - U_2) = 4U_1 - 4U_2 = -3.3535 = -q(0), \\ Q_2^{(4)}\Big|_{\text{def}} &= \frac{du_h}{dx}\Big|_{x=1} = \frac{1}{h}(U_5 - U_4) = 4U_5 - 4U_4 = -3.0585 = q(1). \end{aligned} \tag{17}$$

We note that the values of the secondary variables $Q_1^{(1)}$ and $Q_2^{(4)}$ computed using the definition are in error by $f_1^{(1)}$ and $f_2^{(4)}$, respectively. Of course, the values of $f_1^{(1)}$ and $f_2^{(4)}$ reduce in magnitude as the mesh is refined. In general, the secondary variables calculated from the finite element equations are the most accurate due to the fact that the FEM equations are derived from the equilibrium or governing differential equation.

For the case of a mesh with two quadratic elements, the condensed equations for the unknown primary nodal values are

$$\frac{2}{3}\begin{bmatrix} 16 & -8 & 0 \\ -8 & 14 & -8 \\ 0 & -8 & 16 \end{bmatrix} \begin{Bmatrix} U_2 \\ U_3 \\ U_4 \end{Bmatrix} = \begin{Bmatrix} 3.2096 \\ 1.4806 \\ 2.4238 \end{Bmatrix}. \tag{18}$$

The solution of Eq. (18) is ($U_1 = U_5 = 0$): $U_2 = 0.8381$, $U_3 = 1.0743$, and $U_4 = 0.7644$.

Using the first and last equations of the assembled system in Eq. (18), we obtain the secondary variables at nodes 1 and 5 as (the actual, not truncated to four decimal points, values of U_I are used here)

$$\begin{aligned} Q_1^{(1)} &= \tfrac{2}{3}(7U_1 - 8U_2 + U_3) - 0.8436 = -4.5970, \\ Q_2^{(4)} &= \tfrac{2}{3}(U_3 - 8U_4 + U_5) - 0.4572 = -3.8177. \end{aligned} \tag{19}$$

Using the definition of $Q_1^{(1)}$ and $Q_3^{(2)}$ and noting that (with $h = L/2 = 0.5$)

$$\frac{d\psi_1}{d\bar{x}} = -\frac{3}{h} + \frac{4}{h^2}\bar{x}, \quad \frac{d\psi_2}{d\bar{x}} = \frac{4}{h} - \frac{8}{h^2}\bar{x}, \quad \frac{d\psi_3}{d\bar{x}} = -\frac{1}{h} + \frac{4}{h^2}\bar{x}, \tag{20}$$

we obtain:

$$\begin{aligned} Q_1^{(1)}\Big|_{\text{def}} &= -\frac{du_h}{d\bar{x}}\Big|_{\bar{x}=0} = -(-6U_1 + 8U_2 - 2U_3) = -4.5558 = -q(0), \\ Q_3^{(2)}\Big|_{\text{def}} &= \frac{du_h}{d\bar{x}}\Big|_{\bar{x}=1} = (2U_3 - 8U_4 + 6U_5) = -3.9665 = q(1). \end{aligned} \tag{21}$$

The numerical values obtained with various finite element meshes (4L = four linear elements, 2Q = two quadratic elements, and so on) are compared in Table 4.2.1 with the exact and half-control HFVM solutions (N denotes the number of subdivisions), where the underlined numbers are interpolated values between the nodes. Note that the FEM always gives exact solutions at the nodes for model equations in which the coefficient a is a constant and coefficients b and c are zero [for any $f(x)$, as long as the integral in Eq. (4) is evaluated exactly]. The table also contains heat flux at the left end ($x = 0$) and the right end ($x = 1$) of the domain (the last two rows), computed using the element equations in the FEM and the half-control volume formulation of the FVM (see HFVM of Table 3.3.1). We recall from Chapter 3 discussions that the half-control volume formulation is more accurate than the zero-thickness control volume formulation. It is clear that the FEM solutions are more accurate than the HFVM solutions for this problem.

Table 4.2.1 Comparison of the FEM and HFVM solutions with the exact solution (u) of: $-\frac{d^2u}{dx^2} = 10\cos x;\quad 0 < x < 1;\quad u(0) = u(1) = 0.$

		FEM Solutions*				HFVM Solutions*	
x	Exact	4L	2Q	8L	4Q	$N=4$	$N=8$
0.0625	0.2678	0.2096	0.2659	0.2483	0.2676	—	0.2481
0.1250	0.4966	0.4192	0.4943	0.4966	0.4966	0.4181	0.4963
0.1875	0.6867	0.6288	0.6850	0.6675	0.6868	—	0.6670
0.2500	0.8384	0.8384	0.8381	0.8384	0.8384	0.8362	0.8378
0.3125	0.9522	0.8974	0.9535	0.9337	0.9518	—	0.9330
0.3750	1.0289	0.9563	1.0314	1.0289	1.0289	0.9538	1.0283
0.4375	1.0693	1.0153	1.0717	1.0516	1.0698	—	1.0509
0.5000	1.0743	1.0743	1.0743	1.0743	1.0743	1.0715	1.0736
0.5625	1.0450	0.9969	1.0394	1.0285	1.0443	—	1.0279
0.6250	0.9827	0.9195	0.9762	0.9827	0.9827	0.9171	0.9821
0.6875	0.8888	0.8421	0.8845	0.8737	0.8895	—	0.8731
0.7500	0.7646	0.7646	0.7644	0.7646	0.7646	0.7626	0.7641
0.8125	0.6119	0.5735	0.6159	0.5985	0.6110	—	0.5981
0.8750	0.4323	0.3823	0.4390	0.4323	0.4323	0.3813	0.4320
0.9375	0.2277	0.1912	0.2337	0.2162	0.2287	—	0.2160
$-q(0)$	4.5970	4.5970	4.5970	4.5970	4.5970	4.5898	4.5946
$q(1)$	3.8177	3.8177	3.8177	3.8177	3.8177	3.7888	3.8101

*The underlined numbers are interpolated values.

Case 2 $[u(0) = 0,\ \ du/dx = 0 \text{ at } x = 1]$. For Case 2, the only change from Case 1 is the boundary condition at the right end. The assembled equations in Eq. (9) for the mesh of four linear elements and in Eq. (12) for the mesh of two quadratic elements are still valid here. The Case 2 boundary conditions are $U_1 = 0$ and $Q_2^{(4)} = 0$ or $Q_3^{(2)} = 0$. Thus, the condensed equations for the mesh of four linear elements is

$$\text{Four linear elements:}\quad \begin{bmatrix} 8 & -4 & 0 & 0 \\ -4 & 8 & -4 & 0 \\ 0 & -4 & 8 & -4 \\ 0 & 0 & -4 & 4 \end{bmatrix} \begin{Bmatrix} U_2 \\ U_3 \\ U_4 \\ U_5 \end{Bmatrix} = \begin{Bmatrix} 2.4097 \\ 2.1826 \\ 1.8197 \\ 0.7592 \end{Bmatrix}. \tag{22}$$

The solution of Eq. (22) is: $U_2 = 1.7928$, $U_3 = 2.9832$, $U_4 = 3.6279$, and $U_5 = 0.3.8177$, which coincides with the exact solution.

The condensed equations for the mesh of two quadratic elements is

$$\text{Two quadratic elements:}\quad \begin{matrix} \\ 2 \\ 3 \\ \\ \end{matrix}\begin{bmatrix} 16 & -8 & 0 & 0 \\ -8 & 14 & -8 & -1 \\ 0 & -8 & 16 & -8 \\ 0 & 1 & -8 & 7 \end{bmatrix} \begin{Bmatrix} U_2 \\ U_3 \\ U_4 \\ U_5 \end{Bmatrix} = \begin{Bmatrix} 3.2096 \\ 1.4806 \\ 2.4238 \\ 0.4572 \end{Bmatrix}. \tag{23}$$

The solution of Eq. (23) is $U_2 = 1.7925$, $U_3 = 2.9832$, $U_4 = 3.6277$, and $U_5 = 3.8177$. The FEM solution using quadratic elements is in slight error at $x = 0$. In both cases, the secondary variable at the left end, computed from element equations, is $Q_1^{(1)} = -8.4147$.

Table 4.2.2 contains a comparison of the FEM and the half-control volume FVM solutions against the exact solution for Case 2 boundary conditions. Once again, the finite element solutions are exact at the nodes, and they are more accurate than the FVM solutions.

Table 4.2.2 Comparison of the FEM and FVM solutions with the exact solution (u) of: $-\frac{d^2u}{dx^2} = 10\cos x;\quad 0 < x < 1;\quad u(0) = 0,\quad \left.\frac{du}{dx}\right|_{x=1} = 0.$

		FEM Solutions*				FVM Solutions*	
x	Exact	4L	2Q	8L	4Q	$N = 4$	$N = 8$
0.0625	0.5064	$\underline{0.4482}$	$\underline{0.5045}$	$\underline{0.4869}$	$\underline{0.5063}$	——	$\underline{0.4863}$
0.1250	0.9738	$\underline{0.8964}$	$\underline{0.9715}$	0.9738	0.9738	$\underline{0.8917}$	0.9725
0.1875	1.4025	$\underline{1.3446}$	$\underline{1.4008}$	$\underline{1.3833}$	$\underline{1.4026}$	——	$\underline{1.3814}$
0.2500	1.7928	1.7928	1.7925	1.7928	1.7928	1.7834	1.7904
0.3125	2.1453	$\underline{2.0904}$	$\underline{2.1466}$	$\underline{2.1267}$	$\underline{2.1448}$	——	$\underline{2.1237}$
0.3750	2.4606	$\underline{2.3880}$	$\underline{2.4631}$	2.4606	2.4606	$\underline{2.3746}$	2.4571
0.4375	2.7396	$\underline{2.6856}$	$\underline{2.7419}$	$\underline{2.7219}$	$\underline{2.7400}$	——	$\underline{1.0509}$
0.5000	2.9832	2.9832	2.9832	2.9832	2.9832	2.9659	2.9787
0.5625	3.1925	$\underline{3.1444}$	$\underline{3.1869}$	$\underline{3.1760}$	$\underline{3.1918}$	——	$\underline{3.1710}$
0.6250	3.3688	$\underline{3.3055}$	$\underline{3.3622}$	3.3688	3.3688	$\underline{3.2851}$	3.3634
0.6875	3.5135	$\underline{3.4667}$	$\underline{3.5092}$	$\underline{3.4984}$	$\underline{3.5142}$	——	$\underline{3.4926}$
0.7500	3.6279	3.6279	3.6277	3.6279	3.6279	3.6042	3.6217
0.8125	3.7138	$\underline{3.6754}$	$\underline{3.7178}$	$\underline{3.7004}$	$\underline{3.7129}$	——	$\underline{3.6938}$
0.8750	3.7728	$\underline{3.7228}$	$\underline{3.7795}$	3.7728	3.7728	$\underline{3.6965}$	3.7659
0.9375	3.8068	$\underline{3.7703}$	$\underline{3.8128}$	$\underline{3.7953}$	$\underline{3.8078}$	——	$\underline{3.7880}$
1.0000	3.8177	3.8177	3.8177	3.8177	3.8177	3.7888	3.8101
$-q(0)$	8.4147	8.4147	8.4147	8.4147	8.4147	8.3786	8.4047

*The underlined numbers are interpolated values.

Example 4.2.2

A uniform steel rod of diameter $D = 0.02$ m, length $L = 0.05$ m, and constant thermal conductivity $k = 50$ W/(m· °C) is exposed to ambient air at $T_\infty = 20$°C with a heat transfer coefficient $\beta = 100$ W/(m^2 · °C). The left end of the rod is maintained at temperature $T(0) = T_0 = 320$°C and the other end, $x = L$, is either (1) insulated (i.e., no heat flow across the end) or (2) exposed to the ambient air at $T_\infty = 20$°C with heat transfer coefficient β, as shown in Fig. 4.2.4 (see **Example 3.3.2**). Assuming that there is no internal heat generation (i.e., $f = 0$), determine the temperature distribution and the heat input at the left end of the rod using uniform mesh of five linear elements. Compare the FEM solutions with the FDM, FVM, and exact solutions of the problem with the same number of mesh points.

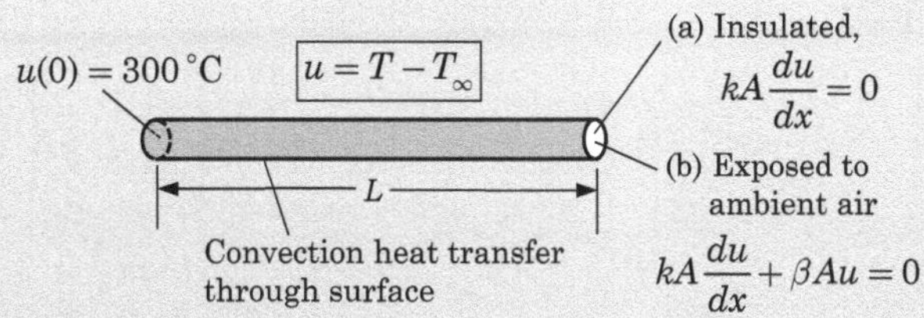

Fig. 4.2.4 Geometry and boundary conditions for heat transfer in an uninsulated rod.

Solution The governing differential equation of the problem is

$$-\frac{d}{dx}\left(kA\frac{dT}{dx}\right) + \beta P(T - T_\infty) = 0, \tag{1}$$

where P denotes the perimeter ($P = \pi D$) and A is the cross-sectional area of the rod ($A = \pi D^2/4$). The boundary conditions for the case of convection at $x = L$ are given by

$$T(0) = T_0, \quad \left[kA\frac{dT}{dx} + \beta A(T - T_\infty)\right]_{x=L} = 0. \tag{2}$$

The boundary conditions for the case in which the right end is insulated can be obtained by setting $\beta = 0$ in Eq. (2).

As in **Example 3.3.2**, we introduce the new variable $u = T - T_\infty$ and divide Eq. (1) throughout by kA (because kA is a constant) to obtain

$$-\frac{d^2u}{dx^2} + m^2u = 0 \quad \text{for} \quad 0 < x < L, \tag{3}$$

where m is given by ($kA = \pi 50 \times 10^{-4}$ and $\beta P = 2\pi$) $m = \sqrt{\beta P/Ak} = 20$.

The boundary conditions on u become

$$\text{Case (1)}: \quad u(0) = u_0 = T(0) - T_\infty = 300°\text{C}, \quad \left[\frac{du}{dx}\right]_{x=L} = 0. \tag{4}$$

$$\text{Case (2)}: \quad u(0) = u_0 = T(0) - T_\infty = 300°\text{C}, \quad \left[\frac{du}{dx} + \frac{\beta}{k}u\right]_{x=L} = 0. \tag{5}$$

Equation (3) is a special case of Eq. (4.2.1), with $a = 1$, $b = 0$, $c = m^2$, and $f = 0$.

Recall from **Example 3.3.2** [see Eq. (3.3.9)] that the exact solution u and heat q for Case (2) boundary conditions is given by

$$u(x) = u_0\left[\frac{\cosh m(L - x) + (\beta/mk)\sinh m(L - x)}{\cosh mL + (\beta/mk)\sinh mL}\right], \tag{6a}$$

$$q(x) \equiv -kA\frac{du}{dx} = kAm\,u_0\left[\frac{\sinh m(L - x) + (\beta/mk)\cosh m(L - x)}{\cosh mL + (\beta/mk)\sinh mL}\right]. \tag{6b}$$

The exact solution for the Case (1) boundary conditions is obtained from Eq. (6a) by setting $\beta = 0$.

Finite Element Solutions

The element equations for a typical linear element are [see Eq. 4.2.29c with $a_e = 1$, $b_e = 0$, $c_e = m^2$, and $f_e = 0$]

$$\left(\frac{1}{h_e}\begin{bmatrix} 1 & -1 \\ -1 & 1 \end{bmatrix} + \frac{m^2h_e}{6}\begin{bmatrix} 2 & 1 \\ 1 & 2 \end{bmatrix}\right)\begin{Bmatrix} u_1^e \\ u_2^e \end{Bmatrix} = \begin{Bmatrix} Q_1^e \\ Q_2^e \end{Bmatrix}, \tag{7}$$

and the element equations for a typical quadratic element are

$$\left(\frac{1}{3h_e}\begin{bmatrix} 7 & -8 & 1 \\ -8 & 16 & -8 \\ 1 & -8 & 7 \end{bmatrix} + \frac{m^2 h_e}{30}\begin{bmatrix} 4 & 2 & -1 \\ 2 & 16 & 2 \\ 1 & 2 & 4 \end{bmatrix}\right)\begin{Bmatrix} u_1^e \\ u_2^e \\ u_3^e \end{Bmatrix} = \begin{Bmatrix} Q_1^e \\ Q_2^e \\ Q_3^e \end{Bmatrix}, \tag{8}$$

where $Q_1^e = -du/dx$ and $Q_n^e = du/dx$ ($n = 2$ for the linear element and $n = 3$ for the quadratic element).

(a) For the uniform mesh (i.e., $h_1 = h_2 = h_3 = h_4 = h_5 = h = 0.01$ m) of five linear elements [see Fig. 4.2.5(a)], the assembled system of equations is ($1/h + m^2h/3 = 304/3$ and $-1/h + m^2h/6 = -298/3$; $u_1^{(1)} = U_1$, $u_2^{(1)} = u_1^{(2)} = U_2$, and so on)

$$\frac{2}{3}\begin{bmatrix} 152 & -149 & 0 & 0 & 0 & 0 \\ -149 & 304 & -149 & 0 & 0 & 0 \\ 0 & -149 & 304 & -149 & 0 & 0 \\ 0 & 0 & -149 & 304 & -149 & 0 \\ 0 & 0 & 0 & -149 & 304 & -149 \\ 0 & 0 & 0 & 0 & -149 & 152 \end{bmatrix}\begin{Bmatrix} U_1 \\ U_2 \\ U_3 \\ U_4 \\ U_5 \\ U_6 \end{Bmatrix} = \begin{Bmatrix} Q_1^{(1)} \\ Q_2^{(1)} + Q_1^{(2)} \\ Q_2^{(2)} + Q_1^{(3)} \\ Q_2^{(3)} + Q_1^{(4)} \\ Q_2^{(4)} + Q_1^{(5)} \\ Q_2^{(5)} \end{Bmatrix}, \tag{9}$$

where U_I denotes the temperature $u(x) = T - T_\infty$ at the Ith global node.

The boundary conditions for Case (1) are

$$U_1 = 300\,^\circ\text{C}, \quad Q_2^{(5)} = 0 \tag{10}$$

and the balance of heats at global nodes 2, 3, 4, and 5 require (because there is no external heat added at the node)

$$Q_2^{(1)} + Q_1^{(2)} = 0, \quad Q_2^{(2)} + Q_1^{(3)} = 0, \quad Q_2^{(3)} + Q_1^{(4)} = 0, \quad Q_2^{(4)} + Q_1^{(5)} = 0. \tag{11}$$

After imposing the conditions in Eqs. (10) and (11), the condensed equations are obtained by omitting the first equation (which can be used to post-compute the heat $Q_1^{(1)}$ at node 1)

$$\frac{2}{3}\begin{bmatrix} 304 & -149 & 0 & 0 & 0 \\ -149 & 304 & -149 & 0 & 0 \\ 0 & -149 & 304 & -149 & 0 \\ 0 & 0 & -149 & 304 & -149 \\ 0 & 0 & 0 & -149 & 152 \end{bmatrix}\begin{Bmatrix} U_2 \\ U_3 \\ U_4 \\ U_5 \\ U_6 \end{Bmatrix} = \frac{2}{3}\begin{Bmatrix} 149\,U_1 \\ 0 \\ 0 \\ 0 \\ 0 \end{Bmatrix}, \tag{12}$$

whose solution is $U_2 = 259.92$, $U_3 = 230.30$, $U_4 = 209.96$, $U_5 = 198.08$, and $U_6 = 194.17$ (the actual temperatures are obtained by adding 20°C to each of the nodal values).

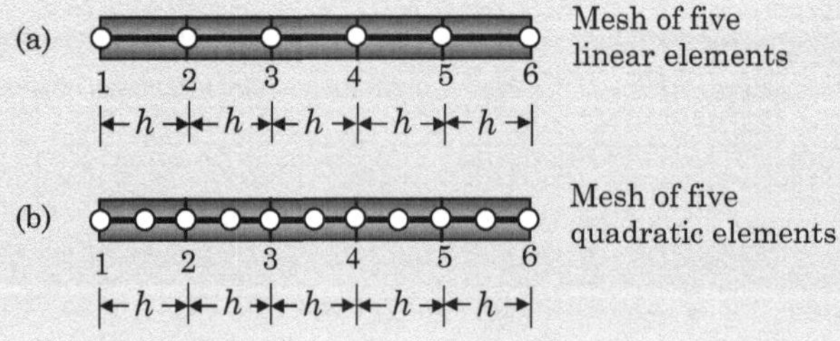

Fig. 4.2.5 (a) Five-element mesh of linear finite elements. (b) Five-element mesh of quadratic elements (an interface between elements is shown with a vertical line).

The heat at node 1 from the element equation is ($kA = 157.0796 \times 10^{-4}$)

$$q(0) = kA\,Q_1^{(1)} = 157.0796 \times 10^{-4}(-4581.4) = 71.96 \text{ W}. \tag{13}$$

Using the definition, we obtain

$$(q(0))_{\text{def}} = kA(Q_1^{(1)})_{\text{def}} = kA\left(\frac{U_2 - U_1}{h}\right) = kA\left(\frac{300 - 259.92}{0.01}\right) = 62.96 \text{ W}. \tag{14}$$

The boundary conditions for Case (2) are

$$U_1 = 300°\text{C}, \quad Q_2^{(5)} = -\frac{\beta}{k}U_6 = -2U_6. \tag{15}$$

Hence, the condensed equations are the same as those in Eq. (9), except that one needs to take $-2U_6$ from the right side to the left side of the matrix equation. Then the diagonal term in location (6,6), $2(152)/3$ is replaced with $2(152)/3 + 2 = (2/3)(155)$:

$$\frac{2}{3}\begin{bmatrix} 304 & -149 & 0 & 0 & 0 \\ -149 & 304 & -149 & 0 & 0 \\ 0 & -149 & 304 & -149 & 0 \\ 0 & 0 & -149 & 304 & -149 \\ 0 & 0 & 0 & -149 & 155 \end{bmatrix} \begin{Bmatrix} U_2 \\ U_3 \\ U_4 \\ U_5 \\ U_6 \end{Bmatrix} = \frac{2}{3}\begin{Bmatrix} 149\,U_1 \\ 0 \\ 0 \\ 0 \\ 0 \end{Bmatrix}. \tag{16}$$

The solution is $U_2 = 257.57$, $U_3 = 225.51$, $U_4 = 202.53$, $U_5 = 187.70$, and $U_6 = 180.44$.

The heat at node 1 from the element equation is ($kA = 157.0796 \times 10^{-4}$)

$$q(0) = kA\,Q_1^{(1)} = 157.0796 \times 10^{-4}(-4815.0) = 75.63 \text{ W}, \tag{17}$$

whereas from the definition we have

$$(q(0))_{\text{def}} = kA(Q_1^{(1)})_{\text{def}} = kA\left(\frac{U_2 - U_1}{h}\right) = kA\left(\frac{300 - 257.57}{0.01}\right) = 66.65 \text{ W}. \tag{18}$$

Finite Difference Solutions

As already discussed in **Example 3.3.1**, the central finite difference formula applied to Eq. (3) yields Eq. (2.3.9) or Eq. (3.3.9):

$$-U_{i-1} + (2 + ch^2)U_i - U_{i+1} = 0, \quad c = m^2 \tag{19}$$

which is valid for any mesh point where u is not specified. By applying the formula in Eq. (19) to nodes $2, 3, \ldots, N$, we obtain the $(N-1)$ relations among the values of u at the mesh points (see Fig. 4.2.6).

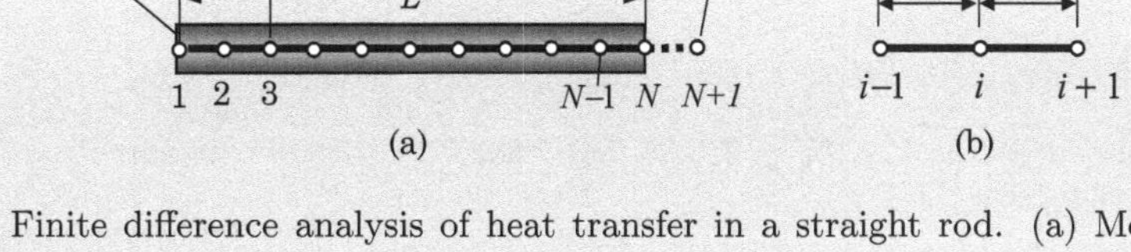

Fig. 4.2.6 Finite difference analysis of heat transfer in a straight rod. (a) Mesh of (base) points and (b) a typical mesh point i with neighboring mesh points $i-1$ and $i+1$.

For a mesh of five subdivisions (i.e., $N = 5$), application of the finite difference formula in Eq. (19) to nodes 2, 3, 4, 5, and 6 yields the following five equations:

$$\begin{aligned} -U_1 + \left(2 + ch^2\right) U_2 - U_3 &= 0, \\ -U_2 + \left(2 + ch^2\right) U_3 - U_4 &= 0, \\ -U_3 + \left(2 + ch^2\right) U_4 - U_5 &= 0, \\ -U_4 + \left(2 + ch^2\right) U_5 - U_6 &= 0, \\ -U_5 + \left(2 + ch^2\right) U_6 - U_7 &= 0, \end{aligned} \tag{20}$$

where $c = 400$ and $U_1 = 300°\text{C}$. Note that U_7 is the value of $u(x)$ at the fictitious node 7.

For Case (1), the boundary condition at mesh point $N + 1$ (i.e., $x = L$) is $(du/dx) = 0$, which can be *approximated* using the central difference formula [error is of the order $O(h^2)$]:

$$\left[\frac{du}{dx}\right]_{x=x_{N+1}} \approx \frac{U_{N+2} - U_N}{2h} = 0 \rightarrow U_{N+2} = U_N. \tag{21}$$

In particular, for a mesh of five subdivisions ($N = 5$), we have

$$\left.\frac{du}{dx}\right|_{x=L} = \frac{U_7 - U_5}{2h} = 0 \rightarrow U_7 = U_5. \tag{22}$$

Then Eq. (20) gives five equations in five unknowns:

$$\begin{bmatrix} 2.04 & -1.00 & 0.00 & 0.00 & 0.00 \\ -1.00 & 2.04 & -1.00 & 0.00 & 0.00 \\ 0.00 & -1.00 & 2.04 & -1.00 & 0.00 \\ 0.00 & 0.00 & -1.00 & 2.04 & -1.00 \\ 0.00 & 0.00 & 0.00 & -2.00 & 2.04 \end{bmatrix} \begin{Bmatrix} U_2 \\ U_3 \\ U_4 \\ U_5 \\ U_6 \end{Bmatrix} = \begin{Bmatrix} U_1 \\ 0 \\ 0 \\ 0 \\ 0 \end{Bmatrix}, \tag{23}$$

whose solution is $U_2 = 260.12$, $U_3 = 230.64$, $U_4 = 210.39$, $U_5 = 198.56$, and $U_6 = 194.66$. The heat at mesh point 1 is (this is the only way to compute the heat/flux in the FDM)

$$q(0) = -kA\left.\frac{du}{dx}\right|_{x=0} = 157.0796 \times 10^{-4}\,\frac{U_1 - U_2}{h} = 62.64\text{ W}. \tag{24}$$

For Case (2), the boundary condition at $x = L$ is $(du/dx) + (\beta/k)u = 0$. Using, again, the the central difference formula, we obtain

$$\frac{U_{N+2} - U_N}{2h} + \frac{\beta}{k}U_{N+1} = 0 \rightarrow \quad U_{N+2} = U_N - 2h\frac{\beta}{k}U_{N+1}. \tag{25}$$

For the mesh of five subdivisions, use of the result from Eq. (23) (i.e., $U_7 = U_5 - 4hU_6 = U_5 - 0.04U_6$) in Eq. (20) yields five equations in five unknowns:

$$\begin{bmatrix} 2.04 & -1.00 & 0.00 & 0.00 & 0.00 \\ -1.00 & 2.04 & -1.00 & 0.00 & 0.00 \\ 0.00 & -1.00 & 2.04 & -1.00 & 0.00 \\ 0.00 & 0.00 & -1.00 & 2.04 & -1.00 \\ 0.00 & 0.00 & 0.00 & -2.00 & 2.08 \end{bmatrix} \begin{Bmatrix} U_2 \\ U_3 \\ U_4 \\ U_5 \\ U_6 \end{Bmatrix} = \begin{Bmatrix} U_1 \\ 0 \\ 0 \\ 0 \\ 0 \end{Bmatrix}, \tag{26}$$

whose solution is $U_2 = 257.77$, $U_3 = 225.85$, $U_4 = 202.97$, $U_5 = 188.20$, and $U_6 = 180.96$. The heat at mesh point 1 is

$$q(0) = -kA\left.\frac{du}{dx}\right|_{x=0} = 157.0796 \times 10^{-4}\,\frac{U_1 - U_2}{h} = 66.33\text{ W}. \tag{27}$$

Example 3.3.2 has a detailed discussion of the HFVM (FVM32) of the problem, and the discussion is not repeated here.

A comparison of the nodal values of $u(x)$ obtained using the FDM, HFVM, and FEM with the exact values for Case (1) and Case (2) is presented in Tables 4.2.3 and 4.2.4, respectively (see Tables 3.3.2 and 3.3.3; note that Table 3.3.3 results are obtained using FVM22 and not FVM32). The numerical solutions are in good agreement, especially for refined meshes, with the exact solutions. We note that the FEM adjusts the nodal values to minimize the error in the approximation over the domain in the weighted-integral sense, whereas the HFVM is based on integral statement of the governing differential equation; the FDM has no such minimization of error, but is based on truncated Taylor series approximation. The FEM solution for u at the nodes matches with the exact solution when a uniform mesh of five quadratic elements is used.

Table 4.2.3 Comparison of the FDM, FVM32, and FEM solutions with the exact solution of:
$$-\frac{d^2u}{dx^2} + 400\,u = 0,\ 0 < x < 0.05;\ \ u(0) = 300,\ \ \left[\frac{du}{dx}\right]\Big|_{x=L} = 0.$$

x	Exact Solution	FDM Solution		FVM32 Solution		FEM Solution		
		$N = 5$	$N = 10$	$N = 5$	$N = 10$	5L	10L	5Q
0.005	278.62	——	278.63	279.98	278.61	279.96	278.60	278.62
0.010	260.02	260.12	260.04	259.97	260.01	259.92	259.99	260.02
0.015	244.03	——	244.06	245.18	244.01	245.11	243.99	244.03
0.020	230.47	230.64	230.52	230.39	230.45	230.30	230.43	230.47
0.025	219.23	——	219.28	220.23	219.20	220.13	219.18	219.23
0.030	210.18	210.39	210.23	210.07	210.15	209.96	210.12	210.18
0.035	203.23	——	203.29	204.13	203.20	204.02	203.17	203.23
0.040	198.32	198.56	198.38	198.20	198.29	198.08	198.26	198.32
0.045	195.39	——	195.45	196.24	195.36	196.12	195.33	195.39
0.050	194.42	194.66	194.48	194.29	194.39	194.17	194.35	194.42

Table 4.2.4 Comparison of the FDM, FVM32, and FEM solutions with the exact solution of:
$$-\frac{d^2u}{dx^2} + 400\,u = 0,\ 0 < x < L;\ \ u(0) = 300,\ \ \left[\frac{du}{dx} + (\beta/k)u\right]\Big|_{x=L} = 0.$$

x	Exact Solution	FDM Solution		FVM32 Solution		FEM Solution		
		$N = 5$	$N = 10$	$N = 5$	$N = 10$	5L	10L	5Q
0.005	277.44	——	277.46	278.81	277.44	278.78	277.43	277.44
0.010	257.66	257.77	257.69	257.62	257.65	257.57	257.64	257.66
0.015	240.46	——	240.50	241.61	240.45	241.54	240.43	240.46
0.020	225.66	225.85	225.71	225.59	225.65	225.51	225.63	225.66
0.025	213.13	——	213.18	214.12	213.11	214.02	213.08	213.13
0.030	202.72	202.97	202.78	202.64	202.70	202.53	202.68	202.72
0.035	194.35	——	194.42	195.23	194.33	195.11	194.30	194.35
0.040	187.92	188.20	187.99	187.83	187.90	187.70	187.87	187.92
0.045	183.37	——	183.44	184.20	183.35	184.07	183.32	183.37
0.050	180.66	180.96	180.73	180.57	180.64	180.44	180.60	180.66

4.2.9 Axisymmetric Problems

Here we consider equations governing axisymmetric problems in one dimension (see **Example 3.3.3** and Fig. 3.3.8). The governing equation for an axisymmetric one-dimensional problem is of the form

$$-\frac{1}{r}\frac{d}{dr}\left[ra(r)\frac{du}{dr}\right] = f(r), \tag{4.2.31}$$

where a and f are given functions of r. Examples of Eq. (4.2.31) are provided by radial heat transfer in a long cylinder, as discussed in **Example 3.3.3** and the axisymmetric deformation of an elastic medium due to the pullout of an embedded rigid rod (see Example 4.4.4 of [8]). In the latter case, a denotes the shear modulus G of the material, $f = 0$, and u is the displacement along the length of the cylinder enclosing the rod.

The weak form of this equation over a typical element, $\Omega^e = (r_a^e, r_b^e)$ (Fig. 4.2.7 shows a hollow cylinder of inner radius a and outer radius b) can be derived using the three-step procedure discussed in Section 4.2.4. We have (area element is $dA = r\,dr\,d\theta$)

$$\begin{aligned} 0 &= \int_0^{2\pi}\int_{r_a^e}^{r_b^e} w_i^e\left\{-\frac{1}{r}\frac{d}{dr}\left[ra(r)\frac{du}{dr}\right] - f(r)\right\} r\,dr\,d\theta \\ &= 2\pi\int_{r_a^e}^{r_b^e}\left(ra\frac{dw_i^e}{dr}\frac{du_h^e}{dr} - rf\right)dr - w_i^e(r_a^e)Q_a^e - w_i^e(r_b^e)Q_b^e, \end{aligned} \tag{4.2.32a}$$

where

$$Q_1^e \equiv -2\pi\left(ra_e\frac{du_h^e}{dr}\right)\bigg|_{r_a^e}, \quad Q_2^e \equiv 2\pi\left(ra_e\frac{du_h^e}{dr}\right)\bigg|_{r_b^e}. \tag{4.2.32b}$$

Substituting the approximation from Eq. (4.2.22) (with x replaced by r) into Eq. (4.2.32a), we obtain

$$\mathbf{K}^e\mathbf{u}^e = \mathbf{f}^e + \mathbf{Q}^e, \tag{4.2.33a}$$

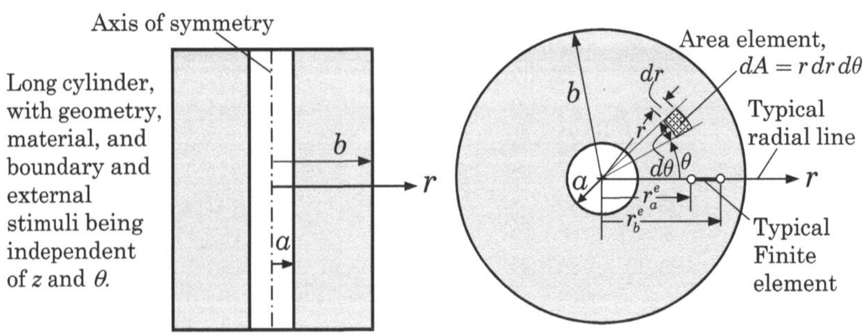

Fig. 4.2.7 Axisymmetric problem involving a long cylinder (i.e., the solution u is independent of the radial coordinate θ and the coordinate z along the length of the cylinder).

where

$$K_{ij}^e = 2\pi \int_{r_a^e}^{r_b^e} a_e \frac{d\psi_i^e}{dr}\frac{d\psi_j^e}{dr}\, r dr, \quad f_i^e = 2\pi \int_{r_a^e}^{r_b^e} \psi_i^e f \, r\, dr. \tag{4.2.33b}$$

The numerical values of the coefficients K_{ij}^e and f_i^e for element-wise constant values of a_e and f_e are given in Eqs. (4.2.34) and (4.2.35), respectively, for the linear and quadratic approximations of $u(r)$. We note that r_a^e denotes the global coordinate of the first node and h_e is the length of the eth element. Note that we can cancel out 2π from all quantities in Eq. (4.2.33a) and (4.2.33b), but one should be careful about how Q_i^e are defined.

Linear element

$$\begin{aligned} \mathbf{K}^e &= \frac{2\pi a_e}{h_e}(r_a^e + \tfrac{1}{2}h_e)\begin{bmatrix} 1 & -1 \\ -1 & 1 \end{bmatrix} \\ \mathbf{f}^e &= \frac{2\pi f_e h_e}{6}\begin{Bmatrix} 3r_a^e + h_e \\ 3r_a^e + 2h_e \end{Bmatrix}. \end{aligned} \tag{4.2.34}$$

Quadratic element

$$\begin{aligned} \mathbf{K}^e &= \frac{2\pi a_e}{6h_e}\begin{bmatrix} 3h_e + 14r_a^e & -(4h_e + 16r_a^e) & h_e + 2r_a^e \\ -(4h_e + 16r_a^e) & 16h_e + 32r_a^e & -(12h_e + 16r_a^e) \\ h_e + 2r_a^e & -(12h_e + 16r_a^e) & 11h_e + 14r_a^e \end{bmatrix} \\ \mathbf{f}^e &= \frac{2\pi f_e h_e}{6}\begin{Bmatrix} r_a^e \\ 4r_a^e + 2h_e \\ r_a^e + h_e \end{Bmatrix}. \end{aligned} \tag{4.2.35}$$

Example 4.2.3

Consider a *long*, homogeneous, isotropic solid circular cylinder of outside radius $R_0 = 0.01$ m, conductivity $k = 20$ W/(m·°C), and a constant rate of internal heat generation $g_0 = 2 \times 10^8$ W/m^3 (see **Example 3.3.3** and Fig. 3.3.8). The boundary surface at $r = R_0$ is maintained at a constant temperature $T_0 = 100$°C. Calculate the temperature distribution $T(r)$ and heat flux $q(r) = -kdT/dr$ in the cylinder using various uniform meshes of linear and quadratic finite elements and compare the solutions against the FDM, FVM, and exact solutions.

Solution The governing equation for this problem is given by Eq. (4.2.31) with $a = k$ and $f = g_0$. The boundary conditions [see Eq. (3.3.13)] are

$$\left[2\pi k r \frac{dT}{dr}\right]_{r=0} = 0, \quad T(R_0) = T_0. \tag{1}$$

The zero-flux boundary condition at $r = 0$ (a result of the radial symmetry) amounts to setting $Q_1^{(1)} = 0$.

The finite element model of the governing equation is given by Eqs. (4.2.33a) and (4.2.33b). For linear and quadratic interpolations of $T(r)$, the element matrices for a typical finite element are given in Eq. (4.2.34) and (4.2.35), respectively.

For the mesh of one linear element, we have ($r_a^1 = 0$ and $r_2^1 = h_1 = R_0$)

$$2\pi k \begin{bmatrix} 1 & -1 \\ -1 & 1 \end{bmatrix}\begin{Bmatrix} U_1 \\ U_2 \end{Bmatrix} = \frac{2\pi g_0 R_0^2}{3}\begin{Bmatrix} 1 \\ 2 \end{Bmatrix} + \begin{Bmatrix} Q_1^{(1)} \\ Q_2^{(1)} \end{Bmatrix}. \tag{2}$$

The boundary conditions are $Q_1^{(1)} = 0$ and $U_2 = T_0$. Hence, the temperature at node 1 is

$$U_1 = \frac{g_0 R_0^2}{3k} + T_0 \tag{3}$$

and the heat at $r = R_0$, using the first element equation, is

$$(Q_2^{(1)})_{equil} = \pi k(U_2 - U_1) - \tfrac{2}{3}\pi g_0 R_0^2 = -\pi g_0 R_0^2. \tag{4}$$

The negative sign indicates that heat is removed from the body. The one-element solution as a function of the radial coordinate r is

$$T_h(r) = U_1\psi_1^1(r) + U_2\psi_2^1(r) = \frac{g_0 R_0^2}{3k}\left(1 - \frac{r}{R_0}\right) + T_0 \tag{5}$$

and the heat flux is

$$q(r) \equiv -k\frac{dT_h}{dr} = \tfrac{1}{3}g_0 R_0. \tag{6}$$

For a mesh of two linear elements ($h_1 = h_2 = \frac{1}{2}R_0$), the assembled equations are

$$\pi k \begin{bmatrix} 1 & -1 & 0 \\ -1 & 1+3 & -3 \\ 0 & -3 & 3 \end{bmatrix} \begin{Bmatrix} U_1 \\ U_2 \\ U_3 \end{Bmatrix} = \frac{\pi g_0 R_0^2}{6} \begin{Bmatrix} \frac{1}{2} \\ 1+2 \\ \frac{1}{2}+2 \end{Bmatrix} + \begin{Bmatrix} Q_1^{(1)} \\ Q_2^{(1)} + Q_1^{(2)} \\ Q_2^{(2)} \end{Bmatrix}. \tag{7}$$

Imposing the boundary and balance conditions $U_3 = T_0$, $Q_1^{(1)} = 0$, and $Q_2^{(1)} + Q_1^{(2)} = 0$, the condensed equations for the unknown temperatures are

$$\pi k \begin{bmatrix} 1 & -1 \\ -1 & 4 \end{bmatrix} \begin{Bmatrix} U_1 \\ U_2 \end{Bmatrix} = \frac{\pi g_0 R_0^2}{12} \begin{Bmatrix} 1 \\ 6 \end{Bmatrix} + \pi k \begin{Bmatrix} 0 \\ 3T_0 \end{Bmatrix}. \tag{8}$$

The nodal values are

$$U_1 = \tfrac{5}{18}\frac{g_0 R_0^2}{k} + T_0, \quad U_2 = \tfrac{7}{36}\frac{g_0 R_0^2}{k} + T_0. \tag{9}$$

From equilibrium, $Q_2^{(2)}$ is computed as

$$Q_2^{(2)} = -\tfrac{5}{12}\pi g_0 R_0^2 + 3\pi k(U_3 - U_2) = -\pi g_0 R_0^2. \tag{10}$$

The FEM solution becomes

$$T_h(r) = \begin{cases} U_1\psi_1^1(r) + U_2\psi_2^1(r) = \left(\frac{5}{18}\frac{g_0R_0^2}{k} + T_0\right)\left(1 - \frac{2r}{R_0}\right) + \frac{2r}{R_0}\left(\frac{7}{36}\frac{g_0R_0^2}{k} + T_0\right) \\ U_2\psi_1^2 + U_3\psi_2^2 = 2\left(\frac{7}{36}\frac{g_0R_0^2}{k} + T_0\right)\left(1 - \frac{r}{R_0}\right) + T_0\left(\frac{2r}{R_0} - 1\right) \end{cases}$$

$$= \begin{cases} \frac{1}{18}\frac{g_0R_0^2}{k}\left(5 - \frac{3r}{R_0}\right) + T_0, & \text{for } 0 \le r \le \frac{1}{2}R_0 \\ \frac{7}{18}\frac{g_0R_0^2}{k}\left(1 - \frac{r}{R_0}\right) + T_0, & \text{for } \frac{1}{2}R_0 \le r \le R_0. \end{cases} \tag{11}$$

The exact solution of the problem is [see Eqs. (3.3.14a) and (3.3.14b)]

$$T(r) = \frac{g_0R_0^2}{4k}\left(1 - \frac{r^2}{R_0^2}\right) + T_0 \ (^\circ\text{C}), \quad q(r) = -k\frac{dT}{dr} = \tfrac{1}{2}g_0 r \ (\text{W/m}^2), \tag{12a}$$

$$Q(R_0) = -\left[2\pi r k\frac{dT}{dr}\right]_{R_0} = \pi g_0 R_0^2 \ (\text{W}). \tag{12b}$$

The exact temperature at the center of the cylinder is $T(0) = g_0R_0^2/4k + T_0$, whereas it is $g_0R_0^2/3k + T_0$ and $5g_0R_0^2/18k + T_0$ according to the one- and two-element models, respectively.

The FEM solutions obtained using four- and eight-element meshes of linear elements and four-element mesh of quadratic elements are compared with the HFVM and exact solution in Table 4.2.5. Convergence of the finite element solutions, $\bar{T} = (T - T_0)k/g_0R_0^2$, to the exact solution with an increasing number of elements is clear.

Table 4.2.5 Comparison of the FEM and HFVM (FVM32) solutions (the underlined terms are the linearly interpolated values) with the exact solutions for temperature distribution in a radially symmetric circular cylinder (L = linear elements, Q = quadratic elements, and N = number of subdivisions in the half-control FVM formulation, HFVM).

	Exact	FVM32 Solution			FEM Solution		
$\frac{r}{R_0}$	Solution	$N=4$	$N=8$	$N=16$	4L	8L	4Q
0.00000	350.00	350.00	350.00	350.00	358.73	352.63	350.00
0.06250	349.02	——	348.05	349.02	353.52	350.03	349.02
0.12500	346.09	342.19	346.09	346.09	348.31	347.42	346.09
0.18750	341.21	——	340.03	314.21	343.11	341.35	341.21
0.25000	334.38	334.38	334.38	334.38	337.90	335.27	334.37
0.31250	325.59	——	324.61	325.59	325.74	325.38	325.59
0.37500	314.84	310.94	314.84	314.84	313.59	315.48	314.84
0.43750	302.15	——	301.17	302.15	301.44	301.71	302.15
0.50000	287.50	287.50	287.50	287.50	289.29	287.95	287.50
0.56250	270.90	——	269.92	270.90	269.49	270.30	270.90
0.62500	252.34	248.44	248.44	248.44	249.70	252.65	252.34
0.68750	231.84	——	230.86	231.84	229.91	231.11	231.84
0.75000	209.37	209.37	209.37	209.37	210.12	209.56	209.37
0.81250	184.96	——	183.98	184.96	182.59	184.12	184.96
0.87500	158.59	154.69	154.69	154.69	155.06	158.68	158.59
0.93750	130.27	——	129.30	130.27	127.53	129.34	130.27
1.00000	100.00	100.00	100.00	100.00	100.00	100.00	100.00

Figures 4.2.8 and 4.2.9 contain plots of the FEM results for $T(r)$ and $\bar{Q}(r) = Q(r)/2\pi R_0 g_0$, respectively, versus $\bar{r} = r/R_0$, where $Q(r) = 2\pi krdT/dr$. The FEM solutions for $Q(r)$ computed at the element centers, not the nodes, are in excellent agreement with the exact solution.

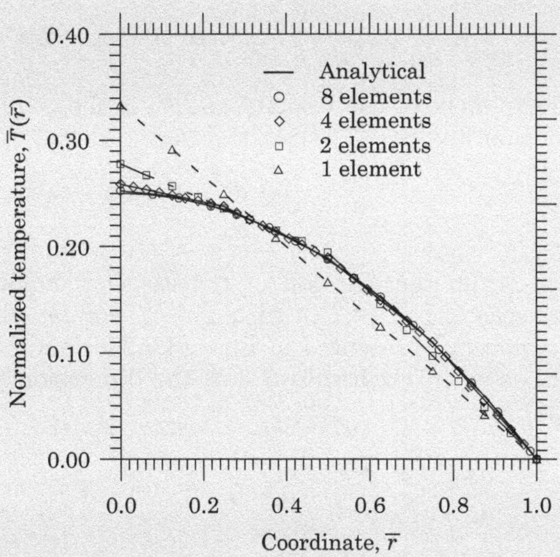

Fig. 4.2.8 Comparison of the finite element solutions for normalized temperature, $\bar{T} = (T - T_0)k/g_0R_0^2$ (obtained with meshes of linear elements), with the exact solution for heat transfer in an axisymmetric problem.

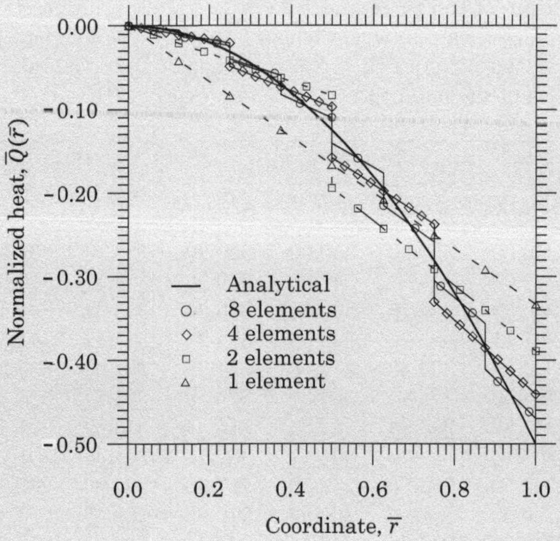

Fig. 4.2.9 Comparison of the finite element solution (obtained with meshes of linear elements) with the exact solution for the normalized heat, $\bar{Q}(\bar{r}) = Q(\bar{r})/2\pi R_0 g_0$, in an axisymmetric problem.

4.2.10 Advection–Diffusion Equation

The steady-state advection–diffusion equation in one dimension is given by [see Eq. (3.3.20)]

$$\rho u \frac{d\phi}{dx} - \frac{d}{dx}\left(k_x \frac{d\phi}{dx}\right) - g = 0, \quad 0 < x < L, \tag{4.2.36}$$

where u is the x-component of the velocity and k_x is the conductivity in the x direction. The diffusion flux is

$$q_n = -k_x \frac{d\phi}{dx}. \tag{4.2.37}$$

When ρu is known (from the solution of the associated flow problem), Eq. (4.2.36) is a special case of Eq. (4.2.1) with $a = k_x$, $b = \rho u$, and $c = 0$. Hence, the finite element equations presented in Eqs. (4.2.29c) and (4.2.30f) are valid with $c = 0$. As discussed in **Example 3.3.4**, the dimensionless forms of Eqs. (4.2.36) and (4.2.37) are

$$\frac{d\phi}{d\bar{x}} - \frac{1}{Pe}\frac{d^2\phi}{d\bar{x}^2} - \bar{g} = 0, \quad 0 < \bar{x} < 1 \tag{4.2.38}$$

and

$$\bar{q}_n = -\frac{1}{Pe}\frac{d\phi}{d\bar{x}}, \tag{4.2.39}$$

where Pe is the Péclet number.

Example 3.3.4 contains a discussion of an advection–diffusion problems with numerical results obtained using the FVM. Here we discuss the same problem in the context of the FEM.

Example 4.2.4

Develop the finite element model of Eq. (4.2.38) and obtain the assembled equations for uniform meshes of (a) four linear finite elements and (b) two quadratic elements. Use the following boundary conditions

$$\phi(0) = 1, \quad \phi(1) = 0. \tag{1}$$

Solution The weak-form Galerkin finite element model of Eq. (4.2.39) is (we note that the coefficient matrix is unsymmetric, requiring the use of unsymmetric but banded equation solvers when direct methods are used)

$$\mathbf{K}^e \boldsymbol{\phi}^e = \mathbf{Q}^e, \tag{2a}$$

$$K_{ij}^e = \int_{x_a^e}^{x_b^e} \left(\psi_i^e \frac{d\psi_j^e}{dx} + \frac{1}{Pe}\frac{d\psi_i^e}{dx}\frac{d\psi_j^e}{dx} \right) dx, \quad Q_1^e \equiv \left[-\frac{1}{Pe}\frac{d\phi}{dx} \right]_{x_a^e}, \quad Q_2^e \equiv \left[\frac{1}{Pe}\frac{d\phi}{dx} \right]_{x_b^e}. \tag{2b}$$

(a) For a linear finite element, the finite element equations are [see Eq. (4.2.29c); set $a = 1/Pe$, $b = 1$, $c = 0$ and $f = 0$]

$$\left(\frac{1}{2}\begin{bmatrix} -1 & 1 \\ -1 & 1 \end{bmatrix} + \frac{1}{h\,Pe}\begin{bmatrix} 1 & -1 \\ -1 & 1 \end{bmatrix} \right) \begin{Bmatrix} u_1^e \\ u_2^e \end{Bmatrix} = \begin{Bmatrix} Q_1^e \\ Q_2^e \end{Bmatrix}. \tag{3}$$

The assembled system of equations for a uniform mesh of four linear elements is

$$\left(\frac{1}{2}\begin{bmatrix} -1 & 1 & 0 & 0 & 0 \\ -1 & 0 & 1 & 0 & 0 \\ 0 & -1 & 0 & 1 & 0 \\ 0 & 0 & -1 & 0 & 1 \\ 0 & 0 & 0 & -1 & 1 \end{bmatrix} + \frac{1}{h\,Pe}\begin{bmatrix} 1 & -1 & 0 & 0 & 0 \\ -1 & 2 & -1 & 0 & 0 \\ 0 & -1 & 2 & -1 & 0 \\ 0 & 0 & -1 & 2 & -1 \\ 0 & 0 & 0 & -1 & 1 \end{bmatrix} \right) \begin{Bmatrix} U_1 \\ U_2 \\ U_3 \\ U_4 \\ U_5 \end{Bmatrix} = \begin{Bmatrix} Q_1^{(1)} \\ 0 \\ 0 \\ 0 \\ Q_2^{(4)} \end{Bmatrix}. \tag{4}$$

From Eq. (4), one can identify the structure of the equation for an interior global node I as

$$\left(-\frac{1}{2} - \frac{1}{h\,Pe} \right) U_{I-1} + \frac{2}{h\,Pe} U_I + \left(\frac{1}{2} - \frac{1}{h\,Pe} \right) U_{I+1} = 0, \tag{5}$$

which is the same as that obtained in the FDM and FVM methods [see Eqs. (3.3.27) and (3.3.36)]. Hence the stability requirements discussed in **Example 3.3.4** are valid for any FEM mesh of linear elements.

(b) For a quadratic finite element, the finite element equations are [see Eq. (4.2.30f)]

$$\left(\frac{1}{6}\begin{bmatrix} -3 & 4 & 1 \\ -4 & 0 & 4 \\ 1 & -4 & 3 \end{bmatrix} + \frac{1}{3h\,Pe}\begin{bmatrix} 7 & -8 & 1 \\ -8 & 16 & -8 \\ 1 & -8 & 7 \end{bmatrix} \right) \begin{Bmatrix} u_1^e \\ u_2^e \\ u_3^e \end{Bmatrix} = \begin{Bmatrix} Q_1^e \\ 0 \\ Q_3^e \end{Bmatrix}. \tag{6}$$

The assembled set of finite element equations for a uniform mesh of two quadratic elements is

$$\left(\frac{1}{6}\begin{bmatrix} -3 & 4 & 1 & 0 & 0 \\ -4 & 0 & 4 & 0 & 0 \\ 1 & -4 & 0 & 4 & 1 \\ 0 & 0 & -4 & 0 & 4 \\ 0 & 0 & 1 & -4 & 3 \end{bmatrix} + \frac{1}{3h\,Pe}\begin{bmatrix} 7 & -8 & 1 & 0 & 0 \\ -8 & 16 & -8 & 0 & 0 \\ 1 & -8 & 14 & -8 & 1 \\ 0 & 0 & -8 & 16 & -8 \\ 0 & 0 & 1 & -8 & 7 \end{bmatrix} \right) \begin{Bmatrix} U_1 \\ U_2 \\ U_3 \\ U_4 \\ U_5 \end{Bmatrix} = \begin{Bmatrix} Q_1^{(1)} \\ 0 \\ 0 \\ 0 \\ Q_3^{(2)} \end{Bmatrix}. \tag{7}$$

Figure 4.2.10 shows a comparison of the exact solution [solid line; see Eq. (3.3.25)] with the finite element solutions for various uniform meshes of linear and quadratic finite elements for $Pe = 100$. Clearly, for four linear elements, the numerical solutions exhibit oscillations. The oscillations disappear when the number of elements is greater than $Pe/2$, as dictated by the stability condition (no stability analysis is available for quadratic elements). Table 4.2.6 shows the convergence with the mesh refinements. The numerical solutions obtained with $N = 100$ linear elements or $N = 50$ quadratic elements are stable and accurate.

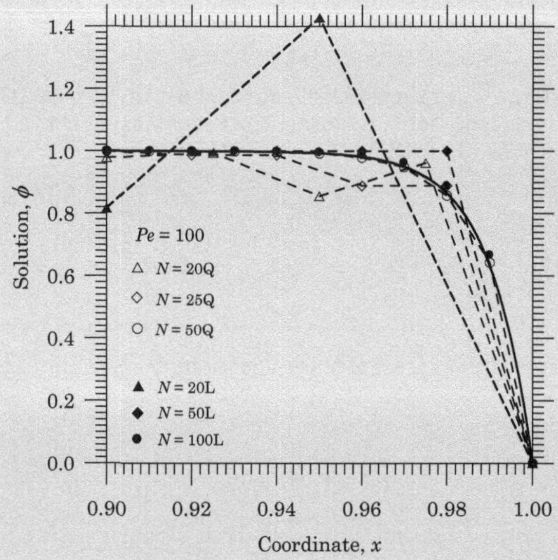

Fig. 4.2.10 A comparison of the FEM solutions against the exact solution for $Pe = 100$.

Table 4.2.6 Comparison of the FEM and exact solutions of the advection–diffusion equation for Péclet number, $Pe = 100$ (L = linear and Q = quadratic elements).

	FEM Solution[a]					Exact
x	20L	50L	100L	25Q	50Q	Solution
0.500	0.9998	1.0000	1.0000	1.0000	1.0000	1.0000
0.550	1.0005	1.0000	1.0000	1.0000	1.0000	1.0000
0.600	0.9989	1.0000	1.0000	1.0000	1.0000	1.0000
0.650	1.0027	1.0000	1.0000	1.0000	1.0000	1.0000
0.700	0.9938	1.0000	1.0000	1.0000	1.0000	1.0000
0.750	1.0145	1.0000	1.0000	1.0000	1.0000	1.0000
0.800	0.9663	1.0000	1.0000	1.0000	1.0000	1.0000
0.850	1.0787	1.0000	1.0000	1.0000	1.0000	1.0000
0.860	——	1.0000	1.0000	1.0000	1.0000	1.0000
0.880	——	1.0000	1.0000	0.9995	1.0000	1.0000
0.900	0.8163	1.0000	1.0000	0.9995	0.9999	0.9999
0.920	——	1.0000	0.9999	0.9941	0.9996	0.9997
0.940	——	1.0000	0.9986	0.9941	0.9971	0.9975
0.960	——	1.0000	0.9877	0.9231	0.9796	0.9817
0.980	——	1.0000	0.8889	0.9231	0.8571	0.8647
0.990	——	0.5000	0.6667	0.5769	0.6429	0.6321

[a]The underlined values are interpolated values.

4.3 Two-Dimensional Problems

4.3.1 Model Differential Equation

In this section, we consider a model differential equation involving a single unknown in two dimensions to present the finite element model and its application to representative problems. Although there is a multitude of engineering problems described by the model equation we consider here (much the same way as the model differential equation in one dimension), it arises in a number of fields, including heat transfer, inviscid flows (e.g., stream function or velocity potential formulations), and solid mechanics (e.g., torsion of cylindrical members and deflections of membranes). Heat transfer in a fluid medium (including the advection–diffusion problem) will be considered in Chapter 5.

We consider the following partial differential equation governing $u(x, y)$:

$$-\left[\frac{\partial}{\partial x}\left(a_{xx}\frac{\partial u}{\partial x}\right)+\frac{\partial}{\partial y}\left(a_{yy}\frac{\partial u}{\partial y}\right)\right]=f(x,y) \quad \text{in } \Omega, \tag{4.3.1}$$

where Ω is a two-dimensional domain with boundary Γ, as shown in Fig. 4.3.1(a). The coefficients a_{xx}, a_{yy}, and $f(x, y)$ are known functions of position (x, y). For example, in the case of two-dimensional heat transfer, u denotes the temperature, $a_{xx} = k_{xx}$ and $a_{yy} = k_{yy}$ are the conductivities of an orthotropic solid medium in the x- and y-directions, respectively (the material coordinate axes are assumed to coincide with the problem coordinates, x and y), and $f(x, y)$ is the known internal heat generation (if any) measured per unit area. For an isotropic medium, we set $k_{xx} = k_{yy} = k$. In the case of groundwater flow problem, u denotes the water head (i.e., velocity potential), a_{xx} and a_{yy} are the permeabilities in the x- and y-directions, respectively, and $f(x, y)$ is the distributed source of water. For other problems where Eq. (4.3.1) arises, see Table 9.1.1 of Reddy [8].

The function u in Eq. (4.3.1) is required to satisfy certain conditions on the boundary Γ of the bounded closed domain Ω. In general, u is required to satisfy one of the following two types of boundary conditions at a boundary point:

$$u = \hat{u}(s) \quad \text{on} \quad \Gamma_u \tag{4.3.2a}$$

$$\left(a_{xx}\frac{\partial u}{\partial x}n_x + a_{yy}\frac{\partial u}{\partial y}n_y\right) + \beta(u - u_\infty) = \hat{q}_n(s) \quad \text{on} \quad \Gamma_q, \tag{4.3.2b}$$

where Γ_u and Γ_q are disjoint portions of the total boundary Γ such that $\Gamma = \Gamma_u \cup \Gamma_q$, $\hat{u}$, β, u_∞, and $\hat{q}_n$ are known quantities (i.e., data of the problem), and (n_x, n_y) are the direction cosines of the unit normal vector $\hat{\mathbf{n}}$ on the boundary Γ:

$$\begin{aligned} n_x &= \cos(x, \hat{\mathbf{n}}), \text{cosine of the angle between positive } x\text{-axis and } \hat{\mathbf{n}} \\ n_y &= \cos(y, \hat{\mathbf{n}}), \text{cosine of the angle between positive } y\text{-axis and } \hat{\mathbf{n}}. \end{aligned} \tag{4.3.3}$$

In most problems of practical interest, the boundary Γ enclosing a bounded domain Ω is not a single continuous line but a collection of connected and non-overlapping lines.

4.3.2 Finite Element Approximation

4.3.2.1 Approximation of the geometry

As in the case of Eq. (4.2.1), we wish to determine an approximation $u_h(x, y)$ of the actual solution u of Eq. (4.3.1) with suitable boundary conditions. In the finite element method, we discretize the domain Ω as a collection of nonoverlapping subdomains (i.e., finite elements) Ω^e such that their union is approximately equal to the domain with the boundary, $\bar{\Omega}$, as illustrated in Fig. 4.3.1(b). This amounts to approximating the geometry of the domain; i.e., a typical point (x, y) of the domain is approximated as

$$x = \sum_{j=1}^{m} x_j^e \hat{\psi}_j^e(\xi, \eta), \quad y = \sum_{j=1}^{m} y_j^e \hat{\psi}_j^e(\xi, \eta), \tag{4.3.4}$$

where (ξ, η) is a local coordinate system, (x_j^e, y_j^e) are the global coordinates of the jth node of the eth finite element Ω^e with respect to the global coordinate system (x, y), and $\hat{\psi}_j^e$ are the approximation functions used to define the element geometry (more details to follow). Geometric shapes that qualify as finite elements are those for which approximation functions ψ_i^e can be derived uniquely. It turns out that only triangles and rectangles meet these requirements, as discussed shortly.

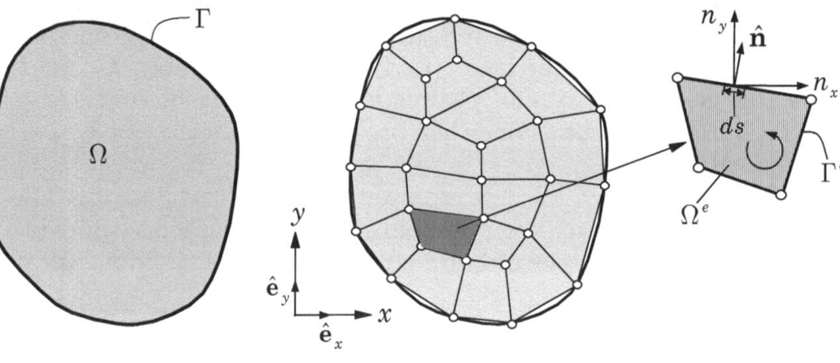

Fig. 4.3.1 (a) A two-dimensional domain Ω with its boundary Γ. (b) Finite element discretization of $\bar{\Omega} \equiv \Omega \cup \Gamma$ and a typical finite element Ω^e with its boundary Γ^e.

The union of all elements $\bar{\Omega}^e$ is called the *finite element mesh* of the domain Ω. We note that the finite element mesh of a domain consists of nonoverlapping finite elements but with overlapping (or coinciding) boundary segments, without leaving gaps between elements (see Fig 4.3.2). In general, the finite element mesh may not equal $\bar{\Omega}$, especially when the boundary Γ is curved. Of course, for polygonal domains, the finite element mesh exactly represents the actual domain.

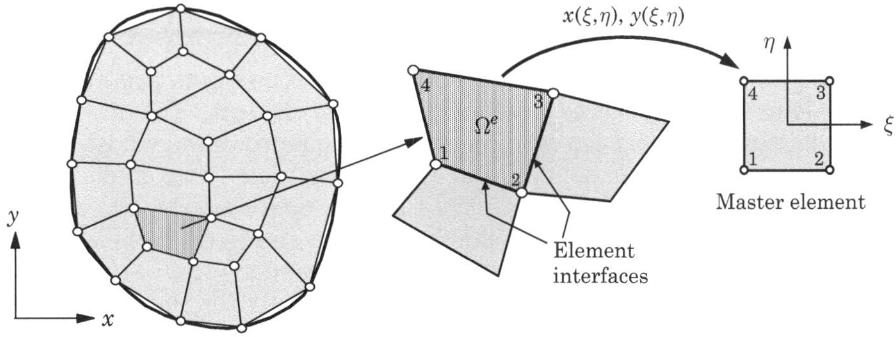

Fig. 4.3.2 Finite element discretization of $\bar{\Omega}$ with a set of nonoverlapping finite elements Ω^e whose boundary segments Γ^e coincide with those of the neighboring elements, without leaving gaps between element interfaces.

4.3.2.2 Approximation of the solution

To keep the formulation steps very general (i.e., not confine the finite element formulation to a specific geometric shape), we develop the weak form of the governing equation over an arbitrary typical element Ω^e and its boundary by Γ^e. The element $\bar{\Omega}^e \equiv \Omega^e \cup \Gamma^e$ can be a triangle or quadrilateral in shape (including the interior and the boundary segments), and the degree of interpolation over it can be linear, quadratic, or higher.

Suppose that the dependent unknown u is approximated over a typical finite element $\bar{\Omega}^e$ by $u_h^e(x, y)$ in the same way as in the one-dimensional case:

$$u(x, y) \approx u_h^e(x, y) = \sum_{j=1}^{n} u_j^e \psi_j^e(x, y), \quad (x, y) \in \bar{\Omega}^e, \tag{4.3.5}$$

where u_j^e $(j = 1, 2, \ldots, n)$ denote the values of the function $u_h^e(x, y)$ at a selected number of points (i.e., element nodes) in the element $\bar{\Omega}^e$, and ψ_j^e are the Lagrange interpolation functions associated with the element.

When the approximation used for the geometry and solution is the same (i.e., $m = n$ and $\hat{\psi}_j^e = \psi_j^e$), the formulation is known as the *isoparametric formulation.* When $m > n$ (i.e., the geometry is approximated using higher-order interpolation than the solution), the formulation is known as the *superparametric formulation.* For $m < n$, it is known as a *subparametric formulation.* The isoparametric formulation is the most common, and there is only one mesh to show the nodes and elements used to approximate the geometry and the solution.

Although it is not necessary to know how the interpolation functions are derived in order to develop finite element models of differential equations, for the sake of completeness and understanding we will briefly visit the topic in Section 4.3.6. For now, few general comments are in order. A triangle is the simplest two-dimensional geometric shape in two dimensions because it is uniquely defined by three points ($n = 3$) that are not collinear in a plane and the approximated function varies between any two points according to the polynomial

$u_h^e(x,y) = c_1^e + c_2^e x + c_3^e y$, which is complete. A triangle with three nodes per side (a total of six nodes in the element) will uniquely define the geometry while representing the quadratic variation $u_h^e(x,y) = c_1^e + c_2^e x + c_3^e y + c_4^e xy + c_5^e x^2 + c_6^e y^2$ uniquely along any of the three sides and inside the element. The other geometric shape of a finite element in two dimensions is a quadrilateral, whose geometry is uniquely defined by four noncollinear points in a plane. A complete polynomial of the form $u_h^e(x,y) = c_1^e + c_2^e x + c_3^e y + c_4^e xy$ is used to derive the associated interpolation functions. When each side has three nodes (a total of eight nodes), the polynomial used is $u_h^e(x,y) = c_1^e + c_2^e x + c_3^e y + c_4^e xy + c_5^e x^2 + c_6^e y^2 + c_7^e x^2 y + c_8^e xy^2$. In general, the interpolation functions are derived only for the so-called *master elements* (see Fig. 4.3.2), whose geometry facilitates easy derivation. Then these functions are transformed to the actual geometry of the element.

Note that any geometric shape in two dimensions that has more than four sides is not qualified as a finite element as there is no simple way to derive interpolation functions. A pentagon, for example, requires five points to define its geometry uniquely, but there is no five-parameter polynomial in (x,y) that uniquely defines the linear variation along any segment of the pentagon. [2] Therefore, a pentagon is not a standard finite element.

4.3.3 Weak Form

As described in Section 4.2.4, we require the governing differential equation to be satisfied only in a weak-form sense, with the weight functions being the same as the approximation functions (i.e., use the Galerkin idea). The resulting finite element model (i.e., set of n algebraic relations among n nodal values u_j^e needed to represent u_h) is called the *weak-form Galerkin finite element model* or the *Ritz finite element model.*

We use the three-step procedure described in Section 4.2.4 to develop the weak form of Eq. (4.3.1) that accounts for the form of the natural boundary condition in Eq. (4.3.2b) over the typical element Ω^e. The weak-form development does not require the knowledge of the interpolation functions. We note that u_h^e, being the dependent variable in the governing equation, is the *primary variable*; and specification of u_h^e constitutes the essential boundary condition.

Step 1. The first step is to move all terms in Eq. (4.3.1) to one side of the equality (say, to the left side), multiply the (left side) expression with a weight function w_i^e from a set of linearly independent functions (which will be replaced with ψ_i^e to evaluate the integrals and obtain the algebraic relations) $\{w_i^e\}_{i=1}^n$, and equate the integral of the weighted expression over the element domain Ω^e to zero:

$$0 = \int_{\Omega^e} w_i^e \left[-\frac{\partial}{\partial x}\left(a_{xx}\frac{\partial u_h^e}{\partial x}\right) - \frac{\partial}{\partial y}\left(a_{yy}\frac{\partial u_h^e}{\partial y}\right) - f(x,y)\right] dxdy. \tag{4.3.6}$$

The expression in the square brackets of Eq. (4.3.6) represents the residual of the approximation of the differential equation (4.3.1) because $u_h^e(x,y)$ is only an approximation of $u(x,y)$.

[2] *Polygonal finite elements* found in the literature are special elements in which higher-order functions are defined inside the element in such a way that they become linear on the edges.

For n independent choices of w_i^e, the weighted-integral statement in Eq. (4.3.6) yields a set of n linearly independent algebraic equations, called the *weighted-residual finite element model.* Such a model requires higher-order interpolation of $u_h(x, y)$ because the residual contains higher-order derivatives with respect to x and y.

Step 2. In the second step, we transfer derivatives from u_h^e to w_i^e so that both u_h^e and w_h^e are required to be differentiable only once with respect to x and y (keeping in mind that both w_i^e and u_h will be replaced in terms of the interpolation functions). This is the step that results in a boundary expression and allows the identification of the secondary variable. If the secondary variable is not a physically meaningful quantity, we should not carry out this trading of differentiation between w_i^e and u_h.

To transfer one derivative from u_h to w_i^e, we use the component form of the gradient (or divergence) theorem:

$$\int_{\Omega^e} \frac{\partial}{\partial x}(w_i^e F_x)\, dxdy = \oint_{\Gamma^e} (w_i^e F_x) n_x\, ds \tag{4.3.7}$$

$$\int_{\Omega^e} \frac{\partial}{\partial y}(w_i^e F_y)\, dxdy = \oint_{\Gamma^e} (w_i^e F_y) n_y\, ds, \tag{4.3.8}$$

where n_x and n_y are the components of the unit normal vector $\hat{\mathbf{n}}$ [see Eq. (4.3.3)]

$$\hat{\mathbf{n}} = n_x \hat{\mathbf{e}}_x + n_y \hat{\mathbf{e}}_y = \cos\alpha\ \hat{\mathbf{e}}_x + \sin\alpha\ \hat{\mathbf{e}}_y \tag{4.3.9}$$

on the boundary Γ^e, ds is the arc length of an infinitesimal line element along the boundary, and F_x and F_y are the components of a vector function $\mathbf{F}$. With

$$F_x = a_{xx}\frac{\partial u_h^e}{\partial x}, \quad F_y = a_{yy}\frac{\partial u_h^e}{\partial y} \tag{4.3.10}$$

and identities

$$-w_i^e \frac{\partial F_x}{\partial x} = -\frac{\partial}{\partial x}(w_i^e F_x) + F_x \frac{\partial w_i^e}{\partial x} \tag{4.3.11}$$

$$-w_i^e \frac{\partial F_y}{\partial y} = -\frac{\partial}{\partial y}(w_i^e F_y) + F_y \frac{\partial w_i^e}{\partial y}, \tag{4.3.12}$$

we obtain

$$\begin{aligned} 0 = &\int_{\Omega^e} \left(a_{xx}\frac{\partial w_i^e}{\partial x}\frac{\partial u_h^e}{\partial x} + a_{yy}\frac{\partial w_i^e}{\partial y}\frac{\partial u_h^e}{\partial y} - w_i^e f \right) dxdy \\ &- \oint_{\Gamma^e} w_i^e \left(a_{xx}\frac{\partial u_h^e}{\partial x} n_x + a_{yy}\frac{\partial u_h^e}{\partial y} n_y \right) ds. \end{aligned} \tag{4.3.13}$$

From an inspection of the boundary term in Eq. (4.3.13), we note that the coefficient of the weight function in the boundary integral is

$$a_{xx}\frac{\partial u_h^e}{\partial x} n_x + a_{yy}\frac{\partial u_h^e}{\partial y} n_y \equiv q_n^e. \tag{4.3.14}$$

Hence, q_n^e is the *secondary variable*; its specification constitutes the natural boundary condition. Clearly, q_n^e is the positive outward flux normal to the surface as one travels counterclockwise along the boundary Γ^e. This is the reason why the element nodes are numbered in the counterclockwise direction and the evaluation of boundary integrals is carried out in the counterclockwise sense. In heat flow problems, the secondary variable q_n^e denotes the *inward* heat flux normal to the boundary of the element because the heat flux vector $\mathbf{q}^e$ is given by

$$\mathbf{q}^e = q_x^e \hat{\mathbf{e}}_x + q_y^e \hat{\mathbf{e}}_y, \quad q_x^e = -a_{xx}\frac{\partial u_h^e}{\partial x}, \quad q_y^e = -a_{yy}\frac{\partial u_h^e}{\partial y}, \tag{4.3.15}$$

and we have

$$\hat{\mathbf{n}} \cdot \mathbf{q}^e = n_x q_x^e + n_y q_y^e = -\left(a_{xx}\frac{\partial u_h^e}{\partial x} n_x + a_{yy}\frac{\partial u_h^e}{\partial y} n_y\right) = -q_n^e. \tag{4.3.16}$$

Step 3. The third and last step involves the use of the boundary condition from Eqs. (4.3.2b) in Eq. (4.3.13), $q_n^e = \hat{q}_n^e - \beta(u - u_\infty)$, and write

$$\begin{aligned} 0 = &\int_{\Omega^e} \left(a_{xx}\frac{\partial w_i^e}{\partial x}\frac{\partial u_h^e}{\partial x} + a_{yy}\frac{\partial w_i^e}{\partial y}\frac{\partial u_h^e}{\partial y} - w_i^e f \right) dA + \oint_{\Gamma^e} \beta w_i^e u_h^e \, ds \\ &- \oint_{\Gamma^e} w_i^e (\hat{q}_n^e + \beta u_\infty) ds. \end{aligned} \tag{4.3.17}$$

Since we have started with the model equation (4.3.1) and made use of Eq. (4.3.2b) in arriving at the weak form (4.3.17), it is obvious that the governing differential equation (4.3.1) and the natural boundary condition (4.3.2a) are equivalent to the weak form (4.3.17). The converse of the statement also holds [i.e., the weak form is equivalent to Eqs. (4.3.1) and (4.3.2b)]; for details, see Reddy [8], Reddy and Gartling [10], Gresho and Sani [9], and Surana and Reddy [14, 15].

4.3.4 Finite Element Model

The weak form in Eq. (4.3.17) requires that u_h^e be at least linear in both x and y so that there are no terms in Eq. (4.3.17) that become identically zero. Suppose that u_h^e is represented over a typical finite element Ω^e by a complete polynomial of the form

$$\begin{aligned} u_h^e(x, y) = &c_1^e + c_2^e x + c_3^e y + c_4^e xy + c_5^e x^2 + c_6^e y^2 \\ &+ c_7^e x^2 y + c_8^e xy^2 + c_9^e x^2 y^2 + \cdots . \end{aligned} \tag{4.3.18}$$

In addition to the property of completeness, we also require *equipresence* of the x- and y-terms in the polynomial. Thus, one must include the first three terms, the first four terms, the first six terms, or the first eight terms and so on, to derive the interpolation functions. More on this topic will be presented in Section 4.3.7.

For now, we assume that the finite element approximation of u_h^e is of the form in Eq. (4.3.5) [i.e., the approximation (4.3.18), for any admissible choice of terms in the polynomial, will reduce to (4.3.5)]. Substituting Eq. (4.3.5) into the weak form (4.3.17), we obtain

$$
\begin{aligned}
0 = \sum_{j=1}^{n} \left[\int_{\Omega^e} \left(a_{xx} \frac{\partial w_i^e}{\partial x} \frac{\partial \psi_j^e}{\partial x} + a_{yy} \frac{\partial w_i^e}{\partial y} \frac{\partial \psi_j^e}{\partial y} \right) dxdy + \oint_{\Gamma^e} \beta w_i^e \psi_j^e \, ds \right] u_j^e \\
- \int_{\Omega^e} w_i^e f \, dxdy - \oint_{\Gamma^e} w_i^e \left(\hat{q}_n^e + \beta u_\infty \right) ds. \qquad (4.3.19)
\end{aligned}
$$

This equation must hold for any weight function w_i^e from the set $\{w_i^e\}_{i=1}^n$. In particular, it must hold for the case $w_i^e = \psi_i^e$ (i.e., the Galerkin idea).

For each choice of w_i^e we obtain an algebraic relation among $(u_1^e, u_2^e, \ldots, u_n^e)$. The ith algebraic equation is obtained by substituting $w_i^e = \psi_i^e$ into Eq. (4.3.19):

$$
\sum_{j=1}^{n} \left(K_{ij}^e + H_{ij}^e \right) u_j^e = f_i^e + P_i^e + q_i^e
$$

or

$$
\left(\mathbf{K}^e + \mathbf{H}^e \right) \mathbf{u}^e = \mathbf{f}^e + \mathbf{P}^e + \mathbf{Q}^e \equiv \mathbf{F}^e, \qquad (4.3.20)
$$

where the coefficients K_{ij}^e, H_{ij}^e ($\mathbf{K}^e$ and $\mathbf{H}^e$ are symmetric matrices), f_i^e, P_i^e, and Q_i^e are defined by

$$
\begin{aligned}
K_{ij}^e &= \int_{\Omega^e} \left(a_{xx} \frac{\partial \psi_i^e}{\partial x} \frac{\partial \psi_j^e}{\partial x} + a_{yy} \frac{\partial \psi_i^e}{\partial y} \frac{\partial \psi_j^e}{\partial y} \right) dxdy, \quad H_{ij}^e = \oint_{\Gamma^e} \beta \psi_i^e \psi_j^e \, ds \\
f_i^e &= \int_{\Omega^e} f \, \psi_i^e \, dxdy, \quad P_i^e = \oint_{\Gamma^e} \beta u_\infty \, \psi_i^e \, ds, \quad Q_i^e = \oint_{\Gamma^e} \hat{q}_n^e \, \psi_i^e \, ds.
\end{aligned}
\qquad (4.3.21)
$$

We note that $\mathbf{H}^e$ and $\mathbf{P}^e$ are nonzero only for elements with convection boundary segments (i.e., they are zero for all interior elements). The evaluations of these coefficients will be discussed in Sections 4.3.6 and 4.3.9.

4.3.5 Axisymmetric Problems

4.3.5.1 Governing equation

As discussed in Section 4.3, certain problems can be reduced from three dimensions to two dimensions or even one dimension, depending on the nature of the geometry, material properties, source term, and boundary conditions. If the cylinder problem discussed in Section 4.3 (see Fig. 4.3.8) is such that the geometry, material properties, heat source, and boundary conditions vary radially and along the length of the cylinder (but not around the circumference; i.e., axisymmetric about the z-axis), then the problem cannot be reduced to one dimension. In such cases, we are required to analyze the resulting two-dimensional problem. Here we consider axisymmetric problems described by a second-order differential equation in the cylindrical coordinate system (r, θ, z); see Fig. 2.3.1 for the cylindrical coordinate system.

Consider the following general differential equation in cylindrical coordinates

$$-\frac{1}{r}\frac{\partial}{\partial r}\left(ra_{rr}\frac{\partial u}{\partial r}\right)-\frac{1}{r^2}\frac{\partial}{\partial \theta}\left(a_{\theta\theta}\frac{\partial u}{\partial \theta}\right)-\frac{\partial}{\partial z}\left(a_{zz}\frac{\partial u}{\partial z}\right)=f(r,\theta,z). \tag{4.3.22}$$

where a_{rr}, $a_{\theta\theta}$, and a_{zz} are material properties (e.g., conductivity in a heat transfer problem) and f is the source term (e.g., the internal heat generation).

When the geometry, loading, and boundary conditions are independent of the circumferential direction (i.e., θ-coordinate direction), the problem is said to be axisymmetric and the governing equation becomes two-dimensional in terms of r and z coordinates:

$$-\frac{1}{r}\frac{\partial}{\partial r}\left(ra_{rr}\frac{\partial u}{\partial r}\right)-\frac{\partial}{\partial z}\left(a_{zz}\frac{\partial u}{\partial z}\right)=f(r,z). \tag{4.3.23}$$

In addition, if the problem geometry and data are independent of z, for example when the cylinder is very long, heat transfer is one-dimensional in the radial coordinate r:

$$-\frac{1}{r}\frac{d}{dr}\left(ra_{rr}\frac{du}{dr}\right)=f(r) \quad \text{for} \quad a<r<b, \tag{4.3.24}$$

which is the one-dimensional equation considered in Section 4.2.9.

4.3.5.2 Finite element model

In this section, we develop the finite element model of Eq. (4.3.23). We begin with the development of the weak form, where the volume element dv is replaced by $dv = rdrd\theta dz$.

Following the three-step procedure, we obtain the weak form (the constant multiplier 2π and element label e are omitted in the following)

$$0=\int_{\Omega^e}\left(a_{rr}\frac{\partial w_i}{\partial r}\frac{\partial u_h}{\partial r}+a_{zz}\frac{\partial w_i}{\partial z}\frac{\partial u_h}{\partial z}-w_i f\right)rdrdz-\oint_{\Gamma_e} w_i q_n^e\, ds, \tag{4.3.25a}$$

where w_i is the ith weight function and q_n^e is the flux normal to the boundary

$$q_n^e = r\left(a_{rr}\frac{\partial u_h}{\partial r}n_r + a_{zz}\frac{\partial u_h}{\partial z}n_z\right) \tag{4.3.25b}$$

and (n_r, n_z) are the direction cosines of the unit normal vector ($n_r = 1$ and $n_z = 0$ on the lateral surface and $n_r = 0$ and $n_z = \pm 1$ on top and bottom surfaces of the cylinder). We note that the domain of the problem is a rectangle in the rz-plane.

We assume that $u(r,z)$ is approximated by the finite element interpolation $u_h^e(r,z)$ over a typical two-dimensional element Ω^e (which can be triangular or quadrilateral in shape and the degree of interpolation is dictated by the number of nodes n in the element)

$$u \approx u_h^e(r,z) = \sum_{j=1}^{n} u_j^e \psi_j^e(r,z), \tag{4.3.26}$$

where interpolation functions $\psi_j^e(r,z)$ are the same as those discussed previously but with $x=r$ and $y=z$. Substitution of Eq. (4.3.26) for u_h^e and ψ_i^e for w_i^e into the weak form gives the ith algebraic equation of the finite element model

$$\begin{aligned}0=&\sum_{j=1}^{n}\left[\int_{\Omega^e}\left(a_{rr}\frac{\partial\psi_i^e}{\partial r}\frac{\partial\psi_j^e}{\partial r}+a_{zz}\frac{\partial\psi_i^e}{\partial z}\frac{\partial\psi_j^e}{\partial z}\right)rdrdz\right]u_j^e\\&-\int_{\Omega^e}\psi_i^e f(r,z)\,r\,dr\,dz-\oint_{\Gamma_e}\psi_i^e q_n^e\,ds,\end{aligned}\tag{4.3.27}$$

or

$$0=\sum_{j=1}^{n}K_{ij}^e\,u_j^e-f_i^e-Q_i^e\quad\text{or}\quad\mathbf{K}^e\mathbf{u}^e=\mathbf{f}^e+\mathbf{Q}^e,\tag{4.3.28}$$

where

$$K_{ij}^e=\int_{\Omega^e}\left(a_{rr}\frac{\partial\psi_i^e}{\partial r}\frac{\partial\psi_j^e}{\partial r}+a_{zz}\frac{\partial\psi_i^e}{\partial z}\frac{\partial\psi_j^e}{\partial z}\right)r\,dr\,dz\tag{4.3.29a}$$

$$f_i^e=\int_{\Omega^e}\psi_i^e f(r,z)\,r\,dr\,dz,\quad Q_i^e=\oint_{\Gamma_e}\psi_i^e q_n^e\,ds.\tag{4.3.29b}$$

When there is convection on the boundary, the heat flux normal to the boundary, q_n^e, in Eq. (4.3.27) is replaced by $\hat{q}_n^e-\beta(u-u_\infty)$ [see Eq. (4.3.2b)]; the finite element model becomes

$$(\mathbf{K}^e+\mathbf{H}^e)\,\mathbf{u}^e=\mathbf{f}^e+\mathbf{P}^e+\mathbf{Q}^e,\tag{4.3.30a}$$

$$\begin{aligned}K_{ij}^e&=\int_{\Omega^e}\left(k_{rr}\frac{\partial\psi_i^e}{\partial r}\frac{\partial\psi_j^e}{\partial r}+k_{zz}\frac{\partial\psi_i^e}{\partial z}\frac{\partial\psi_j^e}{\partial z}\right)r\,dr\,dz,\quad H_{ij}^e=\oint_{\Gamma_e}\beta\psi_i^e\psi_j^e\,ds\\f_i^e&=\int_{\Omega^e}\psi_i^e f(r,z)\,r\,dr\,dz,\quad P_i^e=\oint_{\Gamma_e}\beta u_\infty\psi_i^e\,ds,\quad Q_i^e=\oint_{\Gamma_e}\hat{q}_n^e\psi_i^e\,ds.\end{aligned}\tag{4.3.30b}$$

4.3.6 Advection–Diffusion Equation

4.3.6.1 Governing equation

Consider the two-dimensional, steady-state, advection–diffusion equation in a Cartesian rectangular coordinate system

$$\rho\left(v_x\frac{\partial u}{\partial x}+v_y\frac{\partial u}{\partial y}\right)-\left[\frac{\partial}{\partial x}\left(k_{xx}\frac{\partial u}{\partial x}\right)+\frac{\partial}{\partial y}\left(k_{yy}\frac{\partial u}{\partial y}\right)\right]=g(x,y)\ \text{ in }\ \Omega\tag{4.3.31}$$

where v_x and v_y are the components in the x and y directions, respectively, of the velocity vector $\mathbf{v}$, (k_{xx},k_{yy},g) are the data, and Ω is the domain with closed boundary Γ of the problem. The dependent unknown u is required to satisfy one of the following two types of boundary conditions at a boundary point:

$$u=\hat{u}(s)\quad\text{on}\quad\Gamma_u,\tag{4.3.32a}$$

$$\left(k_{xx}\frac{\partial u}{\partial x}n_x+k_{yy}\frac{\partial u}{\partial y}n_y\right)=\hat{q}_n(s)\quad\text{on}\quad\Gamma_q,\tag{4.3.32b}$$

where Γ_u and Γ_q are disjoint portions of the total boundary Γ such that $\Gamma = \Gamma_u \cup \Gamma_q$, β is the heat transfer coefficient, and (n_x, n_y) are the direction cosines of the unit normal vector $\hat{\mathbf{n}}$ on the boundary Γ.

Assume that k_{xx} and k_{yy} are constants and let

$$\begin{aligned} \bar{x} &= \frac{x}{a}, \quad \bar{y} = \frac{y}{b}, \quad \beta = \frac{v_y}{v_x}, \quad \alpha = \frac{b}{a} \\ \gamma &= \frac{k_{yy}}{k_{xx}}, \quad Pe = \frac{\rho v_x a}{k_{xx}}, \quad \bar{g} = \frac{ag}{v_x \rho} \end{aligned} \tag{4.3.33}$$

Then Eq. (4.3.31) becomes

$$\frac{\partial u}{\partial \bar{x}} + \frac{\beta}{\alpha}\frac{\partial u}{\partial \bar{y}} - \frac{1}{Pe}\left(\frac{\partial^2 u}{\partial \bar{x}^2} + \frac{\gamma}{\alpha^2}\frac{\partial^2 u}{\partial \bar{y}^2}\right) = \bar{g} \quad \text{in } \Omega \tag{4.3.34}$$

For an isotropic material (i.e., $k_{xx} = k_{yy} = k$ and $\gamma = 1$) with $v_x = v_y$ (i.e., $\beta = 1$) and a square domain (i.e., $a = b$ and $\alpha = 1$), Eq. (4.3.34) becomes

$$\frac{\partial u}{\partial \bar{x}} + \frac{\partial u}{\partial \bar{y}} - \frac{1}{Pe}\left(\frac{\partial^2 u}{\partial \bar{x}^2} + \frac{\partial^2 u}{\partial \bar{y}^2}\right) = \bar{g} \quad \text{in } \Omega \tag{4.3.35}$$

In the following, we use Eq. (4.3.34) but omit the over bars on the variables to develop the FEM model.

4.3.6.2 Finite element model

The weak form of Eq. (4.3.35) over a typical finite element Ω^e is

$$\begin{aligned} 0 = &\int_{\Omega^e} \left[w_i \left(\frac{\partial u}{\partial x} + \frac{\partial u}{\partial y}\right) + \frac{1}{Pe}\left(\frac{\partial w_i}{\partial x}\frac{\partial u}{\partial x} + \frac{\partial w_i}{\partial y}\frac{\partial u}{\partial y}\right) - w_i g \right] dxdy \\ &- \int_{\Gamma^e} w_i q_n \, ds, \end{aligned} \tag{4.3.36}$$

where w_i is the weight function and q_n is the diffusion flux

$$q_n = \frac{1}{Pe}\left(n_x \frac{\partial u}{\partial x} + n_y \frac{\partial u}{\partial y}\right). \tag{4.3.37}$$

Substitution of the finite element approximation from Eq. (4.3.5) for u and $w_i = \psi_i^e$ into the weak form in Eq. (4.3.36), we obtain the following finite element model:

$$\left(\mathbf{A}^e + \tfrac{1}{Pe}\mathbf{D}^e\right)\mathbf{u}^e = \mathbf{g}^e + \mathbf{q}^e \tag{4.3.38}$$

where (we note that the coefficient matrix $\mathbf{A}^e$ is not symmetric)

$$\begin{aligned} A_{ij}^e &= \int_{\Omega^e} \psi_i^e \left(\frac{\partial \psi_j^e}{\partial x} + \frac{\partial \psi_j^e}{\partial y}\right) dxdy, \\ D_{ij}^e &= \frac{1}{Pe}\int_{\Omega^e} \left(\frac{\partial \psi_i^e}{\partial x}\frac{\partial \psi_j^e}{\partial x} + \frac{\partial \psi_i^e}{\partial y}\frac{\partial \psi_j^e}{\partial y}\right) dxdy, \end{aligned} \tag{4.3.39}$$

$$g_i^e = \int_{\Omega^e} \psi_i^e g \, dxdy, \quad q_i^e = \int_{\Gamma^e} \psi_i^e q_n \, ds. \tag{4.3.40}$$

4.3.7 Linear Finite Elements and Evaluation of Coefficients

4.3.7.1 Properties of interpolation functions

The finite element approximation $u_h^e(x, y)$ of $u(x, y)$ over an element Ω^e must satisfy the following conditions in order for the approximate solution to converge to the true solution of a problem.

(1) $u_h^e(x, y)$ must be continuous as required in the weak form of the problem; that is, all terms in the weak form are represented as nonzero values.

(2) The polynomials used to represent $u_h^e(x, y)$ must be complete and contain both x and y of the same order (i.e., equipresence of x and y). This means all terms, beginning with a constant term up to the highest order desired, in both x and y, should be included in the polynomial.

(3) All terms in the polynomial should be linearly independent.

The number of linearly independent terms in the representation of u_h^e dictates the geometry and the number of nodes in the element. It turns out that only triangular and quadrilateral shapes meet the interpolation properties stated above. Here we discuss the interpolation functions of triangular and rectangular elements with linear and quadratic approximations. The interpolation functions are readily available in all finite element books, but the intent here is to show how they are derived for the lower-order elements.

4.3.7.2 Linear triangular element

An examination of the statement (4.3.19) and the finite element matrices in Eq. (4.3.21) shows that ψ_i^e $(i = 1, 2, \ldots, n)$ should be, at least, a linear function of x and y. The lowest-order linear polynomial that meets this requirements is

$$u_h^e(x, y) = c_1^e + c_2^e\, x + c_3^e\, y. \tag{4.3.41}$$

The polynomial is complete because the constant and linear terms in x and y are included. To write the three parameters (c_1^e, c_2^e, c_3^e) in terms of three values of u_h^e, we must identify three points, (x_i^e, y_i^e), $i = 1, 2, 3$, that uniquely define the geometry of the element Ω^e and allow the imposition of interelement continuity of the primary variable $u_h^e(x, y)$. Obviously, the geometric shape defined by three noncollinear points in a two-dimensional domain is a triangle. Thus, the polynomial in Eq. (4.3.41) is associated with a triangular element, with the vertices of the triangle being the three nodes [see Fig. 4.3.3(a)].

The linear interpolation functions for an arbitrary three-node triangular element are derived by evaluating Eq. (4.3.41) at (x_i^e, y_i^e):

$$\begin{aligned}
u_1^e &\equiv u_h^e(x_1^e, y_1^e) = c_1^e + c_2^e x_1^e + c_3^e y_1^e \\
u_2^e &\equiv u_h^e(x_2^e, y_2^e) = c_1^e + c_2^e x_2^e + c_3^e y_2^e \\
u_3^e &\equiv u_h^e(x_3^e, y_3^e) = c_1^e + c_2^e x_3^e + c_3^e y_3^e,
\end{aligned}$$

which can be expressed in matrix form as

$$\begin{Bmatrix} u_1^e \\ u_2^e \\ u_3^e \end{Bmatrix} = \begin{bmatrix} 1 & x_1^e & y_1^e \\ 1 & x_2^e & y_2^e \\ 1 & x_3^e & y_3^e \end{bmatrix} \begin{Bmatrix} c_1^e \\ c_2^e \\ c_3^e \end{Bmatrix}. \tag{4.3.42}$$

Solving Eq. (4.3.42) for (c_1^e, c_2^e, c_3^e) and substituting into Eq. (4.3.41), we obtain

$$u_h^e(x,y) = \psi_1^e(x,y)u_1^e + \psi_2^e(x,y)u_2^e + \psi_3^e(x,y)u_3^e = \sum_{j=1}^{n=3} \psi_j^e(x,y)u_j^e, \quad (4.3.43a)$$

where

$$\psi_i^e(x,y) = \frac{1}{2A_e}\left(\alpha_i^e + \beta_i^e x + \gamma_i^e y\right), \qquad (i = 1,2,3). \quad (4.3.43b)$$

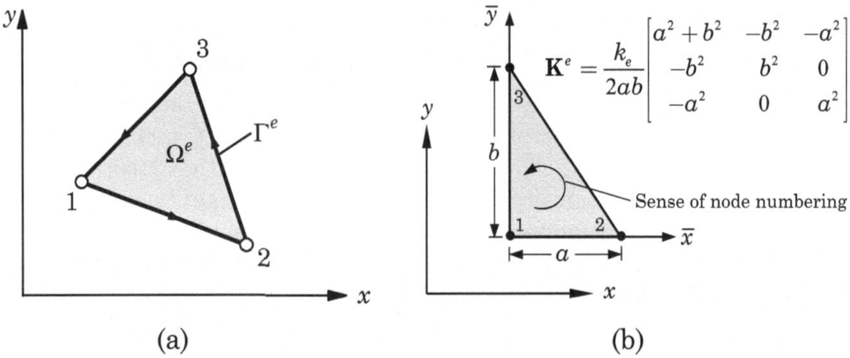

Fig. 4.3.3 (a) General linear triangular element. (b) Right-angled linear triangle element with coefficient matrix $\mathbf{K}^e$ of Eq. (4.3.21) when $a_{xx} = a_{yy} = k^e$.

Here A_e is the area of the triangle [or $2A_e$ is the determinant of the coefficient matrix in Eq. (4.3.42)], and α_i^e, β_i^e, and γ_i^e are the constants which depend only on the nodal global coordinates (x_i^e, y_i^e) of the triangular element Ω^e:

$$\alpha_i^e = x_j^e y_k^e - x_k^e y_j^e; \qquad \beta_i^e = y_j^e - y_k^e; \qquad \gamma_i^e = -(x_j^e - x_k^e) \quad (4.3.43c)$$

for $i \neq j \neq k$, and i, j, and k permute in a natural order. The interpolation functions ψ_i^e $(i = 1, 2, \ldots, n)$ satisfy the following interpolation properties:

$$(1) \quad \psi_i^e(x_j^e, y_j^e) = \delta_{ij}, \quad (i,j = 1,2,3); \qquad (2) \quad \sum_{i=1}^{3} \psi_i^e(x,y) = 1. \quad (4.3.44)$$

The use of linear approximation of the actual function $u(x, y)$, which is possibly a surface, results in a planar shape. For linear triangular element, the derivatives of u_h^e are constant:

$$\frac{\partial \psi_i^e}{\partial x} = \frac{\beta_i^e}{2A_e}, \quad \frac{\partial \psi_i^e}{\partial y} = \frac{\gamma_i^e}{2A_e}. \quad (4.3.45)$$

The integrals in the definition of the coefficients K_{ij}^e and f_i^e can be evaluated for given data: a_{xx}^e, a_{yy}^e, and f_e (all of which can be functions of x and y). For example, for element-wise constant values of the data, the element coefficient matrix and source vector from Eq. (4.3.21) for linear triangular element become

$$K_{ij}^e = \frac{1}{4A_e}\left(a_{xx}^e \beta_i^e \beta_j^e + a_{yy}^e \gamma_i^e \gamma_j^e\right), \quad f_i^e = \frac{f_e A_e}{3}. \tag{4.3.46}$$

In particular, for a right-angled triangular element with base a and height b, and with node 1 at the right angle [see Fig. 4.3.3(b)], $\mathbf{K}^e$ takes the form

$$\mathbf{K}^e = \frac{a_{xx}^e}{2}\begin{bmatrix} \lambda^e & -\lambda^e & 0 \\ -\lambda^e & \lambda^e & 0 \\ 0 & 0 & 0 \end{bmatrix} + \frac{a_{yy}^e}{2}\begin{bmatrix} \mu^e & 0 & -\mu^e \\ 0 & 0 & 0 \\ -\mu^e & 0 & \mu^e \end{bmatrix}, \tag{4.3.47}$$

where $\lambda^e = b/a$ and $\mu^e = a/b$. Of course, for cases in which the conductivity coefficients a_{xx}^e and a_{yy}^e are functions of (x, y), numerical integration (see Section 1.8) can be used to evaluate the coefficients. We note that the first part of the coefficient matrix contains an aspect ratio b/a and the second part contains the reciprocal of the aspect ratio. When a is very large compared with b, the second part contains large numbers compared with the first part, for the same orders of magnitudes of a_{xx} and a_{yy}; hence, one must maintain an aspect ratio that does not unduly affect the physics.

4.3.7.3 Linear rectangular element

The next polynomial that meets the requirements is

$$u_h^e(x, y) = c_1^e + c_2^e x + c_3^e y + c_4^e xy, \tag{4.3.48}$$

which contains four linearly independent terms, $\{1\ x\ y\ xy\}$. In order to represent Eq. (4.3.48) in terms of the values of $u_h^e(x, y)$, an element with four noncollinear points, with linear variation along any side of the element, must be identified. In general, such an element is a quadrilateral, with four corner points being the nodes. It is simpler to derive the approximation functions for a rectangle (see Fig. 4.3.4). When the element is quadrilateral in shape, we use coordinate transformations to represent the integrals posed on the quadrilateral element to those over a square geometry and then use numerical integration to evaluate the integrals (see Section 1.8.4).

For a linear rectangular element (called a *bilinear element* because it contains a term that is linear in both x and y; see Fig. 4.3.4), we have

$$u_h^e(\bar{x}, \bar{y}) = \sum_{i=1}^{4} u_i^e \psi_i^e(\bar{x}, \bar{y}), \tag{4.3.49}$$

where ψ_i^e are the Lagrange interpolation functions

$$\begin{aligned} \psi_1^e &= \left(1 - \frac{\bar{x}}{a}\right)\left(1 - \frac{\bar{y}}{b}\right), \qquad & \psi_2^e &= \frac{\bar{x}}{a}\left(1 - \frac{\bar{y}}{b}\right), \\ \psi_3^e &= \frac{\bar{x}}{a}\frac{\bar{y}}{b}, & \psi_4^e &= \left(1 - \frac{\bar{x}}{a}\right)\frac{\bar{y}}{b}, \end{aligned} \tag{4.3.50}$$

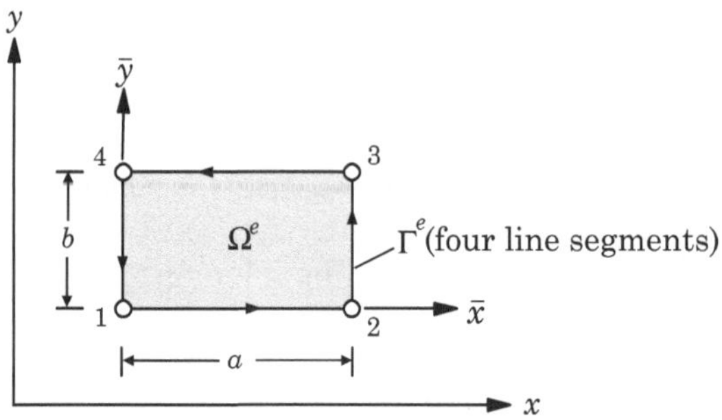

Fig. 4.3.4 The bilinear rectangular finite element.

and $(\bar{x}, \bar{y})$ denote the local coordinates with the origin located at node 1 of the element, and (a, b) denote the horizontal and vertical dimensions of the rectangle, as shown in Fig. 4.3.4.

The integrals in the definition of K_{ij}^e and f_i^e can be easily evaluated over a rectangular element of sides a and b. For example, for element-wise constant values of the data, we have (see Reddy [8]) the following results:

$$\mathbf{K}^e = a_{xx}^e \mathbf{S}^{11} + a_{yy}^e \mathbf{S}^{22}, \qquad f_i^e = \frac{f_e ab}{4}, \tag{4.3.51a}$$

where the matrices $\mathbf{S}^{11}$ and $\mathbf{S}^{22}$ are defined as ($\lambda = b/a$ and $\mu = a/b$; element label e is omitted here):

$$\mathbf{S}^{11} = \frac{1}{6}\begin{bmatrix} 2\lambda & -2\lambda & -\lambda & \lambda \\ -2\lambda & 2\lambda & \lambda & -\lambda \\ -\lambda & \lambda & 2\lambda & -2\lambda \\ \lambda & -\lambda & -2\lambda & 2\lambda \end{bmatrix}, \quad \mathbf{S}^{22} = \frac{1}{6}\begin{bmatrix} 2\mu & \mu & -\mu & -2\mu \\ \mu & 2\mu & -2\mu & -\mu \\ -\mu & -2\mu & 2\mu & \mu \\ -2\mu & -\mu & \mu & 2\mu \end{bmatrix}. \tag{4.3.51b}$$

Because the coefficient matrix $\mathbf{K}^e$ depends on the aspect ratios λ and μ, their values should not be too large or too small (compared with unity). Depending on the problem, values of the order 10 or 0.1 are considered to be reasonable.

We note that the heat flux contributions to a global node come from the line integrals of the distributed fluxes on the sides of all elements connected at the node. In reality, the Q_i^e $(i = 1, 2, \ldots, n)$ of an element which shares all its sides with other elements are not known because q_n^e on its sides is unknown. However, the sum of Q_i^e from all elements at a global node is equal to heat applied (externally) at the node because the contributions of q_n^e from the neighboring elements at the element interfaces inside the domain are canceled. In other words, q_n^e on side (i, j) of element Ω^e cancels with q_n^f on side (p, q) of element Ω^f when sides (i, j) of element Ω^e and (p, q) of element Ω^f are the same (i.e., at the interface of elements Ω^e and Ω^f; see Fig. 4.3.5). This can be viewed as the balance of the internal flux. When an element boundary coincides with the problem boundary, the nodal contribution Q_i^e due to the distributed flux q_n^e on that boundary can be evaluated with the help of Eq. (4.3.30b).

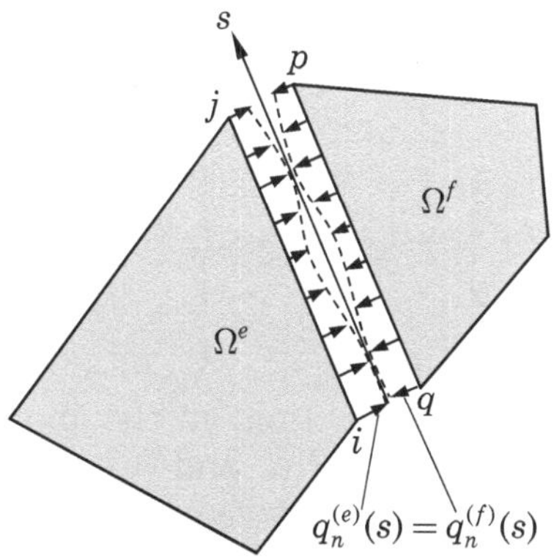

Fig. 4.3.5 Balance of fluxes at the inter-element boundary of elements Ω^e and Ω^f.

4.3.8 Higher-Order Finite Elements

4.3.8.1 Triangular elements

Higher-order triangular elements (i.e., triangular elements with interpolation functions of higher degree) can be systematically developed with the help of the so-called *area coordinates.* For triangular elements, it is possible to construct three natural coordinates L_i ($i = 1, 2, 3$), as in the case of one-dimensional elements, which vary in a direction normal to the sides directly opposite each node, as shown in Fig. 4.3.6(a). The coordinates are defined as

$$\psi_i = L_i = \frac{A_i}{A}\,, \qquad A = \text{ area of the triangle,} \tag{4.3.52}$$

where A_i is the area of the triangle formed by nodes j and k and an arbitrary point P in the element, and A is the total area of the element. For example, A_1 is the area of the shaded triangle that is formed by nodes 2 and 3 and point P. Therefore, A_1 is given by [see Fig. 4.3.6(a)]

$$A_1 = \frac{1}{2}\left(\alpha_1 + \beta_1 x + \gamma_1 y\right); \quad \psi_1 = L_1 = \frac{A_1}{A} = \frac{1}{2A}\left(\alpha_1 + \beta_1 x + \gamma_1 y\right). \tag{4.3.53}$$

Similarly, $2A_2 = \alpha_2 + \beta_2 x + \gamma_2 y$ and $2A_3 = \alpha_3 + \beta_3 x + \gamma_3 y$. Thus, for the linear triangular element, we have

$$\psi_i(x, y) = L_i(x, y). \tag{4.3.54}$$

The area coordinates (L_1, L_2, L_3) can be used to construct interpolation functions for higher-order triangular elements. For example, in the case of quadratic triangular element ψ_1 is given by [see Fig. 4.3.6(b)] $\psi_1 = c_1(L_1 - 0)(L_1 - 0.5)$, where c_1 is determined such that $\psi_1 = 1$ when $L_1 = 1$, giving $c_1 = 2$. Hence,

$$\psi_1 = L_1(2L_1 - 1).$$

The explicit forms of the interpolation functions for the linear and quadratic elements are

$$\mathbf{\Psi}^e = \begin{Bmatrix} L_1 \\ L_2 \\ L_3 \end{Bmatrix}; \qquad \mathbf{\Psi}^e = \begin{Bmatrix} L_1(2L_1 - 1) \\ L_2(2L_2 - 1) \\ L_3(2L_3 - 1) \\ 4L_1L_2 \\ 4L_2L_3 \\ 4L_3L_1 \end{Bmatrix}. \tag{4.3.55}$$

The ordering of the interpolation functions in the above arrays corresponds to the node numbers shown in Figs. 4.3.7(a) and (b). Integrals of the products of the area coordinates L_i over an element can be evaluated using the following formula:

$$\int_A L_1^p \, L_2^q \, L_3^r \, dA = \frac{p!\, q!\, r!}{(p+q+r+1)!} A. \tag{4.3.56}$$

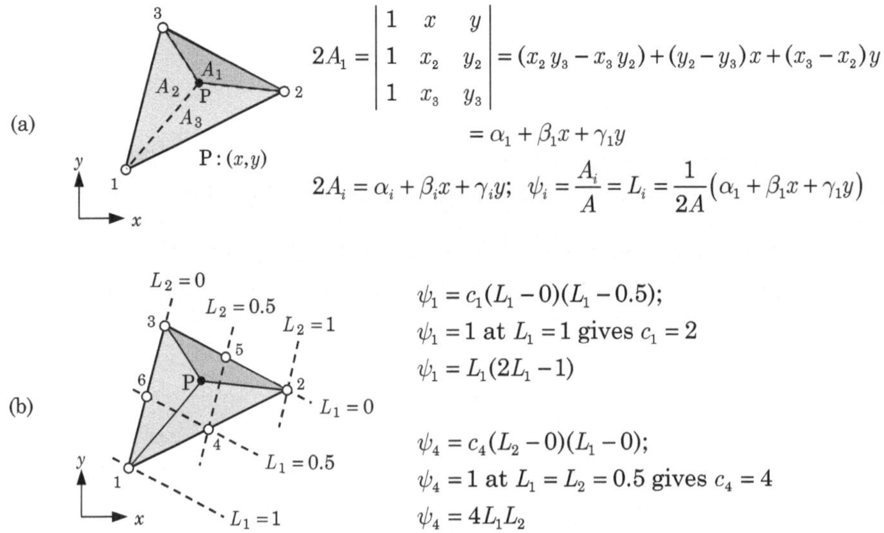

Fig. 4.3.6 (a) Graphical and analytical representations of area coordinates L_i for triangular elements. (b) Determination of ψ_i for the quadratic triangular element.

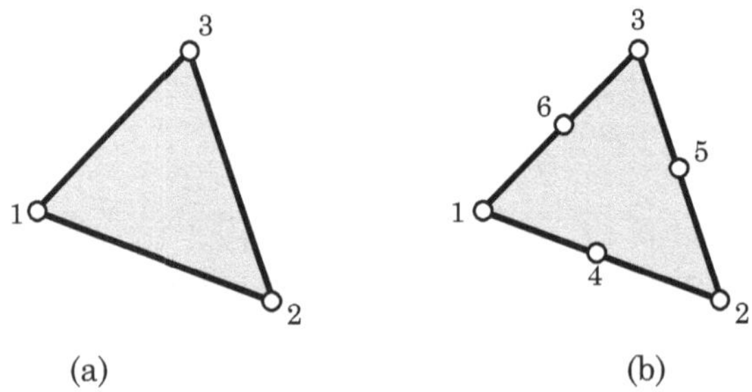

Fig. 4.3.7 (a) Linear and (b) quadratic triangular elements.

4.3.8.2 Rectangular elements

The linear and quadratic Lagrange interpolation functions associated with rectangular family of master finite elements can be obtained from the tensor product of the corresponding one-dimensional Lagrange interpolation functions. The one-dimensional linear and quadratic interpolation functions [see, e.g., Eq. (4.2.29a)] can be expressed in terms of the local coordinates ξ, $-1 \leq \xi \leq 1$, as

$$\begin{aligned} &\text{linear:} && \psi_1 = \tfrac{1}{2}(1-\xi), \quad \psi_2 = \tfrac{1}{2}(1+\xi), \\ &\text{quadratic:} && \psi_1 = -\tfrac{1}{2}\xi(1-\xi), \quad \psi_2 = (1-\xi^2), \quad \psi_3 = \tfrac{1}{2}\xi(1+\xi). \end{aligned} \tag{4.3.57}$$

The approximation functions for the linear rectangular element are obtained by taking the product of the column vector with the row vector:

$$\frac{1}{2}\left\{ \begin{matrix} (1-\xi) \\ (1+\xi) \end{matrix} \right\} \frac{1}{2}\{ (1-\eta) \quad (1+\eta) \}$$

gives four functions, which are placed in the same order as the node numbers shown in Fig. 4.3.8(a):

$$\mathbf{\Psi}^e = \frac{1}{4}\left\{ \begin{matrix} (1-\xi)(1-\eta) \\ (1+\xi)(1-\eta) \\ (1+\xi)(1+\eta) \\ (1-\xi)(1+\eta) \end{matrix} \right\}. \tag{4.3.58}$$

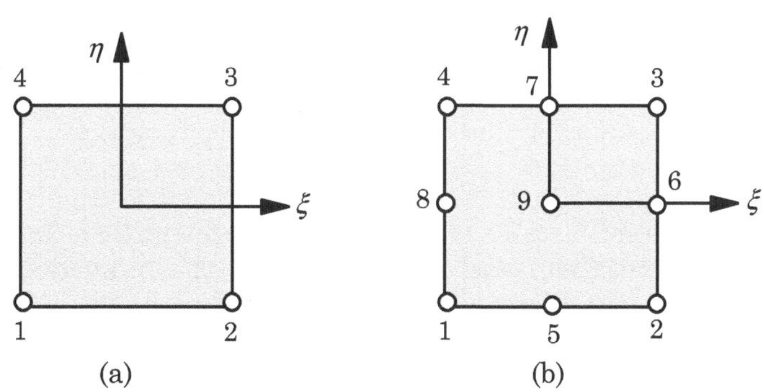

Fig. 4.3.8 (a) Linear and (b) quadratic rectangular elements.

Similarly, the tensor product of the one-dimensional quadratic functions for the two coordinate directions (ξ, η) yields the interpolation functions for the (master) quadratic rectangular element [the resulting quadratic interpolations from the tensor product are renumbered to match the node numbering shown in Fig. 4.3.8(b), which is the common numbering system used in most commercial codes]:

$$\boldsymbol{\Psi}^e = \frac{1}{4}\left\{\begin{array}{c} (1-\xi)(1-\eta)(-\xi-\eta-1)+(1-\xi^2)(1-\eta^2) \\ (1+\xi)(1-\eta)(\xi-\eta-1)+(1-\xi^2)(1-\eta^2) \\ (1+\xi)(1+\eta)(\xi+\eta-1)+(1-\xi^2)(1-\eta^2) \\ (1-\xi)(1+\eta)(-\xi+\eta-1)+(1-\xi^2)(1-\eta^2) \\ 2[(1-\xi^2)(1-\eta)-(1-\xi^2)(1-\eta^2)] \\ 2[(1+\xi)(1-\eta^2)-(1-\xi^2)(1-\eta^2)] \\ 2[(1-\xi^2)(1+\eta)-(1-\xi^2)(1-\eta^2)] \\ 2[(1-\xi)(1-\eta^2)-(1-\xi^2)(1-\eta^2)] \\ 4(1-\xi^2)(1-\eta^2) \end{array}\right\}. \tag{4.3.59}$$

The *serendipity elements* are those elements that have no interior nodes. These elements have fewer nodes compared with the same-order complete Lagrange elements. The approximation functions ψ_i^e of the serendipity elements are not complete (in the sense that they are not complete polynomials), and they cannot be obtained using tensor products of one-dimensional Lagrange interpolation functions. Instead, an alternative procedure must be employed (see Reddy [8] for the details). The lowest-order serendipity element in two dimensions is the eight-node quadratic element shown in Fig. 4.3.9.

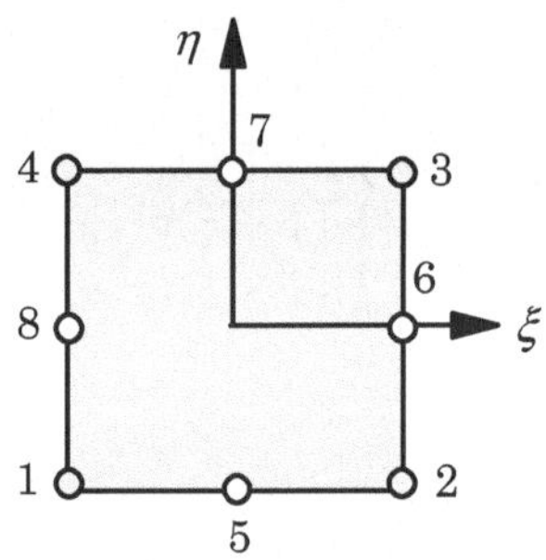

Fig. 4.3.9 Quadratic rectangular serendipity element.

The approximation functions of the eight-node quadratic element, compared with the nine-quadratic element, do not contain the bi-quadratic term $\xi^2\eta^2$ and, therefore, they are incomplete quadratic polynomials. However, ψ_i^e of the serendipity elements do satisfy the Kronecker-delta property [i.e., property (i) in Eq. (4.3.34)]. The interpolation functions for the eight-node quadratic serendipity element are given in Eq. (4.3.50).

$$\boldsymbol{\Psi}^e = \frac{1}{4}\left\{\begin{array}{c} (1-\xi)(1-\eta)(-\xi-\eta-1) \\ (1+\xi)(1-\eta)(\xi-\eta-1) \\ (1+\xi)(1+\eta)(\xi+\eta-1) \\ (1-\xi)(1+\eta)(-\xi+\eta-1) \\ 2(1-\xi^2)(1-\eta) \\ 2(1+\xi)(1-\eta^2) \\ 2(1-\xi^2)(1+\eta) \\ 2(1-\xi)(1-\eta^2) \end{array}\right\}. \tag{4.3.60}$$

4.3.9 Assembly of Elements

In the FEM, the element equations are algebraic relations among the duality pairs of the nodes in that element and do not contain the nodal values of the other elements (even those elements connected to the element). It is necessary to employ the continuity of the primary variables and balance of the secondary variables to eliminate the secondary variables (which are unknown) between elements (which share boundaries that are interior to the domain).

The assembly of finite element equations to obtain the equations of the entire domain is based on the following two physical requirements:

(1) Continuity of the primary variable $u_h^e(\mathbf{x})$ (e.g., displacements and temperature).
(2) Equilibrium of secondary variables Q_i^e (e.g., forces and heats).

We illustrate the assembly procedure by considering a finite element mesh consisting of a triangular element and a quadrilateral element, as shown in Fig. 4.3.10.

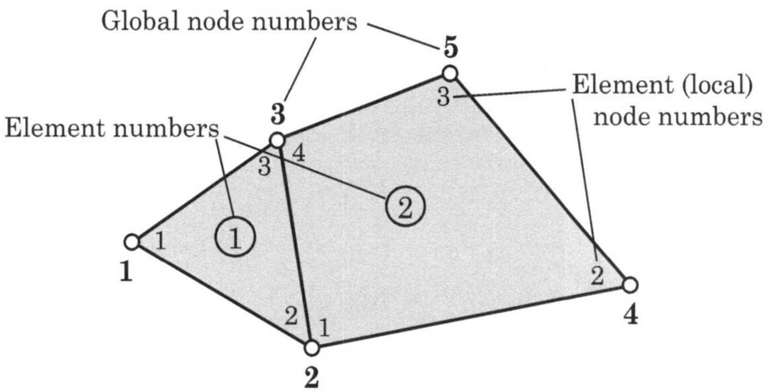

Fig. 4.3.10 Global-local correspondence of nodes for element assembly of two elements.

The nodes of the finite element mesh are called *global nodes*, whereas those of a typical element are called *element nodes*. From the mesh shown in Fig. 4.3.10, it is clear that the following correspondence between global and element nodes exists: nodes 1, 2, and 3 of element 1 correspond to global nodes 1, 2, and 3, respectively. Nodes 1, 2, 3, and 4 of element 2 correspond to global nodes 2, 4, 5, and 3, respectively. Hence, the correspondence between the local (u_i^e) and global (U_I) nodal values of the primary variable u is as follows:

$$u_1^{(1)} = U_1, \;\; u_2^{(1)} = u_1^{(2)} = U_2, \;\; u_3^{(1)} = u_4^{(2)} = U_3, \;\; u_2^{(2)} = U_4, \;\; u_3^{(2)} = U_5, \quad (4.3.61)$$

where the superscripts refer to the element numbers, and the subscripts refer to the element node numbers. This amounts to imposing the continuity of the primary variables at the nodes common to elements 1 and 2. Note that the continuity of the primary variables at the interelement nodes guarantees the continuity of the primary variable along the entire interelement boundary, that is $u^{(1)}(s) = u^{(2)}(s)$, where s denotes a coordinate along the interface of the two elements.

Next, we consider the equilibrium of secondary variables at the interelement boundaries. At the interface between any two elements, the flux from the two elements should be equal in magnitude and opposite in sign. For the two elements shown in Fig. 4.3.10, the interface is along the side connecting global nodes 2 and 3. Hence, the heat flux $q_n^{(1)}$ on side 2–3 of element 1 should balance, in integral sense, the flux $q_n^{(2)}$ on side 4–1 of element 2 (recall the sign convention on $q_n^{(e)}$):

$$\int_{h_{23}^{(1)}} q_n^{(1)} \psi_2^{(1)}\, ds = -\int_{h_{14}^{(2)}} q_n^{(2)} \psi_1^{(2)}\, ds \quad \text{or} \quad \int_{h_{23}^{(1)}} q_n^{(1)} \psi_3^{(1)}\, ds = -\int_{h_{14}^{(2)}} q_n^{(2)} \psi_4^{(2)}\, ds, \tag{4.3.62}$$

where $h_{pq}^{(e)}$ denotes length of the side connecting nodes p and q of element Ω^e.

Now we are ready to assemble the element equations for the two-element mesh. Let $K_{ij}^{(1)}$ $(i, j = 1, 2, 3)$ denote the coefficient matrix corresponding to the triangular element, and let $K_{ij}^{(2)}$ $(i, j = 1, 2, 3, 4)$ denote the coefficient matrix corresponding to the quadrilateral element. The element equations of the two elements are written separately first. The element equations of the triangular element are of the form

$$\begin{aligned}
K_{11}^{(1)} u_1^{(1)} + K_{12}^{(1)} u_2^{(1)} + K_{13}^{(1)} u_3^{(1)} &= f_1^{(1)} + Q_1^{(1)} \\
K_{21}^{(1)} u_1^{(1)} + K_{22}^{(1)} u_2^{(1)} + K_{23}^{(1)} u_3^{(1)} &= f_2^{(1)} + Q_2^{(1)} \\
K_{31}^{(1)} u_1^{(1)} + K_{32}^{(1)} u_2^{(1)} + K_{33}^{(1)} u_3^{(1)} &= f_3^{(1)} + Q_3^{(1)}.
\end{aligned} \tag{4.3.63}$$

Similarly, the finite element equations for the rectangular element are

$$\begin{aligned}
K_{11}^{(2)} u_1^{(2)} + K_{12}^{(2)} u_2^{(2)} + K_{13}^{(2)} u_3^{(2)} + K_{14}^{(2)} u_4^{(2)} &= f_1^{(2)} + Q_1^{(2)} \\
K_{21}^{(2)} u_1^{(2)} + K_{22}^{(2)} u_2^{(2)} + K_{23}^{(2)} u_3^{(2)} + K_{24}^{(2)} u_4^{(2)} &= f_2^{(2)} + Q_2^{(2)} \\
K_{31}^{(2)} u_1^{(2)} + K_{32}^{(2)} u_2^{(2)} + K_{33}^{(2)} u_3^{(2)} + K_{34}^{(2)} u_4^{(2)} &= f_3^{(2)} + Q_3^{(2)} \\
K_{41}^{(2)} u_1^{(2)} + K_{42}^{(2)} u_2^{(2)} + K_{43}^{(2)} u_3^{(2)} + K_{44}^{(2)} u_4^{(2)} &= f_4^{(2)} + Q_4^{(2)}.
\end{aligned} \tag{4.3.64}$$

In order to impose the balance condition in Eq. (4.3.62), it is necessary to add the second equation of element 1 to the first equation of element 2, and also add the third equation of element 1 to the fourth equation of element 2:

$$\begin{aligned}
&\left(K_{21}^{(1)} u_1^{(1)} + K_{22}^{(1)} u_2^{1} + K_{23}^{(1)} u_3^{(1)}\right) + \left(K_{11}^{(2)} u_1^{(2)} + K_{12}^{(2)} u_2^{(2)} + K_{13}^{(2)} u_3^{(2)} + K_{14}^{(2)} u_4^{(2)}\right) \\
&\qquad = \left(f_2^{(1)} + Q_2^{(1)}\right) + \left(f_1^{(2)} + Q_1^{(2)}\right) \\
&\left(K_{31}^{(1)} u_1^{(1)} + K_{32}^{(1)} u_2^{(1)} + K_{33}^{(1)} u_3^{(1)}\right) + \left(K_{41}^{(2)} u_1^{(2)} + K_{42}^{(2)} u_2^{(2)} + K_{43}^{(2)} u_3^{(2)} + K_{44}^{(2)} u_4^{(2)}\right) \\
&\qquad = \left(f_3^{(1)} + Q_3^{(1)}\right) + \left(f_4^{(2)} + Q_4^{(2)}\right).
\end{aligned}$$

Using the local-global nodal variable correspondence in Eq. (4.3.61), we can rewrite the above equations as

$$
\begin{aligned}
&K_{21}^{(1)}U_1 + \left(K_{22}^{(1)} + K_{11}^{(2)}\right)U_2 + \left(K_{23}^{(1)} + K_{14}^{(2)}\right)U_3 + K_{12}^{(2)}U_4 + K_{13}^{(2)}U_5 \\
&\quad = f_2^{(1)} + f_1^{(2)} + \left(Q_2^{(1)} + Q_1^{(2)}\right) \\
&K_{31}^{(1)}U_1 + \left(K_{32}^{(1)} + K_{41}^{(2)}\right)U_2 + \left(K_{33}^{(1)} + K_{44}^{(2)}\right)U_3 + K_{42}^{(2)}U_4 + K_{43}^{(2)}U_5 \\
&\quad = f_3^{(1)} + f_4^{(2)} + \left(Q_3^{(1)} + Q_4^{(2)}\right).
\end{aligned}
\tag{4.3.65}
$$

Now we can impose the conditions in Eq. (4.3.62) by setting appropriate portions of the expressions in parentheses on the right-hand side of Eq. (4.3.65) to zero. This is done by means of the connectivity relations, that is, the correspondence of the element (or local) node number to the global node number.

For elements with sides on the boundary of the computational domain where $q_n^e(s)$ is known, we can evaluate the boundary integral

$$
q_i^e = \oint_{\Gamma^e} \hat{q}_n^e \psi_i^e(s). \tag{4.3.66}
$$

It is necessary to compute such integrals only when Γ^e, or a portion of it, coincides with the boundary Γ_q of the total domain Ω on which the flux q_n^e is specified. When Γ^e falls on the boundary Γ_u of the domain Ω, q_n^e is not known there and can be determined in the post-computation (we note that the primary variable u_h^e is specified on Γ_u).

Another comment is in order on the evaluation of the line integrals. The two-dimensional interpolation functions $\psi_i^e(x, y)$ become one-dimensional function $\psi_i^e(s)$ along the side with coordinate s (see Fig. 4.3.11). Consequently, the evaluation of the line integrals is made simpler, and they are readily available through Eqs. (4.2.29a)–(4.2.30f).

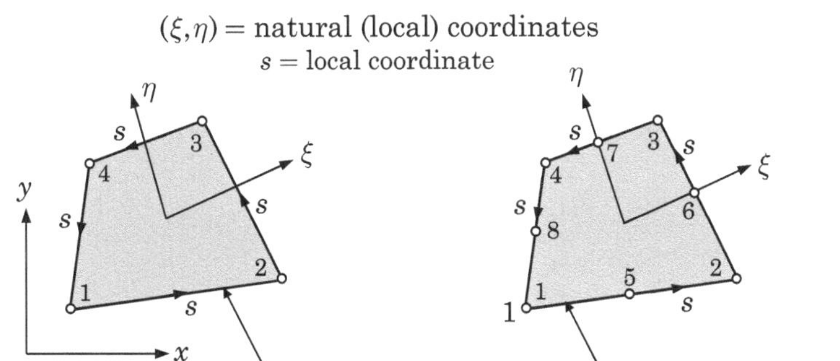

Fig. 4.3.11 Evaluation of line integrals on the boundary of two-dimensional elements.

4.3.10 Numerical Examples

Here we consider three numerical examples to illustrate the assembly and imposition of boundary conditions, including convection. The first example deals with the computation of nodal heats due to distributed boundary flux and contributions due to convection on the boundary.

Example 4.3.1

(a) Give the global coefficients K_{55}, $K_{5(10)}$, $K_{(10)(10)}$, $K_{(10)(15)}$, $K_{(15)(15)}$, F_5, F_{10}, and F_{15}; (b) identify the specified boundary conditions; and (c) compute the nodal contributions of various flux distributions and convection shown for a heat transfer problem with the mesh shown in Fig. 4.3.12. Assume that there is no internal heat generation (i.e., $f = 0$).

Solution (a) From Fig. 4.3.12, the required coefficients are

$$\begin{aligned}
K_{55} &= K_{22}^{(4)} + H_{22}^{(4)}, \quad K_{5(10)} = K_{23}^{(4)} + H_{23}^{(4)} \\
K_{(10)(10)} &= K_{33}^{(4)} + H_{33}^{(4)} + K_{22}^{(8)} + H_{22}^{(8)} \\
K_{(10)(15)} &= K_{23}^{(4)} + H_{23}^{(4)}, \quad K_{(15)(15)} = K_{33}^{(8)} + H_{33}^{(8)} + K_{22}^{(12)} \\
F_5 &= P_2^{(4)}, \quad F_{10} = P_3^{(4)} + P_2^{(8)}, \quad F_{15} = P_3^{(8)} + Q_{15}.
\end{aligned} \tag{1}$$

The numerical values of $K_{ij}^{(e)}$ can be readily obtained from Eqs. (4.3.47) and (4.3.51a). The numerical values of $H_{ij}^{(e)}$ and $P_i^{(e)}$ will be computed in part (c).

(b) From Fig. 4.3.12 it is clear that the known primary nodal degrees of freedom (i.e., temperatures) are:

$$U_1 = U_6 = U_{11} = U_{16} = U_{20} = u_0. \tag{2}$$

The known secondary variables (i.e., heats) are

$$Q_2 = Q_2^{(1)} + Q_1^{(2)} = 0; \quad Q_3 = Q_2^{(2)} + Q_1^{(3)} = 0; \quad Q_4 = Q_2^{(3)} + Q_1^{(4)} = 0. \tag{3}$$

The computation of the known heats at nodes 15, 19, 20, 21, and 22 is discussed next. The contribution of the uniform flux q_0 to nodes 15, 19, and 22 is computed first using Eq. (4.3.21) or (4.3.56) ($h = 2.5$ in.) and utilizing the known result from Eq. (4.2.29c):

$$\begin{aligned}
Q_{15} &= \int_0^h q_0 \psi_1^{(1)}\, ds = \frac{q_0 h}{2}, \quad Q_{22} = \int_0^h q_0 \psi_2^{(2)}\, ds = \frac{q_0 h}{2} \\
Q_{19} &= \int_0^h q_0 \psi_2^{(1)}\, ds + \int_0^h q_0 \psi_1^{(2)}\, ds = q_0 h.
\end{aligned} \tag{4}$$

The contribution of the linearly varying flux $q_1(s/2h_1)$ to nodes 20, 21, and 22 is computed as follows ($h_1 = 2$ in.):

$$Q_{20} = \int_0^{h_1} q_1 \frac{s}{2h_1} \psi_1^{(1)}\, ds = \int_0^{h_1} q_1 \frac{s}{2h_1}\left(1 - \frac{s}{h_1}\right) ds = \frac{q_1 h_1}{12} \tag{5a}$$

$$\begin{aligned}
Q_{21} &= \int_0^{h_1} q_1 \frac{s}{2h_1} \psi_2^{(1)}\, ds + \int_{h_1}^{2h_1} q_1 \frac{s}{2h_1} \psi_1^{(2)}\, ds \\
&= \int_0^{h_1} q_1 \frac{s}{2h_1}\left(\frac{s}{h_1}\right) ds + \int_{h_1}^{2h_1} q_1 \frac{s}{2h_1}\left(2 - \frac{s}{h_1}\right) ds = \frac{q_1 h_1}{6} + \frac{2q_1 h_1}{6} = \frac{q_1 h_1}{2}
\end{aligned} \tag{5b}$$

$$Q_{22} = \int_{h_1}^{2h_1} q_1 \frac{s}{2h_1} \psi_2^{(2)}\, ds = \int_{h_1}^{2h_1} q_1 \frac{s}{2h_1}\left(\frac{s}{h_1} - 1\right) ds = \frac{5q_1 h_1}{12}. \tag{5c}$$

Obviously, Q_{22} just computed should be added to the previously computed value.

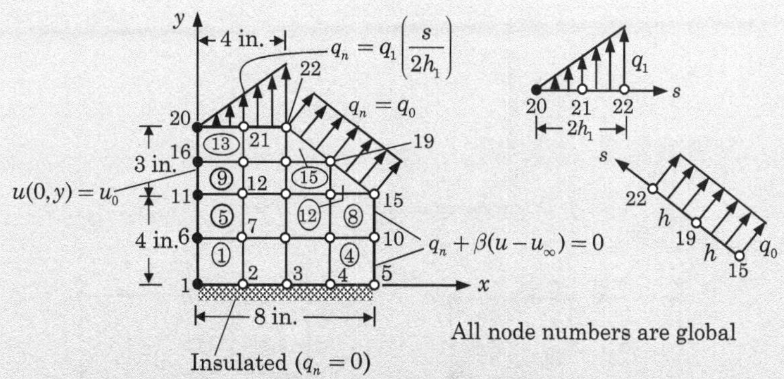

Fig. 4.3.12 A heat transfer problem with various types of boundary conditions.

(c) Lastly, we compute the convective contributions to nodes 5, 10, and 15. The convection contributes to the element coefficients as well as to the right-hand side. Following the discussion of Section 4.3.4, we have

$$\mathbf{H}^{(e)} = \frac{\beta_e h_e}{6}\begin{bmatrix} 2 & 1 \\ 1 & 2 \end{bmatrix}, \quad \mathbf{P}^e = \frac{\beta_e h_e u_\infty}{2}\begin{Bmatrix} 1 \\ 1 \end{Bmatrix}. \tag{6}$$

Since β and $h = 2$ in. is the same for both line elements, we have the following contributions to the global coefficients; note that node 10 has contributions added from the line element below and above:

$$\begin{aligned} H_{22}^{(4)} &= H_{33}^{(4)} = \frac{\beta h}{3}, \quad H_{23}^{(4)} = \frac{\beta h}{6}, \quad P_2^{(4)} = P_3^{(4)} = \frac{\beta u_\infty h}{2} \\ H_{22}^{(8)} &= H_{33}^{(8)} = \frac{\beta h}{3}, \quad H_{23}^{(8)} = \frac{\beta h}{6}, \quad P_2^{(8)} = P_3^{(8)} = \frac{\beta u_\infty h}{2}. \end{aligned} \tag{7}$$

Example 4.3.2

Consider steady-state heat conduction in an isotropic rectangular region of dimensions $3a \times 2a$, as shown in Fig. 4.3.13(a). The origin of the x and y coordinates is taken at the lower left corner such that x is parallel to the side $3a$ and y is parallel to the side $2a$. Boundaries $x = 0$ and $y = 0$ are insulated (i.e., $q_n = 0$), boundary $x = 3a$ is maintained at zero temperature, and boundary $y = 2a$ is maintained at temperature $T = T_0 \cos(\pi x/6a)$. Determine the temperature distribution using (1) mesh of linear triangular elements shown in Fig. 4.3.13(b) and (2) mesh of linear rectangular elements shown in Fig. 4.3.13(c).

Solution The governing equation is a special case of the model equation (4.3.1) with zero internal heat generation $f = 0$ and coefficients $a_{xx} = a_{yy} = k$. Thus, Eq. (4.3.1) takes the form

$$-k\left(\frac{\partial^2 T}{\partial x^2} + \frac{\partial^2 T}{\partial y^2}\right) = 0. \tag{1}$$

The exact solution of Eq. (1) for the boundary conditions shown in Fig. 4.3.13(a) is

$$T(x,y) = T_0 \frac{\cosh(\pi y/6b)\cos(\pi x/6a)}{\cosh(\pi/3)}. \tag{2}$$

The finite element model of Eq. (1) is given by

$$\mathbf{K}^e \mathbf{T}^e = \mathbf{Q}^e, \tag{3}$$

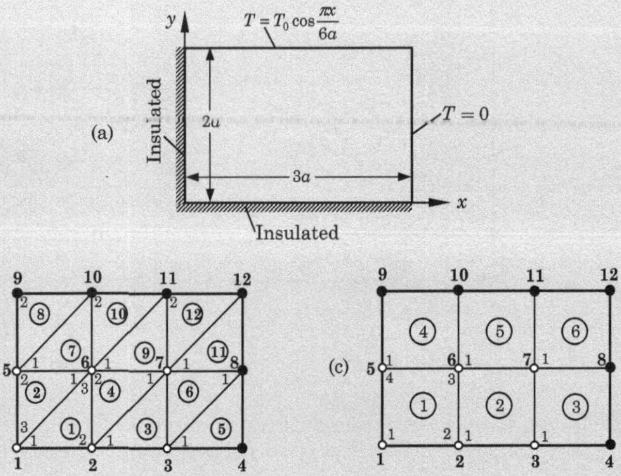

Fig. 4.3.13 Finite element analysis of a heat conduction problem over a rectangular domain: (a) domain; (b) mesh of linear triangular elements; and (c) mesh of linear rectangular elements.

where T_i^e is the temperature at node i of a typical finite element Ω^e, and

$$K_{ij}^e = \int_{\Omega^e} k\left(\frac{\partial\psi_i^{(e)}}{\partial x}\frac{\partial\psi_j^{(e)}}{\partial x} + \frac{\partial\psi_i^{(e)}}{\partial y}\frac{\partial\psi_j^{(e)}}{\partial y}\right) dA, \quad Q_i^e = \oint_{\Gamma_e} q_n\psi_i^{(e)}\, ds. \tag{4}$$

Suppose that we use a 3 × 2 mesh (i.e., three subdivisions along the x-axis and two subdivisions along the y-axis) of linear triangular elements and then a 3 × 2 mesh of linear rectangular elements, as shown in Figs. 4.3.13(b) and 4.3.13(c). Both meshes have the same number of global nodes, namely 12, but differing numbers of elements. There are six unknown nodal values associated with nodes 1, 2, 3, 5, 6, and 7.

Mesh of Triangular Elements. The global node numbers, element numbers, and element node numbers used are shown in Fig. 4.3.13(b). Of course, the global node and element numbering is arbitrary, although the global node numbering dictates the size of the half-bandwidth of the assembled equations, which in turn affects the computational time of Gauss elimination methods used in the solution of algebraic equations in a computer. The element node numbering scheme should be the one that is used in the development of element interpolation functions. By a suitable numbering of the element nodes, all similar elements can be made to have identical element coefficient matrices. Such considerations are important only when hand calculations are carried out.

For a typical element of the mesh of triangles shown in Fig. 4.3.13(b), the element coefficient matrix is given by [see Eq. (4.3.46) and Fig. 4.3.3 for the element matrix and element geometry],

$$\mathbf{K}^e = \frac{k}{2}\begin{bmatrix} 1 & -1 & 0 \\ -1 & 2 & -1 \\ 0 & -1 & 1 \end{bmatrix}, \quad \mathbf{f}^e = \mathbf{0}, \tag{5}$$

where k is the conductivity of the medium. Note that the element matrix is independent of the size of the element, as long as the element is a right-angle triangle with its base equal to its height.

The assembly of the elements follows the logic discussed earlier. From the mesh in Fig. 4.3.13(b), it is clear that global node 1 is connected only to global nodes 2, 5, and 6; hence, the equation associated with node 1, numbered as global equation 1, is of the form

$$K_{11}U_1 + K_{12}U_2 + K_{15}U_5 + K_{16}U_6 = F_6.$$

Similarly, global node 2 is connected only to global nodes 1, 3, 6, and 7, giving the global equation associated with node 2 as

$$K_{21}U_1 + K_{22}U_2 + K_{23}U_3 + K_{26}U_6 + K_{27}U_7 = F_2,$$

and so on. The global coefficients in terms of element coefficients are

$$\begin{aligned} K_{11} &= K_{11}^{(1)} + K_{33}^{(2)} = \tfrac{1}{2}(1+1)k, \quad K_{12} = K_{12}^{(1)} = \tfrac{1}{2}(-1)k \\ K_{15} &= K_{32}^{(2)} = \tfrac{1}{2}(-1)k, \quad K_{16} = K_{13}^{(1)} + K_{31}^{(2)} = 0+0, \quad \text{etc.} \\ K_{21} &= K_{21}^{(1)} = \tfrac{1}{2}(-1)k, \quad K_{22} = K_{22}^{(1)} + K_{11}^{(3)} + K_{33}^{(4)} = \tfrac{1}{2}(2+1+1)k, \quad \text{etc.} \\ F_1 &= Q_1^{(1)} + Q_3^{(2)}, \quad F_2 = Q_2^{(1)} + Q_1^{(3)} + Q_3^{(4)}, \quad \text{etc.} \end{aligned} \tag{6}$$

The boundary conditions on the global nodal temperatures and heats are

$$\begin{aligned} U_4 = U_8 = U_{12} = 0, \quad U_9 = T_0, \quad U_{10} = \tfrac{\sqrt{3}}{2}T_0, \quad U_{11} = \tfrac{1}{2}T_0 \\ F_1 = F_2 = F_3 = F_5 = 0, \end{aligned} \tag{7}$$

and the balance of internal heat flow requires that

$$F_6 = F_7 = 0. \tag{8}$$

Thus, the unknown primary variables and secondary variables are:

$$U_1, \quad U_2, \quad U_3, \quad U_5, \quad U_6, \quad U_7; \quad F_4, \quad F_8, \quad F_9, \quad F_{10}, \quad F_{11}, \quad F_{12}. \tag{9}$$

Condensed Equations for the Unknown Temperatures. We write the six finite element equations for the six unknown primary variables. These equations come from rows 1, 2, 3, 5, 6, and 7 (corresponding to the same global nodes) of the assembled system of matrix equations:

$$\begin{aligned} K_{11}U_1 + K_{12}U_2 + K_{15}U_5 + K_{16}U_6 &= F_1 = 0 \\ K_{21}U_1 + K_{22}U_2 + K_{23}U_3 + K_{26}U_6 + K_{27}U_7 &= F_2 = 0 \\ &\vdots \\ K_{72}U_2 + K_{73}U_3 + K_{76}U_6 + K_{77}U_7 + K_{7(11)}U_{11} &= F_7 = 0. \end{aligned} \tag{10}$$

Using the boundary conditions and the values of K_{IJ}, we obtain

$$\begin{aligned} k(U_1 - \tfrac{1}{2}U_2 - \tfrac{1}{2}U_5) &= 0 \\ k(-\tfrac{1}{2}U_1 + 2U_2 - \tfrac{1}{2}U_3 - U_6) &= 0 \\ k(-\tfrac{1}{2}U_2 + 2U_3 - U_7) &= 0 \\ k(-\tfrac{1}{2}U_1 + 2U_5 - U_6 - \tfrac{1}{2}U_9) &= 0 \quad (U_9 = T_0) \\ k(-U_2 - U_5 + 4U_6 - U_7 - U_{10}) &= 0 \quad (U_{10} = \tfrac{\sqrt{3}}{2}T_0) \\ k(-U_3 - U_6 + 4U_7 - U_{11}) &= 0 \quad (U_{11} = \tfrac{1}{2}T_0), \end{aligned} \tag{11}$$

or, in the matrix form

$$\frac{k}{2}\begin{bmatrix} 2 & -1 & 0 & -1 & 0 & 0 \\ -1 & 4 & -1 & 0 & -2 & 0 \\ 0 & -1 & 4 & 0 & 0 & -2 \\ -1 & 0 & 0 & 4 & -2 & 0 \\ 0 & -2 & 0 & -2 & 8 & -2 \\ 0 & 0 & -2 & 0 & -2 & 8 \end{bmatrix} \begin{Bmatrix} U_1 \\ U_2 \\ U_3 \\ U_5 \\ U_6 \\ U_7 \end{Bmatrix} = \frac{k}{2}\begin{Bmatrix} 0 \\ 0 \\ 0 \\ T_0 \\ \sqrt{3}T_0 \\ T_0 \end{Bmatrix}. \tag{12}$$

The solution of these equations is (in °C)

$$U_1 = 0.6362\,T_0, \quad U_2 = 0.5510\,T_0, \quad U_3 = 0.3181\,T_0$$
$$U_5 = 0.7214\,T_0, \quad U_6 = 0.6248\,T_0, \quad U_7 = 0.3607\,T_0. \tag{13}$$

Evaluating the exact solution in Eq. (2) at the nodes, we have (in °C)

$$T_1 = 0.6249\,T_0, \quad T_2 = 0.5412\,T_0, \quad T_3 = 0.3124\,T_0$$
$$T_5 = 0.7125\,T_0, \quad T_6 = 0.6171\,T_0, \quad T_7 = 0.3563\,T_0. \tag{14}$$

Condensed Equations for the Unknown Secondary Variables. The secondary variable at node 8, for example, can be computed from the eighth finite element equation

$$\begin{aligned} F_8 &= K_{83}U_3 + K_{87}U_7 = \left(K_{13}^{(6)} + K_{31}^{(5)}\right)U_3 + \left(K_{12}^{(6)} + K_{21}^{(11)}\right)U_7 \\ &= -0.3607\,kT_0. \end{aligned} \tag{15}$$

Since q_n is the negative of the heat flux normal to the boundary, F_8 is the negative of the heat at node 8. Thus heat at node 8 is

$$Q_8 = -F_8 = 0.3607\,kT_0. \tag{16}$$

Mesh of Rectangular Elements. For a 3 × 2 mesh of linear rectangular elements [see Fig. 4.3.13(c)], the element coefficient matrix is given by Eqs. (4.3.51a) and (4.3.51b):

$$\mathbf{K}^{(e)} = \frac{k}{6}\begin{bmatrix} 4 & -1 & -2 & -1 \\ -1 & 4 & -1 & -2 \\ -2 & -1 & 4 & -1 \\ -1 & -2 & -1 & 4 \end{bmatrix}, \quad \mathbf{f}^{(e)} = \mathbf{0}. \tag{17}$$

The assembled global coefficients are

$$\begin{aligned} &K_{11} = K_{11}^{(1)}, \quad K_{12} = K_{12}^{(1)}, \quad K_{15} = K_{14}^{(1)}, \quad K_{16} = K_{13}^{(1)}, \quad K_{22} = K_{22}^{(1)} + K_{11}^{(2)} \\ &K_{23} = K_{12}^{(2)}, \quad K_{25} = K_{24}^{(1)}, \quad K_{26} = K_{23}^{(1)} + K_{14}^{(2)}, \quad K_{27} = K_{13}^{(2)}, \quad \text{etc.} \\ &F_1 = Q_1^{(1)}, \quad F_2 = Q_2^{(1)} + Q_1^{(2)}, \quad F_3 = Q_2^{(2)} + q_1^{(3)}, \quad F_4 = q_2^{(3)}, \quad \text{etc.} \end{aligned} \tag{18}$$

The boundary conditions for the present mesh remain the same as those for the mesh of triangles. The condensed equations for the unknown temperatures are

$$\frac{k}{6}\begin{bmatrix} 4 & -1 & 0 & -1 & -2 & 0 \\ -1 & 8 & -1 & -2 & -2 & -2 \\ 0 & -1 & 8 & 0 & -2 & -2 \\ -1 & -2 & 0 & 8 & -2 & 0 \\ -2 & -2 & -2 & -2 & 16 & -2 \\ 0 & -2 & -2 & 0 & -2 & 16 \end{bmatrix}\begin{Bmatrix} U_1 \\ U_2 \\ U_3 \\ U_5 \\ U_6 \\ U_7 \end{Bmatrix} = \frac{k}{6}\begin{Bmatrix} 0 \\ 0 \\ 0 \\ T_0 + \sqrt{3}T_0 \\ 2T_0 + \sqrt{3}T_0 + T_0 \\ \sqrt{3}T_0 + T_0 \end{Bmatrix}. \tag{19}$$

The solution of these equations is

$$U_1 = 0.6128\,T_0, \quad U_2 = 0.5307\,T_0, \quad U_3 = 0.3064\,T_0$$
$$U_5 = 0.7030\,T_0, \quad U_6 = 0.6088\,T_0, \quad U_7 = 0.3515\,T_0. \tag{20}$$

The value of the heat at node 8 is given by ($K_{83} = K_{31}^{(3)}$, $K_{87} = K_{34}^{(3)} + K_{21}^{(6)}$, and $K_{8(11)} = K_{24}^{(6)}$)

$$\begin{aligned} Q_8 &= -F_8 = -K_{83}U_3 - K_{87}U_7 - K_{8(11)}U_{11} \\ &= \tfrac{2}{6}k\,(U_3 + U_7 + U_{11}) = 0.3860\,kT_0 \text{ W}. \end{aligned} \tag{21}$$

Table 4.3.1 contains a comparison of the FEM solutions with the FVM solutions (obtained with 12 × 8 mesh) and analytical solution in Eq. (2) for three different meshes of linear triangular and rectangular elements (results are independent of a and k). We note that there are only half as many elements in the mesh of rectangles compared with the mesh of triangles. It is interesting to note that meshes of triangles provide an upper bound, while the meshes of rectangles provide a lower bound to the exact solution. The FEM solution with triangles is the same as that predicted by the FVM as well as the FDM for 12 × 8 mesh (see Table 3.4.1).

Table 4.3.1 Comparison of the nodal temperatures $T(x, y)/T_0$, obtained using various finite element meshes with the FVM solution (obtained with 12 × 8 mesh) and analytical solution in Eq. (2).

		FEM Solutions							
		Triangles			Rectangles			FVM	Exact
x	y	3 × 2	6 × 4	12 × 8	3 × 2	6 × 4	12 × 8	12 × 8	Solution
0.0	0.0	0.6362	0.6278	0.6256	0.6128	0.6219	0.6242	0.6256	0.6249
0.5	0.0	——	0.6064	0.6043	——	0.6007	0.6029	0.6043	0.6036
1.0	0.0	0.5510	0.5437	0.5418	0.5307	0.5386	0.5405	0.5418	0.5412
1.5	0.0	——	0.4439	0.4424	——	0.4398	0.4413	0.4424	0.4419
2.0	0.0	0.3181	0.3139	0.3128	0.3064	0.3110	0.3121	0.3128	0.3124
2.5	0.0	——	0.1625	0.1619	——	0.1610	0.1615	0.1619	0.1617
0.0	1.0	0.7214	0.7148	0.7131	0.7030	0.7102	0.7119	0.7131	0.7125
0.5	1.0	——	0.6904	0.6888	——	0.6860	0.6877	0.6888	0.6882
1.0	1.0	0.6248	0.6190	0.6176	0.6088	0.6150	0.6166	0.6175	0.6171
1.5	1.0	——	0.5054	0.5042	——	0.5022	0.5034	0.5042	0.5038
2.0	1.0	0.3607	0.3574	0.3565	0.3515	0.3551	0.3560	0.3565	0.3563
2.5	1.0	——	0.1850	0.1846	——	0.1838	0.1843	0.1845	0.1844

The next example deals with heat transfer in a cooling fin with convection boundary condition.

Example 4.3.3

Consider a rectangular cooling fin shown in Fig. 4.3.14 has its base, $x = 0$, maintained at 300 °C and exposed to convection on its remaining three sides. Assume that the dimension normal to the xy-plane (i.e., thickness) is either very small or very long so that the heat flow in that direction is negligible. Analyze the problem using the finite element mesh shown in Fig. 4.3.14.

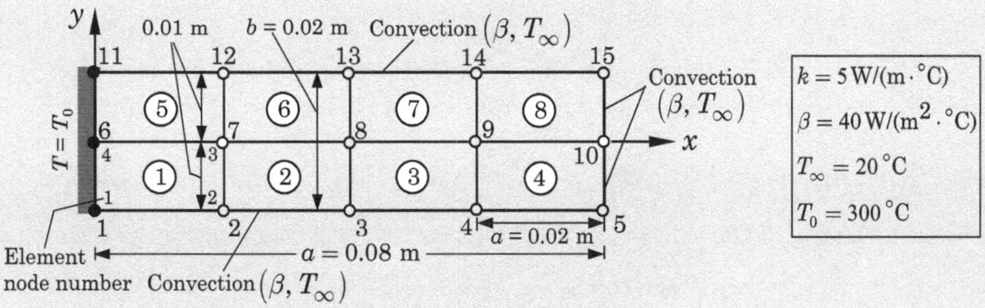

Fig. 4.3.14 Mesh of rectangular elements to discretize a rectangular cooling fin.

Solution Considering the full domain, temperatures at nodes 1, 6, and 11 are specified, and there are 12 unknown nodal temperatures that need to be determined. Thus, there are 12 equations that constitute the condensed equations for the primary unknowns (temperatures). Here we give the global equations associated with a couple of nodes, one interior and another on the boundary. For example, the finite element equation for node 7 is given by

$$\begin{aligned} K_{71}U_1 + K_{72}U_2 + K_{73}U_3 + K_{76}U_6 + K_{77}U_7 + K_{78}U_8 + K_{7(11)}U_{11} \\ + K_{7(12)}U_{12} + K_{7(13)}U_{13} = F_7, \end{aligned} \tag{1}$$

where K_{IJ} and F_I are the global coefficients, which can be expressed in terms of the element coefficients $K_{ij}^{(e)}$ [whose values are readily available from Eqs. (4.3.51a) and (4.3.51b)] and $Q_I^{(e)}$ ($f_i^{(e)} = 0$ for all elements, as there is no internal heat generation specified). Equation (1) becomes

$$\begin{aligned} \left(K_{32}^{(1)} + K_{41}^{(2)}\right)U_2 + K_{42}^{(2)}U_3 + \left(K_{43}^{(2)} + K_{12}^{(6)}\right)U_8 + \left(K_{23}^{(5)} + K_{14}^{(6)}\right)U_{12} + K_{13}^{(6)}U_{13} \\ = -K_{31}^{(1)}U_1 - \left(K_{34}^{(1)} + K_{21}^{(5)}\right)U_6 - K_{24}^{(5)}U_{11} + Q_3^{(1)} + Q_4^{(2)} + Q_1^{(6)} + Q_2^{(5)}, \end{aligned} \tag{2}$$

where $U_1 = U_6 = U_{11} = T_0$ and $Q_3^{(1)} + Q_4^{(2)} + Q_1^{(6)} + Q_2^{(5)} = 0$ by balance of heats at node 7, although individual Q_i^e is not known (and can be determined in the post-computation either by definition or by element equations).

Similarly, the global finite element equation for node 3 can be written as

$$K_{32}U_2 + K_{33}U_3 + K_{34}U_4 + K_{37}U_7 + K_{38}U_8 + K_{39}U_9 = Q_2^{(2)} + Q_1^{(3)}, \tag{3}$$

where $Q_2^{(2)} + Q_1^{(3)}$ is equal to (or balanced by) the heat due to convection

$$\begin{aligned} Q_2^{(2)} + Q_1^{(3)} &= \int_{1-2} q_n^{(2)}\psi_2^{(2)}(s)\,ds + \int_{1-2} q_n^{(3)}\psi_1^{(3)}(s)\,ds \\ &= \int_{1-2} \beta_{12}^2(U_2\psi_1^{(2)} + U_3\psi_2^{(2)} - T_\infty)\psi_2^{(2)}(s)\,ds \\ &\quad + \int_{1-2} \beta_{12}^{(3)}(U_3\psi_1^{(3)} + U_4\psi_2^{(3)} - T_\infty)q_n^{(3)}\psi_1^{(3)}(s)\,ds \\ &= H_{12}^{(2)}U_2 + H_{22}^{(2)}U_3 + H_{11}^{(3)}U_3 + H_{21}^{(2)}U_4 + P_2^{(2)} + P_1^{(3)}. \end{aligned} \tag{4}$$

Here $q_n^{(e)}$ is the flux, and $\beta_{12}^{(e)}$ is the heat transfer coefficient on side 1-2 of element e ($e = 2, 3$), and $H_{ij}^{(e)}$ and $P_i^{(e)}$ are defined by

$$H_{ij}^{(e)} = \int_{p-q} \beta_{pq}^{(e)}\psi_i^{(e)}\psi_j^{(e)}\,ds, \quad P_i^{(e)} = \int_{p-q} T_\infty\beta_{pq}^{(e)}\psi_i^e\,ds. \tag{5}$$

Here p-q refers to the side of the element that is exposed to the ambient temperature T_∞. The values of $H_{ij}^{(e)}$ and $P_i^{(e)}$ are readily available from Eqs. (4.2.29c):

$$H_{11}^{(e)} = H_{22}^{(e)} = \tfrac{1}{3}\beta^{(e)}h_e, \quad H_{12}^{(e)} = H_{21}^{(e)} = \tfrac{1}{6}\beta^e h_e, \quad P_1^{(e)} = P_2^{(e)} = \tfrac{1}{2}\beta^{(e)}h_eT_\infty \tag{6}$$

where h_e is the length of the side 1-2. Thus, in terms of the element coefficients, we have

$$\begin{aligned} \left(K_{12}^{(2)} + H_{12}^{(2)}\right)U_3 + \left(K_{22}^{(2)} + H_{22}^{(2)} + K_{11}^{(3)} + H_{11}^{(3)}\right)U_3 + K_{12}^{(3)}U_4 \\ + K_{37}U_7 + K_{38}U_8 + K_{39}U_9 = P_2^{(2)} + P_2^{(3)}. \end{aligned} \tag{7}$$

It is clear that the element coefficients, due to convection on the boundary of element Ω^e, $H_{ij}^{(e)}$, are added to $K_{ij}^{(e)}$; similarly, the contribution $P_i^{(e)}$ is added to the right-hand side $f_i^{(e)}$ of the element equations.

The problem does have a symmetry about $y = 0$ (see Fig. 4.3.14). In heat transfer problems, a line of symmetry requires that the flux normal to the line of symmetry (i.e., on the surface $y = 0$) is zero. Therefore, it is sufficient to use symmetric half, say upper-half, of the body as the computational domain.

Numerical results obtained (with computer code **FEM2D**) are presented in Table 4.3.2 for four different meshes of rectangular elements in the upper-half domain (using the symmetry). The table also contains the results obtained using the HFVM with 16 × 2 mesh. The convergence of the results with mesh refinement is also clear from the results presented in Table 4.3.2 (L = linear element and Q = nine-node quadratic element). Figure 4.3.15 contains contour plots (i.e., isotherms) of the temperature field.

Table 4.3.2 Nodal temperatures $T(x, 0.01)$ of the cooling fin obtained using various finite element meshes of rectangles (see Fig. 4.3.14 for the geometry and boundary conditions) and the FVM with 16 × 2 mesh in the computational domain (exploiting the symmetry).

	FEM Solutions				FVM
x	4 × 1L	8 × 1L	16 × 2L	8 × 1Q	16 × 2
0.00	300.00	300.00	300.00	300.00	300.00
0.01	——	227.61	227.56	227.65	228.63
0.02	176.31	178.53	178.15	178.07	179.78
0.03	——	141.34	141.28	141.26	143.87
0.04	112.99	113.86	113.95	113.98	117.71
0.05	——	93.84	94.07	94.07	99.23
0.06	78.57	79.69	79.90	79.98	86.98
0.07	——	70.27	70.50	70.58	80.00
0.08	63.97	64.86	65.08	65.15	77.73

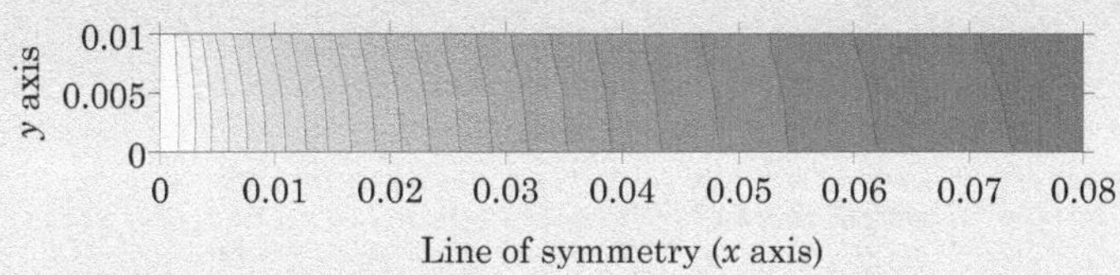

Fig. 4.3.15 Contour plots of the temperature field $T(x, y)$.

The next example is also one of convection (see **Example 3.4.2**) with internal heat generation.

Example 4.3.4

The bus bar shown in Fig. 4.3.16 carries sufficient electrical current to have a heat generation of $g = 10^6$ W/m^3. The bar has a conductivity of $a_{xx} = a_{yy} = k$ W/(m K) and dimensions 0.10 m × 0.05 m (and 0.01 m thick). The left side is maintained at 40 °C and the right side at 10 °C. Assuming that the heat flow is two-dimensional (or one may assume that the front and back faces are insulated), and the bottom edge is insulated and the top edge is exposed to ambient air temperature of $T_\infty = 0$ °C with a heat transfer coefficient of 75 W/(m^2 K), and a uniform conductivity of $k = k_0 = 20$ W/(m· °C), determine the steady-state temperature distribution with the 20 × 10 uniform mesh of bilinear elements.

Solution: Following the discussion of **Example 4.3.3**, this convection heat transfer problem can be analyzed using the FEM. The FVM solution was discussed in **Example 3.4.2**. The FEM and FVM solutions obtained for $T(x,0)$ and $T(x,0.05)$ versus x are presented in Table 4.3.3. A uniform mesh of 20 × 10 subdivisions is used in the FVM and the same mesh of bilinear elements was used in FEM. The solutions predicted by the two methods are very close (they will be indistinguishable if plotted).

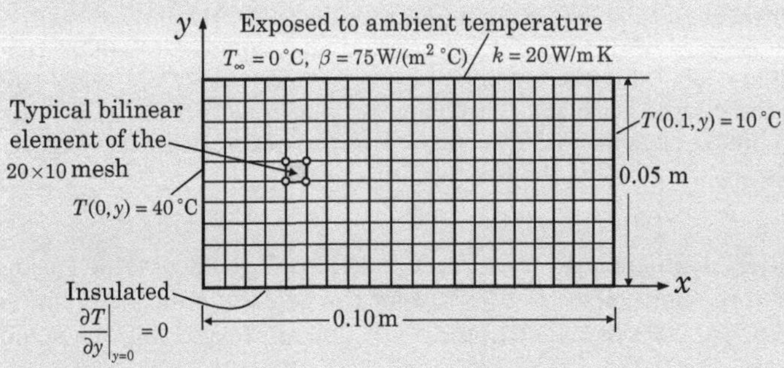

Fig. 4.3.16 Domain, boundary conditions, and 20×10 mesh of bilinear elements for convective heat transfer in a bus bar.

Table 4.3.3 The FEM and FVM solutions (temperature $T(x,y)$ in °C) of a two-dimensional heat conduction problem with convection boundary condition (20 × 10 mesh).

x	$y = 0.0$		$y = 0.05$	
	FEM	FVM	FEM	FVM
0.00	40.000	40.000	40.000	40.000
0.01	58.106	58.099	54.979	55.025
0.02	71.362	71.349	66.553	66.578
0.03	79.894	79.879	74.084	74.103
0.04	83.793	83.777	77.509	77.525
0.05	83.107	83.091	76.811	76.825
0.06	77.845	77.831	71.965	71.978
0.07	67.980	67.967	62.927	62.940
0.08	53.449	53.439	49.631	49.644
0.09	34.160	34.155	31.986	32.003
0.10	10.000	10.000	10.000	10.000

The next example deals with an axisymmetric problem.

Example 4.3.5

The equation governing a diffusion process (e.g., heat flow) in an axisymmetric problem of a cylinder of finite length with internal heat generation f is

$$-\frac{1}{r}\left[\frac{\partial}{\partial r}\left(r\frac{\partial u}{\partial r}\right)+\frac{\partial}{\partial z}\left(r\frac{\partial u}{\partial z}\right)\right]=f(r,z). \tag{1}$$

Determine the finite element solution of Eq. (1) for the following boundary conditions shown in Fig. 4.3.17(b),

$$u(r,b)=g_1(r)=\tfrac{1}{2}\left(3-\frac{r^2}{a^2}\right),\;\; u(a,z)=g_2(z)=\tfrac{1}{2}\left(3-\frac{z^2}{b^2}\right),\;\; \frac{\partial u}{\partial r}\Big|_{r=0}=0,\;\; \frac{\partial u}{\partial z}\Big|_{x=0}=0 \tag{2}$$

The source $f(r,z)$ is given by (see **Example 1.10.2**)

$$f(r,z) = 3\left(\frac{1}{a^2} + \frac{1}{2b^2}\right) - \frac{1}{2b^2}\frac{r^2}{a^2} - \frac{1}{a^2}\frac{z^2}{b^2}. \tag{3}$$

Use $a = 1.0$ and $b = 2.0$ and a uniform mesh of 10×20 linear finite elements to obtain the numerical solutions.

Solution: The exact solution to this problem was developed using the method of manufactured solutions in **Example 1.10.2**. The solution is

$$u(r,z) = g_1(r)g_2(z) = \left[1 + \frac{1}{2}\left(1 - \frac{r^2}{a^2}\right)\right]\left[1 + \frac{1}{2}\left(1 - \frac{z^2}{b^2}\right)\right]. \tag{4}$$

The problem can be analyzed using the procedure discussed in the preceding examples, except that the coefficients take the values $a_{xx} = r$, $a_{yy} = r$, and $f(x,y) = rf(r,z)$.

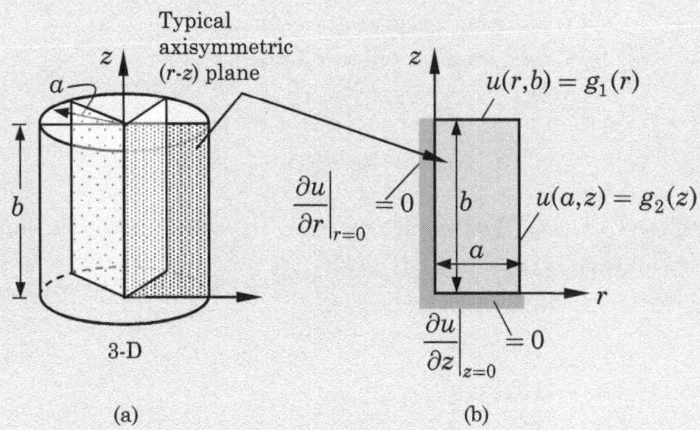

Fig. 4.3.17 (a) Reduction of an axisymmetric cylinder problem to a two-dimensional problem in the rz plane. (b) Analysis domain.

One needs to compute the known values of u on the surfaces at $r = a$ and $z = b$ to input them as the known primary nodal degrees of freedom. For the 10×20 mesh, there will be a total of $11 \times 21 = 231$ nodes. Boundary nodes $11, 22, 33, \ldots, 220$, and $221, 222, \ldots, 231$ (31 of them) have specified primary variables. The values can be computed by evaluating $g_2(r)$ at $r = a$ and $g_1(z)$ at $z = b$. These values are:

1.50000, 1.49875, 1.49500, 1.48875, 1.48000, 1.46875, 1.45500, 1.43875, 1.42000, 1.39875, 1.37500, 1.34875, 1.32000, 1.28875, 1.25500, 1.21875, 1.18000, 1.13875, 1.09500, 1.04875, 1.50000, 1.49500, 1.48000, 1.45500, 1.42000, 1.37500, 1.32000, 1.25500, 1.18000, 1.09500, 1.00000

A revised version (to account for the higher-order nature of the source term) of program **FEM2D** from [8] can be used to analyze the problem [noting that $a_{xx} = r$, $a_{yy} = r$, and $f(x,y) = rf(r,z)$] with (for $a = 1$ and $b = 2$) $f(r,z) = 3.375 - 0.125r^2 - 0.25z^2$.

Figure 4.3.18 contains plots of the FEM solutions (symbols) and the exact solutions (solid lines) for $u(r,z)$ versus r for different values of the coordinate z ($z = 0.0, 1.0$, and 1.5). The FEM solutions are in close agreement with the exact solution developed using the method of manufactured solutions, Eq. (1.10.14). In fact, the finite element solution obtained with the 10×20 mesh of nine quadratic elements coincides (up to the fourth decimal point) with the exact solution. Figure 4.3.19 contains the contour plots of the temperature field $u(r,z)$ obtained with the 10×20 mesh of nine-node quadratic elements.

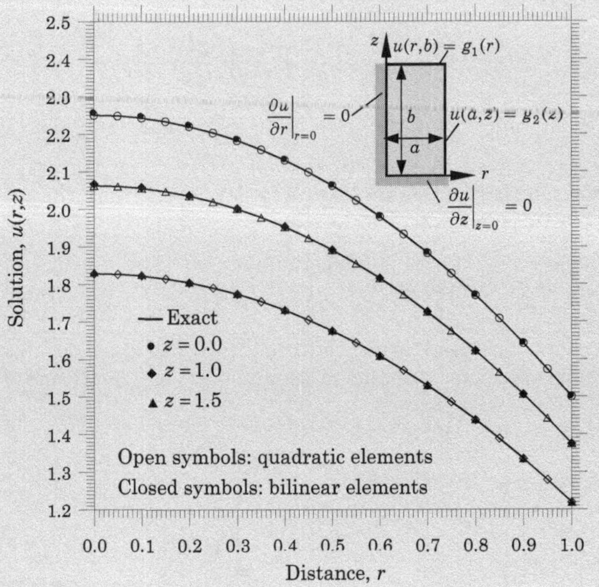

Fig. 4.3.18 Comparison of the FEM solutions with the analytical solution of axisymmetric diffusion problem. The dark symbols correspond to the 10×20 mesh of bilinear elements, while the open symbols correspond to the 10×20 mesh of nine-node quadratic elements.

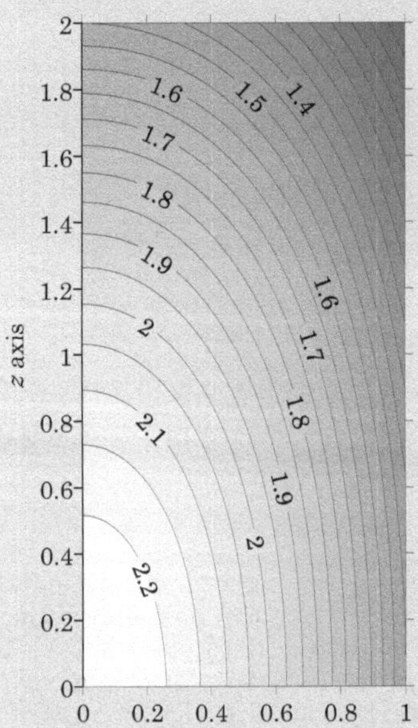

Fig. 4.3.19 Contour plots (isotherms) of $u(r, z)$ obtained with the 10×20 mesh of nine-node quadratic elements.

The last example of this chapter is concerned with an advection–diffusion problem. The problem is one of the two-dimensional version of the one-dimensional problem discussed in **Example 4.2.3**.

Example 4.3.6

Analyze the advection–diffusion problem governed by the differential equation

$$\frac{\partial u}{\partial x}+\frac{\partial u}{\partial y}-\frac{1}{Pe}\left(\frac{\partial^2 u}{\partial x^2}+\frac{\partial^2 u}{\partial y^2}\right)=0 \quad \text{in } \Omega=(1.0)\times(1,0), \tag{4.3.67}$$

subjected to the boundary conditions

$$u(1,y)=0,\;\; 0\le y\le 1;\quad u(x,1)=0,\;\; 0\le x\le 1, \tag{4.3.68}$$

$$u(x,0)=\frac{1-e^{(x-1)Pe}}{1-e^{-Pe}},\;\; 0\le x\le 1;\;\; u(0,y)=\frac{1-e^{(y-1)Pe}}{1-e^{-Pe}},\;\; 0\le y\le 1. \tag{4.3.69}$$

Investigate the performance of the FEM for various values of the Péclet number, Pe, with different uniform meshes of linear finite elements.

Solution: The exact solution of Eqs. (4.3.67)–(4.3.69) is

$$u(x,y)=\frac{\left(1-e^{(x-1)Pe}\right)\left(1-e^{(y-1)Pe}\right)}{(1-e^{-Pe})^2}, \tag{4.3.70}$$

Table 4.3.4 contains the finite element results and exact solution of the advection–diffusion equation for $Pe = 100$ and $Pe = 250$ using, in each case, with two different meshes. The FEM solutions converge to the exact solution only when $N > Pe$ (unlike for the one-dimensional case, where $N > 0.5\,Pe$ gave converged solutions), as can be seen from Fig. 4.3.20.

Table 4.3.4 Comparison of the FEM solutions with the exact solutions of the two-dimensional advection–diffusion equation using various meshes of linear elements for the case of $Pe = 100$ and $Pe = 250$.

	$Pe = 100$			$Pe = 250$		
$x = y$	75×75	150×150	Exact	100×100	150×150	Exact
0.800	1.00000	1.00000	1.00000	1.00000	1.00001	1.00000
0.840	1.00000	1.00000	1.00000	1.00000	1.00000	1.00000
0.860	——	1.00000	1.00000	1.00000	1.00000	1.00000
0.880	1.00000	0.99999	0.99999	1.00000	1.00000	1.00000
0.900	——	0.99994	0.99991	1.00000	1.00000	1.00000
0.910	——	——	0.99975	1.00000	——	1.00000
0.920	0.99987	0.99951	0.99933	1.00000	1.00000	1.00000
0.930	——	——	0.99818	1.00000	——	1.00000
0.940	——	0.99610	0.99505	1.00000	1.00000	1.00000
0.950	——	0.96157	0.98657	1.00000	——	0.99999
0.960	0.98406	0.96899	0.96370	0.99970	1.00000	0.99991
0.970	——	——	0.90290	1.00270	——	0.99889
0.980	——	0.76562	0.74765	0.97546	0.99850	0.98657
0.990	——	——	0.39958	1.23460	——	0.84257

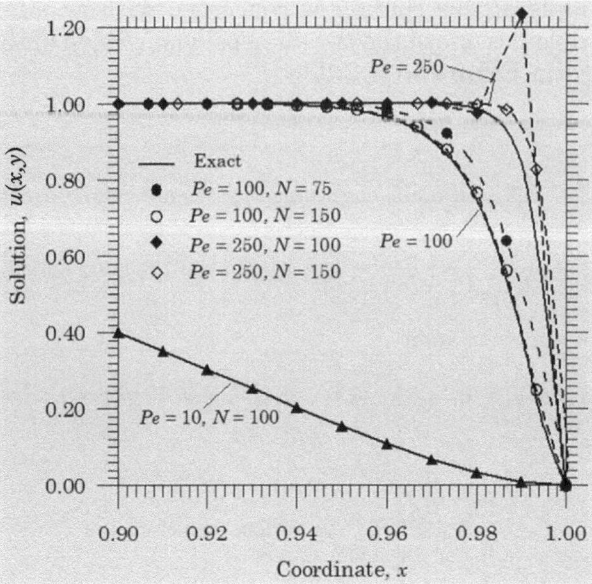

Fig. 4.3.20 A comparison of the FEM solutions, $u(x, y)$ for $x = y$, of the advection–diffusion equation with the exact solution for $Pe = 10$, 100, and 250.

4.4 Summary

The present chapter has been devoted to a study of: (1) the weak-form Galerkin (or Ritz) finite element models of one- and two-dimensional problems involving generalized Poisson's equation (governing plane and axisymmetric problems) typical of heat transfer, (2) a derivation of interpolation functions for basic one- and two-dimensional elements, and (3) finite element models of unsteady problems of one- and two-dimensional heat transfer. A number of numerical examples have been presented to illustrate the ideas presented in the chapter.

The advantages of the weak-form finite element models of second-order BVPs are: (a) identification of the primary and secondary variables – a duality that exists in all phenomena of nature, (b) weakening the differentiability required of approximation functions used for the primary variables and leading to C^0-approximations, and hence simpler finite elements, and (c) possible symmetry of the finite element coefficient matrix for most problems of engineering physics. In general, the derivation of the weak forms from the governing differential equations is subject to the requirement that the resulting secondary variables be physically meaningful in that they can be specified on the portions of the boundary where the corresponding primary variables are unknown. A disadvantage of the C^0-approximation is that the derivatives of the variables being approximated are discontinuous across interelement boundaries.

There are two drawbacks of the FEM. First, representing a system as a collection of connected finite elements often results in a discontinuous representation of the gradients of the solution, unless so-called C^1-continuity is used (which in turn dictates the element type, both in geometry and degrees of freedom per node). Second, the satisfaction of the governing equations in the weak-form or weighted-integral sense tends to smooth the solution and thereby the FEM predicts diffuse solutions when applied to problems with steep gradients. The finite element method has a systematic approximation of the variables and, hence, their derivatives inside the domain. The FEM allows domain-fitted meshes and the satisfaction of gradient boundary conditions in an integral sense.

We remark that families of interpolation functions on various geometries (e.g., triangles, rectangles/quadrilaterals, tetrahedrals, etc.) existed long before the FEM came into existence. Thus, calling these geometries as finite elements is appropriate only when they are used in the FEM; otherwise, they are just what they are geometrically: triangles, quadrilaterals, tetrahedrals, and so on.

The strong point of methods that use global approximations (i.e., no subdivision of the domain into elements, as in the traditional Ritz and Galerkin methods) is that they do not introduce discontinuities in the derivatives of the solution between elements. At the moment, there is no clear consensus on the overall effectiveness and robustness of such computational procedures to compete with or displace the current weak-form finite element technology.

Readers may find additional examples, exercise problems, and computer implementation ideas in the textbook by Reddy [8], which also contains detailed discussions of programs **FEM1D** and **FEM2D** and illustrative examples on how to prepare the input to these programs. The Fortran and MATLAB source codes are available from the website, **http://mechanics.tamu.edu**.

Problems

One-Dimensional Problems

4.1 Consider the differential equation

$$-\frac{d}{dx}\left(a\frac{du}{dx}\right)+\frac{d^2}{dx^2}\left(b\frac{d^2u}{dx^2}\right)+cu=f,$$

where a, b, c, and f are known functions of position x. (a) Develop the weak form over a typical element $\Omega^e = (x_a^e, x_b^e)$ such that the bilinear form is symmetric, (b) identify the bilinear and linear forms and construct the quadratic functional, (c) discuss the type of finite element approximation of u that may be used, and (d) develop the finite element model of the equation.

4.2 Suppose that the one-dimensional Lagrange cubic element with equally spaced nodes has a source of $f(x) = f_0 x/h$, where h is the length of the element (see Fig. P4.2). Compute its contribution to node 2.

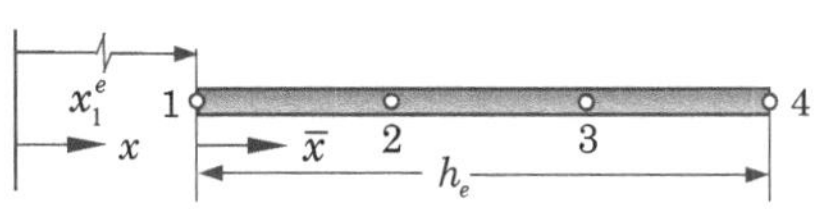

Fig. P4.2

4.3 Derive the finite element model of the differential equation

$$-\frac{d}{dx}\left(a(x)\frac{du}{dx}\right) + cu = f(x) \quad \text{for} \quad 0 < x < L \tag{1}$$

over an element while accounting for the mixed boundary condition of the following form at both ends of the element ($n_x = -1$ at $x = x_a$ and $n_x = 1$ at $x = x_b$)

$$\left[n_x\left(a(x)\frac{du}{dx}\right) + \beta(u - u_c)\right] = Q. \tag{2}$$

4.4 Solve the problem described by the following equations:

$$-\frac{d^2u}{dx^2} = \cos\pi x, \quad 0 < x < 1; \quad u(0) = 0, \quad u(1) = 0.$$

Use the uniform mesh of three linear elements to solve the problem and compare against the exact solution

$$u(x) = \frac{1}{\pi^2}\left(\cos\pi x + 2x - 1\right).$$

4.5 Solve the problem described by the following equations:

$$-\frac{d^2u}{dx^2} = \cos\pi x, \quad 0 < x < 1; \quad u(0) = 0, \quad u(1) = 0.$$

Use the uniform mesh of three linear elements to solve the problem and compare against the exact solution

$$u(x) = \frac{1}{\pi^2}\left(\cos\pi x + 2x - 1\right).$$

4.6 An insulating wall is constructed of three homogeneous layers with conductivities k_1, k_2, and k_3 in intimate contact (see Fig. P4.6). Under steady-state conditions, the temperatures of the media in contact at the left and right surfaces of the wall are at ambient temperatures of T^L_∞ and T^R_∞, respectively, and film coefficients β_L and β_R, respectively. Determine the temperatures when the ambient temperatures T^L_∞ and T^R_∞ are known. Assume that there is no internal heat generation and that the heat flow is one-dimensional ($\partial T/\partial y = 0$).

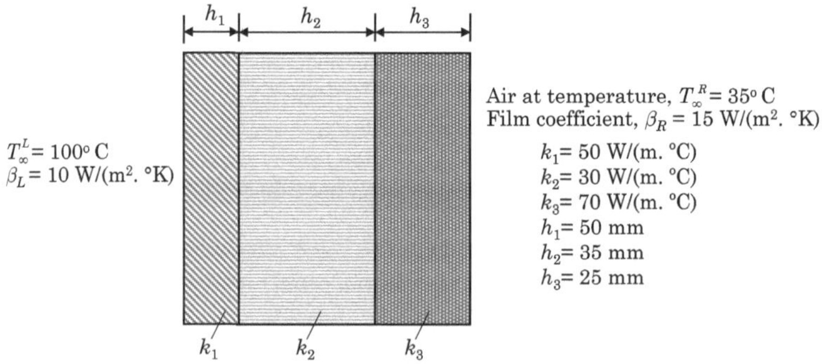

Fig. P4.6

4.7 Consider steady heat conduction in a wire of circular cross-section (radius R_0) with an electrical heat source. Suppose that the radius of the wire is a, its electrical conductivity is K_e (Ω^{-1}/cm), and it is carrying an electric current density of I (A/cm ^{2}). During the transmission of an electric current, some of the electrical energy is converted into thermal energy. The rate of heat generation per unit volume is given by $g = I^2/K_e$. Assume that the temperature rise in the wire is sufficiently small that the dependence of the thermal or electric conductivity on temperature can be neglected. The governing equations of the problem are

$$-\frac{1}{r}\frac{d}{dr}\left(rk\frac{dT}{dr}\right) = g \quad \text{for} \quad 0 \leq r \leq R_0, \quad \left(rk\frac{dT}{dr}\right)\Big|_{r=0} = 0, \quad T(R_0) = T_0.$$

Determine the distribution of temperature in the wire using (a) two linear elements and (b) one quadratic element, and compare the finite element solution with the exact solution (see **Example 4.2.3**),

$$T(r) = T_0 + \frac{ga^2}{4k}\left[1 - \left(\frac{r}{R_0}\right)\right].$$

Also, determine the heat flow, $Q = -2\pi R_0 k(dT/dr)|_{R_0}$, at the surface using (i) the temperature field and (ii) the discrete balance equation.

4.8 Consider the steady laminar flow of a viscous fluid through a long circular cylindrical tube. The governing equation is

$$-\frac{1}{r}\frac{d}{dr}\left(r\mu\frac{dw}{dr}\right) = \frac{P_0 - P_L}{L} \equiv f_0,$$

where w is the axial (i.e., z) component of velocity, μ is the viscosity, and f_0 is the gradient of pressure (which includes the combined effect of static pressure and gravitational force). The boundary conditions are

$$\left(r\frac{dw}{dr}\right)\Big|_{r=0} = 0 \quad w(R_0) = 0.$$

Using the symmetry and (a) two linear elements, (b) one quadratic element, determine the velocity field and compare with the exact solution at the nodes:

$$w_e(r) = \frac{f_0 R_0^2}{4\mu}\left[1 - \left(\frac{r}{R_0}\right)^2\right].$$

4.10 Solve the advection–diffusion problem of **Example 4.2.4** using (a) 50 linear elements, (b) 100 linear elements (c) 25 quadratic elements, and (d) 50 quadratic elements for $Pe = 100$. Plot and tabulate the results.

Two-Dimensional Problems

4.11 Calculate the linear interpolation functions for the linear (a) triangular and (b) rectangular elements shown in Fig. P4.11.

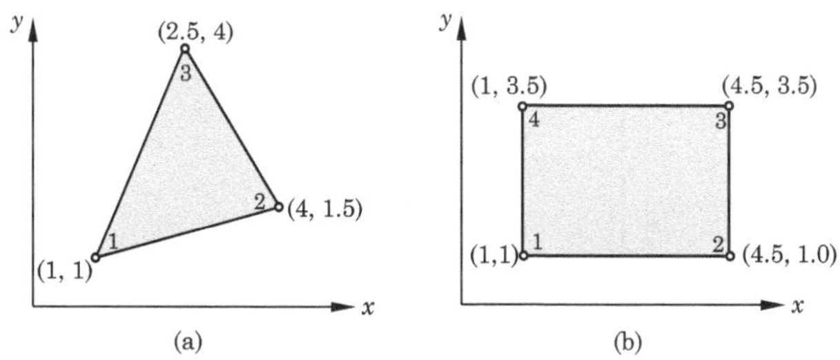

Fig. P4.11

4.12 Evaluate the coefficients K_{ij}^e and F_i^e of Eq. (4.3.20) for a linear triangular element when a_{xx}, a_{yy}, f, h_c and u_c are constants.

4.13 Consider a diffusion process in a square region shown in Fig. P4.13. The governing equation is given by

$$-\left(\frac{\partial^2 u}{\partial x^2} + \frac{\partial^2 u}{\partial y^2}\right) = 0. \tag{1}$$

The boundary conditions for the problem are:

$$u(0, y) = y^2 \ , \ u(x, 0) = x^2 \ , \ u(1, y) = 1 - y \ , \ u(x, 1) = 1 - x. \tag{2}$$

Determine the unknown nodal value of U_5 using the finite element mesh shown in Fig. P4.13.

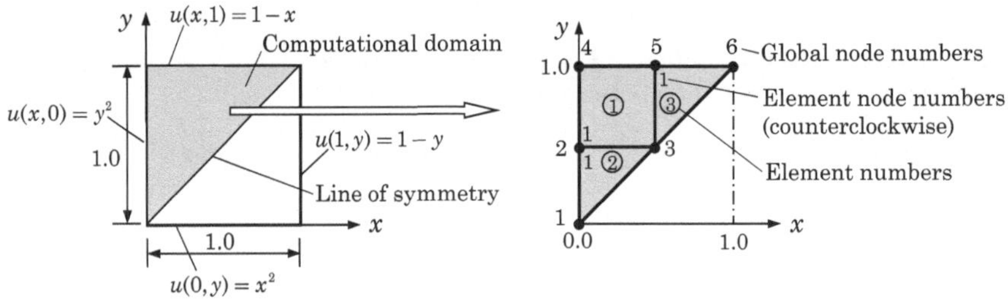

Fig. P4.13

4.14 The nodal values of a linear triangular element Ω^e in the finite-element analysis of the field problem

$$-\nabla^2 u = f_0 \tag{3.1}$$

are $u_1^e = U_{10} = 389.79$, $u_2^e = U_{12} = 337.19$, and $u_3^e = U_{11} = 395.08$.

(a) Find the gradient of the solution,

$$\nabla u \equiv \frac{\partial u}{\partial x}\hat{\mathbf{e}}_x + \frac{\partial u}{\partial y}\hat{\mathbf{e}}_y \tag{3.2}$$

in the element domain Ω^e (see Fig. P4.14).

(b) Determine the places where the $u = 392$ isoline intersects the boundary of the element domain Ω^e shown in Fig. P4.14.

(c) Find the contribution of a point source Q_0 located at $(x = 4.0, y = 1.25)$ to the nodes of the element.

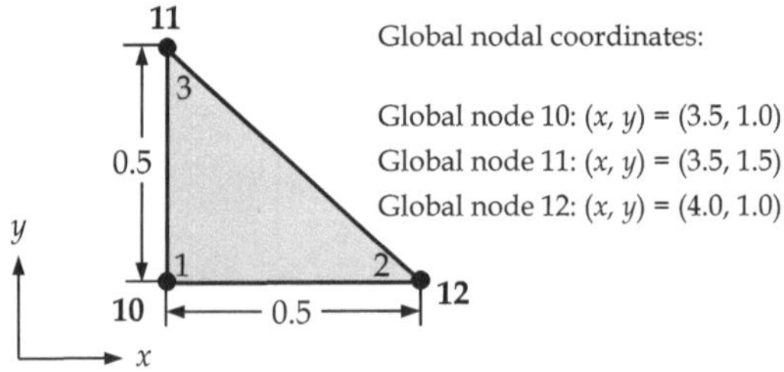

Fig. P4.14

4.15 Consider a diffusion problem governed by the differential equation $-k\nabla^2 u = 0$ in the domain shown in Fig. P4.15. Write the global finite element equation associated with global node 8 and identify all of the coefficients (other than unknown U_I) in the equation in terms of the element coefficients K_{ij}^e (algebraic) and problem data only (i.e., you must give the numerical values of other contributions other than K_{ij}^e).

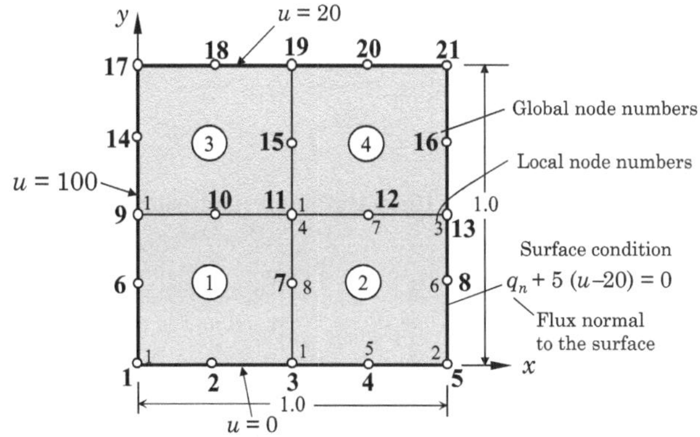

Fig. P4.15

4.16 Consider steady-state heat transfer governed by the differential equation

$$-\frac{\partial}{\partial x}\left(k\frac{\partial u}{\partial x}\right) - \frac{\partial}{\partial y}\left(k\frac{\partial u}{\partial y}\right) = f_0 \tag{1}$$

over the domain shown in Fig. P4.16.

(a) Write the finite element equation associated with global node 1 in terms of element coefficients (include only the nonzero contributions).

(b) Compute the contribution of the flux q_0 to global nodes 1 and 4.

(c) Compute the contribution of the boundary condition $q_n + 5u = 0$ to global node 1 using the coefficients H_{ij}^e and P_i^e due to convection.

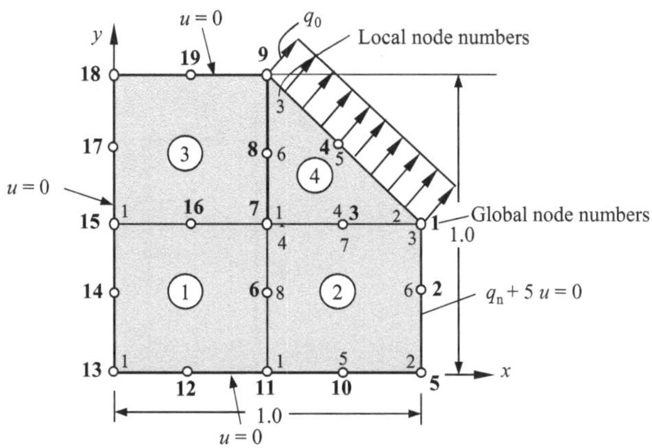

Fig. P4.16

4.17 The Prandtl theory of torsion of a cylindrical member leads to

$$-\nabla^2 u = 2G\theta \quad \text{in } \Omega; \quad u = 0 \quad \text{on } \Gamma$$

where Ω is the cross-section of the cylindrical member being subjected to torsion, Γ is the boundary of Ω, G is the shear modulus of the material of the cylinder, θ is the angle of twist, and u is the Prandtl stress function. Solve the equation for the case in which Ω is a circular section (see Fig. P4.17) using the mesh of linear triangular elements. Compare the finite-element solution with the exact solution (valid for elliptical sections with axes a and b):

$$u_{\text{exact}} = \frac{G\theta a^2 b^2}{a^2 + b^2}\left(1 - \frac{x^2}{a^2} - \frac{y^2}{b^2}\right).$$

Use $a = 1$, $b = 1$, and $f = 2G\theta = 10$ in the numerical calculations.

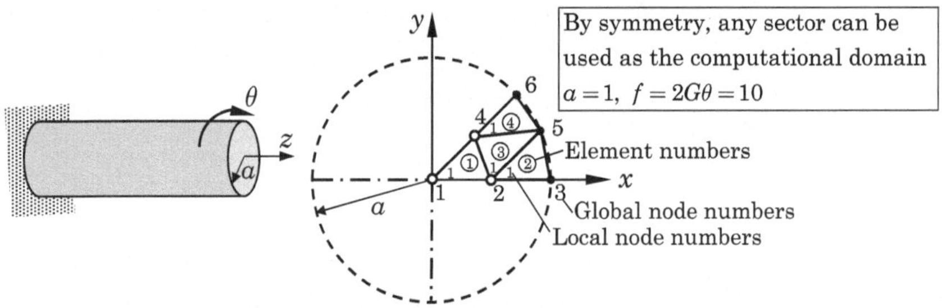

Fig. P4.17

4.18 Consider a diffusion process governed by the differential equation

$$-k\left(\frac{\partial^2 u}{\partial x^2} + \frac{\partial^2 u}{\partial y^2}\right) = 0 \tag{1}$$

over the domain shown in Fig. P4.18. Exploiting the symmetry, analyze the problems using (a) 2×2 mesh of bilinear elements [see Fig. P4.18(a)] and (b) 2×2 mesh of linear triangular elements [see Fig. P4.18(b)]. The exact solution, when $u_0(x) = 1$, is

$$u(x, y) = 4k \sum_{n=0}^{\infty} \frac{\sin(\lambda_n x)\,\sinh(\lambda_n y)}{\lambda_n \sinh(\lambda_n)}, \quad \lambda_n = (2n+1)\pi. \tag{2}$$

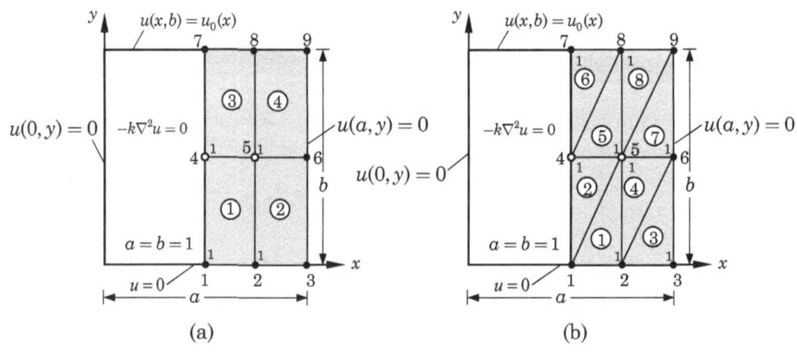

Fig. P4.18

4.19 A series of heating cables has been placed underground, as shown in Fig. P4.19. Assume that the ground (soil) has conductivities of $a_{xx} = k_{xx} = 10$ W/(cm·°C) and $a_{yy} = k_{yy} = 15$ W/(cm·°C), the upper surface is exposed to a temperature of $-5\,°$C [$\beta = 5$ W/(cm^2·K)], and the lower surface has zero heat flow normal to the boundary. Assume that each cable is producing a heat of 250 W/cm. Use a 8×16 mesh of linear triangular (or rectangular) elements in the computational domain (exploit the symmetry available in the problem), and formulate the problem, that is, give (a) element matrices for a typical element, (b) boundary conditions on primary and secondary variables, and (c) convection boundary contributions.

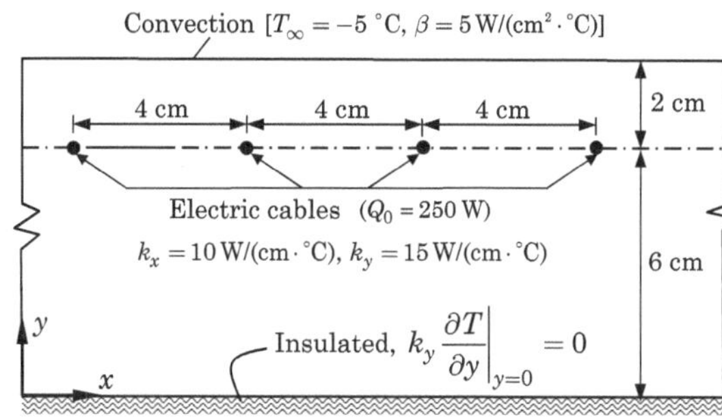

Fig. P4.19

4.20 The torsion of a cylindrical bar is also defined by $-\nabla^2\Psi = 2$ in Ω and $\Psi = 0$ on Γ, where Ω is the cross-section of the cylindrical bar, Γ is the closed surface of Ω, Ψ denotes the stress function, and the components of the gradient of Ψ are the shear stresses on the section due to the twisting (assumed to be small) of the bar, which are of primary interest in the design of shafts:

$$\sigma_{xz} = G\theta\frac{\partial\Psi}{\partial y}, \quad \sigma_{yz} = -G\theta\frac{\partial\Psi}{\partial x}.$$

Here G is the shear modulus of the material of the bar and θ is the angle of twist per unit length. Assuming that the cross-section of the bar is 6×4 in., determine the normalized shear stresses, $\sigma_{xz}/G\theta$ and $\sigma_{yz}/G\theta$ with a uniform 12×8 mesh of linear rectangular elements in a quadrant of the domain (i.e., exploit the biaxial symmetry).

5
Dual Mesh Control Domain Method

5.1 Introduction

Currently, the FEM and FVM are the most commonly used numerical methods for the solution of differential equations with FEM dominating solid and structural mechanics and FVM witnessing popularity in heat transfer and fluid mechanics. Detailed introductions to these two methods as applied to second-order differential equations in a single dependent unknown in one and two dimensions were presented in Chapters 3 and 4.

In 2019, Reddy [16] introduced a numerical approach termed the *dual mesh control domain method* (DMCDM) (originally it was called the *dual mesh finite domain method*), which utilizes the desirable features of the FEM and FVM. In the DMCDM, the domain is represented with a primal mesh of finite elements (defining the approximations used for the dependent variables), and a dual mesh is superimposed on the primal mesh such that the nodes of the primal mesh are at the center of the dual mesh of *control domains*. Then the governing equation is satisfied in an integral sense over the control domain, as in the FVM. The approach does not involve isolating a finite element domain and satisfying the governing equations in a weak sense over it and assembling element equations to obtain the global equations. Instead, the DMCDM results, much like in the FVM, directly in a set of global equations in terms of the nodal values of the primary variables. Thus, the DMCDM brings the desirable features of the FEM and the FVM, and comes closer to the so-called "control volume finite element method" (CVFEM) presented by Voller [35] for linear problems.

Some remarks concerning the DMCDM and CVFEM are in order. The DMCDM brings the best features of the FEM, namely, the interpolation of the variables and imposition of physical boundary conditions, and of the FVM in satisfying the actual balance equations over the control domain. The major merits of the DMCDM are that the method inherits the desirable aspect of the FVM (in satisfying the global form of the governing equations over the control domains) and overcomes the disadvantage of the discontinuity of the secondary variables at the interfaces of the finite elements (primal mesh) by calculating them at the boundaries of the control domains (dual mesh), where they are continuous (i.e., uniquely defined). Readers who are familiar with the CVFEM may draw a parallel between DMCDM described here and the works of Winslow [36], Baliga and Patankar [37, 38], and Voller [35]. Winslow [36] uses the phrases "primary mesh" and "secondary mesh" and triangles to develop a finite difference scheme to solve the generalized Poisson equation. However, the work has nothing to do with either the FEM or the FVM (i.e., the development does not explicitly use finite element interpolation functions or control volumes to satisfy the governing equations). The idea of Baliga and Patankar [37, 38] employs a similar scheme of primal mesh, which the authors never connected to the standard finite elements and no numerical results were presented there. The previous work that comes close to the DMCDM is that of Voller [35], and it does not include the concept of duality and imposition of boundary conditions in the same way as in the DMCDM. The difference between the DMCDM and the ideas presented in the works of Baliga and Patankar [38] and Voller [35] comes from the systematic formulation and implementation of DMCDM that explicitly makes use of the ideas from both the FEM and FVM through finite element interpolation, control domain satisfaction of the governing equations, and the duality concept.

An important point one should be aware of is that the phrase "control volume finite element method" (see [39], [40], among many others) is a misnomer and the CVFEM has nothing to do with the FEM, and the phrase "finite element method" should not have appeared in these titles of many papers on the CVFEM because the features of the FEM are not used in any form. Unfortunately, this error has propagated through the literature for many years and misled people to think that the "control volume finite element method" has something to do with the FEM when it absolutely does not. Mere use of triangle and tetrahedral geometries for meshes and their interpolation functions as approximation functions does not make the discretization procedure as the FEM. The DMCDM (1) explicitly connects the FVM to FEM through finite element interpolation functions and (2) uses the concept of duality in the formulation and implementation of physical boundary conditions, which are not discussed or practiced in the FVM or CVFEM literature. There is a considerable vagueness and arbitrariness in the FVM and CVFEM because of the various ad-hoc and specialized schemes used in the imposition of the gradient boundary conditions, the evaluation of domain integrals, and treatment of nonlinear terms.

In this chapter, the DMCDM is introduced as an alternative formulation of the FVM in which the finite element approximation of the dependent unknowns is used (primal mesh), but the discrete equations are derived using integral statements over the dual mesh of control domains that is different from the primal mesh. In addition, in writing the integral statement, the duality pairs are retained on the boundary of the domain so that one can readily impose specified boundary conditions on either element of the duality pair. Although the DMCDM shares some features of both the FVM and FEM, it is different from them as known today. The DMCDM comes close to or, in some cases, is the same as the half-control FVM presented in Chapter 3.

Following this introduction, the basic idea of the DMCDM is introduced with a model second-order linear differential equation in a single unknown, both in one and two dimensions, like we did in the case of the FDM (Chapter 2), FVM (Chapter 3), and FEM (Chapter 4). The numerical results of examples introduced in the previous chapters are compared with those obtained with the DMCDM. Extension of the DMCDM to multivariable and nonlinear problems will be presented in the forthcoming chapters.

5.2 Dual Mesh Control Domain Method

In the DMCDM, the domain Ω is divided into a collection of nonoverlapping subdomains (i.e., finite elements with their interpolation functions), which we term *primal mesh.* Each nodal point of the primal mesh is enclosed in a *control domain* (we do not call it a "control volume" because such phrase is not meaningful for one- and two-dimensional domains), which may occupy more than one finite element. Figure 5.2.1 shows an arbitrary two-dimensional domain Ω and its discretization into a primal mesh of quadrilateral finite elements and a dual mesh of quadrilateral control domains, each control domain centered around a nodal point. Over each control domain, an integral form (not a weighted-residual statement or a weak form, as in the case of the FEM) of the equation to be solved (or the balance law to be satisfied) is used. Thus, the DMCDM introduced here consists of two different meshes, namely, the mesh of finite elements over which the dependent variables are approximated, in the same way as in the FEM, and a mesh of control domains over which the governing equations are satisfied in an integral sense.

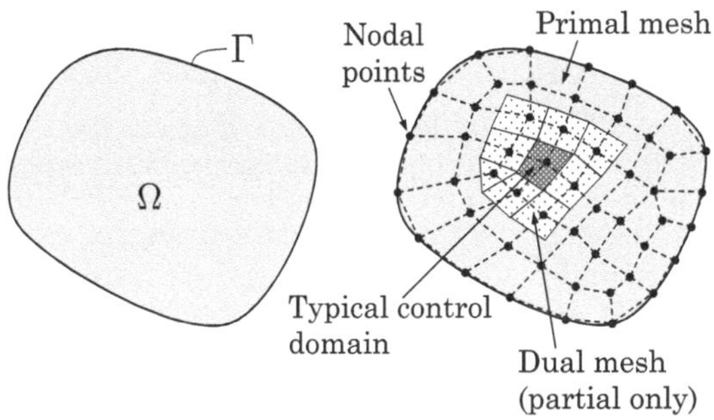

Fig. 5.2.1 Representation of a two-dimensional domain with meshes of quadrilateral finite elements (primal mesh) and node-centered quadrilateral control domains (dual mesh).

The discrete equations in the DMCDM are derived for a typical node centered in a control domain, as opposed to a typical element in the FEM, and therefore no "assembly" of elements is involved in the DMCDM. However, the DMCDM makes use of finite elements and their interpolation functions in deriving discrete equations among nodal values of the finite element mesh. The resulting discrete equations closely resemble finite difference equations applied to a mesh (or nodal) point. The discrete equations contain nodal values of all finite elements that participate in the control domain. The discretized equations derived for a typical node are then applied to all nodal points of the finite element mesh to obtain a system of algebraic equations among the nodal values. Since the control domain associated with a boundary node can only be inside the domain, discrete equations for boundary nodes are developed separately from those inside the domain, as was done in the half-control volume formulation of the FVM. These ideas will be illustrated in the following sections with model differential equations in a single unknown in one and two dimensions.

5.3 One-Dimensional Problems

5.3.1 Model Differential Equation

Consider the differential equation

$$-\frac{d}{dx}\left[a(x)\frac{du}{dx}\right] + c(x)u = f(x), \quad 0 < x < L \tag{5.3.1}$$

subjected to boundary conditions of the form

$$n_x\, a\frac{du}{dx} + \beta(x)\,(u - u_\infty) = \hat{Q}, \qquad \text{or} \qquad u = \hat{u} \tag{5.3.2}$$

at a boundary point. Here $u(x)$ denotes the dependent variable to be determined, a, c, and f are known functions of x, $(u_\infty,\ \hat{u},\ \beta,\ \hat{Q})$ are known (or specified) quantities, and $n_x = -1$ at $x = 0$ and $n_x = 1$ at $x = L$.

5.3.2 Primal and Dual Meshes

In the DMCDM, the domain $(0, L)$ is first divided into a set of N finite elements[1], $\Omega_f^{(1)}$, $\Omega_f^{(2)}$, ..., $\Omega_f^{(N)}$, of lengths h_1, $h_2, \ldots, h_N$. Then a dual mesh of $N+1$ control domains, $\Omega_c^{(1)}$, $\Omega_c^{(2)}$, ..., $\Omega_c^{(N+1)}$, such that the nodes of the finite elements are at the center of each control domain, except for the control domains at node 1 and node $N+1$ (i.e., boundary nodes), where the node is on one side of the control domain, as illustrated in Fig. 5.3.1.

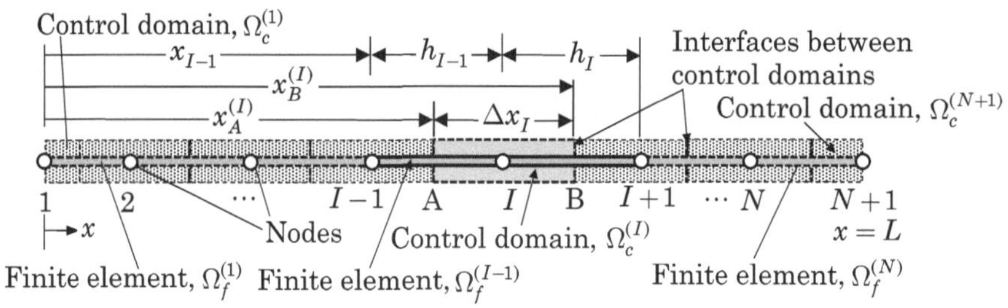

Fig. 5.3.1 A primal mesh of finite elements and dual mesh of control domains used in the discretization of a one-dimensional domain (i.e., a line). We note that the boundary nodes have only half-control domains, whereas the internal nodes have full control domains. Also, each control domain connects two neighboring finite elements, one on the left and the other on the right, and thus automatic assembly of element equations takes place.

[1]Here, the phrase "finite elements" refers to the line segments over which a function can be interpolated uniquely; it has nothing to do with the finite element method of solving a differential equation. In the finite element method, the governing equation is first written as weak form involving the dependent unknown and a weight function, before approximating the dependent variable using finite element interpolation functions.

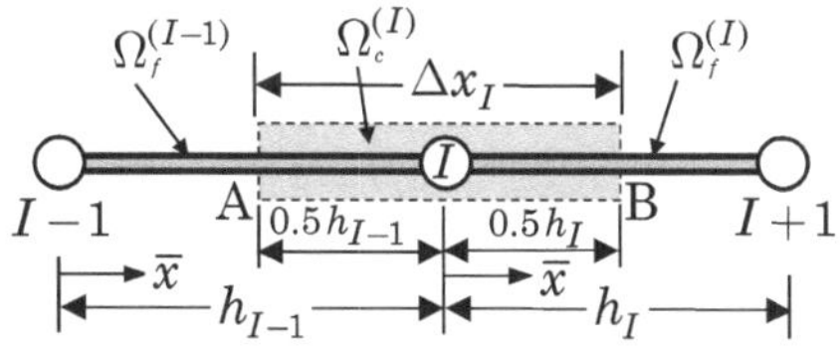

Fig. 5.3.2 Control domain associated with an interior node I. We note that each node has a primary unknown, and the control domain connects three neighboring nodal values of the primary variable (U_{I-1}, U_I, U_{I+1}) through the discretization of the governing equation.

5.3.3 Integral Statement over a Control Domain

The algebraic equations among the nodal values of a typical control domain are obtained by satisfying the governing equation (5.3.1) in an integral sense (with all terms put on one side of the equation, which amounts to requiring the integral of the residual due to the approximation to be zero) over a typical control domain $\Omega(I)_c$ shown in Fig. 5.3.2:

$$0 = \int_{x_A^{(I)}}^{x_B^{(I)}} \left\{ -\frac{d}{dx}\left[a(x)\frac{du}{dx}\right] + c(x)u - f \right\} dx. \tag{5.3.3}$$

The first term (which has two derivatives) is integrated once to reduce the differentiability required of the approximation functions and obtain

$$0 = \int_{x_A^{(I)}}^{x_B^{(I)}} \left[c(x)u - f(x)\right] dx - \left[-a(x)\frac{du}{dx}\right]_{x_A^{(I)}} - \left[a(x)\frac{du}{dx}\right]_{x_B^{(I)}} \tag{5.3.4}$$

or

$$0 = -N_1^{(I)} - N_2^{(I)} + \int_{x_A^{(I)}}^{x_B^{(I)}} \left[c(x)u - f(x)\right] dx, \tag{5.3.5a}$$

where (see Fig. 5.3.3)

$$N_1^{(I)} \equiv \left[-a(x)\frac{du}{dx}\right]_{x_A^{(I)}}, \quad N_2^{(I)} \equiv \left[a(x)\frac{du}{dx}\right]_{x_B^{(I)}}. \tag{5.3.5b}$$

Here $N_1^{(I)}$ and $N_2^{(I)}$ denote the *secondary variables* at the left and right interfaces of the control domain centered at node I. Physically, the secondary variables denote the axial forces or heats, when the model equation is one that describes axial deformation of bars or one-dimensional heat flow. The minus sign in the definition of $N_1^{(I)}$ indicates that it is a compressive force or heat input. The identification of the secondary variables is a significant feature of the DMCDM, and it allows handling of the boundary conditions on the secondary variables in a physically meaningful way, without replacing them in terms of the nodal values of u (as was done in the case of the half-control FVM).

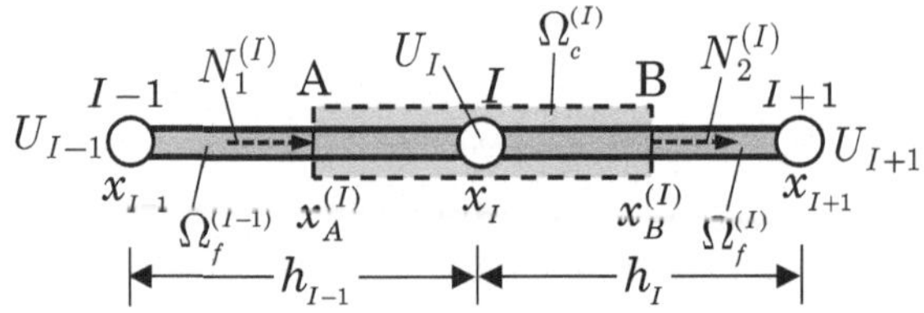

Fig. 5.3.3 A typical control domain $\Omega_c^{(I)}$ for the 1-D model. Here, points A and B refer to the left and right end locations of the control domain associated with node I; these points have the coordinates $x_A^{(I)}$ and $x_B^{(I)}$, respectively; note that point A is in finite element $\Omega_f^{(I-1)}$ and point B is in finite element $\Omega_f^{(I)}$.

5.3.4 Discretized Equations over a Control Domain

We employ, as an example, the linear interpolation of $u(x)$ over a typical finite element, $\Omega_f^{(I)} = (x_I, x_{I+1})$:

$$u(x) \approx u_h(x) \equiv U_I \psi_1^{(I)}(x) + U_{I+1}\psi_2^{(I)}(x), \tag{5.3.6}$$

where U_I is the value of u at node I [i.e., $U_I \approx u(x_I)$] and $\psi_i^{(I)}(x)$ $(i = 1, 2)$ are linear finite element interpolation functions of element $\Omega^{(I)}$ for $I = 1, 2, \ldots, N$:

$$\psi_1^{(I)}(x) = \frac{x_{I+1} - x}{h_I}, \quad \psi_2^{(I)}(x) = \frac{x - x_I}{h_I}. \tag{5.3.7}$$

Substituting the approximation (5.3.6) into Eq. (5.3.5a), we obtain (for $I = 2, 3, \ldots, N$)

$$-N_1^{(I)} - N_2^{(I)} + C_{I-1}U_{I-1} + C_I U_I + C_{I+1}U_{I+1} = F_I, \tag{5.3.8}$$

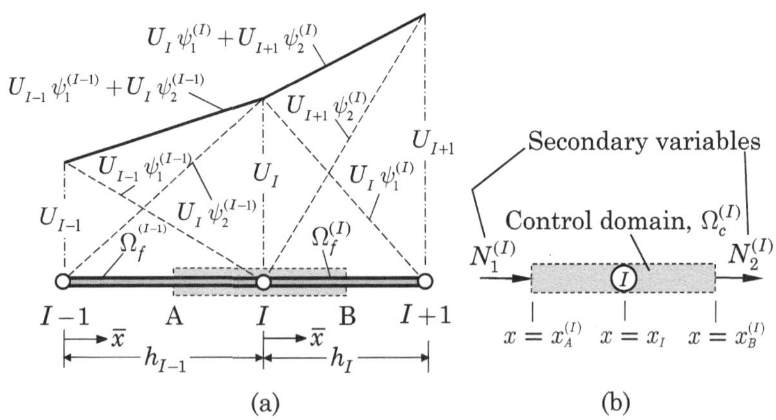

Fig. 5.3.4 (a) Linear approximation of $u(x)$ over two neighboring finite elements. (b) The secondary variables of a typical control domain.

where

$$
\begin{aligned}
a_1^{(I)} &= a(x_A^{(I)}), \quad a_2^{(I)} = a(x_B^{(I)}), \quad F_I = \int_{x_A^{(I)}}^{x_B^{(I)}} f(x)dx \\
C_{I-1} &= \int_{x_A^{(I)}}^{x_I} c(x)\psi_1^{(I-1)}(x)dx, \quad C_{I+1} = \int_{x_I}^{x_B^{(I)}} c(x)\psi_2^{(I)}(x)dx \\
C_I &= \int_{x_A^{(I)}}^{x_I} c(x)\psi_2^{(I-1)}(x)dx + \int_{x_I}^{x_B^{(I)}} c(x)\psi_1^{(I)}(x)dx.
\end{aligned}
\tag{5.3.9}
$$

5.3.4.1 Discretized equations for interior nodes

For interior nodes, we replace the secondary variables $N_1^{(I)} = N_A^{(I)}$ and $N_2^{(I)} = N_B^{(I)}$ in Eq. (5.3.5b) in terms of the nodal values of $u(x)$ using the interpolation in Eq. (5.3.6) (we note that point A is in element $\Omega^{(I-1)}$ and point B is in element $\Omega^{(I)}$; see Fig. 5.3.4). We have

$$
\begin{aligned}
N_1^{(I)} &\equiv \left[-a\frac{du}{dx}\right]_{x_A^{(I)}} = a_1^{(I)}\left(\frac{U_{I-1} - U_I}{h_{I-1}}\right) \\
N_2^{(I)} &\equiv \left[a\frac{du}{dx}\right]_{x_B^{(I)}} = a_2^{(I)}\left(\frac{U_{I+1} - U_I}{h_I}\right).
\end{aligned}
\tag{5.3.10}
$$

Now use Eq. (5.3.10) to replace $N_1^{(I)}$ and $N_2^{(I)}$ in Eq. (5.3.8) and arrive at the relations

$$
a_1^{(I)}\left(\frac{U_I - U_{I-1}}{h_{I-1}}\right) + a_2^{(I)}\left(\frac{U_I - U_{I+1}}{h_I}\right) + C_{I-1}U_{I-1} + C_I U_I + C_{I+1}U_{I+1} = F_I,
\tag{5.3.11}
$$

which can be expressed as

$$
A_{I-1}U_{I-1} + A_I U_I + A_{I+1}U_{I+1} = F_I \tag{5.3.12}
$$

for $I = 2, 3, \ldots, N$. The coefficients A_{I-1}, A_I, and A_{I+1} are defined as

$$
A_{I-1} = C_{I-1} - \frac{a_1^{(I)}}{h_{I-1}}, \quad A_I = C_I + \frac{a_1^{(I)}}{h_{I-1}} + \frac{a_2^{(I)}}{h_I}, \quad A_{I+1} = C_{I+1} - \frac{a_2^{(I)}}{h_I}. \tag{5.3.13}
$$

For the mesh shown in Fig. 5.3.4(b) (i.e., $x_A^{(I)} = x_I - 0.5h_{I-1}$ and $x_B^{(I)} = x_I + 0.5h_I$), the coefficients in Eqs. (5.3.12) can be evaluated for the *linear* expansions of $a(x) = a_0 + a_1 x$, $c(x) = c_0 + c_1 x$, and $f(x) = f_0 + f_1 x$ as follows:

$$
a_1^{(I)} = a(x_A^{(I)}) = a_0 + a_1 x_A^{(I)}, \quad a_2^{(I)} = a(x_B^{(I)}) = a_0 + a_1 x_B^{(I)} \tag{5.3.14a}
$$

$$\begin{aligned}
C_{I-1} &= \tfrac{1}{8}h_{I-1}c_0 + \tfrac{1}{24}\left(2h_{I-1} + 3x_{I-1}\right)h_{I-1}c_1 \\
C_I &= C_{IL} + C_{IR}, \\
C_{IL} &= \tfrac{1}{24}\left[9c_0 + \left(9x_{I-1} + 7h_{I-1}\right)c_1\right]h_{I-1} \qquad (5.3.14b)\\
C_{IR} &= \tfrac{1}{24}\left[9c_0 + \left(9x_{I-1} + 9h_{I-1} + 2h_I\right)c_1\right]h_I, \\
C_{I+1} &= \tfrac{1}{8}h_I c_0 + \tfrac{1}{24}\left(3h_{I-1} + h_I + 3x_{I-1}\right)h_I c_1 \\
F_I &= \left[f_0 + 0.5f_1\left(x_A^{(I)} + x_B^{(I)}\right)\right]\Delta x_I. \qquad (5.3.14c)
\end{aligned}$$

5.3.4.2 Discretized equations for boundary nodes

For a boundary node, the discrete equations will be different, and they have to be formulated as discussed next. For node 1 [see Fig 5.3.5(a)], we have $x_A^{(1)} = 0$, and Eq. (5.3.5a) takes the form

$$0 = -N_1^{(1)} + a_2^{(1)}\left(\frac{U_1 - U_2}{h_1}\right) + \int_0^{x_B^{(1)}}\left[c\left(U_1\psi_1^{(1)} + U_2\psi_2^{(1)}\right) - f\right]dx$$

or

$$-N_1^{(1)} + \bar{A}_1U_1 + \bar{A}_2U_2 - F_1 = 0, \qquad (5.3.15a)$$

where

$$\begin{aligned}
\bar{A}_1 &= \bar{C}_1 + \frac{a_2^{(1)}}{h_1}, \quad \bar{A}_2 = \bar{C}_2 - \frac{a_2^{(1)}}{h_1}, \quad F_1 = \int_0^{x_B^{(1)}} f(x)\,dx \\
\bar{C}_1 &= \int_0^{x_B^{(1)}} c(x)\psi_1^{(1)}(x)\,dx, \quad \bar{C}_2 = \int_0^{x_B^{(1)}} c(x)\psi_2^{(1)}(x)\,dx.
\end{aligned} \qquad (5.3.15b)$$

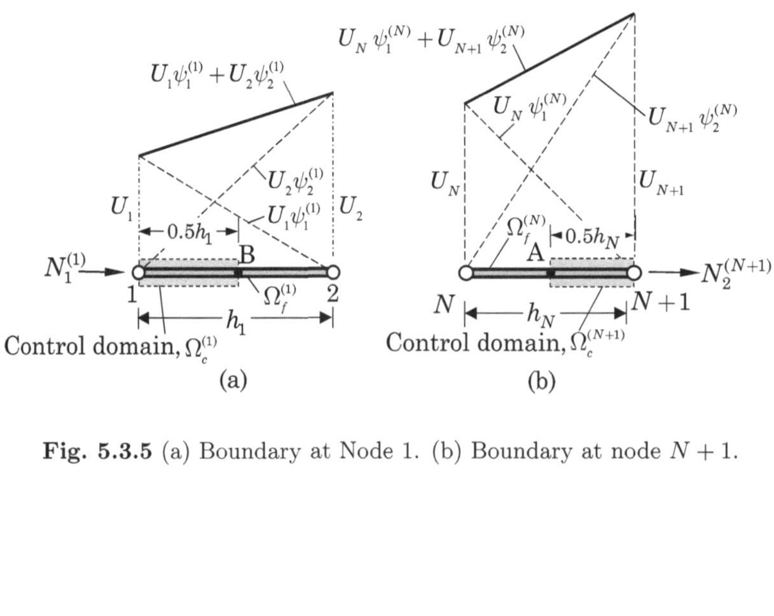

Fig. 5.3.5 (a) Boundary at Node 1. (b) Boundary at node $N+1$.

For linear approximation of $u(x)$ and linear expansions of $c(x)$ and $f(x)$, we obtain

$$\bar{C}_1 = \tfrac{3}{8}c_0h_1 + \tfrac{1}{12}c_1h_1^2, \quad \bar{C}_2 = \tfrac{1}{8}c_0h_1 + \tfrac{1}{24}c_1h_1^2, \quad F_1 = \tfrac{1}{2}h_1f_0 + \tfrac{1}{8}f_1h_1^2. \tag{5.3.16}$$

Similarly, for the right end of the domain [i.e., node $N+1$; see Fig 5.3.5(b)], we have

$$0 = a_1^{(N+1)}\left(\frac{U_{N+1} - U_N}{h_N}\right) - N_2^{(N+1)} + \int_{x^{(N)}}^{x_{N+1}} \left[c\left(U_N\psi_1^{(N)} + U_{N+1}\psi_2^{(N)}\right) - f\right] dx$$

or

$$-N_2^{(N+1)} + \bar{A}_N U_N + \bar{A}_{N+1}U_{N+1} - F_{N+1} = 0, \tag{5.3.17a}$$

where ($x^{(N)} = x_{N+1} - 0.5h_N$)

$$\begin{aligned}
\bar{A}_N &= \bar{C}_N - \frac{a_1^{N+1}}{h_N}, \quad \bar{A}_{N+1} = \bar{C}_{N+1} + \frac{a_1^{N+1}}{h_N}\\
F_{N+1} &= \int_{x^{(N)}}^{x_{N+1}} f(x)\, dx\\
\bar{C}_N &= \int_{x^{(N)}}^{x_{N+1}} c(x)\psi_1^{(N)}(x)dx,\\
\bar{C}_{N+1} &= \int_{x^{(N)}}^{x_{N+1}} c(x)\psi_2^{(N)}(x)dx.
\end{aligned} \tag{5.3.17b}$$

Again, for linear approximation of $u(x)$ and linear expansion of $c(x)$ and $f(x)$ we have

$$\begin{aligned}
\bar{C}_N &= \tfrac{1}{8}h_Nc_0 + \left(\tfrac{1}{12}h_N + \tfrac{1}{8}x_N\right)h_Nc_1,\\
\bar{C}_{N+1} &= \tfrac{3}{8}h_Nc_0 + \left(\tfrac{7}{24}h_N + \tfrac{3}{8}x_N\right)h_Nc_1,\\
F_{N+1} &= \tfrac{1}{2}h_Nf_0 + \left(\tfrac{3}{8}h_N + \tfrac{1}{2}x_N\right)h_Nf_1.
\end{aligned} \tag{5.3.18}$$

5.3.5 Numerical Examples

Here we consider two examples to illustrate the ideas presented in the previous sections. The examples are the same as those analyzed by the FVM and FEM in Chapters 3 and 4, respectively. **Examples 3.3.1** and **4.2.1** are not revisited here because the half-control FVM results are the same as those predicted by the DMCDM. This is due to the fact that the linear interpolation functions used in the DMCDM give the same discrete equations for the first derivatives in the interior of the domain as the FVM [see Eqs. (3.2.12) and (3.2.13)] and the treatment of the boundary conditions in the half-control FVM and the DMCDM is the same. The first example of this section is the same as Case 2 of **Example 3.3.2** and **Example 4.2.2**. The FVM (half-control volume formulation) and FEM results are compared against the results of the DMCDM.

Example 5.3.1

Consider a steel rod of uniform diameter $D = 0.02$ m, length $L = 0.05$ m, and constant thermal conductivity $k = 50$ W/(m·°C) that is exposed to ambient air at 20°C with a heat transfer coefficient $\beta = 100$ W/(m^2·°C). The left end of the rod is maintained at temperature $T(0) = T_0 = 320\,$°C and the other end, $x = L$, is exposed to the ambient air at $T_\infty = 20$°C with heat transfer coefficient β, as shown in Fig. 5.3.6. Assuming that there is no internal heat generation, use the DMCDM to determine the temperature distribution in the rod and the heat input at the left end of the rod and compare with the FVM and FEM results.

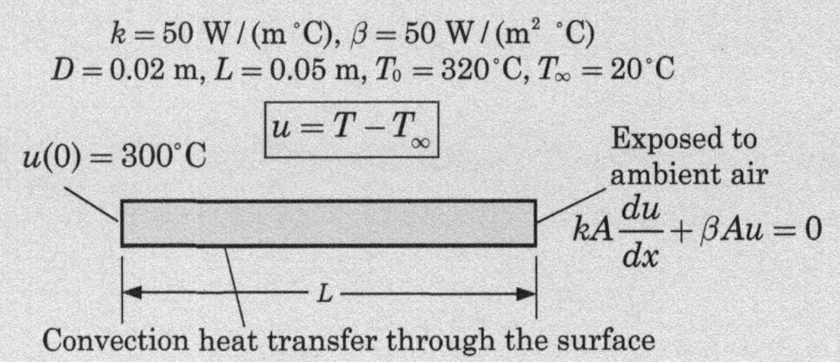

Fig. 5.3.6 Geometry and boundary conditions for heat transfer in an uninsulated rod.

Solution: This is the same problem as that considered in Case 2 of **Examples 3.3.2** and **4.2.2**. The governing differential equation is

$$-\frac{d^2u}{dx^2} + cu = 0 \quad \text{for} \quad 0 < x < L, \tag{1}$$

where $c = \frac{P\beta}{kA}$, P being the perimeter and A being the cross-sectional area of the bar, and u is the temperature above the ambient temperature ($u = T - T_\infty$). In the present case, we have $a = 1$ and $c = c_0 = 400$. The boundary conditions are

$$u_0 = u(0) = T(0) - T_\infty = 300° \text{ C}, \qquad \left[\frac{du}{dx} + \frac{\beta}{k}u\right]_{x=L} = 0. \tag{2}$$

The exact solution of the problem for u is given by

$$u(x) = u_0\left[\frac{\cosh\sqrt{c}(L-x) + (\beta/\sqrt{c}k)\sinh\sqrt{c}(L-x)}{\cosh\sqrt{c}L + (\beta/\sqrt{c}k)\sinh\sqrt{c}L}\right], \tag{3}$$

and the gradient is

$$-a\frac{du}{dx} = u_0\left[\sqrt{c}\frac{\sinh\sqrt{c}(L-x) + (\beta/k)\cosh\sqrt{c}(L-x)}{\cosh\sqrt{c}L + (\beta/\sqrt{c}k)\sinh\sqrt{c}L}\right], \tag{4}$$

where the value of c is given by ($kA = \pi 50 \times 10^{-4}$ and $\beta P = 2\pi$)

$$c = \frac{P\beta}{Ak} = \frac{2\pi}{50 \times 10^{-4}\pi} = 400. \tag{5}$$

As an example, we consider a primal mesh of five linear finite elements and associated dual mesh of six control domains, as shown in Fig. 5.3.7. Figure 5.3.7 also shows the mesh and boundary conditions for the problem. The discrete equations for the mesh shown in Fig. 5.3.7 are obtained using Eqs. (5.3.15a), (5.3.12) (with $I = 2, 3, 4, 5$), and (5.3.17a) (with $a = 1$,

$c = 400$, and $h = 0.01$). We obtain the same discretized equations as those of the half-control FVM with linear approximation of the $u(x)$ [see Eqs. (3.2.8), (3.2.9a), and (3.2.9b); also see Eq. (13) of **Example 3.3.2**]:

$$\begin{bmatrix} 101.5 & -99.5 & 0.0 & 0.0 & 0.0 & 0.0 \\ -99.5 & 203.0 & -99.5 & 0.0 & 0.0 & 0.0 \\ 0.0 & -99.5 & 203.0 & -99.5 & 0.0 & 0.0 \\ 0.0 & 0.0 & -99.5 & 203.0 & -99.5 & 0.0 \\ 0.0 & 0.0 & 0.0 & -99.5 & 203.5 & -99.5 \\ 0.0 & 0.0 & 0.0 & 0.0 & -99.5 & 101.5 \end{bmatrix} \begin{Bmatrix} U_1 \\ U_2 \\ U_3 \\ U_4 \\ U_5 \\ U_6 \end{Bmatrix} = \begin{Bmatrix} Q_1^{(1)} \\ 0 \\ 0 \\ 0 \\ 0 \\ -Q_2^{(5)} \end{Bmatrix}. \tag{6}$$

After imposing the boundary conditions, $U_1 = 300$ and $Q_1^{(5)} = (\beta/k)U_6$, and solving the equations, we obtain [the solution is the same as that obtained using FVM32; see Eq. (40) of **Example 3.3.2**]

$$U_2 = 257.62\,^\circ\text{C},\quad U_3 = 225.59\,^\circ\text{C},\quad U_4 = 202.64\,^\circ\text{C},\quad U_5 = 187.83\,^\circ\text{C},\quad U_6 = 180.57\,^\circ\text{C}.$$

The heat at node 1 is computed from Eq. (5.3.15a) (change of the notation, $N_1^{(1)} = Q_1^{(1)}$): $Q_1^{(1)} = 4817\,\text{W}$. The corresponding exact values are

$$u_2^e = 257.66\,^\circ\text{C},\quad u_3^e = 225.67\,^\circ\text{C},\quad u_4^e = 202.72\,^\circ\text{C},\quad u_5^e = 187.92\,^\circ\text{C},\quad u_6^e = 180.66\,^\circ\text{C}$$

and $Q(0) = 4804$W. The FEM solution using the same mesh (i.e., five linear finite elements) is

$$U_2^{\text{fe}} = 257.57\,^\circ\text{C},\quad U_3^{\text{fe}} = 225.51\,^\circ\text{C},\quad U_4^{\text{fe}} = 202.53\,^\circ\text{C},\quad U_5^{\text{fe}} = 187.70\,^\circ\text{C},\quad U_6^{\text{fe}} = 180.44\,^\circ\text{C}$$

and $Q^{fe}(0) = 4815$ W. The solution obtained with the same mesh and the FVM31 is

$$U_2^{\text{fv}} = 257.77\,^\circ\text{C},\quad U_3^{\text{fv}} = 225.85\,^\circ\text{C},\quad U_4^{\text{fv}} = 202.97\,^\circ\text{C},\quad U_5^{\text{fv}} = 188.20\,^\circ\text{C},\quad U_6^{\text{fv}} = 180.96\,^\circ\text{C}$$

and $Q^{fv}(0) = 4823$ W.

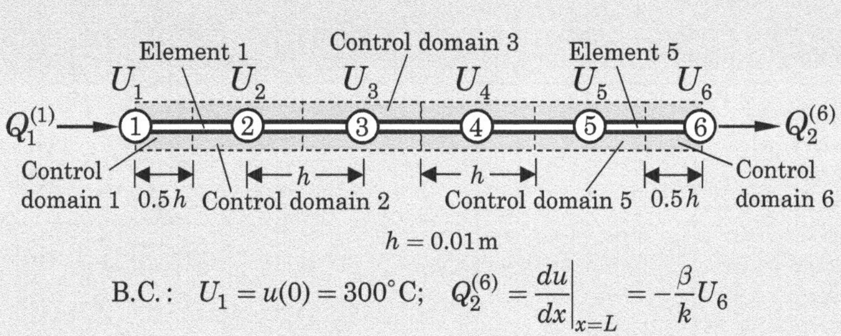

Fig. 5.3.7 The primal mesh of five finite elements and dual mesh of six control domains (six nodes), and boundary conditions for the DMCDM analysis of one-dimensional heat flow in uninsulated rod.

A comparison of the nodal values of u obtained by the FEM, FVM31, and the DMCDM (the FVM32 also gives the same solutions as the DMCDM) with the exact values is presented in Table 5.3.1 for 10 and 20 elements. All numerical solutions are in good agreement with the exact solution, with the DMCDM (and FVM32) being the best for the meshes shown. In addition, the secondary variable predicted by the FVM31 is in more error than any other method. We note that the FVM21 and FVM22 solutions would be even more inaccurate compared to the other solutions.

Table 5.3.1 Comparison of the FEM, FVM31, and DMCDM (the same as that of the FVM32) solutions with the exact solution of Eqs. (1) and (2).

$$-\frac{d^2u}{dx^2} + 400\,u = 0,\ 0 < x < 0.05;\quad u(0) = 300,\quad \left(\frac{du}{dx} + \frac{\beta}{k}u\right)\Big|_{x=L} = 0.$$

x	Exact Solution	FEM Solution		FVM31 Solution		DMCDM Solution	
		$N=10$	$N=20$	$N=10$	$N=20$	$N=10$	$N=20$
0.005	277.44	277.43	277.44	277.46	277.45	277.44	277.44
0.010	257.66	257.64	257.66	257.69	257.67	257.65	257.66
0.015	240.46	240.43	240.45	240.50	240.47	240.45	240.46
0.020	225.66	225.63	225.65	225.71	225.68	225.65	225.66
0.025	213.13	213.08	213.12	213.18	213.14	213.11	213.12
0.030	202.72	202.68	202.71	202.78	202.74	202.70	202.72
0.035	194.35	194.30	194.34	194.42	194.37	194.33	194.34
0.040	187.92	187.87	187.91	187.99	187.94	187.90	187.91
0.045	183.37	183.32	183.36	183.44	183.39	183.35	183.37
0.050	180.66	180.60	180.64	180.73	180.68	180.64	180.65
Q(0)	4803.7	4806.5	4805.0	4808.6	4804.9	4807.0	4804.5

Example 5.3.2

Consider a long, homogeneous, isotropic solid cylinder of outside radius $R_0 = 0.01$ m and conductivity $k = 20$ W/(m·°C), and with a constant rate of internal heat generation $g_0 = 2 \times 10^8$ W/m^3. Suppose that the boundary surface at $r = R_0$ is maintained at $T_0 = 100\,°$C. Calculate the temperature distribution $T(r)$ and heat flux at $r = R_0$ using uniform meshes of four and eight linear elements in the DMCDM and compare with the FEM, FVM, and exact solutions.

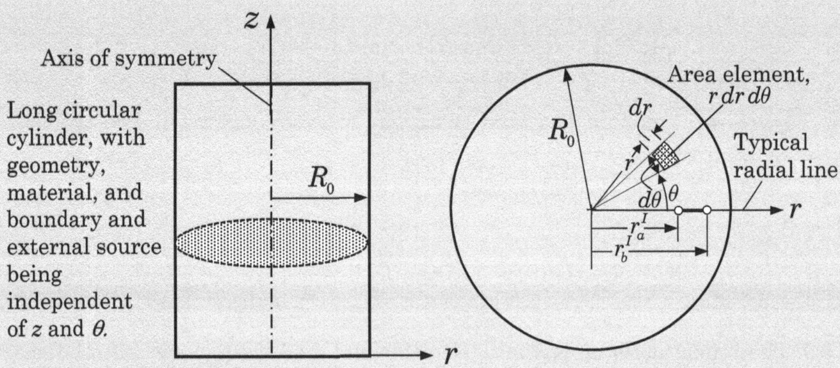

Fig. 5.3.8 Reduction of the three-dimensional heat transfer in a long cylinder to one dimension by using axisymmetry conditions.

Solution: This is the same problem as that considered in **Examples 3.3.3** and **4.2.3**. The governing equation for this problem is given by

$$-\frac{1}{r}\frac{d}{dr}\left(kr\frac{dT}{dr}\right) = g_0. \tag{1}$$

The boundary conditions are

$$\left(2\pi kr\frac{dT}{dr}\right)\Big|_{r=0} = 0, \quad T(R_0) = T_0. \tag{2}$$

The first boundary condition is a statement of the fact that the heat flux at the center of the cylinder (i.e., $r = 0$) is zero due to axisymmetry (see Fig. 5.3.8). The exact solution of the problem is

$$T(r) = T_0 + \frac{f_0}{4k}\left(R_0^2 - r^2\right), \quad Q(r) = -2\pi kr\frac{dT}{dr} = \pi r^2\, g_0. \tag{3}$$

The integral statement of the equation at hand is the same as that (5.3.3) with the fact that $a(r) = 2\pi kr$ (i.e., replace u with T and x with r):

$$\begin{aligned} 0 &= \int_0^{2\pi}\int_{r_a^I}^{r_b^I}\left[-\frac{1}{r}\frac{d}{dr}\left(kr\frac{dT}{dr}\right) - g_0\right] r dr d\theta \\ &= 2\pi\left[-kr\frac{dT}{dr}\right]_{r_b^I} - 2\pi\left[-kr\frac{dT}{dr}\right]_{r_a^I} - 2\pi\int_{r_a^I}^{r_b^I} r g_0\, dr \end{aligned} \tag{4}$$

or

$$-Q_1^{(I)} - Q_2^{(I)} - 2\pi\int_{r_a^I}^{r_b^I} r g_0\, dr = 0, \tag{5}$$

where (linear interpolation of $T(r)$ is assumed)

$$\begin{aligned} Q_1^{(I)} &\equiv 2\pi\left[-kr\frac{dT}{dr}\right]_{r_a^I} = a_1^I\left(\frac{T_{I-1} - T_I}{h_{I-1}}\right) \\ Q_2^{(I)} &\equiv 2\pi\left[kr\frac{dT}{dr}\right]_{r_b^I} = a_2^I\left(\frac{T_{I+1} - T_I}{h_I}\right) \end{aligned} \tag{6}$$

and $a_1^I = 2\pi\, r_a^I\, k(r_a^I)$, $a_2^I = 2\pi\, r_b^I\, k(r_b^I)$; $r_a^I = r_I - 0.5h_{I-1}$, and $r_b^I = r_I + 0.5h_I$.

The coefficients presented in Eqs. (5.3.13), (5.3.14a)–(5.3.14c), (5.3.16), and (5.3.18) can be used (with zero constant coefficients, $a_0 = 0$, $c_0 = 0$, and $f_0 = 0$). The discretized equations for a primal mesh of four ($N = 4$) elements and associated dual mesh of five control domains are given by [note that the factor 2π is included here, whereas it was cancelled out on both sides of Eq. (9) of **Example 3.3.3**]

$$\begin{bmatrix} 62.832 & -62.832 & 0.000 & 0.000 & 0.000 \\ -62.832 & 251.327 & -188.496 & 0.000 & 0.000 \\ 0.000 & -188.496 & 502.655 & -314.159 & 0.000 \\ 0.000 & 0.000 & -314.159 & 753.982 & -439.823 \\ 0.000 & 0.000 & 0.000 & -439.823 & 439.823 \end{bmatrix} \begin{Bmatrix} T_1 \\ T_2 \\ T_3 \\ T_4 \\ T_5 \end{Bmatrix} = \frac{1}{2\pi}\begin{Bmatrix} Q_1^{(1)} \\ 0 \\ 0 \\ 0 \\ Q_2^{(5)} \end{Bmatrix} + 10^3\begin{Bmatrix} 0.982 \\ 7.854 \\ 15.708 \\ 23.562 \\ 14.726 \end{Bmatrix}. \tag{7}$$

The boundary conditions are: $Q_1^{(1)} = 0$ and $T_{N+1} = T_0 = 300\,^\circ\mathrm{C}$. The solution of these equations is [the same as that in Eq. (11) of **Example 3.3.3**]

$$T_1 = 350.0\,^\circ\mathrm{C}, \quad T_2 = 334.375\,^\circ\mathrm{C}, \quad T_3 = 287.5\,^\circ\mathrm{C}, \quad T_4 = 209.375\,^\circ\mathrm{C}. \tag{8}$$

The heat at $r = R_0$ is

$$Q(R_0) = -Q_2^{(5)} = 2\pi\,(70\,T_4 - 70\,T_5 + 2343.75) = 2\pi \times 10^4. \tag{9}$$

The numerical solution obtained with the DMCDM matches with the numerical solutions obtained with the HFVM (FVM31 is the same as the FVM32 in this case, as there is no $c\,u(r)$ term in the governing equation) as well as with the exact solution at the nodes.

A comparison of the FEM solutions for the temperature obtained using four-, and eight-element meshes of linear elements and a four-element mesh of quadratic elements with the HFVM and DMCDM solutions obtained with four- and eight-element meshes, and the exact solution is presented in Table 5.3.2. The FEM mesh of four quadratic elements and the DMCDM mesh of eight linear elements yield exact solutions at the nodes. It is clear that the nodal values for the temperatures predicted by the DMCDM and HFVM are exact for all meshes considered, whereas those predicted by the FEM are inexact but converge to the exact solution with mesh refinements.

Table 5.3.2 Comparison of the FEM, HFVM, and DMCDM solutions with the exact solution for temperature distribution in a axisymmetric circular disc (L - linear and Q - quadratic elements; N denotes the number of subdivisions used in the HFVM; and the underlined numbers are the linearly interpolated values).

	FEM Solution			HFVM Solution		DMCDM Solution		Exact Solution
r/R_0	4L	8L	4Q	$N=4$	$N=8$	4L	8L	
0.000	358.73	352.63	350.00	350.00	350.00	350.00	350.00	350.00
0.125	$\underline{348.32}$	347.42	346.09	$\underline{342.19}$	346.09	$\underline{342.19}$	346.09	346.09
0.250	337.90	335.27	334.37	334.37	334.37	334.37	334.37	334.37
0.375	$\underline{313.60}$	315.48	314.84	$\underline{310.94}$	314.84	$\underline{310.94}$	314.84	314.84
0.500	289.29	287.95	287.50	287.50	287.50	287.50	287.50	287.50
0.625	$\underline{249.71}$	252.65	252.34	$\underline{248.44}$	252.34	$\underline{248.44}$	252.34	252.34
0.750	210.12	209.56	209.37	209.37	209.37	209.37	209.37	209.37
0.875	$\underline{155.06}$	158.68	158.59	$\underline{154.69}$	158.59	$\underline{154.69}$	158.59	158.59

5.4 Two-Dimensional Problems

5.4.1 Preliminary Comments

Here we discuss the DMCDM for two-dimensional problems. Extension of the ideas presented for the dual mesh control domain model of one-dimensional problems to rectangular meshes of two-dimensional problems is straightforward. That is, if the domains of the two-dimensional problems are of rectangular geometry, we can follow the same procedure as in one-dimensional problems to derive the algebraic equations of all nodes in the mesh. In fact, many of the ideas from Section 3.4 on half-control FVM formulation are applicable here, except for the actual values of the coefficients of the discretized equations.

5.4.2 Model Equation

We begin with a model partial differential equation

$$-\left[\frac{\partial}{\partial x}\left(a_{xx}\frac{\partial u}{\partial x}\right)+\frac{\partial}{\partial y}\left(a_{yy}\frac{\partial u}{\partial y}\right)\right]=f(x,y)\quad \text{in } \Omega, \tag{5.4.1}$$

where $u(x,y)$ is the dependent unknown, (a_{xx}, a_{yy}, f) are the data, and Ω is the domain with closed boundary Γ of the problem. In general, u is required to satisfy one of the following two types of boundary conditions at a boundary point:

$$u = \hat{u}(s) \quad \text{on} \quad \Gamma_u, \tag{5.4.2a}$$

$$\left(a_{xx}\frac{\partial u}{\partial x}n_x + a_{yy}\frac{\partial u}{\partial y}n_y\right) + \beta(u - u_\infty) = \hat{q}_n(s) \quad \text{on} \quad \Gamma_q, \tag{5.4.2b}$$

where Γ_u and Γ_q are disjoint portions of the total boundary Γ such that $\Gamma = \Gamma_u \cup \Gamma_q$, β is the heat transfer coefficient, u_∞ the temperature of the medium surrounding the body, $(u_\infty,\ \hat{q}_n(s),\ \hat{u}(s))$ are specified functions on the boundary, and (n_x, n_y) are the direction cosines of the unit normal vector $\hat{\mathbf{n}}$ on the boundary Γ.

We discretize the rectangular domain into a nonuniform primal mesh of N elements along the x coordinate and M elements along the y coordinate, as shown in Fig. 5.4.1. For simplicity, we shall assume that the control domain bisects the neighboring finite elements.

The integral statement of Eq. (5.4.1) over a typical rectangular control domain (see Fig. 5.4.2) is developed as follows

$$\begin{aligned} 0 &= -\int_{x_I-0.5a}^{x_I+0.5a}\int_{y_I-0.5b}^{y_I+0.5b}\left\{\left[\frac{\partial}{\partial x}\left(a_{xx}\frac{\partial u}{\partial x}\right) + \frac{\partial}{\partial y}\left(a_{yy}\frac{\partial u}{\partial y}\right)\right] - f\right\} dxdy \\ &= -\oint_{\Gamma_R}\left[n_x\left(a_{xx}\frac{\partial u}{\partial x}\right) + n_y\left(a_{yy}\frac{\partial u}{\partial y}\right)\right] ds - \int_{x_I-0.5a}^{x_I+0.5a}\int_{y_I-0.5b}^{y_I+0.5b} f\, dxdy, \end{aligned} \tag{5.4.3}$$

where (x_I, y_I) are the global coordinates of the node labelled as I, (n_x, n_y) are the direction cosines of the unit normal vector to the boundary of the control domain, and Γ_R is the boundary of the rectangular control domain. The boundary integration is taken all around the boundary of the control domain in the direction indicated in Fig. 5.4.2.

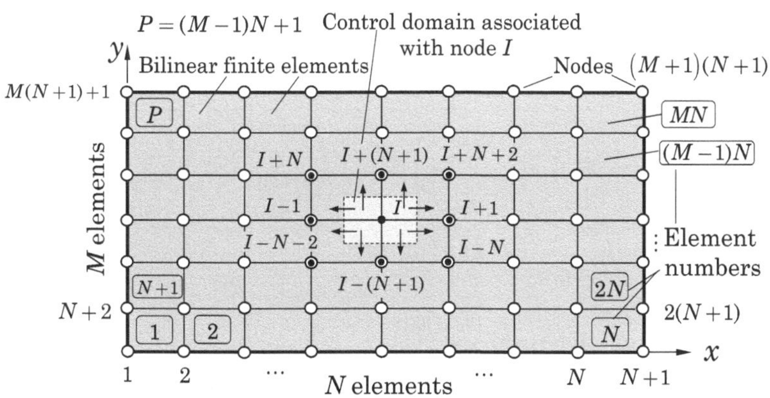

Fig. 5.4.1 $N \times M$ primal mesh of bilinear rectangular finite elements and a dual mesh of control domains is such that the internal nodes of the primal mesh are at the center of the rectangular control domains. The boundary nodes of the primal mesh are not at the center of the boundary control domains.

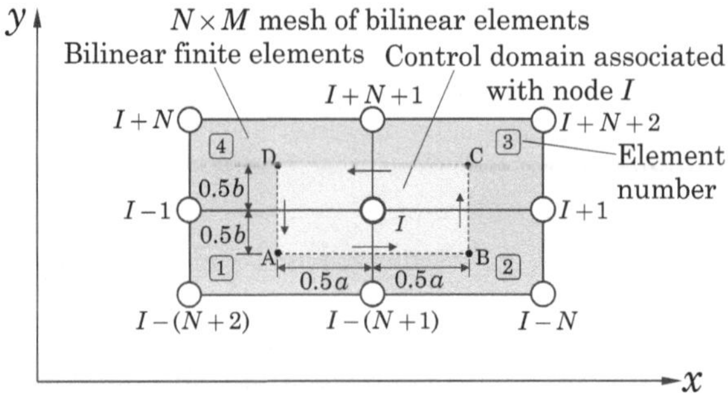

Fig. 5.4.2 Two-dimensional control domain associated with an interior node I when the primal mesh of linear rectangular finite elements is used.

The boundary integrals can be simplified using the values of the direction cosines on each boundary line segment. We have [$\hat{\mathbf{n}} = (n_x, n_y)$; $\hat{\mathbf{n}} = (0, -1)$ and $ds = dx$ on AB; $\hat{\mathbf{n}} = (1, 0)$ and $ds = dy$ on BC; $\hat{\mathbf{n}} = (0, 1)$ and $ds = -dx$ on CD; and $\hat{\mathbf{n}} = (-1, 0)$ and $ds = -dy$ on DA]

$$\begin{aligned}
&\oint_{\Gamma_R} \left[n_x \left(a_{xx} \frac{\partial u}{\partial x} \right) + n_y \left(a_{yy} \frac{\partial u}{\partial y} \right) \right] ds \\
&= -\int_{x_I-0.5a}^{x_I+0.5a} \left[a_{yy} \frac{\partial u}{\partial y} \right]_{y=y_I-0.5b} dx + \int_{y_I-0.5b}^{y_I+0.5b} \left[a_{xx} \frac{\partial u}{\partial x} \right]_{x=x_I+0.5a} dy \\
&\quad + \int_{x_I-0.5a}^{x_I+0.5a} \left[a_{yy} \frac{\partial u}{\partial y} \right]_{y=y_I+0.5b} dx - \int_{y_I-0.5b}^{y_I+0.5b} \left[a_{xx} \frac{\partial u}{\partial x} \right]_{x=x_I-0.5a} dy. \qquad (5.4.4)
\end{aligned}$$

Then the integral form in Eq. (5.4.3) becomes

$$\begin{aligned}
0 &= \int_{x_I-0.5a}^{x_I+0.5a} \left[a_{yy} \frac{\partial u}{\partial y} \right]_{y=y_I-0.5b} dx - \int_{y_I-0.5b}^{y_I+0.5b} \left[a_{xx} \frac{\partial u}{\partial x} \right]_{x=x_I+0.5a} dy \\
&\quad - \int_{x_I-0.5a}^{x_I+0.5a} \left[a_{yy} \frac{\partial u}{\partial y} \right]_{y=y_I+0.5b} dx + \int_{y_I-0.5b}^{y_I+0.5b} \left[a_{xx} \frac{\partial u}{\partial x} \right]_{x=x_I-0.5a} dy \\
&\quad - \int_{x_I-0.5a}^{x_I+0.5a} \int_{y_I-0.5b}^{y_I+0.5b} f(x,y)\, dx dy. \qquad (5.4.5)
\end{aligned}$$

5.4.3 Discretized Equations

As in the finite element method, over each element Ω^e the function u is approximated as

$$u(x,y) \approx u_h^e(x,y) = \sum_{j=1}^{4} u_j^e \psi_j^e(\bar{x}, \bar{y}), \quad \bar{x} = x - x_1^e, \quad \bar{y} = y - y_1^e, \qquad (5.4.6)$$

where u_j^e denote the values of the function u_h^e at element nodes, (x_1^e, y_1^e) are the global coordinates of node 1 of element Ω^e, and ψ_j^e are the Lagrange interpolation functions associated with the element. Here we assume bilinear interpolation of u_h^e (see Fig. 5.4.3). The interpolation functions are given by

$$\begin{aligned} \psi_1^e &= \left(1-\frac{\bar{x}}{a}\right)\left(1-\frac{\bar{y}}{b}\right), \quad & \psi_2^e &= \frac{\bar{x}}{a}\left(1-\frac{\bar{y}}{b}\right) \\ \psi_3^e &= \frac{\bar{x}}{a}\frac{\bar{y}}{b}, & \psi_4^e &= \left(1-\frac{\bar{x}}{a}\right)\frac{\bar{y}}{b} \end{aligned} \tag{5.4.7}$$

Here $(\bar{x}, \bar{y})$ denote the local coordinates with the origin located at node 1 of the element, and (a, b) denote the horizontal and vertical dimensions of the rectangle. Since the element (local) coordinate system $(\bar{x}, \bar{y})$ is a translation of the global coordinate system (x, y), the derivatives in both coordinate systems remain the same. One need only to express functions defined in terms of (x, y) to those in terms of $(\bar{x}, \bar{y})$ and use the local coordinates to evaluate the integrals.

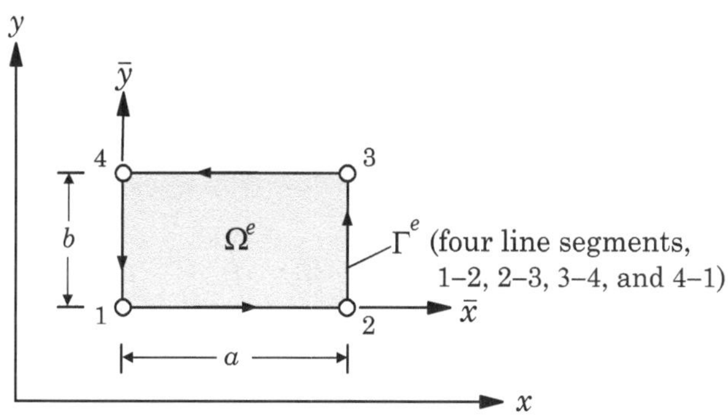

Fig. 5.4.3 Bilinear rectangular finite element.

Since the control domain partially occupies four bilinear elements (see Fig. 5.4.2), the integral statement in Eq. (5.4.5) can be expressed as

$$\begin{aligned} 0 = &\int_{x_I-0.5a}^{x_I}\left[a_{yy}\frac{\partial u}{\partial y}\right]_{y=y_I-0.5b}^{(1)} dx + \int_{x_I}^{x_I+0.5a}\left[a_{yy}\frac{\partial u}{\partial y}\right]_{y=y_I-0.5b}^{(2)} dx \\ &- \int_{y_I-0.5b}^{y_I}\left[a_{xx}\frac{\partial u}{\partial x}\right]_{x=x_I+0.5a}^{(2)} dy - \int_{y_I}^{y_I+0.5b}\left[a_{xx}\frac{\partial u}{\partial x}\right]_{x=x_I+0.5a}^{(3)} dy \\ &- \int_{x_I-0.5a}^{x_I}\left[a_{yy}\frac{\partial u}{\partial y}\right]_{y=y_I+0.5b}^{(4)} dx - \int_{x_I}^{x_I+0.5a}\left[a_{yy}\frac{\partial u}{\partial y}\right]_{y=y_I+0.5b}^{(3)} dx \\ &+ \int_{y_I-0.5b}^{y_I}\left[a_{xx}\frac{\partial u}{\partial x}\right]_{x=x_I-0.5a}^{(1)} dy + \int_{y_I}^{y_I+0.5b}\left[a_{xx}\frac{\partial u}{\partial x}\right]_{x=x_I-0.5a}^{(4)} dy \end{aligned}$$

$$
\begin{aligned}
&- \int_{x_I-0.5a}^{x_I} \int_{y_I-0.5b}^{y_I} f(x,y)\,dxdy - \int_{x_I-0.5a}^{x_I} \int_{y_I-0.5b}^{y_I} f(x,y)\,dxdy \\
&- \int_{x_I}^{x_I+0.5a} \int_{y_I}^{y_I+0.5b} f(x,y)\,dxdy - \int_{x_I-0.5a}^{x_I} \int_{y_I}^{y_I+0.5b} f(x,y)\,dxdy, \qquad (5.4.8)
\end{aligned}
$$

where the superscript on the square brackets denotes the element number of the elements partially covering the control domain. Equation (5.4.8) can be expressed symbolically as

$$
\begin{aligned}
A_{I-N-2}\,U_{I-N-2} &+ A_{I-N-1}\,U_{I-N-1} + A_{I-N}\,U_{I-N} + A_{I-1}\,U_{I-1} \\
&+ A_I\,U_I + A_{I+1}\,U_{I+1} + A_{I+N}\,U_{I+N} + A_{I+N+1}\,U_{I+N+1} \\
&+ A_{I+N+2}\,U_{I+N+2} = F_I, \qquad (5.4.9a)
\end{aligned}
$$

where the coefficients A_K and F_I are defined as follows:

$$
\begin{aligned}
A_{I-N-2} &= \int_{x_I-0.5a}^{x_I} \left[a_{yy}\frac{\partial \psi_1^{(1)}}{\partial y}\right]_{y=y_I-0.5b} dx + \int_{y_I-0.5b}^{y_I} \left[a_{xx}\frac{\partial \psi_1^{(1)}}{\partial x}\right]_{x=x_I-0.5a} dy \\
A_{I-N-1} &= \int_{x_I-0.5a}^{x_I} \left[a_{yy}\frac{\partial \psi_2^{(1)}}{\partial y}\right]_{y=y_I-0.5b} dx + \int_{y_I-0.5b}^{y_I} \left[a_{xx}\frac{\partial \psi_2^{(1)}}{\partial x}\right]_{x=x_I-0.5a} dy \\
&+ \int_{x_I}^{x_I+0.5a} \left[a_{yy}\frac{\partial \psi_1^{(2)}}{\partial y}\right]_{y=y_I-0.5b} dx - \int_{y_I-0.5b}^{y_I} \left[a_{xx}\frac{\partial \psi_1^{(2)}}{\partial x}\right]_{x=x_I+0.5a} dy \\
A_{I-N} &= \int_{x_I}^{x_I+0.5a} \left[a_{yy}\frac{\partial \psi_2^{(2)}}{\partial y}\right]_{y=y_I-0.5b} dx - \int_{y_I-0.5b}^{y_I} \left[a_{xx}\frac{\partial \psi_2^{(2)}}{\partial x}\right]_{x=x_I+0.5a} dy \\
A_{I-1} &= \int_{x_I-0.5a}^{x_I} \left[a_{yy}\frac{\partial \psi_4^{(1)}}{\partial y}\right]_{y=y_I-0.5b} dx + \int_{y_I-0.5b}^{y_I} \left[a_{xx}\frac{\partial \psi_4^{(1)}}{\partial x}\right]_{x=x_I-0.5a} dy \\
&- \int_{x_I-0.5a}^{x_I} \left[a_{yy}\frac{\partial \psi_1^{(4)}}{\partial y}\right]_{y=y_I+0.5b} dx + \int_{y_I}^{y_I+0.5b} \left[a_{xx}\frac{\partial \psi_1^{(4)}}{\partial x}\right]_{x=x_I-0.5a} dy \\
A_I &= \int_{x_I-0.5a}^{x_I} \left[a_{yy}\frac{\partial \psi_3^{(1)}}{\partial y}\right]_{y=y_I-0.5b} dx + \int_{y_I-0.5b}^{y_I} \left[a_{xx}\frac{\partial \psi_3^{(1)}}{\partial x}\right]_{x=x_I-0.5a} dy \\
&+ \int_{x_I}^{x_I+0.5a} \left[a_{yy}\frac{\partial \psi_4^{(2)}}{\partial y}\right]_{y=y_I-0.5b} dx - \int_{y_I-0.5b}^{y_I} \left[a_{xx}\frac{\partial \psi_4^{(2)}}{\partial x}\right]_{x=x_I+0.5a} dy \\
&- \int_{x_I}^{x_I+0.5a} \left[a_{yy}\frac{\partial \psi_1^{(3)}}{\partial y}\right]_{y=y_I+0.5b} dx - \int_{y_I}^{y_I+0.5b} \left[a_{xx}\frac{\partial \psi_1^{(3)}}{\partial x}\right]_{x=x_I+0.5a} dy \\
&- \int_{x_I-0.5a}^{x_I} \left[a_{yy}\frac{\partial \psi_2^{(4)}}{\partial y}\right]_{y=y_I+0.5b} dx + \int_{y_I}^{y_I+0.5b} \left[a_{xx}\frac{\partial \psi_2^{(4)}}{\partial x}\right]_{x=x_I-0.5a} dy \\
A_{I+1} &= \int_{x_I}^{x_I+0.5a} \left[a_{yy}\frac{\partial \psi_3^{(2)}}{\partial y}\right]_{y=y_I-0.5b} dx - \int_{y_I-0.5b}^{y_I} \left[a_{xx}\frac{\partial \psi_3^{(2)}}{\partial x}\right]_{x=x_I+0.5a} dy \\
&- \int_{x_I}^{x_I+0.5a} \left[a_{yy}\frac{\partial \psi_2^{(3)}}{\partial y}\right]_{y=y_I+0.5b} dx - \int_{y_I}^{y_I+0.5b} \left[a_{xx}\frac{\partial \psi_2^{(3)}}{\partial x}\right]_{x=x_I+0.5a} dy
\end{aligned}
$$

$$A_{I+N} = -\int_{x_I-0.5a}^{x_I}\left[a_{yy}\frac{\partial\psi_4^{(4)}}{\partial y}\right]_{y=y_I+0.5b} dx + \int_{y_I}^{y_I+0.5b}\left[a_{xx}\frac{\partial\psi_4^{(4)}}{\partial x}\right]_{x=x_I-0.5a} dy$$

$$A_{I+N+1} = -\int_{x_I}^{x_I+0.5a}\left[a_{yy}\frac{\partial\psi_4^{(3)}}{\partial y}\right]_{y=y_I+0.5b} dx - \int_{y_I}^{y_I+0.5b}\left[a_{xx}\frac{\partial\psi_4^{(3)}}{\partial x}\right]_{x=x_I+0.5a} dy$$

$$-\int_{x_I-0.5a}^{x_I}\left[a_{yy}\frac{\partial\psi_3^{(4)}}{\partial y}\right]_{y=y_I+0.5b} dx + \int_{y_I}^{y_I+0.5b}\left[a_{xx}\frac{\partial\psi_3^{(4)}}{\partial x}\right]_{x=x_I-0.5a} dy$$

$$A_{I+N+2} = -\int_{x_I}^{x_I+0.5a}\left[a_{yy}\frac{\partial\psi_3^{(3)}}{\partial y}\right]_{y=y_I+0.5b} dx - \int_{y_I}^{y_I+0.5b}\left[a_{xx}\frac{\partial\psi_3^{(3)}}{\partial x}\right]_{x=x_I+0.5a} dy \tag{5.4.9b}$$

$$F_I = \int_{x_I-0.5a}^{x_I}\int_{y_I-0.5b}^{y_I} f(x,y)\,dxdy + \int_{x_I-0.5a}^{x_I}\int_{y_I-0.5b}^{y_I} f(x,y)\,dxdy$$

$$+\int_{x_I}^{x_I+0.5a}\int_{y_I}^{y_I+0.5b} f(x,y)\,dxdy + \int_{x_I-0.5a}^{x_I}\int_{y_I}^{y_I+0.5b} f(x,y)\,dxdy, \tag{5.4.9c}$$

where the superscript e on $\psi_i^{(e)}$, $(e = 1, 2, 3, 4)$ denotes the element number that is a part of the Ith control domain (see Fig. 5.4.2).

It is clear from Eqs. (5.4.9b) and (5.4.9c) that we need the first derivatives of the bilinear interpolation functions to evaluate the coefficients. They are

$$\begin{aligned}
&\frac{\partial\psi_i^e}{\partial x} = \frac{\partial\psi_i^e}{\partial\bar{x}}, && \frac{\partial\psi_i^e}{\partial y} = \frac{\partial\psi_i^e}{\partial\bar{y}} \quad (i = 1,2,3,4),\\
&\frac{\partial\psi_1^e}{\partial\bar{x}} = -\frac{1}{a}\left(1-\frac{\bar{y}}{b}\right), && \frac{\partial\psi_1^e}{\partial\bar{y}} = -\frac{1}{b}\left(1-\frac{\bar{x}}{a}\right),\\
&\frac{\partial\psi_2^e}{\partial\bar{x}} = \frac{1}{a}\left(1-\frac{\bar{y}}{b}\right), && \frac{\partial\psi_2^e}{\partial\bar{y}} = -\frac{1}{b}\frac{\bar{x}}{a},\\
&\frac{\partial\psi_3^e}{\partial\bar{x}} = \frac{1}{a}\frac{\bar{y}}{b}, && \frac{\partial\psi_3^e}{\partial\bar{y}} = \frac{1}{b}\frac{\bar{x}}{a},\\
&\frac{\partial\psi_4^e}{\partial\bar{x}} = -\frac{1}{a}\frac{\bar{y}}{b}, && \frac{\partial\psi_4^e}{\partial\bar{y}} = \frac{1}{b}\left(1-\frac{\bar{x}}{a}\right).
\end{aligned} \tag{5.4.10}$$

When the data (a_{xx}, a_{yy}, f) is uniform (i.e., constant) within each element, we can simplify the expressions for the coefficients A_K and F_I using the results in Eqs. (5.4.9b), (5.4.9c), and (5.4.10), and noting that the derivatives of the bilinear interpolations are only functions of one coordinate. Since the finite element interpolation functions are defined in terms of the element coordinates $(\bar{x}, \bar{y})$, it is simpler to evaluate integrals using $(\bar{x}, \bar{y})$. We have, for example,

$$\int_{x_I-0.5a}^{x_I}\frac{\partial\psi_1^{(1)}}{\partial y}\,dx = \int_{0.5a}^{a}\frac{\partial\psi_1^{(1)}}{\partial\bar{y}}\,d\bar{x}$$

$$= -\frac{1}{b}\int_{0.5a}^{a}\left(1-\frac{\bar{x}}{a}\right)d\bar{x} = -\tfrac{1}{8}\frac{a}{b}\,. \tag{5.4.11}$$

For uniform mesh and constant data, the coefficients A_K and F_I are

$$\begin{aligned} A_{I-N-2} &= -a_{xx}\frac{b}{8a} - a_{yy}\frac{a}{8b}, \quad A_{I-N-1} = a_{xx}\frac{b}{4a} - a_{yy}\frac{3a}{4b} \\ A_{I-N} &= -a_{xx}\frac{b}{8a} - a_{yy}\frac{a}{8b}, \quad A_{I-1} = -a_{xx}\frac{3b}{4a} + a_{yy}\frac{a}{4b} \\ A_I &= a_{xx}\frac{3b}{2a} + a_{yy}\frac{3a}{2b}, \quad A_{I+1} = -a_{xx}\frac{3b}{4a} + a_{yy}\frac{a}{4b} \\ A_{I+N} &= -a_{xx}\frac{b}{8a} - a_{yy}\frac{a}{8b}, \quad A_{I+N+1} = a_{xx}\frac{b}{4a} - a_{yy}\frac{3a}{4b} \\ A_{I+N+2} &= -a_{xx}\frac{b}{8a} - a_{yy}\frac{a}{8b}, \quad F_I = f_0 ab. \end{aligned} \tag{5.4.12}$$

We note that the discretized equations of the DMCDM are different from those derived for the half-control volume formulation of the FVM [cf. Eqs. (3.4.8a) and (3.4.8b) with Eqs. (5.4.9a) and (5.4.12)]. In particular, the bilinear interpolation functions used in DMCDM bring all nine nodal values into the discretized equations.

For a rectangular domain, the boundary nodal point locations and their control domains can be classified into several cases as shown in Fig. 5.4.4 (recall the

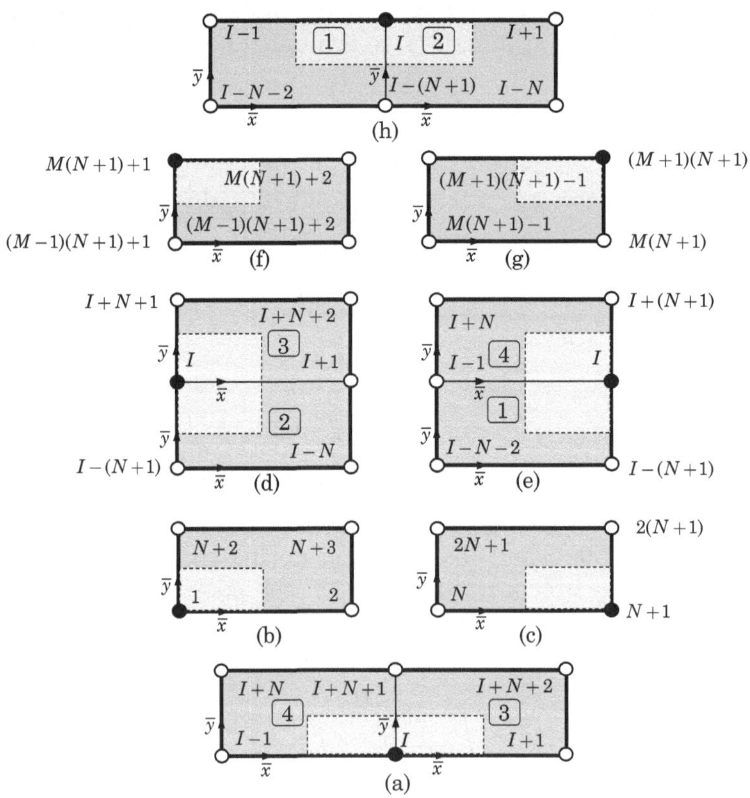

Fig. 5.4.4 Various possible boundary nodes in a rectangular domain.

discussion in Section 3.4.3.2 in connection with the HFVM). The discretized equations associated with a couple of representative boundary cases are presented here. One can follow the logic presented here to develop the relations for the other boundary nodes by considering the control domains shown in Fig. 5.4.4.

Nodes on the bottom boundary [see Fig. 5.4.4(a)]

For this case, the integral form in Eq. (5.4.8) takes the form ($y_I = 0$)

$$\begin{aligned}
0 = &-\int_{x_I-0.5a}^{x_I+0.5a}\left[a_{yy}\frac{\partial u}{\partial y}\right]_{y=y_I} dx - \int_{y_I}^{y_I+0.5b}\left[a_{xx}\frac{\partial u}{\partial x}\right]_{x=x_I+0.5a} dy \\
&- \int_{x_I-0.5a}^{x_I+0.5a}\left[a_{yy}\frac{\partial u}{\partial y}\right]_{y=y_I+0.5b} dx + \int_{y_I}^{y_I+0.5b}\left[a_{xx}\frac{\partial u}{\partial x}\right]_{x=x_I-0.5a} dy \\
&- \int_{x_I-0.5a}^{x_I+0.5a}\int_{y_I}^{y_I+0.5b} f(x,y)\,dxdy \\
= &-\int_{y_I}^{y_I+0.5b}\left[a_{xx}\frac{\partial u}{\partial x}\right]^{(2)}_{x=x_I+0.5a} dy - \int_{x_I}^{x_I+0.5a}\left[a_{yy}\frac{\partial u}{\partial y}\right]^{(2)}_{y=y_I+0.5b} dx \\
&- \int_{x_I-0.5a}^{x_I}\left[a_{yy}\frac{\partial u}{\partial y}\right]^{(1)}_{y=y_I+0.5b} dx + \int_{y_I}^{y_I+0.5b}\left[a_{xx}\frac{\partial u}{\partial x}\right]^{(1)}_{x=x_I-0.5a} dy \\
&- \int_{x_I-0.5a}^{x_I+0.5a} q_y(x,y_I)\,dx, \; - \int_{x_I-0.5a}^{x_I}\int_{y_I}^{y_I+0.5b} f(x,y)\,dxdy \\
&- \int_{x_I}^{x_I+0.5a}\int_{y_I}^{y_I+0.5b} f(x,y)\,dxdy - Q_I. \qquad (5.4.13)
\end{aligned}$$

where

$$Q_I = \int_{x_I-0.5a}^{x_I+0.5a}\left[a_{yy}\frac{\partial u}{\partial y}\right]_{y=y_I} dx$$

Thus, we have

$$\begin{aligned}
A_{I-1}U_{I-1} + A_IU_I + A_{I+1}U_{I+1} + A_{I+N}U_{I+N} + A_{I+N+1}U_{I+N+1} \\
+ A_{I+N+2}U_{I+N+2} = F_I + Q_I \qquad (5.4.14a)
\end{aligned}$$

For uniform mesh and constant data throughout the domain, we obtain

$$\begin{aligned}
A_{I-1} &= -\frac{3b}{8a}a_{xx} + \frac{a}{8b}a_{yy}\,, \quad A_I = \frac{3b}{4a}a_{xx} + \frac{3a}{4b}a_{yy} \\
A_{I+1} &= -\frac{3b}{8a}a_{xx} + \frac{a}{8b}a_{yy}\,, \quad A_{I+N+2} = -\frac{b}{8a}a_{xx} - \frac{a}{8b}a_{yy} \qquad (5.4.14b) \\
A_{I+N+1} &= \frac{b}{4a}a_{xx} - \frac{3a}{4b}a_{yy}\,, \quad A_{I+N} = -\frac{b}{8a}a_{xx} - \frac{a}{8b}a_{yy}\,, \quad F_I = \tfrac{1}{2}f_0ab
\end{aligned}$$

Nodes on the left boundary [see Fig. 5.4.4(d)]

The integral statement for this case is given by ($x_I = 0.0$)

$$
\begin{aligned}
0 =& \int_{x_I}^{x_I+0.5a} \left[a_{yy}\frac{\partial u}{\partial y}\right]_{y=y_I-0.5b} dx - \int_{y_I-0.5b}^{y_I+0.5b} \left[a_{xx}\frac{\partial u}{\partial x}\right]_{x=x_I+0.5a} dy \\
&- \int_{x_I}^{x_I+0.5a} \left[a_{yy}\frac{\partial u}{\partial y}\right]_{y=y_I+0.5b} dx + \int_{y_I-0.5b}^{y_I+0.5b} \left[a_{xx}\frac{\partial u}{\partial x}\right]_{x=x_I} dy \\
&- \int_{x_I}^{x_I+0.5a} \int_{y_I-0.5b}^{y_I+0.5b} f(x,y)\, dxdy - Q_I \\
=& \int_{x_I}^{x_I+0.5a} \left[a_{yy}\frac{\partial u}{\partial y}\right]^{(1)}_{y=y_I-0.5b} dx - \int_{y_I-0.5b}^{y_I} \left[a_{xx}\frac{\partial u}{\partial x}\right]^{(1)}_{x=x_I+0.5a} dy \\
&- \int_{y_I}^{y_I+0.5b} \left[a_{xx}\frac{\partial u}{\partial x}\right]^{(2)}_{x=x_I+0.5a} dy - \int_{x_I}^{x_I+0.5a} \left[a_{yy}\frac{\partial u}{\partial y}\right]^{(2)}_{y=y_I+0.5b} dx \\
&- \int_{y_I-0.5b}^{y_I+0.5b} q_x(x_I,y)dy - \int_{x_I}^{x_I+0.5a} \int_{y_I-0.5b}^{y_I+0.5b} f(x,y)\, dxdy - Q_I
\end{aligned}
\tag{5.4.15}
$$

which can be expressed as

$$
\begin{aligned}
A_{I-N-1}U_{I-N-1} + A_{I-N}U_{I-N} + A_IU_I + A_{I+1}U_{I+1} + A_{I+N+1}U_{I+N+1}& \\
+ A_{I+N+2}U_{I+N+2} = F_I + Q_I&
\end{aligned}
\tag{5.4.16a}
$$

where

$$
Q_I = \int_{y_I-0.5b}^{y_I+0.5b} \left[a_{xx}\frac{\partial u}{\partial xy}\right]_{x=x_I} dy
$$

For uniform mesh and constant data throughout the domain, we obtain

$$
\begin{aligned}
A_{I-N-1} &= \frac{b}{8a}a_{xx} - \frac{3a}{8b}a_{yy}\,, \quad A_{I-N} = -\frac{b}{8a}a_{xx} - \frac{a}{8b}a_{yy} \\
A_I &= \frac{3b}{4a}a_{xx} + \frac{3a}{4b}a_{yy}\,, \quad A_{I+1} = -\frac{3b}{4a}a_{xx} + \frac{a}{4b}a_{yy} \\
A_{I+N+1} &= \frac{b}{8a}a_{xx} - \frac{3a}{8b}a_{yy}\,, \quad A_{I+N+2} = -\frac{b}{8a}a_{xx} - \frac{a}{8b}a_{yy}\,, \quad F_I = \tfrac{1}{2}f_0ab
\end{aligned}
\tag{5.4.16b}
$$

Node on the bottom left corner [see Fig. 5.4.4(b)]

The integral statement for this case can be expressed as

$$
\begin{aligned}
0 =& \int_0^{0.5a} \left[a_{yy}\frac{\partial u}{\partial y}\right]_{y=0} dx + \int_0^{0.5b} \left[a_{xx}\frac{\partial u}{\partial x}\right]_{x=0} dy \\
&- \int_0^{0.5a} \left[a_{yy}\frac{\partial u}{\partial y}\right]_{y=0.5b} dx - \int_0^{0.5b} \left[a_{xx}\frac{\partial u}{\partial x}\right]_{x=0.5a} dy \\
&- \int_0^{0.5a} \int_0^{0.5b} f(x,y)\, dxdy
\end{aligned}
$$

$$= -\int_0^{0.5a} \left[a_{yy}\frac{\partial u}{\partial y}\right]_{y=0.5b} dx - \int_{y_I}^{0.5b} \left[a_{xx}\frac{\partial u}{\partial x}\right]_{x=0.5a} dy$$
$$- \int_0^{0.5a} q_x(x, y_I)\, dx - \int_0^{0.5b} q_x(0, y)\, dy$$
$$- \int_0^{0.5a}\int_0^{0.5b} f(x, y)\, dxdy - Q_I \tag{5.4.17}$$

which can be expressed as

$$A_1U_1 + A_2U_2 + A_{N+2}U_{N+2} + A_{N+3}U_{N+3} = F_1 + Q_1 \tag{5.4.18a}$$

where

$$Q_1 = -\int_0^{0.5a} \left[a_{yy}\frac{\partial u}{\partial y}\right]_{y=0} dx - \int_0^{0.5b} \left[a_{xx}\frac{\partial u}{\partial x}\right]_{x=0} dy$$

For uniform mesh and constant data throughout the domain, we obtain

$$A_1 = \frac{3b}{8a}a_{xx} + \frac{3a}{8b}a_{yy}\,, \quad A_2 = -\frac{3b}{8a}a_{xx} + \frac{a}{8b}a_{yy}$$
$$A_{N+2} = \frac{b}{8a}a_{xx} - \frac{3a}{8b}a_{yy}\,, \quad A_{N+3} = -\frac{b}{8a}a_{xx} - \frac{a}{8b}a_{yy}\,, \quad F_1 = \tfrac{1}{4}f_0ab \tag{5.4.18b}$$

This completes the derivation of control domain equations for Poisson's equation in Eq. (5.4.1) on rectangular domains and uniform meshes.

Extension of the ideas presented for two-dimensional problems with primal mesh of rectangular bilinear elements can be extended to primal mesh of linear triangular elements, while the dual mesh of control domains consists of rectangular elements, which bisect the neighboring triangular finite elements, as shown in Fig. 5.4.5. The control domain now spans partially over six triangular elements of the primal mesh, as shown in Fig. 5.4.6.

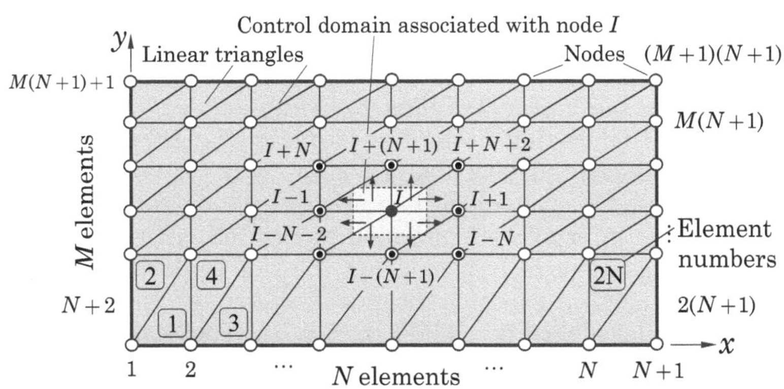

Fig. 5.4.5 A primal mesh of triangular elements with a rectangular control domain.

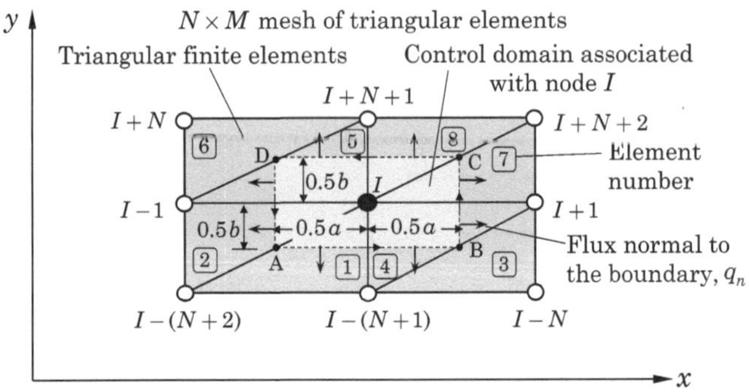

Fig. 5.4.6 Two-dimensional control domain associated with an interior node I when the primal mesh of linear triangular finite elements is used.

The integral statement over the control domain can be written over the elements involved as follows:

$$\begin{aligned} 0 = &\int_{x_I-0.5a_{I-1}}^{a_{I-1}} \left[a_{yy}\frac{\partial u}{\partial y}\right]^{(1)}_{y=y_I-0.5b_{I-1}} dx + \int_0^{x_I+0.5a_I} \left[a_{yy}\frac{\partial u}{\partial y}\right]^{(4)}_{y=y_I-0.5b_{I-1}} dx \\ &- \int_{y_I-0.5b_{I-1}}^{0} \left[a_{xx}\frac{\partial u}{\partial x}\right]^{(4)}_{x=x_I+0.5a_I} dy - \int_0^{y_I+0.5b_I} \left[a_{xx}\frac{\partial u}{\partial x}\right]^{(7)}_{x=x_I+0.5a_I} dy \\ &- \int_0^{x_I+0.5a_I} \left[a_{yy}\frac{\partial u}{\partial y}\right]^{(8)}_{y=y_I+0.5b_I} dx - \int_{x_I-0.5a_{I-1}}^{0} \left[a_{yy}\frac{\partial u}{\partial y}\right]^{(5)}_{y=y_I+0.5b_I} dx \\ &+ \int_0^{y_I+0.5b_I} \left[a_{xx}\frac{\partial u}{\partial x}\right]^{(5)}_{x=x_I-0.5a_{I-1}} dy + \int_{y_I-0.5b_{I-1}}^{0} \left[a_{xx}\frac{\partial u}{\partial x}\right]^{(2)}_{x=x_I-0.5a_{I-1}} dy \\ &- \int_{x_I-0.5a_{I-1}}^{x_I+0.5a_I} \int_{y_I-0.5b_{I-1}}^{y_I+0.5b_I} f(x,y)\, dxdy \end{aligned} \tag{5.4.19}$$

where the subscripts (I and $I-1$) on a and b refer to the subdivision numbers (for a mesh of $N \times M$ subdivisions). The remaining steps are similar to those discussed for primal mesh of rectangular elements.

It is interesting to note that in the case of linear triangular elements, the discretized equations associated with node I contain only the nodal values U_{I-N-1}, U_{I-1}, U_I, U_{I+1}, and U_{I+N+1}, much like in the half-control FVM. This is due to the fact that the interpolation function associated with a node of a linear triangular element is identically zero along the side opposite to the node (see Reddy [8]). Figure 5.4.7(a) contains the stencil associated with the primal mesh of bilinear rectangular elements, while Fig. 5.4.7(b) contains the stencil for the primal mesh of linear triangular elements [cf. Fig. 3.4.7(d) for $a_{xx} = a_{yy} = k$ and $\alpha = \beta = 1$]. However, the coincidence between the discretized equations of the half-control FVM (HFVM) and the primal mesh of triangular elements in the DMCDM will not hold for primal meshes of arbitrary or higher order finite elements.

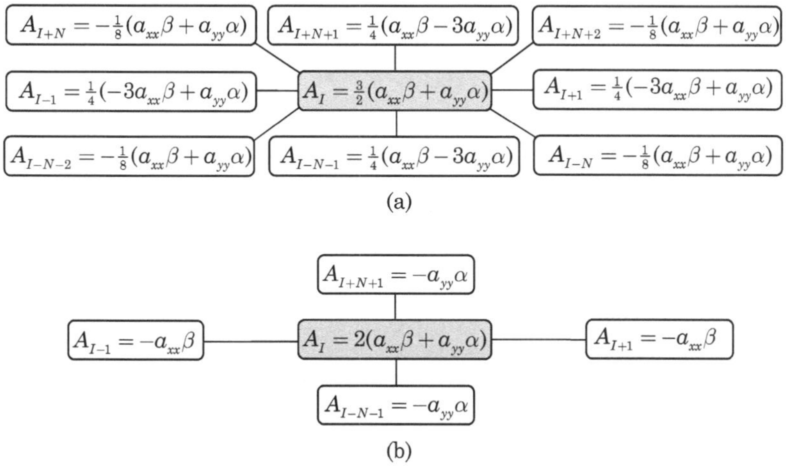

Fig. 5.4.7 Stencils associated with an interior node I for the primal mesh of (a) bilinear rectangular finite elements and (b) linear triangular finite elements for the case of generalized Poisson's operator of Eq. (5.4.1).

5.4.4 Numerical Examples

Example 5.4.1

Consider steady-state heat conduction in an isotropic rectangular region of dimensions $3a \times 2a$, as shown in Fig. 5.4.8(a). The origin of the x and y coordinates is taken at the lower left corner such that x is parallel to the side $3a$, and y is parallel to the side $2a$. Boundaries $x = 0$ and $y = 0$ are insulated (i.e., $q_n = 0$), boundary $x = 3a$ is maintained at zero temperature, and boundary $y = 2a$ is maintained at temperature $T = T_0 \cos(\pi x/6a)$. Determine the temperature distribution using a uniform mesh of linear rectangular elements shown in Fig. 5.4.8(b) and then refine the meshes by doubling the previous mesh, that is 3×2, 6×4, and 12×8.

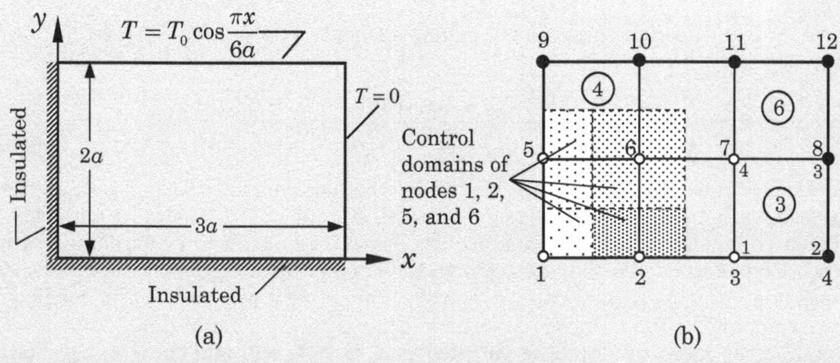

Fig. 5.4.8 The dual mesh control domain analysis of a heat conduction problem over a rectangular domain: (a) domain and (b) 3×2 uniform mesh of linear rectangular elements.

Solution: The governing equation is a special case of the model equation (5.4.1) with zero internal heat generation $f = 0$ and constant coefficients $a_{xx} = a_{yy} = k$. Thus, Eq. (5.4.1) takes the form

$$-k\left(\frac{\partial^2 T}{\partial x^2}+\frac{\partial^2 T}{\partial y^2}\right)=0.$$

The exact solution of Eq. (1) for the boundary conditions shown in Fig. 5.4.8(a) is

$$T(x,y)=T_0\frac{\cosh\frac{\pi y}{6a}\cos\frac{\pi x}{6a}}{\cosh\frac{\pi}{3}}.$$

First, we note that the HFVM formulation gives the same results as the DMCDM with primal mesh of triangular elements. Table 3.4.1 contains the results obtained with the ZFVM and the HFVM for various meshes. For the 3×2 mesh, because of the fact that $f=0$ and the specified boundary conditions, we have $F_I=0$ at all nodes; $Q_I=0$ at nodes 1, 2, 3, and 5; $T_4=0$, $T_8=0$, and $T_{12}=0$; and $T_9=T_0$, $T_{10}=\sqrt{3}/2T_0$, and $T_{11}=0.5T_0$. Thus, we need to determine equations associated with nodes (at which the temperatures are unknown) 1, 2, 3, 5, 6, and 7.

Writing all of the relations (in the order of the sequential node numbers) in matrix form, we obtain (known temperatures are moved to the right-hand side)

$$\frac{k}{4}\begin{bmatrix} 3 & -1 & 0 & -1 & -1 & 0\\ -1 & 6 & -1 & -1 & -2 & -1\\ 0 & -1 & 6 & 0 & -1 & -2\\ -1 & -1 & 0 & 6 & -2 & 0\\ -1 & -2 & -1 & -2 & 12 & -2\\ 0 & -1 & -2 & 0 & -2 & 12\end{bmatrix}\begin{Bmatrix}T_1\\T_2\\T_3\\T_5\\T_6\\T_7\end{Bmatrix}=\frac{k}{4}\begin{Bmatrix}0\\0\\0\\T_0+0.5\sqrt{3}T_0\\T_0+\sqrt{3}T_0+0.5T_0\\0.5\sqrt{3}T_0+T_0\end{Bmatrix}$$

The solution of these equations is (in °C)

$$T_1=0.6190\,T_0,\quad T_2=0.5360\,T_0,\quad T_3=0.3095\,T_0$$
$$T_5=0.7078\,T_0,\quad T_6=0.6130\,T_0,\quad T_7=0.3539\,T_0$$

The corresponding finite element and exact solutions, respectively, are

$$T_1^{fem}=0.6128\,T_0,\quad T_2^{fem}=0.5307\,T_0,\quad T_3^{fem}=0.3064\,T_0$$
$$T_5^{fem}=0.7030\,T_0,\quad T_6^{fem}=0.6088\,T_0,\quad T_7^{fem}=0.3515\,T_0$$
$$T_1^{exct}=0.6249\,T_0,\quad T_2^{exct}=0.5412\,T_0,\quad T_3^{exct}=0.3124\,T_0$$
$$T_5^{exct}=0.7125\,T_0,\quad T_6^{exct}=0.6171\,T_0,\quad T_7^{exct}=0.3563\,T_0$$

The use of 3×2 primal mesh of triangular elements in the DMCDM yields the results (details are not given here)

$$T_1=0.6362\,T_0,\quad T_2=0.5510\,T_0,\quad T_3=0.3181\,T_0$$
$$T_5=0.7214\,T_0,\quad T_6=0.6248\,T_0,\quad T_7=0.3515\,T_0$$

which are exactly the same as those predicted by the half-control FVM as well as the weak-form finite element model. The coincidence of the FEM and DMCDM results when the primal mesh of linear triangular elements is used for the present problem can be explained by the fact that the two formulations give the same global set of equations for the choice of meshes used. This happens only in this special case, where the source term is zero and the coefficients a_{xx} and a_{yy} are constant.

If we refine the mesh by doubling the elements [i.e., $N=6$ and $M=4$; see Fig. 5.4.9], we obtain 24 equations for the 24 unknown nodal temperatures for the 6×4 mesh. These equations can be readily obtained from the representative equations given for the 3×2 mesh. The coefficients remain the same but the node numbers change. Typical control domains are shown in Fig. 5.4.10. We note that the nodes at which the primary variables (i.e., temperatures) are specified, the boundary equations do not come into play. In other words, the equations associated with these nodes will be replaced with the specified temperatures. For example, node 7 equation is replaced with $T_7=0$.

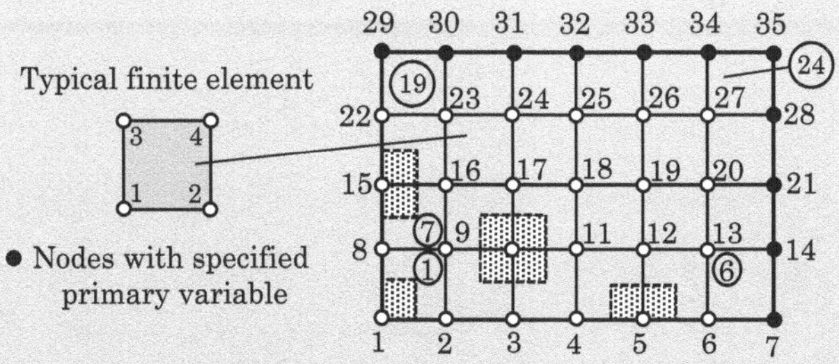

Fig. 5.4.9 6 × 4 uniform mesh of linear rectangular elements for the heat transfer problem of **Example 5.4.1**. Nodes with dark circles have specified temperatures (hence, boundary equations at these nodes are not used).

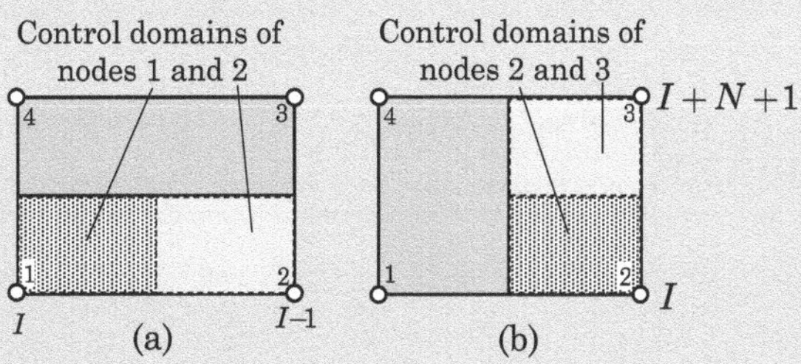

Fig. 5.4.10 Control domains of typical boundary nodes for the 6 × 4 mesh. (a) Control domains for boundary nodes 1 and 2. (b) Control domains for boundary nodes 2 and 3.

Table 5.4.1 contains a comparison of the FEM and DMCDM solutions with the analytical solution for three different meshes of linear triangular and rectangular elements. The DMCDM results with mesh refinement are slightly more accurate than the FEM solutions for primal meshes of rectangular (R) elements. The FEM, HFVM, and DMCDM results are the same for the present problem when primal meshes of triangular (T) elements are used. A contour plot of the temperature field is presented in Fig. 5.4.11.

Table 5.4.1 Comparison of the nodal temperatures $T(x, y)/T_0$, obtained using various meshes (see Figs. 5.4.8(a) and (b) and 5.4.9) in the FEM and DMCDM with the analytical solution.

		FEM Solutions			DMCDM Solutions			Analyt.
x	y	6 × 4R	12 × 8R	12 × 8T	6 × 4R	12 × 8R	12 × 8T	Solution
0.0	0.0	0.6219	0.6242	0.6256	0.6234	0.6245	0.6256	0.6249
0.5	0.0	0.6007	0.6029	0.6043	0.6022	0.6032	0.6043	0.6036
1.0	0.0	0.5386	0.5405	0.5418	0.5399	0.5409	0.5418	0.5412
1.5	0.0	0.4398	0.4413	0.4424	0.4408	0.4416	0.4424	0.4419
2.0	0.0	0.3110	0.3121	0.3128	0.3117	0.3123	0.3128	0.3124
2.5	0.0	0.1610	0.1615	0.1619	0.1614	0.1616	0.1619	0.1617
0.0	1.0	0.7102	0.7119	0.7131	0.7114	0.7122	0.7131	0.7125
0.5	1.0	0.6860	0.6877	0.6888	0.6871	0.6880	0.6888	0.6882
1.0	1.0	0.6150	0.6166	0.6175	0.6161	0.6168	0.6175	0.6171
1.5	1.0	0.5022	0.5034	0.5042	0.5030	0.5036	0.5042	0.5038
2.0	1.0	0.3551	0.3560	0.3565	0.3557	0.3561	0.3565	0.3563
2.5	1.0	0.1838	0.1843	0.1845	0.1841	0.1843	0.1845	0.1844

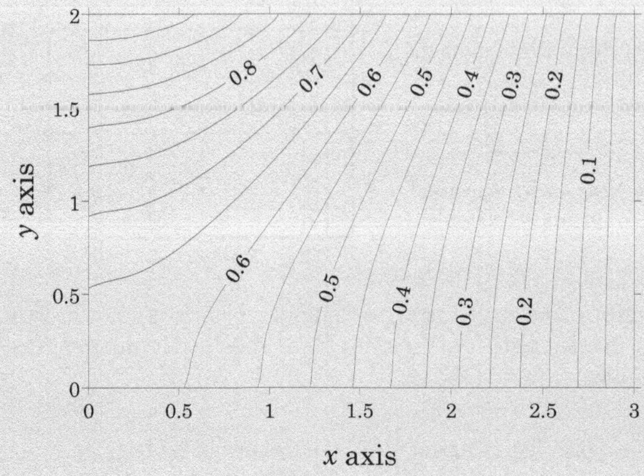

Fig. 5.4.11 A contour plot of the temperature field obtained with the 12×8 primal mesh of bilinear elements in DMCDM.

Example 5.4.2

Consider heat conduction in a rectangular, isotropic medium with conductivity $a_{xx} = a_{yy} = k$. The domain is of dimensions $a \times b$. Use a 8×8 uniform primal mesh of linear rectangular elements and the following data and boundary conditions (see Fig. 5.4.12) in obtaining the numerical solutions:

$$\begin{gathered} a = 0.2\,\text{m}, \quad b = 0.1\,\text{m}, \quad k = 0.2\,\text{W/(m K)}, \\ T(0,y) = 500\,\text{K}\;, \quad T(a,y) = 300\,\text{K}, \quad \left.\frac{\partial T}{\partial y}\right|_{y=0} = 0, \\ T(x,b) = 500(1 - 10x^2)\,\text{K}. \end{gathered} \tag{5.4.20}$$

In addition, we assume that there is no internal heat generation.

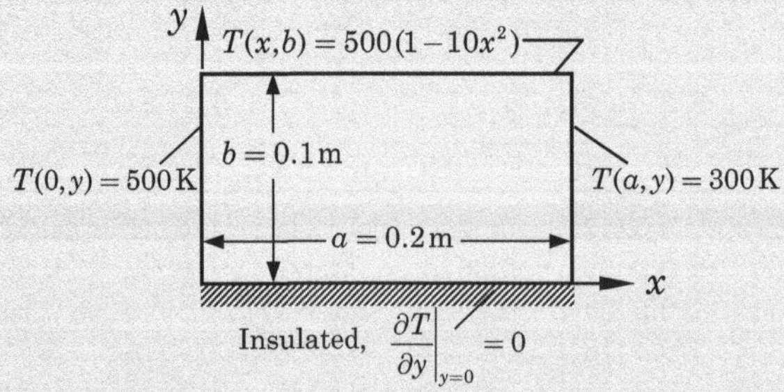

Fig. 5.4.12 Geometry and boundary conditions for heat conduction in a rectangular isotropic medium.

Solution: Other than for the boundary conditions, this problem is similar to the one considered in **Example 5.4.1**. Therefore, details of computing the discretized equations are not presented here. Figure 5.4.13(a) shows the primal 8×8 mesh of bilinear elements (a typical interior control domain is shown with broken lines). In the case of DMCDM (as well as in the FEM and HFVM), the number of nodes is 81, and U_I for $I = 73, 74, \ldots, 81$ have the specified values of 496.875, 487.5, 471.875, 450.0, 421.875, 387.5, and 346.875, respectively. For the ZFVM, there are 100 nodes [see Fig. 5.4.13(b)], and the top boundary node numbers are $92, 93, \ldots, 99$, and the specified values at these nodes are 499.21875, 492.96875, 480.46875, 461.71875, 436.71875, 405.46875, 367.96875, and 324.21875 (the darkened nodes are with specified values of Us). Figure 5.4.14 contains the contour plots (isotherms) of the temperature field obtained.

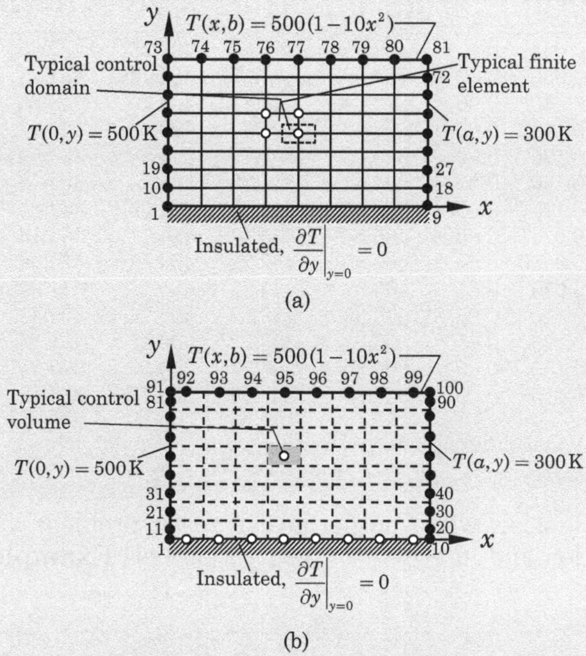

Fig. 5.4.13 (a) A uniform 8×8 (primal) mesh of bilinear finite elements used in the DMCDM (the same mesh applies to FEM and HFVM). (b) A uniform 8×8 mesh of control volumes used in the ZFVM. The dark circles indicate nodes with specified primary variables.

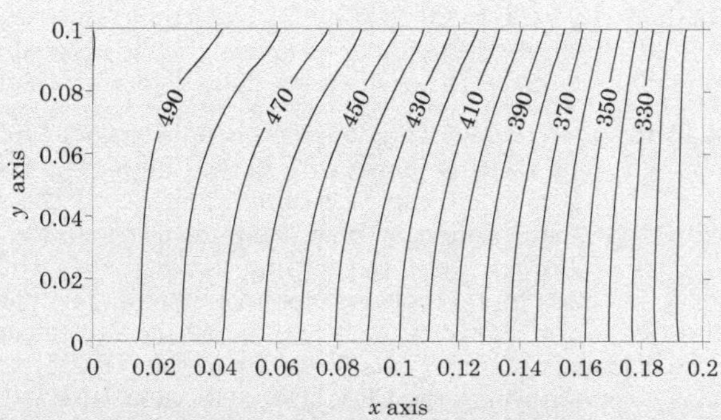

Fig. 5.4.14 Contour plots of the temperature field for the problem of **Example 5.4.2.**

Table 5.4.2 contains solutions $T(x, y)$ as functions of x and y obtained with the DMCDM, FEM, HFVM, and ZFVM. Recall that the HFVM results are the same as those predicted by the primal mesh of triangular elements. In the case of the ZFVM, the actual nodal locations are different from those of the other methods and cannot be compared without interpolating between the nodal values. Clearly, all numerical solutions are very close to each other.

Table 5.4.2 The DMCDM, FEM, and FVM solutions (temperature $T(x, y)$ in K) of a two-dimensional heat conduction problem.

x	y	DMCDM	FEM	HFVM	ZFVM*
0.025	0.00	482.85	482.80	483.00	482.85
0.050	0.00	464.48	464.39	464.74	464.46
0.075	0.00	443.88	443.77	444.20	443.84
0.100	0.00	420.42	420.30	420.76	420.37
0.125	0.00	393.88	393.77	394.20	393.84
0.150	0.00	364.48	364.39	364.74	364.46
0.175	0.00	332.85	332.80	333.00	332.85
0.025	0.05	485.54	485.49	485.71	486.18
0.050	0.05	469.30	469.22	469.54	470.39
0.075	0.05	450.01	449.92	450.27	451.35
0.100	0.05	426.98	426.89	427.24	428.41
0.125	0.05	400.01	399.92	400.27	401.35
0.150	0.05	369.30	369.22	369.54	370.39
0.175	0.05	335.34	335.49	335.71	336.18

* Values linearly interpolated between the nodal values.

The next example deals with another heat transfer problem with a convection boundary condition and internal heat generation (see **Example 4.3.4**).

Example 5.4.3

Consider the bus bar shown in Fig. 5.4.15(a) which carries sufficient electrical current to have a heat generation of $f = 10^6$ W/m^3. The bar has a conductivity of $a_{xx} = a_{yy} = k$ W/(m K) and dimensions 0.10 m × 0.05 m (and 0.01 m thick). The left side is maintained at 40°C and the right side at 10°C. Assuming that the heat flow is two-dimensional (i.e., the heat flow through the thickness is negligible), and the bottom edge is insulated and the top edge is exposed to ambient air temperature of $T_\infty = 0$°C with a heat transfer coefficient of $\beta = 75$ W/(m^2 K), and a conductivity of $k = 20$ W/(m·°C), determine the steady-state solutions with a uniform mesh of 20 × 10 linear rectangular elements. Include the surface convection by adding the contribution $\beta T(x, 0.05)$ to the coefficients associated with the top surface.

Solution: Figure 5.4.15(b) shows the domain, boundary conditions, and 10 × 5 primal mesh of bilinear finite elements. The dual mesh is shown with broken lines. The boundary conditions at nodes on edges $x = 0$ and $x = 0.1$ m can be readily applied and the associated discrete equations can be omitted (which are used to post-compute heats at the nodes). The flux boundary condition ($q_n = -\partial T/\partial y = 0$) at the bottom nodes (i.e., $y = 0$) can be imposed by setting the boundary integrals at the bottom face to be zero. The convection boundary condition at the nodes of the top surface (i.e., $y = 0.05$ m) is implemented by adding the contributions to the coefficients K_{ij}^e and F_i^e (as was done in the HFVM).

Table 5.4.3 contains a comparison of the FEM, DMCDM, and HFVM (half-control FVM; see Table 3.4.2) solutions for $T(x, 0)$ and $T(x, 0.05)$ for different values of x (the results were obtained with a 20 × 10 primal mesh). The solutions are in excellent agreement with each other. The numerical solutions obtained with the ZFVM were presented in Fig. 3.4.8 for $T(x, 0)$ and $T(x, 0.05)$ as functions of x (also see **Examples 3.4.2** and **4.3.4**).

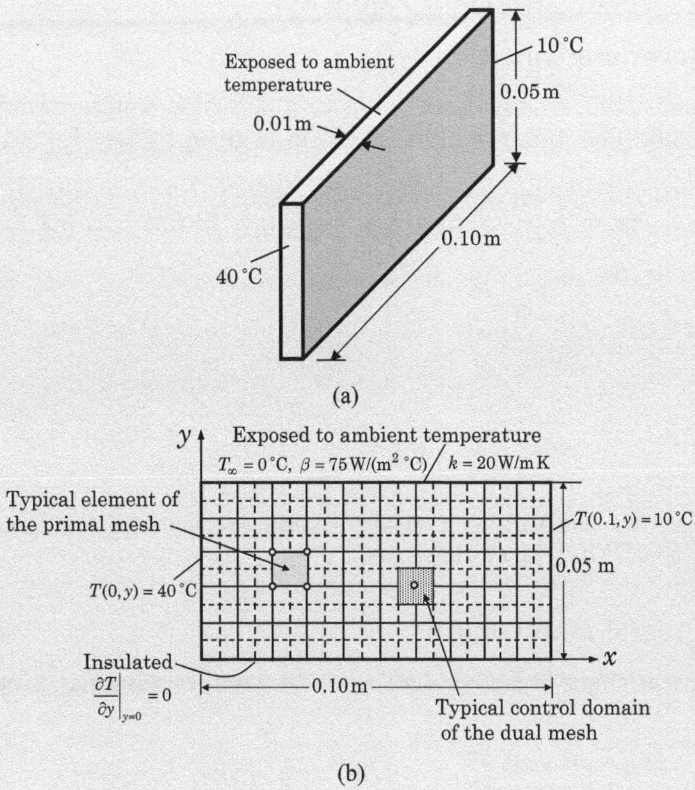

Fig. 5.4.15 Domain, boundary conditions, and uniform primal mesh of 10×5 bilinear finite elements for conductive and convective heat transfer in a bus bar. The bottom is insulated, while the top is exposed to ambient temperature of 0°C; the left side is kept at 40°C, and the right face is maintained at 10°C.

Table 5.4.3 The FEM, DMCDM, and HFVM solutions (temperature $T(x, y)$ in °C) of a two-dimensional heat conduction problem with convection boundary condition (20×10 mesh).

	$y = 0.0$			$y = 0.05$		
x	FEM	DMCDM	HFVM	FEM	DMCDM	HFVM
0.00	40.000	40.000	40.000	40.000	40.000	40.000
0.01	58.129	58.120	58.099	54.975	55.029	55.025
0.02	71.404	71.386	71.349	66.589	66.599	66.578
0.03	79.948	79.926	79.879	74.128	74.132	74.103
0.04	83.852	83.828	83.777	77.559	77.558	77.525
0.05	83.166	83.142	83.091	76.862	76.859	76.825
0.06	77.900	77.878	77.831	72.014	72.010	71.978
0.07	68.027	68.008	67.967	62.972	62.969	62.940
0.08	53.483	53.469	53.439	49.666	49.666	49.644
0.09	34.179	34.171	34.155	32.003	32.014	32.003
0.10	10.000	10.000	10.000	10.000	10.000	10.000

5.4.5 Advection–Diffusion Equation

5.4.5.1 Governing equation

The two-dimensional, steady state, advection–diffusion equation was considered in Section 4.3.6. The dimensionless equation is given by [see Eq. (4.3.34)]

$$\frac{\partial u}{\partial \bar{x}} + \frac{\beta}{\alpha}\frac{\partial u}{\partial \bar{y}} - \frac{1}{Pe}\left(\frac{\partial^2 u}{\partial \bar{x}^2} + \frac{\gamma}{\alpha^2}\frac{\partial^2 u}{\partial \bar{y}^2}\right) = \bar{g} \quad \text{in } \Omega, \tag{5.4.21}$$

where

$$\begin{aligned} \bar{x} &= \frac{x}{a}, \quad \bar{y} = \frac{y}{b}, \quad \beta = \frac{v_y}{v_x}, \quad \alpha = \frac{b}{a} \\ \gamma &= \frac{k_{yy}}{k_{xx}}, \quad Pe = \frac{\rho v_x a}{k_{xx}}, \quad \bar{g} = \frac{ag}{v_x \rho} \end{aligned} \tag{5.4.22}$$

In this section, we use Eq. (5.4.21) (omitting the over bars on the variables) and develop the DMCDM model.

5.4.5.2 Integral statement

The integral statement of Eq. (5.4.21) over a typical rectangular control domain (see Fig. 5.4.2) is

$$\begin{aligned} 0 &= \int_{x_I-0.5a}^{x_I+0.5a}\int_{y_I-0.5b}^{y_I+0.5b}\left[\frac{\partial u}{\partial x} + \frac{\beta}{\alpha}\frac{\partial u}{\partial y} - \frac{1}{Pe}\left(\frac{\partial^2 u}{\partial x^2} + \frac{\gamma}{\alpha^2}\frac{\partial^2 u}{\partial y^2}\right) - g\right] xdy \\ &= -\frac{1}{Pe}\oint_{\Gamma_R}\left(n_x\frac{\partial u}{\partial x} + \frac{\gamma}{\alpha^2}n_y\frac{\partial u}{\partial y}\right) ds \\ &\quad + \int_{x_I-0.5a}^{x_I+0.5a}\int_{y_I-0.5b}^{y_I+0.5b}\left(\frac{\partial u}{\partial x} + \frac{\beta}{\alpha}\frac{\partial u}{\partial y} - g\right) dxdy \end{aligned} \tag{5.4.23}$$

where a and b are the dimensions of the rectangular element, and (x_I, y_I) are the global coordinates of the node labelled as I, (n_x, n_y) are the direction cosines of the unit normal vector, and Γ_R is the boundary of the rectangular control domain.

Since $\hat{\mathbf{n}} = (n_x, n_y)$; $\hat{\mathbf{n}} = (0, -1)$ and $ds = dx$ on AB; $\hat{\mathbf{n}} = (1, 0)$ and $ds = dy$ on BC; $\hat{\mathbf{n}} = (0, 1)$ and $ds = -dx$ on CD; and $\hat{\mathbf{n}} = (-1, 0)$ and $ds = -dy$ on DA, the integral form in Eq. (5.4.23) becomes

$$\begin{aligned} 0 &= \frac{1}{Pe}\int_{x_I-0.5a}^{x_I+0.5a}\left[\frac{\gamma}{\alpha^2}\frac{\partial u}{\partial y}\right]_{y=y_I-0.5b} dx - \frac{1}{Pe}\int_{y_I-0.5b}^{y_I+0.5b}\left[\frac{\partial u}{\partial x}\right]_{x=x_I+0.5a} dy \\ &\quad - \frac{1}{Pe}\int_{x_I-0.5a}^{x_I+0.5a}\left[\frac{\gamma}{\alpha^2}\frac{\partial u}{\partial y}\right]_{y=y_I+0.5b} dx + \frac{1}{Pe}\int_{y_I-0.5b}^{y_I+0.5b}\left[\frac{\partial u}{\partial x}\right]_{x=x_I-0.5a} dy \\ &\quad + \int_{x_I-0.5a}^{x_I+0.5a}\int_{y_I-0.5b}^{y_I+0.5b}\left(\frac{\partial u}{\partial x} + \frac{\beta}{\alpha}\frac{\partial u}{\partial y} - g\right) dxdy. \end{aligned} \tag{5.4.24}$$

5.4.5.3 Discretized equations for the interior nodes

As before, the function u is approximated over each element Ω^e of the primal mesh using Eqs. (5.4.6) and (5.4.7). The derivatives of the bilinear interpolation functions are given in Eqs. (5.4.10).

Since the control domain partially occupies four bilinear elements, the integral statement in Eq. (5.4.24) can be expressed as

$$\begin{aligned}
0 = & \frac{1}{Pe}\int_{x_I-0.5a}^{x_I}\left[\frac{\gamma}{\alpha^2}\frac{\partial u}{\partial y}\right]^{(1)}_{y=y_I-0.5b} dx \\
& + \frac{1}{Pe}\int_{x_I}^{x_I+0.5a}\left[\frac{\gamma}{\alpha^2}\frac{\partial u}{\partial y}\right]^{(2)}_{y=y_I-0.5b} dx \\
& - \frac{1}{Pe}\int_{y_I-0.5b}^{y_I}\left[\frac{\partial u}{\partial x}\right]^{(2)}_{x=x_I+0.5a} dy \\
& - \frac{1}{Pe}\int_{y_I}^{y_I+0.5b}\left[\frac{\partial u}{\partial x}\right]^{(3)}_{x=x_I+0.5a} dy \\
& - \frac{1}{Pe}\int_{x_I-0.5a}^{x_I}\left[\frac{\gamma}{\alpha^2}\frac{\partial u}{\partial y}\right]^{(4)}_{y=y_I+0.5b} dx \\
& - \frac{1}{Pe}\int_{x_I}^{x_I+0.5a}\left[\frac{\gamma}{\alpha^2}\frac{\partial u}{\partial y}\right]^{(3)}_{y=y_I+0.5b} dx \\
& + \frac{1}{Pe}\int_{y_I-0.5b}^{y_I}\left[\frac{\partial u}{\partial x}\right]^{(1)}_{x=x_I-0.5a} dy \\
& + \int_{x_I-0.5a}^{x_I}\int_{y_I-0.5b}^{y_I}\left(\frac{\partial u}{\partial x}+\frac{\beta}{\alpha}\frac{\partial u}{\partial y}-g\right) dxdy \\
& + \int_{x_I}^{x_I+0.5a}\int_{y_I-0.5b}^{y_I}\left(\frac{\partial u}{\partial x}+\frac{\beta}{\alpha}\frac{\partial u}{\partial y}-g\right) dxdy \\
& + \int_{x_I}^{x_I+0.5a}\int_{y_I}^{y_I+0.5b}\left(\frac{\partial u}{\partial x}+\frac{\beta}{\alpha}\frac{\partial u}{\partial y}-g\right) dxdy \\
& + \int_{x_I-0.5a}^{x_I}\int_{y_I}^{y_I+0.5b}\left(\frac{\partial u}{\partial x}+\frac{\beta}{\alpha}\frac{\partial u}{\partial y}-g\right) dxdy
\end{aligned} \tag{5.4.25}$$

where the superscript on the square brackets denotes the element number of the elements partially covering the control domain. Equation (5.4.25) can be expressed symbolically as

$$\begin{aligned}
& \bar{A}_{I-N-2}\,U_{I-N-2} + \bar{A}_{I-N-1}\,U_{I-N-1} + \bar{A}_{I-N}\,U_{I-N} + \bar{A}_{I-1}\,U_{I-1} \\
& + \bar{A}_I\,U_I + \bar{A}_{I+1}\,U_{I+1} + \bar{A}_{I+N}\,U_{I+N} + \bar{A}_{I+N+1}\,U_{I+N+1} \\
& + \bar{A}_{I+N+2}\,U_{I+N+2} = F_I.
\end{aligned} \tag{5.4.26}$$

The coefficients of Eq. (5.4.26) are defined by

$$A_{I-N-2} = \frac{1}{P_e}\int_{x_I-0.5a}^{x_I}\left[\frac{\gamma}{\alpha^2}\frac{\partial\psi_1^{(1)}}{\partial y}\right]_{y=y_I-0.5b} dx + \frac{1}{P_e}\int_{y_I-0.5b}^{y_I}\left[\frac{\partial\psi_1^{(1)}}{\partial x}\right]_{x=x_I-0.5a} dy$$

$$A_{I-N-1} = \frac{1}{P_e}\int_{x_I-0.5a}^{x_I}\left[\frac{\gamma}{\alpha^2}\frac{\partial\psi_2^{(1)}}{\partial y}\right]_{y=y_I-0.5b} dx + \frac{1}{P_e}\int_{y_I-0.5b}^{y_I}\left[\frac{\partial\psi_2^{(1)}}{\partial x}\right]_{x=x_I-0.5a} dy$$
$$+ \frac{1}{P_e}\int_{x_I}^{x_I+0.5a}\left[\frac{\gamma}{\alpha^2}\frac{\partial\psi_1^{(2)}}{\partial y}\right]_{y=y_I-0.5b} dx - \frac{1}{P_e}\int_{y_I-0.5b}^{y_I}\left[\frac{\partial\psi_1^{(2)}}{\partial x}\right]_{x=x_I+0.5a} dy$$

$$A_{I-N} = \frac{1}{P_e}\int_{x_I}^{x_I+0.5a}\left[\frac{\gamma}{\alpha^2}\frac{\partial\psi_2^{(2)}}{\partial y}\right]_{y=y_I-0.5b} dx - \frac{1}{P_e}\int_{y_I-0.5b}^{y_I}\left[\frac{\partial\psi_2^{(2)}}{\partial x}\right]_{x=x_I+0.5a} dy$$

$$A_{I-1} = \frac{1}{P_e}\int_{x_I-0.5a}^{x_I}\left[\frac{\gamma}{\alpha^2}\frac{\partial\psi_4^{(1)}}{\partial y}\right]_{y=y_I-0.5b} dx + \frac{1}{P_e}\int_{y_I-0.5b}^{y_I}\left[\frac{\partial\psi_4^{(1)}}{\partial x}\right]_{x=x_I-0.5a} dy$$
$$- \frac{1}{P_e}\int_{x_I-0.5a}^{x_I}\left[\frac{\gamma}{\alpha^2}\frac{\partial\psi_1^{(4)}}{\partial y}\right]_{y=y_I+0.5b} dx + \frac{1}{P_e}\int_{y_I}^{y_I+0.5b}\left[\frac{\partial\psi_1^{(4)}}{\partial x}\right]_{x=x_I-0.5a} dy$$

$$A_{I} = \frac{1}{P_e}\int_{x_I-0.5a}^{x_I}\left[\frac{\gamma}{\alpha^2}\frac{\partial\psi_3^{(1)}}{\partial y}\right]_{y=y_I-0.5b} dx + \frac{1}{P_e}\int_{y_I-0.5b}^{y_I}\left[\frac{\partial\psi_3^{(1)}}{\partial x}\right]_{x=x_I-0.5a} dy$$
$$+ \frac{1}{P_e}\int_{x_I}^{x_I+0.5a}\left[\frac{\gamma}{\alpha^2}\frac{\partial\psi_4^{(2)}}{\partial y}\right]_{y=y_I-0.5b} dx - \frac{1}{P_e}\int_{y_I-0.5b}^{y_I}\left[\frac{\partial\psi_4^{(2)}}{\partial x}\right]_{x=x_I+0.5a} dy$$
$$- \frac{1}{P_e}\int_{x_I}^{x_I+0.5a}\left[\frac{\gamma}{\alpha^2}\frac{\partial\psi_1^{(3)}}{\partial y}\right]_{y=y_I+0.5b} dx - \frac{1}{P_e}\int_{y_I}^{y_I+0.5b}\left[\frac{\partial\psi_1^{(3)}}{\partial x}\right]_{x=x_I+0.5a} dy$$
$$- \frac{1}{P_e}\int_{x_I-0.5a}^{x_I}\left[\frac{\gamma}{\alpha^2}\frac{\partial\psi_2^{(4)}}{\partial y}\right]_{y=y_I+0.5b} dx + \frac{1}{P_e}\int_{y_I}^{y_I+0.5b}\left[\frac{\partial\psi_2^{(4)}}{\partial x}\right]_{x=x_I-0.5a} dy$$

$$A_{I+1} = \frac{1}{P_e}\int_{x_I}^{x_I+0.5a}\left[\frac{\gamma}{\alpha^2}\frac{\partial\psi_3^{(2)}}{\partial y}\right]_{y=y_I-0.5b} dx - \frac{1}{P_e}\int_{y_I-0.5b}^{y_I}\left[\frac{\partial\psi_3^{(2)}}{\partial x}\right]_{x=x_I+0.5a} dy$$
$$- \frac{1}{P_e}\int_{x_I}^{x_I+0.5a}\left[\frac{\gamma}{\alpha^2}\frac{\partial\psi_2^{(3)}}{\partial y}\right]_{y=y_I+0.5b} dx - \frac{1}{P_e}\int_{y_I}^{y_I+0.5b}\left[\frac{\partial\psi_2^{(3)}}{\partial x}\right]_{x=x_I+0.5a} dy$$

$$A_{I+N} = -\frac{1}{P_e}\int_{x_I-0.5a}^{x_I}\left[\frac{\gamma}{\alpha^2}\frac{\partial\psi_4^{(4)}}{\partial y}\right]_{y=y_I+0.5b} dx + \frac{1}{P_e}\int_{y_I}^{y_I+0.5b}\left[\frac{\partial\psi_4^{(4)}}{\partial x}\right]_{x=x_I-0.5a} dy$$

$$A_{I+N+1} = -\frac{1}{P_e}\int_{x_I}^{x_I+0.5a}\left[\frac{\gamma}{\alpha^2}\frac{\partial\psi_4^{(3)}}{\partial y}\right]_{y=y_I+0.5b} dx - \frac{1}{P_e}\int_{y_I}^{y_I+0.5b}\left[\frac{\partial\psi_4^{(3)}}{\partial x}\right]_{x=x_I+0.5a} dy$$
$$- \frac{1}{P_e}\int_{x_I-0.5a}^{x_I}\left[\frac{\gamma}{\alpha^2}\frac{\partial\psi_3^{(4)}}{\partial y}\right]_{y=y_I+0.5b} dx + \frac{1}{P_e}\int_{y_I}^{y_I+0.5b}\left[\frac{\partial\psi_3^{(4)}}{\partial x}\right]_{x=x_I-0.5a} dy$$

$$A_{I+N+2} = -\frac{1}{P_e}\int_{x_I}^{x_I+0.5a}\left[\frac{\gamma}{\alpha^2}\frac{\partial\psi_3^{(3)}}{\partial y}\right]_{y=y_I+0.5b} dx - \frac{1}{P_e}\int_{y_I}^{y_I+0.5b}\left[\frac{\partial\psi_3^{(3)}}{\partial x}\right]_{x=x_I+0.5a} dy$$

$$C_{I-N-2} = \int_{y_I-0.5b}^{y_I}\int_{x_I-0.5a}^{x_I}\Big(\frac{\partial\psi_1^{(1)}}{\partial x} + \frac{\beta}{\alpha}\frac{\partial\psi_1^{(1)}}{\partial y}\Big)dxdy$$

$$C_{I-N-1} = \int_{y_I-0.5b}^{y_I}\int_{x_I-0.5a}^{x_I}\Big(\frac{\partial\psi_2^{(1)}}{\partial x} + \frac{\beta}{\alpha}\frac{\partial\psi_2^{(1)}}{\partial y}\Big)dxdy + \int_{y_I-0.5b}^{y_I}\int_{x_I}^{x_I+0.5a}\Big(\frac{\partial\psi_1^{(2)}}{\partial x} + \frac{\beta}{\alpha}\frac{\partial\psi_1^{(2)}}{\partial y}\Big)dxdy$$

$$C_{I-N} = \int_{y_I-0.5b}^{y_I}\int_{x_I}^{x_I+0.5a}\Big(\frac{\partial\psi_2^{(2)}}{\partial x} + \frac{\beta}{\alpha}\frac{\partial\psi_2^{(2)}}{\partial y}\Big)dxdy$$

$$C_{I-1} = \int_{y_I-0.5b}^{y_I}\int_{x_I-0.5a}^{x_I}\Big(\frac{\partial\psi_4^{(1)}}{\partial x} + \frac{\beta}{\alpha}\frac{\partial\psi_4^{(1)}}{\partial y}\Big)dxdy + \int_{y_I}^{y_I+0.5b}\int_{x_I-0.5a}^{x_I}\Big(\frac{\partial\psi_1^{(4)}}{\partial x} + \frac{\beta}{\alpha}\frac{\partial\psi_1^{(4)}}{\partial y}\Big)dxdy$$

$$\begin{aligned}C_I = &\int_{y_I-0.5b}^{y_I}\int_{x_I-0.5a}^{x_I}\Big(\frac{\partial\psi_3^{(1)}}{\partial x} + \frac{\beta}{\alpha}\frac{\partial\psi_3^{(1)}}{\partial y}\Big)dxdy + \int_{y_I-0.5b}^{y_I}\int_{x_I}^{x_I+0.5a}\Big(\frac{\partial\psi_4^{(2)}}{\partial x} + \frac{\beta}{\alpha}\frac{\partial\psi_4^{(2)}}{\partial y}\Big)dxdy\\ &+ \int_{y_I}^{y_I+0.5b}\int_{x_I}^{x_I+0.5a}\Big(\frac{\partial\psi_1^{(3)}}{\partial x} + \frac{\beta}{\alpha}\frac{\partial\psi_1^{(3)}}{\partial y}\Big)dxdy + \int_{y_I}^{y_I+0.5b}\int_{x_I-0.5a}^{x_I}\Big(\frac{\partial\psi_2^{(4)}}{\partial x} + \frac{\beta}{\alpha}\frac{\partial\psi_2^{(4)}}{\partial y}\Big)dxdy\end{aligned}$$

$$C_{I+1} = \int_{y_I-0.5b}^{y_I}\int_{x_I}^{x_I+0.5a}\Big(\frac{\partial\psi_3^{(2)}}{\partial x} + \frac{\beta}{\alpha}\frac{\partial\psi_3^{(2)}}{\partial y}\Big)dxdy + \int_{y_I}^{y_I+0.5b}\int_{x_I}^{x_I+0.5a}\Big(\frac{\partial\psi_2^{(3)}}{\partial x} + \frac{\beta}{\alpha}\frac{\partial\psi_2^{(3)}}{\partial y}\Big)dxdy$$

$$C_{I+N} = \int_{y_I}^{y_I+0.5b}\int_{x_I-0.5a}^{x_I}\Big(\frac{\partial\psi_4^{(4)}}{\partial x} + \frac{\beta}{\alpha}\frac{\partial\psi_4^{(4)}}{\partial y}\Big)dxdy$$

$$C_{I+N+1} = \int_{y_I}^{y_I+0.5b}\int_{x_I-0.5a}^{x_I}\Big(\frac{\partial\psi_3^{(4)}}{\partial x} + \frac{\beta}{\alpha}\frac{\partial\psi_3^{(4)}}{\partial y}\Big)dxdy + \int_{y_I}^{y_I+0.5b}\int_{x_I}^{x_I+0.5a}\Big(\frac{\partial\psi_4^{(3)}}{\partial x} + \frac{\beta}{\alpha}\frac{\partial\psi_4^{(3)}}{\partial y}\Big)dxdy$$

$$C_{I+N+2} = \int_{y_I}^{y_I+0.5b}\int_{x_I}^{x_I+0.5a}\Big(\frac{\partial\psi_3^{(3)}}{\partial x} + \frac{\beta}{\alpha}\frac{\partial\psi_3^{(3)}}{\partial y}\Big)dxdy$$

$$\begin{aligned}F_I = &\int_{y_I-0.5b}^{y_I}\int_{x_I-0.5a}^{x_I} g(x,y)\,dxdy + \int_{y_I-0.5b}^{y_I}\int_{x_I}^{x_I+0.5a} g(x,y)\,dxdy\\ &+ \int_{y_I}^{y_I+0.5b}\int_{x_I}^{x_I+0.5a} g(x,y)\,dxdy + \int_{y_I}^{y_I+0.5b}\int_{x_I-0.5a}^{x_I} g(x,y)\,dxdy.\end{aligned} \tag{5.4.27}$$

When the data $(\alpha, \beta, \gamma, g)$ is uniform (i.e., constant) within each element, we can simplify the expressions for the coefficients A_K, C_K, and F_I using the results in Eq. (5.4.26) and noting that the derivatives of the bilinear interpolations are only functions of one coordinate. Since the finite element interpolation functions are defined in element coordinates $(\bar{x}, \bar{y})$, the integrals can be evaluated exactly using the element coordinates. Using the derivatives of the bilinear interpolation functions from Eq. (5.4.10), we obtain the following analytical expressions from the evaluation of the various integrals appearing in the present formulation:

$$\begin{aligned}
\int_{x_I-0.5a}^{x_I} \frac{\partial \psi_1^{(1)}}{\partial y}\, dx &= \int_{0.5a}^{a} \frac{\partial \psi_1^{(1)}}{\partial \bar{y}}\, d\bar{x} = -\frac{1}{b}\int_{0.5a}^{a}\left(1-\frac{\bar{x}}{a}\right) d\bar{x} = -\tfrac{1}{8}\frac{a}{b}\\
\int_{x_I-0.5a}^{x_I} \frac{\partial \psi_3^{(1)}}{\partial y}\, dx &= \int_{0.5a}^{a} \frac{\partial \psi_3^{(1)}}{\partial \bar{y}}\, d\bar{x} = \frac{1}{b}\int_{0.5a}^{a}\frac{\bar{x}}{a}\, d\bar{x} = \tfrac{3}{8}\frac{a}{b}\\
\int_{x_I}^{x_I+0.5a} \frac{\partial \psi_1^{(1)}}{\partial y}\, dx &= \int_{0}^{0.5a} \frac{\partial \psi_1^{(1)}}{\partial \bar{y}}\, d\bar{x} = -\frac{1}{b}\int_{0}^{0.5a}\left(1-\frac{\bar{x}}{a}\right) d\bar{x} = -\tfrac{3}{8}\frac{a}{b}\\
\int_{x_I}^{x_I+0.5a} \frac{\partial \psi_3^{(1)}}{\partial y}\, dx &= \int_{0}^{0.5a} \frac{\partial \psi_3^{(1)}}{\partial \bar{y}}\, d\bar{x} = \frac{1}{b}\int_{0}^{0.5a}\frac{\bar{x}}{a}\, d\bar{x} = \tfrac{1}{8}\frac{a}{b}
\end{aligned} \tag{5.4.28}$$

In view of the analytical expressions in Eq. (5.4.28), the coefficients A_K and F_I can be easily determined. For uniform mesh and constant data, A_K and F_I are given as follows:

$$\begin{aligned}
A_{I-N-2} &= -\frac{1}{8P_e}\left(\frac{b}{a}+\frac{\gamma}{\alpha^2}\frac{a}{b}\right), \quad & A_{I-N-1} &= \frac{1}{4P_e}\left(\frac{b}{a}-\frac{\gamma}{\alpha^2}\frac{3a}{b}\right)\\
A_{I-N} &= -\frac{1}{8P_e}\left(\frac{b}{a}+\frac{\gamma}{\alpha^2}\frac{a}{b}\right), \quad & A_{I-1} &= \frac{1}{4P_e}\left(-\frac{3b}{a}+\frac{\gamma}{\alpha^2}\frac{a}{b}\right)\\
A_{I} &= \frac{3}{2P_e}\left(\frac{b}{a}+\frac{\gamma}{\alpha^2}\frac{a}{b}\right), \quad & A_{I+1} &= \frac{1}{4P_e}\left(-\frac{3b}{a}+\frac{\gamma}{\alpha^2}\frac{a}{b}\right)\\
A_{I+N} &= -\frac{1}{8P_e}\left(\frac{b}{a}+\frac{\gamma}{\alpha^2}\frac{a}{b}\right), \quad & A_{I+N+1} &= \frac{1}{4P_e}\left(\frac{b}{a}-\frac{\gamma}{\alpha^2}\frac{3a}{b}\right)\\
A_{I+N+2} &= -\frac{1}{8P_e}\left(\frac{b}{a}+\frac{\gamma}{\alpha^2}\frac{a}{b}\right), \quad & F_I &= g_0 ab,\\
C_{I-N-2} &= \left(-\frac{b}{16}-\frac{\beta}{\alpha}\frac{a}{16}\right), \quad & C_{I-N-1} &= \left(\frac{b}{16}-\frac{\beta}{\alpha}\frac{3a}{16}\right)+\left(-\frac{b}{16}-\frac{\beta}{\alpha}\frac{3a}{16}\right)\\
C_{I-N} &= \left(\frac{b}{16}-\frac{\beta}{\alpha}\frac{b}{16}\right), \quad & C_{I-1} &= \left(-\frac{3b}{16}+\frac{\beta}{\alpha}\frac{a}{16}\right)+\left(-\frac{3b}{16}-\frac{\beta}{\alpha}\frac{a}{16}\right)
\end{aligned}$$

$$C_I = \left(\frac{3b}{16}+\frac{\beta}{\alpha}\frac{3a}{16}\right)+\left(-\frac{3b}{16}+\frac{\beta}{\alpha}\frac{3a}{16}\right)+\left(-\frac{3b}{16}-\frac{\beta}{\alpha}\frac{3a}{16}\right)+\left(\frac{3b}{16}-\frac{\beta}{\alpha}\frac{3a}{16}\right)$$

$$C_{I+1} = \left(\frac{3b}{16} + \frac{\beta}{\alpha}\frac{a}{16}\right) + \left(\frac{3b}{16} - \frac{\beta}{\alpha}\frac{a}{16}\right), \qquad C_{I+N} = \left(-\frac{b}{16} + \frac{\beta}{\alpha}\frac{a}{16}\right)$$
$$C_{I+N+1} = \left(\frac{b}{16} + \frac{\beta}{\alpha}\frac{3a}{16}\right) + \left(-\frac{b}{16} + \frac{\beta}{\alpha}\frac{3a}{16}\right), \quad C_{I+N+2} = \left(\frac{b}{16} + \frac{\beta}{\alpha}\frac{a}{16}\right) \tag{5.4.29}$$

5.4.5.4 Discretized equations for the boundary nodes

For the nodes on the boundary, we must modify Eq. (5.4.28). For a rectangular domain, the boundary nodal point locations can be classified into several cases as shown in Fig. 5.4.4. The evaluation of coefficients for typical boundary nodes were presented in Eqs. (5.4.13)–(5.4.18b) for the model differential equations in Eq. (5.4.1). The revised coefficients associated with the generalized model equation in Eq. (5.4.25) for a couple of representative boundary cases are derived here. One can follow the logic presented here to develop the relations for the other boundary nodes.

Nodes on the bottom boundary [see Fig. 5.4.4(a)] For this case, the integral form in Eq. (5.4.28) takes the form ($y_I = 0$)

$$\begin{aligned}
0 =& \int_{x_I-0.5a}^{x_I+0.5a} \left[a_{yy}\frac{\partial u}{\partial y}\right]_{y=y_I} dx - \int_{y_I}^{y_I+0.5b} \left[a_{xx}\frac{\partial u}{\partial x}\right]_{x=x_I+0.5a} dy \\
&- \int_{x_I-0.5a}^{x_I+0.5a} \left[a_{yy}\frac{\partial u}{\partial y}\right]_{y=y_I+0.5b} dx + \int_{y_I}^{y_I+0.5b} \left[a_{xx}\frac{\partial u}{\partial x}\right]_{x=x_I-0.5a} dy \\
&- \int_{x_I-0.5a}^{x_I+0.5a}\int_{y_I}^{y_I+0.5b} g(x,y)\, dxdy \\
=& -\int_{y_I}^{y_I+0.5b} \left[a_{xx}\frac{\partial u}{\partial x}\right]^{(2)}_{x=x_I+0.5a} dy - \int_{x_I}^{x_I+0.5a} \left[a_{yy}\frac{\partial u}{\partial y}\right]^{(2)}_{y=y_I+0.5b} dx \\
&- \int_{x_I-0.5a}^{x_I} \left[a_{yy}\frac{\partial u}{\partial y}\right]^{(1)}_{y=y_I+0.5b} dx + \int_{y_I}^{y_I+0.5b} \left[a_{xx}\frac{\partial u}{\partial x}\right]^{(1)}_{x=x_I-0.5a} dy \\
&- \int_{x_I-0.5a}^{x_I+0.5a} q_y(x, y_I)\, dx \\
&- \int_{x_I-0.5a}^{x_I}\int_{y_I}^{y_I+0.5b} g^{(1)}\, dxdy - \int_{x_I}^{x_I+0.5a}\int_{y_I}^{y_I+0.5b} g^{(2)}\, dxdy
\end{aligned} \tag{5.4.30a}$$

or

$$\begin{aligned}
A_{I-1}U_{I-1} + A_I U_I + A_{I+1}U_{I+1} + A_{I+N}U_{I+N} + A_{I+N+1}U_{I+N+1}& \\
+ A_{I+N+2}U_{I+N+2} = F_I + Q_I&
\end{aligned} \tag{5.4.30b}$$

For uniform mesh and constant data throughout the domain, we obtain

$$\begin{aligned} A_{I-1} &= -\frac{3b}{8a}a_{xx} + \frac{a}{8b}a_{yy}\,, \quad A_I = \frac{3b}{4a}a_{xx} + \frac{3a}{4b}a_{yy} \\ A_{I+1} &= -\frac{3b}{8a}a_{xx} + \frac{a}{8b}a_{yy}\,, \quad A_{I+N+2} = -\frac{b}{8a}a_{xx} - \frac{a}{8b}a_{yy} \\ A_{I+N+1} &= \frac{b}{4a}a_{xx} - \frac{3a}{4b}a_{yy}\,, \quad A_{I+N} = -\frac{b}{8a}a_{xx} - \frac{a}{8b}a_{yy}\,, \quad F_I = \tfrac{1}{2}g_0ab \end{aligned} \tag{5.4.30c}$$

Nodes on the left boundary [see Fig. 5.4.4(d)] The integral statement for this case is given by ($x_I = 0.0$)

$$\begin{aligned} 0 &= \int_{x_I}^{x_I+0.5a} \left[a_{yy}\frac{\partial u}{\partial y}\right]_{y=y_I-0.5b} dx - \int_{y_I-0.5b}^{y_I+0.5b} \left[a_{xx}\frac{\partial u}{\partial x}\right]_{x=x_I+0.5a} dy \\ &\quad - \int_{x_I}^{x_I+0.5a} \left[a_{yy}\frac{\partial u}{\partial y}\right]_{y=y_I+0.5b} dx + \int_{y_I-0.5b}^{y_I+0.5b} \left[a_{xx}\frac{\partial u}{\partial x}\right]_{x=x_I} dy \\ &\quad - \int_{x_I}^{x_I+0.5a}\int_{y_I-0.5b}^{y_I+0.5b} g(x,y)\,dxdy \\ &= \int_{x_I}^{x_I+0.5a} \left[a_{yy}\frac{\partial u}{\partial y}\right]^{(1)}_{y=y_I-0.5b} dx - \int_{y_I-0.5b}^{y_I} \left[a_{xx}\frac{\partial u}{\partial x}\right]^{(1)}_{x=x_I+0.5a} dy \\ &\quad - \int_{y_I}^{y_I+0.5b} \left[a_{xx}\frac{\partial u}{\partial x}\right]^{(2)}_{x=x_I+0.5a} dy - \int_{x_I}^{x_I+0.5a} \left[a_{yy}\frac{\partial u}{\partial y}\right]^{(2)}_{y=y_I+0.5b} dx \\ &\quad - \int_{y_I-0.5b}^{y_I+0.5b} q_x(x_I,y)dy - \int_{x_I}^{x_I+0.5a}\int_{y_I-0.5b}^{y_I+0.5b} g(x,y)\,dxdy \end{aligned} \tag{5.4.31a}$$

which can be expressed as

$$\begin{aligned} A_{I-N-1}U_{I-N-1} + A_{I-N}U_{I-N} + A_IU_I + A_{I+1}U_{I+1} + A_{I+N+1}U_{I+N+1} \\ + A_{I+N+2}U_{I+N+2} = F_I + Q_I \end{aligned} \tag{5.4.31b}$$

where, uniform mesh and constant data throughout the domain, we obtain

$$\begin{aligned} A_{I-N-1} &= \frac{b}{8a}a_{xx} - \frac{3a}{8b}a_{yy}\,, \quad A_{I-N} = -\frac{b}{8a}a_{xx} - \frac{a}{8b}a_{yy} \\ A_I &= \frac{3b}{4a}a_{xx} + \frac{3a}{4b}a_{yy}\,, \quad A_{I+1} = -\frac{3b}{4a}a_{xx} + \frac{a}{4b}a_{yy} \\ A_{I+N+1} &= \frac{b}{8a}a_{xx} - \frac{3a}{8b}a_{yy}\,, \quad A_{I+N+2} = -\frac{b}{8a}a_{xx} - \frac{a}{8b}a_{yy}\,, \quad F_I = \tfrac{1}{2}g_0ab \end{aligned} \tag{5.4.31c}$$

Node on the bottom left corner [see Fig. 5.4.4(b)] The integral statement for this case can be expressed as

$$
\begin{aligned}
0 =& \int_0^{0.5a} \left[a_{yy}\frac{\partial u}{\partial y}\right]_{y=0} dx + \int_0^{0.5b} \left[a_{xx}\frac{\partial u}{\partial x}\right]_{x=0} dy \\
&- \int_0^{0.5a} \left[a_{yy}\frac{\partial u}{\partial y}\right]_{y=0.5b} dx - \int_0^{0.5b} \left[a_{xx}\frac{\partial u}{\partial x}\right]_{x=0.5a} dy \\
&- \int_0^{0.5a}\int_0^{0.5b} g(x,y)\, dxdy \\
=& - \int_0^{0.5a} \left[a_{yy}\frac{\partial u}{\partial y}\right]_{y=0.5b} dx - \int_{y_I}^{0.5b} \left[a_{xx}\frac{\partial u}{\partial x}\right]_{x=0.5a} dy \\
&- \int_0^{0.5a} q_x(x, y_I)\, dx - \int_0^{0.5b} q_x(0,y)\, dy \\
&- \int_0^{0.5a}\int_0^{0.5b} g(x,y)\, dxdy
\end{aligned} \tag{5.4.32a}
$$

which can be expressed as

$$
A_1 U_1 + A_2 U_2 + A_{N+2} U_{N+2} + A_{N+3} U_{N+3} = F_1 + Q_1 \tag{5.4.32b}
$$

where, for uniform mesh and constant data throughout the domain, we obtain

$$
\begin{aligned}
A_1 &= \frac{3b}{8a} a_{xx} + \frac{3a}{8b} a_{yy}\,, \quad A_2 = -\frac{3b}{8a} a_{xx} + \frac{a}{8b} a_{yy} \\
A_{N+2} &= \frac{b}{8a} a_{xx} - \frac{3a}{8b} a_{yy}\,, \quad A_{N+3} = -\frac{b}{8a} a_{xx} - \frac{a}{8b} a_{yy}\,, \quad F_1 = \tfrac{1}{4} g_0 ab
\end{aligned} \tag{5.4.32c}
$$

For $\alpha = \beta = \gamma = 1$, Eq. (5.4.25) yields

$$
\begin{aligned}
&- \left[\frac{1}{8P_e}\left(\frac{b}{a}+\frac{a}{b}\right) + \frac{1}{16}(a+b)\right] U_{I-N-2} + \left[\frac{1}{4P_e}\left(\frac{b}{a}-\frac{3a}{b}\right) - \frac{3a}{8}\right] U_{I-N-1} \\
&+ \left[-\frac{1}{8P_e}\left(\frac{b}{a}+\frac{a}{b}\right)\right] U_{I-N} + \left[\frac{1}{4P_e}\left(-\frac{3b}{a}+\frac{a}{b}\right) - \frac{3b}{8}\right] U_{I-1} \\
&+ \left[\frac{3}{2P_e}\left(\frac{b}{a}+\frac{a}{b}\right)\right] U_I + \left[\frac{1}{4P_e}\left(-\frac{3b}{a}+\frac{a}{b}\right) + \frac{3b}{8}\right] U_{I+1} \\
&+ \left[-\frac{1}{8P_e}\left(\frac{b}{a}+\frac{a}{b}\right) + \frac{1}{16}(a-b)\right] U_{I+N} + \left[\frac{1}{4P_e}\left(\frac{b}{a}-\frac{3a}{b}\right) + \frac{3a}{8}\right] U_{I+N+1} \\
&+ \left[-\frac{1}{8P_e}\left(\frac{b}{a}+\frac{a}{b}\right) + \frac{1}{16}(a+b)\right] U_{I+N+2} = g_0 ab
\end{aligned} \tag{5.4.33}
$$

Example 5.4.4

Use Eq. (5.4.33) to investigate the performance of DMCDM for various values of the Péclet number, Pe, with different uniform primal meshes.

Solution: For this two-dimensional case (see **Example 4.3.6**), the exact solution $u(x,y)$ of Eq. (5.4.21) (when $\alpha = \beta = \gamma = 1$) for the boundary conditions

$$u(1,y) = 0, \quad 0 \le y \le 1; \quad u(x,1) = 0, \quad 0 \le x \le 1$$

$$u(x,0) = \frac{1 - e^{(x-1)Pe}}{1 - e^{-Pe}}, \quad 0 \le x \le 1; \quad u(0,y) = \frac{1 - e^{(y-1)Pe}}{1 - e^{-Pe}}, \quad 0 \le y \le 1 \tag{5.4.34}$$

is given in Eq. (4.3.68):

$$u(x,y) = \frac{\left(1 - e^{(x-1)Pe}\right)\left(1 - e^{(y-1)Pe}\right)}{(1 - e^{-Pe})^2}, \tag{5.4.35}$$

Figure 5.4.16 shows the solution $u(x,y)$ of the DMCDM along the diagonal, $x = y$, of the domain [i.e., the results included in the figures are at points on the diagonal that cuts from $(0,0)$ to $(1,1)$, as this will be the line which is least affected by the given boundary conditions and differ most from the exact solution (see [41])]. The DMCDM solution is shown for $Pe = 30$ and $Pe = 10$. The results are markedly similar to the one-dimensional case. It is important to note that in all cases, the DMCDM's error tends to overestimate instead of underestimate the solution. Figure 5.4.17 shows the solution $u(x,y)$ at $x = y$ of the two-dimensional DMCDM for a $N \times N = 100 \times 100$ primal mesh for Péclet numbers 10 and 100, and 150×150 primal mesh for Péclet number 250. Again, the results are shown down the diagonal of the domain, so as to show the points which are least influenced by the boundary conditions.

Table 5.4.4 shows the results of the linear DMCDM solution compared to the exact solution $u(x,y)$ at $x = y$ for $Pe = 75$. This lower Péclet number was chosen because it was found that for the two-dimensional case, the FEM solution is unstable for $Pe\,h \ge 2$ (or $N \le Pe/2$).

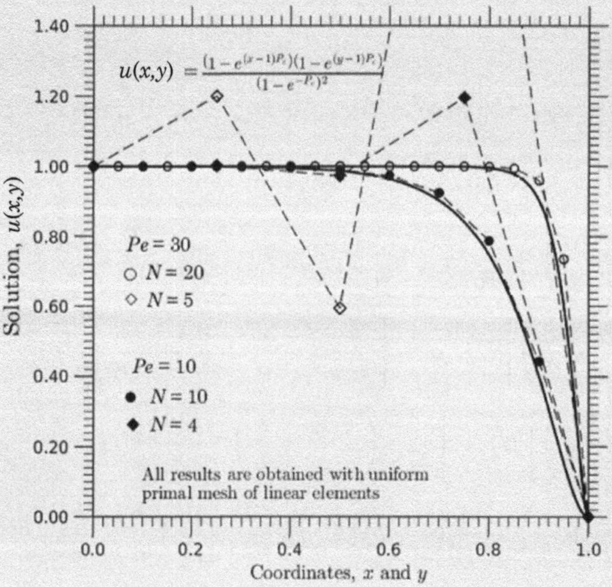

Fig. 5.4.16 Comparison of the two-dimensional numerical solutions, $u(x,y)$, for $x = y$, obtained with various meshes, with the exact solution for $Pe = 10$ and 30.

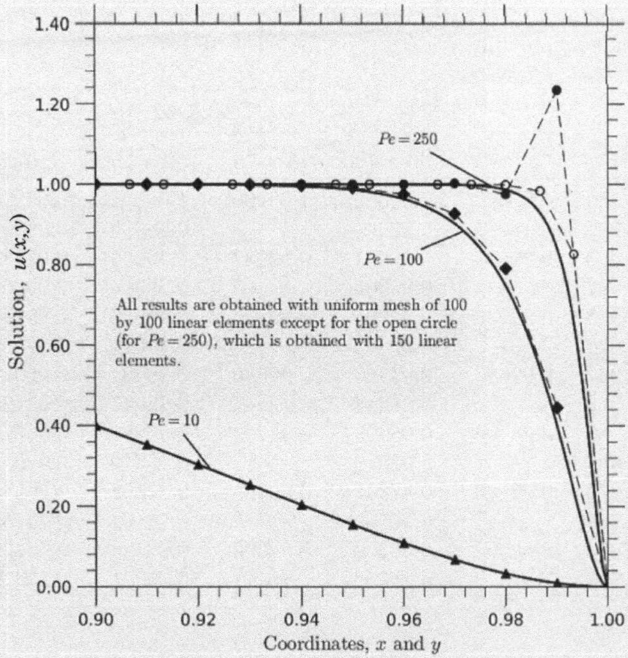

Fig. 5.4.17 Comparison of the two-dimensional numerical solutions, $u(x, y)$ for $x = y$, with the exact solution for $Pe = 10$, 100, and 250.

Table 5.4.4 Comparison of the FEM and DMCDM solutions with the exact solutions of the two-dimensional advection–diffusion equation (using primal meshes of linear elements) for the case of $Pe = 75$.

	FEM		DMCDM		
$x = y$	50×50	100×100	50×50	100×100	Exact
0.800	1.00000	1.00000	1.00000	1.00001	1.00000
0.850	——	0.99999	——	1.00000	0.99997
0.860	1.00000	0.99997	1.00000	0.99998	0.99994
0.880	0.99998	0.99984	0.99999	0.99986	0.99975
0.900	0.99988	0.99925	0.99989	0.99926	0.99889
0.910	——	0.99834	——	0.99836	0.99766
0.920	0.99917	0.99636	0.99917	0.99637	0.99505
0.930	——	0.99200	——	0.99201	0.98953
0.940	0.99418	0.98244	0.99418	0.98245	0.97791
0.950	——	0.96157	——	0.96158	0.95352
0.960	0.95960	0.91645	0.95960	0.91646	0.90290
0.970	——	0.82099	——	0.82100	0.80031
0.980	0.73469	0.62947	0.73470	0.62947	0.60353
0.990	——	0.29752	——	0.29753	0.27840

Table 5.4.5 provides the results for the case of $Pe = 100$. It can be seen that the FEM and DMCDM solutions are nearly identical, but the DMCDM solution shows greater stability.

Table 5.4.5 Comparison of the FEM and DMCDM solutions with the exact solution of the two-dimensional advection–diffusion equation for $Pe = 100$.

	FEM		DMCDM		
$x = y$	75×75	150×150	75×75	150×150	Exact
0.880	1.00000	1.00000	1.00000	1.00001	0.99999
0.900	——	0.99994	——	0.99995	0.99991
0.920	0.99987	0.99951	0.99987	0.99953	0.99933
0.927	——	0.99902	——	0.99904	0.99869
0.933	0.99936	0.99805	0.99936	0.99806	0.99746
0.940	——	0.99610	——	0.99611	0.99505
0.947	0.99680	0.99220	0.99680	0.99222	0.99037
0.953	——	0.98444	——	0.98445	0.98128
0.960	0.98406	0.96899	0.98407	0.96901	0.96370
0.967	——	0.93848	——	0.93849	0.92992
0.973	0.92160	0.87891	0.92160	0.87892	0.86585
0.980	——	0.76562	——	0.76564	0.74765
0.987	0.64000	0.56250	0.64000	0.56521	0.54229
0.993	——	0.25000	——	0.25001	0.23676

5.5 Summary

In summary, the *dual mesh control domain method* (DMCDM) involves the following basic steps:

1. The whole domain Ω is represented as a collection of N nonoverlapping finite elements, Ω^e $(e = 1, 2, \ldots, N)$, called *primal mesh* with a set of associated nodes. The *dual mesh* consists of a set of node-centered control domains (dual mesh) such that all control domains are interconnected, nonoverlapping, and cover the whole domain Ω. Thus, the mesh of control domains is different from the primal mesh of finite elements with their interpolation functions.
2. Over each control domain, discrete (algebraic) relations among the duality pairs (e.g., relations between pairs of dual variables (e.g., "forces" and "displacements"; "heats" and "temperatures") are developed using integral statement(s) of the differential equation(s) to be solved. This results in algebraic relations between the values of the enclosed node (master) as well as nodes (neighbors) in the inter-connected neighboring control domains. Similar relations are also derived for nodes on the boundary.
3. Using the algebraic relations at all nodes (inside as well as on the boundary), a system of algebraic equations among the duality pairs of all nodes in the mesh are obtained. Then boundary conditions are imposed and equations are solved for the unknown variables of the duality pairs at the nodes.
4. Unique values of the secondary variables can be computed at the control domain interfaces (equivalent to using the reduced Gauss point locations in the FEM).

From the formulation presented for the one- and two-dimensional model problems and the numerical examples presented, it is clear that the DMCDM makes use of two different meshes, one for interpolation of the variable and another one to satisfy the integral form of the governing equations. Both meshes cover the domain of the problem, but they do not coincide with each other. Thus, the DMCDM shares some characteristics of the FEM as well as the FVM. The DMCDM is similar to the FEM and FVM (especially the half-control finite volume formulation) in representing the domain into a set of subdomains. The difference between the DMCDM and FEM is not only in the type of integral statement used but in the fact that the (primal) mesh used for the approximation of the variable is different from the (dual) mesh used for the satisfaction of the integral statement. On the other hand, the FVM has no explicit approximation of the variable and replaces the derivatives using the Taylor series approximations, and the domain integrals are often replaced with average integral values. In the FVM and DMCDM, the interfaces of control domains fall inside the mesh of elements used for interpolation, and there is no "assembly of elements." Because of the duality concept used in the DMCDM, the secondary variables are computed at the control domain interfaces, where the derivatives are continuous. The numerical examples show that the DMCDM gives more accurate solution than the FEM and FVM methods.

There is a chance that some researchers of the FVM already practice what is presented here as the DMCDM but failed to articulate their methodology. However, the books and papers found in the literature on the FVM do not contain any such ideas, especially the use of duality concept and finite element approximation of the variables, which makes the DMCDM distinctly different from the ZFVM. Some papers that compare the FVM and FEM also seem to miss to correctly identify the features of the FEM and FVM. The main features of the three methods are summarized here (with problems posed on two-dimensional domains in mind; see Fig. 5.5.1):

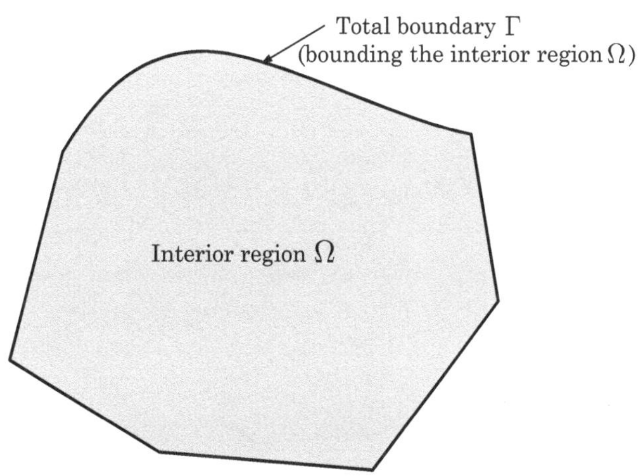

Fig. 5.5.1 Domain and boundary of an arbitrary two-dimensional domain.

(1) *Domain discretization.* In the FEM, the domain Ω is divided into a set of finite elements (triangular or quadrilateral geometry and associated interpolation functions). The nodes are the vertices (and points on the sides and in the interior for higher-order elements) of the elements (this can be viewed as the primal mesh), as shown in Fig. 5.5.2. Domain discretization errors exist when the domain has curved boundary. In the FVM, the domain is divided into a set of rectangular geometry (although some FVM studies suggest finite element geometries, but it is not clear if they used the associated interpolation functions to approximate the dependent variables). Figure 5.5.3(a) illustrates the use of quadrilateral subdomains to represent the geometry in the FVM, leading to control volumes which are arbitrary polygons. When the quadrilaterals are rectangles, the control volumes are also rectangles (as already shown in Fig. 5.4.1). The rectangular meshes are necessitated by the fact that difference formulas to replace derivatives of functions exist only for structured rectangular meshes. When the domain has curved boundary, domain discretization errors exist as in the FEM. Unless one uses finite element type approximations of the dependent unknowns, the FVM has to depend on the Taylor series expansions to replace the derivatives, which in turn dictates the rectangular meshes (i.e., arbitrary domain can have large discretization errors). In the DMCDM, the domain discretization (primal mesh) is exactly like in the FEM, replacing the domain with a set of finite element geometries that have unique interpolation functions. In the DMCDM, the primal mesh is a set of finite elements, and the dual mesh is a set of control domains that are arbitrary polygons, as shown in Fig. 5.5.3(b). Because of the unique finite element interpolation functions used for the dependent unknowns, one can (numerically) evaluate the integrals posed over the control domains and their edges.

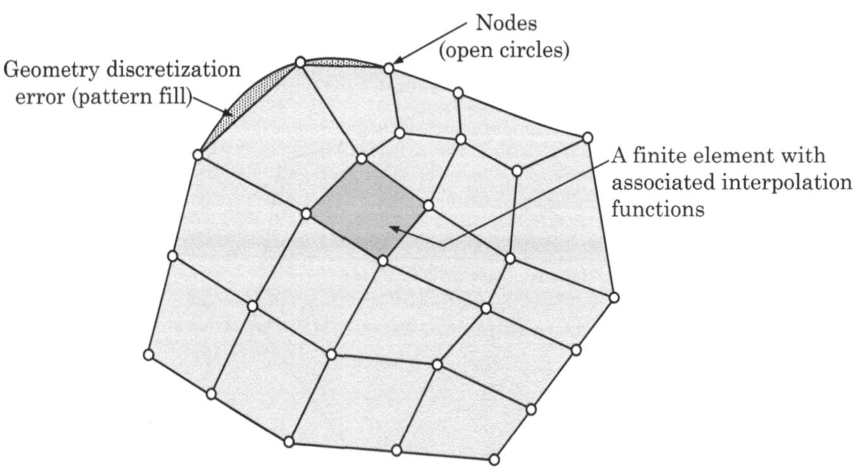

Fig. 5.5.2 Discretization of the domain with a set of quadrilateral finite elements in the FEM.

(2) *Types of differential equations that can be discretized.* In the FEM any order differential equation can be converted to discrete algebraic equations. When the differential equation is higher order (than the second order), the interpolation functions needed will be higher-order type. For example, a fourth-order differential equation requires (dictated by associated weak form) interpolations that are C^1-continuous (i.e., the first-order derivatives of the dependent variables must be continuous) at element interfaces. Alternatively, higher-order equations can be expressed as a set of second-order equations by introducing new unknowns (known as mixed formulations) and the C^1-continuity can be avoided. On the other hand, both the FVM and DMCDM can only be used to discretize first- or second-order differential equations. In reality, this is not a limitation because most physical problems when derived using physical laws are either first- or second-order differential equations.

(3) *Approximation of the functions and their derivatives.* In the FEM, an explicit representation of the dependent variables over each finite element is adopted. Consequently, the derivatives can be readily computed within an element when needed. We note that, in general, the geometry approximation is not the same as the approximation used for the dependent variables. Thus, there is a secondary mesh (not a dual mesh) in the FEM that is associated with the approximation of the dependent variables. When the primal mesh used for the geometry and the secondary mesh used for the dependent variables is the same, it is known as the isoparametric formulation. In the case of the FVM, the derivatives are replaced with difference formulas derived using the Taylor series expansions (of whatever accuracy desired). The DMCDM works exactly like the FEM as far as the representation of the dependent unknowns is concerned. However, in the DMCDM (as introduced in this book) the primal mesh used to discretize the domain is the same as that used for the approximation of the dependent variable.

(4) *Derivation of discretized equations.* The traditional FEM uses the weak-form Galerkin formulation, which requires setting up a weighted-residual statement and carrying out integration-by-parts *over a typical element* to relax the differentiability requirements on the approximation functions used to replace the dependent variables and introduce the secondary variables into the weak form. Consequently, the discretized equations, are valid only over a finite element. In general, the weak-form is only equivalent to the original differential equation (in mathematical sense), and it does not necessarily represent an integral statement of a physical law[2] because of the weight function that is not unity. Since the equations are only valid element-wise, one must assemble them using certain requirements imposed by the physics of the problem. Thus, the overhead involved with the FEM is greater than the FVM or DMCDM, which only use integral statement of the equations to be discretized over a set of control volumes or domains (the dual mesh). In most cases, the integral statements

[2] An exception to this statement is the field of solid and structural mechanics, where the weak forms can be shown to be equivalent to the principles of virtual displacements or the minimum total potential energy.

are the global conservation or balance laws, and therefore the FVM and DMCDM satisfy the local forms of physical laws more directly than the FEM. Also, since the discretized equations in both FVM and DMCDM are derived over a control volume (or control domain), which spans over all neighboring subdomains or finite elements (see Fig 5.5.3), no assembly of discretized equations is required. Unlike in the FVM (especially, the zero-thickness formulation), the DMCDM makes use of the duality concept and introduces the dual variables into the discretized equations associated with the boundary nodes. Thus, equations associated with the boundary nodes are used only in the post-computation.

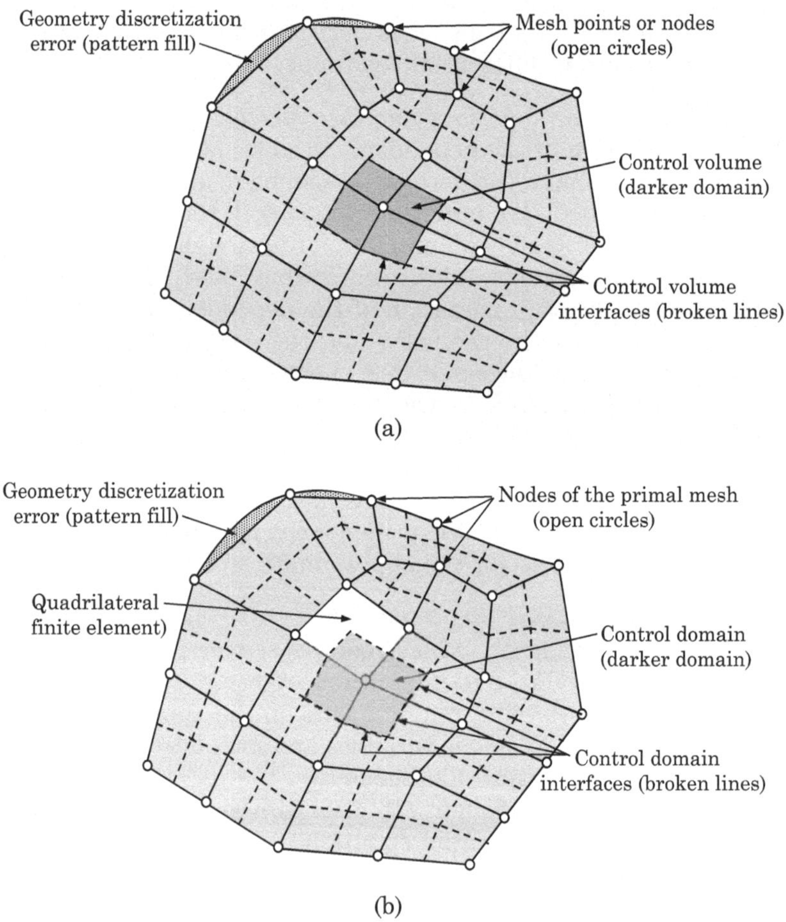

Fig. 5.5.3 Discretization of the domain in the FVM and DMCDM. (a) Typical control volume in the FVM. (b) Typical finite element (a quadrilateral) in the primal mesh and typical control domain (a polygon) in the dual mesh of the DMCDM.

(5) *Solution of discretized equations and computation of the secondary variables.* The solution of discretized equations after the imposition of the boundary conditions is common to all numerical methods. However, in both the FEM and DMCDM, the secondary variables at nodes at which the primary variable is not known can be computed from discretized equations associated with the boundary nodes (this is because the duality concept is exercised in the FEM and DMCDM). In practice, the secondary variables are post-computed in the FEM using the definitions introduced during the weak-form development. The values computed in this manner are not accurate compared to the values computed using the nodal equations. In the FVM and DMCDM, the dual variables are computed at the interfaces of the dual mesh, where they are continuous as well as the most accurate.

In closing this chapter, we conclude that the DMCDM has the desirable features of the FEM (unique representation of the dependent variables over the primal mesh using the finite element interpolation functions and introduction of dual variables into the discretized equations) and the FVM (satisfying the local form of the conservation or balance laws and avoiding assembly of discretized equations). Extensions of the DMCDM to nonlinear and multivariable problems will be presented in the forthcoming chapters. Extending the DMCDM to arbitrary and higher-order primal meshes is a major step toward totally displacing the FEM and FVM to solve real-world problems governed by differential equations. Mathematical aspects such as the existence, uniqueness, and error estimates of the DMCDM are also yet to be studied.

Problems

The readers may find additional examples and problems (involving a single unknown) in the textbook by Reddy [8] which can be solved using the DMCDM method.

ONE-DIMENSIONAL PROBLEMS

5.1 Rewrite Eq. (5.3.1) as a pair of first-order equations

$$-\frac{dv}{dx} + cu - f = 0, \quad \frac{du}{dx} - \frac{v}{a} = 0, \tag{1}$$

and develop the DMCDM model of the pair of equations keeping both u and v as unknowns.

5.2 The following differential equation arises in connection with heat transfer in a plane wall:

$$-\frac{d}{dx}\left(k\frac{dT}{dx}\right) = 0 \quad \text{for} \quad 0 < x < L \tag{1}$$

$$T(0) = T_0, \quad \left[k\frac{dT}{dx} + \beta(T - T_\infty)\right]\Bigg|_{x=L} = 0, \tag{2}$$

where T is the temperature, k the thermal conductivity, and T_∞ is the ambient temperature at $x = L$. Take the following values for the data: $L = 0.1$ m, $k = 0.01$ W m^{-1}°C^{-1}, $\beta = 25$ W m^{-2}°C^{-1}, $T_0 = 50$°C, and $T_\infty = 5$°C. Solve the problem using the primal mesh of two linear finite elements (and three control domains of the dual mesh) for temperature values at $x = 0.5L$ and $x = L$, and heat at $x = 0$.

5.3 An insulating wall is constructed of three homogeneous layers with conductivities k_1, k_2, and k_3 in intimate contact (see Fig. P5.3). Under steady-state conditions, the temperatures at the boundaries of the layers are characterized by the external surface temperatures T_1 and T_4 and the interface temperatures T_2 and T_3. Use the DMCDM to determine the temperatures T_i $(i = 1, 2, 3, 4)$ when the ambient temperatures T_0 and T_5 (at the left and right, respectively) and the (surface) film coefficients β_0 and β_5 are known for the following cases. Case (1) $T_0 = T_5 = 20°\text{C}$ and $\beta_0 = \beta_5 = 500$ W/m^2·C. Case (2) $T_0 = 120°\text{C}$, $T_5 = 20°\text{C}$, $\beta_0 = 500$ W/m^2·C, and $\beta_5 = 560$ W/m^2·C. Case (3) $T_1 = 100°\text{C}$, $T_5 = 20°\text{C}$, and $\beta_5 = 500$ W/m^2·C. Assume that there is no internal heat generation and that the heat flow is one-dimensional in the x-direction.

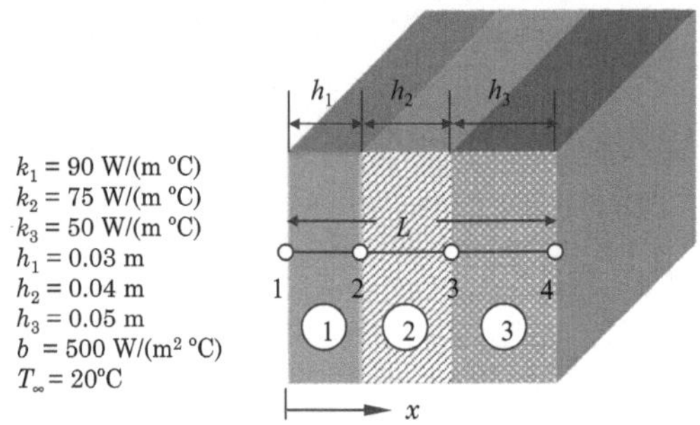

Fig. P5.3

5.4 Consider steady heat conduction in a wire of circular cross-section with an electrical heat source. Suppose that the radius of the wire is a, its electrical conductivity is K_e (Ω^{-1}/cm), and it is carrying an electric current density of I (A/cm ^{2}). During the transmission of an electric current, some of the electrical energy is converted into thermal energy. The rate of heat generation per unit volume is given by $g = I^2/K_e$. Assume that the temperature rise in the wire is sufficiently small that the dependence of the thermal or electric conductivity on temperature can be neglected. The governing equations of the problem are

$$-\frac{1}{r}\frac{d}{dr}\left(rk\frac{dT}{dr}\right) = g \quad \text{for} \quad 0 \le r \le a, \qquad \left(rk\frac{dT}{dr}\right)\bigg|_{r=0} = 0, \quad T(a) = T_0.$$

Determine the distribution of temperature in the wire using a primal mesh of two linear finite elements, and compare the DMCDM solution with the FEM solution and the exact solution,

$$T(r) = T_0 + \frac{ga^2}{4k}\left[1 - \left(\frac{r}{a}\right)\right].$$

Take $a = 0.01$ m, $T_0 = 100\,°\text{C}$, $g = g_0 = 2 \times 10^8$ W/m^3, and $k = 20$ W/(m·°C. Also, determine the heat flow, $Q = -2\pi ak(dT/dr)|_{r=a}$.

5.5 Consider the steady radial heat flow from the inside to the outside of a thick-walled hollow cylinder of internal radius a and external radius b. The inside wall is maintained at temperature T_a and the outside wall is maintained at T_b, as shown in Fig. P5.6. The governing equation for the steady-state case is given by

$$\frac{k}{r}\frac{d}{dr}\left(r\frac{dT}{dr}\right) = 0. \tag{1}$$

Determine the temperature distribution in the wall for a uniform primal mesh of four linear elements. The analytical solution is given by

$$\frac{T(r) - T_a}{T_b - T_a} = \frac{\ln(r/a)}{\ln(b/a)}. \tag{2}$$

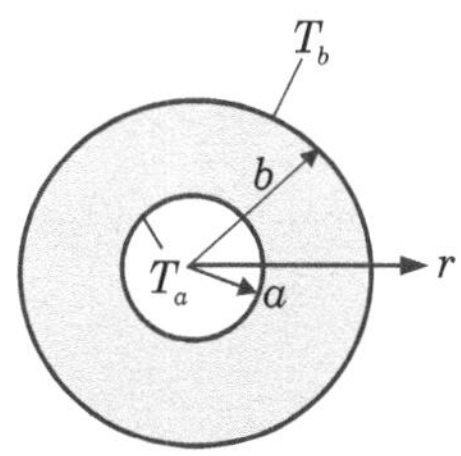

Fig. P5.6

5.6 Formulate the advection-diffusion problem of **Example 4.2.4** [see (4.2.38)] using the DMCDM and solve it using a primal mesh of ten elements for Péclet number of $Pe = 20$.

TWO-DIMENSIONAL PROBLEMS

5.7 Evaluate the coefficients in Eq. (5.4.9a) for the two-dimensional control domain shown in Fig. 5.4.6. Assume that a_{xx}, a_{yy}, and $f(x, y)$ are element-wise constants.

5.8 Consider the steady-state heat transfer in a square region shown in Fig. P5.8. The governing equation is given by

$$-\frac{\partial}{\partial x}\left(k\frac{\partial u}{\partial x}\right) - \frac{\partial}{\partial y}\left(k\frac{\partial u}{\partial y}\right) = f_0. \tag{1}$$

The boundary conditions for the problem are:

$$u(0, y) = y^2\ ,\ u(x, 0) = x^2\ ,\ u(1, y) = 1 - y\ ,\ u(x, 1) = 1 - x. \tag{2}$$

Assuming $k = 1$ and $f_0 = 2$, determine the unknown nodal value U_3 using the primal finite element mesh shown in Fig. P5.8.

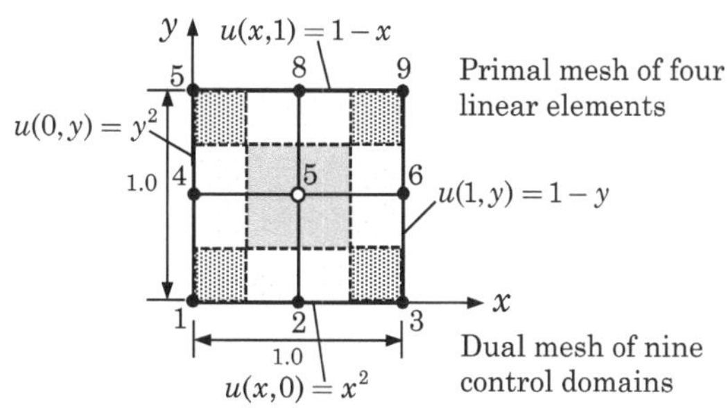

Fig. P5.8

5.9 Consider steady-state diffusion process governed by the differential equation

$$-k\left(\frac{\partial^2 u}{\partial x^2}+\frac{\partial^2 u}{\partial y^2}\right)=0 \tag{1}$$

over the domain shown in Fig. P5.9. Exploiting the symmetry, analyze the problems using DMCDM with the primal mesh of (a) 2×2 mesh of bilinear elements [see Fig. P5.9(a)] and (b) 2×2 mesh of linear triangular elements [see Fig. P5.9(b)]. The exact solution to the problem, when $u_0(x)=1$, is

$$u(x,y)=4k\sum_{n=0}^{\infty}\frac{\sin(\lambda_n x)\,\sinh(\lambda_n y)}{\lambda_n\,\sinh(\lambda_n)},\quad \lambda_n=(2n+1)\pi. \tag{2}$$

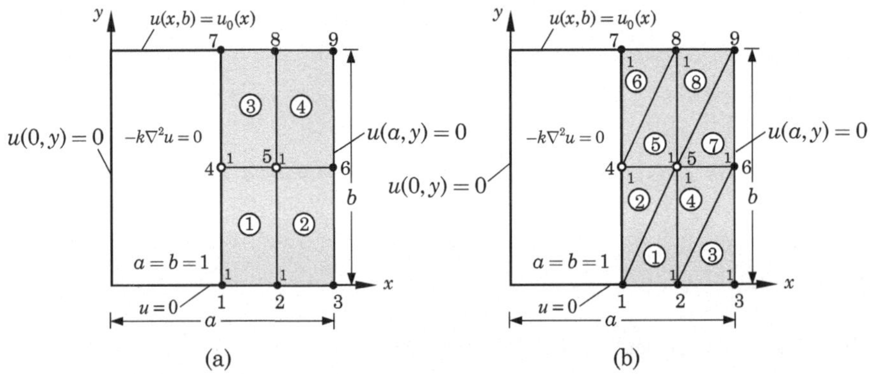

Fig. P5.9

6
Nonlinear Problems with a Single Unknown

6.1 Introduction

The FVM, FEM, and DMCDM were applied in Chapters 3, 4, and 5, respectively, to linear differential equations in one and two dimensions and involving a single dependent unknown. In the present chapter, the FEM and DMCDM methods are extended to nonlinear differential equations with a single dependent unknown in one and two dimensions. Following this introduction, in Section 6.2 the FEM and DMCDM are extended to one-dimensional nonlinear problems governed by model second-order nonlinear differential equation in a single unknown. A number of numerical examples of nonlinear problems in one dimension are presented in Section 6.2.4. Then, in Section 6.3, the FEM and DMCDM are extended to a model second-order nonlinear differential equation in two dimensions involving a single unknown. Numerical results are presented in Section 6.3.4 using example problems from engineering with a particular focus on heat transfer. A summary and concluding remarks are presented in Section 6.4.

6.2 One-Dimensional Problems

6.2.1 Model Differential Equation

Consider the differential equation

$$-\frac{d}{dx}\left[a(x,u)\frac{du}{dx}\right]+b(x,u)\frac{du}{dx}+c(x,u)u=f(x),\quad 0<x<L \tag{6.2.1}$$

subjected to boundary conditions of the form

$$n_x\, a\frac{du}{dx}+\beta(x,u)\,(u-u_\infty)=\hat{Q},\qquad \text{or}\qquad u=\hat{u} \tag{6.2.2}$$

at a boundary point. Here $u(x)$ denotes the dependent variable to be determined, a, b, and c are known functions of x and u (and possibly derivatives of u), f is a known function of x, (u_∞, $\hat{u}$, β, $\hat{Q}$) are known (or specified) quantities (with $\beta=\beta_0+\beta_u\, u$) and $n_x=-1$ at $x=0$ and $n_x=1$ at $x=L$; β denotes a physical parameter (e.g., film conductance).

6.2.2 Finite Element Method

6.2.2.1 Weak formulation

Suppose that the domain $\Omega = (0, L)$ is divided into N line elements. A typical element from the collection of N elements is denoted as $\Omega^e = (x_a^e, x_b^e)$, where x_a^e and x_b^e denote the global coordinates of the end nodes of the line element. The weak form of Eq. (6.2.1) over the element can be developed as follows:

$$
\begin{aligned}
0 &= \int_{x_a^e}^{x_b^e} \left(a\frac{dw_i^e}{dx}\frac{du_h^e}{dx} + bw_i^e\frac{du_h^e}{dx} + cw_i^e u_h^e - w_i^e f \right) dx - \left[w_i^e \left(a\frac{du_h^e}{dx} \right) \right]_{x_a^e}^{x_b^e} \\
&= \int_{x_a^e}^{x_b^e} \left[a(x,u)\frac{dw_i^e}{dx}\frac{du_h^e}{dx} + b(x,u)w_i^e\frac{du_h^e}{dx} + c(x,u)w_i^e u_h^e - w_i^e f(x) \right] dx \\
&\quad - \left\{ Q_a^e - \beta_a\left[u_h^e(x_a^e) - u_\infty^a\right] \right\} w_i^e(x_a^e) - \left\{ Q_b^e - \beta_b\left[u_h^e(x_b^e) - u_\infty^b\right] \right\} w_i^e(x_b^e),
\end{aligned}
\tag{6.2.3}
$$

where $w_i^e(x)$ is the ith weight function. The number of weight functions is equal to the number of unknowns in the approximation of u_h. The boundary expression in the first line of Eq. (6.2.3) suggests that u is the primary variable and $Q = a(du_h^e/dx)$ is the secondary variable of the formulation. Using the mixed boundary condition in Eq. $(6.2.2)_1$, $a(du_h^e/dx)$ is expressed as

$$
\begin{aligned}
-\left[a\frac{du_h^e}{dx} \right]_{x=x_a^e} &= Q_a^e - \beta_a\left[u_h^e(x_a^e) - u_\infty^a\right], \\
\left[a\frac{du_h^e}{dx} \right]_{x=x_b^e} &= Q_b^e - \beta_b\left[u_h^e(x_b^e) - u_\infty^b\right],
\end{aligned}
\tag{6.2.4}
$$

where (Q_a^e, Q_b^e) are the nodal values, (u_∞^a, u_∞^b) denote the values of the variable u_∞, and (β_a, β_b) denote the value of β at the left and right ends of the element, respectively. When a node is in the interior of the element, the corresponding β and u_∞ are zero.

As discussed in Chapter 4, specifying a primary variable constitutes an *essential* (or *geometric*) boundary condition and specifying a secondary variable is a *natural* (or *force*) boundary condition. The first boundary condition in Eq. (6.2.2) is of the *mixed* type since it includes both the primary and secondary variables, and it is nonlinear because of the dependence of β on u; the second boundary condition in Eq. (6.2.2) is of the essential type. In a specific problem, a boundary point may have the essential, natural, or mixed type boundary condition specified.

6.2.2.2 Finite element model

Suppose that the dependent unknown $u(x)$ is approximated over an element Ω^e by the finite element approximation of the form

$$
u(x) \approx u_h^e(x) = \sum_{j=1}^{n} u_j^e \psi_j^e(x) \tag{6.2.5}
$$

Substituting the approximation from Eq. (6.2.5) for u_h^e and $w_i^e = \psi_i^e$ (i.e., using the Galerkin method) into the weak form, Eq. (6.2.3), we obtain the following weak-form Galerkin (or Ritz finite element model):

$$\mathbf{K}^e(\mathbf{u}^e)\,\mathbf{u}^e = \mathbf{F}^e, \tag{6.2.6a}$$

where

$$\begin{aligned}
K_{ij}^e &= \int_{x_a^e}^{x_b^e} \left[a(x,u_h^e)\frac{d\psi_i^e}{dx}\frac{d\psi_j^e}{dx} + b(x,u_h^e)\psi_i^e\frac{d\psi_j^e}{dx} + c(x,u_h^e)\psi_i^e\psi_j^e \right] dx \\
&\quad + \beta_a\psi_i^e(x_a^e)\psi_j^e(x_a^e) + \beta_b\psi_i^e(x_b^e)\psi_j^e(x_b^e), && (6.2.6b)\\
F_i^e &= \int_{x_a^e}^{x_b} f(x)\psi_i^e\,dx + \beta_a u_\infty^a \psi_i^e(x_a^e) + \beta_b u_\infty^b \psi_i^e(x_b^e) \\
&\quad + Q_a\psi_i^e(x_a^e) + Q_b\psi_i^e(x_b^e). && (6.2.6c)
\end{aligned}$$

We note that the coefficient matrix $\mathbf{K}^e$ depends on the unknown nodal values u_j^e, and it is an unsymmetric matrix when $b \neq 0$; that is, when $b = 0$, $\mathbf{K}^e$ is a symmetric matrix. The term involving c is symmetric, independent of whether it depends on u_h and/or du_h/dx. Therefore, it is advisable to include nonlinear terms of the type $(du_h/dx)\,u_h$ in a differential equation as the c-term in the equation by writing it as $(du_h/dx)\,u_h = c\,u_h$, with $c = du_h/dx$; otherwise, the coefficient matrix will be unsymmetric, and the convergence of the solution may become an issue. Of course, one may treat the expression as the b-term by writing it as $(du_h/dx)\,u_h = b\,(du_h/dx)$ with $b = u_h$. The coefficients involving β in $\mathbf{K}^e$ and $\mathbf{F}^e$ should be included only in elements that have end (i.e., boundary) nodes with the convection type boundary condition. **Example 6.2.1** provides more insight into the make-up of the coefficient matrix $\mathbf{K}^e$.

Example 6.2.1

Consider the problem described by Eqs. (6.2.1) and (6.2.2). Suppose that $a(x,u) = a_0 + a_u\,u(x) + a_{ux}\,\frac{du}{dx}$, $b = 0$, and $c(x,u) = c_0 + c_u\,u(x)$, where a_0, a_u, a_{ux}, c_0, and c_u are functions of x only. Determine the explicit form of the element matrices $\mathbf{K}^e$ and $\mathbf{f}^e$ using linear approximation of u and element-wise constant values a_0^e, a_u^e, a_{ux}^e, c_0^e, and c_u^e of a_0, a_u, a_{ux}, c_0, and c_u, respectively.

Solution: For all elements except for the last one, the coefficients β_a and β_b are zero; for the last element we have $\beta_a = 0$ and $\beta_b \neq 0$ when the end $x = L$ is subjected to the convection boundary condition. Thus, for an element interior to the domain (noting that a_0^e, a_u^e, a_{ux}^e, c_0^e, and c_u^e are constant within each element) we have:

$$\begin{aligned}
K_{ij}^e = \int_{x_a^e}^{x_b^e} \Bigg\{ &\left[a_0^e + a_u^e\left(\sum_{k=1}^n u_k^e\psi_k^e\right) + a_{ux}^e\left(\sum_{k=1}^n u_k^e\frac{d\psi_k^e}{dx}\right)\right]\frac{d\psi_i^e}{dx}\frac{d\psi_j^e}{dx} \\
&+ \left[c_0^e + c_u^e\left(\sum_{k=1}^n u_k^e\psi_k^e\right)\right]\psi_i^e\psi_j^e \Bigg\} dx
\end{aligned}$$

$$
\begin{aligned}
&= a_0^e \int_{x_a^e}^{x_b^e} \frac{d\psi_i^e}{dx}\frac{d\psi_j^e}{dx}\,dx + a_u^e \sum_{k=1}^{n} u_k^e \int_{x_a^e}^{x_b^e} \psi_k^e \frac{d\psi_i^e}{dx}\frac{d\psi_j^e}{dx}\,dx \\
&\quad + a_{ux}^e \sum_{k=1}^{n} u_k^e \int_{x_a^e}^{x_b^e} \frac{d\psi_k^e}{dx}\frac{d\psi_i^e}{dx}\frac{d\psi_j^e}{dx}\,dx + c_0^e \int_{x_a^e}^{x_b^e} \psi_i^e \psi_j^e\,dx \\
&\quad + c_u^e \sum_{k=1}^{n} u_k^e \int_{x_a^e}^{x_b^e} \psi_k^e \psi_i^e \psi_j^e\,dx.
\end{aligned} \tag{1}
$$

For the linear ($n = 2$) approximation of $u(x)$, we have [see Eqs. (4.2.14), (4.2.29a), and (4.2.29b)]

$$
\begin{aligned}
u_h^e(x) &= u_1^e \psi_1^e(x) + u_2^e \psi_2^e(x) = u_1^e \Big(\frac{x_b^e - x}{h_e}\Big) + u_2^e \Big(\frac{x - x_a^e}{h_e}\Big) \\
\frac{du_h^e}{dx} &= u_1^e \frac{d\psi_1^e}{dx} + u_2^e \frac{d\psi_2^e}{dx} = u_1^e \Big(-\frac{1}{h_e}\Big) + u_2^e \Big(\frac{1}{h_e}\Big) = \frac{u_2^e - u_1^e}{h_e}
\end{aligned} \tag{2}
$$

and

$$
\begin{aligned}
K_{ij}^e &= a_0^e \int_{x_a^e}^{x_b^e} \frac{d\psi_i^e}{dx}\frac{d\psi_j^e}{dx}\,dx + a_u^e \sum_{k=1}^{2} u_k^e \int_{x_a^e}^{x_b^e} \psi_k^e \frac{d\psi_i^e}{dx}\frac{d\psi_j^e}{dx}\,dx + a_{ux}^e \sum_{k=1}^{2} u_k^e \int_{x_a^e}^{x_b^e} \frac{d\psi_k^e}{dx}\frac{d\psi_i^e}{dx}\frac{d\psi_j^e}{dx}\,dx \\
&\quad + c_0^e \int_{x_a^e}^{x_b^e} \psi_i^e \psi_j^e\,dx + c_u^e \sum_{k=1}^{2} u_k^e \int_{x_a^e}^{x_b^e} \psi_k^e \psi_i^e \psi_j^e\,dx \\
&= (-1)^{i+j} \Big[a_0^e \frac{1}{h_e} + a_u^e \frac{1}{h_e^2} \sum_{k=1}^{2} u_k^e \int_{x_a^e}^{x_b^e} \psi_k^e\,dx + a_{ux}^e \frac{1}{h_e^2} \sum_{k=1}^{n} u_k^e \int_{x_a^e}^{x_b^e} \frac{d\psi_k^e}{dx}\,dx \Big] \\
&\quad + c_0^e \int_{x_a^e}^{x_b^e} \psi_i^e \psi_j^e\,dx + c_u^e \sum_{k=1}^{n} u_k^e \int_{x_a^e}^{x_b^e} \psi_k^e \psi_i^e \psi_j^e\,dx \\
&= (-1)^{i+j} \frac{1}{h_e} \Big[a_0^e + a_u^e \frac{u_1^e + u_2^e}{2} + a_{ux}^e \frac{u_2^e - u_1^e}{h_e} \Big] + c_0^e \int_{x_a^e}^{x_b^e} \psi_i^e \psi_j^e\,dx \\
&\quad + c_u^e \sum_{k=1}^{n} u_k^e \int_{x_a^e}^{x_b^e} \psi_k^e \psi_i^e \psi_j^e\,dx
\end{aligned} \tag{3}
$$

Noting that

$$
\begin{aligned}
&\int_0^{h_e} [\psi_1^e(\bar{x})]^3\,d\bar{x} = \frac{h_e}{4}, \qquad \int_0^{h_e} [\psi_1^e(\bar{x})]^2\,\psi_2^e(\bar{x})\,d\bar{x} = \frac{h_e}{12}, \\
&\int_0^{h_e} \psi_1^e(\bar{x})\,[\psi_2^e(\bar{x})]^2\,d\bar{x} = \frac{h_e}{12}, \qquad \int_0^{h_e} [\psi_2^e(\bar{x})]^3\,d\bar{x} = \frac{h_e}{4},
\end{aligned} \tag{4}
$$

we obtain

$$
\begin{aligned}
\mathbf{K}^e &= \frac{1}{h_e}\Big(a_0^e + a_u^e \frac{u_1^e + u_2^e}{2} + a_{ux}^e \frac{u_2^e - u_1^e}{h_e} \Big) \begin{bmatrix} 1 & -1 \\ -1 & 1 \end{bmatrix} + c_0^e \frac{h_e}{6} \begin{bmatrix} 2 & 1 \\ 1 & 2 \end{bmatrix} \\
&\quad + c_u^e \frac{h_e}{12} \begin{bmatrix} 3u_1^e + u_2^e & u_1^e + u_2^e \\ u_1^e + u_2^e & u_1^e + 3u_2^e \end{bmatrix}, \quad 1 < e < N,
\end{aligned} \tag{5}
$$

The last (i.e., Nth) element coefficient matrix and source vector are given by

$$\mathbf{K}^{(N)} = \frac{1}{h_e}\left(a_0^e + a_u^e \frac{u_1^e + u_2^e}{2} + a_{ux}^e \frac{u_2^e - u_1^e}{h_e}\right)\begin{bmatrix} 1 & -1 \\ -1 & 1 \end{bmatrix} + c_0^e \frac{h_e}{6}\begin{bmatrix} 2 & 1 \\ 1 & 2 \end{bmatrix}$$
$$+ c_u^e \frac{h_e}{12}\begin{bmatrix} 3u_1^e + u_2^e & u_1^e + u_2^e \\ u_1^e + u_2^e & u_1^e + 3u_2^e \end{bmatrix} + \begin{bmatrix} 0 & 0 \\ 0 & \beta_b \end{bmatrix}, \tag{6}$$
$$\mathbf{f}^{(N)} = \begin{Bmatrix} f_1^{(N)} \\ f_2^{(N)} \end{Bmatrix} + \begin{Bmatrix} 0 \\ \beta_b u_\infty^b \end{Bmatrix}.$$

The assembly of element equations follows the same procedure as in the linear finite element analysis. If we denote the global nodal solution vector with $\mathbf{U}$, the assembled system of equations can be expressed as

$$\mathbf{K}(\mathbf{U})\mathbf{U} = \mathbf{F}(\mathbf{U}), \tag{6.2.7}$$

where $\mathbf{K}$ and $\mathbf{F}$ denote the global coefficient matrix and the source vector, respectively. In **Example 6.2.1**, $\mathbf{K}^e$ is a linear function of the nodal values u_i^e and it is a symmetric matrix. Consequently, the resulting global finite element equations are nonlinear in u_i^e; in the present case, the algebraic equations are quadratic in u_i^e (hence, in U_I). The right-hand side vector $\mathbf{F}$ depends on $\mathbf{U}$ only when the boundary conditions are nonlinear.

6.2.2.3 Linearization of nonlinear algebraic equations

Linearization is necessary in order to solve the final algebraic equations resulting from the application of any numerical method to the solution of differential equations. The final algebraic equations (which are nonlinear if the differential equations are nonlinear) obtained with the FEM, FVM, or DMCDM have the form

$$\mathbf{K}(\mathbf{U})\mathbf{U} = \mathbf{F}(\mathbf{U}), \tag{6.2.8}$$

where $\mathbf{K}$ is the coefficient matrix (known in terms of $\mathbf{U}$), $\mathbf{U}$ is the column vector of nodal values of u, and $\mathbf{F}$ is the source vector, which contains the contribution due to the source term $f(x)$ as well as the secondary variables at the boundary nodes. After the imposition of boundary conditions on the elements of $\mathbf{U}$ and $\mathbf{F}$, Eq. (6.2.8) is solved using a successive approximation scheme known as the Picard (or direct) iteration method, which was explained in Section 1.9.4.2.

To summarize the Picard iteration scheme, suppose that we arrived at the end of the rth iteration (i.e., $\mathbf{U}^r$ is known). Then we seek the $(r+1)$st iteration solution $\mathbf{U}^{r+1}$ by solving the algebraic equations

$$\mathbf{K}(\mathbf{U}^r)\mathbf{U}^{r+1} = \mathbf{F}(\mathbf{U}^r) \;\; \rightarrow \;\; \mathbf{U}^{r+1} = (\mathbf{K}^r)^{-1}\mathbf{F}^r, \tag{6.2.9}$$

where $\mathbf{K}^r \equiv \mathbf{K}(\mathbf{U}^r)$ and $\mathbf{F}^r \equiv \mathbf{F}(\mathbf{U}^r)$ are evaluated using known solution $\mathbf{U}^r$ from the rth iteration, which is called a linearization of $\mathbf{K}$ and $\mathbf{F}$. The iteration is continued until the normalized difference between two consecutive solutions, measured with a suitable norm, is within a prescribed tolerance:

$$\sqrt{\frac{\delta\mathbf{U}\cdot\delta\mathbf{U}}{\mathbf{U}^r\cdot\mathbf{U}^r}} \leq \epsilon, \quad \delta\mathbf{U} \equiv \mathbf{U}^{r+1} - \mathbf{U}^r, \tag{6.2.10}$$

where ϵ denotes the value of the error tolerance (say, e.g., $\epsilon = 10^{-3}$). In the beginning of the iteration, one must have a starting guess vector $\mathbf{U}^0$; in the case of problems with a linear term in $\mathbf{K}$, we can take the initial guess vector to be zero so that the first iteration solution corresponds to the linear solution (desirable to know). In the subsequent iterations, the nonlinear terms begin to contribute towards the final solution. These ideas are equally valid for the FVM and DMCDM. When $\mathbf{K}$ is nonlinear without a linear part, one must give a nonzero guess vector based on a qualitative understanding of the solution (which is a challenge).

6.2.3 Dual Mesh Control Domain Method

6.2.3.1 Primal and dual meshes

We recall that in the DMCDM, the domain $(0, L)$ is first divided into a set of N finite elements, $\Omega_f^{(1)}$, $\Omega_f^{(2)}$, ..., $\Omega_f^{(N)}$, of lengths h_1, h_2,..., h_N, respectively. Then a dual mesh of $N+1$ control domains, $\Omega_c^{(1)}$, $\Omega_c^{(2)}$, ..., $\Omega_c^{(N+1)}$, is identified such that the nodes of the finite elements are at the center of each control domain, except for the control domains at node 1 and node $N+1$ (i.e., boundary nodes), where the node is on one side of the control domain, as illustrated in Fig. 6.2.1.

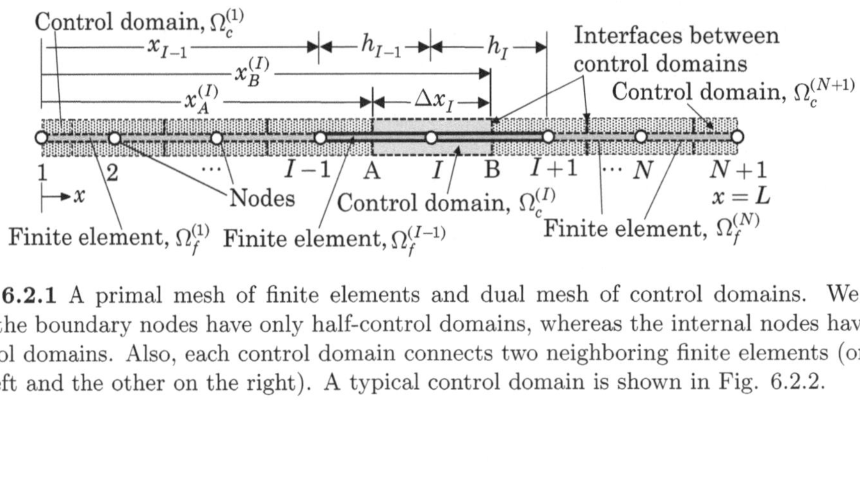

Fig. 6.2.1 A primal mesh of finite elements and dual mesh of control domains. We note that the boundary nodes have only half-control domains, whereas the internal nodes have full control domains. Also, each control domain connects two neighboring finite elements (one on the left and the other on the right). A typical control domain is shown in Fig. 6.2.2.

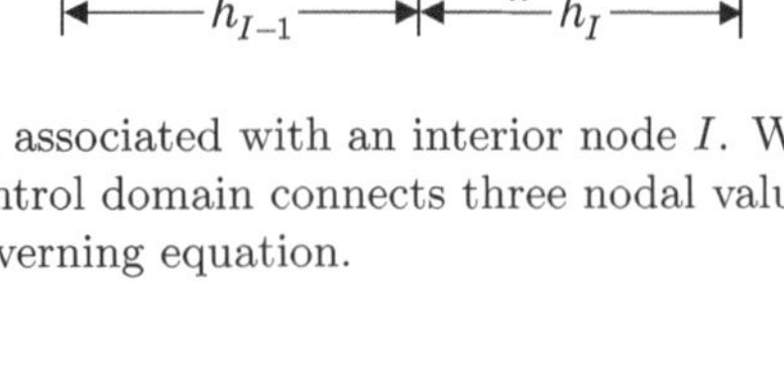

Fig. 6.2.2 Control domain associated with an interior node I. We note that each node has a single unknown and the control domain connects three nodal values (U_{I-1}, U_I, U_{I+1}) through the discretization of the governing equation.

6.2.3.2 Integral statement over a control domain

The algebraic equations among the nodal values of a typical control domain are obtained by satisfying the governing equation (6.2.1) in an integral sense (with all terms put on one side of the equation, which amounts to reducing the integral of the residual due to the approximation to zero) over a typical control domain shown in Fig. 6.2.2:

$$0 = \int_{x_A^{(I)}}^{x_B^{(I)}} \left\{ -\frac{d}{dx}\left[a(x,u)\frac{du}{dx} \right] + b(x,u)\frac{du}{dx} + c(x,u)u - f \right\} dx. \tag{6.2.11}$$

The first term (which has two derivatives) is integrated once to reduce the differentiability required of the approximation functions and obtain

$$\begin{aligned} 0 &= \int_{x_A^{(I)}}^{x_B^{(I)}} \left[b(x,u)\frac{du}{dx} + c(x,u)u - f \right] dx \\ &\quad - \left[-a(x,u)\frac{du}{dx} \right]_{x_A^{(I)}} - \left[a(x,u)\frac{du}{dx} \right]_{x_B^{(I)}} \\ &= -N_1^{(I)} - N_2^{(I)} + \int_{x_A^{(I)}}^{x_B^{(I)}} \left[b(x,u)\frac{du}{dx} + c(x,u)u - f \right] dx, \end{aligned} \tag{6.2.12}$$

where (see Fig. 6.2.3)

$$N_1^{(I)} \equiv \left[-a(x,u)\frac{du}{dx} \right]_{x_A^{(I)}}, \quad N_2^{(I)} \equiv \left[a(x,u)\frac{du}{dx} \right]_{x_B^{(I)}}. \tag{6.2.13}$$

Here $N_1^{(I)}$ and $N_2^{(I)}$ denote the *secondary variables* at the left and right interfaces of the control domain centered at node I, as shown in Fig. 6.2.3. Physically, the secondary variables denote the axial forces or heats, when the model equation is one that describes axial deformation of bars or one-dimensional heat flow, respectively.

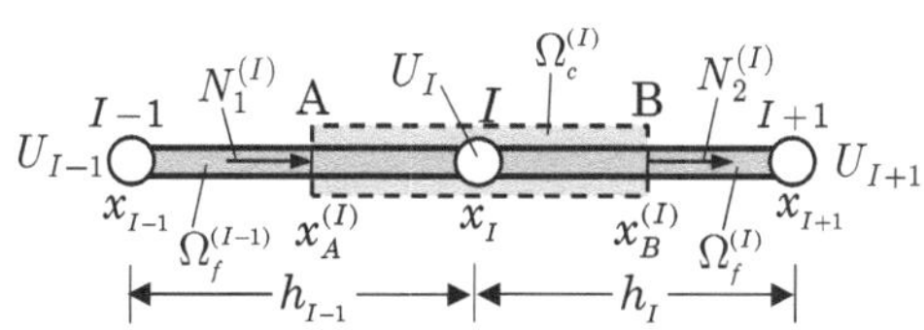

Fig. 6.2.3 A typical control domain $\Omega_c^{(I)}$ for the 1-D model. Points A and B refer to the left and right end locations, respectively, of the control domain associated with node I and have the coordinates $x_A^{(I)}$ and $x_B^{(I)}$, respectively. We note that point A is in element $\Omega_f^{(I-1)}$ and point B is in element $\Omega_f^{(I)}$.

The identification of the secondary variables is a significant feature of the present approach, and it allows handling of the boundary conditions on the secondary variables in a physically meaningful way, without replacing them in terms of the nodal values of u.

6.2.3.3 Discretized equations

To derive the discretized equations, we invoke the approximations of u over a typical finite element $\Omega_f^{(I)} = (x_I, x_{I+1})$. For example, the linear finite element approximation of $u(x)$ over $\Omega_f^{(I)}$ is

$$u(\bar{x}) \approx U_I \psi_1^{(I)}(\bar{x}) + U_{I+1}\psi_2^{(I)}(\bar{x}), \tag{6.2.14a}$$

where U_I is the value of u at node I (i.e., $U_I \approx u(x_I)$) and $\psi_i^{(I)}(\bar{x})$ $(i = 1, 2)$ are linear finite element interpolation functions of element $\Omega^{(I)}$ for $I = 1, 2, \ldots, N$ (see Fig. 6.2.4), expressed in terms of the local coordinate $\bar{x}$ (which has its origin at the left node of each finite element; see Fig. 6.2.4, which is the same as Fig. 5.3.4):

$$\psi_1^{(I)}(\bar{x}) = 1 - \frac{\bar{x}}{h_I}, \quad \psi_2^{(I)}(x) = \frac{\bar{x}}{h_I}. \tag{6.2.14b}$$

We can now express $(N_1^{(I)}, N_2^{(I)})$ in Eq. (6.2.13) for an interior node I in terms of the nodal values (U_{I-1}, U_I, U_{I+1}) of u using the finite element approximation in Eq. (6.2.14a), while linearizing the nonlinear terms.

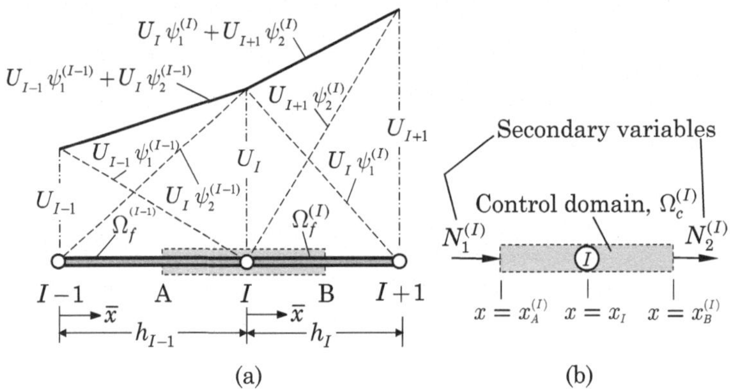

Fig. 6.2.4 (a) Linear finite element approximation over the domains of two neighboring finite elements. (a) Secondary variables on a typical control domain.

As examples of linearization, terms of the form u^2 and $(du/dx)^2$ in coefficients $a(x, u)$, $b(x, u)$, or $c(x, u)$ can be linearized as

$$u^2(x) \approx [u^{(r)}(x)]u(x); \quad \left(\frac{du}{dx}\right)^2 \approx \left[\left(\frac{du}{dx}\right)\right]^{(r)} \frac{du}{dx},$$

where the terms in the square brackets are evaluated using the known solution from the rth (previous) iteration (or a weighted-sum of the previous two iterations to accelerate convergence, which is termed a *relaxation procedure*):

$$\bar{U}_I = (1-\gamma)\,U_I^r + \gamma\,U_I^{r-1}, \quad 0 \le \gamma \le 1. \tag{6.2.15}$$

Here γ is known as the acceleration parameter. Often, the value of γ is assumed to be between 0.0 and 0.5.

Returning to the DMCDM discretization of Eq. (6.2.1) inside the domain, we first express $(N_1^{(I)}, N_2^{(I)})$ in Eq. (6.2.13) in terms of the nodal values of u:

$$N_1^{(I)} = -a_1^{(I)} \frac{U_I - U_{I-1}}{h_{I-1}}, \quad N_2^{(I)} = a_2^{(I)} \frac{U_{I+1} - U_I}{h_I}, \tag{6.2.16}$$

where $a_1^{(I)} = a(x_A^{(I)})$ at the left interface and $a_2^{(I)} = a(x_B^{(I)})$ at the right interface of the control domain centered around node I. We assume that the coefficient a is of the form

$$a(x, u, du/dx) = a_x + a_u \cdot u + a_{u2} \cdot u^2 + a_{du} \cdot \frac{du}{dx} + a_{du2} \cdot \left(\frac{du}{dx}\right)^2 \tag{6.2.17}$$

and the coefficients a_x, a_u, a_{u2}, and so on can be functions of x. Then, we have the following linearized expression in a discretization scheme (where the capital letters U_I denote the discretized values)

$$\begin{aligned} a_1^{(I)} &= a_x(x_A^{(I)}) + 0.5a_u(x_A^{(I)}) \left(\bar{U}_{I-1} + \bar{U}_I\right) + 0.25a_{u2}(x_A^{(I)}) \left(\bar{U}_{I-1} + \bar{U}_I\right)^2 \\ &\quad + a_{du}(x_A^{(I)}) \left(\frac{\bar{U}_I - \bar{U}_{I-1}}{h_{I-1}}\right) + a_{du2}(x_A^{(I)}) \left(\frac{\bar{U}_I - \bar{U}_{I-1}}{h_{I-1}}\right)^2, \end{aligned} \tag{6.2.18a}$$

$$\begin{aligned} a_2^{(I)} &= a_x(x_B^{(I)}) + 0.5a_u(x_B^{(I)}) \left(\bar{U}_{I+1} + \bar{U}_I\right) + 0.25a_{u2}(x_B^{(I)}) \left(\bar{U}_{I+1} + \bar{U}_I\right)^2 \\ &\quad + a_{du}(x_B^{(I)}) \left(\frac{\bar{U}_{I+1} - \bar{U}_I}{h_I}\right) + a_{du2}(x_B^{(I)}) \left(\frac{\bar{U}_{I+1} - \bar{U}_I}{h_I}\right)^2, \end{aligned} \tag{6.2.18b}$$

where $\bar{U}_I$ is the solution known from the previous iteration or a weighted average of the last two iterations, as given in Eq. (6.2.15).

The integral expression in Eq. (6.2.12) can be evaluated using a numerical integration scheme such as Simpson's one-third rule [see Eq. (1.8.4c)]. We assume the coefficients b and c to be of the same form as the coefficient a in Eq. (6.2.17). Then we have

$$\begin{aligned} \int_{x_A^{(I)}}^{x_B^{(I)}} \left(b\frac{du}{dx} + c\,u\right) dx &= \int_{0.5h_{I-1}}^{h_{I-1}} \left(b\frac{du}{dx} + c\,u\right) d\bar{x} + \int_0^{0.5h_I} \left(b\frac{du}{dx} + c\,u\right) d\bar{x} \\ &= \int_{0.5h_{I-1}}^{h_{I-1}} \left[\left(b\frac{d\psi_1^{(I-1)}}{d\bar{x}} + c\psi_1^{(I-1)}\right) U_{I-1} + \left(b\frac{d\psi_2^{(I-1)}}{d\bar{x}} + c\psi_2^{(I-1)}\right) U_I\right] d\bar{x} \\ &\quad + \int_0^{0.5h_I} \left[\left(b\frac{d\psi_1^{(I)}}{d\bar{x}} + c\psi_1^{(I)}\right) U_I + \left(b\frac{d\psi_2^{(I)}}{d\bar{x}} + c\psi_2^{(I)}\right) U_{I+1}\right] d\bar{x} \\ &\equiv C_{I-1}U_{I-1} + C_I U_I + C_{I+1}U_{I+1}, \end{aligned} \tag{6.2.19}$$

where

$$C_{I-1} = \int_{0.5h_{I-1}}^{h_{I-1}} \left(b\,\frac{d\psi_1^{(I-1)}}{d\bar{x}} + c\psi_1^{(I-1)} \right) d\bar{x}, \tag{6.2.20a}$$

$$C_I = \int_{0.5h_{I-1}}^{h_{I-1}} \left(b\,\frac{d\psi_2^{(I-1)}}{d\bar{x}} + c\psi_2^{(I-1)} \right) d\bar{x} + \int_0^{0.5h_I} \left(b\,\frac{d\psi_1^{(I)}}{d\bar{x}} + c\psi_1^{(I)} \right) d\bar{x}, \tag{6.2.20b}$$

$$C_{I+1} = \int_0^{0.5h_I} \left(b\,\frac{d\psi_2^{(I)}}{d\bar{x}} + c\psi_2^{(I)} \right) d\bar{x}. \tag{6.2.20c}$$

The integrals in Eqs. (6.2.20a)–(6.2.20c) are evaluated using Simpson's one-third rule. For example, we have

$$\int_{0.5h_{I-1}}^{h_{I-1}} f(\bar{x})\, d\bar{x} = \frac{0.25h_{I-1}}{3}\left(f_1 + 4f_2 + f_3\right) = \frac{h_{I-1}}{12}\left(f_1 + 4f_2 + f_3\right), \tag{6.2.21a}$$

where

$$f_1 = f(0.5h_{I-1}), \quad f_2 = f(0.75h_{I-1}), \quad f_3 = f(h_{I-1}). \tag{6.2.21b}$$

With relations (6.2.16)–(6.2.21b) in hand, we can now write the discretized equations for the Ith control domain as

$$A_{I-1}U_{I-1} + A_I U_I + A_{I+1}U_{I+1} = F_I \tag{6.2.22a}$$

for $I = 2, 3, \ldots, N$. The coefficients A_{I-1}, A_I, and A_{I+1} are given by

$$A_{I-1} = C_{I-1} - \frac{a_1^{(I)}}{h_{I-1}}, \quad A_I = C_I + \frac{a_1^{(I)}}{h_{I-1}} + \frac{a_2^{(I)}}{h_I}, \quad A_{I+1} = C_{I+1} - \frac{a_2^{(I)}}{h_I}, \tag{6.2.22b}$$

where $a_1^{(I)}$ and $a_2^{(I)}$ were defined in Eqs. (6.2.18a) and (6.2.18b), and F_I is

$$F_I = \int_{x_A^{(I)}}^{x_B^{(I)}} f(x)dx. \tag{6.2.22c}$$

Next, we should obtain the discretized equation for the boundary nodes 1 and $N+1$ (when there are N linear elements in the primal mesh). We note that at node 1, $N_1^{(1)}$ is the dual variable which has the meaning, for example, of being the axial force or heat, which is known or its dual, U_1, is known. Hence, we only evaluate $N_2^{(1)}$ at $h_1/2$. For node 1 [see Fig 6.2.5(a)], we have $x_a^{(1)} = 0$, and Eq. (6.2.22a) takes the form

$$-N_1^{(1)} + A_1U_1 + A_2U_2 - F_1 = 0, \tag{6.2.23a}$$

where

$$\bar{A}_1 = \bar{C}_1 + \frac{a_1^{(1)}}{h_1}, \quad \bar{A}_2 = \bar{C}_2 - \frac{a_2^{(1)}}{h_1}, \quad F_1 = \int_0^{0.5h_1} f(x)\, dx$$
$$\bar{C}_1 = \int_0^{0.5h_1} \left(b\frac{d\psi_1^{(1)}}{dx} + c\,\psi_1^{(1)} \right) d\bar{x}, \quad \bar{C}_2 = \int_0^{0.5h_1} \left(b\frac{d\psi_2^{(1)}}{dx} + c\,\psi_2^{(1)} \right) d\bar{x}. \tag{6.2.23b}$$

For the boundary point at node $N+1$, considering the $N+1$st control domain (with length $0.5h_N$), Eq. (6.2.22a) takes the form

$$-N_2^{(N+1)} + A_N U_N + A_{N+1} U_{N+1} - F_N = 0, \tag{6.2.24a}$$

where

$$\bar{A}_N = \bar{C}_N - \frac{a_1^{(N+1)}}{h_N}, \quad \bar{A}_{N+1} = \bar{C}_{N+1} + \frac{a_2^{(N+1)}}{h_N}, \quad F_1 = \int_{0.5h_N}^{h_N} f\, d\bar{x}$$
$$\bar{C}_N = \int_{0.5h_N}^{h_N} \left(b\frac{d\psi_1^{(N)}}{dx} + c\,\psi_1^{(N)} \right) d\bar{x}, \quad \bar{C}_{N+1} = \int_{0.5h_N}^{h_N} \left(b\frac{d\psi_2^{(N)}}{dx} + c\,\psi_2^{(N)} \right) d\bar{x}. \tag{6.2.24b}$$

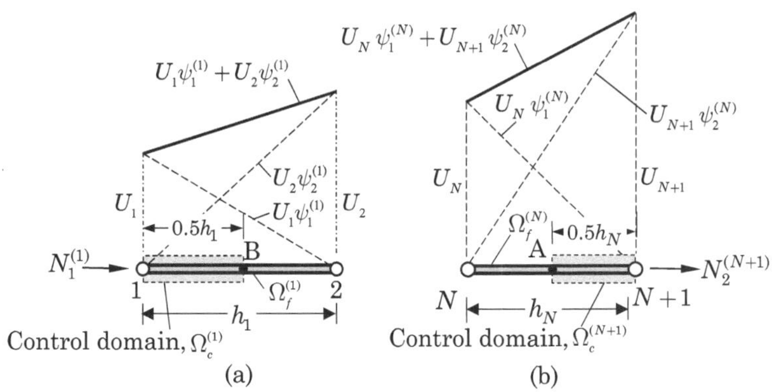

Fig. 6.2.5 (a) Boundary at node 1. (b) Boundary at node $N+1$.

6.2.4 Numerical Examples

In this section, we consider example problems that utilize the methodology developed in the preceding sections. A computer program based on the developments of the previous sections is used to solve these problems. Numerical results obtained with the FEM and DMCDM are compared in all cases.

Example 6.2.2

Consider a nonlinear problem described by the following equations:

$$-\frac{d^2u}{dx^2} + 2u^3 = 0, \quad 1 < x < 2; \quad u(1) = 1, \qquad \left[\frac{du}{dx} + u^2\right]_{x=2} = 0 \tag{6.2.25}$$

Determine the solution $u(x)$ of Eq. (6.2.25) using the FEM and DMCDM.

Solution: Comparing with the model equation in Eq. (6.2.1), we have $a = 1$, $b = 0$, and $c = 2u^2$. The boundary condition at $x = 2$ is a mixed nonlinear boundary condition with $\beta = \beta_0 + \beta_u u$ ($\beta_0 = 0$, $\beta_u = 1$) and $u_\infty = 0$ [see Eq. (6.2.2)]. The problem becomes linear when c and β are zero, and the solution of the linear problem is $u(x) = 1$.

The initial guess vector of $(0.0, 0.0, \ldots, 0.0)$ takes 15 iterations to converge within a convergence tolerance of $\epsilon = 10^{-3}$ and acceleration parameter $\gamma = 0$ (even though we took $U_1^0 = 0.0$, it is replaced by the boundary condition $U_1 = 1.0$ at the beginning of each iteration). A different guess vector and acceleration parameter may take more or less iterations, or may not even converge. For example, a guess vector of $(1.0, 0.5, \ldots, 0.5)$ takes 11 iterations to converge within a convergence tolerance of $\epsilon = 10^{-3}$.

For a mesh of four linear elements, the (linear) coefficient matrix at the beginning of the iteration process, with the initial guess vector of $(0.0, 0.0, \ldots, 0.0)$ is (i.e., $r = 1$)

```
 0.40000E+01 -0.40000E+01  0.00000E+00  0.00000E+00  0.00000E+00
-0.40000E+01  0.80000E+01 -0.40000E+01  0.00000E+00  0.00000E+00
 0.00000E+00 -0.40000E+01  0.80000E+01 -0.40000E+01  0.00000E+00
 0.00000E+00  0.00000E+00 -0.40000E+01  0.80000E+01 -0.40000E+01
 0.00000E+00  0.00000E+00  0.00000E+00 -0.40000E+01  0.40000E+01
```

The linear solution obtained at the nodes is {1.0000 1.0000 1.0000 1.0000 1.0000}. The coefficient matrix for the second iteration (with the linear solution as the guess vector) is

```
 0.41875E+01 -0.39375E+01  0.00000E+00  0.00000E+00  0.00000E+00
-0.39375E+01  0.83750E+01 -0.39375E+01  0.00000E+00  0.00000E+00
 0.00000E+00 -0.39375E+01  0.83750E+01 -0.39375E+01  0.00000E+00
 0.00000E+00  0.00000E+00 -0.39375E+01  0.83750E+01 -0.39375E+01
 0.00000E+00  0.00000E+00  0.00000E+00 -0.39375E+01  0.41875E+01
```

The solution at the end of iteration 2 is {1.00000 0.70890 0.50782 0.37123 0.28178} with an error of 0.7913. At the end of 15 iterations, the error is 0.6429×10^{-3} and the solution $u_h(x)$ at the nodes is {1.0000 0.8001 0.6669 0.5719 0.5006}.

The results obtained with the FEM and DMCDM (both using direct iteration) for three different meshes of four, eight, and 16 elements are compared with the exact solution $u(x) = 1/x$ in Table 6.2.1. In all cases, the number of iterations taken was 15 (with zero guess vector, tolerance $\epsilon = 10^{-3}$, and acceleration parameter $\gamma = 0$). The solution $u(x)$ and flux $q(x) = -du/dx$ predicted by the DMCDM agrees very well with FEM and the exact solutions.

Table 6.2.1 Comparison of the FEM and DMCDM solutions $u(x)$ with the exact solution, $u(x) = 1/x$, of the problem described by Eq. (6.2.25).

	4 Elements		8 Elements		16 Elements		
x	FEM	DMCDM	FEM	DMCDM	FEM	DMCDM	Exact
1.125	—	—	0.8887	0.8889	0.8889	0.8889	0.8889
1.250	0.7987	0.8001	0.7998	0.8001	0.8000	0.8001	0.8000
1.375	—	—	0.7271	0.7274	0.7273	0.7274	0.7273
1.500	0.6653	0.6669	0.6665	0.6669	0.6667	0.6668	0.6667
1.625	—	—	0.6153	0.6156	0.6155	0.6156	0.6154
1.750	0.5703	0.5719	0.5713	0.5717	0.5716	0.5717	0.5714
1.875	—	—	0.5333	0.5336	0.5335	0.5336	0.5333
2.000	0.4991	0.5006	0.5000	0.5003	0.5002	0.5003	0.5000

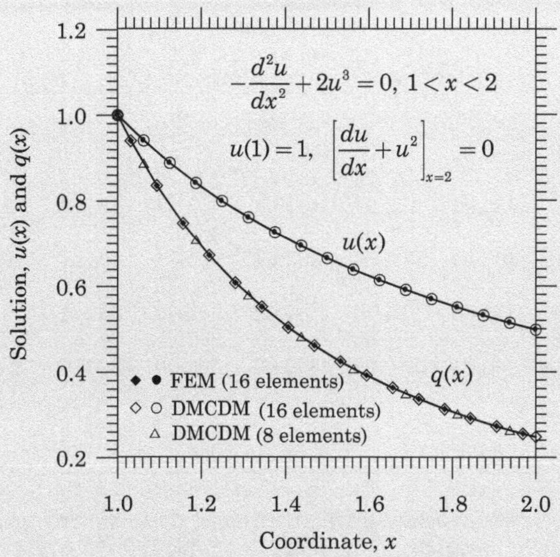

Fig. 6.2.6 Graphical comparison of the FEM and DMCDM solutions with the exact solutions $u(x)$ and $q(x)$ as functions of x.

Example 6.2.3

Next, we consider heat transfer in an isotropic bar of length $L = 0.18$ m whose left end is maintained at a temperature of 500 K and the right end at 300 K. There is no internal heat generation ($f = 0$) and the surface of the bar is insulated so that there is no convection from the surface. The governing differential equation and boundary conditions of the problem are

$$-\frac{d}{dx}\left[k(T)\,\frac{dT}{dx}\right] = 0, \quad 0 < x < L,$$
$$T(0) = 500\,\mathrm{K}, \quad T(L) = 300\,\mathrm{K}, \tag{6.2.26}$$

where the conductivity $k(T)$ varies according to the relation

$$k(T) = k_0\,(1 + k_1\,\Delta T)\,, \quad \Delta T = T - T_0. \tag{6.2.27}$$

Here k_0 is the thermal conductivity [$k_0 = 0.2$ W/(m K)], k_1 is the temperature coefficient of thermal conductivity [$k_1 = 2 \times 10^{-3}$ (K^{-1})], and $T_0 = 0$ (K) is the reference temperature. Determine the numerical solution using the FEM, FVM, and DMCDM.

Solution: This problem does have a linear solution when we set $k_1 = 0$; the linear solution $T(x)$ is a linear variation of the temperature between 500 K at one end to 300 K at the other end. Therefore, a zero guess vector would yield the linear numerical solution (which coincides with the exact solution) after the first iteration. The convergence is reached after three iterations (with $\epsilon = 10^{-3}$ and $\gamma = 0$) for this problem.

Table 6.2.2 contains nonlinear solutions $T(x)$ obtained with the DMCDM, FEM, and FVM for different values of x and two different meshes. Clearly, all three methods for this case give almost the same solution, with FEM and DMCDM being closer to each other.

Table 6.2.2 FEM and DMCDM solutions (temperature, $T(x)$ in K) of a nonlinear heat conduction problem described by Eqs. (6.2.26) and (6.2.27).

	DMCDM		FEM		FVM	
x	4L	8L	4L	8L	4L	8L
0.0000	500.00	500.00	500.00	500.00	500.00	500.00
0.0225	——	477.24	——	477.24	——	477.24
0.0450	453.94	453.94	453.94	453.94	453.95	453.94
0.0675	——	430.06	——	430.06	——	430.06
0.0900	405.54	405.54	405.54	405.54	405.56	405.55
0.1125	——	380.35	——	380.35	——	380.35
0.1350	354.40	354.40	354.40	354.40	354.42	354.41
0.1575	——	327.65	——	327.65	——	327.65
0.1800	300.00	300.00	300.00	300.00	300.00	300.00

Example 6.2.4

Consider the large-deformation analysis of a bar of length L, uniform cross-sectional area A, and made of an isotropic linear elastic material (with modulus E). Suppose that the bar is fixed at the left end and subjected to axial load P at the right end. As discussed in the book by Reddy [13], the governing equation is (the governing equation is obtained using the principle of virtual displacements, which is the same as the weak form)

$$-\frac{d}{dx}\left[a(x,u)\frac{du}{dx}\right] = 0, \quad 0 < x < L; \qquad u(0) = 0, \quad \left[a\frac{du}{dx}\right]_{x=L} = P, \tag{6.2.28}$$

where u is the axial displacement, P is the point load at the right end (i.e., $x = L$) of the bar, and

$$a(x,u) = EA\Big(1 + \frac{du}{dx}\Big)\Big(1 + \tfrac{1}{2}\frac{du}{dx}\Big) = EA\Big[1 + 1.5\frac{du}{dx} + 0.5\Big(\frac{du}{dx}\Big)^2\Big]. \tag{6.2.29}$$

Solution: For this problem, we have $a = a(x,u)$, $b = 0$, $c = 0$, and $f = 0$. This problem has a linear solution [i.e., $u(x) = Px/EA$] and, hence, the zero initial guess vector for the solution can be used. The axial stiffness EA (assumed to be a constant) is taken as unity in the analysis (amounts to normalizing the solution with EA).

The problem is different from those considered in the previous examples in the sense that the present problem is to be solved for different values of the load P. The total load is divided into 25 load steps of magnitude $\Delta P = 0.2$ to achieve convergence. No load step took more than seven iterations (with $\epsilon = 10^{-3}$) to yield a converged solution. The converged solution from the previous load step is used as the guess vector for the next load step. For example, the converged nonlinear solution using the DMCDM as well as the FEM (when a eight-element primal mesh is used) for $P = 0.2$ is {0.0000 0.0120 0.0399 0.0599 0.0798 0.0998 0.1198 0.1397 0.1597}; for $P = 1.0$ (after five load steps) is {0.0000 0.0652 0.1303 0.1955 0.2606 0.3258 0.3910 0.4561 0.5213}; and for $P = 5$ is {0.0000 0.1636 0.3272 0.4908 0.6544 0.8179 0.9815 1.1451 1.3087}. The nodal values obtained with the FEM and DMCDM agree with each other.

Figure 6.2.7 shows plots of $u(L)/L$ versus $\bar{P} = P/AE$. The figure also contains results for the case in which the right end of the bar is attached to a nonlinear spring, with spring constant equal to $k = k_0 + k_u\, u$. Thus, the free end boundary condition becomes

$$\left[a\frac{du}{dx} + k\,u\right]_{x=L} = P. \tag{6.2.30}$$

In obtaining the numerical results, we have used two cases: (a) $k_0 = 1$ and $k_u = -1$ (i.e., spring weakens as the bar deforms) and (b) $k_0 = 1$ and $k_u = 1$ (i.e., spring stiffens as it is compressed).

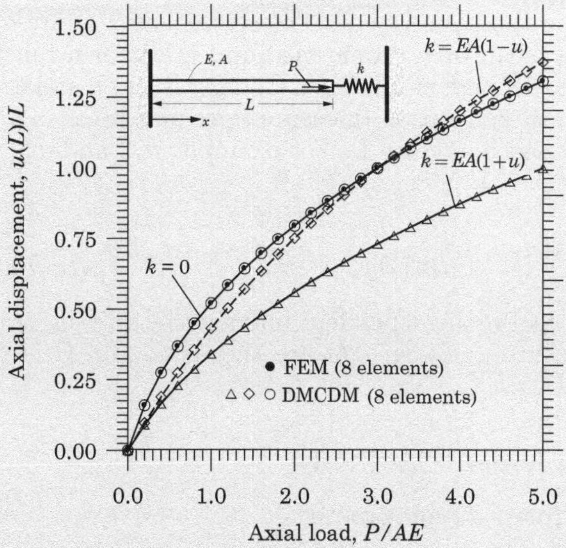

Fig. 6.2.7 Plots of $u(L)/L$ versus the load parameter P/AE. The numerical solutions predicted by FEM and DMCDM are in close agreement with each other and are indistinguishable from each other in the plots.

6.3 Two-Dimensional Problems

6.3.1 Model Differential Equation

In this section, we discuss the FEM and DMCDM models of two-dimensional problems governed by PDEs involving a single unknown. Extension of the ideas from the previous section for one-dimensional problems to two-dimensional problems is presented here.

Consider the problem of finding the solution $u(x, y)$ to the following second-order partial differential equation governing, for example, heat transfer in a solid medium:

$$-\frac{\partial}{\partial x}\left(a_{xx}\frac{\partial u}{\partial x}\right) - \frac{\partial}{\partial y}\left(a_{yy}\frac{\partial u}{\partial x}\right) = f(x, y) \quad \text{in } \Omega \tag{6.3.1}$$

subjected to the following types of boundary conditions:

$$u = \hat{u}(s) \quad \text{on} \quad \Gamma_u \tag{6.3.2a}$$

and

$$\left(a_{xx}\frac{\partial u}{\partial x}n_x + a_{yy}\frac{\partial u}{\partial y}n_y\right) + \beta(u - u_\infty) = \hat{q}_n(s) \quad \text{on} \quad \Gamma_q, \tag{6.3.2b}$$

where Γ_u and Γ_q denote disjoint portions of the total boundary Γ such that $\Gamma = \Gamma_u \cup \Gamma_q$, β is the heat transfer coefficient, u_∞ is the temperature of the medium surrounding the body, and (n_x, n_y) are the direction cosines of the unit normal vector $\hat{\mathbf{n}}$ on the boundary Γ. The coefficients a_{xx} are a_{yy}, which denote the conductivities of an orthotropic medium, are known functions of position (x, y) as well as the dependent unknown (temperature) u and its derivatives, and f is the internal heat generation (measured per unit area) in a two-dimensional domain Ω with closed boundary Γ. For example, a_{xx} and a_{yy} may be assumed to be of the form

$$a_{xx} = a_{xx}\left(x, y, u, \frac{\partial u}{\partial x}, \frac{\partial u}{\partial y}\right), \quad a_{yy} = a_{yy}\left(x, y, u, \frac{\partial u}{\partial x}, \frac{\partial u}{\partial y}\right) \tag{6.3.3}$$

In Eq. (6.3.1), $u(x, y)$ is the dependent unknown (e.g., temperature above some reference temperature), (a_{xx}, a_{yy}, f) are the data, and Ω is the domain with closed boundary Γ.

6.3.2 Finite Element Method

6.3.2.1 Weak form development

In the finite element method, the domain $\bar{\Omega}$ is discretized into a mesh of finite elements Ω^e (with associated interpolation functions), as shown in Figs. 4.3.1 and 4.3.2:

$$\bar{\Omega} \approx \bar{\Omega}_h = \bigcup_{e=1}^{N} \bar{\Omega}^e, \quad \bar{\Omega} = \Omega \cup \Gamma, \quad \bar{\Omega}^e = \Omega^e \cup \Gamma^e, \tag{6.3.4}$$

where N is the total number of elements in the mesh. The residual due to the approximation of u by u_h^e over Ω^e is

$$\mathbb{R}^e(u_h^e) = -\frac{\partial}{\partial x}\left(a_{xx}^e \frac{\partial u_h^e}{\partial x}\right) - \frac{\partial}{\partial y}\left(a_{yy}^e \frac{\partial u_h^e}{\partial y}\right) - f^e(x, y) \tag{6.3.5}$$

We use a representative element Ω^e to derive the weak form of the model equation, Eq. (6.3.1). Following the three steps of Section 4.3.3, the first step is to multiply the residual $\mathbb{R}^e$ with the ith weight function $w_i(x, y)$, which is assumed to be differentiable once with respect to x and y, and then set the integral of the product $w_i\,\mathbb{R}^e$ over the element domain Ω^e to zero:

$$0 = \int_{\Omega^e} w_i \left[-\frac{\partial}{\partial x}\left(a_{xx}^e \frac{\partial u_h^e}{\partial x}\right) - \frac{\partial}{\partial y}\left(a_{yy}^e \frac{\partial u_h^e}{\partial y}\right) - f^e\right] dx\, dy. \tag{6.3.6}$$

In the second step, we distribute the differentiation among u_h and w_i equally in the first two terms of Eq. (6.3.6) using the Green–Gauss theorem:

$$\begin{aligned} 0 = &\int_{\Omega^e} \left[\frac{\partial w_i}{\partial x}\left(a_{xx}^e \frac{\partial u_h^e}{\partial x}\right) + \frac{\partial w_i}{\partial y}\left(a_{yy}^e \frac{\partial u_h^e}{\partial y}\right) - w_i f^e\right] dxdy \\ &- \oint_{\Gamma^e} w_i \left[\left(a_{xx}^e \frac{\partial u_h^e}{\partial x}\right) n_x + \left(a_{yy}^e \frac{\partial u_h^e}{\partial y}\right) n_y\right] ds, \end{aligned} \tag{6.3.7}$$

where n_x and n_y are the components (i.e., the direction cosines) of the unit normal vector $\hat{\mathbf{n}}$ on the boundary Γ^e, and ds is the arc length of an infinitesimal line element along the boundary. The circle on the boundary integral denotes integration over the closed boundary Γ^e. From an inspection of the boundary term in Eq. (6.3.7), we note that u_h^e is the primary variable. The coefficient of the weight function w_i in the boundary expression is

$$\left(a_{xx}^e \frac{\partial u_h^e}{\partial x}\right) n_x + \left(a_{yy}^e \frac{\partial u_h^e}{\partial y}\right) n_y \equiv q_n^e, \tag{6.3.8}$$

and it constitutes the secondary variable. Thus, the weak form of Eq. (6.3.1) is

$$\begin{aligned} 0 = &\int_{\Omega^e} \left[\frac{\partial w_i}{\partial x}\left(a_{xx}^e \frac{\partial u_h^e}{\partial x}\right) + \frac{\partial w_i}{\partial y}\left(a_{yy}^e \frac{\partial u_h^e}{\partial y}\right) - w_i f^e\right] dxdy \\ &- \oint_{\Gamma^e} w_i\, q_n^e\, ds. \end{aligned} \tag{6.3.9}$$

The function $q_n^e = q_n^e(s)$ denotes the outward flux normal to the boundary Γ^e as we move counterclockwise along the boundary. The secondary variable q_n^e is of physical interest in most problems. For example, in the case of the heat transfer through an anisotropic medium, q_n denotes the heat influx normal to the boundary of the element. The weak form in Eq. (6.3.9) will be the basis of the weak-form Galerkin finite element model, in which w_i will be replaced by the approximation functions used to represent u_h^e.

6.3.2.2 Finite element model

The weak form in Eq. (6.3.9) requires that the approximation chosen for u should be at least linear in both x and y so that every term in Eq. (6.3.9) has a nonzero contribution to the integral. Since the primary variable is just u, which must be made continuous between elements, the Lagrange family of interpolation functions is admissible. Hence, the approximation u_h of u is represented over a typical finite element Ω^e by the expression

$$u(x,y) \approx u_h^e(x,y) = \sum_{j=1}^{n} u_j^e \psi_j^e(x,y), \tag{6.3.10}$$

where u_j^e is the value of u_h^e at the jth node of the element, and $\psi_j^e(x,y)$ are the Lagrange interpolation functions derived in Sections 4.3.6 and 4.3.7, and they have the following properties:

$$\psi_i^e(x_j^e, y_j^e) = \delta_{ij}, \quad \sum_{j=1}^{n} \psi_j^e(x,y) = 1, \tag{6.3.11}$$

where (x_j^e, y_j^e) denote the global coordinates of the jth node of element Ω^e. In reality, ψ^e are derived only for the so-called master elements using a normalized local coordinate system (ξ, η).

Substituting the finite element approximation in Eq. (6.3.10) for u_h^e into the weak form, Eq. (6.3.9), we obtain

$$
\begin{aligned}
0 = \sum_{j=1}^{n} u_j^e \int_{\Omega^e} \left[\frac{\partial w_i}{\partial x} \left(a_{xx}^e \frac{\partial \psi_j^e}{\partial x} \right) + \frac{\partial w_i}{\partial y} \left(a_{yy}^e \frac{\partial \psi_j^e}{\partial y} \right) \right] dxdy \\
- \int_{\Omega^e} w_i f^e \, dxdy - \oint_{\Gamma^e} w_i \, q_n^e \, ds.
\end{aligned} \tag{6.3.12}
$$

For the weak-form Galerkin model, we replace the weight function w_i with ψ_i^e and obtain

$$
\sum_{j=1}^{n} K_{ij}^e u_j^e - f_i^e - Q_i^e = 0 \quad \text{or} \quad \mathbf{K}^e \mathbf{u}^e = \mathbf{f}^e + \mathbf{Q}^e, \tag{6.3.13}
$$

where

$$
K_{ij}^e = \int_{\Omega^e} \left(a_{xx}^e \frac{\partial \psi_i^e}{\partial x} \frac{\partial \psi_j^e}{\partial x} + a_{yy}^e \frac{\partial \psi_i^e}{\partial y} \frac{\partial \psi_j^e}{\partial y} \right) dxdy, \tag{6.3.14a}
$$

$$
f_i^e = \int_{\Omega^e} \psi_i^e f^e \, dx \, dy, \quad Q^e = \oint_{\Gamma^e} \psi_i^e q_n^e \, ds. \tag{6.3.14b}
$$

Note that $K_{ij}^e = K_{ji}^e$ (i.e., $\mathbf{K}^e$ is symmetric). Equation (6.3.13) represents a set of n nonlinear algebraic equations.

This completes the finite element model development for a second-order nonlinear equation in two dimensions. The usual tasks of assembly of element equations, imposition of boundary conditions, and solution of linear algebraic equations (after linearization and an iterative method is applied) are standard, which were discussed in Chapter 4. Therefore, they are not discussed here again.

6.3.3 Dual Mesh Control Domain Formulation

6.3.3.1 Primal and dual meshes

Extension of the DMCDM ideas presented for one-dimensional problems in Section 6.2.3 to two-dimensional problems governed by a nonlinear partial differential equation is discussed here. That is, if the domains of the two-dimensional problems are of rectangular geometry, we can follow the same procedure as in one-dimensional problems to derive the algebraic equations of all nodes in the mesh. The primal mesh is one of bilinear finite elements, and the dual mesh is one of rectangular control domains, as shown in Fig. 6.3.1.

Here we consider a rectangular domain discretized by an $N \times M$ mesh of bilinear finite elements (i.e., N elements along the x-axis and M elements along the y-axis), as shown in Fig. 6.3.2. The dual mesh of rectangles, called control domains, are placed such that the control domain bisects four rectangular elements of the primal mesh in the interior and bisects two or one element on the boundary. The discussion presented in Section 5.4 applies here with the understanding that the coefficients a_{xx} and a_{yy} are functions of the unknown u and possibly its first derivatives with respect to x and y.

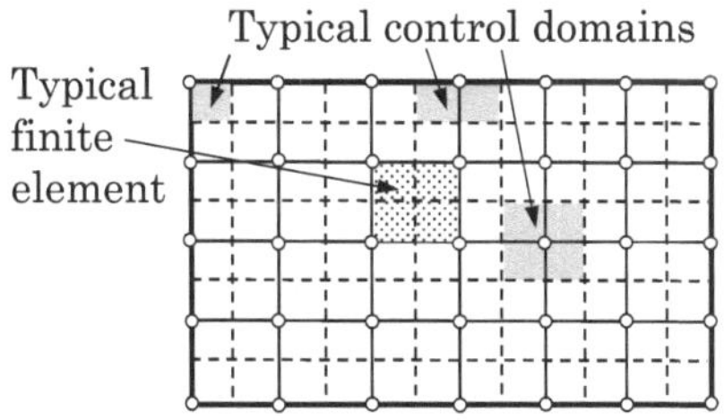

Fig. 6.3.1 Two-dimensional domain with a primal mesh of bilinear rectangular elements and a dual mesh of rectangular control domains that bisect four elements in the interior and occupy two or single finite elements on the boundary.

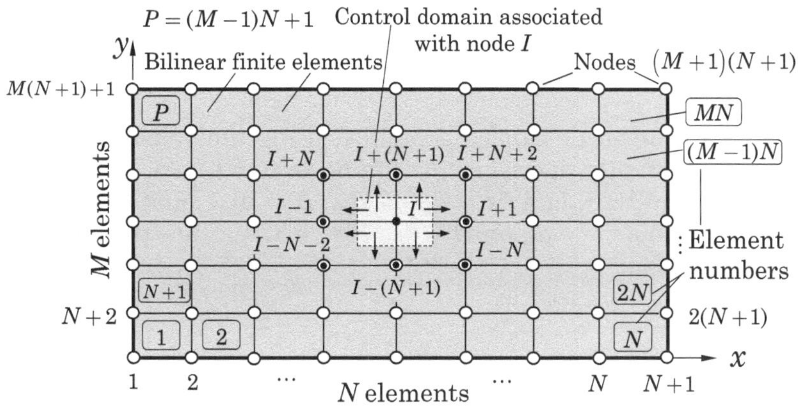

Fig. 6.3.2 Rectangular domain with a $N \times M$ primal mesh of bilinear rectangular elements and a dual mesh of rectangular control domains with the element and node numbering schemes.

6.3.3.2 Integral statements over a control domain

The integral statement of Eq. (6.3.1) over a typical interior rectangular control domain (see Fig. 6.3.3) is developed as follows [we note that the local coordinate system $(\bar{x}, \bar{y})$ used for each element is a translation of the global coordinate system (x, y)]:

$$
\begin{aligned}
0 &= -\int_{x_I - 0.5a_{I-1}}^{x_I + 0.5a_I} \int_{y_I - 0.5b_{I-1}}^{y_I + 0.5b_I} \left\{ \left[\frac{\partial}{\partial x}\left(a_{xx} \frac{\partial u}{\partial x} \right) + \frac{\partial}{\partial y}\left(a_{yy} \frac{\partial u}{\partial y} \right) \right] - f \right\} dxdy \\
&= -\oint_{\Gamma_R} \left[\left(a_{xx} \frac{\partial u}{\partial x} \right) n_x + \left(a_{yy} \frac{\partial u}{\partial y} \right) n_y \right] ds - \int_{x_I - 0.5a_{I-1}}^{x_I + 0.5a_I} \int_{y_I - 0.5b_{I-1}}^{y_I + 0.5b_I} f \, dxdy,
\end{aligned}
\tag{6.3.15}
$$

where (x_I, y_I) are the global coordinates of the node labelled as I, (n_x, n_y) are the direction cosines of the unit normal vector (see Fig. 6.3.3), and Γ_R is the boundary of the rectangular control domain. The boundary integration is taken all around the boundary of the control domain in the direction indicated.

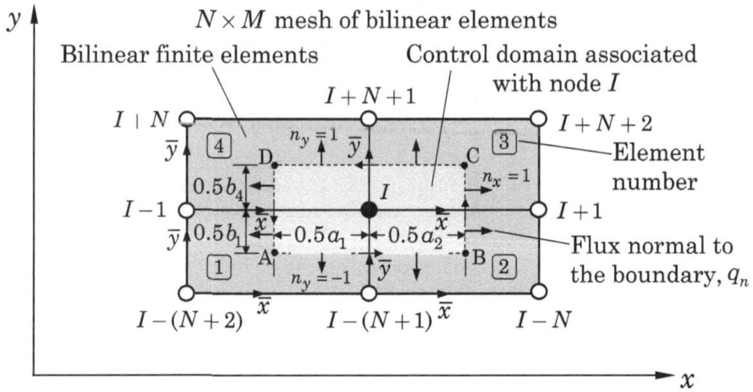

Fig. 6.3.3 A typical interior control domain, which bisects four rectangular finite elements. Each finite element has its own local coordinate system $(\bar{x}, \bar{y})$.

The boundary integrals can be simplified using the values of the direction cosines on each boundary line segment. We have [$\hat{\mathbf{n}} = (n_x, n_y)$; $\hat{\mathbf{n}} = (0, -1)$ and $ds = dx$ on AB; $\hat{\mathbf{n}} = (1, 0)$ and $ds = dy$ on BC; $\hat{\mathbf{n}} = (0, 1)$ and $ds = -dx$ on CD; and $\hat{\mathbf{n}} = (-1, 0)$ and $ds = -dy$ on DA; see Fig. 6.3.3]. Since the control domain partially occupies four bilinear elements (see Fig. 6.3.3), the line integrals in Eq. (6.3.15) can be expressed as

$$\oint_\Gamma \left(a_{xx} \frac{\partial u}{\partial x} \right) n_x \, ds = -\int_{0.5b_1}^{b_1} \left[a_{xx} \frac{\partial u}{\partial x} \right]^{(1)}_{\bar{x}=0.5a_1} d\bar{y} + \int_{0.5b_2}^{b_2} \left[a_{xx} \frac{\partial u}{\partial x} \right]^{(2)}_{\bar{x}=0.5a_2} d\bar{y}$$
$$+ \int_0^{0.5b_3} \left[a_{xx} \frac{\partial u}{\partial x} \right]^{(3)}_{\bar{x}=0.5a_3} d\bar{y} - \int_0^{0.5b_4} \left[a_{xx} \frac{\partial u}{\partial x} \right]^{(4)}_{\bar{x}=0.5a_4} d\bar{y}. \tag{6.3.16}$$

Here the superscript (e) $(e = 1, 2, 3, 4)$ refers to the element number shown in Fig. 6.3.3 and $(\bar{x}, \bar{y})$ are the local coordinates with the origin at node 1 of each finite element. Similar relation holds for the closed contour (line) integral of $(\cdot)\, n_y$:

$$\oint_\Gamma \left(a_{yy} \frac{\partial u}{\partial y} \right) n_y \, ds = -\int_{0.5a_1}^{a_1} \left[a_{yy} \frac{\partial u}{\partial y} \right]^{(1)}_{\bar{y}=0.5b_1} d\bar{x} - \int_0^{0.5a_2} \left[a_{yy} \frac{\partial u}{\partial y} \right]^{(2)}_{\bar{y}=0.5b_2} d\bar{x}$$
$$+ \int_0^{0.5a_3} \left[a_{yy} \frac{\partial u}{\partial y} \right]^{(3)}_{\bar{y}=0.5b_3} d\bar{x} + \int_{0.5a_4}^{a_4} \left[a_{yy} \frac{\partial u}{\partial y} \right]^{(4)}_{\bar{y}=0.5b_4} d\bar{x}. \tag{6.3.17}$$

Using Eqs. (6.3.16) and (6.3.17) in Eq. (6.3.15), we can numerically evaluate, as in one-dimensional case, the line integrals appearing in Eq. (6.3.15). Then the integral form in Eq. (6.3.15) becomes ($a_i \times b_i$ are the dimensions of the ith element of the primal mesh with $a_4 = a_1$, $a_3 = a_2$, $b_2 = b_1$, and $b_3 = b_4$):

$$\begin{aligned}
0 = & \int_{0.5a_1}^{a_1} \left[a_{yy} \frac{\partial u}{\partial y} \right]^{(1)}_{\bar{y}=0.5b_1} d\bar{x} + \int_{0.5b_1}^{b_1} \left[a_{xx} \frac{\partial u}{\partial x} \right]^{(1)}_{\bar{x}=0.5a_1} d\bar{y} \\
& + \int_{0}^{0.5a_2} \left[a_{yy} \frac{\partial u}{\partial y} \right]^{(2)}_{\bar{y}=0.5b_1} d\bar{x} - \int_{0.5b_1}^{b_1} \left[a_{xx} \frac{\partial u}{\partial x} \right]^{(2)}_{\bar{x}=0.5a_2} d\bar{y} \\
& - \int_{0}^{0.5a_2} \left[a_{yy} \frac{\partial u}{\partial y} \right]^{(3)}_{\bar{y}=0.5b_4} d\bar{x} - \int_{0}^{0.5b_4} \left[a_{xx} \frac{\partial u}{\partial x} \right]^{(3)}_{\bar{x}=0.5a_2} d\bar{y} \\
& - \int_{0.5a_1}^{a_1} \left[a_{yy} \frac{\partial u}{\partial y} \right]^{(4)}_{\bar{y}=0.5b_4} d\bar{x} + \int_{0}^{0.5b_4} \left[a_{xx} \frac{\partial u}{\partial x} \right]^{(4)}_{\bar{x}=0.5a_1} d\bar{y} \\
& - \int_{x_I-0.5a_1}^{x_I+0.5a_2} \int_{y_I-0.5b_1}^{y_I+0.5b_4} f(x,y)\, dxdy.
\end{aligned} \tag{6.3.18}$$

The numerical integration is necessary because the coefficients a_{xx} and a_{yy} are functions of u and its derivatives; see Section 1.8.4 for a discussion of numerical integration in two dimensions.

6.3.3.3 Finite element approximation and discretized equations

As in the FEM, over each finite element Ω_f^e $(e = 1, 2, \ldots, N)$ of the primal mesh, the function u is approximated as

$$u(x,y) \approx u_h(x,y) = \sum_{j=1}^{4} u_j^e \psi_j^e(\bar{x}, \bar{y}), \quad \bar{x} = x - x_1^e, \quad \bar{y} = y - y_1^e \tag{6.3.19}$$

where u_j^e denote the values of the function u^e at element nodes, (x_1^e, y_1^e) are the global coordinates of node 1 of element Ω_f^e and ψ_j^e are the Lagrange interpolation functions associated with the element Ω_f^e:

$$\begin{aligned}
\psi_1^e &= \left(1 - \frac{\bar{x}}{a}\right)\left(1 - \frac{\bar{y}}{b}\right), & \psi_2^e &= \frac{\bar{x}}{a}\left(1 - \frac{\bar{y}}{b}\right) \\
\psi_3^e &= \frac{\bar{x}}{a}\frac{\bar{y}}{b}, & \psi_4^e &= \left(1 - \frac{\bar{x}}{a}\right)\frac{\bar{y}}{b}
\end{aligned} \tag{6.3.20}$$

Here $(\bar{x}, \bar{y})$ denote the local coordinates with the origin located at node 1 of the element, and (a, b) denote the horizontal and vertical dimensions of a typical rectangle. Since the element (local) coordinate system $(\bar{x}, \bar{y})$ is a translation of the global coordinate system (x, y), the derivatives in both coordinate systems remain the same. One need only to express functions defined in terms of (x, y) to those in terms of $(\bar{x}, \bar{y})$ and use the local coordinates to evaluate the integrals. As explained in Section 5.4, one may use a primal mesh of linear triangles.

After the substitution of the approximation functions and evaluating the integrals, Eq. (6.3.18) can be expressed symbolically as [see Eq. (5.4.9a)]

$$\begin{aligned} A_{I-N-2}\,U_{I-N-2} &+ A_{I-N-1}\,U_{I-N-1} + A_{I-N}\,U_{I-N} + A_{I-1}\,U_{I-1} \\ &+ A_I\,U_I + A_{I+1}\,U_{I+1} + A_{I+N}\,U_{I+N} + A_{I+N+1}\,U_{I+N+1} \\ &+ A_{I+N+2}\,U_{I+N+2} = F_I, \end{aligned} \tag{6.3.21}$$

where the coefficients A_K, which contain the nonlinear contributions, and F_I can be determined using numerical evaluation of the integrals appearing in Eqs. (5.4.9b) and (5.4.9c).

For the nodes on the boundary, we must modify Eq. (6.3.18). For a rectangular domain, the nodal point locations can be classified into several cases, as shown in Fig. 5.4.4 (see Section 5.4.3 for the integral expressions of the coefficients A_K for typical boundary nodes; the integrals have to be numerically evaluated due to the fact that the coefficients a_{xx} and a_{yy} appearing in these integrals are functions of u). This completes the derivation of DMCDM equations for the nonlinear two-dimensional Poisson equation on rectangular domains.

6.3.4 Numerical Examples

Here we extend the linear problems considered in Chapters 4 and 5 to nonlinear problems. A computer program based on the developments of the previous sections is developed to solve these problems.

Example 6.3.1

Consider nonlinear heat conduction in a rectangular, isotropic medium with conductivity $a_{xx} = a_{yy} = k$. The domain is of dimensions $a \times b$. The conductivity k is assumed to vary according to the relation

$$k = k_0\left[1 + k_1(T - T_0)\right], \tag{6.3.22}$$

where T denotes the temperature, k_0 is the constant thermal conductivity, k_1 is the temperature coefficient of thermal conductivity, and T_0 is a reference temperature (taken to be zero), and T is the temperature. Use the following data and boundary conditions in obtaining the numerical solutions:

$$\begin{aligned} a = 0.2\text{ m},\quad b = 0.1\text{ m},\quad T_0 = 0\,^\circ\text{C},\quad k_0 = 0.2\text{ W/(m K)},\quad k_1 = 100\,\text{K}^{-1} \\ T(0, y) = 500^\circ\text{ C},\quad T(a, y) = 300\,^\circ\text{C} \\ \frac{\partial T}{\partial y} = 0 \text{ at } y = 0,\quad T(x, b) = 500(1 - 10x^2)\,^\circ\text{C}. \end{aligned} \tag{6.3.23}$$

Solution: A uniform mesh of 8×8 is used to analyze the problem. The error tolerance is taken to be $\epsilon = 10^{-3}$. The nonlinear convergence is achieved with three iterations. Table 6.3.1 contains linear and nonlinear solutions $T(x, y)$ obtained with the DMCDM, FEM, and FVM as functions of x and y. Clearly, both the FEM and DMCDM for this problem give very close solutions (and indistinguishable if plotted), and the FVM solutions are slightly different from the solutions predicted by DMCDM and FEM. In fact, doubling the mesh did not improve the FVM solution, indicating that the gradient boundary conditions in the FVM are not satisfied accurately.

Table 6.3.1 The FEM and DMCDM solutions [$T(x, y)$ is in K] of a nonlinear, two-dimensional heat conduction problem.

		DMCDM		FEM		FVM	
x	y	Linear	Nonlin.	Linear	Nonlin.	Linear	Nonlin.
0.025	0.00	482.85	485.05	482.80	485.00	484.01	486.76
0.050	0.00	464.48	468.64	464.39	468.57	466.85	471.93
0.075	0.00	443.88	449.85	443.77	449.76	447.47	454.31
0.100	0.00	420.42	427.95	420.30	427.86	425.05	433.06
0.125	0.00	393.88	402.40	393.77	402.32	399.01	407.53
0.150	0.00	364.48	372.81	364.39	372.76	369.13	377.12
0.175	0.00	332.85	338.84	332.80	338.81	335.71	341.19
0.025	0.05	485.54	487.18	485.49	487.12	486.00	488.24
0.050	0.05	469.30	472.45	469.22	472.37	470.65	474.80
0.075	0.05	450.01	454.60	449.92	454.52	452.97	458.54
0.100	0.05	426.98	432.88	426.89	432.81	432.07	438.60
0.125	0.05	400.01	406.86	399.92	406.81	407.16	414.16
0.150	0.05	369.30	376.26	369.22	376.22	377.36	384.18
0.175	0.05	335.34	340.80	335.49	340.78	341.88	346.96

Example 6.3.2

The bus bar shown in Fig. 6.3.4(a) carries sufficient electrical current to have a heat generation of $f = 10^6$ W/m^3. The bar has a conductivity of $a_{xx} = a_{yy} = k$ W/(m K) and dimensions 0.10 m × 0.05 m (and 0.01 m thick). The left side is maintained at $T(0, y) = 40°$C, the right side at $T(0.1, y) = 10°$C, the bottom edge is insulated, and the top edge is exposed to ambient air temperature of $T_\infty = 0°$C with a heat transfer coefficient of $\beta = 75$ W/(m^2 K). Assuming that the heat flow is two-dimensional (or one may assume that the front and back faces are insulated) and a temperature-dependent conductivity of $k(T) = 20 + 0.2(T - T_0)$ (with $T_0 = 0$), determine the linear and nonlinear steady-state solutions with a uniform primal mesh of 10 × 5 bilinear rectangular elements (i.e., $\Delta x = \Delta y = 0.01$ m). Use a tolerance of $\epsilon = 10^{-3}$ for nonlinear convergence check.

Solution: The surface convection is included by adding the contribution $\beta T(x, 0.05)$ to the coefficients associated with the top surface. For the primal 10 × 5 mesh of bilinear elements shown in Fig. 6.3.4(b), the matrix coefficients associated with nodes 56 to 66 will have additional contributions from edges exposed to convection. For example, global node 56 will have convection contribution from the side on its right, while node 56 will have contributions from sides on its left as well as right.

The numerical solutions obtained (convergence was achieved with five iterations) with the FEM and DMCDM (both methods with the 10 × 5 mesh of bilinear elements) for $T(x, 0)$ and $T(x, 0.05)$ as functions of x are shown in Fig. 6.3.5. For example, the linear solution using DMCDM at nodes 57–65 is

55.029, 66.599, 74.132, 77.558, 76.859, 72.010, 62.969, 49.666, 32.014.

The nonlinear solution at the same nodes is

50.233, 57.599, 62.088, 63.871, 63.016, 59.464, 53.013, 43.251, 29.405.

The corresponding linear and nonlinear solutions obtained with the FEM are, respectively,

54.975, 66.589, 74.128, 77.559, 76.862, 72.014, 62.972, 49.666, 32.003

and

50.196, 57.592, 62.085, 63.871, 63.017, 59.466, 53.014, 43.250, 29.395.

Thus, the solutions predicted by the DMCDM and FEM are in excellent agreement with each other.

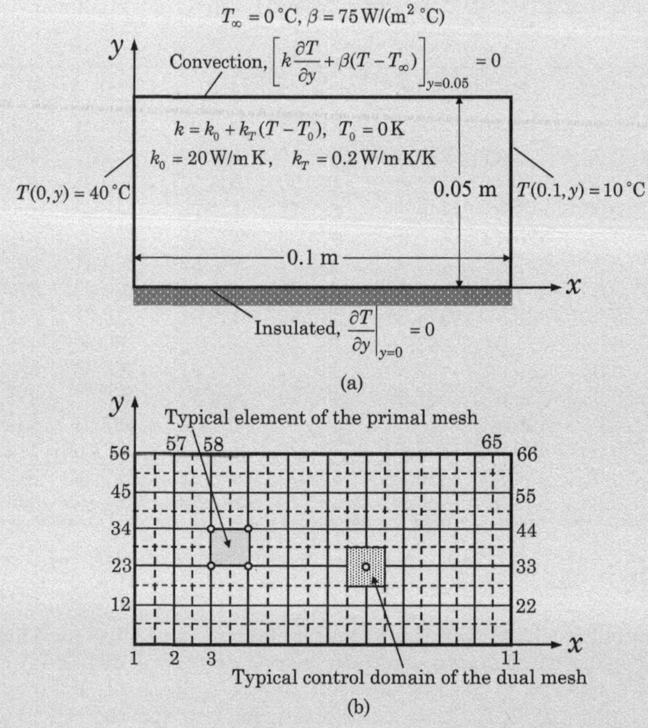

Fig. 6.3.4 (a) Domain and boundary conditions for conductive and convective heat transfer in a bus bar. The bottom is insulated, while the top is exposed to ambient temperature of 0°C; the left side is kept at 40°C and the right face is maintained at 10°C. (b) A primal mesh of 10 × 5 bilinear elements.

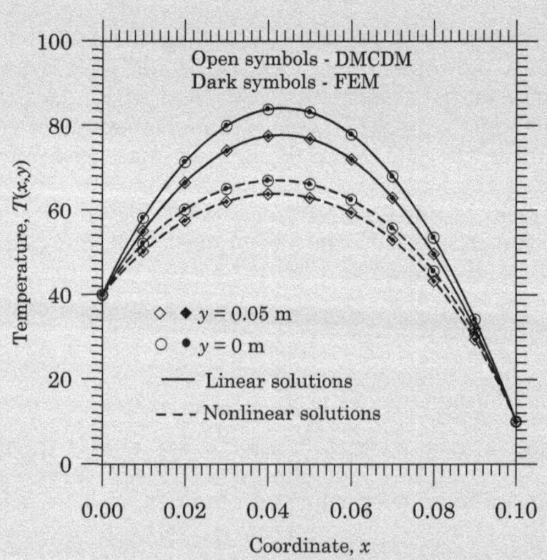

Fig. 6.3.5 Plots of linear (solid lines) and nonlinear (broken lines) temperature distributions at the bottom and top surfaces of the bar. The FEM solutions (shown with dark symbols) and dual mesh DMCDM solutions (shown with open symbols) are indistinguishable in the plots.

6.4 Summary

In this chapter, the FEM and DMCDM methods are extended to solve typical one- and two-dimensional nonlinear differential equations involving a single unknown, with numerical examples drawn primarily from heat transfer; however, any field problem described by the model equations can be analyzed. The numerical results of one- and two-dimensional nonlinear problems indicate that the DMCDM gives very accurate results, as good as the FEM but with relatively very low overhead compared to the FEM (which requires the development of weak forms, construction of element equations, and assembly of the element equations to obtain the final global equations). In general, the DMCDM and FEM do not give the same set of final algebraic equations. They will yield the same equations for isolated linear problems in one dimension in which the coefficients b and c of the model equation [see Eq. (6.2.1)] are zero. The numerical results show that the exact imposition of boundary conditions makes a difference in the accuracy of the solutions (i.e., the FVM solutions are not as accurate as the FEM or DMCDM).

Problems

The readers may find additional examples and problems (involving a single unknown) in the textbook by Reddy [13], which can be solved using the DMCDM method.

ONE-DIMENSIONAL PROBLEMS

6.1 Consider the nonlinear differential equation

$$-(\sqrt{2}+u)\frac{d^2u}{dx^2} - \left(\frac{du}{dx}\right)^2 = 1, \quad 0 < x < 1, \tag{1}$$

$$\frac{du}{dx}(0) = 0, \qquad u(1) = 1. \tag{2}$$

Develop the FEM and DMCDM formulations of the equation and solve the nonlinear problem using the Picard iteration procedure. Tabulate the nodal values of $u(x)$ for four and eight linear elements. Use an error tolerance of $\epsilon = 10^{-3}$.

6.2 Consider the nonlinear differential equation

$$-\frac{d^2u}{dx^2} - 2u\frac{du}{dx} = 0, \quad 0 < x < 1, \tag{1}$$

$$u(0) = 1, \qquad u(1) = 0.5. \tag{2}$$

Develop the FEM and DMCDM formulations of the equation and solve the problem using the Picard iteration procedure. Tabulate the nodal values of $u(x)$ for four and eight linear elements. The exact solution is given by $u(x) = 1/(1+x)$.

6.3 Formulate the nonlinear differential equation in **Problem 6.2** subject to the boundary conditions

$$\left[\frac{du}{dx}\right]_{x=0} = -1, \qquad \left[\frac{du}{dx} + u^2\right]_{x=1} = 0 \tag{1}$$

and solve it with the FEM and DMCDM with a convergence tolerance of 10^{-4}. Tabulate the nodal values of $u(x)$ for four and eight linear elements, along with the exact solution $u(x) = 1/(1+x)$.

6.4 Formulate the nonlinear differential equation in **Problem 6.2** subject to the boundary conditions

$$u(0) = 1, \qquad \left[\frac{du}{dx} + u^2\right]_{x=1} = 0 \tag{1}$$

and solve the problem using the FEM and DMCDM (use an error tolerance of $\epsilon = 10^{-3}$). Tabulate the nodal values of $u(x)$ for four and eight linear elements. The exact solution is given by $u(x) = 1/(1+x)$.

6.5 Solve the nonlinear differential equation in **Problem 6.2** subject to the boundary conditions

$$\left[\frac{du}{dx} + 2u\right]_{x=0} = 1, \qquad \left[\frac{du}{dx} + u^2\right]_{x=1} = 0. \tag{1}$$

Use an error tolerance of $\epsilon = 10^{-4}$, and tabulate the nodal values of $u(x)$ for four and eight linear elements. The exact solution is given by $u(x) = 1/(1+x)$.

6.6 Develop the FEM and DMCDM formulations of the nonlinear differential equation

$$-\frac{d^2u}{dx^2} - \left(\frac{du}{dx}\right)^3 = 0, \quad 0 < x < 1, \tag{1}$$

$$\left[\frac{du}{dx} + u\right]_{x=0} = \frac{3}{\sqrt{2}}, \qquad \left[\frac{du}{dx}\right]_{x=1} = 0.5 \tag{2}$$

and solve the problem using the Picard iteration procedure. Tabulate the nodal values of $u(x)$ for four and eight linear elements. The exact solution is given by $u(x) = \sqrt{2(1+x)}$.

Two-Dimensional Problems

6.7 Solve the problem of **Example 6.3.1** for the following data:

$$a = 0.2 \text{ m}, \quad b = 0.1 \text{ m}, \tag{1}$$

$$T(0,y) = 500°\text{C}\,, \; T(a,y) = 300°\text{C}\,, \; \frac{\partial T}{\partial y} = 0 \text{ at } y = 0, \quad T(x,b) = 500(1-2x)°\text{C}. \tag{2}$$

All other data remain the same. Use the uniform 8×8 linear element mesh to analyze the problem using the FEM and DMCDM. Use a convergence tolerance of $\varepsilon = 10^{-3}$ and maximum allowable iterations to be 10.

6.8 Consider the nonlinear problem of **Example 6.3.1**. Use the uniform 8×8 mesh to analyze the problem using the following data and boundary conditions and the Picard (direct) iteration procedure:

$$a = 0.18 \text{ m}, \quad b = 0.1 \text{ m}, \quad f_0 = 0 \text{ W/m}^3, \tag{1}$$

$$k = k_0\,(1 + k_1 T)\,, \quad k_0 = 25 \text{ W/(m °C)}, \tag{2}$$

$$T(0,y) = 100°\text{C}\,, \quad T(a,y) = 50°\text{C}, \quad k\frac{\partial T}{\partial n} = 0 \text{ at } y = 0,$$

$$k\frac{\partial T}{\partial n} + h_c(T - T_\infty) = 0 \text{ at } y = b. \tag{3}$$

Use $k_1 = 0.2$, $T_\infty = 10°$C, and $h_c = 50$ W/(m^2 °C), and $\epsilon = 10^{-3}$ and ten maximum iterations. Plot linear and nonlinear solutions $T(x,y)$ as functions of x for fixed $y = 0$ and $y = 0.05$.

7
Bending of Straight Beams

7.1 Introduction

7.1.1 Background

The branch of mechanics that deals with the deformation and stress in solid bodies subjected to forces is called *solid mechanics*. A subset of solid mechanics is *structural mechanics*, which deals with deformation and stress state in solid bodies of specific type of geometry that allows reduction of a three-dimensional problem to a two- or one-dimensional problem, and the resulting solids are known as *structural elements*. Many commonly known structural elements are broadly classified as bars, beams, frames, plates, and shells. Most real-world structures, microscale or macroscale (e.g., buildings, automobiles, aircraft, ships, bridges, MEMS and so on; as an example, Fig. 7.1.1 shows a nanobeam in a nano-indentation setup) are mostly composed of structural elements, and only few parts are geometrically arbitrary solids with all three dimensions comparable in size. All of the structural elements have one of the three (geometric) dimensions very small (at least one-twentieth to one-hundredth) compared to the remaining two dimensions, and the smallest dimension is termed the thickness. Thus, plates and shells are thin solids with the two dimensions comparable to each other but large compared to their thickness. The difference between plates and shells is the fact that shells are thin objects with curvature, while plates are thin flat bodies.

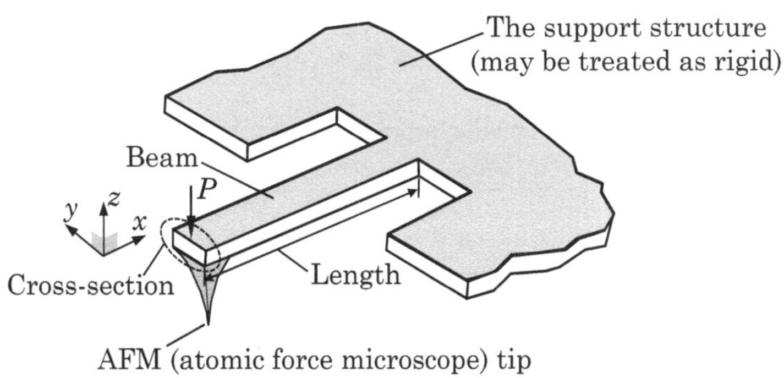

Fig. 7.1.1 Beam element in a nano-indentation setup. Force P can be a mechanical or electromagnetic force.

Beams are structural elements that have a ratio of length-to-cross-sectional dimensions very large, say, 20 to 100 or more, and subjected to forces both along and transverse to the length, and moments that tend to rotate and bend them about an axis perpendicular to their length. When all applied loads are along the length only, they are often called *bars* (i.e., bars experience only tensile or compressive strains). All solid bodies subjected to forces can be analyzed for deformation and stress using the elasticity equations; certain simple geometries, such as those of beams, allow us to develop theories that are simple and yet yield results that are accurate enough for engineering analysis and design. These simplified continuum theories, known as *structural theories*, have been developed for centuries using simplified kinematics (see Reddy [17, 42]).

The most commonly used structural theories of beams are: (a) the classical beam theory in which transverse shear strain is assumed to be zero, (b) the first-order shear deformation beam theory, which accounts for transverse shear strains as a constant through the beam thickness (or height), and (c) the third-order shear deformation beam theory, which accommodates quadratic variation of the transverse shear strains through the thickness. The commonly used names for these theories, respectively, are the *Euler–Bernoulli beam theory* (EBT), the *Timoshenko beam theory* (TBT), and the *Reddy third-order beam theory* (RBT).

7.1.2 Functionally Graded Structures

Structural elements can be made of many thin layers bonded together (called a laminate), each layer having different material property, or they can be made up of different microscopic architectures. In both cases, the objective is to achieve certain functionalities (e.g., stiffness, strength, fracture toughness, thermal resistance, and other properties). If two dissimilar materials are bonded together, there is a very high chance that debonding will occur at interface due to some extreme loading conditions, may it be mechanical or thermal. Another problem in a laminated structure is the presence of residual stresses due to the difference in coefficients of thermal expansion between different material layers. These problems can be resolved by gradually varying the volume fraction of the material constituents rather than abruptly changing them over an interface. A gradual variation of the material results in a very efficient structure tailored to suit the needs.

Functionally graded structural elements are a class of heterogeneous structures that have a gradual variation of material properties through the thickness. Functionally graded materials (FGMs) are often used in thermal barrier structures for aerospace applications as well as in space structures. In fact, functionally graded material characteristics are present in most structures found in nature (including the human body), and perhaps a better understanding of the highly complex form of materials in nature will help us in synthesizing new materials.

Two-constituent functionally graded through-thickness materials are characterized by a power-law variation, among other types of variations, of the modulus of elasticity while the Poisson ratio is kept constant. For example, if the z-coordinate is taken along the thickness (or height) of the beam (see Fig. 7.1.2), the modulus $E(z)$ of an FGM beam along the thickness coordinate z is assumed to be represented by a simple power-law, such as (see Reddy [43])

$$E(z) = (E_1 - E_2)\, f(z) + E_2, \quad f(z) = \left(\frac{1}{2} + \frac{z}{h}\right)^n, \tag{7.1.1}$$

where E_1 and E_2 are the material properties of Material 1 (top face, $z = +h/2$) and Material 2 (bottom face, $z = -h/2$) of the beam, respectively, and n is known as the power-law index. Note that when $n = 0$, we obtain the single-material structure (with modulus E_1).

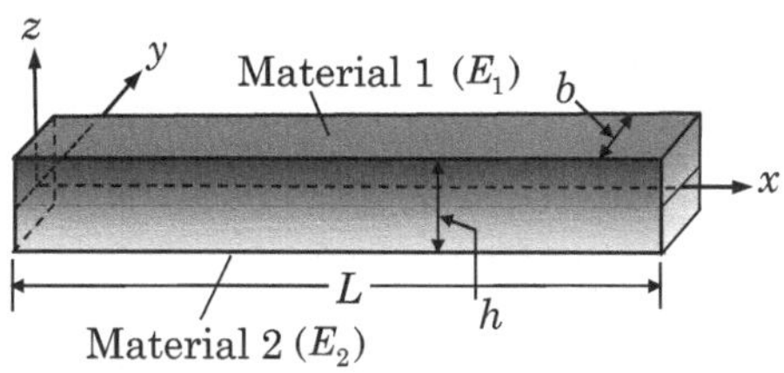

Fig. 7.1.2 FGM beam of rectangular cross-section. The bottom ($z = -h/2$) is 100% Material 2, and the top ($z = h/2$) is 100% Material 1.

7.1.3 Present Study

In the present chapter, the FEM and DMCDM are extended to study the response of functionally graded beams under different types of loads and boundary conditions. As a special case of FGM beams, one can obtain the results for homogeneous isotropic beams. The governing differential equations of the EBT include a fourth-order equation, while those of the TBT are second-order. For FGM beams, both theories require the solution of multiple coupled differential equations. Thus, in the present chapter, displacement and mixed formulations involving several dependent variables are developed, and numerical results are presented for both linear and geometrically nonlinear bending of beams.

Following this introduction, a review of the governing equations of the EBT and TBT (see Reddy [17, 42]), under the assumption of small strains and rotations, as applied to functionally graded beams, is presented in Section 7.2. Since the present book is more concerned with the numerical methods, only a review of the governing equations is presented without detailed derivations. The reader may find additional details in the references cited above. Various finite element models of the governing equations are presented in Section 7.3. Discrete models based on the dual mesh control domain method for linear problems are presented in Section 7.4. Then, in Section 7.5, numerical solutions are presented for beams with various boundary conditions, and the numerical solutions obtained with FEM and DMCDM models are compared with the exact solutions to assess the relative accuracy in predicting the displacements and moments. Section 7.6 is dedicated to various nonlinear finite element models (using the EBT and TBT) as well as the discrete models using the DMCDM. Lastly, Section 7.7 contains a summary of the chapter.

7.2 Linear Theories of FGM Beams

7.2.1 Euler–Bernoulli Beam Theory

7.2.1.1 Kinematics and equilibrium equations

Consider a straight, through-thickness functionally graded beam. The geometry and the coordinate system are shown in Fig. 7.2.1(a). The displacement field of the classical beam theory (CBT) is constructed assuming that transverse lines perpendicular to the beam axis (x) remain: (1) straight, (2) inextensible, and (2) perpendicular to the tangents of the deflected x-axis [see Fig. 7.2.1(b)]. These three assumptions together are known as the *Euler–Bernoulli hypothesis.* For bending in the xz-plane (i.e., bending about the y-axis), the Euler–Bernoulli hypothesis results in the following displacement field in a rectangular Cartesian coordinate system (x, y, z) (see Reddy [42] for details):

$$\mathbf{u}(x,z) = [u(x) + z\,\theta_x(x)]\,\hat{\mathbf{e}}_x + w(x)\,\hat{\mathbf{e}}_z,\ \theta_x \equiv -\frac{dw}{dx}, \tag{7.2.1}$$

where (u, w) denote the axial and transverse displacements, respectively, of a point on the midplane of the beam, and $(\hat{\mathbf{e}}_x, \hat{\mathbf{e}}_z)$ are unit base vectors in the x- and y-coordinate directions, respectively.

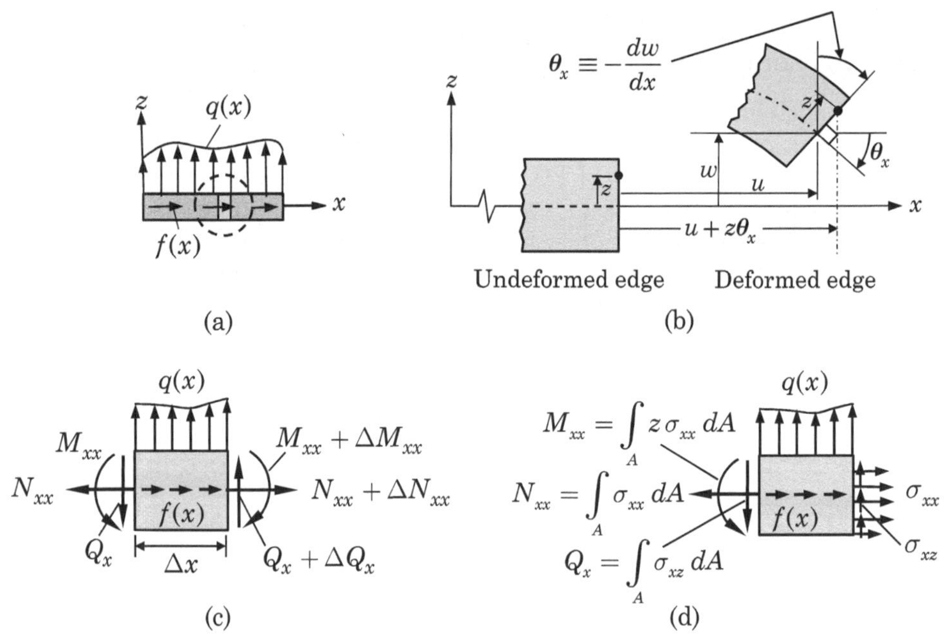

Fig. 7.2.1 (a) Straight beam with distributed axial ($f(x)$) and transverse ($q(x)$) forces. (b) Deformed (i.e., elongated and bent) beam; the deformation is exaggerated. (c) Element of length Δx of the beam (presuming that the geometry did not change significantly after deformation) with stress resultants. (d) Point-equilibrium of internal stress resultants (N_{xx}, M_{xx}, Q_x) with stresses.

The only nonzero strain is the axial strain $\varepsilon_{xx}(x,z)$, which includes both stretching and bending strains,

$$\varepsilon_{xx} = \frac{du}{dx} + z\,\frac{d\theta_x}{dx} = \frac{du}{dx} - z\,\frac{d^2w}{dx^2}. \tag{7.2.2}$$

The equilibrium of forces [see Fig. 7.2.1(c)] in the EBT require

$$-\frac{dN_{xx}}{dx} - f = 0, \tag{7.2.3a}$$

$$-\frac{d^2M_{xx}}{dx^2} + c_f\,w - q = 0, \tag{7.2.3b}$$

where $f(x)$ and $q(x)$ are distributed loads in the axial and transverse directions, respectively, c_f is the modulus of the elastic foundation (if none, set $c_f = 0$) on which the beam rests, and N_{xx} and M_{xx} are the stress resultants [which are in equilibrium with the internally developed stresses, as shown in Fig. 7.2.1(d)] defined by

$$N_{xx} = \int_A \sigma_{xx}\,dA, \quad M_{xx} = \int_A z\sigma_{xx}\,dA, \tag{7.2.4}$$

where $\sigma_{xx}(x,z)$ is the axial stress at a point (x,z) in the beam.

7.2.1.2 Equations in terms of displacements

Next, we express the equilibrium equations in Eqs. (7.2.3a) and (7.2.3b) in terms of the displacements u and w. This requires us to assume a constitutive relation between the axial stress σ_{xx} and axial strain ε_{xx}. Following the one-dimensional Hooke's law, we write

$$\sigma_{xx} = E(x,z)\,\varepsilon_{xx}, \tag{7.2.5}$$

where $E(x,z)$ is Young's modulus, which can be a function of x and z. In the present study, we assume that E is only a function of z, as indicated in Eq. (7.1.1). Then, the stress resultants can be expressed in terms of u and w as

$$N_{xx} = A_{xx}\,\frac{du}{dx} - B_{xx}\,\frac{d^2w}{dx^2}, \tag{7.2.6a}$$

$$M_{xx} = B_{xx}\,\frac{du}{dx} - D_{xx}\frac{d^2w}{dx^2}, \tag{7.2.6b}$$

where A_{xx}, B_{xx}, and D_{xx} are the extensional, extensional-bending, and bending stiffness coefficients

$$(A_{xx}, B_{xx}, D_{xx}) = \int_A (1, z, z^2)E(z)\,dA. \tag{7.2.7}$$

In arriving at the relations in Eqs. (7.2.6a) and (7.2.6b), we assumed that the x-axis passes through the geometric centroid of the beam cross-section so that

$$\int_A z\,dA = 0. \tag{7.2.8}$$

The values of A_{xx}, B_{xx}, and D_{xx} for rectangular cross-section beams, when $E(z)$ varies according to Eq. (7.1.1), are presented in Appendix 7 at the end of the chapter.

Substitution of Eqs. (7.2.6a) and (7.2.6b) into Eqs. (7.2.3a) and (7.2.3b) gives the governing equilibrium equations in terms of the displacements:

$$-\frac{d}{dx}\left(A_{xx}\frac{du}{dx} - B_{xx}\frac{d^2w}{dx^2}\right) - f = 0, \tag{7.2.9a}$$

$$-\frac{d^2}{dx^2}\left(B_{xx}\frac{du}{dx} - D_{xx}\frac{d^2w}{dx^2}\right) + c_f w - q = 0. \tag{7.2.9b}$$

We note that Eqs. (7.2.9a) and (7.2.9b) are coupled differential equations involving u and w due to the presence of the coupling coefficient B_{xx}. When E is either a constant or a symmetric function of z, we have $B_{xx} = 0$, and Eqs. (7.2.9a) and (7.2.9b) are uncoupled [i.e., Eq. (7.2.9a) can be used to solve for u and Eq. (7.2.9b) can be used to solve for w]. Another point to note is (independent of whether B_{xx} is zero or not) that the equation governing w is fourth order, preventing us from constructing discretized equations using the DMCDM. Next, we discuss a formulation in which a different choice of dependent unknowns reduces the equations to second order so that DMCDM can be used.

7.2.1.3 Equations in terms of displacements and bending moment

To express the governing equations in terms of u, w, and M_{xx}, we first rewrite Eqs. (7.2.6a) and (7.2.6b) in the form (i.e., displacement gradients in terms of the stress resultants)

$$\frac{du}{dx} = \frac{1}{D^*_{xx}}\left(D_{xx}N_{xx} - B_{xx}M_{xx}\right), \tag{7.2.10a}$$

$$-\frac{d^2w}{dx^2} = \frac{1}{D^*_{xx}}\left(-B_{xx}N_{xx} + A_{xx}M_{xx}\right), \tag{7.2.10b}$$

where

$$D^*_{xx} \equiv D_{xx}A_{xx} - B^2_{xx}. \tag{7.2.10c}$$

Solving Eq. (7.2.10a) for N_{xx} in terms of du/dx and M_{xx}, we obtain

$$N_{xx} = \bar{A}_{xx}\frac{du}{dx} + \bar{B}_{xx}M_{xx}, \tag{7.2.11a}$$

where

$$\bar{A}_{xx} \equiv \frac{D^*_{xx}}{D_{xx}}, \quad \bar{B}_{xx} \equiv \frac{B_{xx}}{D_{xx}}. \tag{7.2.11b}$$

Substituting Eq. (7.2.11a) for N_{xx} in Eq. (7.2.10b), we obtain

$$-\frac{d^2w}{dx^2} = -\bar{B}_{xx}\frac{du}{dx} + \frac{1}{D_{xx}}M_{xx}. \tag{7.2.11c}$$

Substituting Eq. (7.2.11a) in Eq. (7.2.3a) yields

$$-\frac{d}{dx}\left(\bar{A}_{xx}\frac{du}{dx}+\bar{B}_{xx}\,M_{xx}\right)=f. \tag{7.2.11d}$$

Equations (7.2.11d), (7.2.3b), and (7.2.11c) provide the governing second-order differential equations in terms of u, w, and M_{xx}:

$$-\frac{d}{dx}\left(\bar{A}_{xx}\frac{du}{dx}+\bar{B}_{xx}\,M_{xx}\right)=f, \tag{7.2.12a}$$

$$-\frac{d^2M_{xx}}{dx^2}+c_f w=q, \tag{7.2.12b}$$

$$-\frac{d^2w}{dx^2}-\frac{1}{D_{xx}}M_{xx}+\bar{B}_{xx}\frac{du}{dx}=0. \tag{7.2.12c}$$

7.2.2 Timoshenko Beam Theory

7.2.2.1 Kinematics and equilibrium equations

The Timoshenko beam theory (TBT) is based on the displacement field

$$\mathbf{u}(x,z)=\left[u(x)+z\,\phi_x(x)\right]\hat{\mathbf{e}}_x+w(x)\,\hat{\mathbf{e}}_z, \tag{7.2.13}$$

where ϕ_x denotes the rotation of the cross-section about the y-axis. In the TBT the normality assumption of the EBT is relaxed (i.e., the rotation ϕ_x is not equal to the slope θ_x) and a constant state of transverse shear strain (and thus constant shear stress computed from the constitutive equation) with respect to the thickness coordinate z is included.

The strains in this case are

$$\varepsilon_{xx}(x,z)=\frac{du}{dx}+z\frac{d\phi_x}{dx},\quad \gamma_{xz}(x)\equiv 2\,\varepsilon_{xz}(x)=\phi_x+\frac{dw}{dx}. \tag{7.2.14}$$

We note that the strains involve only the first derivatives of the dependent unknowns (u,w,ϕ_x). In addition, the transverse shear strain γ_{xz} is only a function of x (i.e., constant through the beam height).

In addition to the axial stress–strain relation in Eq. (7.2.5), we assume a linear relation between the shear stress σ_{xz} and shear strain γ_{xz}

$$\sigma_{xz}=G(x,z)\,\gamma_{xz},\quad G(x,z)=\frac{1}{2(1+\nu)}\,E(x,z), \tag{7.2.15}$$

where G is the shear modulus and ν is the Poisson ratio.

The equations of equilibrium of the TBT are

$$-\frac{dN_{xx}}{dx}-f=0, \tag{7.2.16a}$$

$$-\frac{dQ_x}{dx}+c_f\,w-q=0, \tag{7.2.16b}$$

$$-\frac{dM_{xx}}{dx}+Q_x=0, \tag{7.2.16c}$$

where (N_{xx}, M_{xx}) are the stress resultants defined in Eq. (7.2.4), and Q_x is the shear force on a beam cross-section

$$Q_x = \int_A \sigma_{xz}\, dA. \tag{7.2.17}$$

The TBT requires a shear correction factor to compensate for the error introduced due to the constant state of shear stress. The correction is made to the shear force Q_x:

$$Q_x = K_s \int_A \sigma_{xz}\, dA, \tag{7.2.18}$$

where K_s is the shear correction coefficient. For rectangular cross-sections, the shear correction coefficient K_s is taken to be $K_s = 5/6$.

The stress resultants (N_{xx}, M_{xx}, Q_x) in the TBT can be expressed in terms of the generalized displacements (u, w, ϕ_x) as

$$N_{xx} = A_{xx}\frac{du}{dx} + B_{xx}\frac{d\phi_x}{dx}, \tag{7.2.19a}$$

$$M_{xx} = B_{xx}\frac{du}{dx} + D_{xx}\frac{d\phi_x}{dx}, \tag{7.2.19b}$$

$$Q_x = K_s \int_A \sigma_{xz}\, dA = S_{xz}\left(\phi_x + \frac{dw}{dx}\right), \tag{7.2.19c}$$

$$S_{xz} = \frac{K_s}{2(1+\nu)} \int_A E(z)\, dA, \tag{7.2.19d}$$

where S_{xz} denotes the shear stiffness. The word "generalized" is used here because the rotation ϕ_x can also be viewed as a "displacement." Substitution of Eqs. (7.2.19a)–(7.2.19c) into Eqs. (7.2.16a)–(7.2.16c) yields the following governing equations in terms of the displacements:

$$-\frac{d}{dx}\left(A_{xx}\frac{du}{dx} + B_{xx}\frac{d\phi_x}{dx}\right) - f = 0, \tag{7.2.20a}$$

$$-\frac{d}{dx}\left[S_{xz}\left(\phi_x + \frac{dw}{dx}\right)\right] + c_f w - q = 0, \tag{7.2.20b}$$

$$-\frac{d}{dx}\left(B_{xx}\frac{du}{dx} + D_{xx}\frac{d\phi_x}{dx}\right) + S_{xz}\left(\phi_x + \frac{dw}{dx}\right) = 0. \tag{7.2.20c}$$

7.2.2.2 Equations in terms of displacements and bending moment

As in the case of EBT, we can also develop a mixed formulation of the TBT equations. First rewrite Eqs. (7.2.16a)–(7.2.16c) by replacing Q_x from Eq. (7.2.19c) in terms of M_{xx} as

$$-\frac{dN_{xx}}{dx} - f = 0, \tag{7.2.21a}$$

$$-\frac{d^2 M_{xx}}{dx^2} + c_f w - q = 0, \tag{7.2.21b}$$

$$-\frac{dM_{xx}}{dx} + S_{xz}\left(\phi_x + \frac{dw}{dx}\right) = 0. \tag{7.2.21c}$$

Then we rewrite Eqs. (7.2.19a) and (7.2.19b) for displacement gradients in terms of stress resultants,

$$\frac{du}{dx} = \frac{1}{D^*_{xx}}\left(D_{xx}N_{xx} - B_{xx}M_{xx}\right), \tag{7.2.22a}$$

$$\frac{d\phi_x}{dx} = \frac{1}{D^*_{xx}}\left(-B_{xx}N_{xx} + A_{xx}M_{xx}\right). \tag{7.2.22b}$$

Solving Eq. (7.2.22a) for N_{xx} in terms of du/dx and M_{xx} [see Eq. (7.2.11a)]

$$N_{xx} = \bar{A}_{xx}\,\frac{du}{dx} + \bar{B}_{xx}\,M_{xx}, \tag{7.2.23a}$$

where D^*_{xx}, $\bar{A}_{xx}$, and $\bar{B}_{xx}$ are defined in Eqs. (7.2.10c) and (7.2.11b). Substituting for N_{xx} from Eq. (7.2.23a) in Eq. (7.2.22b), we obtain

$$\frac{d\phi_x}{dx} = -\bar{B}_{xx}\frac{du}{dx} + \frac{1}{D_{xx}}M_{xx}. \tag{7.2.23b}$$

Differentiating Eq. (7.2.21c) with respect to x, and solving for $d\phi_x/dx$, we obtain

$$\frac{d\phi_x}{dx} = -\frac{d^2w}{dx^2} + \frac{1}{S_{xz}}\frac{d^2M_{xx}}{dx^2}. \tag{7.2.23c}$$

From Eqs. (7.2.23b) and (7.2.23c), we obtain

$$-\frac{d^2w}{dx^2} = -\frac{1}{S_{xz}}\frac{d^2M_{xx}}{dx^2} - \bar{B}_{xx}\frac{du}{dx} + \frac{1}{D_{xx}}M_{xx}. \tag{7.2.24}$$

Equations (7.2.21a) [with N_{xx} replaced using Eq. (7.2.23a)], (7.2.21b), and (7.2.24) constitute the governing second-order differential equations in terms of u, w, and M_{xx} for the TBT:

$$-\frac{d}{dx}\left(\bar{A}_{xx}\frac{du}{dx} + \bar{B}_{xx}M_{xx}\right) - f = 0, \tag{7.2.25a}$$

$$-\frac{d^2M_{xx}}{dx^2} + c_f w - q = 0, \tag{7.2.25b}$$

$$-\frac{d}{dx}\left(\frac{dw}{dx} - \frac{1}{S_{xz}}\frac{dM_{xx}}{dx}\right) + \bar{B}_{xx}\frac{du}{dx} - \frac{1}{D_{xx}}M_{xx} = 0. \tag{7.2.25c}$$

The effective rotation ϕ_x is [see Eq. (7.2.23c)]

$$\phi_x = -\frac{dw}{dx} + \frac{1}{S_{xz}}\frac{dM_{xx}}{dx}. \tag{7.2.26}$$

7.3 Linear Finite Element Models

7.3.1 Euler–Bernoulli Beam Theory

7.3.1.1 Displacement finite element model

The *displacement* finite element model of Eqs. (7.2.9a) and (7.2.9b) is based on their weak forms [see Section 4.2.4 for the details of the weak-form development] or the principle of virtual displacements (see Reddy [17, 42]), where the role of a weight function is played by the virtual displacements (i.e., $w_1 \sim \delta u$ and $w_2 \sim \delta w$). The three-step procedure of developing weak forms, applied to Eqs. (7.2.9a) and (7.2.9b), results in the following weak forms:

$$0 = \int_{x_a^e}^{x_b^e} \left[\frac{d\delta u}{dx} \left(A_{xx}^e \frac{du}{dx} - B_{xx}^e \frac{d^2 w}{dx^2} \right) - \delta u\, f \right] dx - \delta u(x_a^e)\, Q_1^e - \delta u(x_b^e)\, Q_4^e, \tag{7.3.1a}$$

$$\begin{aligned} 0 = &\int_{x_a^e}^{x_b^e} \left[-\frac{d^2 \delta w}{dx^2} \left(B_{xx}^e \frac{du}{dx} - D_{xx}^e \frac{d^2 w}{dx^2} \right) + c_f^e\, \delta w\, w - \delta w\, q^e(x) \right] dx \\ &- \delta w(x_a^e)\, Q_2^e - \left[-\frac{d\delta w}{dx} \right]_{x_a^e} Q_3^e - \delta w(x_b^e)\, Q_5^e - \left[-\frac{d\delta w}{dx} \right]_{x_b^e} Q_6^e, \end{aligned} \tag{7.3.1b}$$

where Q_i^e are the generalized forces at the element nodes (see Fig. 7.3.1)

$$\begin{aligned} Q_1^e &= -N_{xx}(x_a^e), \quad Q_2^e = -\left[\frac{dM_{xx}}{dx} \right]_{x_a^e}, \quad Q_3^e = -M_{xx}(x_a^e), \\ Q_4^e &= N_{xx}(x_b^e), \quad Q_5^e = \left[\frac{dM_{xx}}{dx} \right]_{x_b^e}, \quad Q_6^e = M_{xx}(x_b^e). \end{aligned} \tag{7.3.2}$$

The superscript e over quantities refers to the fact that they are defined over a typical line element (x_a^e, x_b^e)].

The duality pair of the weak form in Eq. (7.3.1a) is (u, N_{xx}), whereas the weak form in Eq. (7.3.1b) has two duality pairs, namely, (w, Q_x) and $(-dw/dx, M_{xx})$. The product of the variables in each duality pair has the meaning of work done (see Reddy [17, 42]). We note that there are two primary variables associated with w, requiring an approximation that interpolates both w

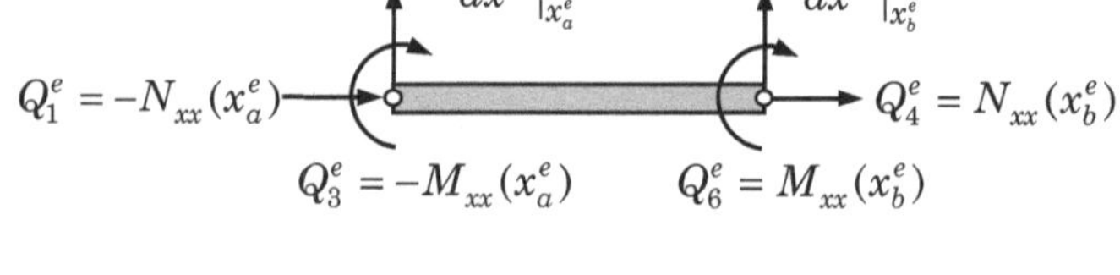

Fig. 7.3.1 Generalized forces in the displacement finite element model of the EBT.

and $\theta_x = -dw/dx$ (so that both w and θ_x can be made continuous at the nodes between elements; this is known as the C^1-continuity), leading, at the minimum, to the Hermite cubic interpolations functions, whose derivation can be found in Reddy [8].

We use the Lagrange linear interpolation of $u(x)$ and Hermite cubic interpolation of $w(x)$:

$$u_h^e(x) = u_1^e\psi_1^e(x) + u_2^e\psi_2^e(x), \tag{7.3.3a}$$

$$w_h^e(x) = \Delta_1^e\varphi_1^e(x) + \Delta_2^e\varphi_2^e(x) + \Delta_3^e\varphi_3^e(x) + \Delta_4^e\varphi_4^e(x), \tag{7.3.3b}$$

where ψ_i^e $(i = 1, 2)$ are the linear interpolation functions [see Eqs. (4.2.13) and (4.2.14)]

$$\psi_1^e(x) = \frac{x_b^e - x}{h_e} = 1 - \frac{\bar{x}}{h_e}, \qquad \psi_2^e(x) = \frac{x - x_a^e}{h_e} = \frac{\bar{x}}{h_e} \tag{7.3.4}$$

and φ_i^e are the Hermite cubic interpolation functions

$$\begin{aligned} \varphi_1^e(\bar{x}) &= 1 - 3\left(\frac{\bar{x}}{h_e}\right)^2 + 2\left(\frac{\bar{x}}{h_e}\right)^3, \quad &\varphi_2^e(\bar{x}) &= -\bar{x}\left(1 - \frac{\bar{x}}{h_e}\right)^2, \\ \varphi_3^e(\bar{x}) &= 3\left(\frac{\bar{x}}{h_e}\right)^2 - 2\left(\frac{\bar{x}}{h_e}\right)^3, &\varphi_4^e(\bar{x}) &= -\bar{x}\left[\left(\frac{\bar{x}}{h_e}\right)^2 - \frac{\bar{x}}{h_e}\right]. \end{aligned} \tag{7.3.5}$$

Here $\bar{x}$ denotes the element (or local) coordinate, $\bar{x} = x - x_a^e$. Note that each node has three primary degrees of freedom, namely, axial displacement, u_i^e, transverse displacement w_i^e, and rotation θ_i^e $(i = 1, 2)$. We have used the notation $\Delta_1 = w_1$, $\Delta_2 = \theta_1$, $\Delta_3 = w_2$, and $\Delta_4 = \theta_2$.

The *displacement finite element model* of the EBT is obtained by substituting the approximations in Eqs. (7.3.4) and (7.3.5) for u and w, respectively, and $\delta u = \psi_i^e$ and $\delta w = \varphi_i^e$ into the weak forms in Eqs. (7.3.1a) and (7.3.1b). We obtain

$$\begin{bmatrix} \mathbf{K}^{11} & \mathbf{K}^{12} \\ \mathbf{K}^{21} & \mathbf{K}^{22} \end{bmatrix} \begin{Bmatrix} \mathbf{u} \\ \boldsymbol{\Delta} \end{Bmatrix} = \begin{Bmatrix} \mathbf{F}^1 \\ \mathbf{F}^2 \end{Bmatrix} \tag{7.3.6a}$$

where

$$\begin{aligned} K_{ij}^{11} &= \int_{x_a^e}^{x_b^e} A_{xx}^e \frac{d\psi_i^e}{dx}\frac{d\psi_j^e}{dx}\, dx, \quad K_{iJ}^{12} = -\int_{x_a^e}^{x_b^e} B_{xx}^e \frac{d\psi_i}{dx}\frac{d^2\varphi_J^e}{dx^2}\, dx \\ F_i^1 &= \psi_i(x_a)\, Q_1^e + \psi_i(x_b^e)\, Q_4^e \\ K_{Ij}^{21} &= -\int_{x_a^e}^{x_b^e} B_{xx}^e \frac{d^2\varphi_I^e}{dx^2}\frac{d\psi_j}{dx}\, dx, \quad K_{IJ}^{22} = \int_{x_a^e}^{x_b^e} \left(D_{xx}^e \frac{d^2\varphi_I^e}{dx^2}\frac{d^2\varphi_J^e}{dx^2} + c_f^e\varphi_I^e\varphi_J^e \right) dx \\ F_I^2 &= \int_{x_a^e}^{x_b^e} \varphi_I q^e(x)\, dx + \varphi_I^e(x_a^e)\, Q_2^e - \left.\frac{d\varphi_I^e}{dx}\right|_{x_a^e} Q_3^e + \varphi_I^e(x_b^e)\, Q_5^e - \left.\frac{d\varphi_I^e}{dx}\right|_{x_b} Q_6^e \end{aligned} \tag{7.3.6b}$$

and Q_i^e are the generalized nodal forces, as shown in Fig. 7.3.1; $\mathbf{u}$ denotes the vector of nodal displacements associated with the linear approximation of

$u(x)$; and $\mathbf{\Delta}$ denotes the nodal displacements (transverse deflection and rotation at each node) associated with the Hermite cubic interpolation of $w(x)$. The element coefficient matrix size is 6×6 (i.e., with three nodal degrees of freedom per node).

7.3.1.2 Mixed finite element model

The finite element formulation based on the set of equations in Eqs. (7.2.12a)–(7.2.12c) is known as a *mixed* model because displacement variables (u, w) and a force variable (M_{xx}) are used as the nodal values. The weak forms of Eqs. (7.2.12a)–(7.2.12c) over a typical finite element (x_a^e, x_b^e) are:

$$\begin{aligned} 0 &= \int_{x_a^e}^{x_b^e} \left(\bar{A}_{xx}^e \frac{d\delta u}{dx}\frac{du}{dx} + \bar{B}_{xx}^e \frac{d\delta u}{dx} M_{xx} - \delta u\, f \right) dx \\ &\quad - \delta u(x_a^e)\, Q_1^e - \delta u(x_b)\, Q_4^e \end{aligned} \tag{7.3.7a}$$

$$0 = \int_{x_a^e}^{x_b^e} \left(\frac{d\delta w}{dx}\frac{dM_{xx}}{dx} + c_f^e\, \delta w\, w - \delta w\, q \right) - \delta w(x_a^e)\, Q_2^e - \delta w(x_b)\, Q_5^e \tag{7.3.7b}$$

$$\begin{aligned} 0 &= \int_{x_a^e}^{x_b^e} \left(\frac{d\delta M_{xx}}{dx}\frac{dw}{dx} - \frac{1}{D_{xx}^e}\delta M_{xx}\, M_{xx} + \frac{B_{xx}^e}{D_{xx}^e}\,\delta M_{xx}\frac{du}{dx} \right) dx \\ &\quad + Q_3^e\, [-\delta M_{xx}(x_a^e)] + Q_6^e\, \delta M_{xx}(x_b^e) \end{aligned} \tag{7.3.7c}$$

where (see Fig. 7.3.2)

$$\begin{aligned} Q_1^e &= -N_{xx}(x_a^e), \quad Q_2^e = -\left[\frac{dM_{xx}}{dx}\right]_{x_a^e}, \quad Q_3^e = -\left.\frac{dw}{dx}\right|_{x_a^e} \\ Q_4^e &= N_{xx}(x_b^e), \quad Q_5^e = \left[\frac{dM_{xx}}{dx}\right]_{x_b^e}, \quad Q_6^e = -\left.\frac{dw}{dx}\right|_{x_b^e} \end{aligned} \tag{7.3.8}$$

Here (Q_1^e, Q_2^e, Q_3^e) and (Q_4^e, Q_6^e, Q_6^e) denote the axial force, transverse shear force, and rotation at nodes 1 and 2, respectively (see Fig 7.3.2). We note that there are three duality pairs in the mixed formulation: $(u, N_{xx}), (w, dM_{xx}/dx)$, and (M_{xx}, θ_x). Clearly, the weak forms in Eqs. (7.3.7a)–(7.3.7c) admit Lagrange interpolations of degree as low as linear for u, w, and M_{xx}.

The *mixed finite element model* of the EBT is developed using the Lagrange interpolation of all variables (u, w, M_{xx}),

$$\begin{aligned} u_h^e(x) &= \sum_{j=1}^{m} u_j^e\, \psi_j^{(1)}(x), \\ w_h^e(x) &= \sum_{j=1}^{n} w_j^e\, \psi_j^{(2)}(x), \\ M_{xx}^e(x) &= \sum_{j=1}^{p} M_j^e\, \psi_j^{(3)}(x), \end{aligned} \tag{7.3.9}$$

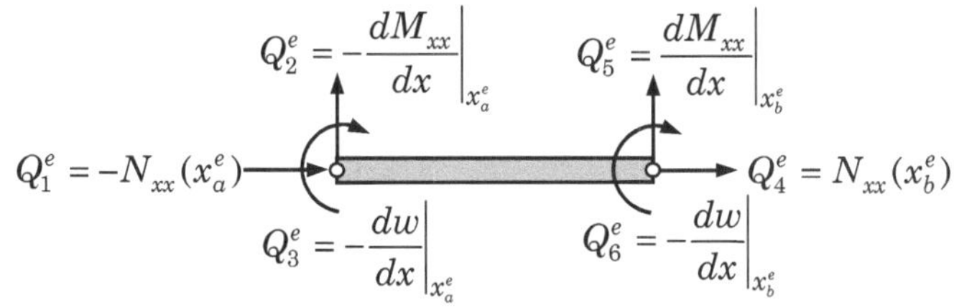

Fig. 7.3.2 Generalized forces in the mixed finite element model of the EBT.

where $\psi_i^{(\alpha)}$ ($\alpha = 1, 2, 3$) are the Lagrange interpolation functions of possibly different degree. However, in the present study we use the same (i.e., $m = n = p$) degree of interpolation for all three variables.

Substitution of the expansions from Eq. (7.3.4) into the weak forms of Eqs. (7.3.7a)–(7.3.7c), we obtain (the label e is omitted when the quantities already have superscript)

$$\begin{bmatrix} \mathbf{K}^{11} & \mathbf{K}^{12} & \mathbf{K}^{12} \\ \mathbf{K}^{21} & \mathbf{K}^{22} & \mathbf{K}^{23} \\ \mathbf{K}^{31} & \mathbf{K}^{32} & \mathbf{K}^{33} \end{bmatrix} \begin{Bmatrix} \mathbf{u} \\ \mathbf{w} \\ \mathbf{M} \end{Bmatrix} = \begin{Bmatrix} \mathbf{F}^1 \\ \mathbf{F}^2 \\ \mathbf{F}^3 \end{Bmatrix}, \tag{7.3.10a}$$

where

$$K_{ij}^{11} = \int_{x_a^e}^{x_b^e} \bar{A}_{xx}^e \frac{d\psi_i^{(1)}}{dx}\frac{d\psi_j^{(1)}}{dx}\, dx, \quad K_{ij}^{12} = 0 \quad K_{ij}^{13} = \int_{x_a^e}^{x_b^e} \bar{B}_{xx}^e \frac{d\psi_i^{(1)}}{dx}\psi_j^{(3)}\, dx,$$

$$F_i^1 = \psi_i^{(1)}(x_a^e)Q_1^e + \psi_i^{(1)}(x_b^e)Q_4^e, \quad K_{ij}^{21} = 0, \quad K_{ij}^{22} = \int_{x_a^e}^{x_b^e} c_f\, \psi_i^{(2)}\psi_j^{(2)}\, dx,$$

$$K_{ij}^{23} = \int_{x_a^e}^{x_b^e} \frac{d\psi_i^{(2)}}{dx}\frac{d\psi_j^{(3)}}{dx}\, dx, \quad F_I^2 = \int_{x_a^e}^{x_b^e} \psi_i^{(2)} q\, dx + \psi_i^{(2)}(x_a^e)\, Q_2^e + \psi_i^{(2)}(x_b^e)\, Q_5^e,$$

$$K_{ij}^{31} = \int_{x_a^e}^{x_b^e} \bar{B}_{xx}^e\, \psi_i^{(3)} \frac{d\psi_j^{(1)}}{dx}\, dx, \quad K_{ij}^{32} = \int_{x_a^e}^{x_b^e} \frac{d\psi_i^{(3)}}{dx}\frac{d\psi_j^{(2)}}{dx} dx,$$

$$K_{ij}^{33} = -\int_{x_a^e}^{x_b^e} \frac{1}{D_{xx}^e}\psi_i^{(3)}\psi_j^{(3)}\, dx, \quad F_I^3 = \psi_i^{(3)}(x_a^e)\, Q_3^e + \psi_i^{(3)}(x_b^e)\, Q_6^e. \tag{7.3.10b}$$

7.3.2 Timoshenko Beam Theory

7.3.2.1 Displacement finite element model

The weak forms of the equilibrium equations in terms of the displacements, Eqs. (7.2.20a)–(7.2.20c), are (again, the weak-form development follows the same ideas as discussed in Section 4.2.4):

$$0 = \int_{x_a^e}^{x_b^e} \left[\frac{d\delta u}{dx} \left(A_{xx}^e \frac{du}{dx} + B_{xx}^e \frac{d\phi_x}{dx} \right) - \delta u\, f \right] dx - \delta u(x_a^e)\, Q_1^e - \delta u(x_b^e)\, Q_4, \tag{7.3.11a}$$

$$0 = \int_{x_a^e}^{x_b^e} \left\{ \frac{d\delta w}{dx} \left[S_{xz}^e \left(\phi_x + \frac{dw}{dx} \right) \right] + c_f^e\, \delta w\, w - \delta w\, q^e(x) \right\} dx - \delta w(x_a^e)\, Q_2^e - \delta w(x_b^e)\, Q_5^e, \tag{7.3.11b}$$

$$0 = \int_{x_a^e}^{x_b^e} \left[\frac{d\delta \phi_x}{dx} \left(B_{xx}^e \frac{du}{dx} + D_{xx}^e \frac{d\phi_x}{dx} \right) + S_{xz}^e\, \delta\phi_x \left(\phi_x + \frac{dw}{dx} \right) \right] dx - \delta\phi_x(x_a^e)\, Q_3^e - \delta\phi_x(x_b^e)\, Q_6^e, \tag{7.3.11c}$$

where Q_i^e are the generalized forces at the element nodes (see Fig. 7.3.3)

$$\begin{aligned} Q_1^e &= -N_{xx}(x_a^e), \quad Q_2^e = -Q_x(x_a^e), \quad Q_3^e = -M_{xx}(x_a^e) \\ Q_4^e &= N_{xx}(x_b^e), \qquad Q_5^e = Q_x(x_b^e), \qquad Q_6^e = M_{xx}(x_b^e). \end{aligned} \tag{7.3.12}$$

The three duality pairs of the displacement model of the TBT are: (u, N_{xx}), (w, Q_x), and (ϕ_x, M_{xx}). The weak forms indicate that Lagrange interpolation of $(u,\ w,\ \phi_x)$ are required.

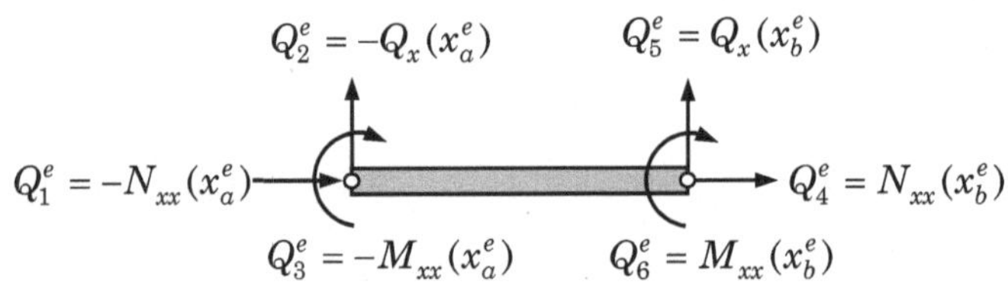

Fig. 7.3.3 Generalized nodal forces in the displacement model of the TBT.

The *displacement finite element model* of the TBT is obtained by substituting finite element approximations of the form,

$$u_h^e(x) = \sum_{j=1}^{m} u_j^e \psi_j^{(1)}(x), \quad w_h^e(x) = \sum_{j=1}^{n} w_j^e \psi_j^{(2)}(x), \quad \phi_x^e(x) = \sum_{j=1}^{p} s_j^e \psi_j^{(3)}(x). \tag{7.3.13}$$

Substituting Eq. (7.3.13) into the weak forms of Eqs. (7.3.11a)–(7.3.11c), we obtain

$$\begin{bmatrix} \mathbf{K}^{11} & \mathbf{K}^{12} & \mathbf{K}^{13} \\ \mathbf{K}^{21} & \mathbf{K}^{22} & \mathbf{K}^{23} \\ \mathbf{K}^{31} & \mathbf{K}^{32} & \mathbf{K}^{33} \end{bmatrix} \begin{Bmatrix} \mathbf{u} \\ \mathbf{w} \\ \mathbf{s} \end{Bmatrix} = \begin{Bmatrix} \mathbf{F}^1 \\ \mathbf{F}^2 \\ \mathbf{F}^3, \end{Bmatrix} \tag{7.3.14a}$$

where

$$K_{ij}^{11} = \int_{x_a^e}^{x_b^e} A_{xx}^e \frac{d\psi_i^{(1)}}{dx}\frac{d\psi_j^{(1)}}{dx}\,dx, \quad K_{ij}^{12} = 0,$$

$$K_{ij}^{13} = \int_{x_a^e}^{x_b^e} B_{xx}^e \frac{d\psi_i^{(1)}}{dx}\frac{d\psi_j^{(3)}}{dx}\,dx, \quad K_{ij}^{21} = 0,$$

$$F_i^1 = \int_{x_a^e}^{x_b} f\,\psi_i^{(1)}\,dx + \psi_i^{(1)}(x_a^e)Q_1^e + \psi_i^{(1)}(x_b^e)Q_4^e$$

$$K_{ij}^{22} = \int_{x_a^e}^{x_b^e} \left(S_{xz}^e \frac{d\psi_i^{(2)}}{dx}\frac{d\psi_j^{(2)}}{dx} + c_f^e\,\psi_i^{(2)}\psi_j^{(2)}\right)dx, \quad K_{ij}^{23} = \int_{x_a^e}^{x_b^e} S_{xz}^e \frac{d\psi^{(2)}}{dx}\psi_j^{(3)}\,dx,$$

$$F_i^2 = \int_{x_a^e}^{x_b^e} q\psi_i^{(2)}\,dx + \psi_i^{(2)}(x_a^e)Q_2^e + \psi_i^{(2)}(x_b^e)\,Q_5^e,$$

$$K_{ij}^{31} = \int_{x_a^e}^{x_b^e} B_{xx}^e \frac{d\psi_i^{(3)}}{dx}\frac{d\psi_j^{(1)}}{dx}\,dx, \quad K_{ij}^{32} = \int_{x_a^e}^{x_b^e} S_{xz}^e\,\psi_i^{(3)}\frac{d\psi_j^{(2)}}{dx}\,dx,$$

$$K_{ij}^{33} = \int_{x_a^e}^{x_b^e} \left(S_{xz}^e\,\psi_i^{(3)}\psi_j^{(3)} + D_{xx}^e \frac{d\psi_i^{(3)}}{dx}\frac{d\psi_j^{(3)}}{dx}\right)dx,$$

$$F_i^3 = \psi_i^{(3)}(x_a^e)\,Q_3^e + \psi_i^{(3)}(x_b^e)\,Q_6^e. \tag{7.3.14b}$$

We use the same degree of interpolation for all variables: $\psi_i^{(1)} = \psi_i^{(2)} = \psi_i^{(3)}$.

It is well-known that equal and lower-order interpolation of w and ϕ_x results in overly constrained finite element equations because the shear strain $\gamma_{xz} = \phi_x + dw/dx$ cannot go to zero when the beam becomes thin (because thin beams have vanishing shear strain). This is known as the *shear locking*, which is alleviated by using selective reduced integration in which integrals of shear terms involving ϕ_x are evaluated as if ϕ_x is interpolated using a lower-order polynomial (see Reddy [8, 13] for details). Shear locking goes away as we increase the degree of interpolation used for w and ϕ_x (especially when both are interpolated using cubic polynomials).

7.3.2.2 Mixed finite element model

The mixed finite element model of the TBT is based on the weak forms of Eqs. (7.2.25a)–(7.2.25c), and it has the same form as the mixed finite element model of the EBT because of the similarity of the equations in terms of (u, w, M_{xx}) in the two theories, and the coefficients $K_{ij}^{\alpha\beta}$ and F_i^{α} $(\alpha, \beta = 1, 2, 3)$ also remain the same as those in Eq. (7.3.10b), except for the following coefficient:

$$K_{ij}^{33} = -\int_{x_a^e}^{x_b^e} \left(\frac{1}{D_{xx}^e}\psi_i^{(3)}\psi_j^{(3)} + \frac{1}{S_{xz}^e}\frac{d\psi_i^{(3)}}{dx}\frac{d\psi_j^{(3)}}{dx}\right)dx. \tag{7.3.15}$$

The mixed model does not suffer from shear locking.

7.4 Linear Dual Mesh Control Domain Model

7.4.1 Euler–Bernoulli Beam Theory

The DMCDM is applicable to second-order differential equations. Therefore, we can only consider the mixed model of the EBT. In the DMCDM, we divide the domain $\Omega = (0, L)$ into a primal mesh of N finite elements (with their interpolation), as shown in Fig. 7.4.1, with each node having its own control domain (a dual mesh). The finite element nodes as well as the control domains are numbered sequentially from the left to the right. Except for the control domains associated with the boundary nodes, all other control domains are node-centered. We consider two adjacent elements of the primal mesh connected at a typical interior node I and control domain associated with that node (see Fig. 7.4.2). We note that the primal mesh of finite elements does not have to be uniform.

Next, we derive the discretized equations associated with Eqs. (7.2.12a)–(7.2.12c). The complete steps of the DMCDM are presented by considering Eq. (7.2.12a) first, and then the results are summarized for Eqs. (7.2.12b) and (7.2.12c).

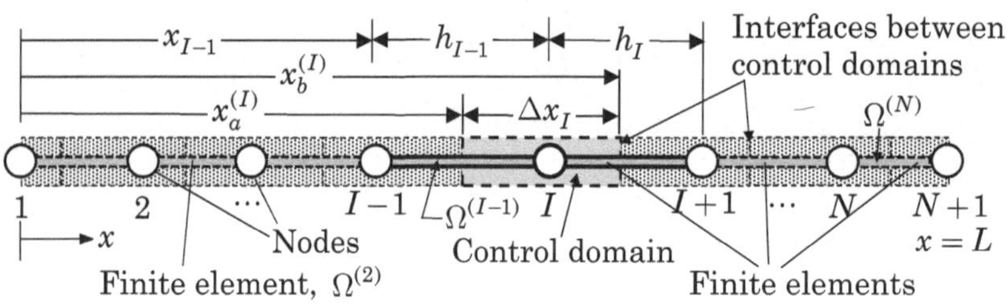

Fig. 7.4.1 Primal mesh of finite elements and a dual mesh of control domains.

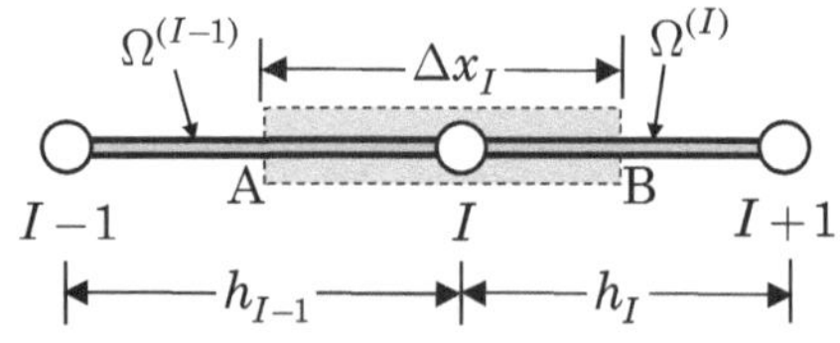

Fig. 7.4.2 Control domain associated with an interior node I.

7.4.1.1 Control domain statement

The first step is to set up the integral statement of Eq. (7.2.12a) over a typical control domain:

$$0 = \int_A^B \left[-\frac{d}{dx}\left(\bar{A}_{xx}\frac{du}{dx} + \bar{B}_{xx}M_{xx} \right) - f \right] dx, \tag{7.4.1}$$

where A and B refer to the left and right end locations of the control domain (associated with node I), which have the coordinates $x_a^{(I)}$ and $x_b^{(I)}$, respectively, as shown in Fig. 7.4.1. Unlike in the FEM, we weaken the differentiability on u by carrying out the indicated integration and obtain

$$\begin{aligned} 0 &= \int_{x_a^{(I)}}^{x_b^{(I)}} \left[-\frac{d}{dx}\left(\bar{A}_{xx}\frac{du}{dx} + \bar{B}_{xx}M_{xx} \right) - f \right] dx \\ &= \left[\bar{A}_{xx}\frac{du}{dx} + \bar{B}_{xx}M_{xx} \right]_{x_a^{(I)}} - \left[\bar{A}_{xx}\frac{du}{dx} + \bar{B}_{xx}M_{xx} \right]_{x_b^{(I)}} - \int_{x_a^{(I)}}^{x_b^{(I)}} f\,dx \end{aligned} \quad (7.4.2)$$

or

$$0 = -N_1^{(I)} - N_2^{(I)} - F_I, \quad (7.4.3a)$$

where

$$N_1^{(I)} \equiv -\left[\bar{A}_{xx}\frac{du}{dx} + \bar{B}_{xx}M_{xx} \right]_{x_a^{(I)}}, \quad N_2^{(I)} \equiv \left[\bar{A}_{xx}\frac{du}{dx} + \bar{B}_{xx}M_{xx} \right]_{x_b^{(I)}}, \quad (7.4.3b)$$

$$F_I = \int_{x_a^{(I)}}^{x_b^{(I)}} f\,dx. \quad (7.4.3c)$$

Here $N_1^{(I)}$ and $N_2^{(I)}$ denote the secondary variables (axial forces) at the left and right interfaces of the control domain centered at node I. The minus sign in the definition of $N_1^{(I)}$ indicates that it is a compressive force, and both $N_1^{(I)}$ and $N_2^{(I)}$ are axial forces in the positive x-direction.

7.4.1.2 Discretized equations

Next, we use finite element approximations of $u(x)$, $w(x)$, and $M_{xx}(x)$ over a typical finite element, $\Omega^{(I)} = (x_I, x_{I+1})$. For example, $u(x)$ is approximated using linear interpolation

$$u(x) \approx u_h(x) \equiv U_I\psi_1^{(I)}(x) + U_{I+1}\psi_2^{(I)}(x), \quad (7.4.4)$$

where U_I is the value of u at node I (i.e., $U_I \approx u(x_I)$) and $\psi_i^{(I)}(x)$ $(i = 1, 2)$ are linear finite element interpolation functions of element $\Omega^{(I)}$ for $I = 1, 2, \ldots, N$ (see Fig. 7.4.3 for the nodal degrees of freedom):

$$\psi_1^{(I)}(x) = \frac{x_{I+1} - x}{h_I}, \quad \psi_2^{(I)}(x) = \frac{x - x_I}{h_I}. \quad (7.4.5)$$

Hence, we can calculate parts (because of the nonlinearity involving w) of $N_1^{(I)}$ and $N_2^{(I)}$ in Eq. (7.4.3b) using the linear interpolation for each of the dependent variables (u, w, M_{xx}) of the formulation. We note that point A is in element $\Omega^{(I-1)}$, and point B is in element $\Omega^{(I)}$; see Figs. 7.4.3 and 7.4.4.

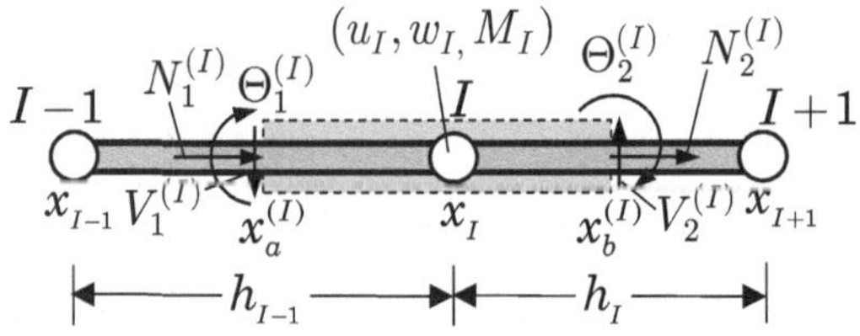

Fig. 7.4.3 Typical control domain for the mixed model of the EBT.

The following formulas are employed in the development of the DMCDM discretized equations, where the generic coefficients a and c are assumed to be element-wise constants:

$$\int_{x_a^{(I)}}^{x_b^{(I)}} c\, u\, dx = \tfrac{1}{8}\Big[C_{I-1}h_{I-1}U_{I-1} + 3\big(C_{I-1}h_{I-1} + C_I h_I\big)U_I + C_I h_I U_{I+1}\Big] \tag{7.4.6a}$$

$$\int_{0}^{0.5h_1} a\, u(x)\, dx = \tfrac{1}{8}A_1 h_1\,(U_2 + 3U_1) \tag{7.4.6b}$$

$$\int_{x_b^{(N)}-0.5h_N}^{x_b^{(N)}} a\, u(x)\, dx = \tfrac{1}{8}A_N h_N\,(3U_{N+1} + U_N) \tag{7.4.6c}$$

$$\int_{x_a^{(I)}}^{x_b^{(I)}} a\,\frac{du}{dx}\, dx = \tfrac{1}{2}\left[-A_{I-1}U_{I-1} + (A_{I-1} - A_I)\,U_I + A_I U_{I+1}\right] \tag{7.4.6d}$$

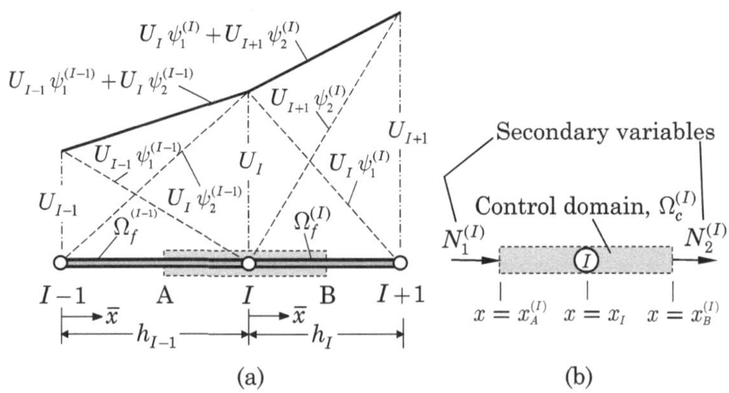

Fig. 7.4.4 (a) Linear finite element approximation over two neighboring finite elements. (b) Secondary variables over a typical control domain.

$$\int_0^{0.5h_1} a\,\frac{du}{dx}\,dx = \tfrac{1}{2}\,A_1\,(U_2 - U_1) \tag{7.4.6e}$$

$$\int_{x_b^{(N)}-0.5h_N}^{x_b^{(N)}} a\,\frac{du}{dx}\,dx = \tfrac{1}{2}\,A_N\,(U_{N+1} - U_N) \tag{7.4.6f}$$

$$\left[a\,\frac{du}{dx}\right]_{x_a^{(I)}}^{x_b^{(I)}} = \frac{A_{I-1}}{h_{I-1}}\,U_{I-1} - \left(\frac{A_{I-1}}{h_{I-1}} + \frac{A_I}{h_I}\right)U_I + \frac{A_I}{h_I}\,U_{I+1} \tag{7.4.6g}$$

$$\left[a\,\frac{du}{dx}\right]_0 = \frac{A_1}{h_1}\,U_2 - \frac{A_1}{h_1}\,U_1, \tag{7.4.6h}$$

$$\left[a\,\frac{du}{dx}\right]_{x_b^{(I)}+0.5h_N} = \frac{A_N}{h_N}\,U_{N+1} - \frac{A_N}{h_N}\,U_N \tag{7.4.6i}$$

$$[a\,u(x)]_{x_a^{(I)}}^{x_b^{(I)}} = \tfrac{1}{2}\left[-A_{I-1}U_{I-1} - (A_{I-1} - A_I)\,U_I + A_I U_{I+1}\right] \tag{7.4.6j}$$

$$[a\,u(x)]_0^{0.5h_1} = \tfrac{1}{2}\,A_1\,(U_2 - U_1)\,, \quad [a\,u(x)]_{0.5h_1} = \tfrac{1}{2}\,A_1\,(U_1 + U_2) \tag{7.4.6k}$$

$$[a\,u(x)]_{x_b^{(I)}-0.5h_N}^{x_b^{(N)}} = \tfrac{1}{2}\,A_N\,(U_{N+1} - U_N)\,, \tag{7.4.6l}$$

$$[a\,u(x)]_{0.5h_N} = \tfrac{1}{2}\,A_N\,(U_N + U_{N+1}) \tag{7.4.6m}$$

$$\begin{aligned}\left[\bar{A}_{xx}\,\frac{du}{dx} + \bar{B}_{xx}\,M_{xx}\right]_{x_a^{(I)}}^{x_b^{(I)}} &= \frac{\bar{A}_{I-1}}{h_{I-1}}\,U_{I-1} - \left(\frac{\bar{A}_{I-1}}{h_{I-1}} + \frac{\bar{A}_I}{h_I}\right)U_I + \frac{\bar{A}_I}{h_I}\,U_{I+1}\\ &\quad + \tfrac{1}{2}\left[-\bar{B}_{I-1}M_{I-1} + \left(\bar{B}_{I-1} - \bar{B}_I\right)M_I + \bar{B}_I M_{I+1}\right]\end{aligned} \tag{7.4.6n}$$

where $\bar{A}_{I-1} = \bar{A}_{xx}(x_a^{(I)})$ at the left interface and $\bar{A}_I = \bar{A}_{xx}(x_b^{(I)})$ at the right interface of the control domain centered around node I. Similar meaning applies to $\bar{B}_{I-1}$ and $\bar{B}_I$; M_I denotes the nodal value of M_{xx} at node I.

Substituting the representations in Eqs. (7.4.6n) into Eq. (7.4.3a), we obtain (for $I = 2, 3, \ldots, N$)

$$\begin{aligned}-\frac{\bar{A}_{I-1}}{h_{I-1}}U_{I-1} + \left(\frac{\bar{A}_{I-1}}{h_{I-1}} + \frac{\bar{A}_I}{h_I}\right)U_I - \frac{\bar{A}_I}{h_I}U_{I+1} + \tfrac{1}{2}\bar{B}_{I-1}M_{I-1}\\ - \tfrac{1}{2}\left(\bar{B}_{I-1} - \bar{B}_I\right)M_I - \tfrac{1}{2}\bar{B}_I M_{I+1} - F_I = 0\end{aligned} \tag{7.4.7}$$

where

$$\bar{A}_{I-1} = \left.\frac{D^*_{xx}}{D_{xx}}\right|_{x_a^{(I)}}, \quad \bar{A}_I = \left.\frac{D^*_{xx}}{D_{xx}}\right|_{x_b^{(I)}}, \quad \bar{B}_{I-1} = \left.\frac{B_{xx}}{D_{xx}}\right|_{x_a^{(I)}}, \quad \bar{B}_I = \left.\frac{B_{xx}}{D_{xx}}\right|_{x_b^{(I)}} \tag{7.4.8}$$

The integral of a function $f(x)$ over the control domain $(x_a^{(I)}, x_b^{(I)})$ can be evaluated using either exact integration or numerical integration (e.g., trapezoidal rule, Simpson's one-third rule, and so on).

Lastly, we write the discretized equations for the boundary nodes [see Fig. 7.4.5 and Eqs. (7.4.6a)–(7.4.6n)]:

$$N_1^{(1)} = \frac{\bar{A}_1}{h_1}U_1 - \frac{\bar{A}_1}{h_1}U_2 - \tfrac{1}{2}\bar{B}_1\, M_1 - \tfrac{1}{2}\bar{B}_1\, M_2 - F_1 \tag{7.4.9}$$

$$N_2^{(N+1)} = -\frac{\bar{A}_{N+1}}{h_N}U_N + \frac{\bar{A}_{N+1}}{h_N}U_{N+1} + \tfrac{1}{2}\bar{B}_{N+1}\, M_N + \tfrac{1}{2}\bar{B}_{N+1}\, M_{N+1} - F_{N+1}. \tag{7.4.10}$$

where $\bar{x}$ is the local coordinate with origin at node 1 of element N and $N_1^{(1)}$ and $N_2^{(N+1)}$ are the boundary forces (at nodes 1 and $N+1$, respectively), which are either specified or their duality counterparts, namely, the displacements U_1 and $U_{(N+1)}$, are specified. This completes the discretization of Eq. (7.2.12a).

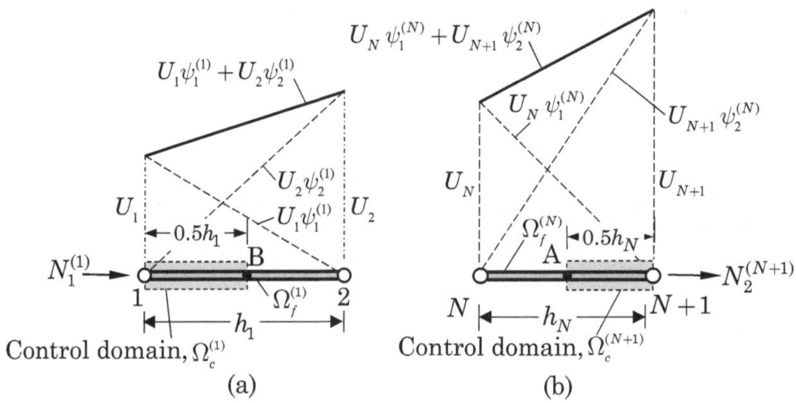

Fig. 7.4.5 (a) Half-control domain at the left boundary node. (b) Half-control domain at the right boundary node.

The same procedure can be applied to Eqs. (7.2.12b) and (7.2.12c) to obtain the discretized equations for the interior and boundary nodes. We have the following integral statements of Eqs. (7.2.12b) and (7.2.12c):

$$0 = -V_1^{(I)} - V_2^{(I)} + \int_{x_a^{(I)}}^{x_b^{(I)}} (c_f w - q)\, dx, \tag{7.4.11}$$

$$0 = -\Theta_1^{(I)} - \Theta_2^{(I)} + \int_{x_a^{(I)}}^{x_b^{(I)}} \left(-\frac{1}{D_{xx}}M_{xx} + \bar{B}_{xx}\frac{du}{dx}\right) dx, \tag{7.4.12}$$

where $V_1^{(I)}$ and $V_2^{(I)}$ denote the secondary variables (shear forces acting upward positive) at the left and right interfaces of the control domain centered at node I,

$$V_1^{(I)} \equiv -\left[\frac{dM}{dx}\right]_{x_a^{(I)}} = -\frac{M_I - M_{I-1}}{h_{I-1}}, \quad V_2^{(I)} \equiv \left[\frac{dM}{dx}\right]_{x_b^{(I)}} = \frac{M_{I+1} - M_I}{h_I}, \tag{7.4.13}$$

and $\Theta_1^{(I)}$ and $\Theta_2^{(I)}$ denote the secondary variables (rotations in counterclockwise direction) at the left and right interfaces of the control domain centered at node I,

$$\Theta_1^{(I)} \equiv -\left[\frac{dw}{dx}\right]_{x_a^{(I)}} = \frac{W_{I-1} - W_I}{h_{I-1}}, \quad \Theta_2^{(I)} \equiv \left[\frac{dw}{dx}\right]_{x_b^{(I)}} = \frac{W_{I+1} - W_I}{h_I}. \tag{7.4.14}$$

The discretized equations associated with Eq. (7.2.12b) at an interior node are:

$$-\frac{1}{h_{I-1}} M_{I-1} + \left(\frac{1}{h_{I-1}} + \frac{1}{h_I}\right) M_I - \frac{1}{h_I} M_{I+1} + \tfrac{1}{8} C_{I-1} h_{I-1} W_{I-1} \\ + \tfrac{3}{8}\left(C_{I-1} h_{I-1} + C_I h_I\right) W_I + \tfrac{1}{8} C_I h_I W_{I+1} = Q_I, \tag{7.4.15}$$

where $C^{(I)}$ denotes the value of c_f in element I and Q_I has the same form as Eq. (7.4.3c). For the boundary nodes 1 and $N+1$, we have

$$V_1^{(1)} = \frac{1}{h_1} M_1 - \frac{1}{h_1} M_2 + \tfrac{3}{8} h_1 C_1 W_1 + \tfrac{1}{8} h_1 C_1 W_2 - Q_1 \tag{7.4.16}$$

$$V_2^{(N+1)} = -\frac{1}{h_N} M_N + \frac{1}{h_N} M_{N+1} + \tfrac{1}{8} h_N C_N W_N + \tfrac{3}{8} h_N C_N W_{N+1} - Q_{N+1}. \tag{7.4.17}$$

The discretized equations associated with Eq. (7.2.12c) are

$$-\frac{1}{h_{I-1}} W_{I-1} + \left(\frac{1}{h_{I-1}} + \frac{1}{h_I}\right) W_I - \frac{1}{h_I} W_{I+1} - \frac{1}{8}\frac{h_{I-1}}{D_{I-1}} M_{I-1} \\ - \frac{3}{8}\left(\frac{h_{I-1}}{D_{I-1}} + \frac{h_I}{D_I}\right) M_I - \frac{1}{8}\frac{h_I}{D_I} M_{I+1} \\ - 0.5\bar{B}_{I-1} U_{I-1} + 0.5\left(\bar{B}_{I-1} - \bar{B}_I\right) U_I + 0.5\bar{B}_I U_{I+1} = 0 \tag{7.4.18}$$

for an interior node. Here D_I denotes the value of D_{xx} in element I, and $\bar{B}_I$ denotes the value of B_{xx}/D_{xx} in element I. For the boundary nodes 1 and $N+1$, we have

$$\Theta_1^{(1)} = \frac{1}{h_1} W_1 - \frac{1}{h_1} W_2 - \tfrac{3}{8}\frac{h_1}{D_1} M_1 - \tfrac{1}{8}\frac{h_1}{D_1} M_2 + \tfrac{1}{2}\bar{B}_1\left(U_2 - U_1\right) \tag{7.4.19}$$

$$\Theta_2^{(N+1)} = -\frac{1}{h_N} W_N + \frac{1}{h_N} W_{N+1} - \tfrac{1}{8}\frac{h_N}{D_N} M_N - \tfrac{3}{8}\frac{h_N}{D_N} M_{N+1} \\ + \tfrac{1}{2}\bar{B}_N\left(U_{N+1} - U_N\right). \tag{7.4.20}$$

This completes the development of the discretized equations using the DMCDM for the mixed formulation of the Euler–Bernoulli beam theory.

7.4.2 Timoshenko Beams

7.4.2.1 Displacement model

In order to derive the discretized equations associated with Eqs. (7.2.20a) (7.2.20c), we follow the same procedure as described for the EBT. The integral statements over the Ith control domain centered around node I, occupying the domain between points A and B (see Fig. 7.4.6 for the nodal degrees of freedom) for each of these three equations are:

$$0 = -N_1^{(I)} - N_2^{(I)} - F_I, \tag{7.4.21a}$$

$$N_1^{(I)} \equiv -\left[A_{xx}\frac{du}{dx} + B_{xx}\frac{d\phi_x}{dx}\right]_{x_a^{(I)}}, \quad N_2^{(I)} \equiv \left[A_{xx}\frac{du}{dx} + B_{xx}\frac{d\phi_x}{dx}\right]_{x_b^{(I)}}, \tag{7.4.21b}$$

$$0 = -V_1^{(I)} - V_2^{(I)} + \int_{x_a^{(I)}}^{x_b^{(I)}} (c_f w - q)dx, \tag{7.4.22a}$$

$$V_1^{(I)} \equiv -\left[S_{xz}\left(\phi_x + \frac{dw}{dx}\right)\right]_{x_a^{(I)}}, \quad V_2^{(I)} \equiv \left[S_{xz}\left(\phi_x + \frac{dw}{dx}\right)\right]_{x_b^{(I)}}, \tag{7.4.22b}$$

$$0 = -M_1^{(I)} - M_2^{(I)} + \int_{x_a^{(I)}}^{x_b^{(I)}} S_{xz}\left(\phi_x + \frac{dw}{dx}\right)dx, \tag{7.4.23a}$$

$$M_1^{(I)} \equiv -\left[B_{xx}\frac{du}{dx} + D_{xx}\frac{d\phi_x}{dx}\right]_{x_a^{(I)}}, \quad M_2^{(I)} \equiv \left[B_{xx}\frac{du}{dx} + D_{xx}\frac{d\phi_x}{dx}\right]_{x_b^{(I)}}. \tag{7.4.23b}$$

Here $(N_1^{(I)}, V_1^{(I)}, M_1^{(I)})$ and $(N_2^{(I)}, V_2^{(I)}, M_2^{(I)})$ denote the axial forces, shear forces, and bending moments at the left and right interfaces, respectively, of the control domain centered at node I (see Fig. 7.4.6). Since the displacement model of the TBT suffers from shear locking, we evaluate the integral appearing in Eq. (7.4.23a) (i.e., the shear force) as a constant to avoid shear locking:

$$\begin{aligned}\int_{x_a^{(I)}}^{x_b^{(I)}} S_{xz}\left(\phi_x + \frac{dw}{dx}\right)dx &= \tfrac{1}{2}S_{I-1}\left(\Phi_{I-1} + \Phi_I\right)\frac{h_{I-1}}{2} + \tfrac{1}{2}S_I\left(\Phi_I + \Phi_{I+1}\right)\frac{h_I}{2}\\ &+ S_{I-1}\frac{W_I - W_{I-1}}{h_{I-1}}\frac{h_{I-1}}{2}\\ &+ S_I\frac{W_{I+1} - W_I}{h_I}\frac{h_I}{2}\end{aligned} \tag{7.4.24}$$

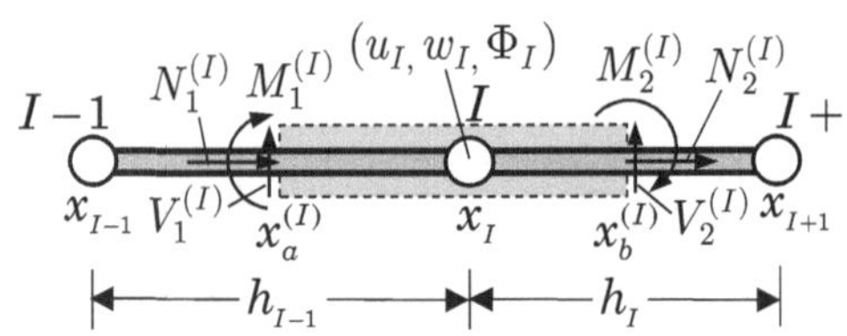

Fig. 7.4.6 Typical control domain for the displacement model of the TBT.

The discretized equations associated with Eqs. (7.4.21a), (7.4.22a), and (7.4.23a) for an interior node I are

$$\begin{aligned}0 = &-\frac{A_{I-1}}{h_{I-1}}U_{I-1} + \left(\frac{A_{I-1}}{h_{I-1}} + \frac{A_I}{h_I}\right)U_I - \frac{A_I}{h_I}U_{I+1}\\ &- \frac{B_{I-1}}{h_{I-1}}\Phi_{I-1} + \left(\frac{B_{I-1}}{h_{I-1}} + \frac{B_I}{h_I}\right)\Phi_I - \frac{B_I}{h_I}\Phi_{I+1} - F_I\end{aligned} \tag{7.4.25a}$$

$$\begin{aligned}0 = &-\frac{S_{I-1}}{h_{I-1}}W_{I-1} + \left(\frac{S_{I-1}}{h_{I-1}} + \frac{S_I}{h_I}\right)W_I - \frac{S_I}{h_I}W_{I+1} + \tfrac{1}{8}C_{I-1}W_{I-1}h_{I-1}\\ &+ \tfrac{3}{8}\left(C_{I-1}h_{I-1} + C_I h_I\right)W_I + \tfrac{1}{8}C_I W_{I+1}h_I + \tfrac{1}{2}S_{I-1}\Phi_{I-1}\\ &+ \tfrac{1}{2}\left(S_{I-1} - S_I\right)\Phi_I - \tfrac{1}{2}S_I\Phi_{I+1} - Q_I\end{aligned} \tag{7.4.25b}$$

$$\begin{aligned}0 = &-\frac{B_{I-1}}{h_{I-1}}U_{I-1} + \left(\frac{B_{I-1}}{h_{I-1}} + \frac{B_I}{h_I}\right)U_I - \frac{B_I}{h_I}U_{I+1} - \tfrac{1}{2}S_{I-1}W_{I-1}\\ &+ \tfrac{1}{2}\left(S_{I-1} - S_I\right)W_I + \tfrac{1}{2}S_I W_{I+1} - \frac{D_{I-1}}{h_{I-1}}\Phi_{I-1} + \left(\frac{D_{I-1}}{h_{I-1}} + \frac{D_I}{h_I}\right)\Phi_I\\ &- \frac{D_I}{h_I}\Phi_{I+1} + \tfrac{1}{4}S_{I-1}h_{I-1}\Phi_{I-1} + \tfrac{1}{4}\left(S_{I-1}h_{I-1} + S_I h_I\right)\Phi_I\\ &+ \tfrac{1}{4}S_I h_I\Phi_{I+1}\end{aligned} \tag{7.4.25c}$$

where

$$F_I = \int_{x_a^{(I)}}^{x_b^{(I)}} f\,dx, \quad Q_I = \int_{x_a^{(I)}}^{x_b^{(I)}} q\,dx, \tag{7.4.25d}$$

The discretized equations of the left boundary node are

$$0 = -N_1^{(1)} + \frac{A_1}{h_1}U_1 - \frac{A_1}{h_1}U_2 + \frac{B_1}{h_1}\Phi_1 - \frac{B_1}{h_1}\Phi_2 - F_1 \tag{7.4.26a}$$

$$0 = -V_1^{(1)} + \frac{S_1}{h_1}W_1 - \frac{S_1}{h_1}W_2 - \tfrac{1}{2}S_1\Phi_1 - \tfrac{1}{2}S_1\Phi_2 + \frac{h_1}{8}C_1\left(3W_1 + W_2\right) - Q_1 \tag{7.4.26b}$$

$$\begin{aligned}0 = &-M_1^{(1)} - \frac{B_1}{h_1}\left(U_2 - U_1\right) + \tfrac{1}{2}S_1(W_2 - W_1) - \frac{D_1}{h_1}\left(\Phi_2 - \Phi_1\right)\\ &+ \tfrac{1}{4}S_1 h_1\left(\Phi_1 + \Phi_2\right)\end{aligned} \tag{7.4.26c}$$

For the node on the right boundary, we have

$$0 = -N_1^{(N+1)} + \frac{A_N}{h_N}U_{N+1} - \frac{A_N}{h_1}U_N + \frac{B_N}{h_N}\Phi_{N+1} - \frac{B_N}{h_N}\Phi_N - F_{N+1} \quad (7.4.27a)$$

$$0 = -V_2^{(N+1)} + \frac{S_N}{h_N}\left(W_{N+1} - W_N\right) + \frac{h_N}{8}C_N\left(W_N + 3W_{N+1}\right)$$
$$+ \tfrac{1}{2}S_N(\Phi_{N+1} + \Phi_N) - Q_{N+1} \quad (7.4.27b)$$

$$0 = -M_2^{(N+1)} + \frac{B_N}{h_N}\left(U_{N+1} - U_N\right) + \tfrac{1}{2}S_N(W_{N+1} - W_N)$$
$$+ \frac{D_N}{h_N}\left(\Phi_{N+1} - \Phi_N\right) + \tfrac{1}{4}S_N h_N\left(\Phi_N + \Phi_{N+1}\right) \quad (7.4.27c)$$

where

$$F_1 = \int_0^{0.5h_1} f(x)\,dx, \qquad Q_1 = \int_0^{0.5h_1} q(x)\,dx,$$
$$F_{N+1} = \int_{0.5h_N}^{h_N} f(\bar{x})\,d\bar{x}, \quad Q_{N+1} = \int_{0.5h_N}^{h_N} q(\bar{x})\,d\bar{x}. \quad (7.4.27d)$$

7.4.2.2 Mixed model

Lastly, we develop the mixed DMCDM model of Eqs. (7.2.25a)–(7.2.25c), without providing details of the steps which were amply illustrated in the preceding sections. The complete steps of the DMCDM are presented by considering Eq. (7.2.25a) first and then the results are summarized for Eqs. (7.2.25b) and (7.2.25c).

The discretized equations associated with Eq. (7.2.25a), which represents equilibrium of forces in the x direction, at a typical interior Ith node are given by

$$0 = -\frac{\bar{A}_{I-1}}{h_{I-1}}U_{I-1} + \left(\frac{\bar{A}_{I-1}}{h_{I-1}} + \frac{\bar{A}_I}{h_I}\right)U_I - \frac{\bar{A}_I}{h_I}U_{I+1} + \tfrac{1}{2}\bar{B}_{I-1}M_{I-1}$$
$$+ \tfrac{1}{2}\left(\bar{B}_{I-1} - \bar{B}_I\right)M_I - \tfrac{1}{2}\bar{B}_I M_{I+1} - F_I. \quad (7.4.28a)$$

The discretized equations associated with Eq. (7.2.25a) at boundary node 1 and boundary node $N+1$ are given by

$$0 = -N_1^{(1)} + \frac{\bar{A}_1}{h_1}U_1 - \frac{\bar{A}_1}{h_1}U_2 + \tfrac{1}{2}\bar{B}_1\,M_1 + \tfrac{1}{2}\bar{B}_1\,M_2 - F_1, \quad (7.4.28b)$$

$$0 = -N_2^{(N+1)} - \frac{\bar{A}_N}{h_N}U_N + \frac{\bar{A}_N}{h_N}U_{N+1} + \tfrac{1}{2}\bar{B}_N\,M_N + \tfrac{1}{2}\bar{B}_N\,M_{N+1}$$
$$- F_{N+1}. \quad (7.4.28c)$$

The discretized equations associated with Eq. (7.2.25b) are:

$$0 = -\frac{M_{I+1} - M_I}{h_I} + \frac{M_I - M_{I-1}}{h_{I-1}} + \tfrac{3}{8} W_I \left(C_{I-1} h_{I-1} + C_I h_I \right) + \tfrac{1}{8} C_{I-1} W_{I-1} h_{I-1} + \tfrac{1}{8} C_I W_{I+1} h_I - Q_I, \tag{7.4.29a}$$

$$0 = -V_1^{(1)} + \frac{1}{h_1} M_1 - \frac{1}{h_1} M_2 + \tfrac{1}{8} h_1 C_1 \left(3W_1 + W_2 \right) - Q_1, \tag{7.4.29b}$$

$$0 = -V_2^{(N+1)} + \frac{1}{h_N} M_{N+1} - \frac{1}{h_N} M_N + \tfrac{1}{8} h_N C_N \left(W_N + 3W_{N+1} \right) - Q_{N+1}. \tag{7.4.29c}$$

Finally, the discretized equations associated with Eq. (7.2.20c) at the Ith node, node 1, and node $N + 1$ are:

$$\begin{aligned} 0 = & -\tfrac{1}{2} \bar{B}_{I-1} U_{I-1} + \tfrac{1}{2} \left(\bar{B}_{I-1} - \bar{B}_I \right) U_I + \tfrac{1}{2} \bar{B}_I U_{I+1} \\ & - \frac{1}{h_{I-1}} W_{I-1} + \left(\frac{1}{h_{I-1}} + \frac{1}{h_I} \right) W_I - \frac{1}{h_I} W_{I+1} \\ & + \frac{1}{h_{I-1}} \frac{1}{S_{I-1}} M_{I-1} - \left(\frac{1}{h_{I-1}} \frac{1}{S_{I-1}} + \frac{1}{h_I} \frac{1}{S_I} \right) M_I + \frac{1}{h_I} \frac{1}{S_I} M_{I+1} \\ & - \tfrac{1}{8} \frac{h_{I-1}}{D_{I-1}} M_{I-1} - \tfrac{3}{8} \left(\frac{h_{I-1}}{D_{I-1}} + \frac{h_I}{D_I} \right) M_I - \tfrac{1}{8} \frac{h_I}{D_I} M_{I+1}, \end{aligned} \tag{7.4.30a}$$

$$0 = -\Theta_1^{(1)} + \tfrac{1}{2} \bar{B}_1 \left(U_2 - U_1 \right) - \frac{W_2 - W_1}{h_1} + \frac{1}{S_1} \frac{M_2 - M_1}{h_1} - \tfrac{1}{8} \frac{h_1}{D_1} \left(3M_1 + M_2 \right), \tag{7.4.30b}$$

$$\begin{aligned} 0 = & -\Theta_2^{(N+1)} + \tfrac{1}{2} \bar{B}_N U_{N+1} - \tfrac{1}{2} \bar{B}_N U_N \\ & - \frac{1}{h_N} W_N + \frac{1}{h_N} W_{N+1} + \frac{1}{S_N h_N} M_N - \frac{1}{S_N h_N} M_{N+1} \\ & - \tfrac{1}{8} \frac{h_N}{D_N} \left(M_N + 3M_{N+1} \right). \end{aligned} \tag{7.4.30c}$$

Various discrete DMCDM models developed in this section will be evaluated in comparison with the FEM models.

7.5 Numerical Results for Linear Problems

Here we consider homogeneous and FGM beams to illustrate the ideas presented in the previous sections. Two different examples, namely, pinned-pinned and clamped-clamped beams with uniformly distributed load (UDL) of intensity q_0 are considered. Due to symmetry about the center of the beam, only a half-beam, $0 \leq x \leq L/2$, is considered. Numerical results obtained with the FEM and DMCDM are compared. In addition, the effect of the power-law index n [see Eq. (7.1.1)], which dictates the material distribution through the beam thickness, on deflections is investigated. There are four models of FEM and three models DMCDM, as summarized here:

- **FE-EB(D)** - Displacement finite element model of the EBT
- **FE-EB(M)** - Mixed finite element model of the EBT
- **FE-TB(D)** - Displacement finite element model of the TBT
- **FE-TB(M)** - Mixed finite element model of the TBT
- **DM-EB(M)** - Mixed dual mesh control domain model of the EBT
- **DM-TB(D)** - Displacement dual mesh control domain model of the TBT
- **DM-TB(M)** - Mixed dual mesh control domain model of the TBT

Few remarks are in order on various elements. The displacement-based EBT element, FE-EB(D), uses Hermite cubic interpolation of $w(x)$, and it always gives exact solution at the nodes for homogeneous beams when $D_{xx} = EI$ is a constant, independent of the mesh and the load $q(x)$. All other models are based on Lagrange interpolation of the variables involved. All finite element models other than FE-EB(D) may also use quadratic or higher order interpolation of the variables, whereas the dual mesh control domain formulations presented herein are based on linear interpolation. Therefore, all numerical results presented herein, with the exception of FE-EB(D), are obtained with linear approximations of all field variables.

We investigate the effect of mesh and the power-law index n on the deflections and stresses. We consider a beam of length $L = 100$ in (254 mm), $b \times h = 1 \times 1$ in^2 (2.54^2 mm^2) cross-sectional dimensions, functionally graded through the height (h) [$E_1 = 30 \times 10^6$ psi (210 GPa), $E_2 = 3 \times 10^6$ psi (21 GPa), and $\nu = 0.3$], and subjected to uniformly distributed transverse load of intensity q_0 lb/in (175 N/m). For the pinned-pinned and clamped-clamped boundary conditions, we can exploit the symmetry about $x = L/2$, and use the left half of the beam as the computational domain. In every problem, *one element of each of the duality pairs must be specified at a boundary node.*

Based on the numerical studies, the following observations are made:

(1) The nodal generalized displacements predicted by FE-EB(D) match the exact EBT solutions.

(2) The nodal generalized displacements predicted by FE-TB(D) and DM-TB(D) are the same.

(3) The nodal transverse displacements predicted by FE-EB(D) and FE-EB(M) are the same.
(4) The nodal generalized displacements and post-computed bending moments predicted by FE-TB(D) and DM-TB(D) are identical.
(5) The nodal bending moments predicted by FE-EB(M), FE-TB(M), DM-EB(M), and DM-TB(M) are the same and match the exact solution.
(6) The nodal transverse displacements predicted by FE-TB(M) match the exact TBT solutions.
(7) The post-computed slopes in FE-TB(M), FE-EB(M), EB(M), and TB(M) and the nodal slopes in FE-EB(D) and FE-TB(D) are the same.
(8) The post-computed bending moments in FE-EB(D), FE-TB(D), FE-EB(M), and FE-TB(M) are the same.

Example 7.5.1

Consider a beam of modulus $E = 30\times 10^6$ psi, rectangular cross-section (1 in.×1 in.), with both ends pinned (i.e., each pinned end will have $u = w = M_{xx} = 0$) and subjected to uniformly distributed load $q(x) = q_0$. Exploiting the symmetry, determine the deflections using the EBT and TBT with FEM and DMCDM.

Solution: The boundary conditions on the *primary variables* in various models for half-beam, using the left half of the beam (in the TBT, ϕ_x is used in place of dw/dx), are:

$$\begin{aligned}\text{Displacement models: } & u(0) = w(0) = u(0.5L) = 0, \quad \left.\frac{dw}{dx}\right|_{x=0.5L} = 0 \text{ or } \phi_x(0.5L) = 0.\\ \text{Mixed models: } & u(0) = w(0) = M(0) = 0, \quad u(0.5L) = 0.\end{aligned} \tag{7.5.1}$$

The boundary conditions on the *secondary variables* (in an integral sense) are:

$$\begin{aligned}\text{Displacement models: } & M(0) = 0, \quad V(0.5L) = 0.\\ \text{Mixed models: } & V(0.5L) = 0, \quad \left.\frac{dw}{dx}\right|_{x=0.5L} = 0. \text{ or } \phi_x(0.5L) = 0.\end{aligned} \tag{7.5.2}$$

When the boundary conditions on the secondary variables are homogeneous, we do not impose them in the finite element analysis, as the right-hand side is already zero.

The exact solutions of pinned-pinned functionally graded beams according to the TBT, with the power-law given in Eq. (7.1.1), are given by

$$\begin{aligned}\bar{D}_{xx}\, u(x) &= \frac{q_0 L^3}{12}\left(\xi - 3\xi^2 + 2\xi^3\right),\\ \hat{D}_{xx}\, w(x) &= \frac{q_0 L^4}{24}\left(\xi - 2\xi^3 + \xi^4\right) + \tilde{D}_{xx}\frac{q_0 L^4}{2}\xi(1-\xi) - \hat{B}_{xx}\frac{q_0 L^4}{24}\xi(1-\xi),\\ \hat{D}_{xx}\, \phi_x(x) &= -\frac{q_0 L^3}{24}\left(1 - 6\xi^2 + 4\xi^3\right) + \hat{B}_{xx}\frac{q_0 L^3}{24}(1-2\xi),\\ M_{xx}(x) &= \frac{q_0 L^2}{2}\xi(1-\xi), \quad Q_x(x) = \frac{dM_{xx}}{dx} = \frac{q_0 L}{2}(1-2\xi),\\ \sigma(x, 0.5h) &= \frac{M_{xx}(x)h}{2I} = \frac{hq_0 L^2}{4I}\xi(1-\xi),\end{aligned} \tag{7.5.3}$$

where $\xi = x/L$ and

$$\hat{D}_{xx} = \frac{D^*_{xx}}{A_{xx}}, \quad \bar{D}_{xx} = \frac{D^*_{xx}}{B_{xx}}, \quad \hat{B}_{xx} = \frac{B^2_{xx}}{D_{xx}A_{xx}}, \quad \tilde{D}_{xx} = \frac{D^*_{xx}}{A_{xx}S_{xz}}. \tag{7.5.4}$$

We note that for homogeneous beams $u(x) = 0$ everywhere. The EBT solutions are obtained from Eq. (7.5.3) by setting $\tilde{D}_{xx} = 0$ and replacing ϕ_x with $-dw/dx$. The bending stress, $\sigma(x, z)$, is computed at $x = L/2$ (where the bending moment is the maximum) and $z = h/2$, h being the beam height. The stress is post-computed in the displacement models at the element center using the relation

$$\sigma(x,z) = -E(z)\, z \frac{d^2 w}{dx^2} \ \text{for EBT;}\ \ \sigma(x,z) = E(z)\, z \frac{d\phi_x}{dx} \ \text{for TBT.} \tag{7.5.5}$$

On the other hand, the stress in the mixed models is computed using the bending moment $M_{xx}(x)$ according to the formula

$$\sigma(x,z) = \frac{M_{xx}(x)\, z}{I}, \tag{7.5.6}$$

where $M_{xx}(x)$ is calculated from the finite element interpolation (i.e., $\sigma(L/2, h/2) = M(L/2)h/2I$, and $M(L/2)$ is the nodal value).

The shear locking is alleviated in the TBT displacement models of the FEM and DMCDM by the use of reduced integration. No such trick is used in the mixed FEM and mixed DMCDM models. We also note that for this slender beam ($L/h = 100$), the effect of shear deformation is negligible, and the EBT and TBT solutions for $\bar{w}$ are the same up to the fourth decimal point.

Tables 7.5.1 and 7.5.2 contain the dimensional center deflection and bending moment, respectively, for homogeneous ($D_{xx} = EI$ and $B_{xx} = 0$) pinned-pinned (P-P) beams for different number of elements in the half-beam (for $q_0 = 0.5$ lb/in). From the results it is clear that the mixed DMCDM models are the most accurate in predicting the displacements and stresses. As already noted, all mixed and displacement models of the DMCDM give exact stresses for any number of elements (the DMCDM gives the exact value of the bending moment at $x = L/2$).

Table 7.5.1 The center deflection $w(L/2)$ (in.) of homogeneous P-P beams predicted by various models for $q_0 = 0.5$ lb/in.

Mesh	FE-EB Displ.	FE-EB Mixed	FE-TB Displ.	FE-TB Mixed	DM-EB Mixed	DM-TB Displ.	DM-TB Mixed
4	0.2604	0.2572	0.2540	0.2572	0.2588	0.2540	0.2588
8	0.2604	0.2596	0.2588	0.2596	0.2600	0.2588	0.2600
16	0.2604	0.2602	0.2601	0.2602	0.2603	0.2600	0.2604
32	0.2604	0.2604	0.2604	0.2604	0.2604	0.2604	0.2604
64	0.2604	0.2604	0.2604	0.2604	0.2604	0.2604	0.2604
Exact	0.2604	0.2604	0.2604	0.2604	0.2604	0.2604	0.2604

Table 7.5.2 The center bending moment $M_{xx}(L/2) \times 10^{-3}$ (lb-in.) for homogeneous P-P beams predicted by various models for $q_0 = 0.5$ lb/in.

Mesh	FE-EB Displ.	FE-EB Mixed	FE-TB Displ.	FE-TB Mixed	DM-EB Mixed	DM-TB Displ.	DM-TB Mixed
4	0.6120	0.6250	0.6054	0.6250	0.6250	0.6250	0.6250
8	0.6218	0.6250	0.6202	0.6250	0.6250	0.6250	0.6250
16	0.6242	0.6250	0.6238	0.6250	0.6250	0.6250	0.6250
32	0.6248	0.6250	0.6248	0.6250	0.6250	0.6250	0.6250
64	0.6250	0.6250	0.6250	0.6250	0.6250	0.6250	0.6250
Exact	0.6250	0.6250	0.6250	0.6250	0.6250	0.6250	0.6250

Table 7.5.3 contains the normalized center deflection for functionally graded pinned-pinned (P-P) beams for different values of the power-law index n. All of the results are obtained using

16 elements in the half-beam. All models predict solutions that match the exact solutions very closely (the difference between the EBT and TBT is negligibly small). The stresses in the FGM beams are exactly the same as those in the homogeneous beams, because the bending moment is independent of the stiffness coefficients.

Table 7.5.3 The center transverse deflection $\bar{w}(L/2) \times 10 = w(L/2)\frac{\hat{D}_{xx}}{q_0 L^4} \times 10$ of FGM P-P beams predicted by various models (16 elements).

n	FE-EB Displ.	FE-TB Displ.	FE-EB Mixed	FE-TB Mixed	DM-EB Mixed	DM-TB Displ.	DM-TB Mixed	Exact
0.0	0.1304	0.1300	0.1300	0.1301	0.1302	0.1300	0.1302	0.1302
1.0	0.1069	0.1068	0.1067	0.1069	0.1069	0.1068	0.1069	0.1070
2.0	0.0919	0.0918	0.0919	0.0919	0.0919	0.0918	0.0919	0.0920
3.0	0.0879	0.0878	0.0879	0.0879	0.0879	0.0878	0.0879	0.0880
5.0	0.0900	0.0899	0.0899	0.0900	0.0900	0.0899	0.0900	0.0900
7.5	0.0959	0.0958	0.0958	0.0958	0.0958	0.0958	0.0959	0.0959
10.0	0.1012	0.1011	0.1012	0.1012	0.1012	0.1011	0.1012	0.1013
12.0	0.1048	0.1047	0.1047	0.1048	0.1048	0.1047	0.1048	0.1048
15.0	0.1091	0.1090	0.1090	0.1090	0.1090	0.1090	0.1091	0.1091
20.0	0.1142	0.1140	0.1141	0.1141	0.1141	0.1140	0.1142	0.1142

It is interesting to note that the effect of the power-law index n on the deflections is *not* monotonic. As n goes from zero to a value of about $n = 2$, the deflection decreases and then increases for $n > 2$, as shown in Fig. 7.5.1. This is due to the fact that B_{xx} is not a monotonically increasing or decreasing function of n (where M denotes the moduli ratio, $M = E_1/E_2$ and $B_0 = bh^2$), as can be seen from Fig. 7.5.2 (see [13]). Figure 7.5.3 shows the variation of the normalized center deflection $\bar{w}$ with the power-law index n.

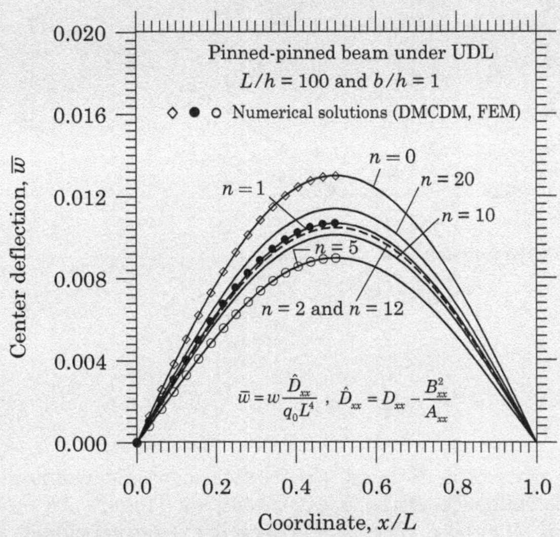

Fig. 7.5.1 Normalized deflection $\bar{w}$ versus x/L for different values of n.

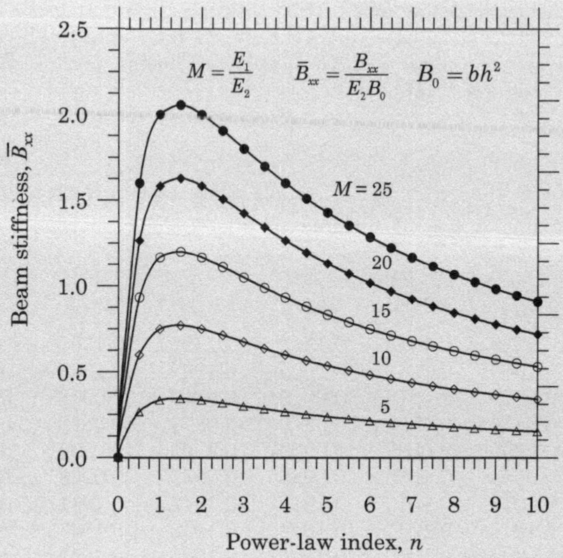

Fig. 7.5.2 Variation of the coupling stiffness $\bar{B}_{xx}$ with n ($M = E_1/E_2$).

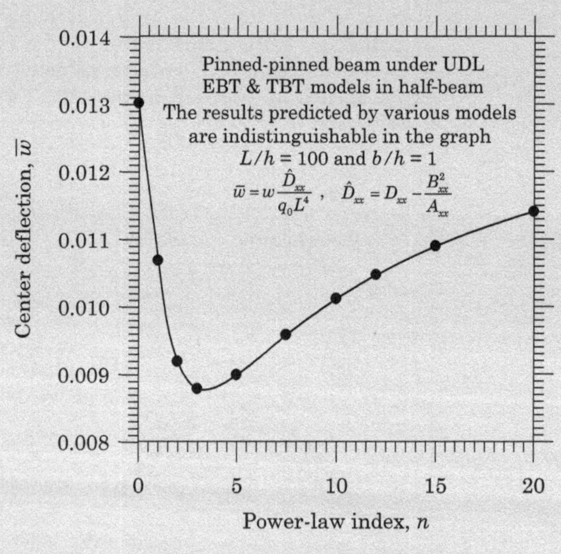

Fig. 7.5.3 Variation of the normalized center deflection $\bar{w}$ with n for P-P beams.

Next, we present plots of deflections, slopes/rotations, and bending moments in dimensional form to see the influence of the power-law index. Figure 7.5.4 contains plots of the dimensional deflections $w(x)$ predicted for P-P beams by various models as a function of x/L (for the half-beam) for a number of values of the power-law index, $n = 0, 1, 2, 5, 10$, and 20. Since increasing value of n means that the FGM beam becomes increasingly more flexible (because the percentage of Material 2 increases).

Figure 7.5.5 contains plots of the dimensional bending moment M_{xx} as a function of x/L for the half-beam (independent of the power-law index n). For the displacement model, the bending moment is computed at the center of each element, while for the mixed model it is obtained as a nodal value. The agreement between the mixed models (of FEM and DMCDM) and the displacement models (EBT and TBT finite element models) is excellent.

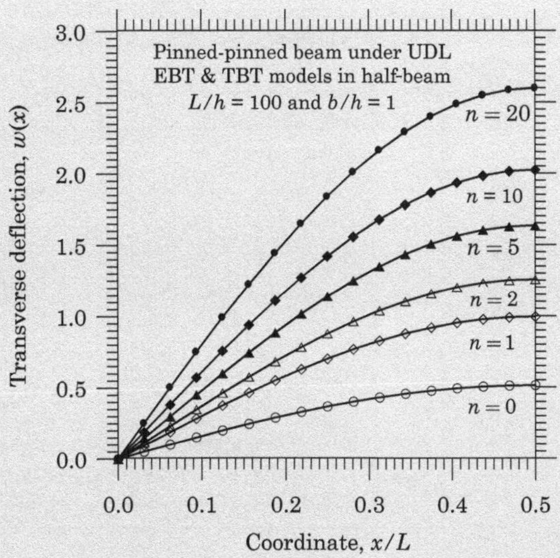

Fig. 7.5.4 Deflection w versus x/L for different values of n.

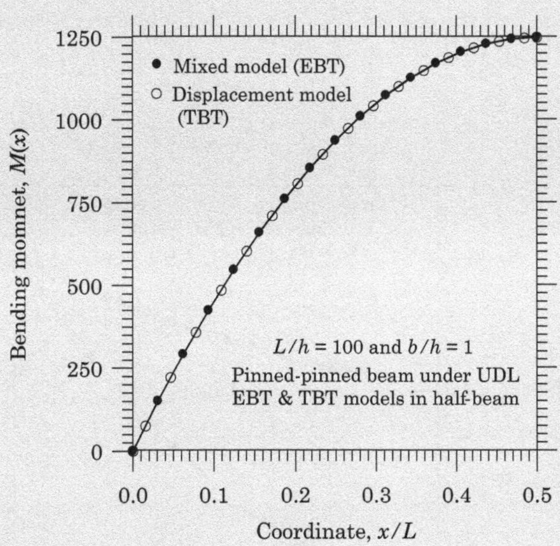

Fig. 7.5.5 Bending moment M_{xx} versus x/L obtained with the displacement and mixed models of FEM and DMCDM for $n = 0$.

Figure 7.5.6 contains plots of the slope $(-dw/dx)$ or rotation (ϕ_x) as functions of x/L for $n = 0$; we note that slope or rotation are the nodal variables for the displacement models, whereas they are post-computed (secondary variables) in the mixed models. Figure 7.5.7 contains plots of the slope $(-dw/dx)$ or rotation (ϕ_x) as functions of x/L for $n = 0$ and $n = 3$ in the half-beam (note that the scale in Fig. 7.5.7 is smaller than that in Fig. 7.5.6). Once again, we see excellent agreement between the mixed and displacement models. The results obtained with the mixed model of the EBT and TBT and the displacement model of the TBT of the DMCDM are the same as those presented in Figs. 7.5.5–7.5.7.

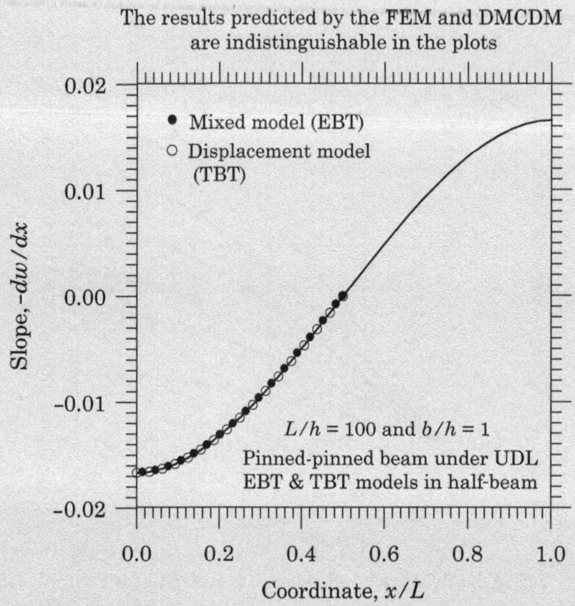

Fig. 7.5.6 Slope ($-dw/dx$) or rotation (ϕ_x) versus x/L obtained with the displacement and mixed models of FEM and DMCDM for $n = 0$.

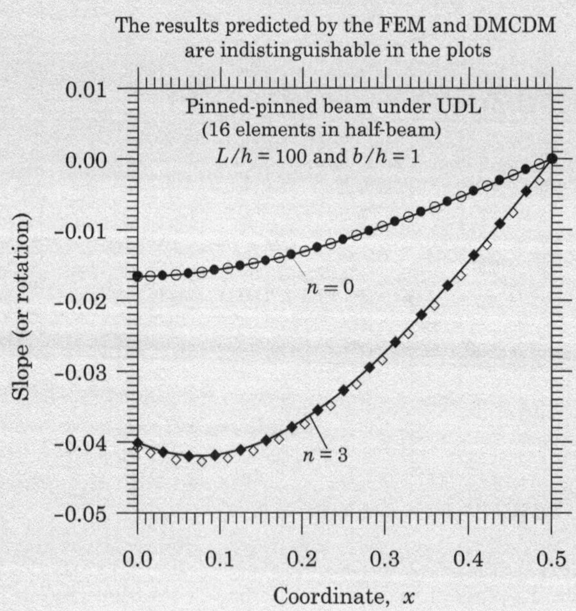

Fig. 7.5.7 Slope ($-dw/dx$) or rotation (ϕ_x) versus x/L obtained with the displacement and mixed models of FEM and DMCDM for $n = 0$ and $n = 3$.

Example 7.5.2

Consider a beam clamped (C-C) at both ends (i.e., $u = w = \phi_x = -dw/dx = 0$ at the clamped end), and subjected to uniformly distributed load of intensity q_0. Exploiting the symmetry, determine the deflections using the FEM and DMCDM.

Solution: The boundary conditions on the primary variables in various models for this problem are as follows (replace dw/dx with ϕ_x for the TBT):

$$\begin{aligned}\text{Displacement models}: \quad & u(0) = w(0) = 0, \quad \left.\frac{dw}{dx}\right|_{x=0} = 0, \quad u(0.5L) = \left.\frac{dw}{dx}\right|_{x=0.5L} = 0.\\ \text{Mixed models}: \quad & u(0) = w(0) = 0, \quad u(0.5L) = 0.\end{aligned} \tag{7.5.7}$$

The boundary conditions on the secondary variables in various models for this problem (satisfied in an integral sense) are as follows:

$$\begin{aligned}\text{Displacement models}: \quad & V(0.5L) = 0.\\ \text{Mixed models}: \quad & \left.\frac{dw}{dx}\right|_{x=0} = 0, \quad \left.\frac{dw}{dx}\right|_{x=0.5L} = 0.\end{aligned} \tag{7.5.8}$$

The exact solutions for clamped-clamped beams according to the TBT are ($\xi = x/L$)

$$\begin{aligned}\bar{D}_{xx}u(x) &= \frac{q_0L^3}{12}\left(\xi - 3\xi^2 + 2\xi^3\right),\\ \hat{D}_{xx}w(x) &= \frac{q_0L^4}{24}\xi^2(1-\xi)^2 + \tilde{D}_{xx}\frac{q_0L^2}{2}\left(\xi - \xi^2\right),\\ \hat{D}_{xx}\phi_x(x) &= -\frac{q_0L^3}{12}\left(\xi - 3\xi^2 + 2\xi^3\right),\\ M_{xx}(x) &= -\frac{q_0L^2}{12}\left(1 - 6\xi + 6\xi^2\right),\\ Q_x(x) &= \frac{q_0L}{2}(1 - 2\xi),\\ \sigma(x, 0.5h) &= -\frac{q_0hL^2}{24I}\left(1 - 6\xi + 6\xi^2\right).\end{aligned} \tag{7.5.9}$$

Tables 7.5.4 and 7.5.5 contain the normalized center deflection and stress, respectively, for the clamped-clamped (C-C) homogeneous beam. From the results it is clear that the mixed DMCDM models and the FEM results are very close, if not identical. The displacement finite element model is the most accurate by virtue of the higher (Hermite cubic) approximation of the deflection. Table 7.5.6 contains the normalized center deflection for the functionally graded clamped-clamped beam for different values of n. All of the results were obtained with 16 elements in half-beam. For the C-C beams, the normalized deflections differ only in the fourth or fifth decimal point.

Table 7.5.4 The center transverse deflection $\bar{w}(L/2) \times 10^2 = w(L/2)\frac{\hat{D}_{xx}}{q_0L^4} \times 10^2$ predicted by various models for homogeneous beams.

Mesh	FE-EB Displ.	FE-EB Mixed	FE-TB Displ.	FE-TB Mixed	DM-EB Mixed	DM-TB Displ.	DM-TB Mixed
4	0.2604	0.2604	0.2445	0.2607	0.2685	0.2445	0.2688
8	0.2604	0.2604	0.2567	0.2607	0.2624	0.2567	0.2628
16	0.2604	0.2604	0.2597	0.2607	0.2609	0.2597	0.2612
32	0.2604	0.2604	0.2605	0.2607	0.2605	0.2605	0.2609
64	0.2604	0.2604	0.2607	0.2607	0.2604	0.2607	0.2608
Exact	0.2604	0.2604	0.2607	0.2607	0.2604	0.2607	0.2607

Table 7.5.5 The center stress $\bar{\sigma} \times 10 = \sigma \, \frac{I}{h \, q_0 L^2} \times 10$ predicted by various models for C-C beams.

Mesh	FE-EB Displ.	FE-EB Mixed	FE-TB Displ.	FE-TB Mixed	DM-EB Mixed	DM-TB Displ.	DM-TB Mixed
4	0.1953	0.2148	0.1953	0.2148	0.2148	0.2148	0.2148
8	0.2051	0.2100	0.2051	0.2100	0.2100	0.2100	0.2100
16	0.2075	0.2087	0.2075	0.2087	0.2087	0.2087	0.2087
32	0.2081	0.2084	0.2081	0.2084	0.2084	0.2084	0.2084
64	0.2083	0.2083	0.2083	0.2083	0.2083	0.2083	0.2083
Exact	0.2083	0.2083	0.2083	0.2083	0.2083	0.2083	0.2083

Table 7.5.6 The center deflection $\bar{w}(L/2) \times 10^2 = w(L/2) \frac{\hat{D}_{xx}}{q_0 L^4} \times 10^2$ of FGM C-C beams predicted by various models.

n	FE-EB Displ.	FE-TB Displ.	FE-EB Mixed	FE-TB Mixed	DM-EB Mixed	DM-TB Displ.	DM-TB Mixed
0.0	0.26042	0.25972	0.26042	0.26074	0.26093	0.25972	0.26125
1.0	0.26019	0.25965	0.26019	0.26044	0.26058	0.25965	0.26084
2.0	0.26004	0.25964	0.26004	0.26028	0.26037	0.25964	0.26060
3.0	0.26000	0.25965	0.26000	0.26025	0.26031	0.25965	0.26055
5.0	0.26002	0.25968	0.26002	0.26031	0.26034	0.25968	0.26062
10.0	0.26013	0.25976	0.26013	0.26049	0.26050	0.25976	0.26086
15.0	0.26021	0.25979	0.26021	0.26060	0.26062	0.25979	0.26100
20.0	0.26026	0.25980	0.26026	0.26066	0.26069	0.25980	0.26109

Figure 7.5.8 shows the variation of the deflection $w(x)$ versus x/L for various values of the power-law index n. Plot of the bending moment M_{xx} versus x/L, computed in the displacement and mixed models, is presented in Fig. 7.5.9. Both FEM and DMCDM models give the same results. Figure 7.5.10 contains plots of the nodal and post-computed values of the slope or rotation for three different values of the power-law index $n = 0, 1$, and 3.

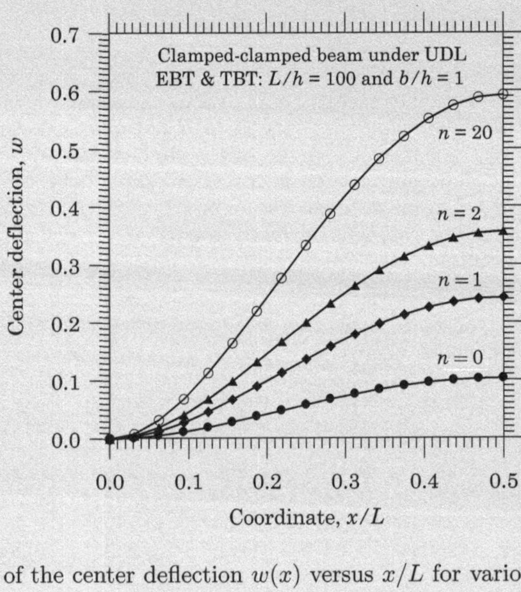

Fig. 7.5.8 Variation of the center deflection $w(x)$ versus x/L for various values of n for C-C beams.

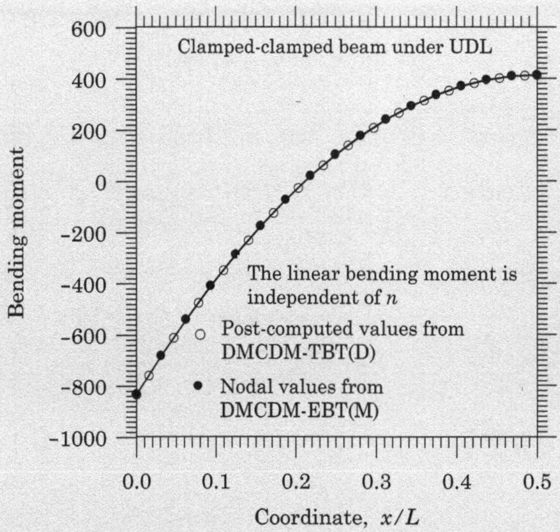

Fig. 7.5.9 Variation of the bending moment $M_{xx}(x)$ versus x/L for C-C beams under uniform load.

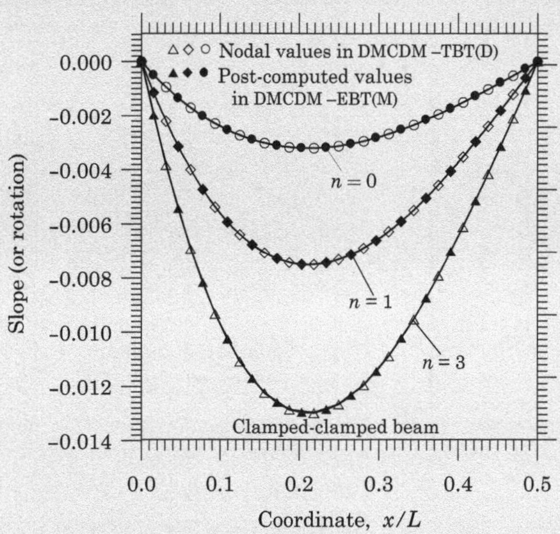

Fig. 7.5.10 Variation of the slope $\theta_x = -dw/dx$ or rotation ϕ_x versus x/L for C-C beams under uniform load.

Figure 7.5.11 shows the variation of the normalized deflection $\bar{w}$ versus n, which has the same variation as that for the P-P beams; Fig. 7.5.12 contains plots of dimensional deflection versus n for both the P-P and C-C beams. It is interesting to note that the rate of increase in the deflection has two different regions; the first region has a rapid increase of the deflection, while the second region is marked with a slow increase. This is primarily because of the fact that the coupling coefficient B_{xx} varies with n rapidly for the lower values of n followed by a slow decay after $n > 3$.

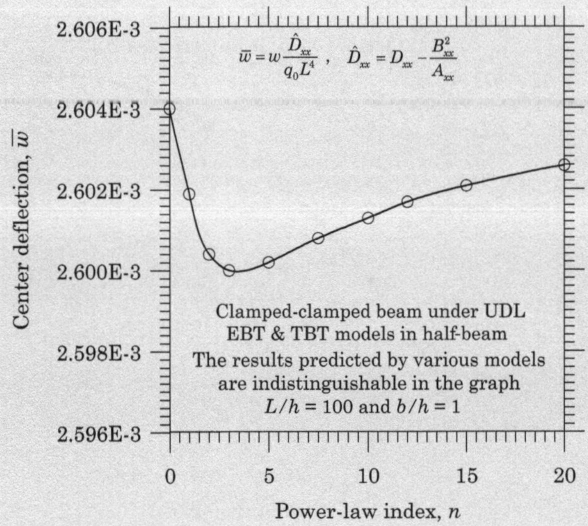

Fig. 7.5.11 Variation of the normalized center deflection $\bar{w}(L/2) = w(L/2)\frac{\hat{D}_{xx}}{q_0L^4}$ with n for C-C beams.

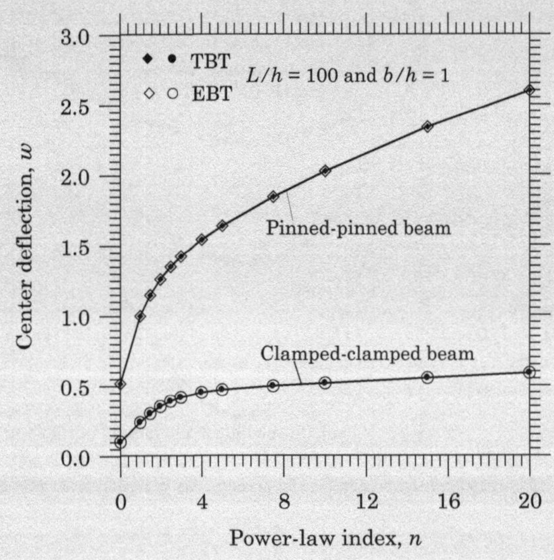

Fig. 7.5.12 Variation of the dimensional center deflection $w(L/2)$ with n for P-P and C-C beams.

We close this section with a comment that the FEM and DMCDM both give results that are very close to each other (indistinguishable in the plots). In addition, the mixed and displacement models give solutions that are essentially the same, although the mixed model of the TBT does not exhibit shear locking.

7.6 Nonlinear Analysis of Beams

7.6.1 Euler–Bernoulli Beam Theory

If one presumes that the strains are small but rotations about the y-axis of the material lines transverse to the x-axis are moderately large (i.e., the squares and products of dw/dx are not negligible, but the squares of du/dx are negligible), the axial strain resulting from the Green strain tensor components (see Reddy [23] for details) are nonlinear only in dw/dx, and they are known as the von Kármán strains. We assume that the axial strain du/dx and the curvature d^2w/dx^2 are of order ϵ, while dw/dx is of order $\sqrt{\epsilon}$, where $\epsilon << 1$ is a small parameter (compared to the ratio of the beam height h to the length L). Then the von Kármán strain (for beams, membrane strain is the only strain component that is affected) is

$$\varepsilon_{xx}^{(0)} = \frac{du}{dx} + \underline{\frac{1}{2}\left(\frac{dw}{dx}\right)^2} \tag{7.6.1}$$

The underlined term in Eq. (7.6.1) is nonlinear in w. We note that the geometric nonlinearity appearing in Eq. (7.6.1) involves only the derivative of the transverse deflection w.

7.6.1.1 Equations in terms of displacements

The equations of equilibrium of the EBT, accounting for the von Kármán nonlinear strain, are (see Reddy [13, 17, 42] for the details):

$$-\frac{dN_{xx}}{dx} - f = 0, \quad -\frac{d^2 M_{xx}}{dx^2} - \frac{d}{dx}\left(N_{xx}\frac{dw}{dx}\right) + c_f w - q = 0, \tag{7.6.2}$$

where $f(x)$ and $q(x)$ axial and transverse distributed loads (measured as force per unit length), respectively, on the beam; c_f is the modulus of the foundation on which the beam rests; and N_{xx} and M_{xx} are the stress resultants defined by (and expressed in terms of the generalized displacements u and w, with $\theta_x = -dw/dx$):

$$N_{xx} = \int_A \sigma_{xx}\, dA = A_{xx}\varepsilon_{xx}^{(0)} + B_{xx}\varepsilon_{xx}^{(1)} \tag{7.6.3a}$$

$$M_{xx} = \int_A z\sigma_{xx}\, dA = B_{xx}\varepsilon_{xx}^{(0)} + D_{xx}\varepsilon_{xx}^{(1)} \tag{7.6.3b}$$

Here $\varepsilon_{xx}^{(0)}$ and $\varepsilon_{xx}^{(1)}$ denote the membrane and bending strains, respectively,

$$\varepsilon_{xx}^{(0)}(x) = \frac{du}{dx} + \frac{1}{2}\left(\frac{dw}{dx}\right)^2, \quad \varepsilon_{xx}^{(1)}(x) = -\frac{d^2 w}{dx^2} \tag{7.6.4}$$

and A_{xx}, B_{xx}, and D_{xx} are the extensional, extensional-bending, and bending stiffness coefficients

$$(A_{xx}, B_{xx}, D_{xx}) = \int_A (1, z, z^2)E(z)\, dA. \tag{7.6.5}$$

Substitution of the stress resultant-displacement relation in Eqs. (7.6.3a), (7.6.3b), and (7.6.4) into the equilibrium equations in Eqs. $(7.6.2)_1$ and $(7.6.2)_2$ gives the governing equations in terms of the displacements for FGM beams. These equations are coupled; there are two sources of coupling in FGM beams: first, the coupling is due to the extensional-bending coefficient B_{xx}, and it is independent of the von Kármán nonlinearity; second, the coupling is due to the von Kármán nonlinearity, which is independent of the coupling coefficient B_{xx}. Of course, the coefficient B_{xx} has a stronger coupling in the presence of the von Kármán nonlinearity. The duality pairs for the EBT in the nonlinear case are: (u, N_{xx}), (w, V_x), and (θ_x, M_{xx}), where

$$V_x \equiv \frac{dM_{xx}}{dx} + N_{xx}\frac{dw}{dx}. \tag{7.6.6}$$

We note that the products of the elements in each pair have the meaning of work done.

7.6.1.2 Equations in terms of displacements and bending moment

The second equation in Eq. (7.6.2), when expressed in terms of the displacement w, results in a fourth-order differential equation, which requires Hermite cubic interpolation of w in constructing the displacement finite element model, as already discussed in Sections 4.3.1 and 7.3.1. However, discrete models using the DMCDM are restricted to only second- or lower-order differential equations. Therefore, we reformulate the governing equations as second-order differential equations in terms of u, w, and M_{xx}, as already presented in Section 7.2.1.3. Here we extend the formulation to the nonlinear case.

Substitution of Eqs. (7.6.3a) and (7.6.3b) into Eqs. (7.6.1) and (7.6.2) gives the following governing equations in terms of the displacements:

$$-\frac{d}{dx}\left\{\bar{A}_{xx}\left[\frac{du}{dx}+\frac{1}{2}\left(\frac{dw}{dx}\right)^2\right]+\bar{B}_{xx}M_{xx}\right\}=f \tag{7.6.7a}$$

$$-\frac{d^2M_{xx}}{dx^2}-\frac{d}{dx}\left\{\bar{A}_{xx}\frac{dw}{dx}\left[\frac{du}{dx}+\frac{1}{2}\left(\frac{dw}{dx}\right)^2\right]+\bar{B}_{xx}\frac{dw}{dx}M_{xx}\right\} +c_f w = q \tag{7.6.7b}$$

$$-\frac{d^2w}{dx^2}-\frac{M_{xx}}{D_{xx}}+\bar{B}_{xx}\left[\frac{du}{dx}+\frac{1}{2}\left(\frac{dw}{dx}\right)^2\right]=0 \tag{7.6.7c}$$

where

$$D^*_{xx} \equiv D_{xx}A_{xx} - B^2_{xx}, \quad \bar{A}_{xx} \equiv \frac{D^*_{xx}}{D_{xx}}, \quad \bar{B}_{xx} \equiv \frac{B_{xx}}{D_{xx}} \tag{7.6.8}$$

We note that, for the nonlinear FGM beams, the axial force N_{xx} is expressed in terms of the displacements (u, w) and moment (M_{xx}) as

$$N_{xx} = \bar{A}_{xx}\left[\frac{du}{dx}+\frac{1}{2}\left(\frac{dw}{dx}\right)^2\right]+\bar{B}_{xx}M_{xx} \tag{7.6.9}$$

7.6.1.3 Displacement finite element model

The displacement finite element model of the EBT can be developed using the same ideas as those discussed in Section 7.3.1.1 but accounting for the nonlinear terms in the weak-form development [i.e., Eqs. (7.3.1a) and (7.3.1b) are modified to include the nonlinear terms]. The resulting finite element model is of the form

$$\begin{bmatrix} \mathbf{K}^{11} & \mathbf{K}^{12} \\ \mathbf{K}^{21} & \mathbf{K}^{22} \end{bmatrix} \begin{Bmatrix} \mathbf{u} \\ \mathbf{\Delta} \end{Bmatrix} = \begin{Bmatrix} \mathbf{F}^1 \\ \mathbf{F}^2 \end{Bmatrix}, \tag{7.6.10}$$

where the element stiffness coefficients for the nonlinear case are

$$\begin{aligned}
K_{ij}^{11} &= \int_{x_a^e}^{x_b^e} A_{xx}^e \frac{d\psi_i^e}{dx}\frac{d\psi_j^e}{dx}\,dx, \quad F_i^1 = \psi_i(x_a^e)\,Q_1^e + \psi_i^e(x_b^e)\,Q_4^e \\
K_{iJ}^{12} &= -\int_{x_a^e}^{x_b^e} B_{xx}^e \frac{d\psi_i^e}{dx}\frac{d^2\varphi_J^e}{dx^2}\,dx + \frac{1}{2}\int_{x_a^e}^{x_b^e} A_{xx}^e \frac{dw}{dx}\frac{d\psi_i^e}{dx}\frac{d\varphi_J}{dx}\,dx \\
K_{Ij}^{21} &= -\int_{x_a^e}^{x_b^e} B_{xx}^e \frac{d^2\varphi_I^e}{dx^2}\frac{d\psi_j^e}{dx}\,dx + \int_{x_a^e}^{x_b^e} A_{xx}^e \frac{dw}{dx}\frac{d\varphi_I^e}{dx}\frac{d\psi_j^e}{dx}\,dx \qquad (7.6.11)\\
K_{IJ}^{22} &= \int_{x_a^e}^{x_b^e} \left(D_{xx}^e \frac{d^2\varphi_I^e}{dx^2}\frac{d^2\varphi_J^e}{dx^2} + c_f^e \varphi_I^e \varphi_J^e \right) dx \\
&\quad + \int_{x_a^e}^{x_b^e} \left[\frac{1}{2} A_{xx}^e \left(\frac{dw}{dx}\right)^2 \frac{d\varphi_I^e}{dx}\frac{d\varphi_J^e}{dx} - B_{xx}^e \frac{dw}{dx}\left(\frac{1}{2}\frac{d^2\varphi_I^e}{dx^2}\frac{d\varphi_J^e}{dx} + \frac{d\varphi_I^e}{dx}\frac{d^2\varphi_J^e}{dx^2} \right) \right] dx \\
F_I^2 &= \int_{x_a^e}^{x_b^e} \varphi_I^e\, q(s)\,dx + \varphi_I^e(x_a^e)\,Q_2^e - \left.\frac{d\varphi_I^e}{dx}\right|_{x_a^e} Q_3^e + \varphi_I^e(x_b^e)\,Q_5^e - \left.\frac{d\varphi_I^e}{dx}\right|_{x_b^e} Q_6^e
\end{aligned}$$

and Q_i^e are the generalized nodal forces, as shown in Fig. 7.6.1, where we replace ϕ_x with $-dw/dx$; φ_I^e are the Hermite cubic interpolation functions and ψ_i^e are the linear Lagrange interpolation functions; $\mathbf{u}^e$ denotes the vector of nodal

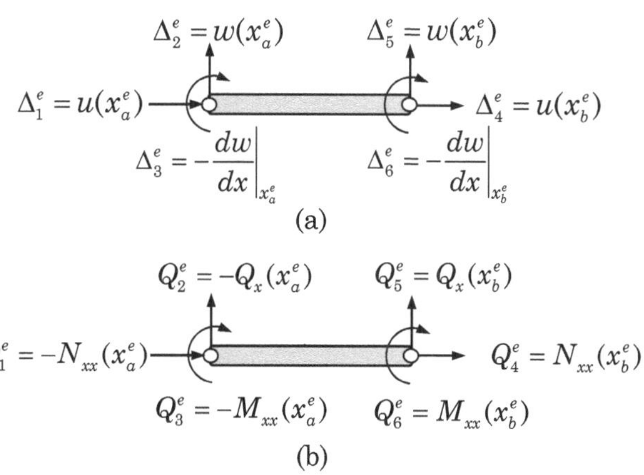

Fig. 7.6.1 Generalized nodal (a) displacements and (b) forces in the displacement model of the EBT.

displacements associated with the linear approximation of $u(x)$; and $\mathbf{\Delta}^e$ denotes the nodal displacements (i.e., transverse deflection and rotation at each node) associated with the Hermite cubic interpolation of $w(x)$.

Similar to shear locking, inconsistent interpolation of u and w leads to *membrane locking* in all models with the nonlinearity (because the membrane strain in Eq. (7.6.1) is not represented correctly when equal interpolation is used). Again, it is cured using reduced integration.

7.6.1.4 Mixed finite element model

The *mixed finite element model* of the EBT, based on the Lagrange interpolation of all variables, is constructed using modified weak forms [i.e., Eqs. (7.3.7a)–(7.3.7c) are modified according to the governing nonlinear equations (7.6.7a)–(7.6.7c)]. The nonlinear mixed finite element model is given by

$$\begin{bmatrix} \mathbf{K}^{11} & \mathbf{K}^{12} & \mathbf{K}^{12} \\ \mathbf{K}^{21} & \mathbf{K}^{22} & \mathbf{K}^{23} \\ \mathbf{K}^{31} & \mathbf{K}^{32} & \mathbf{K}^{33} \end{bmatrix} \begin{Bmatrix} \mathbf{u} \\ \mathbf{w} \\ \mathbf{M} \end{Bmatrix} = \begin{Bmatrix} \mathbf{F}^1 \\ \mathbf{F}^2 \\ \mathbf{F}^3 \end{Bmatrix} \tag{7.6.12}$$

where

$$\begin{aligned}
K_{ij}^{11} &= \int_{x_a^e}^{x_b^e} \frac{D_{xx}^*}{D_{xx}} \frac{d\psi_i^{(1)}}{dx} \frac{d\psi_j^{(1)}}{dx}\, dx, \quad F_i^1 = \psi_i^{(1)}(x_a^e) Q_1^e + \psi_i^{(1)}(x_b^e) Q_4^e \\
K_{ij}^{12} &= \frac{1}{2} \int_{x_a^e}^{x_b^e} \frac{D_{xx}^*}{D_{xx}} \frac{dw}{dx} \frac{d\psi_i^{(1)}}{dx} \frac{d\psi_j^{(2)}}{dx}\, dx, \quad K_{ij}^{13} = \int_{x_a^e}^{x_b^e} \frac{B_{xx}}{D_{xx}} \frac{d\psi_i^{(1)}}{dx} \psi_j^{(3)}\, dx \\
K_{ij}^{21} &= \int_{x_a^e}^{x_b^e} \frac{D_{xx}^*}{D_{xx}} \frac{dw}{dx} \frac{d\psi_i^{(2)}}{dx} \frac{d\psi_j^{(1)}}{dx}\, dx \\
K_{ij}^{22} &= \int_{x_a^e}^{x_b^e} \left[\frac{1}{2} \frac{D_{xx}^*}{D_{xx}} \left(\frac{dw}{dx} \right)^2 \frac{d\psi_i^{(2)}}{dx} \frac{d\psi_j^{(2)}}{dx} + c_f \psi_i^{(2)} \psi_j^{(2)} \right] dx \\
K_{ij}^{23} &= \int_{x_a^e}^{x_b^e} \left(\frac{d\psi_i^{(2)}}{dx} \frac{d\psi_j^{(3)}}{dx} + \frac{B_{xx}}{D_{xx}} \frac{dw}{dx} \frac{d\psi_i^{(2)}}{dx} \psi_j^{(3)} \right) dx \\
F_I^2 &= \int_{x_a^e}^{x_b^e} \psi_i^{(2)} q\, dx + \psi_i^{(2)}(x_a^e) Q_2^e + \psi_i^{(2)}(x_b^e) Q_4^e, \quad K_{ij}^{31} = \int_{x_a^e}^{x_b^e} \frac{B_{xx}}{D_{xx}} \psi_i^{(3)} \frac{d\psi_j^{(1)}}{dx}\, dx \\
K_{ij}^{32} &= \int_{x_a^e}^{x_b^e} \left(\frac{d\psi_i^{(3)}}{dx} \frac{d\psi_j^{(2)}}{dx} + \frac{1}{2} \frac{B_{xx}}{D_{xx}} \frac{dw}{dx} \psi_i^{(3)} \frac{d\psi_j^{(2)}}{dx} \right) dx \\
K_{ij}^{33} &= -\int_{x_a^e}^{x_b^e} \frac{1}{D_{xx}} \psi_i^{(3)} \psi_j^{(3)}\, dx, \quad F_I^3 = \psi_i^{(3)}(x_a^e) Q_3^e - \psi_i^{(3)}(x_b^e) Q_6^e
\end{aligned} \tag{7.6.13}$$

Here $(\psi_i^{(1)}, \psi_i^{(2)}, \psi_i^{(3)})$ are the Lagrange interpolation functions used for (u, w, M_{xx}), respectively. In general, they are different from each other, but in this study they are taken to be the same for all variables. The nodal secondary variables, Q_i^e, for the mixed model are shown in Fig. 7.6.2.

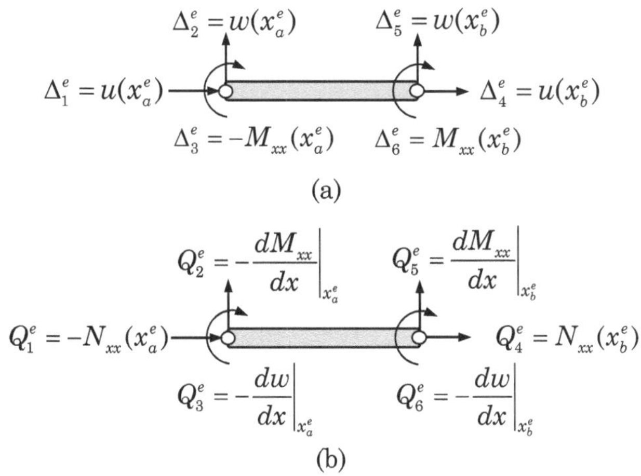

Fig. 7.6.2 (a) Primary and (b) secondary variables at the nodes in the *mixed* model of the EBT.

7.6.2 Timoshenko Beam Theory

7.6.2.1 Equations in terms of displacements

The nonlinear equations of equilibrium of the Timoshenko beam theory are (see Reddy [13, 42])

$$-\frac{dN_{xx}}{dx} - f = 0 \tag{7.6.14a}$$

$$-\frac{dQ_x}{dx} - \frac{d}{dx}\left(N_{xx}\frac{dw}{dx}\right) + c_f w - q = 0 \tag{7.6.14b}$$

$$-\frac{dM_{xx}}{dx} + Q_x = 0 \tag{7.6.14c}$$

The stress resultants (N_{xx}, M_{xx}, Q_x) in the TBT can be expressed in terms of the displacements as

$$N_{xx} = A_{xx}\left[\frac{du}{dx} + \frac{1}{2}\left(\frac{dw}{dx}\right)^2\right] + B_{xx}\frac{d\phi_x}{dx} \tag{7.6.15a}$$

$$M_{xx} = B_{xx}\left[\frac{du}{dx} + \frac{1}{2}\left(\frac{dw}{dx}\right)^2\right] + D_{xx}\frac{d\phi_x}{dx} \tag{7.6.15b}$$

$$Q_x = K_s \int_A \sigma_{xz}\, dA = S_{xz}\left(\phi_x + \frac{dw}{dx}\right) \tag{7.6.15c}$$

where ϕ_x denotes the rotation of the cross-section about the y-axis, K_s is the shear correction factor, and S_{xz} is the shear stiffness coefficient

$$S_{xz} = \frac{K_s}{2(1+\nu)} \int_A E(z)\, dA. \tag{7.6.16}$$

The substitution of the relations from Eqs. (7.6.15a)–(7.6.15c) into Eqs. (7.6.14a)–(7.6.14c) yields the governing equations in terms of the generalized displacements $(u,\, w,\, \phi_x)$.

7.6.2.2 Equations in terms of displacements and bending moment

To develop a mixed model of the TBT, the governing equations of the TBT are reformulated in terms of (u, w, M_{xx}) (i.e., eliminate ϕ_x), as discussed in Section 7.2.2.2. The resulting nonlinear differential equations are

$$-\frac{d}{dx}\left\{\bar{A}_{xx}\left[\frac{du}{dx}+\tfrac{1}{2}\left(\frac{dw}{dx}\right)^2\right]+\bar{B}_{xx}M_{xx}\right\} = f \tag{7.6.17a}$$

$$-\frac{d^2M_{xx}}{dx^2} - \frac{d}{dx}\left\{\bar{A}_{xx}\frac{dw}{dx}\left[\frac{du}{dx}+\tfrac{1}{2}\left(\frac{dw}{dx}\right)^2\right]+\bar{B}_{xx}\frac{dw}{dx}M_{xx}\right\} + c_f w = q \tag{7.6.17b}$$

$$-\frac{d}{dx}\left[\frac{dw}{dx} - \frac{1}{S_{xz}}\frac{dM_{xx}}{dx}\right] + \bar{B}_{xx}\left[\frac{du}{dx}+\tfrac{1}{2}\left(\frac{dw}{dx}\right)^2\right] - \frac{1}{D_{xx}}M_{xx} = 0 \tag{7.6.17c}$$

where the effective rotation $\hat{\phi}_x$ is

$$\hat{\phi}_x = -\frac{dw}{dx} + \frac{1}{S_{xz}}\frac{dM_{xx}}{dx} \tag{7.6.18}$$

7.6.2.3 Displacement finite element model

The displacement finite element model of the TBT is of the form (see Section 7.3.2.1. The only difference is to account for the nonlinear terms; see Reddy [13] for details)

$$\begin{bmatrix} \mathbf{K}^{11} & \mathbf{K}^{12} & \mathbf{K}^{13} \\ \mathbf{K}^{21} & \mathbf{K}^{22} & \mathbf{K}^{23} \\ \mathbf{K}^{31} & \mathbf{K}^{32} & \mathbf{K}^{33} \end{bmatrix} \begin{Bmatrix} \mathbf{u} \\ \bar{\mathbf{w}} \\ \mathbf{s} \end{Bmatrix} = \begin{Bmatrix} \mathbf{F}^1 \\ \mathbf{F}^2 \\ \mathbf{F}^3 \end{Bmatrix}, \tag{7.6.19}$$

where (see Fig. 7.6.3)

$$K_{ij}^{11} = \int_{x_a^e}^{x_b^e} A_{xx}^e \frac{d\psi_i^{(1)}}{dx}\frac{d\psi_j^{(1)}}{dx}\, dx, \quad K_{ij}^{12} = \frac{1}{2}\int_{x_a}^{x_b} A_{xx}^e \frac{dw}{dx}\frac{d\psi_i^{(1)}}{dx}\frac{d\psi_j^{(2)}}{dx}\, dx,$$

$$K_{ij}^{13} = \int_{x_a^e}^{x_b^e} B_{xx}^e \frac{d\psi_i^{(1)}}{dx}\frac{d\psi_j^{(3)}}{dx}\, dx, \quad K_{ij}^{21} = \int_{x_a^e}^{x_b^e} A_{xx}^e \frac{dw}{dx}\frac{d\psi_i^{(2)}}{dx}\frac{d\psi_j^{(1)}}{dx}\, dx$$

$$
\begin{aligned}
F_i^1 &= \int_{x_a^e}^{x_b^e} f\psi_i^{(1)}\, dx + \psi_i^{(1)}(x_a^e)Q_1^e + \psi_i^{(1)}(x_b^e)Q_4^e \\
K_{ij}^{22} &= \int_{x_a^e}^{x_b^e} \left[S_{xz}\frac{d^2\psi_i^{(2)}}{dx}\frac{d\psi_j^{(2)}}{dx} + c_f^e\psi_i^{(2)}\psi_j^{(2)} + \frac{1}{2}A_{xx}^e\left(\frac{dw}{dx}\right)^2\frac{d\psi_i^{(2)}}{dx}\frac{d\psi_j^{(2)}}{dx}\right] dx \\
K_{ij}^{23} &= \int_{x_a^e}^{x_b^e} \left(S_{xz}^e\frac{d\psi^{(2)}}{dx}\psi_j^{(3)} + B_{xx}^e\frac{dw}{dx}\frac{d\psi_i^{(2)}}{dx}\frac{d\psi_j^{(3)}}{dx}\right) dx \\
F_i^2 &= \int_{x_a^e}^{x_b^e} q\psi_i^{(2)}\, dx + \psi_i^{(2)}(x_a^e)Q_2^e + \psi_i^{(2)}(x_b^e)Q_5^e \\
K_{ij}^{31} &= \int_{x_a^e}^{x_b^e} B_{xx}^e\frac{d\psi_i^{(3)}}{dx}\frac{d\psi_j^{(1)}}{dx}\, dx \\
K_{ij}^{32} &= \int_{x_a^e}^{x_b^e} \left(S_{xz}^e\psi_i^{(3)}\frac{d\psi_j^{(2)}}{dx} + \frac{B_{xx}^e}{2}\frac{dw}{dx}\frac{d\psi_i^{(3)}}{dx}\frac{d\psi_j^{(2)}}{dx}\right) dx \\
K_{ij}^{33} &= \int_{x_a^e}^{x_b^e} \left(S_{xz}^e\psi_i^{(3)}\psi_j^{(3)} + D_{xx}^e\frac{d\psi_i^{(3)}}{dx}\frac{d\psi_j^{(3)}}{dx}\right) dx \\
F_i^3 &= \psi_i^{(3)}(x_a^e)Q_3^e + \psi_i^{(3)}(x_b^e)Q_6^e
\end{aligned}
\tag{7.6.20}
$$

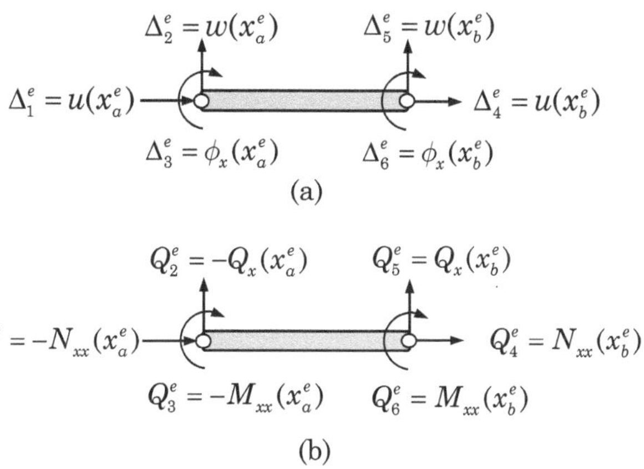

Fig. 7.6.3 Generalized nodal (a) displacements and (b) forces in the displacement model of the TBT.

7.6.2.4 Mixed finite element model

The mixed finite element mode of the TBT has the same form as the mixed model of the EBT, Eq. (7.6.12), and the coefficients $K_{ij}^{\alpha\beta}$ and F_i^{α} $(\alpha, \beta = 1, 2, 3)$ also remain the same as those in Eq. (7.6.13), except for the following coefficient (see Fig. 7.6.4):

$$K_{ij}^{33} = -\int_{x_a}^{x_b} \left(\frac{1}{D_{xx}} \psi_i^{(3)} \psi_j^{(3)} + \frac{1}{S_{xz}} \frac{d\psi_i^{(3)}}{dx} \frac{d\psi_j^{(3)}}{dx} \right) dx \tag{7.6.21}$$

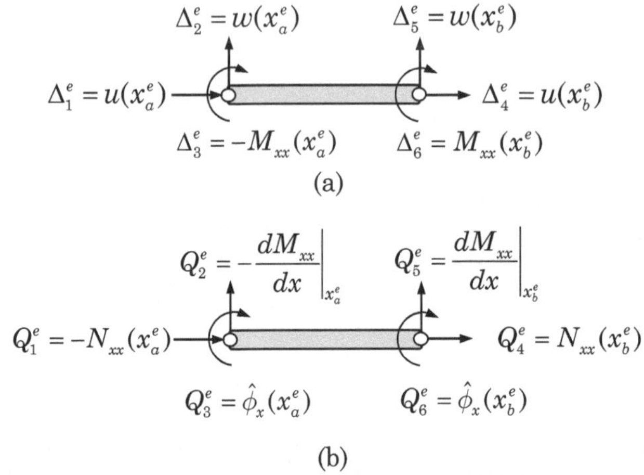

Fig. 7.6.4 (a) Primary and (b) secondary variables at the nodes in the *mixed* model of the TBT.

7.6.3 Dual Mesh Control Domain Models

7.6.3.1 Mixed Euler–Bernoulli beam model

The DMCDM is best suited to discretize second-order equations. Therefore, we can only consider the mixed model of the EBT using Eqs. (7.6.7a)–(7.6.7c). The domain $\Omega = (0, L)$ divided into a set of N finite elements (can be a nonuniform mesh) separated by nodes, as shown in Fig. 7.6.5, with each node having its own control domain (a dual mesh) around it. The first and last nodes have half-control domains. The nodes and elements are numbered sequentially from the left to the right. We consider a typical interior node I and the control domain associated with that node (see Fig. 7.6.6) to discretize the equations.

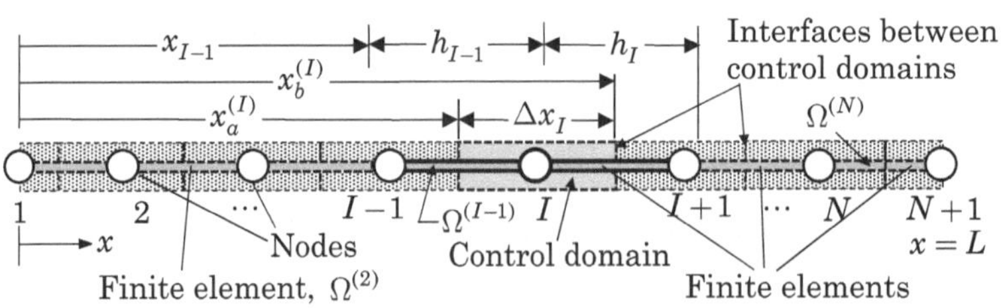

Fig. 7.6.5 A primal mesh of finite elements and dual mesh of control domains. We note that the boundary nodes have only half-control domains, whereas the internal nodes have full control domains. Also, each control domain connects two neighboring finite elements.

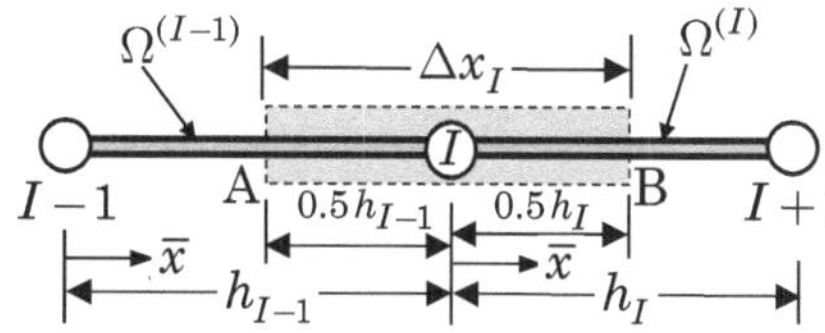

Fig. 7.6.6 Control domain associated with an interior node I. We note that each node has three unknowns and the control domain connects nine nodal values $(U_{I-1}, W_{I-1}, M_{I-1})$, (U_I, W_I, M_I), and $(U_{I+1}, W_{I+1}, M_{I+1})$ through the discretization of three governing equations.

In order to derive the discretized equations, we write the integral statements of Eqs. (7.6.7a)–(7.6.7c) over the control domain and carry out integration of expressions which contain the second differential; that is, unlike in a weighted-residual method (or weak form), we weaken the differentiability by carrying out the integration of second-order derivatives. For example, considering Eq. (7.6.7a), we obtain

$$0 = \int_A^B \left\{ -\frac{d}{dx}\left[\bar{A}_{xx}\left(\frac{du}{dx} + \frac{1}{2}\left(\frac{dw}{dx}\right)^2 \right) + \bar{B}_{xx}M_{xx} \right] - f \right\} dx$$

$$= \left[\bar{A}_{xx}\left(\frac{du}{dx} + \frac{1}{2}\left(\frac{dw}{dx}\right)^2 \right) + \bar{B}_{xx}\,M_{xx} \right]_{x_a^{(I)}}$$

$$- \left[\bar{A}_{xx}\left(\frac{du}{dx} + \frac{1}{2}\left(\frac{dw}{dx}\right)^2 \right) + \bar{B}_{xx}\,M_{xx} \right]_{x_b^{(I)}} - \int_{x_a^{(I)}}^{x_b^{(I)}} f\,dx \qquad (7.6.22)$$

or

$$0 = -N_1^{(I)} - N_2^{(I)} - \int_{x_a^{(I)}}^{x_b^{(I)}} f\,dx \qquad (7.6.23a)$$

where

$$N_1^{(I)} \equiv - \left[\bar{A}_{xx}\left(\frac{du}{dx} + \frac{1}{2}\left(\frac{dw}{dx}\right)^2 \right) + \bar{B}_{xx}\,M_{xx} \right]_{x_a^{(I)}} \qquad (7.6.23b)$$

$$N_2^{(I)} \equiv \left[\bar{A}_{xx}\left(\frac{du}{dx} + \frac{1}{2}\left(\frac{dw}{dx}\right)^2 \right) + \bar{B}_{xx}\,M_{xx} \right]_{x_b^{(I)}} \qquad (7.6.23c)$$

As before, points A and B refer to the left and right end locations of the control domain (associated with node I), which have the coordinates $x_a^{(I)}$ and $x_b^{(I)}$, respectively (note that point A is in element $\Omega^{(I-1)}$ and point B is in element $\Omega^{(I)}$;

see Fig. 7.6.6); $N_1^{(I)}$ and $N_2^{(I)}$ denote the secondary variables (in the present case, they are the axial forces) at the left and right interfaces of the control domain centered at node I (see Fig. 7.6.7 for the nodal degrees of freedom). The minus sign in the definition of $N_1^{(I)}$ indicates that it is a compressive force; both $N_1^{(I)}$ and $N_2^{(I)}$ are axial forces in the positive x direction.

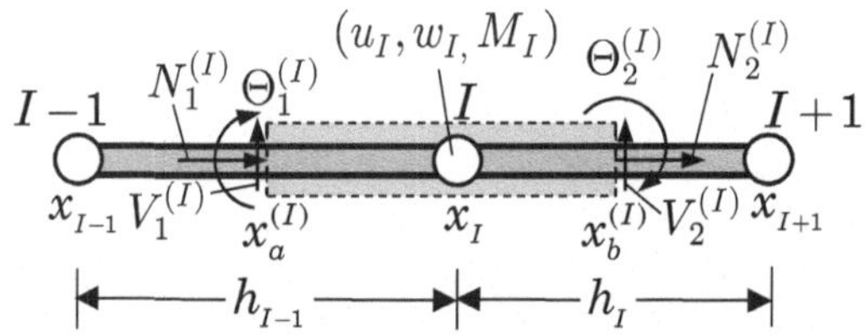

Fig. 7.6.7 Typical control domain for the mixed model of the EBT.

Similarly, Eq. (7.6.23b) takes the form

$$0 = -V_1^{(I)} - V_2^{(I)} + \int_{x_a^{(I)}}^{x_b^{(I)}} (c_f w - q)\, dx \tag{7.6.24a}$$

where $V_1^{(I)}$ and $V_2^{(I)}$ denote the secondary variables (shear forces acting upward positive) at the left and right interfaces of the control domain centered at node I,

$$V_1^{(I)} \equiv -\left[\frac{dM}{dx} + N_{xx}\frac{dw}{dx}\right]_{x_a^{(I)}}, \quad V_2^{(I)} \equiv \left[\frac{dM}{dx} + N_{xx}\frac{dw}{dx}\right]_{x_b^{(I)}} \tag{7.6.24b}$$

and N_{xx} is known in terms of the displacements (u, w) and bending moment M_{xx} through Eq. (7.2.11a).

The integral statement associated with Eq. (7.6.7c) is

$$0 = -\Theta_1^{(I)} + \Theta_2^{(I)} + \int_{x_a^{(I)}}^{x_b^{(I)}} \left[-\frac{1}{D_{xx}} M_{xx} + \bar{B}_{xx}\left(\frac{du}{dx} + \frac{1}{2}\left(\frac{dw}{dx}\right)^2\right)\right] dx \tag{7.6.25a}$$

where $\Theta_1^{(I)}$ and $\Theta_2^{(I)}$ denote the secondary variables (rotations in counterclockwise direction) at the left and right interfaces of the control domain centered at node I,

$$\Theta_1^{(I)} \equiv \left[-\frac{dw}{dx}\right]_{x_a^{(I)}}, \quad \Theta_2^{(I)} \equiv \left[-\frac{dw}{dx}\right]_{x_b^{(I)}} \tag{7.6.25b}$$

To complete the discretization, we invoke the finite element approximations of (u, w, M_{xx}) over a typical finite element $\Omega^{(I)} = (x_I, x_{I+1})$. Here we use equal degree (Lagrange) interpolation of all three variables. For example, the finite element approximation of $u(x)$ over $\Omega^{(I)}$ is

$$u(\bar{x}) \approx U_I \psi_1^{(I)}(\bar{x}) + U_{I+1} \psi_2^{(I)}(\bar{x}) \tag{7.6.26}$$

where U_I is the value of u at node I (i.e., $U_I \approx u(x_I)$) and $\psi_i^{(I)}(\bar{x})$ $(i = 1, 2)$ are linear finite element interpolation functions of element $\Omega^{(I)}$ for $I = 1, 2, \ldots, N$ written in terms of the local coordinate $\bar{x}$ ($\bar{x}$ has its origin at the left node of each finite element):

$$\psi_1^{(I)}(\bar{x}) = 1 - \frac{\bar{x}}{h_I}, \quad \psi_2^{(I)}(x) = \frac{\bar{x}}{h_I} \tag{7.6.27}$$

Hence, we can calculate the $(N_1^{(I)}, N_2^{(I)})$ in Eqs. (7.6.23b) and (7.6.23c), $(V_1^{(I)}, V_2^{(I)})$ in Eq. (7.6.24b), and $(\Theta_1^{(I)}, \Theta_2^{(I)})$ in Eq. (7.6.25b) in terms of the nodal values of (u, w, M_{xx}) using the interpolation of the type in Eq. (7.6.26) for each of the dependent variables of the formulation, while linearizing the nonlinear terms.

Discretization of Eq. (7.6.7a)

Returning to the DMCDM discretization of Eq. (7.6.7a), we first express ($N_1^{(I)}$ and $N_2^{(I)}$) in Eqs. (7.6.23b) and (7.6.23c) in terms of the nodal values of the primary variables:

$$N_1^{(I)} = -\bar{A}_{I-1} \frac{U_I - U_{I-1}}{h_{I-1}} - \tfrac{1}{2}\bar{A}_{I-1}\, \Delta W_{I-1} \frac{W_I - W_{I-1}}{h_{I-1}} - \bar{B}_{I-1} \frac{M_{I-1} + M_I}{2} \tag{7.6.28a}$$

$$N_2^{(I)} = \bar{A}_I \frac{U_{I+1} - U_I}{h_I} + \tfrac{1}{2}\bar{A}_I\, \Delta W_I \frac{W_{I+1} - W_I}{h_I} + \bar{B}_I \frac{M_I + M_{I+1}}{2} \tag{7.6.28b}$$

where $\bar{A}_{I-1} = \bar{A}_{xx}(x_a^{(I)})$ at the left interface and $\bar{A}_I = \bar{A}_{xx}(x_b^{(I)})$ at the right interface of the control domain centered around node I. Similar meaning applies to $\bar{B}_{I-1}$ and $\bar{B}_I$; W_I and M_I denote the nodal values of w and M_{xx}, respectively, at node I, while ΔW_I denotes the value of dw/dx in element $\Omega^{(I)}$, based on the previous iteration solution.

Substituting the approximations (7.6.28a) and (7.6.28b) into Eq. (7.6.23a), we obtain (for $I = 2, 3, \ldots, N$)

$$\begin{aligned}
&-\frac{\bar{A}_{I-1}}{h_{I-1}} U_{I-1} + \left(\frac{\bar{A}_{I-1}}{h_{I-1}} + \frac{\bar{A}_I}{h_I}\right) U_I - \frac{\bar{A}_I}{h_I} U_{I+1} \\
&+ \tfrac{1}{2}\left[-\frac{\bar{A}_{I-1}\Delta W_{I-1}}{h_{I-1}} W_{I-1} + \left(\frac{\bar{A}_{I-1}\Delta W_{I-1}}{h_{I-1}} + \frac{\bar{A}_I \Delta W_I}{h_I}\right) W_I - \frac{\bar{A}_I \Delta W_I}{h_I} W_{I+1}\right] \\
&+ \tfrac{1}{2}\bar{B}_{I-1} M_{I-1} + \tfrac{1}{2}\left(\bar{B}_{I-1} - \bar{B}_I\right) M_I - \tfrac{1}{2}\bar{B}_I M_{I+1} - F_I = 0
\end{aligned} \tag{7.6.29}$$

where

$$\bar{A}_{I-1} = \left.\frac{D_{xx}^*}{D_{xx}}\right|_{x_a^{(I)}}, \quad \bar{A}_I = \left.\frac{D_{xx}^*}{D_{xx}}\right|_{x_b^{(I)}}, \quad \bar{B}_{I-1} = \left.\frac{B_{xx}}{D_{xx}}\right|_{x_a^{(I)}}, \quad \bar{B}_I = \left.\frac{B_{xx}}{D_{xx}}\right|_{x_b^{(I)}}$$

$$\Delta W_{I-1} = \frac{\bar{W}_I - \bar{W}_{I-1}}{h_{I-1}}, \quad \Delta W_I = \frac{\bar{W}_{I+1} - \bar{W}_I}{h_I}, \quad F_I = \int_{x_a^{(I)}}^{x_b^{(I)}} f(x)\, dx \tag{7.6.30}$$

Here $\bar{W}_I$ denotes the value of w at node I from the previous iteration to solve the nonlinear algebraic equations. The integral of a function f over the control domain $(x_a^{(I)}, x_b^{(I)})$ can be evaluated using either exact integration or numerical integration (e.g., the one-third Simpson rule).

Next, we should obtain the discretized equation for the boundary nodes, node 1 and node $N+1$ (when there are N linear elements in the primal mesh). We note that at node 1, $N_1^{(1)}$ is the boundary axial force, which is known or its dual, U_1, is known. Hence, we only evaluate $N_2^{(1)}$ at $h_1/2$. The discretized equations of the left boundary node is

$$\begin{aligned} 0 &= -N_1^{(1)} - \bar{A}_1 \frac{U_2 - U_1}{h_1} - \tfrac{1}{2}\bar{A}_1\, \Delta W_1 \frac{W_2 - W_1}{h_1} - \bar{B}_1 \frac{M_2 + M_1}{2} - \int_0^{0.5h_1} f\, dx \\ &= -N_1^{(1)} + \frac{\bar{A}_1}{h_1} U_1 - \frac{\bar{A}_1}{h_1} U_2 + \tfrac{1}{2} \frac{\bar{A}_1\, \Delta W_1}{h_1} W_1 - \tfrac{1}{2} \frac{\bar{A}_1\, \Delta W_1}{h_1} W_2 \\ &\quad - \tfrac{1}{2}\bar{B}_1\, M_1 - \tfrac{1}{2}\bar{B}_1\, M_2 - \int_0^{0.5h_1} f\, dx \end{aligned} \tag{7.6.31}$$

Similarly, for the node on the right boundary, we have

$$\begin{aligned} 0 &= -N_2^{(N+1)} - \frac{\bar{A}_N}{h_N} U_N + \frac{\bar{A}_N}{h_N} U_{N+1} - \tfrac{1}{2} \frac{\bar{A}_N\, \Delta W_N}{h_N} W_N + \tfrac{1}{2} \frac{\bar{A}_N\, \Delta W_N}{h_N} U_{N+1} \\ &\quad + \tfrac{1}{2}\bar{B}_N\, M_N + \tfrac{1}{2}\bar{B}_N\, M_{N+1} - \int_0^{0.5h_N} f\, d\bar{x} \end{aligned} \tag{7.6.32}$$

This completes the discretization of Eq. (7.6.7a).

Discretization of Eq. (7.6.7b)

The same procedure can be applied to Eq. (7.6.7b) to obtain the discretized equations for the interior and boundary nodes. Discretized values of $V_1^{(I)}$ and $V_2^{(I)}$ are

$$\begin{aligned} V_1^{(I)} &= -\frac{M_I - M_{I-1}}{h_{I-1}} - \bar{A}_{I-1} \Delta W_{I-1} \frac{U_I - U_{I-1}}{h_{I-1}} - 0.5 \bar{A}_{I-1} (\Delta W)^2_{I-1} \frac{W_I - W_{I-1}}{h_{I-1}} \\ &\quad - \tfrac{1}{2} \left[\bar{B}_{I-1}\, \Delta W_{I-1} \left(M_{I-1} + M_{I+1} \right) \right] \end{aligned} \tag{7.6.33a}$$

$$\begin{aligned} V_2^{(I)} &= \frac{M_{I+1} - M_I}{h_I} + \bar{A}_I\, \Delta W_I \frac{U_{I+1} - U_I}{h_I} + 0.5 \bar{A}_I\, (\Delta W)^2_I \frac{W_{I+1} - W_I}{h_I} \\ &\quad + \tfrac{1}{2} \left[\bar{B}_I \Delta W_I \left(M_I + M_{I+1} \right) \right] \end{aligned} \tag{7.6.33b}$$

The integral of $c_f\, w$ over the control domain $(x_a^{(I)}, x_b^{(I)})$, for the linear interpolation used, is

$$\int_{x_a^{(I)}}^{x_b^{(I)}} c_f w\, dx = \tfrac{1}{8} \left[C_{I-1} W_{I-1} h_{I-1} + 3 W_I \left(C_{I-1} h_{I-1} + C_I h_I \right) + C_I W_{I+1} h_I \right] \tag{7.6.34}$$

where C_I is the value of c_f in element I.

Substitution of the expressions from Eqs. (7.6.33a), (7.6.33b), and (7.6.34) into Eq. (7.6.24a), we obtain

$$\begin{aligned}
&-\frac{1}{h_{I-1}}M_{I-1}+\left(\frac{1}{h_{I-1}}+\frac{1}{h_I}\right)M_I-\frac{1}{h_I}M_{I+1}+\tfrac{1}{8}C_{I-1}h_{I-1}W_{I-1}\\
&+\tfrac{3}{8}\left(C_{I-1}h_{I-1}+C_Ih_I\right)W_I+\tfrac{1}{8}C_Ih_IW_{I+1}-\frac{\bar{A}_{I-1}\,\Delta W_{I-1}}{h_{I-1}}U_{I-1}\\
&+\left(\frac{\bar{A}_{I-1}\,\Delta W_{I-1}}{h_{I-1}}+\frac{\bar{A}_I\,\Delta W_I}{h_I}\right)U_I-\frac{\bar{A}_I\,\Delta W_I}{h_I}U_{I+1}-\tfrac{1}{2}\frac{\bar{A}_{I-1}\,(\Delta W)^2_{I-1}}{h_{I-1}}W_{I-1}\\
&+\tfrac{1}{2}\left(\frac{\bar{A}_{I-1}\,(\Delta W)^2_{I-1}}{h_{I-1}}+\frac{\bar{A}_I\,(\Delta W)^2_I}{h_I}\right)W_I-\tfrac{1}{2}\frac{\bar{A}_I\,(\Delta W)^2_I}{h_I}W_{I+1}-Q_I\\
&+\tfrac{1}{2}\left[\bar{B}_{I-1}\,\Delta W_{I-1}\,M_{I-1}+\left(\bar{B}_{I-1}\,\Delta W_{I-1}-\bar{B}_I\,\Delta W_I\right)M_I-\bar{B}_I\,\Delta W_I\,M_{I+1}\right]
\end{aligned}\tag{7.6.35}$$

For the boundary nodes 1 and $N+1$, we have

$$\begin{aligned}
0=&-V_1^{(1)}+\frac{1}{h_1}M_1-\frac{1}{h_1}M_2+\frac{3h_1}{8}C_1W_1+\frac{h_1}{8}C_1W_2\\
&+\frac{\bar{A}_1\,\Delta W_1}{h_1}\left(U_1-U_2\right)+\tfrac{1}{2}\frac{\bar{A}_1\,(\Delta W)^2_1}{h_1}\left(W_1-W_2\right)\\
&-\tfrac{1}{2}\bar{B}_1\,\Delta W_1\,M_1-\tfrac{1}{2}\bar{B}_1\,\Delta W_1\,M_2-Q_1
\end{aligned}\tag{7.6.36a}$$

$$\begin{aligned}
0=&-V_2^{(N+1)}-\frac{1}{h_N}M_N+\frac{1}{h_N}M_{N+1}+\frac{h_N}{8}C_NW_N+\frac{3h_N}{8}C_NW_{N+1}\\
&+\frac{\bar{A}_N\,\Delta W_N}{h_N}\left(U_{N+1}-U_N\right)+\tfrac{1}{2}\frac{\bar{A}_N\,(\Delta W)^2_N}{h_N}\left(W_{N+1}-W_N\right)\\
&+\tfrac{1}{2}\bar{B}_N\,\Delta W_N\,M_N+\tfrac{1}{2}\bar{B}_N\,\Delta W_N\,M_{N+1}-Q_{N+1}
\end{aligned}\tag{7.6.36b}$$

where

$$Q_I=\int_{0.5h_{I-1}}^{h_{I-1}}q\,d\bar{x}+\int_0^{h_I}q\,d\bar{x},\quad Q_1=\int_0^{0.5h_1}q\,d\bar{x},\quad Q_{N+1}=\int_{0.5h_N}^{h_N}q\,d\bar{x},\tag{7.6.37}$$

Discretization of Eq. (7.6.7c)

The discretized equations associated with Eq. (7.6.7c) are obtained substituting the following approximations for $\Theta_1^{(I)}$ and $\Theta_2^{(I)}$ and replacing other terms as before,

$$\Theta_1^{(I)}=\frac{W_{I-1}-W_I}{h_{I-1}},\quad \Theta_2^{(I)}=\frac{W_I-W_{I+1}}{h_I}\tag{7.6.38}$$

For an interior node, we obtain

$$
\begin{aligned}
&-\frac{1}{h_{I-1}}W_{I-1}+\left(\frac{1}{h_{I-1}}+\frac{1}{h_I}\right)W_I-\frac{1}{h_I}W_{I+1}\\
&\quad-\frac{1}{8}\frac{h_{I-1}}{D_{I-1}}M_{I-1}-\frac{3}{8}\left(\frac{h_{I-1}}{D_{I-1}}+\frac{h_I}{D_I}\right)M_I-\frac{1}{8}\frac{h_I}{D_I}M_{I+1}\\
&\quad-0.5\bar{B}_{I-1}U_{I-1}+0.5\left(\bar{B}_{I-1}-\bar{B}_I\right)U_I+0.5\bar{B}_I U_{I+1}\\
&\quad-0.25\bar{B}_{I-1}\,\Delta W_{I-1}\,W_{I-1}+0.25\left(\bar{B}_{I-1}\,\Delta W_{I-1}-\bar{B}_I\,\Delta W_I\right)W_I\\
&\quad+0.25\bar{B}_I\,\Delta W_I\,W_{I+1}=0
\end{aligned}
\tag{7.6.39}
$$

Here D_I denotes the value of D_{xx} in element I and $\bar{B}_I$ denotes the value of B_{xx}/D_{xx} in element I. For the boundary nodes 1 and $N+1$, we have

$$
\begin{aligned}
0&=-\Theta_1^{(1)}+\frac{1}{h_1}W_1-\frac{1}{h_1}W_2-\tfrac{3}{8}\frac{h_1}{D_1}M_1-\tfrac{1}{8}\frac{h_1}{D_1}M_2\\
&\quad-\tfrac{1}{2}\bar{B}_1U_1+\tfrac{1}{2}\bar{B}_1U_2-\tfrac{1}{4}\bar{B}_1\,\Delta W_1\,W_1+\tfrac{1}{4}\bar{B}_1\,\Delta W_1\,W_2
\end{aligned}
\tag{7.6.40a}
$$

$$
\begin{aligned}
0&=\Theta_2^{(N+1)}-\frac{1}{h_N}W_N+\frac{1}{h_N}W_{N+1}-\tfrac{1}{8}\frac{h_N}{D_N}M_N-\tfrac{3}{8}\frac{h_N}{D_N}M_{N+1}-\tfrac{1}{2}\bar{B}_NU_N\\
&\quad+\tfrac{1}{2}\bar{B}_NU_{N+1}-\tfrac{1}{4}\bar{B}_N\,\Delta W_N\,W_N+\tfrac{1}{4}\bar{B}_N\,\Delta W_N\,W_{N+1}
\end{aligned}
\tag{7.6.40b}
$$

This completes the development of the discretized equations based on the DM-CDM for the mixed formulation of the EBT.

7.6.3.2 Displacement model of the Timoshenko beam theory

We present discretized equations associated with Eqs. (7.6.14a)–(7.6.14c), with (N_{xx}, M_{xx}, Q_x) replaced in terms of the displacements using Eqs. (7.6.15a)–(7.6.15c). Figure 7.6.8 shows the nodal degrees of freedom for the displacement model of the TBT. The integral statements of Eqs. (7.6.14a)–(7.6.14c) are:

$$0=-N_1^{(I)}-N_2^{(I)}-\int_{x_a^{(I)}}^{x_b^{(I)}} f\,dx \tag{7.6.41a}$$

$$0=-V_1^{(I)}-V_2^{(I)}+\int_{x_a^{(I)}}^{x_b^{(I)}} (c_f w-q)dx \tag{7.6.41b}$$

$$0=-M_1^{(I)}-M_2^{(I)}+\int_{x_a^{(I)}}^{x_b^{(I)}} S_{xz}\left(\phi_x+\frac{dw}{dx}\right)dx \tag{7.6.41c}$$

where

$$N_1^{(I)}\equiv-\left[A_{xx}\left(\frac{du}{dx}+\tfrac{1}{2}\left(\frac{dw}{dx}\right)^2\right)+B_{xx}\frac{d\phi_x}{dx}\right]_{x_a^{(I)}} \tag{7.6.42a}$$

$$N_2^{(I)}\equiv\left[A_{xx}\left(\frac{du}{dx}+\tfrac{1}{2}\left(\frac{dw}{dx}\right)^2\right)+B_{xx}\frac{d\phi_x}{dx}\right]_{x_b^{(I)}} \tag{7.6.42b}$$

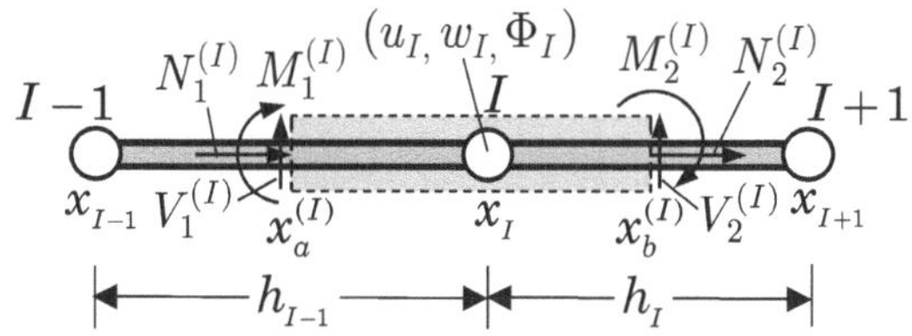

Fig. 7.6.8 A typical control domain for the displacement model of the TBT.

$$V_1^{(I)} \equiv -\left[S_{xz}\left(\phi_x + \frac{dw}{dx}\right) + N_{xx}\frac{dw}{dx}\right]_{x_a^{(I)}} \tag{7.6.42c}$$

$$V_2^{(I)} \equiv \left[S_{xz}\left(\phi_x + \frac{dw}{dx}\right) + N_{xx}\frac{dw}{dx}\right]_{x_b^{(I)}} \tag{7.6.42d}$$

$$M_1^{(I)} \equiv -\left[B_{xx}\left(\frac{du}{dx} + \frac{1}{2}\left(\frac{dw}{dx}\right)^2\right) + D_{xx}\frac{d\phi_x}{dx}\right]_{x_a^{(I)}} \tag{7.6.42e}$$

$$M_2^{(I)} \equiv \left[B_{xx}\left(\frac{du}{dx} + \frac{1}{2}\left(\frac{dw}{dx}\right)^2\right) + D_{xx}\frac{d\phi_x}{dx}\right]_{x_b^{(I)}} \tag{7.6.42f}$$

The values of $N_i^{(I)}$, $V_i^{(I)}$, and $M_i^{(I)}$ $(i = 1, 2)$ in Eqs. (7.6.42a)–(7.6.42f) can be expressed in terms of the nodal values (U_I, W_I, Φ_I) of ($u(x)$, $w(x)$, $\phi_x(x)$), respectively, as follows:

$$\begin{aligned}
N_1^{(I)} &= -A_{I-1}\,\frac{U_I - U_{I-1}}{h_{I-1}} - \tfrac{1}{2}A_{I-1}\,\Delta W_{I-1}\,\frac{W_I - W_{I-1}}{h_{I-1}} - B_{I-1}\frac{\Phi_I - \Phi_{I-1}}{h_{I-1}}\\
N_2^{(I)} &= A_I\,\frac{U_{I+1} - U_I}{h_I} + \tfrac{1}{2}A_I\,\Delta W_I\,\frac{W_{I+1} - W_I}{h_I} + B_I\frac{\Phi_{I+1} - \Phi_I}{h_I}\\
V_1^{(I)} &= -S_{I-1}\left[\frac{\Phi_{I-1} + \Phi_I}{2} + \frac{W_I - W_{I-1}}{h_{I-1}}\right]\\
V_2^{(I)} &= S_I\left[\frac{\Phi_I + \Phi_{I+1}}{2} + \frac{W_{I+1} - W_I}{h_I}\right]\\
M_1^{(I)} &= -B_{I-1}\,\frac{U_I - U_{I-1}}{h_{I-1}} - \tfrac{1}{2}B_{I-1}\,\Delta W_{I-1}\,\frac{W_I - W_{I-1}}{h_{I-1}} - D_{I-1}\frac{\Phi_I - \Phi_{I-1}}{h_{I-1}}\\
M_2^{(I)} &= B_I\,\frac{U_{I+1} - U_I}{h_I} + \tfrac{1}{2}B_I\,\Delta W_I\,\frac{W_{I+1} - W_I}{h_I} + D_I\frac{\Phi_{I+1} - \Phi_I}{h_I}\\
\int_{x_a^{(I)}}^{x_b^{(I)}} S_{xz}\left(\phi_x + \frac{dw}{dx}\right)dx &= \tfrac{1}{4}S_{I-1}\,(\Phi_{I-1} + \Phi_I)\,h_{I-1} + \tfrac{1}{4}S_I\,(\Phi_I + \Phi_{I+1})\,h_I\\
&\quad + \tfrac{1}{2}S_{I-1}\,(W_I - W_{I-1}) + \tfrac{1}{2}S_I\,(W_{I+1} - W_I)\,.
\end{aligned} \tag{7.6.43}$$

We note that in evaluating the integral of $\phi_x + dw/dx$ in Eq. (7.6.43), ϕ_x is treated as a constant to avoid shear locking.

With the relations in Eq. (7.6.43), Eqs. (7.6.41a)–(7.6.41c) can be expressed as

$$
\begin{aligned}
0 = &-\frac{A_{I-1}}{h_{I-1}} U_{I-1} + \left(\frac{A_{I-1}}{h_{I-1}} + \frac{A_I}{h_I}\right) U_I - \frac{A_I}{h_I} U_{I+1} + \tfrac{1}{2}\left[-\frac{A_{I-1}\Delta W_{I-1}}{h_{I-1}} W_{I-1}\right. \\
&\left. + \left(\frac{A_{I-1}\Delta W_{I-1}}{h_{I-1}} + \frac{A_I \Delta W_I}{h_I}\right) W_I - \frac{A_I \Delta W_I}{h_I} W_{I+1}\right] \\
&- \frac{B_{I-1}}{h_{I-1}} \Phi_{I-1} + \left(\frac{B_{I-1}}{h_{I-1}} + \frac{B_I}{h_I}\right) \Phi_I - \frac{B_I}{h_I} \Phi_{I+1} - F_I
\end{aligned} \tag{7.6.44a}
$$

$$
\begin{aligned}
0 = &-\frac{S_{I-1}}{h_{I-1}} W_{I-1} + \left(\frac{S_{I-1}}{h_{I-1}} + \frac{S_I}{h_I}\right) W_I - \frac{S_I}{h_I} W_{I+1} + \tfrac{1}{8} C_{I-1} W_{I-1} h_{I-1} \\
&+ \tfrac{3}{8}\left(C_{I-1} h_{I-1} + C_I h_I\right) W_I + 0.125 C_I W_{I+1} h_I \\
&+ \tfrac{1}{2} S_{I-1} \Phi_{I-1} + \tfrac{1}{2}\left(S_{I-1} - S_I\right) \Phi_I - \tfrac{1}{2} S_I \Phi_{I+1} - Q_I
\end{aligned} \tag{7.6.44b}
$$

$$
\begin{aligned}
0 = &-\frac{B_{I-1}}{h_{I-1}} U_{I-1} + \left(\frac{B_{I-1}}{h_{I-1}} + \frac{B_I}{h_I}\right) U_I - \frac{B_I}{h_I} U_{I+1} \\
&- \tfrac{1}{2} \frac{B_{I-1}\, \Delta W_{I-1}}{h_{I-1}} W_{I-1} + \tfrac{1}{2}\left(\frac{B_{I-1}\, \Delta W_{I-1}}{h_{I-1}} + \frac{B_I\, \Delta W_I}{h_I}\right) W_I \\
&- \tfrac{1}{2} \frac{B_I\, \Delta W_I}{h_I} W_{I+1} - 0.5 S_{I-1} W_{I-1} + \tfrac{1}{2}\left(S_{I-1} - S_I\right) W_I + \tfrac{1}{2} S_I W_{I+1} \\
&- \frac{D_{I-1}}{h_{I-1}} \Phi_{I-1} + \left(\frac{D_{I-1}}{h_{I-1}} + \frac{D_I}{h_I}\right) \Phi_I - \frac{D_I}{h_I} \Phi_{I+1} + 0.25 S_{I-1} h_{I-1} \Phi_{I-1} \\
&+ \tfrac{1}{4}\left(S_{I-1} h_{I-1} + S_I h_I\right) \Phi_I + \tfrac{1}{4} S_I h_I \Phi_{I+1}.
\end{aligned} \tag{7.6.44c}
$$

Next, we should obtain the discretized equations for the boundary nodes. The discretized equations of the left boundary node are

$$
0 = -N_1^{(1)} + \frac{A_1}{h_1}\left(U_1 - U_2\right) + \frac{B_1}{h_1}\left(\Phi_1 - \Phi_2\right) + \tfrac{1}{2} \frac{A_1\, \Delta W_1}{h_1}\left(W_1 - W_2\right) - F_1, \tag{7.6.45a}
$$

$$
0 = -V_1^{(1)} + \frac{S_1}{h_1}\left(W_1 - W_2\right) - \tfrac{1}{2} S_1\left(\Phi_1 + \Phi_2\right) + \frac{h_1}{8} C_1\left(3W_1 + W_2\right) - Q_1, \tag{7.6.45b}
$$

$$
\begin{aligned}
0 = &-M_1^{(1)} - \frac{B_1}{h_1}\left(U_2 - U_1\right) + \tfrac{1}{2} S_1 (W_2 - W_1) - \frac{D_1}{h_1}\left(\Phi_2 - \Phi_1\right) \\
&+ \tfrac{1}{4} S_1 h_1\left(\Phi_1 + \Phi_2\right) - \tfrac{1}{2} \frac{B_1\, \Delta W_1}{h_1}\left(U_2 - U_1\right).
\end{aligned} \tag{7.6.45c}
$$

For the node $N+1$ on the right boundary, we have

$$0 = -N_1^{(N+1)} + \frac{A_N}{h_N}(U_{N+1} - U_N) + \frac{B_N}{h_N}(\Phi_{N+1} - \Phi_N) + \tfrac{1}{2}\frac{A_N\,\Delta W_N}{h_N}(W_{N+1} - W_N) - F_{N+1}, \tag{7.6.46a}$$

$$0 = -V_2^{(N+1)} + \frac{S_N}{h_N}(W_{N+1} - W_N) + \frac{h_N}{8}C_N(W_N + 3W_{N+1}) + \tfrac{1}{2}S_N(\Phi_{N+1} + \Phi_N) - Q_{N+1}, \tag{7.6.46b}$$

$$0 = -M_2^{(N+1)} + \frac{B_N}{h_N}(U_{N+1} - U_N) + \tfrac{1}{2}S_N(W_{N+1} - W_N) + \frac{D_N}{h_N}(\Phi_{N+1} - \Phi_N) + \tfrac{1}{4}S_N h_N(\Phi_N + \Phi_{N+1}) + \tfrac{1}{2}\frac{B_N\,\Delta W_N}{h_N}(W_{N+1} - W_N). \tag{7.6.46c}$$

7.6.3.3 Mixed model of the Timoshenko beam theory

Lastly, we develop the mixed DMCDM model of Eqs. (7.6.17a)–(7.6.17c). Due to the close similarity between Eqs. (7.6.17a)–(7.6.17c) and Eqs. (7.6.7a)–(7.6.7c), the discretized equations in Eqs. (7.6.29), (7.6.31), (7.6.32), (7.6.35), (7.6.36a), (7.6.36b), (7.6.39), (7.6.40a), and (7.6.40b) are valid here, with the additional contributions to Eqs. (7.6.39), (7.6.40a), and (7.6.40b) due to the expression [see Eq. (7.6.17c)]

$$\left[\frac{1}{S_{xz}}\frac{dM_{xx}}{dx}\right]_{x_a^{(I)}}^{x_b^{(I)}} \tag{7.6.47}$$

The additional terms are:

Node I:
$$\frac{1}{h_{I-1}}\frac{1}{S_{I-1}}M_{I-1} - \left(\frac{1}{h_{I-1}}\frac{1}{S_{I-1}} + \frac{1}{h_I}\frac{1}{S_I}\right)M_I + \frac{1}{h_I}\frac{1}{S_I}M_{I+1} \tag{7.6.48a}$$

Node 1:
$$\frac{1}{S_1}\frac{M_2 - M_1}{h_1} \tag{7.6.48b}$$

Node $N+1$:
$$\frac{1}{S_N h_N}M_N - \frac{1}{S_N h_N}M_{N+1} \tag{7.6.48c}$$

where S_I is the value of S_{xz} in element $\Omega^{(I)}$.

7.6.4 Linearization of Equations

We note that the stiffness coefficients K_{iJ}^{12}, K_{Ji}^{21}, and K_{IJ}^{22} depend on the unknown displacement function w. Consequently, the finite element equations in Eq. (7.6.10) are nonlinear, and they are linearized as discussed in Section 1.9.4 before assembly. The linearization is a necessary step to be able to solve the final algebraic equations resulting from the application of a numerical method.

The final algebraic equations obtained with the FEM, FVM, or DMCDM has the form

$$\mathbf{K}(\boldsymbol{\Delta})\boldsymbol{\Delta} = \mathbf{F} \tag{7.6.49}$$

where $\mathbf{K}$ is the coefficient matrix (known in terms of $\boldsymbol{\Delta}$), $\boldsymbol{\Delta}$ is the column vector of nodal unknowns, and $\mathbf{F}$ is the source vector (known). Equation (7.6.49) is solved using a successive approximation known as the direct iteration method or the Picard iteration method. Suppose that we are at the end of the rth iteration and seeking the $(r+1)$st iteration solution. Then we have

$$\mathbf{K}(\boldsymbol{\Delta}^r)\boldsymbol{\Delta}^{r+1} = \mathbf{F} \quad \rightarrow \quad \boldsymbol{\Delta}^{r+1} = (\mathbf{K}^r)^{-1}\mathbf{F} \tag{7.6.50}$$

where $\mathbf{K}^r \equiv \mathbf{K}(\boldsymbol{\Delta}^r)$. The iteration is continued until the difference between two consecutive solutions (measured with a suitable measure) is within a prescribed tolerance:

$$\sqrt{\frac{\delta\boldsymbol{\Delta}\cdot\delta\boldsymbol{\Delta}}{\boldsymbol{\Delta}^r\cdot\boldsymbol{\Delta}^r}} \leq \epsilon, \quad \delta\boldsymbol{\Delta} \equiv \boldsymbol{\Delta}^{r+1} - \boldsymbol{\Delta}^r \tag{7.6.51}$$

where ϵ denotes a preselected value of the error tolerance (say, $\epsilon = 10^{-3}$). In the beginning of the iteration, one must have a starting guess vector $\boldsymbol{\Delta}^0$; in the case of structural problems, we can take the initial guess vector to be zero so that the first iteration solution is the linear solution.

In the present case, the linearization amounts to calculating the nonlinear terms using the previous iteration solution. For example, we can linearize $(dw/dx)^2$ and $(dw/dx)^3$ as

$$\left(\frac{dw}{dx}\right)^2 \approx \left[\frac{dw}{dx}\right]^{(r)} \frac{dw}{dx}; \quad \left(\frac{dw}{dx}\right)^3 \approx \left[\left(\frac{dw}{dx}\right)^2\right]^{(r)} \frac{dw}{dx}, \tag{7.6.52}$$

where the term in the square bracket is evaluated using the known solution from the rth iteration.

The direct iteration scheme does not converge unless an acceleration parameter, β, is used to evaluate the stiffness matrix, $\mathbf{K}^r = \mathbf{K}(\bar{\boldsymbol{\Delta}}^r)$, at each iteration:

$$\bar{\boldsymbol{\Delta}}^r = (1-\beta)\boldsymbol{\Delta}^r + \beta\boldsymbol{\Delta}^{r-1}, \quad 0 \leq \beta \leq 1, \tag{7.6.53}$$

where r denotes the iteration number. Thus, using a weighted average of the last two iteration solutions to update the stiffness matrix accelerates the convergence. In the present case, a value of $\beta = 0.25 - 0.35$ is used (after a parametric study, starting with $\beta = 0$).

7.6.5 Numerical Results

In this section, we consider applications of the methodology developed in the preceding sections. Numerical results obtained with the FEM and DMCDM are compared in all cases. As stated in Section 7.5, we use four beam models of the FEM and three beam models of DMCDM. The FE-EB(D) model uses Hermite cubic interpolation of $w(x)$ and linear interpolation of $u(x)$, whereas all other elements are based on Lagrange interpolations of all variables. All finite element models other than FE-EB(D) can also use quadratic or higher order interpolations, whereas the dual mesh control domain formulations presented herein are based on linear interpolations. Thus, for consistency, all numerical results presented herein, with the exception of FE-EB(D), are obtained with linear approximations of all field variables.

Here, we shall consider a functionally graded beam of length $L = 100$ in (254 cm), height $h = 1$ in (2.54 cm), and width $b = 1$ in (2.54 cm), and subjected to a uniformly distributed load of intensity q_0 lb/in (1 lb/in = 175 N/m). The FGM beam is made of two materials with the following values of the moduli, Poisson's ratio, and shear correction coefficient:

$$E_1 = 30\times10^6 \text{ psi (210 GPa)}, \quad E_2 = 3\times10^6 \text{ psi (21 GPa)}, \quad \nu = 0.3, \quad K_s = \frac{5}{6}.$$

We shall investigate the parametric effects of the power-law index n and boundary conditions on the transverse deflections and stresses.

Numerical studies have been carried out with various models, using the value of the acceleration parameter on the convergence, effect of the power-law index, and post-computation of the secondary variables (either bending moments or rotations). In all cases, both the DMCDM and FEM models, 16 linear elements in the half-beam are used. We consider functionally graded beams which are either pin-supported at both ends (P-P) or clamped at both ends (C-C). Using the symmetry about $x = L/2$, we use the left half of the beam as the computational domain and investigate the effect of the power-law index on the transverse deflections and bending moments. The boundary conditions on the primary variables in the displacement and mixed models for P-P beams are given in Eqs. (7.5.1) and (7.5.2). The exact (or analytical) solutions for the linear case of pinned-pinned functionally graded beams according to the TBT, with the power-law given in Eq. (7.1.1), are given in Eqs. (7.5.3) and (7.5.4) (see Reddy [42]). We note that the bending moment and shear force for the linear case do not depend on B_{xx}. However, in the nonlinear case, the bending moment does depend on B_{xx}.

The boundary conditions on the primary variables in the displacement and mixed models for the C-C beams are presented in Eqs. (7.5.7) and (7.5.8). The exact solutions for the linear case of clamped-clamped beams according to the TBT are given in Eq. (7.5.9).

Load increments of $\Delta q_0 = 1.0$ lb/in (175 N/m) and a tolerance of $\epsilon = 10^{-3}$ are used in the nonlinear analysis. The initial solution vector is chosen to be $\mathbf{\Delta}^0 = \mathbf{0}$ so that the first iteration is the linear solution for the first load step.

The nonlinear analysis shows that all models yield nonlinear solutions that are indistinguishable in the graphs of dimensionless center deflection, $\bar{w} = w(0.5L)\hat{D}_{xx}/L^4$ and bending moment $\bar{M}_{xx} = M_{xx}(0.5L)/L^2$ versus the intensity of the uniformly distributed load, q_0. Table 7.6.1 contains the results obtained with various models with a uniform mesh (16 elements) of linear approximations of all variables in the half-beam (with the direct iteration scheme). Convergence is achieved with almost the same number of iterations for different load steps (see Reddy et al. [44]) when an acceleration parameter β [see Eq. (7.6.53)] of $\beta = 0.35$ is used. Convergence was not achieved with $\beta = 0.0$ (when 100% of the previous iteration solution is used as the guess vector for the current iteration) and the direct iteration method.

Table 7.6.1 Numerical results obtained by various models for the (nonlinear) center deflection, $w(0.5L) \times 10$, of a *pinned–pinned* beam ($n = 0$) under uniform distributed load of intensity q_0.

	FEM				DMCDM		
q_0	EB(D)	TB(D)	EB(M)	TB(M)	TB(D)	EB(M)	TB(M)
0.5	0.0564	0.0563	0.0564	0.0564	0.0563	0.0564	0.0563 (7)[a]
1	0.0922	0.0921	0.0921	0.0921	0.0921	0.0921	0.0921 (8)
2	0.1364	0.1364	0.1364	0.1364	0.1364	0.1364	0.1364 (13)
3	0.1664	0.1663	0.1664	0.1663	0.1663	0.1664	0.1664 (10)
4	0.1888	0.1888	0.1888	0.1888	0.1888	0.1888	0.1888 (14)
5	0.2078	0.2078	0.2078	0.2078	0.2078	0.2078	0.2078 (14)
6	0.2238	0.2237	0.2237	0.2237	0.2237	0.2237	0.2237 (11)
7	0.2387	0.2386	0.2386	0.2387	0.2386	0.2387	0.2387 (15)
8	0.2117	0.2516	0.2516	0.2516	0.2516	0.2516	0.2516 (15)
9	0.2635	0.2634	0.2634	0.2634	0.2634	0.2634	0.2634 (15)
10	0.2744	0.2743	0.2743	0.2743	0.2743	0.2743	0.2743 (15)
⋮							
20	0.3550	0.3547	0.3547	0.3547	0.3547	0.3547	0.3547 (16)

[a]Number of iterations taken to converge (when 16 linear elements are used in the half-beam); all models took about the same number of iterations for a given load step when the acceleration parameter of $\beta = 0.35$ [see Eq. (7.6.53)] is used.

Figures 7.6.9 and 7.6.10 contain plots of the center deflection $\bar{w}$ versus q_0 and the center bending moment $\bar{M}_{xx}$ vs. q_0, respectively, for P-P beams and for different values of n. For FGM beams, the material distribution through the thickness follows the relation in Eq. (7.1.1):

$$E(z) = (E_1 - E_2)\, f(z) + E_2, \quad f(z) = \left(\tfrac{1}{2} + \tfrac{z}{h}\right)^n .$$

The stiffening of the beam with an increased load is due to the fact that the von Kármán nonlinear strain is quadratic in the gradient of the deflection w. Also, as n increases, the beam becomes more flexible (i.e., as $n \to \infty$ modulus $E \to E_1$) and experiences greater bending, which contributes to the stiffening effect. It is interesting to note that the dimensionless bending moment of FGM beams has a cross-over of the bending moment of the homogeneous beam for $n = 1$ with an increase in load, although this is not exhibited for $n > 2$ because of the variation of B_{xx} with n (see Fig. 7.5.2). Figures 7.6.11 and 7.6.12 contain plots of w vs. x/L for $n = 0$ and $n = 5$, respectively, for different values of q_0.

Figures 7.6.13 and 7.6.14 contain plots of the center deflection $\bar{w}$ vs. q_0 and the center bending moment $\bar{M}_{xx}$ versus q_0, respectively, for C-C beams and for different values of n. As in the case of P-P beams, the beams become stiffer but with less rate of increase of nonlinearity because C-C beams are relative stiffer due to the fixed ends. Unlike the P-P beams, the C-C beams do not exhibit the cross over of the bending moment. Figures 7.6.15 and 7.6.16 contain plots of w vs. x/L for $n = 0$ and $n = 5$, respectively, for different values of q_0.

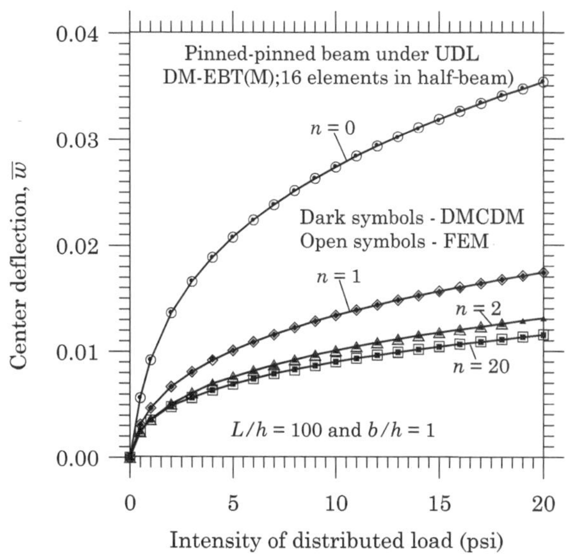

Fig. 7.6.9 Nonlinear dimensionless center deflection $\bar{w}$ versus q_0 curves for P-P beams. Different power-law index (n) values are used to show its effect on the deflection. The solutions predicted by various models of the FEM and DMCDM coincide with each other in the graphs.

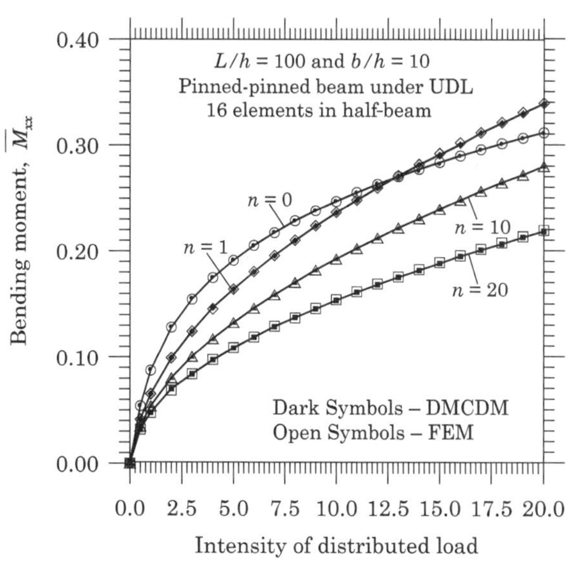

Fig. 7.6.10 Nonlinear dimensionless center bending moment $\bar{M}_{xx}$ versus q_0 curves for P-P beams. Different power-law index (n) values are used to show its effect on the bending moment. The solutions predicted by various models of the FEM and DMCDM coincide with each other in the graphs.

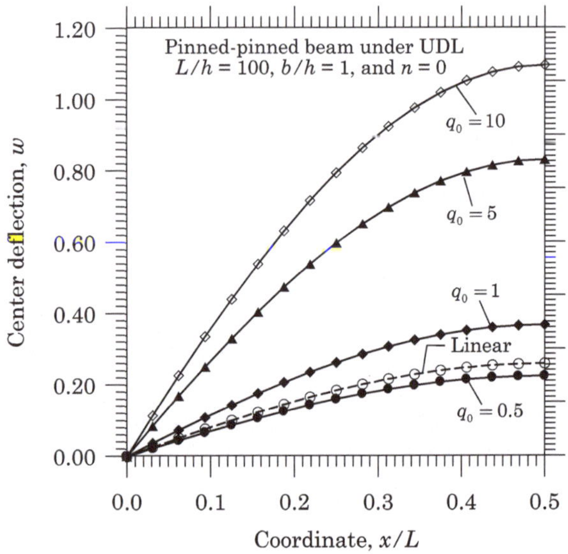

Fig. 7.6.11 Variation of the center deflection w (in.) versus x/L for P-P beams ($n = 0$) under uniform load (UDL) of different magnitudes of q_0 (lb/in).

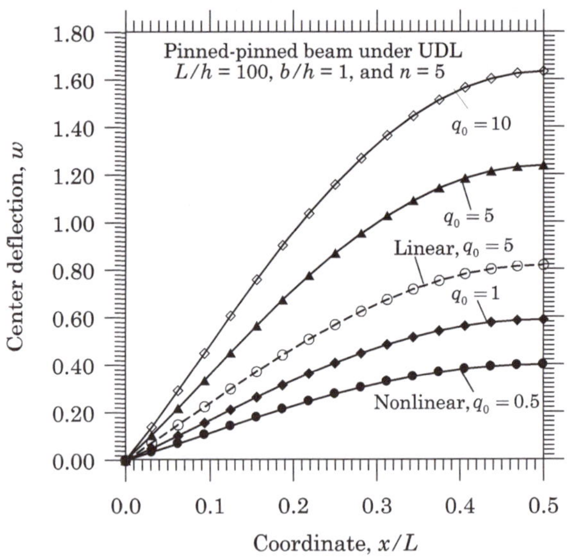

Fig. 7.6.12 Variation of the center deflection w (in.) versus x/L for P-P beams ($n = 5$) under uniform load (UDL) of different magnitudes of q_0 (lb/in).

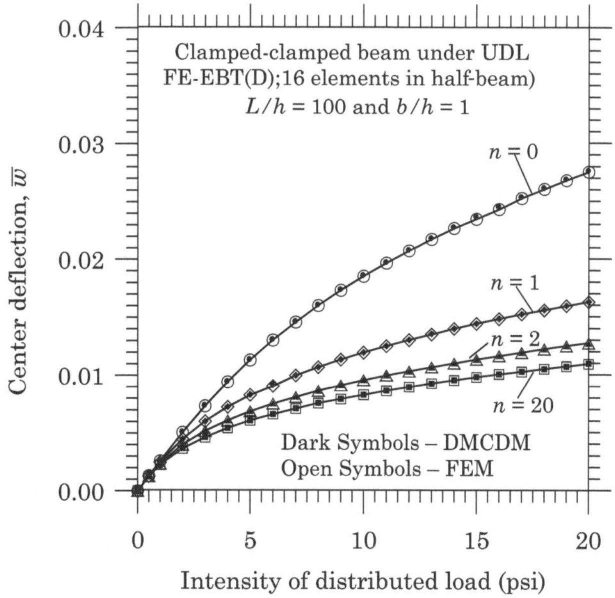

Fig. 7.6.13 Nonlinear dimensionless center deflection $\bar{w}$ versus q_0 curves for C-C beams. Different power-law index (n) values are used to show its effect on the deflection. The solutions predicted by various models of the FEM and DMCDM coincide with each other in the graphs.

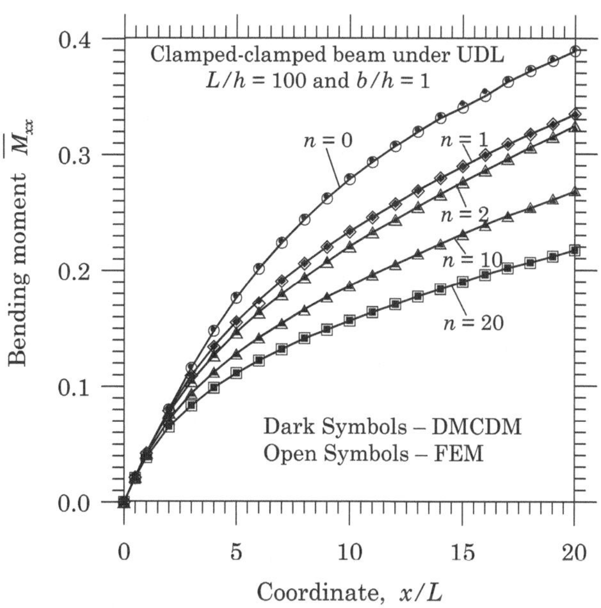

Fig. 7.6.14 Nonlinear dimensionless center bending moment $\bar{M}_{xx}$ versus q_0 curves for C-C beams. Different power-law index (n) values are used to show its effect on the bending moment. The solutions predicted by various models of the FEM and DMCDM coincide with each other in the graphs.

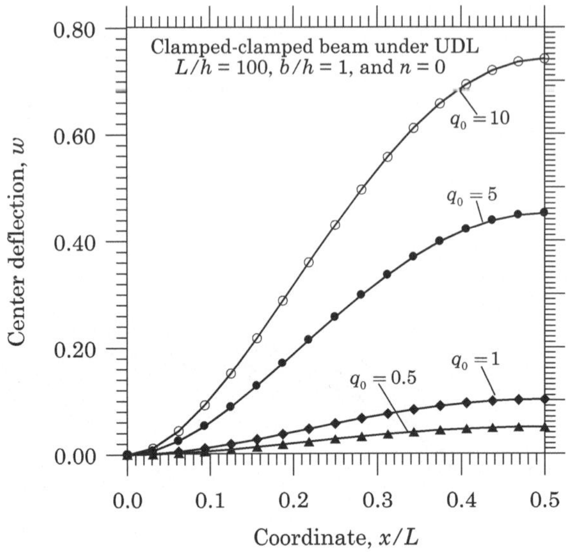

Fig. 7.6.15 Center deflection w (in.) versus x/L for C-C beams ($n = 0$) under uniform distributed load (UDL) of different magnitudes of q_0 (lb/in).

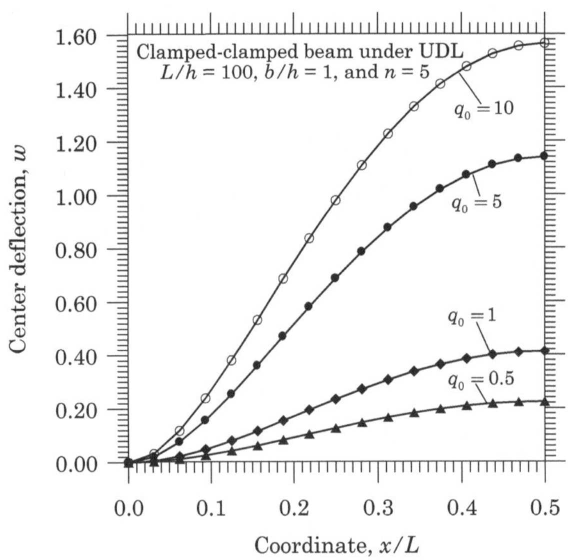

Fig. 7.6.16 Center deflection w (in.) versus x/L for C-C beams ($n = 5$) under uniform distributed load (UDL) of different magnitudes of q_0 (lb/in).

We note that in the numerical results presented in this section, thin beams (length-to-height ratio is $L/h = 100$) are considered. For thicker (hence stiffer) beams, the geometric nonlinearity exhibited, for the same load range, is not significant.

7.7 Summary

In this chapter, the linear and nonlinear bending of functionally graded material (FGM) beams using the FEM and DMCDM are studied. Mixed models (i.e., models that involve displacements and moments) of the Euler–Bernoulli and Timoshenko beams are formulated using the FEM and DMCDM for linear and nonlinear bending analysis of beams. Numerical results are presented to show the effect of the power-law index on the deflections. The mixed models have the advantage of enabling stress calculation at the boundary nodes, instead of Gauss points in the interior of the domain. Numerical results indicate that the DMCDM gives very accurate results, especially for the bending moments, which are used to compute stresses. A study of the FGM beams also revealed that the dimensionless bending deflections ($\bar{w} = w\,\hat{D}_{xx}/q_0L^4$) are not monotonic functions of the power-law index n because the coupling stiffness B_{xx} is not a monotonically increasing or decreasing function of the modulus ratio, E_1/E_2, of two constituent, through-thickness, functionally graded beams.

Problems

7.1 Use the stress resultant-displacement relation in Eqs. (7.6.3a), (7.6.3b), and (7.6.4) to express the equilibrium equations in Eq. (7.6.2) in terms of the displacements u and w.

7.2 Modify the weak forms in Eqs. (7.3.1a) and (7.3.1b) to account for the von Kármán nonlinearity and develop the finite element model given in Eq. (7.6.10) and verify the stiffness coefficients given in Eq. (7.6.11).

7.3 Verify the mixed finite element model of the EBT and the stiffness coefficients given in Eqs. (7.6.12) and (7.6.13).

7.4 Use the stress resultant-displacement relations in Eqs. (7.6.15a)–(7.6.15c), to express the equilibrium equations in Eqs. (7.6.14a)–(7.6.14c) in terms of the displacements u, w, and ϕ_x.

7.5 Modify the weak forms in Eqs. (7.3.11a)–(7.3.11c) to account for the von Kármán nonlinearity and develop the finite element model given in Eq. (7.6.19) and verify the stiffness coefficients given in Eq. (7.6.20).

7.6 Verify the discrete equations, Eqs. (7.6.29), (7.6.35), and (7.6.39) of the mixed DMCDM model of the EBT.

7.7 Verify the discrete equations, Eqs. (7.6.44a)–(7.6.44c) of the displacement DMCDM model of the TBT.

7.8 The finite element equations derived for the EBT and TBT beam theories have the form

$$\begin{bmatrix} \mathbf{K}^{11} & \mathbf{K}^{12} & \mathbf{K}^{13} \\ \mathbf{K}^{21} & \mathbf{K}^{22} & \mathbf{K}^{23} \\ \mathbf{K}^{31} & \mathbf{K}^{32} & \mathbf{K}^{33} \end{bmatrix} \begin{Bmatrix} \mathbf{\Delta}^{(1)} \\ \mathbf{\Delta}^{(2)} \\ \mathbf{\Delta}^{(3)} \end{Bmatrix} = \begin{Bmatrix} \mathbf{F}^{1} \\ \mathbf{F}^{2} \\ \mathbf{F}^{2} \end{Bmatrix}. \tag{1}$$

In the case of the displacement model of the EBT [see Eq. (7.6.10)], the 4×4 matrix $\mathbf{K}^{22}$ is split two 2×2 submatrices [i.e., $\mathbf{\Delta}$ in Eq. (7.6.10) is split into $\mathbf{\Delta}^{(2)}$ and $\mathbf{\Delta}^{(3)}$].

In Newton's scheme of solving nonlinear equations (see Section 1.9.4.3), the matrix equations associated with Eq. (1) is also of the form,

$$\begin{bmatrix} \mathbf{T}^{11} & \mathbf{T}^{12} & \mathbf{K}^{13} \\ \mathbf{T}^{21} & \mathbf{T}^{22} & \mathbf{T}^{23} \\ \mathbf{T}^{31} & \mathbf{T}^{32} & \mathbf{T}^{33} \end{bmatrix}^{(r-1)} \begin{Bmatrix} \delta\boldsymbol{\Delta}^{(1)} \\ \delta\boldsymbol{\Delta}^{(2)} \\ \delta\boldsymbol{\Delta}^{(3)} \end{Bmatrix}^{(r)} = - \begin{Bmatrix} \mathbf{R}^1 \\ \mathbf{R}^2 \\ \mathbf{R}^3 \end{Bmatrix}^{(r-1)}, \tag{2}$$

where R_i^α are the components of the residual vector [see Eq. (7.6.10)],

$$R_i^\alpha = \sum_{\gamma=1}^{2}\sum_{p=1}^{N_\gamma} K_{ip}^{\alpha\gamma}\Delta_p^\gamma - F_i^\alpha = \sum_{p=1}^{2} K_{ip}^{\alpha 1}\Delta_p^1 + \sum_{P=1}^{4} K_{iP}^{\alpha 2}\Delta_P^2 - F_i^\alpha. \tag{3}$$

Here N_γ $(\gamma = 1, 2)$ denotes the number of element degrees of freedom (in the present case, we have $N_1 = 2$ and $N_2 = 4$), $\Delta_p^1 = u_p$, and $\Delta_P^2 = \Delta_P$.

The specific forms of $T_{ij}^{\alpha\beta}$ for each finite element model is different because $K_{ij}^{\alpha\beta}$ are different. Show that the only nonzero tangent stiffness coefficients for the displacement model of the EBT are given by

$$T_{ij}^{\alpha 2} = \frac{\partial R_i^\alpha}{\partial \bar{\Delta}_j} = K_{ij}^{\alpha 2} + \sum_{p=1}^{2} \frac{\partial}{\partial \bar{\Delta}_j}\left(K_{ip}^{\alpha 1}\right) u_p + \sum_{P=1}^{4} \frac{\partial}{\partial \bar{\Delta}_j}\left(K_{iP}^{\alpha 2}\right)\bar{\Delta}_P - \frac{\partial F_i^\alpha}{\partial \bar{\Delta}_j} \tag{4}$$

for $\alpha = 1, 2$. In particular, show that

$$T_{iJ}^{12} = K_{iJ}^{12} + \frac{1}{2}\int_{x_a^e}^{x_b^e} A_{xx}\frac{\partial w}{\partial x}\frac{d\psi_i}{dx}\frac{d\varphi_J}{dx}\,dx, \tag{5a}$$

$$T_{IJ}^{22} = K_{IJ}^{22} + \int_{x_a^e}^{x_b^e} A_{xx}\frac{\partial u}{\partial x}\frac{d\varphi_I}{dx}\frac{d\varphi_J}{dx}\,dx + \int_{x_a^e}^{x_b^e} A_{xx}\left(\frac{\partial w}{\partial x}\right)^2\frac{d\varphi_I}{dx}\frac{d\varphi_J}{dx}\,dx$$
$$- \int_{x_a^e}^{x_b^e} B_{xx}\left(\frac{1}{2}\frac{\partial w}{\partial x}\frac{d^2\varphi_I}{dx^2} + \frac{\partial^2 w}{\partial x^2}\frac{d\varphi_I}{dx}\right)\frac{d\varphi_J}{dx}\,dx. \tag{5b}$$

Hint: Note the following identities for the EBT:

$$\frac{\partial}{\partial \bar{\Delta}_I}\left(\frac{\partial w}{\partial x}\right) = \frac{d\varphi_I}{dx}, \quad \frac{\partial}{\partial \bar{\Delta}_I}\left(\frac{\partial w}{\partial x}\right)^2 = 2\left(\frac{\partial w}{\partial x}\right)\frac{d\varphi_I}{dx}. \tag{6}$$

7.9 Using the information provided in **Problem 7.8**, show that the tangent stiffness coefficients associated with the mixed model of the EBT are

$$T_{ij}^{12} = K_{ij}^{12} + \frac{1}{2}\int_{x_a^e}^{x_b^e} \bar{A}_{xx}\frac{\partial w}{\partial x}\frac{d\psi_i^{(1)}}{dx}\frac{d\psi_j^{(2)}}{dx}\,dx, \tag{1a}$$

$$T_{ij}^{22} = K_{ij}^{22} + \int_{x_a}^{x_b} \bar{A}_{xx}\left[\frac{\partial u}{\partial x} + \left(\frac{\partial w}{\partial x}\right)^2\right]\frac{d\psi_i^{(2)}}{dx}\frac{d\psi_j^{(2)}}{dx}\,dx$$
$$+ \int_{x_a^e}^{x_b^e} \bar{B}_{xx}M_{xx}\frac{d\psi_i^{(2)}}{dx}\frac{d\psi_j^{(2)}}{dx}\,dx, \tag{1b}$$

$$T_{ij}^{32} = K_{ij}^{32} + \frac{1}{2}\int_{x_a^e}^{x_b^e} \bar{B}_{xx}\frac{\partial w}{\partial x}\psi_i^{(3)}\frac{d\psi_j^{(2)}}{dx}\,dx. \tag{1c}$$

Hint: Note the following identities for the TBT:

$$\frac{\partial}{\partial w_j}\left(\frac{\partial w}{\partial x}\right) = \frac{d\psi_j}{dx}, \quad \frac{\partial}{\partial w_j}\left(\frac{\partial w}{\partial x}\right)^2 = 2\left(\frac{\partial w}{\partial x}\right)\frac{d\psi_j}{dx} \tag{2}$$

7.10 Using the information provided in **Problem 7.8**, show that the tangent stiffness coefficients associated with the displacement model of the TBT are

$$T_{ij}^{12} = K_{ij}^{12} + \frac{1}{2}\int_{x_a^e}^{x_b^e} A_{xx}\frac{\partial w}{\partial x}\frac{d\psi_i^{(1)}}{dx}\frac{d\psi_j^{(2)}}{dx}\,dx, \tag{1a}$$

$$T_{ij}^{22} = K_{ij}^{22} + \int_{x_a^e}^{x_b^e} A_{xx}\left[\frac{\partial u}{\partial x} + \left(\frac{\partial w}{\partial x}\right)^2\right]\frac{d\psi_i^{(2)}}{dx}\frac{d\psi_j^{(2)}}{dx}\,dx$$
$$+ \int_{x_a^e}^{x_b^e} B_{xx}\frac{\partial \phi_x}{\partial x}\frac{d\psi_i^{(2)}}{dx}\frac{d\psi_j^{(2)}}{dx}\,dx, \tag{1b}$$

$$T_{ij}^{32} = K_{ij}^{32} + \frac{1}{2}\int_{x_a^e}^{x_b^e} B_{xx}\frac{\partial w}{\partial x}\frac{d\psi_j^{(3)}}{dx}\frac{d\psi_j^{(2)}}{dx}\,dx. \tag{1c}$$

7.11 Show that the tangent stiffness coefficients associated with the mixed model of the TBT are the same as those in the mixed model of the EBT:

$$T_{ij}^{12} = K_{ij}^{12} + \frac{1}{2}\int_{x_a^e}^{x_b^e} \bar{A}_{xx}\frac{\partial w}{\partial x}\frac{d\psi_i^{(1)}}{dx}\frac{d\psi_j^{(2)}}{dx}\,dx, \tag{1a}$$

$$T_{ij}^{22} = K_{ij}^{22} + \int_{x_a^e}^{x_b^e} \bar{A}_{xx}\left[\frac{\partial u}{\partial x} + \left(\frac{\partial w}{\partial x}\right)^2\right]\frac{d\psi_i^{(2)}}{dx}\frac{d\psi_j^{(2)}}{dx}\,dx$$
$$+ \int_{x_a^e}^{x_b^e} \bar{B}_{xx}\,M_{xx}\frac{d\psi_i^{(2)}}{dx}\frac{d\psi_j^{(2)}}{dx}\,dx, \tag{1b}$$

$$T_{ij}^{32} = K_{ij}^{32} + \frac{1}{2}\int_{x_a^e}^{x_b^e} \bar{B}_{xx}\frac{\partial w}{\partial x}\psi_i^{(3)}\frac{d\psi_j^{(2)}}{dx}\,dx. \tag{1c}$$

Appendix 7: Evaluation of integrals through beam height

The following integrals of $f(z) = (1/2 + z/h)^n$ appearing in Eq. (7.1.1) are useful in evaluating the integrals for A_{xx}, B_{xx}, and D_{xx} for functionally graded beams when the modulus of elasticity $E(z)$ is given by Eq. (7.1.1):

$$\int_{-\frac{h}{2}}^{\frac{h}{2}} \left(\frac{1}{2} + \frac{z}{h}\right)^n \; dz = \frac{h}{1+n}, \tag{A7.1}$$

$$\int_{-\frac{h}{2}}^{\frac{h}{2}} \left(\frac{1}{2} + \frac{z}{h}\right)^n z\, dz = \frac{h^2 n}{2(1+n)(2+n)}, \tag{A7.2}$$

$$\int_{-\frac{h}{2}}^{\frac{h}{2}} \left(\frac{1}{2} + \frac{z}{h}\right)^n z^2\, dz = \frac{h^3(2+n+n^2)}{4(1+n)(2+n)(3+n)}, \tag{A.3}$$

$$\int_{-\frac{h}{2}}^{\frac{h}{2}} \left(\frac{1}{2} + \frac{z}{h}\right)^n z^3\, dz = \frac{h^4 n(8+3n+n^2)}{8(1+n)(2+n)(3+n)(4+n)}, \tag{A7.4}$$

$$\int_{-\frac{h}{2}}^{\frac{h}{2}} \left(\frac{1}{2} + \frac{z}{h}\right)^n z^4\, dz = \frac{h^5(24+18n+23n^2+6n^3+n^4)}{16(1+n)(2+n)(3+n)(4+n)(5+n)}, \tag{A7.5}$$

$$\int_{-\frac{h}{2}}^{\frac{h}{2}} \left(\frac{1}{2} + \frac{z}{h}\right)^n z^5\, dz = \frac{h^6 n(184+110n+55n^2+10n^3+n^4)}{32(1+n)(2+n)(3+n)(4+n)(5+n)(6+n)}. \tag{A7.6}$$

These integral values can be used to determine the structural stiffness coefficients A_{xx}, B_{xx}, and D_{xx} for an FGM beam with the material variation in Eq. (7.1.1). We have

$$\begin{aligned} A_{xx} &= b\int_{-\frac{h}{2}}^{\frac{h}{2}} E(z)\, dz = E_2 b\left[(M-1)\int_{-\frac{h}{2}}^{\frac{h}{2}} \left(\frac{1}{2} + \frac{z}{h}\right)^n dz + \int_{-\frac{h}{2}}^{\frac{h}{2}} dz\right] \\ &= E_2 b\left[(M-1)\frac{h}{(1+n)} + h\right] = E_2 bh\frac{M+n}{1+n}, \end{aligned} \tag{A7.7}$$

$$\begin{aligned} B_{xx} &= b\int_{-\frac{h}{2}}^{\frac{h}{2}} z\, E(z)\, dz = E_2 b\left[(M-1)\int_{-\frac{h}{2}}^{\frac{h}{2}} \left(\frac{1}{2} + \frac{z}{h}\right)^n z\, dz + \int_{-\frac{h}{2}}^{\frac{h}{2}} z\, dz\right] \\ &= E_2 b\left[(M-1)\frac{nh^2}{2(1+n)(2+n)} + 0\right] = E_2\frac{bh^2}{2}\frac{(M-1)n}{(1+n)(2+n)}, \end{aligned} \tag{A7.8}$$

$$\begin{aligned} D_{xx} &= b\int_{-\frac{h}{2}}^{\frac{h}{2}} z^2\, E(z)\, dz = E_2 b\left[(M-1)\int_{-\frac{h}{2}}^{\frac{h}{2}} \left(\frac{1}{2} + \frac{z}{h}\right)^n z^2\, dz + \int_{-\frac{h}{2}}^{\frac{h}{2}} z^2\, dz\right] \\ &= E_2 b\left[(M-1)\frac{h^3(2+n+n^2)}{4(1+n)(2+n)(3+n)} + \frac{h^3}{12}\right] \\ &= E_2\frac{bh^3}{12}\left[\frac{3M(2+n+n^2)+8n+3n^2+n^3}{(1+n)(2+n)(3+n)}\right]. \end{aligned} \tag{A7.9}$$

8
Bending of Axisymmetric Circular Plates

8.1 General Kinematic and Constitutive Relations

8.1.1 Geometry and Coordinate System

The present chapter deals with the extension and application of the FEM and DMCDM to functionally graded circular plates under axisymmetric conditions. Circular plates are thin (i.e., thickness h is very small compared to the outer radius a of the plate) discs subjected to transverse loads which tend to bend and stretch them. We use the cylindrical coordinate system (r, θ, z) to describe the deformation and stress state in circular plates. The word "axisymmetry" refers to the case in which the solution (i.e., displacements as well as stresses) is independent of the angular coordinate θ (see Fig. 8.1.1). This is possible if and only if the (a) geometry, (b) material properties, (c) loads, and (d) boundary conditions are also independent of θ. We assume such is the case in dealing with the bending of circular plates. In this case, the governing equations can be described in terms of the radial coordinate r alone.

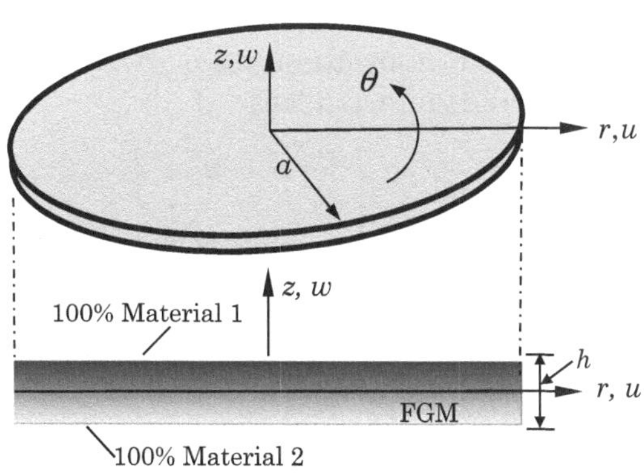

Fig. 8.1.1 Geometry and coordinate system used for circular plates.

8.1.2 Kinematic Relations

Let $\mathbf{u}$ denote the displacement vector with components (u_r, u_θ, u_z) in the (r, θ, z) coordinate directions, respectively. Due to the assumed axisymmetry, we have $u_\theta = 0$, and u_r and u_z are independent of θ. In addition, if we assume the inextensibility of the transverse normal lines, then u_z is only a function of the radial coordinate r. When the inside radius $b = 0$, we have solid circular plates of radius a.

The modified Green–Lagrange strain tensor that accounts for small strains but moderate rotations of normal lines perpendicular to the plane of the plate (i.e., the von Kármán strain) is given by (see Reddy [17, 42])

$$\begin{aligned}\mathbf{E} &= \frac{1}{2}\left[\nabla\mathbf{u} + (\nabla\mathbf{u})^{\mathrm{T}} + \nabla\mathbf{u}\cdot(\nabla\mathbf{u})^{\mathrm{T}}\right]\\ &\approx \frac{1}{2}\left[\nabla\mathbf{u} + (\nabla\mathbf{u})^{\mathrm{T}} + \frac{\partial u_z}{\partial r}\frac{\partial u_z}{\partial r}\hat{\mathbf{e}}_r\hat{\mathbf{e}}_r\right] \equiv \boldsymbol{\varepsilon},\end{aligned} \tag{8.1.1}$$

where $(\hat{\mathbf{e}}_r, \hat{\mathbf{e}}_\theta, \hat{\mathbf{e}}_z)$ are the basis vectors in the cylindrical coordinate system (r, θ, z). Thus, the nonzero strain components in the cylindrical coordinate system for the axisymmetric case are:

$$\varepsilon_{rr} = \frac{\partial u_r}{\partial r} + \frac{1}{2}\left(\frac{\partial u_z}{\partial r}\right)^2, \quad \varepsilon_{rz} = \frac{1}{2}\left(\frac{\partial u_r}{\partial z} + \frac{\partial u_z}{\partial r}\right), \quad \varepsilon_{\theta\theta} = \frac{u_r}{r}. \tag{8.1.2}$$

Here $\varepsilon_{\theta\theta}$ denotes the hoop strain.

8.1.3 Constitutive Equations

As in the case of beams, we consider the general case in which the plate is graded through the thickness. We use the power-law variation (among other variations) of modulus of elasticity, while keeping the Poisson's ratio constant. If the z-coordinate is taken along the thickness of the plate, then for a two-constituent functionally graded linear elastic material, the plane stress constitutive equations relating the nonzero stresses $(\sigma_{rr}, \sigma_{\theta\theta}, \sigma_{rz})$ to the nonzero strains $(\varepsilon_{rr}, \varepsilon_{\theta\theta}, \varepsilon_{rz})$ of the axisymmetric case are

$$\begin{Bmatrix}\sigma_{rr}\\ \sigma_{\theta\theta}\\ \sigma_{rz}\end{Bmatrix} = \frac{E(z)}{1-\nu^2}\begin{bmatrix}1 & \nu & 0\\ \nu & 1 & 0\\ 0 & 0 & \frac{1-\nu}{2}\end{bmatrix}\begin{Bmatrix}\varepsilon_{rr}\\ \varepsilon_{\theta\theta}\\ 2\varepsilon_{rz}\end{Bmatrix}, \tag{8.1.3}$$

where Young's modulus E varies with z, for a two-material composition, according to Eq. (7.1.1)

$$E(z) = (E_1 - E_2)\,f(z) + E_2, \quad f(z) = \left(\frac{1}{2} + \frac{z}{h}\right)^n, \tag{8.1.4}$$

where E_1 and E_2 are Young's moduli of material 1 and material 2, respectively, and ν is Poisson's ratio, which is assumed to be a constant.

8.2 Classical Theory of Plates

8.2.1 Displacements and Strains

Consider a through-thickness functionally graded circular plate of thickness h and outer radius a subjected to an axisymmetric distributed load $q(r)$ (i.e., independent of the angular coordinate) on the top face. If, further, the boundary conditions are also selected to be axisymmetric, then the plate can be considered as a functionally graded axisymmetric circular plate (i.e., every radial line of the plate has exactly the same deformation).

The classical plate theory (CPT) is an extension of the EBT to plates. The Euler–Bernoulli hypothesis of beams extended to plates is termed the *Kirchhoff hypothesis*. The hypothesis assumes that straight lines normal to the mid-plane of the plate before deformation remain: (1) inextensible, (2) straight, and (3) normal to the midsurface of the plate after deformation. The hypothesis amounts to neglecting both transverse shear and transverse normal strains, i.e., $\varepsilon_{rz} = 0$ and $\varepsilon_{zz} = 0$. The total displacements (u_r, u_θ, u_z) along the coordinate directions (r, θ, z), as implied by the Kirchhoff hypothesis for axisymmetric bending of circular plates are assumed in the form

$$\begin{aligned} \mathbf{u}(r,\theta,z) &= u_r\,\hat{\mathbf{e}}_r + u_\theta\,\hat{\mathbf{e}}_\theta + u_z\,\hat{\mathbf{e}}_z,\\ u_r &= u(r) - z\frac{dw}{dr}, \quad u_\theta = 0, \quad u_z = w(r), \end{aligned} \tag{8.2.1}$$

where $(u,\, w)$ are the displacements in the radial (r) and transverse (z) directions, respectively, of a point on the midplane $(z = 0)$ of the plate.

The von Kármán strains in (8.1.2) for the classical plate theory take the form

$$\varepsilon_{rr} = \varepsilon_{rr}^{(0)} + z\varepsilon_{rr}^{(1)}, \quad \varepsilon_{\theta\theta} = \varepsilon_{\theta\theta}^{(0)} + z\varepsilon_{\theta\theta}^{(1)} \tag{8.2.2}$$

where

$$\begin{aligned} \varepsilon_{rr}^{(0)} &= \frac{du}{dr} + \frac{1}{2}\left(\frac{dw}{dr}\right)^2, & \varepsilon_{rr}^{(1)} &= -\frac{d^2w}{dr^2}\\ \varepsilon_{\theta\theta}^{(0)} &= \frac{u}{r}, & \varepsilon_{\theta\theta}^{(1)} &= -\frac{1}{r}\frac{dw}{dr} \end{aligned} \tag{8.2.3}$$

We note that the transverse shear strain is zero, $\varepsilon_{rz} = 0$, in the CPT; however, the transverse shear force is not zero, and it is calculated using the equilibrium equations, as was done in the case of the classical (or the Euler–Bernoulli) beam theory, EBT.

8.2.2 Equilibrium Equations

The equations of equilibrium of the functionally graded axisymmetric circular plate can be obtained using the principle of virtual displacements (i.e., the energy approach) or by considering the equilibrium of a typical element of the plate (i.e., the vector approach). Here we use the former approach because of its utility in developing the finite element models.

The statement of the principle of virtual displacements for an axisymmetric circular plate is (not including any externally applied point loads)

$$0 = \int_0^a \int_{-h/2}^{h/2} (\sigma_{rr}\delta\varepsilon_{rr} + \sigma_{\theta\theta}\delta\varepsilon_{\theta\theta})\, r dz dr d\theta - \int_0^{2\pi}\int_0^a q\delta w r dr d\theta$$
$$= \left[N_{rr}\left(\frac{d\delta u}{dr} + \frac{dw}{dr}\frac{d\delta w}{dr}\right) - M_{rr}\left(\frac{d^2\delta w}{dr^2}\right) + N_{\theta\theta}\frac{\delta u}{r} - M_{\theta\theta}\frac{1}{r}\frac{dw}{dr} - q\delta w\right] r dr, \tag{8.2.4}$$

where the stress resultants N_{rr}, $N_{\theta\theta}$, M_{rr}, and $M_{\theta\theta}$ of the CPT are defined in terms of the stresses and expressed in terms of the displacements using the constitutive relations in Eq. (8.1.3) as (see Fig. 8.2.1)

$$N_{rr} = \int_{-\frac{h}{2}}^{\frac{h}{2}} \sigma_{rr}\, dz = A_{rr}\left[\frac{du}{dr} + \frac{1}{2}\left(\frac{dw}{dr}\right)^2 + \nu\frac{u}{r}\right] - B_{rr}\left(\frac{d^2w}{dr^2} + \frac{\nu}{r}\frac{dw}{dr}\right) \tag{8.2.5a}$$

$$N_{\theta\theta} = \int_{-\frac{h}{2}}^{\frac{h}{2}} \sigma_{\theta\theta}\, dz = A_{rr}\left[\frac{u}{r} + \nu\frac{du}{dr} + \frac{\nu}{2}\left(\frac{dw}{dr}\right)^2\right] - B_{rr}\left(\nu\frac{d^2w}{dr^2} + \frac{1}{r}\frac{dw}{dr}\right) \tag{8.2.5b}$$

$$M_{rr} = \int_{-\frac{h}{2}}^{\frac{h}{2}} \sigma_{rr} z\, dz = B_{rr}\left[\frac{du}{dr} + \frac{1}{2}\left(\frac{dw}{dr}\right)^2 + \nu\frac{u}{r}\right] - D_{rr}\left(\frac{d^2w}{dr^2} + \frac{\nu}{r}\frac{dw}{dr}\right) \tag{8.2.5c}$$

$$M_{\theta\theta} = \int_{-\frac{h}{2}}^{\frac{h}{2}} \sigma_{\theta\theta} z\, dz = B_{rr}\left[\frac{u}{r} + \nu\frac{du}{dr} + \frac{\nu}{2}\left(\frac{dw}{dr}\right)^2\right] - D_{rr}\left(\nu\frac{d^2w}{dr^2} + \frac{1}{r}\frac{dw}{dr}\right), \tag{8.2.5d}$$

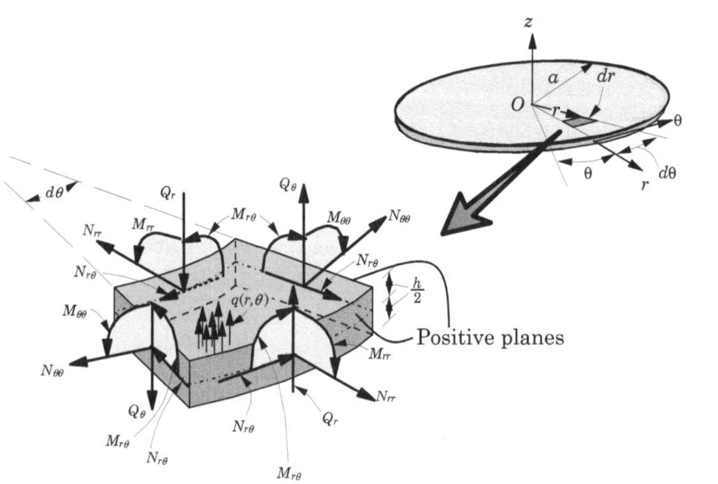

Fig. 8.2.1 Geometry and coordinate system used for circular plates.

where

$$A_{rr} = \frac{1}{1-\nu^2}\int_{-h/2}^{h/2} E(z)\,dz, \qquad B_{rr} = \frac{1}{1-\nu^2}\int_{-h/2}^{h/2} E(z)z\,dz,$$
$$D_{rr} = \frac{1}{1-\nu^2}\int_{-h/2}^{h/2} E(z)z^2\,dz. \tag{8.2.6}$$

The explicit expressions for A_{rr}, B_{rr}, and D_{rr} when $E(z)$ varies according to Eq. (8.1.4) are given as

$$A_{rr} = \frac{E_2 h}{1-\nu^2}\frac{M+n}{1+n}, \quad B_{rr} = \frac{E_2 h^2}{2(1-\nu^2)}\frac{(M-1)n}{(1+n)(2+n)},$$
$$D_{rr} = \frac{E_2 h^3}{12(1-\nu^2)}\left[\frac{3M(2+n+n^2)+8n+3n^2+n^3}{(1+n)(2+n)(3+n)}\right], \quad M = \frac{E_1}{E_2}. \tag{8.2.7}$$

The governing differential equations of the axisymmetric circular plate are obtained by taking the Euler equations of the variational statement in Eq. (8.2.4) (see Reddy [42]). The resulting governing equations read

$$\frac{1}{r}\left[\frac{d}{dr}(rN_{rr}) - N_{\theta\theta}\right] = 0, \tag{8.2.8}$$

$$\frac{1}{r}\frac{d}{dr}(rV_r) + q = 0, \tag{8.2.9}$$

where V_r denotes the *effective* shear force acting on the rz-plane and Q_r is the transverse shear force,

$$V_r = Q_r + N_{rr}\frac{dw}{dr}, \quad Q_r = \frac{1}{r}\left[\frac{d}{dr}(rM_{rr}) - M_{\theta\theta}\right]. \tag{8.2.10}$$

In the case of linear theory of circular plates, we have $V_r = Q_r$.

The boundary conditions involve specifying one element of each of the following duality pairs:

$$(u \text{ or } rN_{rr}), \quad (w \text{ or } rV_r), \quad \left(-\frac{dw}{dr} \text{ or } rM_{rr}\right). \tag{8.2.11}$$

8.2.3 Governing Equations in Terms of Displacements

The governing differential equations of axisymmetric circular plate based on the CPT can be expressed in terms of the displacements u and w by first replacing the stress resultants N_{rr}, $N_{\theta\theta}$, M_{rr}, and $M_{\theta\theta}$ in Eqs. (8.2.8) and (8.2.9) in terms of u and w [i.e., use Eqs. (8.2.5a)–(8.2.5d)]. The resulting differential equations are second-order in u and fourth-order in w.

The equations of equilibrium in terms of u and w are

$$
\begin{aligned}
&-\frac{1}{r}\frac{d}{dr}\left\{rA_{rr}\left[\frac{du}{dr}+\frac{1}{2}\left(\frac{dw}{dr}\right)^2+\nu\frac{u}{r}\right]-rB_{rr}\left(\frac{d^2w}{dr^2}+\frac{\nu}{r}\frac{dw}{dr}\right)\right\}\\
&+A_{rr}\frac{1}{r}\left[\frac{u}{r}+\nu\frac{du}{dr}+\frac{\nu}{2}\left(\frac{dw}{dr}\right)^2\right]-B_{rr}\frac{1}{r}\left(\nu\frac{d^2w}{dr^2}+\frac{1}{r}\frac{dw}{dr}\right)=0,
\end{aligned}
\tag{8.2.12}
$$

$$
\begin{aligned}
&-\frac{1}{r}\frac{d}{dr}\left[\frac{d}{dr}\left\{rB_{rr}\left[\frac{du}{dr}+\frac{1}{2}\left(\frac{dw}{dr}\right)^2+\nu\frac{u}{r}\right]-rD_{rr}\left(\frac{d^2w}{dr^2}+\frac{\nu}{r}\frac{dw}{dr}\right)\right\}\right.\\
&\left.-B_{rr}\left[\frac{u}{r}+\nu\frac{du}{dr}+\frac{\nu}{2}\left(\frac{dw}{dr}\right)^2\right]+D_{rr}\left(\nu\frac{d^2w}{dr^2}+\frac{1}{r}\frac{dw}{dr}\right)\right]\\
&-\frac{1}{r}\frac{d}{dr}\left\{A_{rr}\,r\frac{dw}{dr}\left[\frac{du}{dr}+\frac{1}{2}\left(\frac{dw}{dr}\right)^2+\nu\frac{u}{r}\right]\right.\\
&\left.-B_{rr}\,r\frac{dw}{dr}\left(\frac{d^2w}{dr^2}+\frac{\nu}{r}\frac{dw}{dr}\right)\right\}-q=0.
\end{aligned}
\tag{8.2.13}
$$

8.2.4 Equations in Terms of Displacements and Bending Moment

Since the dual mesh control domain method (DMCDM) is applicable only to first- or second-order differential equations, Eqs. (8.2.12) and (8.2.13), are not suitable for the application of DMCDM. Therefore, we express the governing equations, (8.2.8)–(8.2.10), as second-order differential equations in terms of the three variables (u, w, M_{rr}). Here we follow the same type of algebraic manipulations as in the case of beams.

The constitutive relations between the stress resultants and strain components, Eqs. (8.2.5a) –(8.2.5d), can be rewritten in the form

$$
\begin{aligned}
N_{rr} &= A_{rr}\,\hat{\varepsilon}_{rr}^{(0)} + B_{rr}\,\hat{\varepsilon}_{rr}^{(1)}, \quad & N_{\theta\theta} &= A_{rr}\,\hat{\varepsilon}_{\theta\theta}^{(0)} + B_{rr}\,\hat{\varepsilon}_{\theta\theta}^{(1)},\\
M_{rr} &= B_{rr}\,\hat{\varepsilon}_{rr}^{(0)} + D_{rr}\,\hat{\varepsilon}_{rr}^{(1)}, \quad & M_{\theta\theta} &= B_{rr}\,\hat{\varepsilon}_{\theta\theta}^{(0)} + D_{rr}\,\hat{\varepsilon}_{\theta\theta}^{(1)},
\end{aligned}
\tag{8.2.14}
$$

where

$$\hat{\varepsilon}_{rr}^{(0)} = \left[\frac{du}{dr}+\frac{1}{2}\left(\frac{dw}{dr}\right)^2+\nu\frac{u}{r}\right], \tag{8.2.15a}$$

$$\hat{\varepsilon}_{rr}^{(1)} = -\left(\frac{d^2w}{dr^2}+\frac{\nu}{r}\frac{dw}{dr}\right), \tag{8.2.15b}$$

$$\hat{\varepsilon}_{\theta\theta}^{(0)} = \left[\frac{u}{r}+\nu\frac{du}{dr}+\frac{\nu}{2}\left(\frac{dw}{dr}\right)^2\right], \tag{8.2.15c}$$

$$\hat{\varepsilon}_{\theta\theta}^{(1)} = -\left(\nu\frac{d^2w}{dr^2}+\frac{1}{r}\frac{dw}{dr}\right). \tag{8.2.15d}$$

Inverting Eqs. $(8.2.14)_1$–$(8.2.14)_4$ for the effective strains $\hat{\varepsilon}_{rr}^{(0)}$, $\hat{\varepsilon}_{rr}^{(1)}$, $\hat{\varepsilon}_{\theta\theta}^{(0)}$, and $\hat{\varepsilon}_{\theta\theta}^{(1)}$ in terms of the stress resultants N_{rr}, M_{rr}, $N_{\theta\theta}$, and $M_{\theta\theta}$, we obtain

$$\hat{\varepsilon}_{rr}^{(0)} = \frac{1}{D_{rr}^*}\left(D_{rr}\, N_{rr} - B_{rr}\, M_{rr}\right), \tag{8.2.16a}$$

$$\hat{\varepsilon}_{rr}^{(1)} = \frac{1}{D_{rr}^*}\left(-B_{rr}\, N_{rr} + A_{rr}\, M_{rr}\right), \tag{8.2.16b}$$

$$\hat{\varepsilon}_{\theta\theta}^{(0)} = \frac{1}{D_{rr}^*}\left(D_{rr}\, N_{\theta\theta} - B_{rr}\, M_{\theta\theta}\right), \tag{8.2.16c}$$

$$\hat{\varepsilon}_{\theta\theta}^{(1)} = \frac{1}{D_{rr}^*}\left(-B_{rr}\, N_{\theta\theta} + A_{rr}\, M_{\theta\theta}\right), \tag{8.2.16d}$$

where

$$D_{rr}^* = D_{rr}\, A_{rr} - B_{rr}\, B_{rr}. \tag{8.2.17}$$

We can solve Eqs. (8.2.16a) and (8.2.16b) for N_{rr} in terms of $\hat{\varepsilon}_{rr}^{(0)}$ and bending moment M_{rr} and Eqs. (8.2.16c) and (8.2.16d) for $N_{\theta\theta}$ in terms of $\hat{\varepsilon}_{\theta\theta}^{(0)}$ and bending moment $M_{\theta\theta}$:

$$N_{rr} = \bar{A}_{rr}\, \hat{\varepsilon}_{rr}^{(0)} + \bar{B}_{rr}\, M_{rr}, \quad N_{\theta\theta} = \bar{A}_{rr}\, \hat{\varepsilon}_{\theta\theta}^{(0)} + \bar{B}_{rr}\, M_{\theta\theta}, \tag{8.2.18}$$

where

$$\bar{A}_{rr} = \frac{D_{rr}^*}{D_{rr}}, \quad \bar{B}_{rr} = \frac{B_{rr}}{D_{rr}}, \tag{8.2.19}$$

and $\hat{\varepsilon}_{rr}^{(0)}$ and $\hat{\varepsilon}_{\theta\theta}^{(0)}$ are known in terms of u and w through Eqs. (8.2.15a) and (8.2.15c). Using Eq. $(8.2.18)_1$, we rewrite the Eq. (8.2.16b) as

$$\hat{\varepsilon}_{rr}^{(1)} = -\bar{B}_{rr}\hat{\varepsilon}_{rr}^{(0)} + \frac{1}{D_{rr}} M_{rr}, \tag{8.2.20}$$

where $\hat{\varepsilon}_{rr}^{(1)}$ and $\hat{\varepsilon}_{rr}^{(0)}$ are known in terms of u and w through Eqs. (8.2.15a) and (8.2.15b).

Next, we write all stress resultants N_{rr}, $N_{\theta\theta}$, and $M_{\theta\theta}$ in terms of (u, w, M_{rr}). We already have N_{rr} in terms of (u, w, M_{rr}) through Eq. $(8.2.18)_1$. We use the last two equations of Eq. (8.2.14) to write $M_{\theta\theta}$ in terms of (u, w, M_{rr}) as

$$M_{\theta\theta} = \nu M_{rr} + (1 - \nu^2)\frac{1}{r}\left(B_{rr} u - D_{rr}\frac{dw}{dr}\right). \tag{8.2.21}$$

Similarly, we use the first two equations of Eq. (8.2.14) to write $N_{\theta\theta}$ in terms of (u, w, N_{rr}) and finally in terms of (u, w, M_{rr}). The result is similar in form to that for $M_{\theta\theta}$ in Eq. (8.2.21), where we make use of the identity $A_{rr} =$

$\bar{A}_{rr} + \bar{B}_{rr} B_{rr}$. We have

$$\begin{aligned}
N_{\theta\theta} &= \nu\, N_{rr} + (1-\nu^2)\frac{1}{r}\left(A_{rr} u - B_{rr}\frac{dw}{dr}\right) \\
&= \nu \bar{A}_{rr}\left[\frac{du}{dr} + \left(\frac{dw}{dr}\right)^2 + \nu\frac{u}{r}\right] + (1-\nu^2)\frac{1}{r}\left(A_{rr}\, u - B_{rr}\frac{dw}{dr}\right) + \nu \bar{B}_{rr}\, M_{rr} \\
&= \bar{A}_{rr}\left[\nu\frac{du}{dr} + \nu\left(\frac{dw}{dr}\right)^2 + \frac{u}{r}\right] + (1-\nu^2)\frac{B_{rr}}{r}\left(\bar{B}_{rr}\, u - \frac{dw}{dr}\right) + \nu \bar{B}_{rr}\, M_{rr}.
\end{aligned} \tag{8.2.22}$$

The two equilibrium equations, Eqs. (8.2.8) and (8.2.9), with N_{rr}, $N_{\theta\theta}$, and $M_{\theta\theta}$ expressed in terms of u, w, and M_{rr} [through Eqs. $(8.2.18)_1$, (8.2.22), and (8.2.21), respectively], and Eq. (8.2.20) with $\hat{\varepsilon}_{rr}^{(0)}$ and $\hat{\varepsilon}_{rr}^{(1)}$ expressed in terms of u and w [through Eqs. (8.2.15a) and (8.2.15b), respectively], provide the three required equations of the mixed formulation (additional algebraic identities are used in arriving at these equations):

$$\begin{aligned}
&-\frac{1}{r}\frac{d}{dr}\left\{r\bar{A}_{rr}\left[\frac{du}{dr} + \frac{1}{2}\left(\frac{dw}{dr}\right)^2 + \nu\frac{u}{r}\right] + r\bar{B}_{rr}\, M_{rr}\right\} \\
&\quad + \frac{1}{r}\bar{A}_{rr}\left[\nu\frac{du}{dr} + \frac{\nu}{2}\left(\frac{dw}{dr}\right)^2 + \frac{u}{r}\right] + \frac{1}{r}\nu\bar{B}_{rr}\, M_{rr} \\
&\quad + \frac{1}{r}(1-\nu^2)B_{rr}\left(\bar{B}_{rr}\frac{u}{r} - \frac{1}{r}\frac{dw}{dr}\right) = 0,
\end{aligned} \tag{8.2.23}$$

$$\begin{aligned}
&-\frac{1}{r}\frac{d}{dr}\left[r\frac{dM_{rr}}{dr} + (1-\nu)M_{rr} - (1-\nu^2)D_{rr}\left(\bar{B}_{rr}\frac{u}{r} - \frac{1}{r}\frac{dw}{dr}\right)\right] - q \\
&-\frac{1}{r}\frac{d}{dr}\left\{r\bar{A}_{rr}\left[\frac{dw}{dr}\frac{du}{dr} + \frac{1}{2}\left(\frac{dw}{dr}\right)^3 + \nu\frac{dw}{dr}\frac{u}{r}\right] + r\bar{B}_{rr}\frac{dw}{dr}M_{rr}\right\} = 0,
\end{aligned} \tag{8.2.24}$$

$$\begin{aligned}
&-\frac{1}{r}\frac{d}{dr}\left(r\frac{dw}{dr}\right) + (1-\nu)\frac{1}{r}\frac{dw}{dr} + \bar{B}_{rr}\left[\frac{du}{dr} + \frac{1}{2}\left(\frac{dw}{dr}\right)^2 + \nu\frac{u}{r}\right] \\
&\quad - \frac{1}{D_{rr}}M_{rr} = 0.
\end{aligned} \tag{8.2.25}$$

The governing equations of equilibrium for homogeneous plates can be obtained from Eqs. (8.2.23)–(8.2.25) by setting $B_{rr} = 0$ and $\bar{A}_{rr} = A_{rr}$.

8.3 First-Order Shear Deformation Plate Theory

8.3.1 Displacements and Strains

The *first-order shear deformation theory* (FST) is the simplest plate theory that accounts for nonzero transverse shear strain [17, 42]. The FST is based on the same kinematics as the TBT for beams in Chapter 7. The assumed displacement field is

$$\mathbf{u} = u_r\,\hat{\mathbf{e}}_r + u_z\,\hat{\mathbf{e}}_z, \quad u_r(r,z) = u(r) + z\phi_r(r), \quad u_z(r,z) = w(r), \tag{8.3.1}$$

where ϕ_r denotes the rotation of a transverse normal in the plane θ =constant. The FST includes a constant state of transverse shear strain with respect to the thickness coordinate and, hence, requires the use of a shear correction factor, (as in the case of the TBT) which depends, in general, not only on the material and geometric parameters but also on the loads and boundary conditions.

The nonzero von Kármán strains of the FST are

$$\varepsilon_{rr} = \varepsilon_{rr}^{(0)} + z\varepsilon_{rr}^{(1)}, \quad \varepsilon_{\theta\theta} = \varepsilon_{\theta\theta}^{(0)} + z\varepsilon_{\theta\theta}^{(1)}, \quad \varepsilon_{rz} = \varepsilon_{rz}^{(0)}, \tag{8.3.2}$$

where

$$\begin{aligned} \varepsilon_{rr}^{(0)} &= \frac{du}{dr} + \frac{1}{2}\left(\frac{dw}{dr}\right)^2, \qquad \varepsilon_{rr}^{(1)} = \frac{d\phi_r}{dr}, \\ \varepsilon_{\theta\theta}^{(0)} &= \frac{u}{r}, \qquad \varepsilon_{\theta\theta}^{(1)} = \frac{\phi_r}{r}, \qquad 2\varepsilon_{rz}^{(0)} = \phi_r + \frac{dw}{dr}. \end{aligned} \tag{8.3.3}$$

8.3.2 Equations of Equilibrium

The principle of virtual displacements for the FST can be expressed as

$$\begin{aligned} 0 &= \int_0^a \int_{-\frac{h}{2}}^{\frac{h}{2}} \left(\sigma_{rr}\,\delta\varepsilon_{rr} + \sigma_{\theta\theta}\,\delta\varepsilon_{\theta\theta} + 2K_s\sigma_{rz}\,\delta\varepsilon_{rz}\right) r\,dr\,dz - \int_0^a q\,\delta w\,r\,dr \\ &= \int_0^a \left[N_{rr}\left(\frac{d\delta u}{dr} + \frac{dw}{dr}\frac{d\delta w}{dr}\right) + N_{\theta\theta}\left(\frac{\delta u}{r}\right) + M_{rr}\frac{d\delta\phi_r}{dr} \right. \\ &\qquad \left. + \frac{1}{r}M_{\theta\theta}\delta\phi_r + Q_r\left(\delta\phi_r + \frac{d\delta w}{dr}\right)\right] r\,dr, \end{aligned} \tag{8.3.4}$$

where the various stress resultants are defined by

$$\begin{aligned} (N_{rr}, M_{rr}) &= \int_{-\frac{h}{2}}^{\frac{h}{2}} (1, z)\sigma_{rr}\,dz, \\ Q_r &= K_s \int_{-\frac{h}{2}}^{\frac{h}{2}} \sigma_{rz}\,dz \\ (N_{\theta\theta}, M_{\theta\theta}) &= \int_{-\frac{h}{2}}^{\frac{h}{2}} (1, z)\sigma_{\theta\theta}\,dz, \end{aligned} \tag{8.3.5}$$

and K_s denotes the shear correction coefficient.

The Euler equations obtained from the principle of virtual displacements in Eq. (8.3.4) are the equations of equilibrium of the FST:

$$\frac{1}{r}\left[\frac{d}{dr}(rN_{rr}) - N_{\theta\theta}\right] = 0, \tag{8.3.6}$$

$$\frac{1}{r}\frac{d}{dr}(rV_r) + q = 0, \tag{8.3.7}$$

$$\frac{1}{r}\left[\frac{d}{dr}(rM_{rr}) - M_{\theta\theta}\right] - Q_r = 0, \tag{8.3.8}$$

where

$$V_r = Q_r + N_{rr}\frac{dw}{dr}. \tag{8.3.9}$$

The boundary conditions obtained from the principle of virtual work consist of specifying one element of each of the following duality pairs:

$$(u \text{ or } rN_{rr}); \quad (w \text{ or } rV_r); \quad (\phi_r \text{ or } rM_{rr}). \tag{8.3.10}$$

8.3.3 Equations of Equilibrium in Terms of Displacements

The stress resultants defined in Eqs. (8.3.5) for the FST can be expressed in terms of the generalized displacements (u, w, ϕ_r) as

$$\begin{aligned}
N_{rr} &= A_{rr}\left[\frac{du}{dr} + \frac{1}{2}\left(\frac{dw}{dr}\right)^2 + \nu\frac{u}{r}\right] + B_{rr}\left(\frac{d\phi_r}{dr} + \frac{\nu}{r}\phi_r\right),\\
N_{\theta\theta} &= A_{rr}\left[\frac{u}{r} + \nu\frac{du}{dr} + \frac{\nu}{2}\left(\frac{dw}{dr}\right)^2\right] + B_{rr}\left(\nu\frac{d\phi_r}{dr} + \frac{1}{r}\phi_r\right),\\
M_{rr} &= B_{rr}\left[\frac{du}{dr} + \frac{1}{2}\left(\frac{dw}{dr}\right)^2 + \nu\frac{u}{r}\right] + D_{rr}\left(\frac{d\phi_r}{dr} + \frac{\nu}{r}\phi_r\right),\\
M_{\theta\theta} &= B_{rr}\left[\frac{u}{r} + \nu\frac{du}{dr} + \frac{\nu}{2}\left(\frac{dw}{dr}\right)^2\right] + D_{rr}\left(\nu\frac{d\phi_r}{dr} + \frac{1}{r}\phi_r\right),\\
Q_r &= S_{rz}\left(\phi_r + \frac{dw}{dr}\right).
\end{aligned} \tag{8.3.11}$$

Here A_{rr}, B_{rr}, D_{rr}, and $S_{rz} = K_s A_{rz}$ are the extensional, extensional-bending, bending, and shear stiffness coefficients, respectively [see Eqs. (8.2.6) and (8.2.7)]

$$S_{rz} = \frac{K_s}{2(1+\nu)}\int_{-\frac{h}{2}}^{\frac{h}{2}} E(z)\,dz. \tag{8.3.12}$$

Substitution of Eqs. (8.3.11) for the stress resultants into the equations of equilibrium, Eqs. (8.3.7)–(8.3.9) give the equilibrium equations in terms of the displacements:

$$-\frac{1}{r}\frac{d}{dr}\left\{rA_{rr}\left[\frac{du}{dr}+\frac{1}{2}\left(\frac{dw}{dr}\right)^2+\nu\frac{u}{r}\right]+rB_{rr}\left(\frac{d\phi_r}{dr}+\frac{\nu}{r}\phi_r\right)\right\}$$
$$+\frac{1}{r}\left\{A_{rr}\left[\frac{u}{r}+\nu\frac{du}{dr}+\frac{\nu}{2}\left(\frac{dw}{dr}\right)^2\right]+B_{rr}\left(\nu\frac{d\phi_r}{dr}+\frac{1}{r}\,\phi_r\right)\right\}=0, \quad (8.3.13)$$

$$-\frac{1}{r}\frac{d}{dr}\left\{rS_{rz}\left(\phi_r+\frac{dw}{dr}\right)+rA_{rr}\frac{dw}{dr}\left[\frac{du}{dr}+\frac{1}{2}\left(\frac{dw}{dr}\right)^2+\nu\frac{u}{r}\right]\right.$$
$$\left.+rB_{rr}\frac{dw}{dr}\left(\frac{d\phi_r}{dr}+\frac{\nu}{r}\,\phi_r\right)\right\}-q=0, \quad (8.3.14)$$

$$-\frac{1}{r}\frac{d}{dr}\left\{rB_{rr}\left[\frac{du}{dr}+\frac{1}{2}\left(\frac{dw}{dr}\right)^2+\nu\frac{u}{r}\right]+rD_{rr}\left(\frac{d\phi_r}{dr}+\frac{\nu}{r}\,\phi_r\right)\right\}$$
$$+\frac{1}{r}\left\{B_{rr}\left[\frac{u}{r}+\nu\frac{du}{dr}+\frac{\nu}{2}\left(\frac{dw}{dr}\right)^2\right]+D_{rr}\left(\nu\frac{d\phi_r}{dr}+\frac{1}{r}\,\phi_r\right)\right\}$$
$$+S_{rz}\left(\phi_r+\frac{dw}{dr}\right)=0. \quad (8.3.15)$$

8.4 Finite Element Models

8.4.1 Introduction

In this section, finite element models of axisymmetric circular plates based on the CPT and FST are presented. Displacement (CPT-D) and mixed (CPT-M) models of the CPT and the displacement model (FST-D) of the FST, all accounting for the von Kármán nonlinearity as well as two-material gradation through the thickness (see Fig. 8.4.1), are presented.

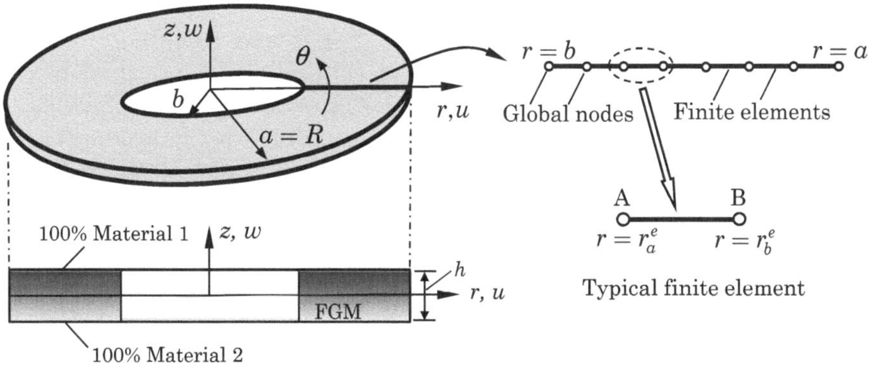

Fig. 8.4.1 Axisymmetric bending of an annular plate of inner radius b and outer radius a; the computational domain, finite element mesh, and a typical element are shown.

As is made clear in the previous sections, the domain of analysis for all axisymmetric circular plate problems is a radial line, as shown in Fig. 8.4.1. In the polar coordinate system, the domain can be represented as a line between points $r = b$ and $r = a$, with $b = 0$ for solid circular plates. In the finite element analysis, a typical finite element occupies the domain between two points A and B with coordinates $r = r_a$ and $r = r_b$, respectively [i.e., $\Omega^e = (r_a, r_b)$]. The generalized primary and secondary variables are not shown on the element because they are different for different finite element models.

8.4.2 Displacement Model of the CPT

8.4.2.1 Weak forms

From statement of the principle of virtual displacements in Eq. (8.2.4), with the stress resultants N_{rr}, $N_{\theta\theta}$, M_{rr}, and $M_{\theta\theta}$ expressed in terms of u and w via Eqs. (8.2.5a)–(8.2.5d), we have the following weak forms for the CPT over a typical finite element $\Omega^e = (r_a^e, r_b^e)$:

$$0 = \int_{r_a^e}^{r_b^e} \left[N_{rr} \frac{d\delta u}{dr} + N_{\theta\theta} \left(\frac{\delta u}{r} \right) \right] r\, dr - Q_1^e \delta u(r_a^e) - Q_4^e \delta u(r_b^e), \tag{8.4.1}$$

$$\begin{aligned} 0 = &\int_{r_a^e}^{r_b^e} \left[N_{rr} \frac{dw}{dr}\frac{d\delta w}{dr} - M_{rr} \frac{d^2 \delta w}{dr^2} - \frac{1}{r} M_{\theta\theta} \frac{d\delta w}{dr} - q\delta w \right] r\, dr \\ &- \left[Q_2^e\, \delta w(r_a^e) + Q_5^e\, \delta w(r_b^e) + Q_3^e\, \delta\theta(r_a^e) + Q_6^e\, \delta\theta(r_b^e) \right], \end{aligned} \tag{8.4.2}$$

where θ denotes the slope $\theta = -(dw/dr)$, Q_i^e are the generalized forces at the nodes of the element for a circular plate [see Fig. 8.4.2(a)]

$$\begin{aligned} Q_1^e &\equiv - \left[r N_{rr} \right]_{r_a^e}, & Q_4^e &\equiv \left[r N_{rr} \right]_{r_b^e} \\ Q_2^e &\equiv - \left[r V_r \right]_{r_a^e}, & Q_5^e &\equiv \left[r V_r \right]_{r_b^e} \\ Q_3^e &\equiv - \left[r M_{rr} \right]_{r_a^e}, & Q_6^e &\equiv \left[r M_{rr} \right]_{r_b^e}, \end{aligned} \tag{8.4.3}$$

and the corresponding generalized displacements are: $(u,\, w,\, \theta)$ at the two nodes of the element [see Fig. 8.4.2(b)].

The two variational statements in Eqs. (8.4.1) and (8.4.2) form the basis of the displacement finite element model of the CPT, as described next. We note that the two statements are coupled (i.e., the dependent unknowns u and w appear in both equations) when geometric nonlinearity is accounted or the beam is functionally graded.

8.4.2.2 Finite element model

An examination of the dual variables listed in Eq. (8.2.11) shows that the Lagrange interpolation of u and Hermite interpolation of w (i.e., C^1-continuity) is necessary. Let

$$u(r) = \sum_{j=1}^{2} \Delta_j^1\, \psi_j^e(r), \quad w(r) = \sum_{J=1}^{4} \Delta_J^2\, \varphi_J^e(r) \tag{8.4.4}$$

where $\psi_j^e(r)$ are the linear polynomials, $\varphi_J^e(r)$ are the Hermite cubic polynomials (they are the same as those listed in Eqs. (7.3.4) and (7.3.5) with x replaced with r), (Δ_1^1, Δ_2^1) are the nodal values of $u(r)$ at r_a^e and r_b^e, respectively, and Δ_J^2 $(J = 1, 2, 3, 4)$ are the nodal values associated with $w(r)$ and its first derivative; the superscript on Δ refers to the variable number, with 1 being for u and 2 for w; and the subscript refers to the degree-of-freedom number:

$$
\begin{gathered}
\Delta_1^1 = u(r_a^e), \quad \Delta_2^1 = u(r_b^e), \\
\Delta_1^2 = w(r_a^e), \quad \Delta_2^2 = -\left.\frac{dw}{dr}\right|_{r_a^e}, \quad \Delta_3^2 = w(r_b^e), \quad \Delta_4^2 = -\left.\frac{dw}{dr}\right|_{r_b^e}.
\end{gathered}
\tag{8.4.5}
$$

The generalized forces and generalized displacements for the CPT are shown in Figs. 8.4.2(a) and 8.4.2(b), respectively.

Substitution of the approximations in Eq.(8.4.4) into Eqs. (8.4.2) and (8.4.3), we obtain the following displacement finite element model of the CPT:

$$
\begin{bmatrix} \mathbf{K}^{11} & \mathbf{K}^{12} \\ \mathbf{K}^{21} & \mathbf{K}^{22} \end{bmatrix} \begin{Bmatrix} \mathbf{\Delta}^1 \\ \mathbf{\Delta}^2 \end{Bmatrix} = \begin{Bmatrix} \mathbf{F}^1 \\ \mathbf{F}^2 \end{Bmatrix}, \tag{8.4.6}
$$

where the stiffness coefficients $K_{ij}^{\alpha\beta}$ and force coefficients F_i^α $(\alpha, \beta = 1, 2)$ are defined in Eq. (8.4.7).

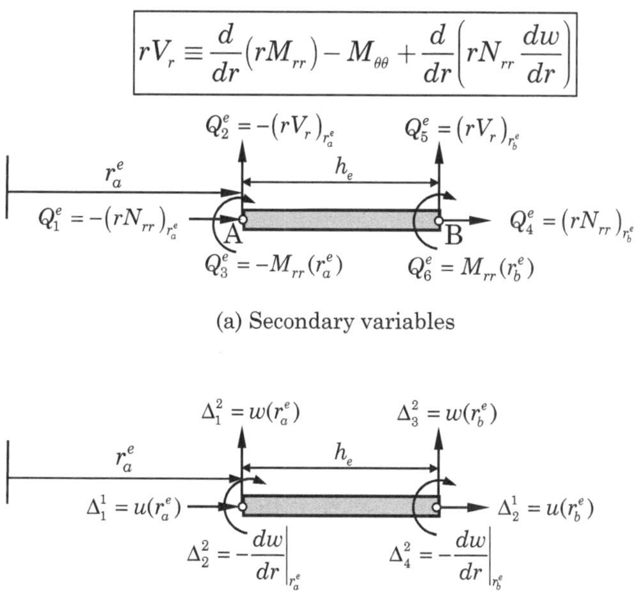

Fig. 8.4.2 (a) Secondary and (b) primary variables of a typical displacement finite element model for the CPT when linear interpolation of u and Hermite cubic interpolation of w is assumed.

$$
\begin{aligned}
K_{ij}^{11} &= \int_{r_a}^{r_b} A_{rr}^e \left[\frac{d\psi_i^e}{dr}\left(\frac{d\psi_j^e}{dr} + \frac{\nu}{r}\psi_j^e\right) + \frac{1}{r}\psi_i^e\left(\frac{1}{r}\psi_j^e + \nu\frac{d\psi_j^e}{dr}\right)\right] r dr,\\
K_{iJ}^{12} &= \int_{r_a^e}^{r_b^e} \left\{ \frac{d\psi_i^e}{dr}\left[A_{rr}^e \frac{1}{2}\frac{dw}{dr}\frac{d\varphi_J^e}{dr} - B_{rr}^e\left(\frac{d^2\varphi_J^e}{dr^2} + \frac{\nu}{r}\frac{d\varphi_J^e}{dr}\right)\right],\right.\\
&\quad \left. + \frac{1}{r}\psi_i^e\left[\frac{1}{2}A_{rr}^e \nu \frac{dw}{dr}\frac{d\varphi_J^e}{dr} - B_{rr}^e\left(\nu\frac{d^2\varphi_J^e}{dr^2} + \frac{1}{r}\frac{d\varphi_J^e}{dr}\right)\right]\right\} r dr,\\
K_{Ij}^{21} &= \int_{r_a^e}^{r_b^e} \left[\left(A_{rr}^e \frac{dw}{dr}\frac{d\varphi_I^e}{dr} - B_{rr}^e \frac{d^2\varphi_I^e}{dr^2}\right)\left(\frac{d\psi_j^e}{dr} + \frac{\nu}{r}\psi_j^e\right)\right.\\
&\quad \left. - B_{rr}^e \frac{1}{r}\frac{d\varphi_I^e}{dr}\left(\frac{1}{r}\psi_j^e + \nu\frac{d\psi_j^e}{dr}\right)\right] r dr,\\
K_{IJ}^{22} &= \int_{r_a^e}^{r_b^e} \left\{ \frac{d\varphi_I^e}{dr}\left[\frac{1}{2}A_{rr}^e\left(\frac{dw}{dr}\right)^2 \frac{d\varphi_J^e}{dr} - B_{rr}^e \frac{dw}{dr}\left(\frac{d^2\varphi_J^e}{dr^2} + \frac{\nu}{r}\frac{d\varphi_J^e}{dr}\right)\right]\right.\\
&\quad - \frac{d^2\varphi_I^e}{dr^2}\left[\frac{1}{2}B_{rr}^e \frac{dw}{dr}\frac{d\varphi_J^e}{dr} - D_{rr}^e\left(\frac{d^2\varphi_J^e}{dr^2} + \frac{\nu}{r}\frac{d\varphi_J^e}{dr}\right)\right]\\
&\quad \left. - \frac{1}{r}\frac{d\varphi_I^e}{dr}\left[\frac{1}{2}B_{rr}^e \nu \frac{dw}{dr}\frac{d\varphi_J^e}{dr} - D_{rr}^e\left(\nu\frac{d^2\varphi_J^e}{dr^2} + \frac{1}{r}\frac{d\varphi_J^e}{dr}\right)\right]\right\} r dr,\\
F_i^1 &= Q_1^e \psi_i^e(r_a^e) + Q_4^e \psi_i(r_b^e)\\
F_I^2 &= \int_{r_a^e}^{r_b^e} q\varphi_I^e \, r dr + Q_2^e \varphi_I^e(r_a^e) + Q_5^e \varphi_I^e(r_b^e) + Q_3^e\left[-\frac{d\varphi_I^e}{dx}\right]_{r_a^e} + Q_6^e\left[-\frac{d\varphi_I^e}{dx}\right]_{r_b^e}.
\end{aligned}
\tag{8.4.7}
$$

As is clear from the coefficients, the nonlinear finite element model of circular plates contains many more terms compared to the corresponding beam finite element model. However, the logical units of the computer programs developed for beams and circular plates remain the same (including the linearization, assembly, imposition of boundary conditions, solution of equations, and post-computations).

8.4.2.3 Tangent stiffness coefficients

For the linearized finite element equations, the tangent stiffness matrix $\mathbf{T}$ is assumed to be of the same form as the direct stiffness matrix $\mathbf{K}$. Then the coefficients of the submatrices $\mathbf{T}^{\alpha\beta}$ of the tangent stiffness matrix $\mathbf{T}$ can be computed using the definition

$$
\begin{aligned}
T_{ij}^{\alpha\beta} &\equiv \frac{\partial R_i^\alpha}{\partial \Delta_j^\beta} = K_{ij}^{\alpha\beta} + \sum_{\gamma=1}^{2}\sum_{k=1}^{n_\gamma} \frac{\partial K_{ik}^{\alpha\gamma}}{\partial \Delta_j^\beta}\Delta_k^\gamma - \frac{\partial F_i^\alpha}{\partial \Delta_j^\beta}\\
&= K_{ij}^{\alpha\beta} + \sum_{k=1}^{n_\gamma} \frac{\partial K_{ik}^{\alpha 1}}{\partial \Delta_j^\beta}\Delta_k^1 + \sum_{k=1}^{n_\gamma} \frac{\partial K_{ik}^{\alpha 2}}{\partial \Delta_j^\beta}\Delta_k^2 - \frac{\partial F_i^\alpha}{\partial \Delta_j^\beta},
\end{aligned}
\tag{8.4.8}
$$

The tangent stiffness coefficients for displacement model of the CPT are

$$
\begin{aligned}
T_{ij}^{11} &= K_{ij}^{11}, \quad T_{Ij}^{21} = K_{Ij}^{21}, \\
T_{iJ}^{12} &= K_{iJ}^{12} + \int_{r_a^e}^{r_b^e} \frac{1}{2} A_{rr}^e \frac{dw}{dr}\left(\frac{d\psi_i^e}{dr}\frac{d\varphi_J^e}{dr} + \frac{\nu}{r}\psi_i \frac{d\varphi_J}{dr}\right) r dr = T_{Ji}^{21}, \\
T_{IJ}^{22} &= K_{IJ}^{22} + \int_{r_a^e}^{r_b^e} \left[A_{rr}^e \left\{ \frac{d\varphi_I^e}{dr}\frac{d\varphi_J^e}{dr}\left[\left(\frac{dw}{dr}\right)^2 + \frac{du}{dr} + \nu\frac{u}{r}\right]\right\}\right. \\
&\qquad - B_{rr}^e \frac{d\varphi_I^e}{dr}\frac{d\varphi_J^e}{dr}\left(\frac{d^2w}{dr^2} + \frac{\nu}{r}\frac{dw}{dr}\right) \\
&\qquad \left. - \frac{1}{2} B_{rr}^e \frac{dw}{dr}\left(\frac{d^2\varphi_I^e}{dr^2}\frac{d\varphi_J^e}{dr} + \frac{\nu}{r}\frac{d\varphi_I^e}{dr}\frac{d\varphi_J^e}{dr}\right)\right] r dr. \qquad (8.4.9)
\end{aligned}
$$

The tangent stiffness matrix is symmetric (i.e., $T_{ij}^{\alpha\beta} = T_{ji}^{\beta\alpha}$).

8.4.3 Mixed Model of the CPT

8.4.3.1 Weak forms

The weak forms of Eqs. (8.2.23)–(8.2.25) can be developed using the three-step procedure discussed in Chapter 4. The weak forms of these three equations are:

$$
\begin{aligned}
0 = &\int_{r_a^e}^{r_b^e} \frac{d\delta u}{dr}\left\{\bar{A}_{rr}^e \left[r\frac{du}{dr} + \frac{r}{2}\left(\frac{dw}{dr}\right)^2 + \nu\, u\right] + r\bar{B}_{rr}^e M_{rr}\right\} dr \\
&+ \int_{r_a^e}^{r_b^e} \left\{\bar{A}_{rr}^e \left[\nu\,\delta u\frac{du}{dr} + \nu\frac{1}{2}\delta u\left(\frac{dw}{dr}\right)^2 + \frac{1}{r}\delta u\, u\right] + \nu\bar{B}_{rr}^e \delta u\, M_{rr}\right\} dr \\
&+ \int_{r_a^e}^{r_b^e} (1-\nu^2) B_{rr}^e \frac{1}{r}\left(\bar{B}_{rr}^e \delta u\, u - \delta u\,\frac{dw}{dr}\right) dr - Q_1^e\,\delta u(r_a^e) - Q_4^e\,\delta u(r_b^e),
\end{aligned}
\tag{8.4.10}
$$

$$
\begin{aligned}
0 = &\int_{r_a^e}^{r_b^e} \frac{d\delta w}{dr}\left[r\frac{dM_{rr}}{dr} + (1-\nu)M_{rr} - (1-\nu^2)\frac{D_{rr}^e}{r}\left(\bar{B}_{rr}^e\, u - \frac{dw}{dr}\right)\right] dr \\
&+ \int_{r_a^e}^{r_b^e} \left\{\bar{A}_{rr}^e \frac{dw}{dr}\frac{d\delta w}{dr}\left[r\frac{du}{dr} + \frac{r}{2}\left(\frac{dw}{dr}\right)^2 + \nu u\right] + r\bar{B}_{rr}^e \frac{d\delta w}{dr}\frac{dw}{dr} M_{rr}\right\} dr \\
&- \int_{r_a^e}^{r_b^e} \delta w\, r\, q - Q_2^e\,\delta w(r_a^e) - Q_5^e\,\delta w(r_b^e),
\end{aligned}
\tag{8.4.11}
$$

$$
\begin{aligned}
0 = &\int_{r_a}^{r_b} \left\{ r\frac{\partial \delta M_{rr}}{dr}\frac{dw}{dr} + (1-\nu)\delta M_{rr}\frac{dw}{dr} + \bar{B}_{rr}^e\,\delta M_{rr}\left[r\frac{du}{dr} + \frac{r}{2}\left(\frac{dw}{dr}\right)^2 + \nu\, u\right]\right. \\
&\left. - \frac{r}{D_{rr}^e}\delta M_{rr}\, M_{rr}\right\} dr - Q_3^e\,\delta M_{rr}(r_a^e) - Q_6^e\,\delta M_{rr}(r_b^e),
\end{aligned}
\tag{8.4.12}
$$

where [see Eq. (8.2.6) for the definition of A^e_{rr}, B^e_{rr}, and D^e_{rr}; they are assumed to be element-wise constant; the label e is omitted in the following equation]

$$D^*_{rr} = D_{rr}\,A_{rr} - B^2_{rr}, \quad \bar{A}_{rr} = \frac{D^*_{rr}}{D_{rr}}, \quad \bar{B}_{rr} = \frac{B_{rr}}{D_{rr}}, \tag{8.4.13}$$

and Q^e_i are the generalized forces [i.e., secondary variables; see Fig. 8.4.3(a)] at the two nodes of the element and they can be expressed in terms of the stress resultants as

$$\begin{aligned}
Q^e_1 &= -[rN_{rr}]_{r=r^e_a}, \qquad Q^e_4 = [rN_{rr}]_{r=r^e_b},\\
Q^e_2 &= -\left[\frac{d}{dr}(rM_{rr}) - M_{\theta\theta} + \frac{d}{dr}\left(rN_{rr}\frac{dw}{dr}\right)\right]_{r=r^e_a},\\
Q^e_5 &= \left[\frac{d}{dr}(rM_{rr}) - M_{\theta\theta} + \frac{d}{dr}\left(rN_{rr}\frac{dw}{dr}\right)\right]_{r=r^e_b},\\
Q^e_3 &= \left[r\frac{dw}{dr}\right]_{r=r^e_a}, \qquad Q^e_6 = \left[-r\frac{dw}{dr}\right]_{r=r^e_b}.
\end{aligned} \tag{8.4.14}$$

8.4.3.2 Finite element model

An examination of Eqs. (8.4.10)–(8.4.12) indicates that the Lagrange interpolation of all three variables (u, w, M_{rr}) is admissible. Thus, one may use linear or higher-order Lagrange interpolation of these variables. The general form of the finite element interpolation is given by

$$u(r) = \sum_{j=1}^{m} u^e_j\,\psi^{(1)}_j(r), \quad w(r) = \sum_{j=1}^{n} w^e_j\,\psi^{(2)}_j(r), \quad M_{rr}(r) = \sum_{j=1}^{p} m^e_j\,\psi^{(3)}_j(r), \tag{8.4.15}$$

where $\psi^{(1)}_j$, $\psi^{(2)}_j$, and $\psi^{(3)}_j$ are the Lagrange polynomials of different degree (i.e., $m \neq n \neq p$) used for u, w, and M_{rr}, respectively. However, when using different degree of interpolation, one must be careful in selecting the values of m, n, and p for compatibility of these fields with each other. For simplicity, we take equal order interpolation (i.e., $m = n = p$) in implementing the mixed finite element model in a computer program. Figure 8.4.3 shows the secondary and primary degrees of freedom of the mixed finite element model when all variables are approximated using linear interpolation.

Substitution of the approximations in Eq. (8.4.15) into the weak forms in Eqs. (8.4.10)–(8.4.12) gives the following mixed finite element model of the CPT:

$$\begin{bmatrix} \mathbf{K}^{11} & \mathbf{K}^{12} & \mathbf{K}^{13}\\ \mathbf{K}^{21} & \mathbf{K}^{22} & \mathbf{K}^{23}\\ \mathbf{K}^{31} & \mathbf{K}^{32} & \mathbf{K}^{33}\end{bmatrix}^{(e)} \begin{Bmatrix}\mathbf{u}\\ \mathbf{w}\\ \mathbf{m}\end{Bmatrix}^{(e)} = \begin{Bmatrix}\mathbf{F}^1\\ \mathbf{F}^2\\ \mathbf{F}^3\end{Bmatrix}^{(e)}, \tag{8.4.16}$$

where $(\mathbf{u}, \mathbf{w}, \mathbf{m})$ denote the vectors of displacements u and w and moment M_{rr} at the nodes of the element.

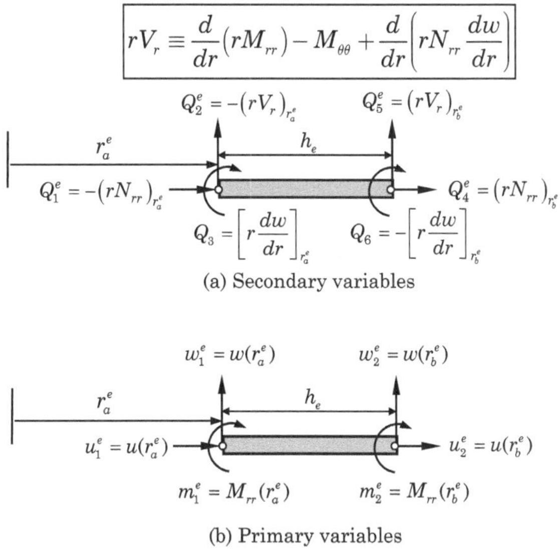

Fig. 8.4.3 (a) Secondary and (b) primary variables of a typical mixed finite element model of the classical plate theory (CPT) when linear interpolation of u, w, and M_{rr} is assumed; $rV_r = d(rM_{rr})/dr - M_{\theta\theta} + d(rN_{rr}(dw/dr))/dr$.

The non-zero coefficients of Eq. (8.4.16) are given by

$$K_{ij}^{11} = \int_{r_a^e}^{r_b^e} \left[\bar{A}_{rr}^e r \frac{d\psi_i^{(1)}}{dr}\frac{d\psi_j^{(1)}}{dr} + \bar{A}_{rr}^e \nu \left(\frac{d\psi_i^{(1)}}{dr}\psi_j^{(1)} + \psi_i^{(1)}\frac{d\psi_j^{(1)}}{dr} + \frac{1}{r}\psi_i^{(1)}\psi_j^{(1)} \right) \right.$$
$$\left. + (1-\nu^2) B_{rr}^e \bar{B}_{rr}^e \frac{1}{r}\psi_i^{(1)}\psi_j^{(1)} \right] dr,$$

$$K_{ij}^{12} = \tfrac{1}{2}\int_{r_a^e}^{r_b^e} \bar{A}_{rr}^e \frac{dw}{dr}\left(r\frac{d\psi_i^{(1)}}{dr}\frac{d\psi_j^{(2)}}{dr} + \nu\psi_i^{(1)}\frac{d\psi_j^{(2)}}{dr} \right) dr$$
$$- \int_{r_a}^{r_b} B_{rr}^e (1-\nu^2)\, \frac{1}{r}\psi_i^{(1)}\frac{d\psi_j^{(2)}}{dr}\, dr,$$

$$K_{ij}^{13} = \int_{r_a}^{r_b} \bar{B}_{rr}^e \left(r\frac{d\psi_i^{(1)}}{dr}\psi_j^{(3)} + \nu\psi_i^{(1)}\psi_j^{(3)} \right) dr,$$

$$K_{ij}^{21} = -\int_{r_a^e}^{r_b^e} B_{rr}^e (1-\nu^2)\, \frac{1}{r}\frac{d\psi_i^{(2)}}{dr}\psi_j^{(1)} dr$$
$$+ \int_{r_a^e}^{r_b^e} \bar{A}_{rr}^e \frac{dw}{dr}\left(r\frac{d\psi_i^{(2)}}{dr}\frac{d\psi_j^{(1)}}{dr} + \nu\frac{d\psi_i^{(2)}}{dr}\psi_j^{(1)} \right) dr,$$

$$K_{ij}^{22} = \int_{r_a^e}^{r_b^e} D_{rr}^e (1-\nu^2)\frac{1}{r}\frac{d\psi_i^{(2)}}{dr}\frac{d\psi_j^{(2)}}{dr} dr + \frac{1}{2}\int_{r_a^e}^{r_b^e} \bar{A}_{rr}^e \left(\frac{dw}{dr}\right)^2 \frac{d\psi_i^{(2)}}{dr}\frac{d\psi_j^{(2)}}{dr} r\, dr,$$

$$
\begin{aligned}
K_{ij}^{23} &= \int_{r_a^e}^{r_b^e} \left[r\frac{d\psi_i^{(2)}}{dr}\frac{d\psi_j^{(3)}}{dr} + (1-\nu)\frac{d\psi_i^{(2)}}{dr}\psi_j^{(3)} \right] dr + \int_{r_a}^{r_b} \bar{B}_{rr}^e \frac{dw}{dr}\frac{d\psi_i^{(2)}}{dr}\psi_j^{(3)}\, rdr,\\
K_{ij}^{31} &= \int_{r_a^e}^{r_b^e} \bar{B}_{rr}^e \left(r\psi_i^{(3)}\frac{d\psi_j^{(1)}}{dr} + \nu\psi_i^{(3)}\psi_j^{(1)} \right) dr,\\
K_{ij}^{32} &= \int_{r_a^e}^{r_b^e} \left[r\frac{d\psi_i^{(3)}}{dr}\frac{d\psi_j^{(2)}}{dr} + (1-\nu)\psi_i^{(3)}\frac{d\psi_j^{(2)}}{dr} \right] dr + \frac{1}{2}\int_{r_a}^{r_b} \bar{B}_{rr}^e \frac{dw}{dr}\psi_i^{(3)}\frac{d\psi_j^{(2)}}{dr}\, rdr,\\
K_{ij}^{33} &= -\int_{r_a^e}^{r_b^e} \frac{1}{D_{rr}^e}\psi_i^{(3)}\psi_j^{(3)}\, rdr, \quad F_i^1 = \psi_i^{(1)}(r_a)Q_1^e + \psi_i^{(1)}(r_b^e)Q_4^e,\\
F_i^2 &= \int_{r_a^e}^{r_b^e} qr\, dr + \psi_i^{(2)}(r_a)Q_2^e + \psi_i^{(2)}(r_b^e)Q_5^e,\\
F_i^3 &= \psi_i^{(3)}(r_a^e)Q_3^e + \psi_i^{(3)}(r_b^e)Q_6^e. \qquad (8.4.17)
\end{aligned}
$$

Here i and j take suitable ranges of values when different degrees of interpolation are used. When $\psi_i^{(1)} = \psi_i^{(2)} = \psi_i^{(3)}$, i, j take the values of $1, 2, \ldots, n$, n being the number of nodes in the element.

8.4.3.3 Mixed model of the CPT

The tangent stiffness coefficients for this case are given by

$$
\begin{aligned}
T_{ij}^{11} &= K_{ij}^{11}, \quad T_{ij}^{21} = K_{ij}^{21}, \quad T_{ij}^{13} = K_{ij}^{13}, \quad T_{ij}^{23} = K_{ij}^{23},\\
T_{ij}^{12} &= K_{ij}^{12} + \int_{r_a}^{r_b} \frac{1}{2}A_{rr}\frac{dw}{dr}\left(r\frac{d\psi_i^{(1)}}{dr}\frac{d\psi_j^{(2)}}{dr} + \nu\psi_i^{(1)}\frac{d\psi_j^{(2)}}{dr} \right) dr = T_{ji}^{21},\\
T_{ij}^{22} &= K_{ij}^{22} + \int_{r_a^e}^{r_b^e} \left\{ \bar{A}_{rr}^e \left[\left(\frac{dw}{dr}\right)^2 + \frac{du}{dr} + \frac{\nu}{r}u \right] + \bar{B}_{rr}^e M_{rr} \right\} \frac{d\psi_i^{(2)}}{dr}\frac{d\psi_j^{(2)}}{dr} rdr,\\
T_{ij}^{32} &= K_{ij}^{32} + \frac{1}{2}\int_{r_a^e}^{r_b^e} \bar{B}_{rr}^e \frac{dw}{dr}\psi_i^{(3)}\frac{d\psi_j^{(2)}}{dr}\, rdr,\\
T_{ij}^{31} &= K_{ij}^{31}, \quad T_{ij}^{33} = K_{ij}^{33}. \qquad (8.4.18)
\end{aligned}
$$

The tangent stiffness matrix is symmetric (i.e., $T_{ij}^{\alpha\beta} = T_{ji}^{\beta\alpha}$).

Because the same coefficients that are also in the corresponding beam finite element models, one can use the beam finite element programs with slight modification (to include the Poisson effect and additional terms) to analyze axisymmetric circular plates. One may use a parameter (must be read as an input) to analyze a beam or circular plate problem. We also note that shear locking and membrane locking issues with the circular plates are the same as those for the beams, and we use the same remedy (i.e., use reduced integration) to alleviate locking.

8.4.4 Displacement Model of the FST

8.4.4.1 Weak forms

The virtual work statement in Eq. (8.3.4) for the FST is equivalent to the following three integral statements over a typical finite element $\Omega^e = (r_a^e, r_b^e)$:

$$0 = \int_{r_a^e}^{r_b^e} \left(N_{rr} \frac{d\delta u}{dr} + N_{\theta\theta} \frac{\delta u}{r} \right) r\, dr - Q_1^e\, \delta u(r_a^e) - Q_4^e\, \delta u(r_b^e), \tag{8.4.19}$$

$$0 = \int_{r_a^e}^{r_b^e} \left[N_{rr} \frac{dw}{dr}\frac{d\delta w}{dr} + Q_r \frac{d\delta w}{dr} - q\delta w \right] r\, dr - Q_2^e\, \delta w(r_a^e) - Q_5^e\, \delta w(r_b^e), \tag{8.4.20}$$

$$0 = \int_{r_a^e}^{r_b^e} \left[M_{rr} \frac{d\delta\phi_r}{dr} + \frac{1}{r} M_{\theta\theta}\delta\phi_r + Q_r\delta\phi_r \right] r\, dr - Q_3^e\, \delta\phi_r(r_a^e) - Q_6^e\, \delta\phi_r(r_b^e), \tag{8.4.21}$$

where Q_i^e are the generalized forces at the nodes of the element for a circular plate [see Fig. 8.4.4(b)]:

$$\begin{aligned} Q_1^e &\equiv -\left[rN_{rr}\right]_{r_a^e}, & Q_4^e &\equiv \left[rN_{rr}\right]_{r_b^e}, \\ Q_2^e &\equiv -\left[rV_r\right]_{r_a^e}, & Q_5^e &\equiv \left[rV_r\right]_{r_b^e}, \\ Q_3^e &\equiv -\left[r\bar{M}_{rr}\right]_{r_a^e}, & Q_6^e &\equiv \left[rM_{rr}\right]_{r_b^e}, \end{aligned} \tag{8.4.22}$$

and N_{rr}, $N_{\theta\theta}$, M_{rr}, $M_{\theta\theta}$, and Q_r are known in terms of u and w through Eq. (8.3.11); V_r is known in terms of Q_r and N_{rr} by Eq. (8.3.9).

8.4.4.2 Finite element model

An examination of the weak forms in Eqs. (8.4.19)–(8.4.21) shows that the Lagrange interpolation of (u, w, ϕ_r) is required. Let

$$\begin{aligned} u(r) &= \sum_{j=1}^{m} u_j^e\, \psi_j^{(1)}(r), \quad w(r) = \sum_{j=1}^{n} w_j^e\, \psi_j^{(2)}(r) \\ \phi_r(r) &= \sum_{j=1}^{p} s_j^e\, \psi_j^{(3)}(r), \end{aligned} \tag{8.4.23}$$

where $\psi_j^{(1)}$, $\psi_j^{(2)}$, and $\psi_j^{(3)}$ are the Lagrange polynomials of different degree used for u, w, and ϕ_r, respectively. We can use equal degree interpolation of all variables, $\psi_j^{(1)} = \psi_j^{(2)} = \psi_j^{(3)}$, and they can be linear or higher order. Figure 8.4.4 shows the secondary and primary variables of the displacement finite element model of the FST when linear interpolation of u, w, and ϕ_r is used.

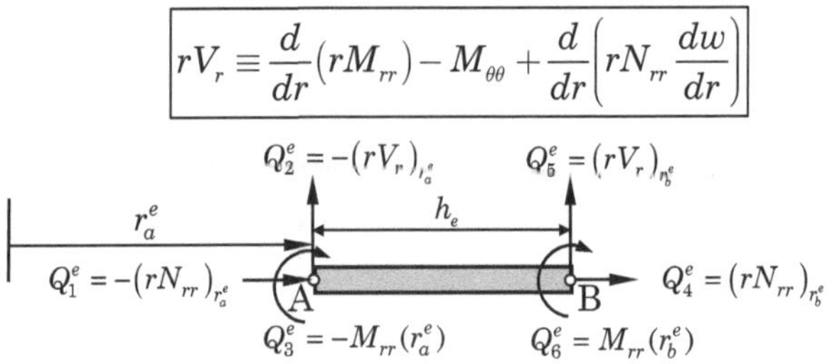

(a) Secondary variables

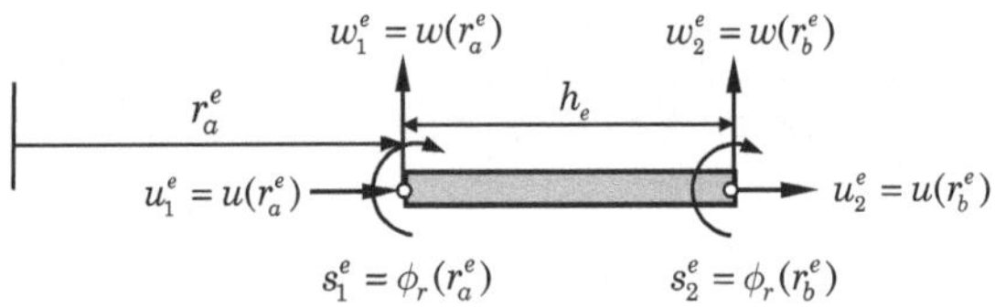

(b) Primary variables

Fig. 8.4.4 (a) Secondary and (b) primary variables of a typical displacement finite element model of the FST when linear interpolation of u and ϕ_r and Hermite cubic approximation of w are assumed.

Substitution of the approximations from Eq. (8.4.23) into the weak forms in Eqs. (8.4.19)–(8.4.21), we obtain the following finite element equations:

$$\begin{bmatrix} \mathbf{K}^{11} & \mathbf{K}^{12} & \mathbf{K}^{13} \\ \mathbf{K}^{21} & \mathbf{K}^{22} & \mathbf{K}^{23} \\ \mathbf{K}^{31} & \mathbf{K}^{32} & \mathbf{K}^{33} \end{bmatrix} \begin{Bmatrix} \mathbf{u} \\ \mathbf{\Delta} \\ \mathbf{s} \end{Bmatrix} = \begin{Bmatrix} \mathbf{F}^1 \\ \mathbf{F}^2 \\ \mathbf{F}^3 \end{Bmatrix} \tag{8.4.24}$$

The nonzero stiffness coefficients $K_{ij}^{\alpha\beta}$ and force coefficients F_i^α $(\alpha, \beta = 1, 2, 3)$ are defined as follows:

$$K_{ij}^{11} = \int_{r_a^e}^{r_b^e} A_{rr}^e \left[\frac{d\psi_i^{(1)}}{dr} \left(\frac{d\psi_j^{(1)}}{dr} + \frac{\nu}{r}\psi_j^{(1)} \right) + \frac{1}{r}\psi_i^{(1)} \left(\frac{1}{r}\psi_j^{(1)} + \nu\frac{d\psi_j^{(1)}}{dr} \right) \right] rdr$$

$$K_{ij}^{12} = \tfrac{1}{2} \int_{r_a^e}^{r_b^e} A_{rr}^e \frac{dw}{dr} \left(\frac{d\psi_i^{(1)}}{dr} + \frac{\nu}{r}\psi_i^{(1)} \right) \frac{d\psi_j^{(2)}}{dr}\, rdr$$

$$K_{ij}^{21} = \int_{r_a^e}^{r_b^e} A_{rr}^e \frac{dw}{dr}\frac{d\psi_i^{(2)}}{dr} \left(\frac{d\psi_j^{(1)}}{dr} + \frac{\nu}{r}\psi_j^{(1)} \right) rdr,$$

$$K_{ij}^{13} = \int_{r_a^e}^{r_b^e} B_{rr}^e \left[\frac{d\psi_i^{(1)}}{dr} \left(\frac{d\psi_j^{(3)}}{dr} + \frac{\nu}{r}\psi_j^{(3)} \right) + \frac{1}{r}\psi_i^{(1)} \left(\nu\frac{d\psi_j^{(3)}}{dr} + \frac{1}{r}\psi_j^{(3)} \right) \right] rdr,$$

$$K_{ij}^{22} = \int_{r_a^e}^{r_b^e} \left[\frac{1}{2} A_{rr}^e \left(\frac{dw}{dr} \right)^2 \frac{d\psi_i^{(2)}}{dr} \frac{d\psi_j^{(2)}}{dr} + S_{rz}^e \frac{d\psi_i^{(2)}}{dr} \frac{d\psi_j^{(2)}}{dr} \right] rdr,$$

$$K_{ij}^{23} = \int_{r_a^e}^{r_b^e} \left[B_{rr}^e \frac{dw}{dr} \frac{d\psi_i^{(2)}}{dr} \left(\frac{d\psi_j^{(3)}}{dr} + \frac{\nu}{r} \psi_j^{(3)} \right) + S_{rz}^e \frac{d\psi_i^{(2)}}{dr} \psi_j^{(3)} \right] rdr$$

$$K_{ij}^{31} = \int_{r_a^e}^{r_b^e} B_{rr}^e \left[\frac{d\psi_i^{(3)}}{dr} \left(\frac{d\psi_j^{(1)}}{dr} + \frac{\nu}{r} \psi_j^{(1)} \right) + \frac{1}{r} \psi_i^{(3)} \left(\nu \frac{d\psi_j^{(1)}}{dr} + \frac{1}{r} \psi_j^{(1)} \right) \right] rdr,$$

$$K_{ij}^{32} = \int_{r_a^e}^{r_b^e} \left[\frac{1}{2} B_{rr}^e \frac{dw}{dr} \left(\frac{d\psi_i^{(3)}}{dr} + \frac{\nu}{r} \psi_i^{(3)} \right) \frac{d\psi_j^{(2)}}{dr} + S_{rz}^e \psi_i^{(3)} \frac{d\psi_j^{(2)}}{dr} \right] rdr,$$

$$K_{ij}^{33} = \int_{r_a^e}^{r_b^e} \left\{ D_{rr}^e \left[\frac{d\psi_i^{(3)}}{dr} \left(\frac{d\psi_j^{(3)}}{dr} + \frac{\nu}{r} \psi_j^{(3)} \right) + \frac{1}{r} \psi_i^{(3)} \left(\nu \frac{d\psi_j^{(3)}}{dr} + \frac{1}{r} \psi_j^{(3)} \right) \right] \right.$$
$$\left. + S_{rz}^e \psi_i^{(3)} \psi_j^{(3)} \right\} rdr,$$

$$F_i^1 = Q_1^e \psi_i^{(1)}(r_a^e) + Q_4^e \psi_i^{(1)}(r_b^e), \quad F_i^3 = Q_3^e \psi_i^{(3)}(r_a^e) + Q_6^e \psi_i^{(3)}(r_b^e),$$

$$F_i^2 = \int_{r_a}^{r_b^e} q \psi_i^{(2)} \, rdr + Q_2^e \psi_i^{(2)}(r_a^e) + Q_5^e \psi_i^{(2)}(r_b^e). \tag{8.4.25}$$

8.4.4.3 Tangent stiffness coefficients

The tangent stiffness coefficients for the displacement model of the FST are given by

$$T_{ij}^{11} = K_{ij}^{11}, \quad T_{ij}^{12} = 2K_{ij}^{12}, \quad T_{ij}^{13} = K_{ij}^{13}, \quad T_{ij}^{21} = K_{ij}^{21},$$
$$T_{ij}^{31} = K_{ij}^{31} = T_{ji}^{13}, \quad T_{ij}^{23} = K_{ij}^{23}, \quad T_{ij}^{33} = K_{ij}^{33},$$
$$T_{ij}^{22} = K_{ij}^{22} + \int_{r_a^e}^{r_b^e} \left\{ A_{rr}^e \left[\left(\frac{dw}{dr} \right)^2 + \frac{du}{dr} + \nu \frac{u}{r} \right] \right.$$
$$\left. + B_{rr}^e \left(\frac{d\phi_r}{dr} + \frac{\nu}{r} \phi_r \right) \right\} \frac{d\psi_i^{(2)}}{dr} \frac{d\psi_j^{(2)}}{dr} rdr,$$

$$T_{ij}^{32} = K_{ij}^{32} + \frac{1}{2} \int_{r_a^e}^{r_b^e} B_{rr}^e \frac{dw}{dr} \left(\frac{d\psi_i^{(3)}}{dr} + \frac{\nu}{r} \psi_i^{(3)} \right) \frac{d\psi_j^{(2)}}{dr} rdr. \tag{8.4.26}$$

The tangent stiffness matrix of the FST element is also symmetric.

8.5 Dual Mesh Control Domain Models

8.5.1 Preliminary Comments

As discussed in Chapters 5 and 7, in the DMCDM, the domain is represented with a primal mesh of finite elements, and a dual mesh is superimposed on the primal mesh such that the nodes of the primal mesh are inside the dual mesh (if the mesh is uniform, they will be at the center of the control domains), except for the nodes on the boundary. Then the governing equation is required to be satisfied in an integral sense over each control domain of the dual mesh. The second-order terms in the differential equation are integrated and expressed as dual variables on the interfaces of the dual mesh. When the interfaces fall on the boundary, either the dual variables or their counterparts (i.e., primary variables) are known and thereby, the derivatives are not replaced at the boundary nodes, eliminating the need for the so-called zero-thickness control volumes or fictitious control volumes.

In the following discussion, we consider DMCDM formulations of circular plates with axisymmetric conditions (i.e., material properties, boundary conditions, and loads are independent of the circumferential coordinate). As in the case of the FEM, we use a typical radial line as the domain (see Fig. 8.5.1).

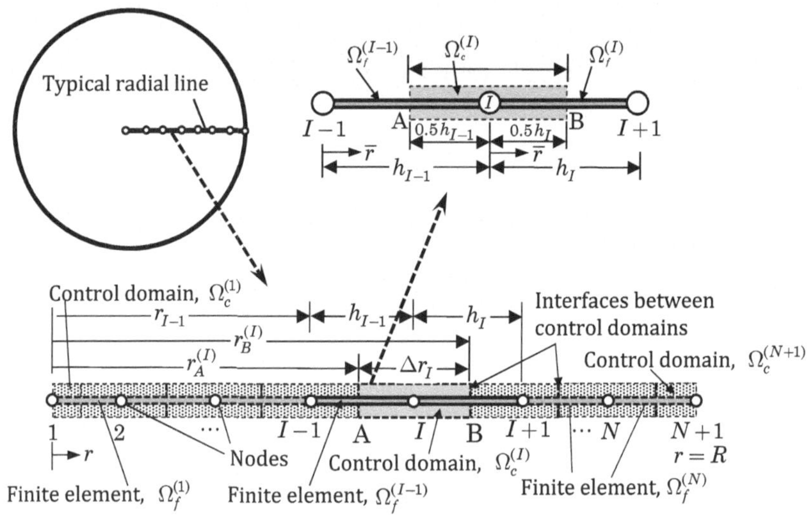

Fig. 8.5.1 A primal mesh of finite elements and dual mesh of control domains shown for a typical radial line of a circular plate. The Ith finite element is denoted by $\Omega_f^{(I)}$. We note that the boundary nodes have only half-control domains, whereas the internal nodes have full control domains. Also, each control domain connects two neighboring finite elements, one on the left and the other on the right. The Ith control domain, which houses the Ith node, is denoted by $\Omega_c^{(I)}$. The control domain $\Omega_c^{(I)}$ associated with an interior node I is isolated to discuss the discretization. Every interior control domain connects three nodes (nine nodal degrees of freedom) through the discretization of the governing equation, whereas the boundary nodes connect two adjacent nodes.

The domain is divided into a set (the primal mesh) of N line finite elements with linear interpolation functions (i.e., we use finite element interpolation but not the finite element method to obtain the discretized equations). For now, we assume that each element has two nodes (i.e., linear approximation), positioned at the ends of the line element. Then, the dual mesh of line elements covers the whole domain and bisects the elements of the primal mesh on either side of the nodes (i.e., the interfaces of the line elements of the dual mesh are at the center of the finite elements of the primal mesh), as shown in Fig. 8.5.1. The line elements of the dual mesh are called *control domains*. Every control domain contains a node of the primal mesh. Then the integral statements of the governing equations are satisfied on every line element of the dual mesh. Since the control domain spans two adjacent elements, the satisfaction of the governing equations automatically relates the nodal values of the dependent unknowns at three consecutive nodes, leading to tridiagonal system of discretized equations as in the finite element method.

In the following subsections, we detail the discretization process for the mixed model of the CPT and the displacement model of the FST. For the purpose of readily seeing the meaning of the dual variables, we write the governing equations in terms of the strains $\hat{\varepsilon}_{rr}^{(0)}$, $\hat{\varepsilon}_{\theta\theta}^{(0)}$, $\hat{\varepsilon}_{\theta\theta}^{(1)}$, and M_{rr}. Ultimately, all equations expressed in terms of the unknowns of the model and the unknowns are approximated using linear finite element interpolation functions. For example, a typical unknown u is approximated over a typical finite element $\Omega_f^{(J)} = (r_J, r_{J+1})$ (the element $\Omega_f^{(J)}$ is on the right side of the node J) by

$$u(r) \approx U_J \, \psi_1^{(J)}(r) + U_{J+1} \, \psi_2^{(J)}(r) \tag{8.5.1}$$

where U_J is the value of u at node J (i.e., $U_J \approx u(r_J)$) and $\psi_i^{(J)}(r)$ $(i = 1, 2)$ are linear finite element interpolation functions of element $\Omega_f^{(J)}$ for $J = 1, 2, \ldots, N$, expressed in terms of the coordinate r (r has its origin at the center of the plate):

$$\psi_1^{(J)}(r) = \frac{r_{J+1} - r}{h_J}, \quad \psi_2^{(J)}(r) = \frac{r - r_J}{h_J} \tag{8.5.2}$$

Similar approximations are used for other dependent unknowns.

When the integral statements are evaluated at a boundary node, either the secondary variable or the corresponding primary variable is known at the node and, therefore, one need not express the secondary variables in terms of the gradients of the primary variables and approximate them in terms of the nodal values of the primary variables. In the interior of the domain, the secondary variables appearing in the integral statements for an interior node I are replaced in terms of the nodal values of the dependent unknowns using the finite element approximation of the form in Eq. (8.5.1), while linearizing the nonlinear terms. In this study, we assume that all of the stiffness coefficients A_{rr}, B_{rr}, and D_{rr} are constant (i.e., independent of position). The details are presented next.

8.5.2 Mixed Model of the Classical Plate Theory

The dual mesh finite domain statements of Eqs. (8.2.23)–(8.2.25) are obtained as follows:

$$
\begin{aligned}
0 &= -\int_{r_A^{(J)}}^{r_B^{(J)}} \frac{1}{r}\left[\frac{d}{dr}\left(r\bar{A}_{rr}\hat{\varepsilon}_{rr}^{(0)} + r\bar{B}_{rr}M_{rr}\right) - A_{rr}\,\hat{\varepsilon}_{\theta\theta}^{(0)} - B_{rr}\hat{\varepsilon}_{\theta\theta}^{(1)}\right] r dr \\
&= -\left[r\bar{A}_{rr}\hat{\varepsilon}_{rr}^{(0)} + r\bar{B}_{rr}M_{rr}\right]_{r_A^{(J)}}^{r_B^{(J)}} - \int_{r_A^{(J)}}^{r_B^{(J)}} \left(A_{rr}\,\hat{\varepsilon}_{\theta\theta}^{(0)} + B_{rr}\hat{\varepsilon}_{\theta\theta}^{(1)}\right) dr \\
&= -N_1^{(J)} - N_2^{(J)} - \int_{r_A^{(J)}}^{r_B^{(I)}} \left(A_{rr}\,\hat{\varepsilon}_{\theta\theta}^{(0)} + B_{rr}\hat{\varepsilon}_{\theta\theta}^{(1)}\right) dr
\end{aligned}
\tag{8.5.3}
$$

$$
\begin{aligned}
0 &= -\int_{r_A^{(J)}}^{r_B^{(J)}} \left\{\frac{1}{r}\frac{d}{dr}\left[\frac{d}{dr}(rM_{rr}) - B_{rr}\,\hat{\varepsilon}_{\theta\theta}^{(0)} - D_{rr}\,\hat{\varepsilon}_{\theta\theta}^{(1)} + \bar{A}_{rr}\,r\frac{dw}{dr}\hat{\varepsilon}_{rr}^{(0)}\right.\right. \\
&\qquad \left.\left. + \bar{B}_{rr}r\frac{dw}{dr}M_{rr}\right] + q\right\} r dr \\
&= -\left[\frac{d}{dr}(rM_{rr}) - B\,\hat{\varepsilon}_{\theta\theta}^{(0)} - D_{rr}\,\hat{\varepsilon}_{\theta\theta}^{(1)} + \bar{A}_{rr}\,r\frac{dw}{dr}\hat{\varepsilon}_{rr}^{(0)} + \bar{B}_{rr}\,r\frac{dw}{dr}M_{rr}\right]_{r_A^{(J)}}^{r_B^{(J)}} \\
&\qquad - \int_{r_A^{(J)}}^{r_B^{(J)}} q\, r dr \\
&= -V_1^{(J)} - V_2^{(J)} - \int_{r_A^{(J)}}^{r_B^{(J)}} q\, r dr
\end{aligned}
\tag{8.5.4}
$$

$$
\begin{aligned}
0 &= \int_{r_A^{(J)}}^{r_B^{(J)}} \left\{-\frac{1}{r}\frac{d}{dr}\left(r\frac{dw}{dr}\right) + (1-\nu)\frac{1}{r}\frac{dw}{dr} + \bar{B}_{rr}\left[\frac{du}{dr} + \frac{1}{2}\left(\frac{dw}{dr}\right)^2 + \nu\frac{u}{r}\right]\right. \\
&\qquad \left. - \frac{1}{D_{rr}}M_{rr}\right\} r dr \\
&= -\left[r\frac{dw}{dr}\right]_{r_A^{(I)}}^{r_B^{(I)}} + \int_{r_A^{(I)}}^{r_B^{(I)}} \left\{(1-\nu)\frac{dw}{dr} + r\bar{B}\left[\frac{du}{dr} + \frac{1}{2}\left(\frac{dw}{dr}\right)^2 + \nu\, u\right]\right. \\
&\qquad \left. - \frac{r}{D_{rr}}M_{rr}\right\} dr \\
&= -\Theta_1^{(J)} - \Theta_2^{(J)} + \int_{r_A^{(J)}}^{r_B^{(J)}} \left\{(1-\nu)\frac{dw}{dr} + r\bar{B}_{rr}\left[\frac{du}{dr} + \frac{1}{2}\left(\frac{dw}{dr}\right)^2 + \nu\, u\right]\right. \\
&\qquad \left. - \frac{r}{D_{rr}}M_{rr}\right\} dr
\end{aligned}
\tag{8.5.5}
$$

where the secondary variables N, V, and Θ are defined by

$$N_1^{(J)} = -(rN_{rr})\Big|_{r_A^{(J)}}, \qquad N_2^{(J)} = (rN_{rr})\Big|_{r_B^{(J)}} \tag{8.5.6a}$$

$$V_1^{(J)} = -(rV_r)\Big|_{r_A^{(J)}}, \qquad V_2^{(J)} = (rV_r)\Big|_{r_B^{(J)}} \tag{8.5.6b}$$

$$\Theta_1^{(J)} = -\left(r\frac{dw}{dr}\right)\Big|_{r_A^{(J)}}, \qquad \Theta_2^{(J)} = \left(r\frac{dw}{dr}\right)\Big|_{r_B^{(J)}} \tag{8.5.6c}$$

For a typical interior node I, the secondary variables in Eqs. (8.5.6a)–(8.5.6c) can be expressed in terms of the nodal values of u, w, and M_{rr}:

$$\begin{aligned} N_1^{(I)} &= -\bar{A}_{rr}\left(r_A^{(I)}\frac{U_I - U_{I-1}}{h_{I-1}} + 0.5r_A^{(I)}\,\delta W_{I-1}\frac{W_I - W_{I-1}}{h_{I-1}} + \nu\frac{U_I + U_{I-1}}{2}\right)\\ &\quad - r_A^{(I)}\,\bar{B}_{rr}\frac{M_I + M_{I-1}}{2} \end{aligned} \tag{8.5.7a}$$

$$\begin{aligned} N_2^{(I)} &= \bar{A}_{rr}\left(r_B^{(I)}\frac{U_{I+1} - U_I}{h_I} + 0.5r_B^{(I)}\,\delta W_I\frac{W_{I+1} - W_I}{h_I} + \nu\frac{U_{I+1} + U_I}{2}\right)\\ &\quad + r_B^{(I)}\,\bar{B}_{rr}\frac{M_{I+1} + M_I}{2} \end{aligned} \tag{8.5.7b}$$

$$\begin{aligned} V_1^{(I)} &= -\left[r_A^{(I)}\frac{M_I - M_{I-1}}{h_{I-1}} + (1-\nu)\frac{M_I + M_{I-1}}{2} + r_A^{(I)}\,\bar{B}_{rr}\,\delta W_{I-1}\frac{M_I + M_{I-1}}{2}\right.\\ &\quad \left. - (1-\nu^2)D_{rr}\left(\bar{B}_{rr}\frac{U_I + U_{I-1}}{2r_A^{(I)}} - \frac{W_I - W_{I-1}}{h_{I-1}r_A^{(I)}}\right)\right]\\ &\quad - \bar{A}_{rr}\,\delta W_{I-1}\left(r_A^{(I)}\frac{U_I - U_{I-1}}{h_{I-1}} + 0.5r_A^{(I)}\,\delta W_{I-1}\frac{W_I - W_{I-1}}{h_{I-1}} + \nu\frac{U_I + U_{I-1}}{2}\right) \end{aligned} \tag{8.5.7c}$$

$$\begin{aligned} V_2^{(I)} &= \left[r_B^{(I)}\frac{M_{I+1} - M_I}{h_I} + (1-\nu)\frac{M_{I+1} + M_I}{2}\right.\\ &\quad \left. - (1-\nu^2)D_{rr}\left(\bar{B}_{rr}\frac{U_{I+1} + U_I}{2r_B^{(I)}} - \frac{W_{I+1} - W_I}{h_I r_B^{(I)}}\right)\right]\\ &\quad + \bar{A}_{rr}\,\delta W_I\left(r_B^{(I)}\frac{U_{I+1} - U_I}{h_I} + 0.5r_B^{(I)}\,\delta W_I\frac{W_{I+1} - W_I}{h_I} + \nu\frac{U_{I+1} + U_I}{2}\right)\\ &\quad + r_B^{(I)}\,\bar{B}_{rr}\,\delta W_I\frac{M_{I+1} + M_I}{2} \end{aligned} \tag{8.5.7d}$$

$$\Theta_1^{(I)} = -r_A^{(I)}\frac{W_I - W_{I-1}}{h_{I-1}}, \quad \Theta_2^{(I)} = r_B^{(I)}\frac{W_{I+1} - W_I}{h_I} \tag{8.5.7e}$$

where the superscript (I) refers to the control domain $\Omega_c^{(I)}$, the subscript I

refers to the element number $\Omega_f^{(I)}$ (the control domain $\Omega_c^{(I)}$ partially occupies finite element $\Omega_f^{(I-1)}$ on the left and finite element $\Omega_f^{(I)}$ on the right), and

$$\delta W_{I-1} = \frac{\bar{W}_I - \bar{W}_{I-1}}{h_{I-1}}, \quad \delta W_I = \frac{\bar{W}_{I+1} - \bar{W}_I}{h_I} \tag{8.5.7f}$$

and $\bar{W}_{I-1}$ and $\bar{W}_I$ are the nodal values of w from the previous iteration (because of the linearization).

The following formulas are used in the development of the discretized equations:

$$[u(r)]_{r_A^{(I)}}^{r_B^{(I)}} = \tfrac{1}{2}\left(U_{I+1} - U_{I-1}\right) \tag{8.5.8a}$$

$$[u(r)]_{r=0.5h_1} = \tfrac{1}{2}(U_1 + U_2), \qquad [u(r)]_{r=R-0.5h_N} = \tfrac{1}{2}(U_N + U_{N+1}) \tag{8.5.8b}$$

$$\left[r\frac{du}{dr}\right]_{r_A^{(I)}}^{r_B^{(I)}} = \left(\frac{r_A^{(I)}}{h_{I-1}}\right)U_{I-1} - \left(\frac{r_A^{(I)}}{h_{I-1}} + \frac{r_B^{(I)}}{h_I}\right)U_I + \left(\frac{r_B^{(I)}}{h_I}\right)U_{I+1} \tag{8.5.8c}$$

$$\left[r\frac{du}{dr}\right]_{r=0.5h_1} = \tfrac{1}{2}(U_2 - U_1), \qquad [ru(r)]_{r=0.5h_1} = \tfrac{1}{4}h_1(U_1 + U_2), \tag{8.5.8d}$$

$$\left[r\frac{du}{dr}\right]_{r=R-0.5h_N} = \frac{(R-0.5h_N)}{h_N}(U_{N+1} - U_N) \tag{8.5.8e}$$

$$[ru(r)]_{r_A^{(I)}}^{r_B^{(I)}} = \tfrac{1}{2}\left[-r_A^{(I)}U_{I-1} + \left(r_B^{(I)} - r_A^{(I)}\right)U_I + r_B^{(I)}U_{I+1}\right] \tag{8.5.8f}$$

$$[ru(r)]_{r=R-0.5h_N} = \frac{(R-0.5h_N)}{2}(U_{N+1} + U_N) \tag{8.5.8g}$$

$$[r^2u(r)]_{r_A^{(I)}}^{r_B^{(I)}} = \tfrac{1}{2}\left\{-\left(r_A^{(I)}\right)^2 U_{I-1} + \left[\left(r_B^{(I)}\right)^2 - \left(r_A^{(I)}\right)^2\right]U_I + \left(r_B^{(I)}\right)^2 U_{I+1}\right\} \tag{8.5.8h}$$

$$[r^2u(r)]_{r=0.5h_1} = \tfrac{1}{8}h_1^2(U_1 + U_2), \quad \left[r^2\frac{du}{dr}\right]_{r=0.5h_1} = \tfrac{1}{4}h_1(U_2 - U_1), \tag{8.5.8i}$$

$$[r^2u(r)]_{r=R-0.5h_N} = \tfrac{1}{2}(R-0.5h_N)^2(U_N + U_{N+1}) \tag{8.5.8j}$$

$$\left[r^2\frac{du}{dr}\right]_{r_A^{(I)}}^{r_B^{(I)}} = \frac{\left(r_A^{(I)}\right)^2}{h_{I-1}}U_{I-1} - \frac{\left(r_B^{(I)}\right)^2}{h_I}U_I - \frac{\left(r_A^{(I)}\right)^2}{h_{I-1}}U_I + \frac{\left(r_B^{(I)}\right)^2}{h_I}U_{I+1} \tag{8.5.8k}$$

$$\left[r^2\frac{du}{dr}\right]_{r=R-0.5h_N} = \frac{(R-0.5h_N)^2}{h_N}(U_{N+1} - U_N) \tag{8.5.8l}$$

To obtain the discretized equations associated with Eqs. (8.5.3)–(8.5.5), we use the relations in Eqs. (8.5.7a)–(8.5.7e). With the relations in Eqs. (8.5.8a)–(8.5.8l), the discretized form of Eqs. (8.5.3)–(8.5.5) can be expressed in the form (we note that these are global equations and contain coefficients of nine variables associated with three consecutive nodes because the number of degrees of freedom per node is 3):

$$
\begin{aligned}
0 &= K^I_{I-3}\,\Delta_{I-3} + K^I_I\,\Delta_I + K^I_{I+3}\,\Delta_{I+3} + K^I_{I-2}\,\Delta_{I-2} + K^I_{I+1}\,\Delta_{I+1} \\
&\quad + K^I_{I+4}\,\Delta_{I+4} + K^I_{I-1}\,\Delta_{I-1} + K^I_{I+2}\,\Delta_{I+2} + K^I_{I+5}\,\Delta_{I+5}
\end{aligned} \tag{8.5.9}
$$

$$
\begin{aligned}
0 &= K^{I+1}_{I-3}\,\Delta_{I-3} + K^{I+1}_I\,\Delta_I + K^{I+1}_{I+3}\,\Delta_{I+3} + K^{I+1}_{I-2}\,\Delta_{I-2} + K^{I+1}_{I+1}\,\Delta_{I+1} \\
&\quad + K^{I+1}_{I+4}\,\Delta_{I+4} + K^{I+1}_{I-1}\,\Delta_{I-1} + K^{I+1}_{I+2}\,\Delta_{I+2} + K^{I+1}_{I+5}\,\Delta_{I+5} - F^{I+1}_{I+1}
\end{aligned} \tag{8.5.10}
$$

$$
\begin{aligned}
0 &= K^{I+2}_{I-3}\,\Delta_{I-3} + K^{I+2}_I\,\Delta_I + K^{I+2}_{I+3}\,\Delta_{I+3} + K^{I+2}_{I-2}\,\Delta_{I-2} + K^{I+2}_{I+1}\,\Delta_{I+1} \\
&\quad + K^{I+2}_{I+4}\,\Delta_{I+4} + K^{I+2}_{I-1}\,\Delta_{I-1} + K^{I+2}_{I+2}\,\Delta_{I+2} + K^{I+2}_{I+5}\,\Delta_{I+5}
\end{aligned} \tag{8.5.11}
$$

for $I = 4, 7, 10, \ldots, 3N$, where N is the number of elements (or $N+1$ is the number of nodes) in the mesh. Here $\boldsymbol{\Delta}$ is the vector of global primary degrees of freedom (see Fig. 8.5.2):

$$
\begin{aligned}
\Delta_{I-3} &= U_{J-1}, & \Delta_I &= U_J, & \Delta_{I+3} &= U_{J+1} \\
\Delta_{I-2} &= W_{J-1}, & \Delta_{I+1} &= W_J, & \Delta_{I+4} &= W_{J+1} \\
\Delta_{I-1} &= M_{J-1}, & \Delta_{I+2} &= M_J, & \Delta_{I+5} &= M_{J+1}
\end{aligned} \tag{8.5.12}
$$

and $K^I_K = K_{IK}$ are the stiffness coefficients defined by

$$
\begin{aligned}
K^I_{I-3} &= \bar{A}_{rr}\left[r\frac{d\psi_1^{(J-1)}}{dr} + \nu\psi_1^{(J-1)}\right]_{r=r_A^{(I)}} + \bar{A}_{rr}\int_{r_A^{(I)}}^{r_I}\left(\nu\frac{d\psi_1^{(J-1)}}{dr} + \frac{1}{r}\psi_1^{(J-1)}\right)dr \\
&\quad + \bar{B}_{rr}\,B_{rr}(1-\nu^2)\int_{r_A^{(I)}}^{r_I}\frac{1}{r}\psi_1^{(J-1)}\,dr
\end{aligned}
$$

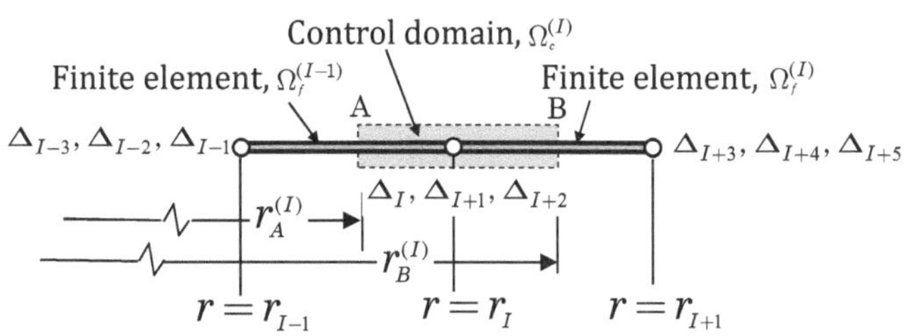

Fig. 8.5.2 The control domain $\Omega_c^{(I)}$ associated with an interior node I, the nodal coordinates, coordinates of the control volume interfaces, and the nodal degrees of freedom (3 degrees of freedom per node).

$$K_I^I = \bar{A}_{rr}\left[r\frac{d\psi_2^{(J-1)}}{dr} + \nu\psi_2^{(J-1)}\right]_{r=r_A^{(I)}} + \bar{A}_{rr}\int_{r_A^{(I)}}^{r_I}\left(\nu\frac{d\psi_2^{(J-1)}}{dr} + \frac{1}{r}\psi_2^{(J-1)}\right)dr$$

$$+ \bar{B}_{rr}\,B_{rr}(1-\nu^2)\int_{r_A^{(I)}}^{r_I}\frac{1}{r}\psi_2^{(J-1)}\,dr - \bar{A}_{rr}\left[r\frac{d\psi_1^{(J)}}{dr} + \nu\psi_1^{(J)}\right]_{r=r_B^{(I)}}$$

$$+ \bar{A}_{rr}\int_{r_I}^{r_B^{(I)}}\left(\nu\frac{d\psi_2^{(J)}}{dr} + \frac{1}{r}\psi_2^{(J)}\right)dr + \bar{B}_{rr}\,B(1-\nu^2)\int_{r_I}^{r_B^{(I)}}\frac{1}{r}\psi_2^{(J)}\,dr$$

$$K_{I+3}^I = -\bar{A}_{rr}\left[r\frac{d\psi_2^{(J)}}{dr} + \nu\psi_2^{(J)}\right]_{r=r_B^{(I)}} + \bar{A}\int_{r_I}^{r_B^{(I)}}\left(\nu\frac{d\psi_2^{(J)}}{dr} + \frac{1}{r}\psi_2^{(J)}\right)dr$$

$$+ \bar{B}_{rr}\,B(1-\nu^2)\int_{r_I}^{r_B^{(I)}}\frac{1}{r}\psi_2^{(J)}\,dr$$

$$K_{I-2}^I = \frac{1}{2}\bar{A}_{rr}\left[r\delta W_{J-1}\frac{d\psi_1^{(J-1)}}{dr}\right]_{r=r_A^{(I)}} + \frac{\nu}{2}\bar{A}\int_{r_A^{(I)}}^{r_I}\delta W_{J-1}\frac{d\psi_1^{(J-1)}}{dr}\,dr$$

$$- B_{rr}(1-\nu^2)\int_{r_A^{(I)}}^{r_I}\frac{1}{r}\frac{d\psi_1^{(J-1)}}{dr}\,dr$$

$$K_{I+1}^I = \frac{1}{2}\bar{A}_{rr}\left[r\delta W_{J-1}\frac{d\psi_2^{(J-1)}}{dr}\right]_{r=r_A^{(I)}} + \frac{\nu}{2}\bar{A}_{rr}\int_{r_A^{(I)}}^{r_I}\delta W_{J-1}\frac{d\psi_2^{(J-1)}}{dr}\,dr$$

$$- B_{rr}(1-\nu^2)\int_{r_A^{(I)}}^{r_I}\frac{1}{r}\frac{d\psi_2^{(J-1)}}{dr}\,dr - \frac{1}{2}\bar{A}_{rr}\left[r\delta W_J\frac{d\psi_1^{(J)}}{dr}\right]_{r=r_B^{(I)}}$$

$$+ \frac{\nu}{2}\bar{A}_{rr}\int_{r_I}^{r_B^{(I)}}\delta W_J\frac{d\psi_1^{(J)}}{dr}\,dr - B_{rr}(1-\nu^2)\int_{r_I}^{r_B^{(I)}}\frac{1}{r}\frac{d\psi_1^{(J)}}{dr}\,dr$$

$$K_{I+4}^I = -\frac{1}{2}\bar{A}_{rr}\left[r\delta W_J\frac{d\psi_2^{(J)}}{dr}\right]_{r=r_B^{(I)}} + \frac{\nu}{2}\bar{A}_{rr}\int_{r_I}^{r_B^{(I)}}\delta W_J\frac{d\psi_2^{(J)}}{dr}\,dr$$

$$- B_{rr}(1-\nu^2)\int_{r_I}^{r_B^{(I)}}\frac{1}{r}\frac{d\psi_2^{(J)}}{dr}\,dr$$

$$K_{I-1}^I = \bar{B}_{rr}\left\{\left[r\psi_1^{(J-1)}\right]_{r=r_A^{(I)}} + \nu\int_{r_A^{(I)}}^{r_I}\psi_1^{(J-1)}\,dr\right\}$$

$$K^I_{I+2} = \bar{B}_{rr}\left\{\left[r\psi_2^{(J-1)}\right]_{r=r_A^{(I)}} + \nu\int_{r_A^{(I)}}^{r_I}\psi_2^{(J-1)}\,dr - \left[r\psi_1^{(J)}\right]_{r=r_B^{(I)}}\right.$$
$$\left. + \nu\int_{r_I}^{r_B^{(I)}}\psi_1^{(J)}\,dr\right\}$$

$$K^I_{I+5} = \bar{B}_{rr}\left\{-\left[r\psi_2^{(J)}\right]_{r=r_B^{(I)}} + \nu\int_{r_I}^{r_B^{(I)}}\psi_2^{(J)}\,dr\right\}$$

$$K^{I+1}_{I-3} = -(1-\nu^2)D_{rr}\,\bar{B}_{rr}\left[\frac{1}{r}\psi_1^{(J-1)}\right]_{r=r_A^{(I)}}$$
$$+ \bar{A}_{rr}\left[\delta W_{J-1}\left(r\frac{d\psi_1^{(J-1)}}{dr} + \nu\psi_1^{(J-1)}\right)\right]_{r=r_A^{(I)}}$$

$$K^{I+1}_{I} = -(1-\nu^2)D_{rr}\,\bar{B}_{rr}\left[\frac{1}{r}\psi_2^{(J-1)}\right]_{r=r_A^{(I)}}$$
$$+ \bar{A}_{rr}\left[\delta W_{J-1}\left(r\frac{d\psi_2^{(J-1)}}{dr} + \nu\psi_2^{(J-1)}\right)\right]_{r=r_A^{(I)}}$$
$$+ (1-\nu^2)D_{rr}\,\bar{B}_{rr}\left[\frac{1}{r}\psi_1^{(J)}\right]_{r=r_B^{(I)}} - \bar{A}_{rr}\left[\delta W_J\left(r\frac{d\psi_1^{(J)}}{dr} + \nu\psi_1^{(J)}\right)\right]_{r=r_B^{(I)}}$$

$$K^{I+1}_{I+3} = (1-\nu^2)D_{rr}\,\bar{B}_{rr}\left[\frac{1}{r}\psi_2^{(J)}\right]_{r=r_B^{(I)}} - \bar{A}_{rr}\left[\delta W_J\left(r\frac{d\psi_2^{(J)}}{dr} + \nu\psi_2^{(J)}\right)\right]_{r=r_B^{(I)}}$$

$$K^{I+1}_{I-2} = (1-\nu^2)D_{rr}\left[\frac{1}{r}\frac{d\psi_1^{(J-1)}}{dr}\right]_{r=r_A^{(I)}} - \frac{1}{2}\bar{A}_{rr}\left[\left(\delta W_{J-1}\right)^2 r\frac{d\psi_1^{(J-1)}}{dr}\right]_{r=r_A^{(I)}}$$

$$K^{I+1}_{I+1} = (1-\nu^2)D_{rr}\left[\frac{1}{r}\frac{d\psi_2^{(J-1)}}{dr}\right]_{r=r_A^{(I)}} - \frac{1}{2}\bar{A}_{rr}\left[\left(\delta W_{J-1}\right)^2 r\frac{d\psi_2^{(J-1)}}{dr}\right]_{r=r_A^{(I)}}$$
$$- (1-\nu^2)D_{rr}\left[\frac{1}{r}\frac{d\psi_1^{(J)}}{dr}\right]_{r=r_B^{(I)}} - \frac{1}{2}\bar{A}_{rr}\left[\left(\delta W_J\right)^2 r\frac{d\psi_1^{(J)}}{dr}\right]_{r=r_B^{(I)}}$$

$$K^{I+1}_{I+4} = -(1-\nu^2)D_{rr}\left[\frac{1}{r}\frac{d\psi_2^{(J)}}{dr}\right]_{r=r_B^{(I)}} - \frac{1}{2}\bar{A}_{rr}\left[\left(\delta W_J\right)^2 r\frac{d\psi_2^{(J)}}{dr}\right]_{r=r_B^{(I)}}$$

$$K^{I+1}_{I-1} = \left[r\frac{d\psi_1^{(J-1)}}{dr} + (1-\nu)\psi_1^{(J-1)} - \bar{B}_{rr}\,r\delta W_{J-1}\,\psi_1^{(J-1)}\right]_{r=r_A^{(I)}}$$

$$
\begin{aligned}
K_{I+2}^{I+1} &= \left[r\frac{d\psi_2^{(J-1)}}{dr} + (1-\nu)\psi_2^{(J-1)} - \bar{B}_{rr}\, r\delta W_{J-1}\, \psi_2^{(J-1)} \right]_{r=r_A^{(I)}} \\
&\quad - \left[r\frac{d\psi_1^{(J)}}{dr} + (1-\nu)\psi_1^{(J)} + \bar{B}_{rr}\, r\delta W_J\, \psi_1^{(J)} \right]_{r=r_B^{(I)}} \\
K_{I+5}^{I+1} &= -\left[r\frac{d\psi_2^{(J)}}{dr} + (1-\nu)\psi_2^{(J)} + \bar{B}_{rr}\, r\delta W_J\, \psi_2^{(J)} \right]_{r=r_B^{(I)}} \\
K_{I-3}^{I+2} &= \bar{B}_{rr} \int_{r_A^{(I)}}^{r_I} \left[r\frac{d\psi_1^{(J-1)}}{dr} + \nu\psi_1^{(J-1)} \right] dr \\
K_{I}^{I+2} &= \bar{B}_{rr} \int_{r_A^{(I)}}^{r_I} \left[r\frac{d\psi_2^{(J-1)}}{dr} + \nu\psi_2^{(J-1)} \right] dr + \bar{B} \int_{r_I}^{r_B^{(I)}} \left[r\frac{d\psi_1^{(J)}}{dr} + \nu\psi_1^{(J)} \right] dr \\
K_{I+3}^{I+2} &= \bar{B}_{rr} \int_{r_I}^{r_B^{(I)}} \left[r\frac{d\psi_2^{(J)}}{dr} + \nu\psi_2^{(J)} \right] dr \\
K_{I-2}^{I+2} &= \left[r\frac{d\psi_1^{(J-1)}}{dr} \right]_{r=r_A^{(I)}} + \int_{r_A^{(I)}}^{r_I} \left[(1-\nu)\frac{d\psi_1^{(J-1)}}{dr} + \frac{1}{2}\bar{B}_{rr} r\delta W_{J-1}\, \frac{d\psi_1^{(J-1)}}{dr} \right] dr \\
K_{I+1}^{I+2} &= \left[r\frac{d\psi_2^{(J-1)}}{dr} \right]_{r=r_A^{(I)}} + \int_{r_A^{(I)}}^{r_I} \left[(1-\nu)\frac{d\psi_2^{(J-1)}}{dr} + \frac{1}{2}\bar{B}_{rr}\, r\delta W_{J-1}\, \frac{d\psi_2^{(J-1)}}{dr} \right] dr \\
&\quad - \left[r\frac{d\psi_1^{(J)}}{dr} \right]_{r=r_B^{(I)}} + \int_{r_I}^{r_B^{(I)}} \left[(1-\nu)\frac{d\psi_1^{(J)}}{dr} + \frac{1}{2}\bar{B}_{rr}\, r\delta W_J\, \frac{d\psi_1^{(J)}}{dr} \right] dr \\
K_{I+4}^{I+2} &= -\left[r\frac{d\psi_2^{(J)}}{dr} \right]_{r=r_B^{(I)}} + \int_{r_I}^{r_B^{(I)}} \left[(1-\nu)\frac{d\psi_2^{(J)}}{dr} + \frac{1}{2}\bar{B}_{rr} r\delta W_J\, \frac{d\psi_2^{(J)}}{dr} \right] dr \\
K_{I-1}^{I+2} &= -\frac{1}{D_{rr}} \int_{r_A^{(I)}}^{r_I} r\psi_1^{(J-1)}\, dr, \quad K_{I+5}^{I+2} = -\frac{1}{D_{rr}} \int_{r_I}^{r_B^{(I)}} r\psi_2^{(J)}\, dr \\
K_{I+2}^{I+2} &= -\frac{1}{D_{rr}} \left(\int_{r_A^{(I)}}^{r_I} r\psi_2^{(J-1)}\, dr + \int_{r_I}^{r_B^{(I)}} r\psi_1^{(J)}\, dr \right) \\
F^{I+1} &= \int_{r_A^{(I)}}^{r_I} rq(r)\, dr + \int_{r_I}^{r_B^{(I)}} rq(r)\, dr. \qquad (8.5.13)
\end{aligned}
$$

Here the superscripts and subscripts $(J-1)$ and J $(J = 2, 3, \ldots, N)$ refer to the element numbers on the left and right, respectively, of the node number J.

For boundary node 1, the discretized equations are

$$0 = -N_1^{(1)} + K_1^1\,\Delta_1 + K_4^1\,\Delta_4 + K_2^1\,\Delta_2 + K_5^1\,\Delta_5 + K_3^1\,\Delta_3 + K_6^1\,\Delta_6 \tag{8.5.14}$$

$$0 = -V_1^{(1)} + K_1^2\,\Delta_1 + K_4^2\,\Delta_4 + K_2^2\,\Delta_2 + K_5^2\,\Delta_5 + K_3^2\,\Delta_3 + K_6^2\,\Delta_6 - F^2 \tag{8.5.15}$$

$$0 = -\Theta_1^{(1)} + K_1^3\,\Delta_1 + K_4^3\,\Delta_I + K_2^3\,\Delta_2 + K_5^3\,\Delta_5 + K_3^3\,\Delta_3 + K_6^3\,\Delta_6 \tag{8.5.16}$$

where the coefficients K_J^I are defined by

$$K_1^1 = -\bar{A}_{rr}\left[r\frac{d\psi_1^{(1)}}{dr} + \nu\psi_1^{(1)}\right]_{r=0.5h_1} + \bar{A}^e\int_0^{0.5h_1}\left(\nu\frac{d\psi_1^{(1)}}{dr} + \frac{1}{r}\psi_1^{(1)}\right)dr$$
$$+ \bar{B}_{rr}\,B_{rr}(1-\nu^2)\int_0^{0.5h_1}\frac{1}{r}\psi_1^{(1)}\,dr$$

$$K_4^1 = -\bar{A}_{rr}\left[r\frac{d\psi_2^{(1)}}{dr} + \nu\psi_1^{(1)}\right]_{r=0.5h_1} + \bar{A}^e\int_0^{0.5h_1}\left(\nu\frac{d\psi_2^{(1)}}{dr} + \frac{1}{r}\psi_2^{(1)}\right)dr$$
$$+ \bar{B}_{rr}\,B_{rr}(1-\nu^2)\int_0^{0.5h_1}\frac{1}{r}\psi_2^{(1)}\,dr$$

$$K_2^1 = -\tfrac{1}{2}\bar{A}_{rr}\left[r\delta W\frac{d\psi_1^{(1)}}{dr}\right]_{r=0.5h_1} + \frac{\nu}{2}\bar{A}^e\int_0^{0.5h_1}\delta W\frac{d\psi_1^{(1)}}{dr}\,dr$$
$$- B_{rr}(1-\nu^2)\int_0^{0.5h_1}\frac{1}{r}\frac{d\psi_1^{(1)}}{dr}\,dr$$

$$K_5^1 = -\frac{1}{2}\bar{A}_{rr}^e\left[r\delta W\frac{d\psi_2^{(1)}}{dr}\right]_{r=0.5h_1} + \frac{\nu}{2}\bar{A}_{rr}^e\int_0^{0.5h_1}\delta W\frac{d\psi_2^{(1)}}{dr}\,dr$$
$$- B_{rr}(1-\nu^2)\int_0^{0.5h_1}\frac{1}{r}\frac{d\psi_2^{(1)}}{dr}\,dr$$

$$K_3^1 = -\bar{B}_{rr}\left\{\left[r\psi_1^{(1)}\right]_{r=0.5h_1} + \nu\int_0^{0.5h_1}\psi_1^{(1)}\,dr\right\}$$

$$K_6^1 = -\bar{B}_{rr}\left\{\left[r\psi_2^{(1)}\right]_{r=0.5h_1} + \nu\int_0^{0.5h_1}\psi_2^{(1)}\,dr\right\}$$

$$K_1^2 = (1-\nu^2)D_{rr}^e\,\bar{B}_{rr}\left[\frac{1}{r}\psi_1^{(1)}\right]_{r=0.5h_1} - \bar{A}_{rr}\left[\delta W\left(r\frac{d\psi_1^{(1)}}{dr} + \nu\psi_1^{(1)}\right)\right]_{r=0.5h_1}$$

$$
\begin{aligned}
K_4^2 &= (1-\nu^2)D_{rr}^e\,\bar{B}_{rr}\left[\frac{1}{r}\psi_2^{(1)}\right]_{r=0.5h_1} - \bar{A}_{rr}\left[\delta W\left(r\frac{d\psi_2^{(1)}}{dr}+\nu\psi_2^{(1)}\right)\right]_{r=0.5h_1}\\
K_2^2 &= -(1-\nu^2)D_{rr}\left[\frac{1}{r}\frac{d\psi_1^{(1)}}{dr}\right]_{r=0.5h_1} - \bar{A}_{rr}\left[\left(\delta W\right)^2 r\frac{d\psi_1^{(1)}}{dr}\right]_{r=0.5h_1}\\
K_5^2 &= -(1-\nu^2)D_{rr}\left[\frac{1}{r}\frac{d\psi_2^{(1)}}{dr}\right]_{r=0.5h_1} - \bar{A}_{rr}\left[\left(\delta W\right)^2 r\frac{d\psi_2^{(1)}}{dr}\right]_{r=0.5h_1}\\
K_3^2 &= -\left[r\frac{d\psi_1^{(1)}}{dr}+(1-\nu)\psi_1^{(1)}+\bar{B}_{rr}\,r\delta W\,\psi_1^{(1)}\right]_{r=0.5h_1}\\
K_6^2 &= -\left[r\frac{d\psi_2^{(1)}}{dr}+(1-\nu)\psi_2^{(1)}+\bar{B}_{rr}\,r\delta W\,\psi_2^{(1)}\right]_{r=0.5h_1}\\
K_1^3 &= \bar{B}_{rr}\int_0^{0.5h_1}\left[r\frac{d\psi_1^{(1)}}{dr}+\nu\psi_1^{(1)}\right]dr,\quad K_4^3 = \bar{B}_{rr}\int_0^{0.5h_1}\left[r\frac{d\psi_2^{(1)}}{dr}+\nu\psi_2^{(1)}\right]dr\\
K_2^3 &= -\left[r\frac{d\psi_1^{(1)}}{dr}\right]_{r=0.5h_1}+\int_0^{0.5h_1}\left[(1-\nu)\frac{d\psi_1^{(1)}}{dr}+\frac{1}{2}\bar{B}_{rr}\,r\delta W\,\frac{d\psi_1^{(1)}}{dr}\right]dr\\
K_5^3 &= \left[r\frac{d\psi_2^{(1)}}{dr}\right]_{r=0.5h_1}+\int_0^{0.5h_1}\left[(1-\nu)\frac{d\psi_2^{(1)}}{dr}+\frac{1}{2}\bar{B}_{rr}\,r\delta W\,\frac{d\psi_2^{(1)}}{dr}\right]dr\\
K_3^3 &= -\frac{1}{D_{rr}}\int_0^{0.5h_1} r\psi_1^{(1)}\,dr,\quad K_6^3 = -\frac{1}{D_{rr}}\int_0^{0.5h_1} r\psi_2^{(1)}\,dr\\
F^2 &= \int_0^{0.5h_1} rq(r)\,dr+\int_0^{0.5h_1} rq(r)\,dr,\quad \delta W=\frac{W_2-W_1}{h_1}
\end{aligned}
\tag{8.5.17}
$$

8.5.3 Displacement Model of the FST

The dual mesh finite domain statements of Eqs. (8.3.13)–(8.3.15) are obtained as follows:

$$
0=\int_{r_A^{(J)}}^{r_B^{(J)}}\left\{\frac{1}{r}\left[-\frac{d}{dr}\left(rN_{rr}\right)+N_{\theta\theta}\right]\right\}rdr=-N_1^{(J)}-N_2^{(J)}+\int_{r_A^{(J)}}^{r_B^{(J)}}N_{\theta\theta}\,dr
\tag{8.5.18}
$$

$$
0=\int_{r_A^{(J)}}^{r_B^{(J)}}\left[-\frac{1}{r}\frac{d}{dr}\left(rV_r\right)-q\right]rdr=-V_1^{(J)}-V_2^{(J)}-\int_{r_A^{(J)}}^{r_B^{(J)}}rq(r)\,dr
\tag{8.5.19}
$$

$$
\begin{aligned}
0&=\int_{r_A^{(J)}}^{r_B^{(J)}}\left\{\frac{1}{r}\left[-\frac{d}{dr}\left(rM_{rr}\right)+M_{\theta\theta}\right]+Q_r\right\}rdr\\
&=-M_1^{(J)}-M_2^{(J)}+\int_{r_A^{(J)}}^{r_B^{(J)}}\left(M_{\theta\theta}+rQ_r\right)dr
\end{aligned}
\tag{8.5.20}
$$

where the secondary variables N, V, and M are defined by

$$N_1^{(J)} = -(rN_{rr})\Big|_{r_A^{(J)}}, \qquad N_2^{(J)} = (rN_{rr})\Big|_{r_B^{(J)}} \tag{8.5.21a}$$

$$V_1^{(J)} = -(rV_r)\Big|_{r_A^{(J)}}, \qquad V_2^{(J)} = (rV_r)\Big|_{r_B^{(J)}} \tag{8.5.21b}$$

$$M_1^{(J)} = -\left(rM_{rr}\right)\Big|_{r_A^{(J)}}, \qquad M_2^{(J)} = \left(rM_{rr}\right)\Big|_{r_B^{(J)}} \tag{8.5.21c}$$

For a typical interior node I, the secondary variables in Eqs. (8.5.21a)–(8.5.21c) can be expressed in terms of the nodal values of u, w, and ϕ_r. In discretizing the expressions for $V_1^{(I)}$ and $V_2^{(I)}$, the coefficients of ϕ_r corresponding to the integral

$$\int_{r_A^{(I)}}^{r_B^{(I)}} r\, S_{rz}\left(\phi_r + \frac{dw}{dr}\right) dr$$

are evaluated by considering ϕ_r to be constant within each finite element (of the primal mesh). In other words, ϕ_r in element $\Omega^{(I-1)}$ is replaced with $\phi_r = \frac{\Phi_I+\Phi_{I-1}}{2}$, while in element $\Omega^{(I)}$ it is replaced with $\phi_r = \frac{\Phi_{I+1}+\Phi_I}{2}$. This is done to remedy the shear locking [42]. Thus, we have

$$\begin{aligned} N_1^{(I)} &= -A_{rr}\left(r_A^{(I)}\frac{U_I - U_{I-1}}{h_{I-1}} + 0.5r_A^{(I)}\,\delta W_{I-1}\frac{W_I - W_{I-1}}{h_{I-1}} + \nu\frac{U_I + U_{I-1}}{2}\right)\\ &\quad - B_{rr}\left(r_A^{(I)}\frac{\Phi_I - \Phi_{I-1}}{h_{I-1}} + \nu\frac{\Phi_I + \Phi_{I-1}}{2}\right) \end{aligned} \tag{8.5.22a}$$

$$\begin{aligned} N_2^{(I)} &= A_{rr}\left(r_B^{(I)}\frac{U_{I+1} - U_I}{h_I} + 0.5r_B^{(I)}\,\delta W_I\frac{W_{I+1} - W_I}{h_I} + \nu\frac{U_{I+1} + U_I}{2}\right)\\ &\quad + B_{rr}\left(r_B^{(I)}\frac{\Phi_{I+1} - \Phi_I}{h_I} + \nu\frac{\Phi_{I+1} + \Phi_I}{2}\right) \end{aligned} \tag{8.5.22b}$$

$$\begin{aligned} V_1^{(I)} &= -\Bigg\{r_A^{(I)}\,S_{rz}\left(\frac{\Phi_I + \Phi_{I-1}}{2} + \frac{W_I - W_{I-1}}{h_{I-1}}\right)\\ &\quad + A_{rr}\,\delta W_{I-1}\left[r_A^{(I)}\frac{U_I - U_{I-1}}{h_{I-1}} + 0.5r_A^{(I)}\,\delta W_{I-1}\,\frac{W_I - W_{I-1}}{h_{I-1}} + \nu\frac{U_I + U_{I-1}}{2}\right]\\ &\quad + B_{rr}\,\delta W_{I-1}\left(r_A^{(I)}\frac{\Phi_I - \Phi_{I-1}}{h_{I-1}} + \nu\frac{\Phi_I + \Phi_{I-1}}{2}\right)\Bigg\} \end{aligned} \tag{8.5.22c}$$

$$\begin{aligned} V_2^{(I)} &= \Bigg\{r_B^{(I)}\,S_{rz}\left(\frac{\Phi_I + \Phi_{I+1}}{2} + \frac{W_{I+1} - W_I}{h_I}\right)\\ &\quad + A_{rr}\,\delta W_I\left[r_B^{(I)}\frac{U_{I+1} - U_I}{h_I} + 0.5r_B^{(I)}\,\delta W_I\,\frac{W_{I+1} - W_I}{h_I} + \nu\frac{U_I + U_{I+1}}{2}\right]\\ &\quad + B_{rr}\,\delta W_I\left(r_B^{(I)}\frac{\Phi_{I+1} - \Phi_I}{h_I} + \nu\frac{\Phi_I + \Phi_{I+1}}{2}\right)\Bigg\} \end{aligned} \tag{8.5.22d}$$

$$
\begin{aligned}
M_1^{(I)} &= -B_{rr}\left(r_A^{(I)}\frac{U_I - U_{I-1}}{h_{I-1}} + 0.5r_A^{(I)}\,\delta W_{I-1}\frac{W_I - W_{I-1}}{h_{I-1}} + \nu\frac{U_I + U_{I-1}}{2}\right)\\
&\quad - D_{rr}\left(r_A^{(I)}\frac{\Phi_I - \Phi_{I-1}}{h_{I-1}} + \nu\frac{\Phi_I + \Phi_{I-1}}{2}\right) \qquad (8.5.22e)\\
M_2^{(I)} &= B_{rr}\left(r_B^{(I)}\frac{U_{I+1} - U_I}{h_I} + 0.5r_B^{(I)}\,\delta W_I\frac{W_{I+1} - W_I}{h_I} + \nu\frac{U_{I+1} + U_I}{2}\right)\\
&\quad + D_{rr}\left(r_B^{(I)}\frac{\Phi_{I+1} - \Phi_I}{h_I} + \nu\frac{\Phi_{I+1} + \Phi_I}{2}\right) \qquad (8.5.22f)
\end{aligned}
$$

The discretized form of Eqs. (8.5.18)–(8.5.20) can be expressed in the same form as Eqs. (8.5.9)–(8.5.11), with $\mathbf{\Delta}$ defined by

$$
\begin{aligned}
\Delta_{I-3} &= U_{J-1}, & \Delta_I &= U_J, & \Delta_{I+3} &= U_{J+1}\\
\Delta_{I-2} &= W_{J-1}, & \Delta_{I+1} &= W_J, & \Delta_{I+4} &= W_{J+1}\\
\Delta_{I-1} &= \Phi_{J-1}, & \Delta_{I+2} &= \Phi_J, & \Delta_{I+5} &= \Phi_{J+1}
\end{aligned} \qquad (8.5.23)
$$

The coefficients $K_K^I = K_{IK}$ for this model are defined by

$$
\begin{aligned}
K_{I-3}^I &= A_{rr}\left[r\frac{d\psi_1^{(J-1)}}{dr} + \nu\psi_1^{(J-1)}\right]_{r=r_A^{(I)}} + A_{rr}\int_{r_A^{(I)}}^{r_I}\left(\nu\frac{d\psi_1^{(J-1)}}{dr} + \frac{1}{r}\psi_1^{(J-1)}\right)dr\\
K_I^I &= A_{rr}\left[r\frac{d\psi_2^{(J-1)}}{dr} + \nu\psi_2^{(J-1)}\right]_{r=r_A^{(I)}} + A_{rr}\int_{r_A^{(I)}}^{r_I}\left(\nu\frac{d\psi_2^{(J-1)}}{dr} + \frac{1}{r}\psi_2^{(J-1)}\right)dr\\
&\quad - A_{rr}\left[r\frac{d\psi_1^{(J)}}{dr} + \nu\psi_1^{(J)}\right]_{r=r_B^{(I)}} + A_{rr}\int_{r_I}^{r_B^{(I)}}\left(\nu\frac{d\psi_1^{(J)}}{dr} + \frac{1}{r}\psi_1^{(J)}\right)dr\\
K_{I+3}^I &= -A_{rr}\left[r\frac{d\psi_2^{(J)}}{dr} + \nu\psi_2^{(J)}\right]_{r=r_B^{(I)}} + A_{rr}\int_{r_I}^{r_B^{(I)}}\left(\nu\frac{d\psi_2^{(J)}}{dr} + \frac{1}{r}\psi_2^{(J)}\right)dr\\
K_{I-2}^I &= \frac{1}{2}A_{rr}\left[r\delta W_{J-1}\frac{d\psi_1^{(J-1)}}{dr}\right]_{r=r_A^{(I)}} + \frac{\nu}{2}A_{rr}\int_{r_A^{(I)}}^{r_I}\delta W_{J-1}\frac{d\psi_1^{(J-1)}}{dr}\,dr\\
K_{I+1}^I &= \frac{1}{2}A_{rr}\left[r\delta W_{J-1}\frac{d\psi_2^{(J-1)}}{dr}\right]_{r=r_A^{(I)}} + \frac{\nu}{2}A_{rr}\int_{r_A^{(I)}}^{r_I}\delta W_{J-1}\frac{d\psi_2^{(J-1)}}{dr}\,dr\\
&\quad - \frac{1}{2}A_{rr}\left[r\delta W_J\frac{d\psi_1^{(J)}}{dr}\right]_{r=r_B^{(I)}} + \frac{\nu}{2}A_{rr}\int_{r_I}^{r_B^{(I)}}\delta W_J\frac{d\psi_1^{(J)}}{dr}\,dr\\
K_{I+4}^I &= -\frac{1}{2}A_{rr}\left[r\delta W_J\frac{d\psi_2^{(J)}}{dr}\right]_{r=r_B^{(I)}} + \frac{\nu}{2}A_{rr}\int_{r_I}^{r_B^{(I)}}\delta W_J\frac{d\psi_2^{(J)}}{dr}\,dr
\end{aligned}
$$

$$K_{I-1}^{I} = B_{rr}\left[r\frac{d\psi_1^{(J-1)}}{dr} + \nu\psi_1^{(J-1)}\right]_{r=r_A^{(I)}} + B_{rr}\int_{r_A^{(I)}}^{r_I}\left(\nu\frac{d\psi_1^{(J-1)}}{dr} + \frac{1}{r}\psi_1^{(J-1)}\right)dr$$

$$\begin{aligned}K_{I+2}^{I} = {} & B_{rr}\left[r\frac{d\psi_2^{(J-1)}}{dr} + \nu\psi_2^{(J-1)}\right]_{r=r_A^{(I)}} + B_{rr}\int_{r_A^{(I)}}^{r_I}\left(\nu\frac{d\psi_2^{(J-1)}}{dr} + \frac{1}{r}\psi_2^{(J-1)}\right)dr \\ & - B_{rr}\left[r\frac{d\psi_1^{(J)}}{dr} + \nu\psi_1^{(J)}\right]_{r=r_B^{(I)}} + B_{rr}\int_{r_I}^{r_B^{(I)}}\left(\nu\frac{d\psi_1^{(J)}}{dr} + \frac{1}{r}\psi_1^{(J)}\right)dr\end{aligned}$$

$$K_{I+5}^{I} = -B_{rr}\left[r\frac{d\psi_2^{(J)}}{dr} + \nu\psi_2^{(J)}\right]_{r=r_B^{(I)}} + B_{rr}\int_{r_I}^{r_B^{(I)}}\left(\nu\frac{d\psi_2^{(J)}}{dr} + \frac{1}{r}\psi_2^{(J)}\right)dr$$

$$K_{I-3}^{I+1} = A_{rr}\left[\delta W_{J-1}\left(r\frac{d\psi_1^{(J-1)}}{dr} + \nu\psi_1^{(J-1)}\right)\right]_{r=r_A^{(I)}}$$

$$\begin{aligned}K_{I}^{I+1} = {} & A_{rr}\left[\delta W_{J-1}\left(r\frac{d\psi_2^{(J-1)}}{dr} + \nu\psi_2^{(J-1)}\right)\right]_{r=r_A^{(I)}} \\ & - A_{rr}\left[\delta W_{J}\left(r\frac{d\psi_1^{(J)}}{dr} + \nu\psi_1^{(J)}\right)\right]_{r=r_B^{(I)}}\end{aligned}$$

$$K_{I+3}^{I+1} = -A_{rr}\left[\delta W_{J}\left(r\frac{d\psi_2^{(J)}}{dr} + \nu\psi_2^{(J)}\right)\right]_{r=r_B^{(I)}}$$

$$K_{I-2}^{I+1} = S_{rz}\left[r\frac{d\psi_1^{(J-1)}}{dr}\right]_{r=r_A^{(I)}} + \frac{1}{2}A_{rr}\left[\left(\delta W_{J-1}\right)^2 r\frac{d\psi_1^{(J-1)}}{dr}\right]_{r=r_A^{(I)}}$$

$$\begin{aligned}K_{I+1}^{I+1} = {} & S_{rz}\left[r\frac{d\psi_2^{(J-1)}}{dr}\right]_{r=r_A^{(I)}} + \frac{1}{2}A_{rr}\left[\left(\delta W_{J-1}\right)^2 r\frac{d\psi_2^{(J-1)}}{dr}\right]_{r=r_A^{(I)}} \\ & - S_{rz}\left[r\frac{d\psi_1^{(J)}}{dr}\right]_{r=r_B^{(I)}} - \frac{1}{2}A_{rr}\left[\left(\delta W_{J}\right)^2 r\frac{d\psi_1^{(J)}}{dr}\right]_{r=r_B^{(I)}}\end{aligned}$$

$$K_{I+4}^{I+1} = -S_{rz}\left[r\frac{d\psi_2^{(J)}}{dr}\right]_{r=r_B^{(I)}} - \frac{1}{2}A_{rr}\left[\left(\delta W_{J}\right)^2 r\frac{d\psi_2^{(J)}}{dr}\right]_{r=r_B^{(I)}}$$

$$K_{I-1}^{I+1} = \left[S_{rz}r\psi_1^{(J-1)} + B_{rr}\,\delta W_{J-1}\left(r\frac{d\psi_1^{(J-1)}}{dr} + \nu\psi_1^{(J-1)}\right)\right]_{r=r_A^{(I)}}$$

$$\begin{aligned}K_{I+2}^{I+1} = {} & \left[S_{rz}r\psi_2^{(J-1)} + B_{rr}\,\delta W_{J-1}\left(r\frac{d\psi_2^{(J-1)}}{dr} + \nu\psi_2^{(J-1)}\right)\right]_{r=r_A^{(I)}} \\ & - \left[S_{rz}r\psi_1^{(J)} + B_{rr}\,\delta W_{J}\left(r\frac{d\psi_1^{(J)}}{dr} + \nu\psi_1^{(J)}\right)\right]_{r=r_B^{(I)}}\end{aligned}$$

$$K_{I+5}^{I+1} = -\left[S_{rz} r\psi_2^{(J)} + B_{rr}\,\delta W_J\left(r\frac{d\psi_2^{(J)}}{dr} + \nu\psi_2^{(J)}\right)\right]_{r=r_B^{(I)}}$$

$$K_{I-3}^{I+2} = B_{rr}\left[r\frac{d\psi_1^{(J-1)}}{dr} + \nu\psi_1^{(J-1)}\right]_{r=r_A^{(I)}} + B_{rr}\int_{r_A^{(I)}}^{r_I}\left(\nu\frac{d\psi_1^{(J-1)}}{dr} + \frac{1}{r}\psi_1^{(J-1)}\right)dr$$

$$K_{I}^{I+2} = B_{rr}\left[r\frac{d\psi_2^{(J-1)}}{dr} + \nu\psi_2^{(J-1)}\right]_{r=r_A^{(I)}} + B_{rr}\int_{r_A^{(I)}}^{r_I}\left(\nu\frac{d\psi_2^{(J-1)}}{dr} + \frac{1}{r}\psi_2^{(J-1)}\right)dr$$

$$- B_{rr}\left[r\frac{d\psi_1^{(J)}}{dr} + \nu\psi_1^{(J)}\right]_{r=r_B^{(I)}} + B\int_{r_I}^{r_B^{(I)}}\left(\nu\frac{d\psi_2^{(J)}}{dr} + \frac{1}{r}\psi_2^{(J)}\right)dr$$

$$K_{I+3}^{I+2} = -B_{rr}\left[r\frac{d\psi_2^{(J)}}{dr} + \nu\psi_2^{(J)}\right]_{r=r_B^{(I)}} + B_{rr}\int_{r_I}^{r_B^{(I)}}\left(\nu\frac{d\psi_2^{(J)}}{dr} + \frac{1}{r}\psi_2^{(J)}\right)dr$$

$$K_{I-2}^{I+2} = \frac{1}{2}B_{rr}\left[r\delta W_{J-1}\frac{d\psi_1^{(J-1)}}{dr}\right]_{r=r_A^{(I)}} + \frac{\nu}{2}B_{rr}\int_{r_A^{(I)}}^{r_I}\delta W_{J-1}\frac{d\psi_1^{(J-1)}}{dr}\,dr$$

$$+ S_{rz}\int_{r_A^{(I)}}^{r_I}\frac{d\psi_1^{(J-1)}}{dr}\,rdr$$

$$K_{I+1}^{I+2} = \frac{1}{2}B_{rr}\left[r\delta W_{J-1}\frac{d\psi_2^{(J-1)}}{dr}\right]_{r=r_A^{(I)}} + \frac{\nu}{2}B_{rr}\int_{r_A^{(I)}}^{r_I}\delta W_{J-1}\frac{d\psi_2^{(J-1)}}{dr}\,dr$$

$$- \frac{1}{2}B_{rr}\left[r\delta W_J\frac{d\psi_1^{(J)}}{dr}\right]_{r=r_B^{(I)}} + \frac{\nu}{2}B_{rr}\int_{r_I}^{r_B^{(I)}}\delta W_J\frac{d\psi_1^{(J)}}{dr}\,dr$$

$$+ S_{rz}\left(\int_{r_A^{(I)}}^{r_I}\frac{d\psi_2^{(J-1)}}{dr}\,rdr + \int_{r_I}^{r_A^{(I)}}\frac{d\psi_1^{(J)}}{dr}\,rdr\right)$$

$$K_{I+4}^{I+2} = -\frac{1}{2}B_{rr}\left[r\delta W_J\frac{d\psi_2^{(J)}}{dr}\right]_{r=r_B^{(I)}} + \frac{\nu}{2}B_{rr}\int_{r_I}^{r_B^{(I)}}\delta W_J\frac{d\psi_2^{(J)}}{dr}\,dr$$

$$+ S_{rz}\int_{r_I}^{r_A^{(I)}}\frac{d\psi_2^{(J)}}{dr}\,rdr$$

$$K_{I-1}^{I+2} = D_{rr}\left[r\frac{d\psi_1^{(J-1)}}{dr} + \nu\psi_1^{(J-1)}\right]_{r=r_A^{(I)}} + D_{rr}\int_{r_A^{(I)}}^{r_I}\left(\nu\frac{d\psi_1^{(J-1)}}{dr} + \frac{1}{r}\psi_1^{(J-1)}\right)dr$$

$$+ S_{rz}\int_{r_A^{(I)}}^{r_I}\psi_1^{(J-1)}\,rdr$$

$$K_{I+2}^{I+2} = D_{rr}\left[r\frac{d\psi_2^{(J-1)}}{dr} + \nu\psi_2^{(J-1)}\right]_{r=r_A^{(I)}} + D_{rr}\int_{r_A^{(I)}}^{r_I}\left(\nu\frac{d\psi_2^{(J-1)}}{dr} + \frac{1}{r}\psi_2^{(J-1)}\right)dr$$

$$- D_{rr}\left[r\frac{d\psi_1^{(J)}}{dr} + \nu\psi_1^{(J)}\right]_{r=r_B^{(I)}} + D_{rr}\int_{r_I}^{r_B^{(I)}}\left(\nu\frac{d\psi_1^{(J)}}{dr} + \frac{1}{r}\psi_1^{(J)}\right)dr$$

$$+ S_{rz}\left(\int_{r_A^{(I)}}^{r_I}\psi_2^{(J-1)}\,rdr + \int_{r_I}^{r_A^{(I)}}\psi_1^{(J)}\,rdr\right)$$

$$K_{I+5}^{I+2} = -D_{rr}\left[r\frac{d\psi_2^{(J)}}{dr} + \nu\psi_2^{(J)}\right]_{r=r_B^{(I)}} + D_{rr}\int_{r_I}^{r_B^{(I)}}\left(\nu\frac{d\psi_2^{(J)}}{dr} + \frac{1}{r}\psi_2^{(J)}\right)dr$$

$$+ S_{rz}\int_{r_I}^{r_A^{(I)}}\psi_2^{(J)}\,rdr$$

$$F^{I+1} = \int_{r_A^{(I)}}^{r_I} rq(r)\,dr + \int_{r_I}^{r_B^{(I)}} rq(r)\,dr \tag{8.5.24}$$

Here the superscripts and subscripts $(J-1)$ and J $(J = 2, 3, \ldots, N)$ refer to the element numbers on the left and right, respectively, of the node number J.

For boundary node 1, the discretized equations are

$$0 = -N_1^{(1)} + K_1^1\,\Delta_1 + K_4^1\,\Delta_4 + K_2^1\,\Delta_2 + K_5^1\,\Delta_5 + K_3^1\,\Delta_3 + K_6^1\,\Delta_6 \tag{8.5.25}$$

$$0 = -V_1^{(1)} + K_1^2\,\Delta_1 + K_4^2\,\Delta_4 + K_2^2\,\Delta_2 + K_5^2\,\Delta_5 + K_3^2\,\Delta_3 + K_6^2\,\Delta_6 - F^2 \tag{8.5.26}$$

$$0 = -M_1^{(1)} + K_1^3\,\Delta_1 + K_4^3\,\Delta_I + K_2^3\,\Delta_2 + K_5^3\,\Delta_5 + K_3^3\,\Delta_3 + K_6^3\,\Delta_6 \tag{8.5.27}$$

where K_J^I are defined by

$$K_1^1 = -A_{rr}\left[r\frac{d\psi_1^{(1)}}{dr} + \nu\psi_1^{(1)}\right]_{r=0.5h_1} + A_{rr}\int_0^{0.5h_1}\left(\nu\frac{d\psi_1^{(1)}}{dr} + \frac{1}{r}\psi_1^{(1)}\right)dr$$

$$K_4^1 = -A_{rr}\left[r\frac{d\psi_2^{(1)}}{dr} + \nu\psi_1^{(1)}\right]_{r=0.5h_1} + A_{rr}\int_0^{0.5h_1}\left(\nu\frac{d\psi_2^{(1)}}{dr} + \frac{1}{r}\psi_2^{(1)}\right)dr$$

$$K_2^1 = -\tfrac{1}{2}A_{rr}\left[r\delta W\frac{d\psi_1^{(1)}}{dr}\right]_{r=0.5h_1} + \tfrac{1}{2}\nu A_{rr}\int_0^{0.5h_1}\delta W\frac{d\psi_1^{(1)}}{dr}\,dr$$

$$K_5^1 = -\tfrac{1}{2}A_{rr}\left[r\delta W\frac{d\psi_2^{(1)}}{dr}\right]_{r=0.5h_1} + \tfrac{1}{2}\nu A_{rr}\int_0^{0.5h_1}\delta W\frac{d\psi_2^{(1)}}{dr}\,dr$$

$$K_3^1 = -B_{rr}\left[r\frac{d\psi_1^{(1)}}{dr} + \nu\psi_1^{(1)}\right]_{r=0.5h_1} + B_{rr}\int_0^{0.5h_1}\left(\nu\frac{d\psi_1^{(1)}}{dr} + \frac{1}{r}\psi_1^{(1)}\right)dr$$

$$
\begin{aligned}
K_6^1 &= -B_{rr}\left[r\frac{d\psi_2^{(1)}}{dr} + \nu\psi_1^{(1)}\right]_{r=0.5h_1} + B_{rr}\int_0^{0.5h_1}\left(\nu\frac{d\psi_2^{(1)}}{dr} + \frac{1}{r}\psi_2^{(1)}\right)dr\\
K_1^2 &= -A_{rr}\left[\delta W\left(r\frac{d\psi_1^{(1)}}{dr} + \nu\psi_1^{(1)}\right)\right]_{r=0.5h_1}\\
K_4^2 &= -A_{rr}\left[\delta W\left(r\frac{d\psi_2^{(1)}}{dr} + \nu\psi_2^{(1)}\right)\right]_{r=0.5h_1}\\
K_2^2 &= -\left[S_{rz}r\frac{d\psi_1^{(1)}}{dr} + \tfrac{1}{2}A_{rr}\left(\delta W\right)^2 r\frac{d\psi_1^{(1)}}{dr}\right]_{r=0.5h_1}\\
K_5^2 &= \left[-S_{rz}r\frac{d\psi_2^{(1)}}{dr} + \tfrac{1}{2}A_{rr}\left(\delta W\right)^2 r\frac{d\psi_2^{(1)}}{dr}\right]_{r=0.5h_1}\\
K_3^2 &= -\left[S_{rz}r\psi_1^{(1)} + B_{rr}\,\delta W\left(r\frac{d\psi_1^{(1)}}{dr} + \nu\psi_1^{(1)}\right)\right]_{r=0.5h_1}\\
K_6^2 &= -\left[S_{rz}r\psi_2^{(1)} + B_{rr}\,\delta W\left(r\frac{d\psi_2^{(1)}}{dr} + \nu\psi_2^{(1)}\right)\right]_{r=0.5h_1}\\
K_1^3 &= -B_{rr}\left[r\frac{d\psi_1^{(1)}}{dr} + \nu\psi_1^{(1)}\right]_{r=0.5h_1} + B_{rr}\int_0^{0.5h_1}\left(\nu\frac{d\psi_1^{(1)}}{dr} + \frac{1}{r}\psi_1^{(1)}\right)dr\\
K_4^3 &= -B_{rr}\left[r\frac{d\psi_2^{(1)}}{dr} + \nu\psi_1^{(1)}\right]_{r=0.5h_1} + B_{rr}\int_0^{0.5h_1}\left(\nu\frac{d\psi_2^{(1)}}{dr} + \frac{1}{r}\psi_2^{(1)}\right)dr\\
K_2^3 &= -\tfrac{1}{2}B_{rr}\left[r\delta W\frac{d\psi_1^{(1)}}{dr}\right]_{r=0.5h_1} + \tfrac{1}{2}\nu B_{rr}\int_0^{0.5h_1}\delta W\frac{d\psi_1^{(1)}}{dr}\,dr\\
K_5^3 &= -\tfrac{1}{2}B_{rr}\left[r\delta W\frac{d\psi_2^{(1)}}{dr}\right]_{r=0.5h_1} + \tfrac{1}{2}\nu B_{rr}\int_0^{0.5h_1}\delta W\frac{d\psi_2^{(1)}}{dr}\,dr\\
K_3^3 &= S_{rz}\int_0^{0.5h_1}\psi_1^{(1)}\,rdr - D_{rr}\left[r\frac{d\psi_1^{(1)}}{dr} + \nu\psi_1^{(1)}\right]_{r=0.5h_1}\\
&\quad + D_{rr}\int_0^{0.5h_1}\left(\nu\frac{d\psi_1^{(1)}}{dr} + \frac{1}{r}\psi_1^{(1)}\right)dr\\
K_6^3 &= S_{rz}\int_0^{0.5h_1}\psi_2^{(1)}\,rdr - D_{rr}\left[r\frac{d\psi_2^{(1)}}{dr} + \nu\psi_1^{(1)}\right]_{r=0.5h_1}\\
&\quad + D_{rr}\int_0^{0.5h_1}\left(\nu\frac{d\psi_2^{(1)}}{dr} + \frac{1}{r}\psi_2^{(1)}\right)dr\\
F^2 &= \int_0^{0.5h_1} rq(r)\,dr, \qquad \delta W = \frac{W_2 - W_1}{h_1}
\end{aligned}
\tag{8.5.28}
$$

Similar expressions can be written for node $N+1$.

8.6 Numerical Results

8.6.1 Preliminary Comments

Here we present numerical results obtained with various FEM and DMCDM models developed in the preceding sections. Numerical results obtained with the FEM and DMCDM are compared in all cases [45, 46]. We use three models of the FEM and two models of the DMCDM, as designated here:

- FE-CP(D) - Displacement finite element model of the CPT
- FE-CP(M) - Mixed finite element model of the CPT
- FE-FS(D) - Displacement finite element model of the FST
- DM-CP(M) - Mixed dual mesh finite domain model of the CPT
- DM-FS(D) - Displacement dual mesh finite domain model of the FST

The FE-CP(D) model uses Hermite cubic interpolation of $w(r)$ and linear interpolation of $u(r)$, whereas all other elements are based on Lagrange interpolations of all variables. All finite element models other than FE-CP(D) can also use quadratic or higher order interpolations, whereas the dual mesh control domain formulations presented herein are based on linear interpolations. Thus, for consistency, all numerical results presented herein, with the exception of FE-CP(D), are obtained with linear approximations of all field variables.

In obtaining the numerical solutions, we shall consider functionally graded circular plates of radius $a = R = 10$ in (25.4 cm) and thickness $h = 0.1$ in (0.254 cm), and subjected to uniformly distributed load of intensity q_0 lb/in (1 lb/in = 175 N/m). The FGM beam is made of two materials with the following values of the moduli, Poisson's ratio, and shear correction coefficient:

$$E_1 = 30 \times 10^6 \text{ psi (207 GPa)}, \quad E_2 = 3 \times 10^6 \text{ psi (20.7 GPa)}, \\ \nu = 0.3, \quad K_s = \frac{5}{6}. \tag{8.6.1}$$

We shall investigate the parametric effects of the power-law index, n, and boundary conditions on the linear and nonlinear transverse deflections and bending moments.

Extensive numerical studies have been carried out with various models, including mesh independence and value of the acceleration parameter on the nonlinear convergence, effect of the power-law index, and post-computation of the stress resultants (either the bending moments or the rotations). In all cases, both the DMCDM and FEM models, using 16 linear elements, yield results that are indistinguishable in a graphical presentation. Most interestingly, it is found that the post-computed rotations (in mixed models) and bending moments (in displacement models) are very accurate (one cannot distinguish between the exact and numerical solutions), except at $r = 0$. Based on the numerical studies, the following observations are made concerning the linear solutions:

(1) The nodal generalized displacements predicted by FE-CP(D) match the exact CPT solutions for the pinned and clamped plates.

(2) The post-computed shear forces in DM-FS(D) and DM-CP(M) match the exact solution for the pinned and clamped plates.

8.6.2 Linear Analysis

The discrete equations of the DMCDM can be specialized to linear analysis by setting $\delta W = 0$ in the DMCDM models developed for the CPT and FST. To illustrate the working of the dual mesh finite domain models presented in the previous sections, we consider two different examples, namely, hinged and clamped functionally graded axisymmetric circular plates subjected to uniformly distributed load of intensity q_0. A comparison between the numerical results obtained with the FEM and DMCDM models for each boundary condition considered is presented.

For the mixed models, the second derivative of the transverse deflection needs to be calculated from the moment M_{rr}, which is determined as a nodal variable. The second derivative of transverse deflection w in the CPT can be obtained using the following equation [see Eq. (8.2.25); set the nonlinear term to zero]:

$$\frac{d^2w}{dr^2} = -\frac{\nu}{r}\frac{dw}{dr} + \bar{B}_{rr}\left(\frac{du}{dr} + \nu\frac{u}{r}\right) - \frac{1}{D_{rr}}M_{rr} \tag{8.6.2}$$

Once the required derivatives of the primary variables are calculated, the stresses can be computed using Eqs. (8.1.3), while the stress resultants can be computed using Eqs. (8.2.5a)–(8.2.5d).

Here we consider functionally graded circular plates which are either pinned or clamped at the (outer) edge. The boundary conditions on the primary variables in various models for the pinned case (for the linear analysis, there is no difference between the pinned and hinged boundary condition) are as follows:

Displacement Models:

$$u(0) = 0, \quad \text{and} \quad \frac{dw}{dr} = 0 \ \text{ or } \ \phi_r(0) = 0; \quad u(a) = w(a) = 0. \tag{8.6.3a}$$

Mixed Models:

$$u(0) = 0, \quad u(a) = w(a) = M_{rr}(a) = 0. \tag{8.6.3b}$$

The exact solution for the transverse deflection of pinned functionally graded circular plates according to the FST, with the power-law given in Eq. (8.1.4) [$B_{rr} \neq 0$; $D^*_{rr} = D_{rr}A_{rr} - B^2_{rr}$; see Eqs. (8.2.6) and (8.2.7)], is given by (see Reddy [42])

$$u(r) = \frac{B_{rr}\, q_0 a^3}{16 D^*_{rr}} \left(-\xi + \xi^3\right) \tag{8.6.4a}$$

$$w(r) = \frac{A_{rr}\, q_0 R^4}{64 D^*_{rr}} \left[\xi^4 - 2\left(\frac{3+\nu}{1+\nu}\right)\xi^2 + \frac{5+\nu}{1+\nu} + \frac{4B^2_{rr}}{D_{rr}\, A_{rr}(1+\nu)}\left(\xi^2 - 1\right)\right] + \frac{q_0 a^2}{4 S_{rz}}\left(1 - \xi^2\right) \tag{8.6.4b}$$

$$\phi_r(r) = \frac{A_{rr}\, q_0 R^3}{16 D^*_{rr}} \left[-\frac{2B^2_{rr}}{D_{rr}\, A_{rr}(1+\nu)}\xi + \frac{(3+\nu)}{(1+\nu)}\xi - \xi^3\right] \tag{8.6.4c}$$

$$M_{rr}(r) = \frac{(3+\nu) q_0 a^2}{16}\left(1 - \xi^2\right) \tag{8.6.4d}$$

where $\xi = r/a$ and D^*_{rr} is defined in Eq. (8.4.13). The CPT solutions are given by setting $1/S_{rz}$ to zero, and the solutions for homogeneous plates are obtained by setting $B_{rr} = 0$.

The boundary conditions on the primarily variables in various models for the clamped circular plate are as follows (replace dw/dr with ϕ_r for the FST):

Displacement Models:

$$u(0) = 0, \quad \frac{dw}{dr}(0) = 0, \quad u(a) = w(a) = \frac{dw}{dr}(a) = 0. \tag{8.6.5a}$$

Mixed Models:

$$u(0) = 0, \quad u(a) = w(a) = 0. \tag{8.6.5b}$$

The exact solution for the transverse deflection of a functionally graded clamped circular plate according to the FST is

$$u(r) = \frac{B_{rr}\, q_0 a^3}{16 D^*_{rr}} \left(-\xi + \xi^3\right) \tag{8.6.6a}$$

$$w(r) = \frac{A_{rr}\, q_0 a^4}{64 D^*_{rr}} \left(1 - \xi^2\right)^2 + \frac{q_0 a^2}{4 S_{rz}}\left(1 - \xi^2\right) \tag{8.6.6b}$$

$$\phi_r(r) = \frac{A_{rr}\, q_0 a^3}{16 D^*_{rr}} \left(\xi - \xi^3\right) \tag{8.6.6c}$$

$$M_{rr}(r) = \frac{q_0 a^2}{16}\left[(1+\nu) - (3+\nu)\xi^2\right] \tag{8.6.6d}$$

A comparison between the transverse deflections at the center of a hinged homogeneous circular plate obtained from various models is presented in Table 8.6.1 (see [45]). When 32 elements are used, all the models give the same results as the exact solution up to the fourth decimal point. The mixed finite element model has a slow mesh convergence rate at the origin for transverse deflection, w, and hence a nonuniform finite element mesh with a finer mesh at the origin is used in reporting the values of $w(0)$. The nonuniform mesh is chosen such that for a two-element mesh, the lengths of the elements are in the ratio of 1:9, with the shortest element being at the origin. All the subsequent refinements are made by breaking the finite elements into half. Thus, for a four-element mesh the element lengths would be $\{0.5a, 0.5a, 4.5a, 4.5a\}$, and further refinement is made by breaking these elements into their halves and so on. Since the radius to thickness ratio is $a/h = 100$, both the CPT and FST predict almost the same results.

Table 8.6.1 Center transverse deflection, $w(0)$, of a hinged homogeneous axisymmetric circular plate as predicted by various models. The results for the transverse deflection are in inches.

Mesh	FE-CP(D)	FE-CP(M)	FE-FS(D)	DM-CP(M)	DM-FS(D)
2	0.1159	0.1120	0.1076	0.1149	0.1052
4	0.1159	0.1136	0.1143	0.1156	0.1134
8	0.1159	0.1155	0.1156	0.1158	0.1153
16	0.1159	0.1159	0.1159	0.1159	0.1158
32	0.1159	0.1159	0.1159	0.1159	0.1159
64	0.1159	0.1159	0.1159	0.1159	0.1159
Exact	0.1159	0.1159	0.1159	0.1159	0.1159

To assess the accuracy of the DMCDM, we compare the transverse deflections at the center, $w(0)$, obtained from the analytical solution of the FST and various finite element models described earlier with those obtained from the DMCDM. Various values of the power-law index, n, are considered to bring out the effect of n on the bending response of the plate. A comparison between the solutions of various finite element models with the analytical solution (of the FST) is presented in Table 8.6.2. Since $a/h = 100$, both the CPT and FST predict close to each other; 32 elements are used in reporting the transverse deflections given in Table 8.6.2. Uniform mesh is used in all the numerical models except for the mixed finite element model of the CPT. A nonuniform mesh described earlier is used for FE-CP(M) to achieve faster mesh convergence. It can be seen that as the power-law index, n, increases the plate stiffness decreases.

Table 8.6.2 Center transverse deflection, $w(0)$, of a clamped FGM axisymmetric circular plate for various values of n, as predicted by various models; 32 elements are used in all the models. The results are given in inches.

n	FE-CP(D)	FE-CP(M)	DM-CP(M)	FE-FP(D)	DM-FP(D)	FST Exact
0.0	0.0284	0.0285	0.0284	0.0285	0.0284	0.0284
1.0	0.0665	0.0666	0.0665	0.0666	0.0665	0.0666
2.0	0.0976	0.0976	0.0975	0.0977	0.0976	0.0976
3.0	0.1152	0.1152	0.1151	0.1153	0.1152	0.1152
4.0	0.1244	0.1244	0.1242	0.1245	0.1244	0.1244
5.0	0.1296	0.1296	0.1294	0.1297	0.1296	0.1296
7.5	0.1370	0.1370	0.1369	0.1371	0.1370	0.1370
10.0	0.1425	0.1425	0.1424	0.1426	0.1425	0.1425
12.0	0.1467	0.1467	0.1466	0.1468	0.1469	0.1467
15.0	0.1527	0.1528	0.1527	0.1529	0.1527	0.1523
20.0	0.1622	0.1622	0.1621	0.1623	0.1622	0.1622

Figure 8.6.1 contains plots of the deflections $w(0)$ predicted for the pinned plates and clamped plates as a function of the normalized radial coordinate, r/a ($a = R$). The deflections predicted (shown by symbols) by all FEM and DMCDM models are essentially the same (i.e., the differences cannot be seen in the graph) and match with the analytical solutions (shown as lines); this also indicates that the effect of shear deformation is negligible (because $a/h = 100$, a thin plate). The post-computed and nodal values of the bending moment M_{rr} are plotted as functions of the normalized radial coordinate, r/a in Fig. 8.6.2. Except for the value at $r = 0$ the results match with the exact solution for both pinned and clamped plates.

Figure 8.6.3 shows the center deflection $w(0)$ and rotation $-dw/dr$ at $r = 0.5265\,a$ as functions of the power-law index n for the pinned and clamped circular plates ($a = R$). We note that the rate of increase of the deflection has two different regions; the first region has a rapid increase of the deflection, while the second region is marked with a relatively slow increase. This is primarily because of the fact that the coupling coefficient B_{xx} varies with n rapidly for the smaller values of n followed by a slow decay after $n > 3$. The rate of increase in the deflection or slope in the second part is less for clamped plates than for the pinned plates. The reason is the fact that the clamped plate is relatively stiffer than the pinned plate.

Although the numerical results for the bending moment M_{rr} are not presented here, few remarks are in order. While the moments obtained from all the models are very close, the mixed FEM and mixed DMCDM models deviate from the displacement FEM model near the origin. The mixed models of both the FEM and DMCDM are prone to give erroneous moments at the origin. However, the displacement models do not exhibit such a behavior.

8.6.3 Nonlinear Analysis

The resulting nonlinear algebraic equations can be solved using either direct iteration or Newton's iteration schemes, which amounts to linearization of these equations (see Reddy [13]). It is found that the direct iteration scheme does not converge even after 50 iterations in some cases, especially for FGM plates. On the other hand, the Newton iteration scheme converges for less than ten iterations (most often for less than four iterations).

Load increments of $\Delta q_0 = 1.0$ lb/in (175 N/m) and a tolerance of $\epsilon = 10^{-3}$ are used in the nonlinear analysis. The initial solution vector is chosen to be $\mathbf{\Delta}^0 = \mathbf{0}$ so that the first iteration of the first load step yields the linear solution. The direct iteration scheme does not converge in most cases unless an acceleration parameter, β, is used to evaluate the stiffness matrix, $\mathbf{K}^r = \mathbf{K}(\bar{\mathbf{\Delta}}^r)$, at each iteration:

$$\bar{\mathbf{\Delta}}^r = (1-\beta)\mathbf{\Delta}^r + \beta\mathbf{\Delta}^{r-1}, \quad 0 \leq \beta \leq 1, \tag{8.6.7}$$

where r denotes the iteration number. Thus, using a weighted average of the last two iteration solutions to update the stiffness matrix accelerates the convergence. In the present case, a value of $\beta = 0.25 - 0.35$ is used (after some study with varying β, starting with $\beta = 0$). In some cases, even this strategy does not help to achieve convergence, forcing us to use the Newton iteration scheme.

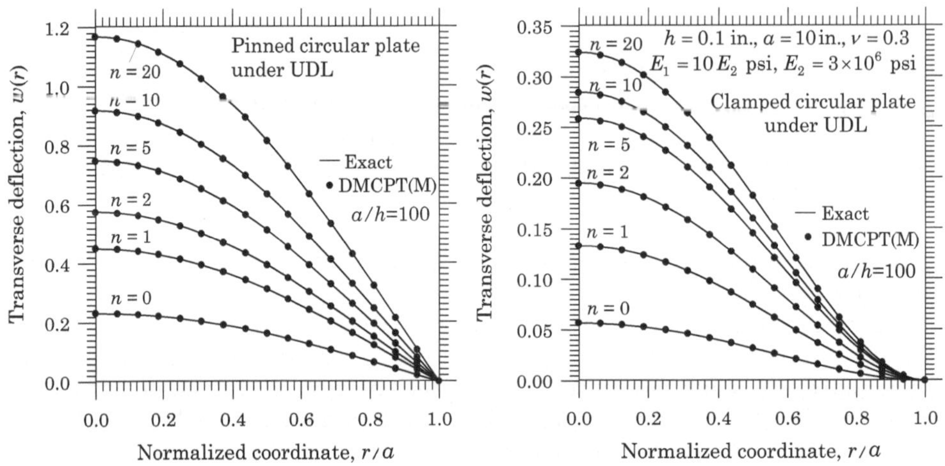

Fig. 8.6.1 Plots of the center deflection (in.) of pinned circular plates as functions of the normalized radial coordinate, r/a, for various value of the power-law index, n. The solid lines correspond to the analytical solutions and symbols to the numerical solutions.

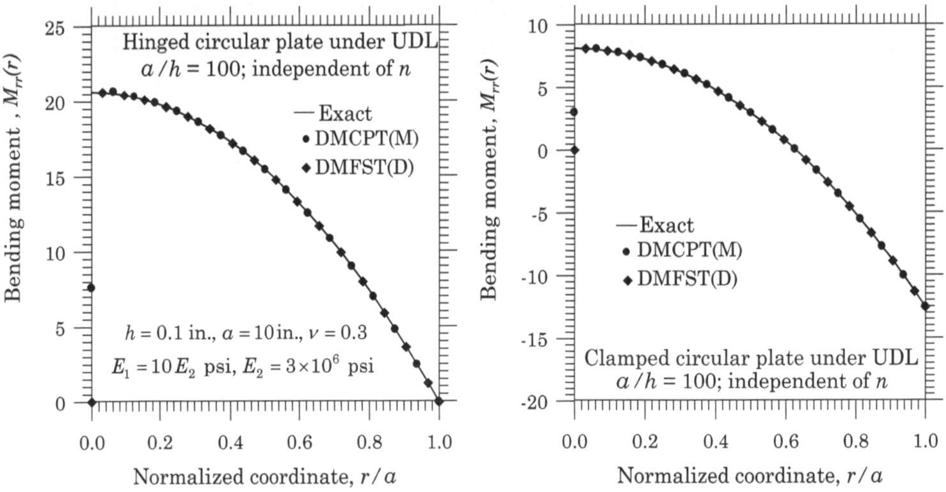

Fig. 8.6.2 Plots of the post-computed and nodal values of the bending moment M_{rr} (lb-in) of pinned and clamped circular plates as functions of the normalized radial coordinate, r/a. The solid lines correspond to the analytical solutions and symbols to the numerical solutions. The results are independent of n.

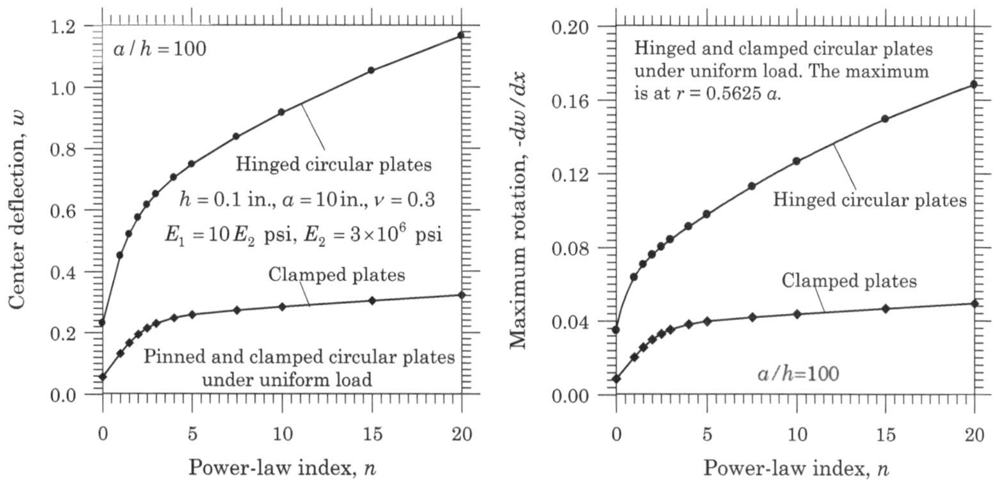

Fig. 8.6.3 Plots of the center deflection $w(0)$ (in.) and $-dw/dr$ (radians) at $r = 0.5625\,a$ of pinned and clamped circular plates as a function of the power-law index, n.

The nonlinear analysis results for deflections and bending moment obtained by various models are also indistinguishable in the graphs of dimensionless center deflection, $\bar{w} = w(0)/h$ versus the load parameter, $P = q_0 a^4/E_1 h^4$, as shown in Figs. 8.6.4 and 8.6.5 for pinned and clamped plates, respectively. The dual mesh control domain method gives essentially the same results as the finite element method, except that the former method has less computational effort due to the computation of the global stiffness matrix and force vector directly, without computing local matrices and assembling.

Plots of the normalized deflection $\bar{w} = w(a)/h$ versus the load parameter $P = q_0 a^4/E_1 h^4$ for pinned (at the outer rim) annular plates ($b = 0.2\,a$) are presented in Fig. 8.6.6, while Fig. 8.6.7 contains plots of the normalized deflection $\bar{w} = w(a)/h$ for annular plates with internal edge clamped. Again, all models yield solutions that are indistinguishable in plots. The deflections of the annular plates will be higher than the solid circular plates because annular plates have less stiffness compared to the solid circular plates for the same boundary conditions. The difference between the deflections of annular plates and solid plates is bigger than that is shown in the plots because the deflections of the solid circular plates (shown in solid lines) are at the center of the plate, while those of the annular plate (shown in broken lines) are at $r/a = 0.2$. As the internal radius of the annular plate increases from $b = 0.2\,a$ to $b = 0.5\,a$, the deflection of the outer edge decreases substantially because of the reduction in free span of the plate.

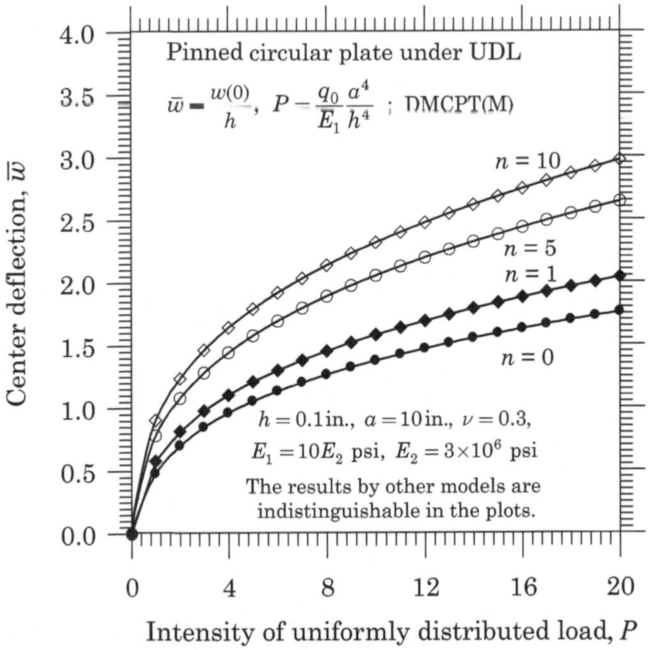

Fig. 8.6.4 Plots of the normalized center deflection $\bar{w} = w(0)/h$ versus the load parameter $P = q_0a^4/E_1h^4$ for pinned circular plates, for various values of the power-law index, n.

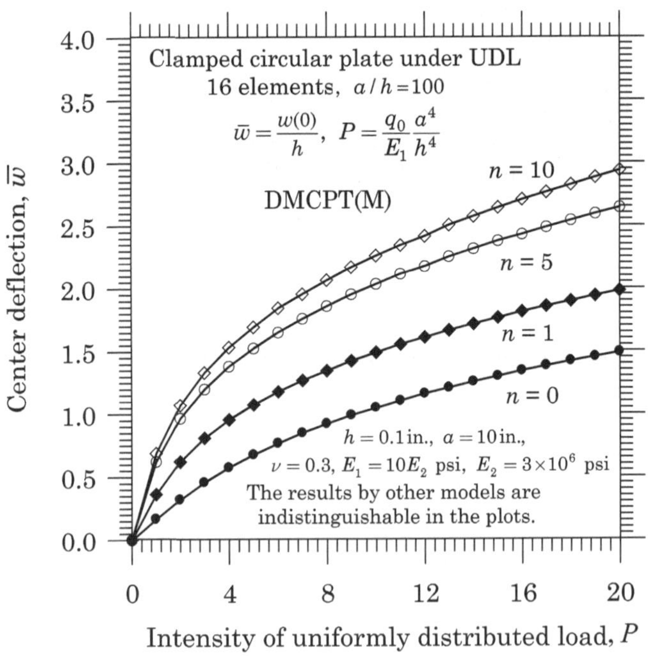

Fig. 8.6.5 Plots of the normalized center deflection $\bar{w} = w(0)/h$ versus the load parameter $P = q_0a^4/E_1h^4$ for clamped circular plates, for various values of the power-law index, n.

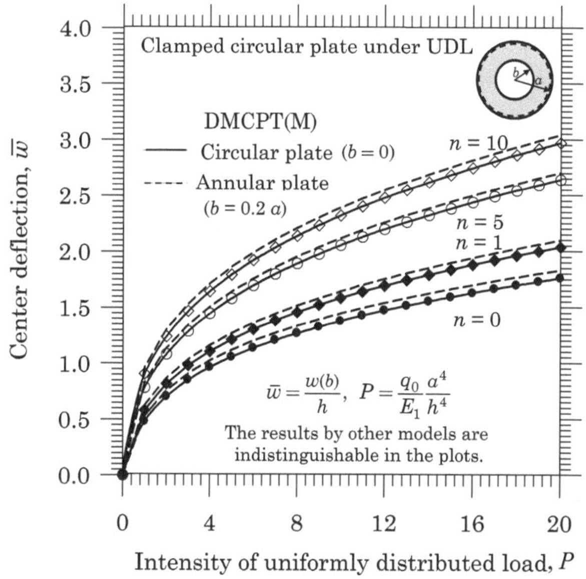

Fig. 8.6.6 Plots of the normalized deflection $\bar{w} = w(b)/h$ versus the load parameter $P = q_0 a^4/E_1 h^4$ for pinned (at the outer rim) annular plates, for various values of the power-law index, n; $b = 0$ for solid circular plates and $b = 0.2\,a$ for annular plates.

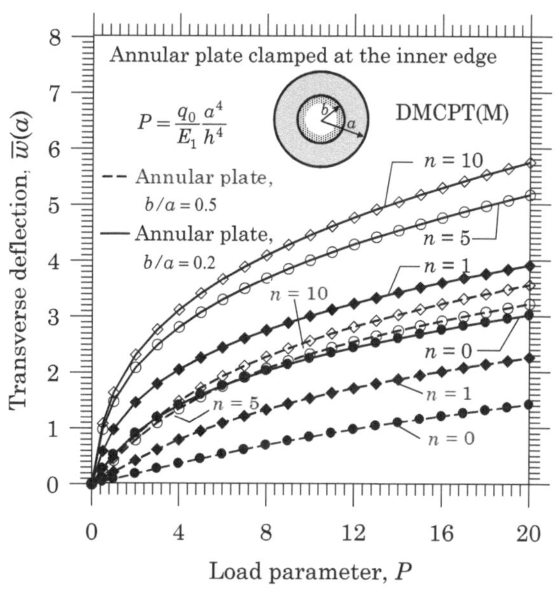

Fig. 8.6.7 Plots of the normalized deflection $\bar{w} = w(a)/h$ versus the load parameter $P = q_0 a^4/E_1 h^4$ for annular plates clamped at the inner edge ($r = b$, $b = 0.2a, 0.5a$), for various values of the power-law index, n.

8.7 Summary

The dual mesh control domain method (DMCDM) is presented for the linear, axisymmetric bending analysis of two-constituent through thickness functionally graded circular plates. Both classical (CPT) and shear deformation (FST) theories are used. The displacement and mixed finite element models of CPT and the displacement finite element model of the FST are developed. The mixed model of the CPT and displacement model of the FST are developed using the DMCDM. The mixed model includes the bending moment as a dependent variable in addition to the radial and transverse displacements. Numerical results are presented for pinned and clamped solid circular and pinned annular plates, showing the effect of the power-law index on the load-deflection behavior. In all cases, the results predicted by all computational models gave essentially the same results.

Numerical results indicate that the mixed DMCDM model has a better mesh convergence characteristic than the mixed FEM model. In general, the DMCDM models give as accurate results as the FEM, with the DMCDM giving the bending moments directly at the nodes. While both FEM and DMCDM have comparable accuracy, the DMCDM has less formulative steps and computational expense because there is no element concept and, hence, no assembly of element equations; these features will have significant savings in multidimensional problems.

With the developed programs for the five models presented here, a number of other problems with a variety of boundary conditions may be analyzed. For example, circular plates with spring (extensional as well as torsional) supports and annular plates with a combination of boundary conditions at the inner and outer edges, and for different material distributions, can be readily analyzed.

Problems

8.1 Verify the tangent stiffness coefficients in Eq. (8.4.9) for the displacement model of CPT.

8.2 Verify the tangent stiffness coefficients in Eq. (8.4.18) for the mixed model of the CPT.

8.3 Verify the tangent stiffness coefficients in Eq. (8.4.26) for the displacement model of the FST.

8.4 Show that the discretized equations for the first control domain of the mixed formulation of the CPT, for the linear case, using the DMCDM are

$$-N_1^{(1)} + U_1\left[\frac{\bar{A}_{rr}(1-\nu)}{2} + \int_0^{0.5h_1}\left(\bar{A}_{rr}\nu\frac{d\psi_1^{(1)}}{dr} + \frac{A_{rr}}{r}\psi_1^{(1)} + \frac{B_{rr}^2(1-\nu)}{rD_{rr}}\psi_1^{(1)}\right)dr\right]$$

$$-W_1B_{rr}(1-\nu^2)\left(\int_0^{0.5h_1}\frac{1}{r}\frac{d\psi_1^{(1)}}{dr}\,dr\right) + M_{rr(1)}\bar{B}_{rr}\left(\int_0^{0.5h_1}\nu\psi_1^{(1)}dr - \frac{h_1}{4}\right)$$

$$+ U_2\left[-\frac{\bar{A}_{rr}(1+\nu)}{2} + \int_0^{0.5h_1}\left(\bar{A}_{rr}\nu\frac{d\psi_2^{(1)}}{dr} + \frac{\bar{A}_{rr}}{r}\psi_2^{(1)} + \frac{B_{rr}^2(1-\nu^2)}{rD_{rr}}\psi_2^{(1)}\right)dr\right]$$

$$- W_2B_{rr}(1-\nu^2)\left(\int_0^{0.5h_1}\frac{1}{r}\frac{d\psi_2^{(1)}}{dr}\,dr\right) + M_{rr(2)}\bar{B}_{rr}\left(\int_0^{0.5h_1}\nu\psi_2^{(1)}\,dr - \frac{h_1}{4}\right) = 0. \quad (1)$$

$$-V_1^{(1)} + U_1B_{rr}(1-\nu^2)\frac{1}{h_1} + 2W_1D_{rr}(1-\nu^2)\frac{1}{h_1^2} + \frac{1}{2}\nu M_{rr(1)}$$

$$+ U_2\left[B_{rr}(1-\nu^2)\frac{1}{h_1}\right] - 2W_2D_{rr}(1-\nu^2)\frac{1}{h_1^2} - \frac{1}{2}M_{rr(2)}(2-\nu) - \int_0^{0.5h_1} qr\,dr = 0 \quad (2)$$

$$-\Phi_1^{(1)} + \bar{B}_{rr}U_1\int_0^{0.5h_1}\left(r\frac{d\psi_1^{(1)}}{dr} + \nu\psi_1^{(1)}\right)dr + W_1\left[\frac{1}{2} + \int_0^{0.5h_1}(1-\nu)\frac{d\psi_1^{(1)}}{dr}\,dr\right]$$

$$+ \bar{B}_{rr}U_2\int_0^{0.5h_1}\left(r\frac{d\psi_2^{(1)}}{dr} + \nu\psi_2^{(1)}\right)dr + W_2\left[-\frac{1}{2} + \int_0^{0.5h_1}(1-\nu)\frac{d\psi_2^{(1)}}{dr}\,dr\right]$$

$$- \frac{1}{D_{rr}}\left(M_{rr(1)}\int_0^{0.5h_1} r\psi_1^{(1)}\,dr\right.$$

$$\left. - M_{rr(2)}\int_0^{0.5h_1} r\psi_2^{(1)}\,dr\right) = 0. \quad (3)$$

8.5 Show that the discretized equations for the last control domain (i.e., $N+1$th control domain) of the mixed formulation of the CPT, for the linear case, using the DMCDM are

$$-N_2^{(N+1)} + \bar{A}_{rr}U_N\left[-\frac{(R-0.5h_N)}{h_N} + \frac{\nu}{2}\right.$$

$$\left. + \int_{R-0.5h_N}^{R}\left(\nu\frac{d\psi_1^{(N)}}{dr} + \frac{1}{r}\psi_1^{(N)} + \frac{B_{rr}^2(1-\nu^2)}{\bar{A}_{rr}D_{rr}}\frac{1}{r}\psi_1^{(N)}\right)dr\right]$$

$$- B_{rr}(1-\nu^2)W_N\int_{R-0.5h_N}^{R}\frac{1}{r}\frac{d\psi_1^{(N)}}{dr}\,dr$$

$$+ \bar{B}_{rr}M_{rr(N)}\left[\int_{R-0.5h_N}^{R}\nu\psi_1^{(N)}\,dr + \frac{(R-0.5h_N)}{2}\right] + \bar{A}_{rr}U_{N+1}\left[\frac{(R-0.5h_N)}{h_N}\right.$$

$$\left. + \frac{\nu}{2} + \int_{R-0.5h_N}^{R}\left(\nu\frac{d\psi_2^{(N)}}{dr} + \frac{1}{r}\psi_2^{(N)} + \frac{B_{rr}^2(1-\nu^2)}{\bar{A}_{rr}D_{rr}}\frac{1}{r}\psi_2^{(N)}\right)dr\right]$$

$$- B_{rr}(1-\nu^2)W_{N+1}\int_{R-0.5h_N}^{R}\frac{1}{r}\frac{d\psi_2^{(N)}}{dr}\,dr$$

$$+ \bar{B}_{rr}M_{rr(N+1)}\left[\int_{R-0.5h_N}^{R}\nu\psi_2^{(N)}\,dr + \frac{(R-0.5h_N)}{2}\right] = 0 \quad (1)$$

$$-V_2^{(N+1)} + U_N\left[-\frac{B_{rr}(1-\nu^2)}{2(R-0.5h_N)}\right] + W_N\left[-\frac{D_{rr}(1-\nu^2)}{h_N(R-0.5h_N)}\right]$$
$$+ M_{rr(N)}\left[1-\frac{R}{h_N}-\frac{\nu}{2}\right] + U_{N+1}\left[-\frac{B_{rr}(1-\nu^2)}{2(R-0.5h_N)}\right] + W_{N+1}\left[\frac{D_{rr}(1-\nu^2)}{h_N(R-0.5h_N)}\right]$$
$$+ M_{rr(N+1)}\left[\frac{R}{h_N}-\frac{\nu}{2}\right] - \int_{R-0.5h_N}^{R} qr\,dr = 0. \tag{2}$$

$$-\Phi_2^{(N+1)} + \bar{B}_{rr}U_N\left[\int_{R-0.5h_N}^{R}\left(r\frac{d\psi_1^{(N)}}{dr} + \nu_N\psi_1^{(N)}\right)dr\right]$$
$$+ W_N\left[-\frac{R-0.5h_N}{h_N} + \int_{R-0.5h_N}^{R}(1-\nu_N)\frac{d\psi_1^{(N)}}{dr}dr\right]$$
$$-\frac{1}{D_{rr}}M_{rr(N)}\int_{R-0.5h_N}^{R} r\psi_1^{(N)}dr + \bar{B}_{rr}U_{N+1}\int_{R-0.5h_N}^{R}\left(r\frac{d\psi_2^{(N)}}{dr} + \nu\psi_2^{(N)}\right)dr$$
$$+ W_{N+1}\left[\frac{R-0.5h_N}{h_N} + \int_{R-0.5h_N}^{R}(1-\nu)\frac{d\psi_2^{(N)}}{dr}dr\right]$$
$$-\frac{1}{D_{rr}}M_{rr(N+1)}\int_{R-0.5h_N}^{R} r\psi_2^{(N)}dr = 0. \tag{3}$$

8.6 Show that the discretized equations for the first control domain of the displacement formulation of the FST, for the linear case, using the DMCDM are

$$-N_1^{(1)} + U_1\left[\frac{A_{rr}(1-\nu)}{2} + \int_0^{0.5h_1} A_{rr}\left(\nu\frac{d\psi_1^{(1)}}{dr} + \frac{\psi_1^{(1)}}{r}\right)dr\right]$$
$$+ \Phi_1\left[\frac{B_{rr}(1-\nu)}{2} + \int_0^{0.5h_1} B_{rr}\left(\nu\frac{d\psi_1^{(1)}}{dr} + \frac{\psi_1^{(1)}}{r}\right)dr\right]$$
$$+ U_2\left[-\frac{A_{rr}(1+\nu)}{2} + \int_0^{0.5h_1} A_{rr}\left(\nu\frac{d\psi_2^{(1)}}{dr} + \frac{\psi_2^{(1)}}{r}\right)dr\right]$$
$$+ \Phi_2\left[-\frac{B_{rr}(1+\nu)}{2} + \int_0^{0.5h_1} B_{rr}\left(\nu\frac{d\psi_2^{(1)}}{dr} + \frac{\psi_2^{(1)}}{r}\right)dr\right] = 0 \tag{1}$$

$$-V_1^{(1)} + W_1\left[\frac{S_{rz}}{2}\right] + \Phi_1\left[-\frac{h_1S_{rz}}{4}\right] + W_2\left[\frac{-S_{rz}}{2}\right] + \Phi_2\left[-\frac{h_1S_{rz}}{4}\right]$$
$$-\int_0^{0.5h_1} qr\,dr = 0 \tag{2}$$

$$-M_1^{(1)} + U_1\left[\frac{B_{rr}(1-\nu)}{2} + \int_0^{0.5h_1} B_{rr}\left(\nu_1\frac{d\psi_1^{(1)}}{dr} + \frac{\psi_1^{(1)}}{r}\right)dr\right]$$
$$+ \Phi_1\left[\frac{D_1(1-\nu)}{2} + \int_0^{0.5h_1} D_{rr}\left(\nu\frac{d\psi_1^{(1)}}{dr} + \frac{\psi_1^{(1)}}{r}\right)dr + \int_0^{0.5h_1}\frac{S_{rz}r}{2}dr\right]$$
$$+ U_2\left[-\frac{B_{rr}(1+\nu)}{2} + \int_0^{0.5h_1} B_{rr}\left(\nu\frac{d\psi_2^{(1)}}{dr} + \frac{\psi_2^{(1)}}{r}\right)dr\right]$$
$$+ W_1\left[\int_0^{0.5h_1} S_{rz}r\frac{d\psi_1^{(1)}}{dr}dr\right] + W_2\left[\int_0^{0.5h_1} S_{rz}r\frac{d\psi_2^{(1)}}{dr}dr\right]$$

$$+\Phi_2\left[-\frac{D_{rr}(1+\nu)}{2}+\int_0^{0.5h_1} D_{rr}\left(\nu\frac{d\psi_2^{(1)}}{dr}+\frac{\psi_2^{(1)}}{r}\right)dr+\int_0^{0.5h_1}\frac{S_{rz}r}{2}dr\right]=0 \quad (3)$$

8.7 Show that the discretized equations for the last control domain (i.e., $N+1$th control domain) of the displacement formulation of the FST, for the linear case, using the DMCDM are

$$\begin{aligned}
&-N_2^{(N+1)}+A_{rr}U_N\left[\frac{\nu}{2}-\frac{(R-0.5h_N)}{h_N}+\int_{R-0.5h_N}^{R}\left(\nu\frac{d\psi_1^{(N)}}{dr}+\frac{\psi_1^{(N)}}{r}\right)dr\right]\\
&+B_{rr}\Phi_N\left[\frac{\nu_N}{2}-\frac{(R-0.5h_N)}{h_N}+\int_{R-0.5h_N}^{R}\left(\nu\frac{d\psi_1^{(N)}}{dr}+\frac{\psi_1^{(N)}}{r}\right)dr\right]\\
&+A_{rr}U_{N+1}\left[\frac{(R-0.5h_N)}{h_N}+\frac{\nu}{2}+\int_{R-0.5h_N}^{R}\left(\nu\frac{d\psi_2^{(N)}}{dr}+\frac{\psi_2^{(N)}}{r}\right)dr\right]\\
&+B_{rr}\Phi_{N+1}\left[\frac{(R-0.5h_N)}{h_N}+\frac{\nu}{2}+\int_{R-0.5h_N}^{R}\left(\nu\frac{d\psi_2^{(N)}}{dr}+\frac{\psi_2^{(N)}}{r}\right)dr\right]=0 \qquad (1)
\end{aligned}$$

$$\begin{aligned}
&-V_2^{(N+1)}-S_{rz}W_N\frac{(R-0.5h_N)}{h_N}+S_{rz}\Phi_N\frac{(R-0.5h_N)}{2}\\
&+S_{rz}W_{N+1}\frac{(R-0.5h_N)}{h_N}+S_{rz}\Phi_{N+1}\frac{(R-0.5h_N)}{2}-\int_{R-0.5h_N}^{R} qr\,dr=0 \qquad (2)
\end{aligned}$$

$$\begin{aligned}
&-M_2^{(N+1)}+B_{rr}U_N\left[\frac{\nu}{2}-\frac{(R-0.5h_N)}{h_N}+\int_{R-0.5h_N}^{R}\left(\nu\frac{d\psi_1^{(N)}}{dr}+\frac{\psi_1^{(N)}}{r}\right)dr\right]\\
&+D_{rr}\Phi_N\left[\frac{\nu}{2}-\frac{(R-0.5h_N)}{h_N}+\int_{R-0.5h_N}^{R}\left(\nu\frac{d\psi_1^{(N)}}{dr}+\frac{\psi_1^{(N)}}{r}\right)dr\right]\\
&+B_{rr}U_{N+1}\left[\frac{(R-0.5h_N)}{h_N}+\frac{\nu}{2}+\int_{R-0.5h_N}^{R}\left(\nu\frac{d\psi_2^{(N)}}{dr}+\frac{\psi_2^{(N)}}{r}\right)dr\right]\\
&+D_{rr}\Phi_{N+1}\left[\frac{(R-0.5h_N)}{h_N}+\frac{\nu}{2}+\int_{R-0.5h_N}^{R}\left(\nu\frac{d\psi_2^{(N)}}{dr}+\frac{\psi_2^{(N)}}{r}\right)dr\right]\\
&+S_{rz}W_N\int_{R-0.5h_N}^{R} r\frac{d\psi_1^{(N)}}{dr}dr+S_{rz}W_{N+1}\int_{R-0.5h_N}^{R} r\frac{d\psi_2^{(N)}}{dr}dr\\
&+\frac{1}{2}S_{rz}\left(\Phi_I+\Phi_{N+1}\right)\int_{R-0.5h_N}^{R} r\,dr=0. \qquad (3)
\end{aligned}$$

9
Plane Elasticity and Viscous Incompressible Flows

9.1 Introduction

9.1.1 Two-Dimensional Elasticity

Elasticity is a subject of solid mechanics that deals with stress and deformation of solid continua. A material is said to be elastic if it regains its original shape and form when the forces causing the deformation are removed (i.e., without leaving permanent strains). A linearly elastic material is one for which the stress-strain behavior is linear. Linearized elasticity is concerned with small deformations (i.e., strains and displacements are very small compared to unity) in linear elastic solids (i.e., obeys Hooke's law). A class of problems in elasticity, due to geometry, boundary conditions, and external applied loads, have their solutions (i.e., displacements and stresses) not dependent on one of the dimensions; the smaller dimension is termed thickness. Such problems are called *plane elasticity* problems.

The plane elasticity problems considered here are grouped into *plane strain* and *plane stress* problems. Both classes of problems are described by a pair of *coupled* partial differential equations expressed in terms of the two components (u_x, u_y) of the displacement vector $\mathbf{u}$ referred to a rectangular Cartesian coordinate system (x, y). The governing equations of plane strain problems differ from those of the plane stress problems only in the coefficients of the differential equations but not in the form of the equations.

Most linear elasticity problems are solved by the finite element method; use of the FDM or FVM to solve elasticity problems is a rarity. In solving the plane elasticity problems, the governing equations are expressed in terms of the displacement components, and the displacement finite element models are based on the principle of virtual displacements, which is entirely equivalent to the weak-form Galerkin formulation discussed in Chapter 4. In this chapter, finite element models and the dual mesh control domain models of the linearized two-dimensional elasticity equations are developed. As will be seen shortly, the governing equations of plane elasticity and those of the penalty function formulation of the flows of viscous incompressible fluids are remarkably similar and, therefore, the DMCDM developments of one set of equations benefit the developments for the other set.

9.1.2 Flows of Viscous Fluids

Flows of viscous incompressible fluids can be classified into two major types: a smooth, orderly motion is called *laminar flow*, and a random fluctuating motion is called *turbulent flow*. Flows of viscous fluids can be characterized by a non-dimensional parameter known as the *Reynolds number*, $Re = \rho VL/\mu$, which is defined as the ratio of inertial forces ρV^2 to viscous forces $\mu V/L$. Here ρ denotes the density of the fluid, μ the fluid viscosity, V the characteristic flow velocity, and L is a characteristic dimension of the flow region. High viscosity fluids and/or small velocities produce relatively small Reynolds numbers and a laminar flow. The case of $Re << 1$ corresponds to the flow, called *Stokes flow*, in which the inertial effects are small compared to the viscous effects and are therefore neglected. The flow of less viscous fluids and/or higher velocities lead to higher Reynolds numbers and a turbulent flow. When the change in the mass density of a fluid particle is negligible, the continuum is termed *incompressible*. In this study, we assume the fluid under consideration to be Newtonian (i.e., the constitutive relations are linear). Further, the fluids are assumed to be incompressible, and the flow to be laminar.

Most numerical simulations of the Navier–Stokes equations for incompressible fluids are based on low-order finite difference and finite volume technologies. Although the finite element method has become the dominant method of choice in the numerical analysis of solids, it has yet to receive such widespread acceptance when applied to fluid flow problems. However, the FEM can naturally deal with complex regions and complicated boundary conditions, and it possesses a rich mathematical foundation compared to the FDM and FVM.

In this chapter, we present the FEM and DMCDM models for the analysis of linearized plane elasticity and two-dimensional flows of viscous incompressible fluids. We begin with a review of the governing equations (see Chapter 2 of [13]) of plane elasticity and two-dimensional fluid flows. The equations are presented here in two-dimensional rectangular Cartesian component forms. The equations for these two cases (when the pressure variable is eliminated from the fluid flow equations) are very similar, although they represent two different physical problems, and their respective FEM and DMCDM models share the same structure, making it easier for the reader to follow.

9.2 Governing Equations

9.2.1 Plane Elasticity

9.2.1.1 Equilibrium equations

The equations of solid continua are formulated using the Lagrangian description (see Reddy [23]), in which one considers a fixed body of material and observes its deformation (i.e., geometric changes) under the action of external loads. The geometric changes are referred to a coordinate system (x, y, z) fixed in the body. In linearized elasticity, the mass of the body remains unaltered and, thus, the principle of conservation of mass is trivially satisfied.

The balance of linear momentum for a two-dimensional elastic solid continua in equilibrium (plane stress or plane strain), under the assumption of small strains, yields the following stress-equilibrium equations (see, e.g., Reddy [23]):

$$\frac{\partial \sigma_{xx}}{\partial x} + \frac{\partial \sigma_{xy}}{\partial y} + f_x = 0 \tag{9.2.1}$$

$$\frac{\partial \sigma_{xy}}{\partial x} + \frac{\partial \sigma_{yy}}{\partial y} + f_y = 0 \tag{9.2.2}$$

where $(\sigma_{xx}, \sigma_{yy}, \sigma_{xy})$ are the second Piola–Kirchhoff stress tensor components, which are (approximately) the same as the Cauchy stress components for linearized elasticity (i.e., for the case of small strains); (f_x, f_y) are the body force components per unit volume of the undeformed body.

9.2.1.2 Constitutive relations

The constitutive relations for an isotropic plane elastic body are given by

$$\begin{Bmatrix} \sigma_{xx} \\ \sigma_{yy} \\ \sigma_{xy} \end{Bmatrix} = \begin{bmatrix} c_{11} & c_{12} & 0 \\ c_{12} & c_{11} & 0 \\ 0 & 0 & c_{66} \end{bmatrix} \begin{Bmatrix} \varepsilon_{xx} \\ \varepsilon_{yy} \\ 2\varepsilon_{xy} \end{Bmatrix}, \tag{9.2.3}$$

where the strains $(\varepsilon_{xx}, \varepsilon_{yy}, 2\varepsilon_{xy})$ are related to the displacements by

$$\varepsilon_{xx} = \frac{\partial u_x}{\partial x}, \quad \varepsilon_{yy} = \frac{\partial u_y}{\partial y}, \quad 2\varepsilon_{xy} = \left(\frac{\partial u_x}{\partial y} + \frac{\partial u_y}{\partial x} \right). \tag{9.2.4}$$

The elastic constants c_{ij} for plane strain and plane stress differ. The elastic constants for plane strain are (for the isotropic case)

$$c_{11} = \frac{E(1-\nu)}{(1+\nu)(1-2\nu)}, \quad c_{12} = \frac{\nu}{1-\nu} c_{11}, \quad c_{66} = G = \frac{E}{2(1+\nu)} \tag{9.2.5}$$

and for plane stress, they are given by

$$c_{11} = \frac{E}{(1-\nu^2)}, \quad c_{12} = \nu\, c_{11}, \quad c_{66} = G = \frac{E}{2(1+\nu)}, \tag{9.2.6}$$

where E is Young's modulus, G is the shear modulus, and ν is Poisson's ratio defined by $\nu \equiv -\varepsilon_{22}/\varepsilon_{11}$.

If the media is incompressible, then in addition to the governing equations given by Eqs. (9.2.1) and (9.2.2), the conservation of mass requires the volume change to be zero:

$$\frac{\partial u_x}{\partial x} + \frac{\partial u_y}{\partial y} = 0. \tag{9.2.7}$$

In this case, the governing equations can be reformulated using the penalty function method, as discussed in Section 9.5 on flows of viscous incompressible fluids.

9.2.1.3 Boundary conditions

The boundary conditions for a plane elasticity problem involve specifying one element of the duality pair $(\mathbf{u}, \mathbf{t})$ at every point on the boundary:

$$u_x = \hat{u}_x, \quad u_y = \hat{u}_y \quad \text{on} \quad \Gamma_u, \tag{9.2.8a}$$

$$t_x \equiv n_x\, \sigma_{xx} + n_y\, \sigma_{xy} = \hat{t}_x, \quad t_y \equiv n_x\, \sigma_{xy} + n_y\, \sigma_{yy} = \hat{t}_y, \quad \text{on} \quad \Gamma_\sigma, \tag{9.2.8b}$$

where $\hat{\mathbf{n}}$ is the unit normal to the boundary (n_x and n_y are the direction cosines) and Γ_u and Γ_σ are the boundary portions on which the displacement $\mathbf{u}$ and tractions (t_x, t_y), respectively, are specified.

The governing equations in terms of the displacements for a compressible solid media are obtained using Eqs. (9.2.1)–(9.2.4):

$$\frac{\partial}{\partial x}\left(c_{11}\frac{\partial u_x}{\partial x} + c_{12}\frac{\partial u_y}{\partial y}\right) + \frac{\partial}{\partial y}\left[c_{66}\left(\frac{\partial u_x}{\partial y} + \frac{\partial u_y}{\partial x}\right)\right] + f_x = 0, \tag{9.2.9}$$

$$\frac{\partial}{\partial x}\left[c_{66}\left(\frac{\partial u_x}{\partial y} + \frac{\partial u_y}{\partial x}\right)\right] + \frac{\partial}{\partial y}\left(c_{12}\frac{\partial u_x}{\partial x} + c_{22}\frac{\partial u_y}{\partial y}\right) + f_y = 0. \tag{9.2.10}$$

9.2.2 Two-Dimensional Flows of Viscous Incompressible Fluids

9.2.2.1 Equation of mass continuity

Application of the principle of conservation of mass to an element of an incompressible fluid region results in the following equation, known as the *continuity equation*:

$$\frac{\partial v_x}{\partial x} + \frac{\partial v_y}{\partial y} = 0, \tag{9.2.11}$$

where ρ is the density (kg/m^3) of the medium, $\mathbf{v}$ is the velocity vector (m/s) with rectangular components v_x and v_y, and $\boldsymbol{\nabla}$ is the vector differential operator. The condition in Eq. (9.2.11) expresses the fact that the volume change is zero for an incompressible fluid during its deformation.

9.2.2.2 Equations of equilibrium

The principle of conservation of linear momentum applied to steady flows of an incompressible medium yields the following form of vector equation (two equations in two dimensions):

$$\rho\left(v_x\frac{\partial v_x}{\partial x} + v_y\frac{\partial v_x}{\partial y}\right) + \frac{\partial \sigma_{xx}}{\partial x} - \frac{\partial \sigma_{xy}}{\partial y} - f_x = 0, \tag{9.2.12}$$

$$\rho\left(v_x\frac{\partial v_y}{\partial x} + v_y\frac{\partial v_y}{\partial y}\right) - \frac{\partial \sigma_{xy}}{\partial x} + \frac{\partial \sigma_{yy}}{\partial y} - f_y = 0, \tag{9.2.13}$$

where $(\sigma_{xx}, \sigma_{yy}, \sigma_{xy})$ are the rectangular Cartesian components of the Cauchy stress tensor $\boldsymbol{\sigma}$ (N/m^2), ρ is the mass density, and (f_x, f_y) are the rectangular Cartesian components of the body force vector $\mathbf{f}$ measured per unit volume.

Equations (9.2.12) and (9.2.13) are known as the Navier–Stokes equations. The principle of conservation of angular momentum, in the absence of distributed couples, leads to the symmetry of the stress tensor components: $\sigma_{ij} = \sigma_{ji}$ (i.e., $\sigma_{xy} = \sigma_{yx}$).

9.2.2.3 Constitutive relations

For viscous incompressible fluids, the total stress $\boldsymbol{\sigma}$ can be decomposed into hydrostatic and viscous parts:

$$\sigma_{xx} = \tau_{xx} - P, \quad \sigma_{yy} = \tau_{yy} - P, \quad \sigma_{xy} = \tau_{xy}, \tag{9.2.14}$$

where P is the hydrostatic pressure and $(\tau_{xx}, \tau_{yy}, \tau_{xy})$ are the components of the viscous stress tensor $\boldsymbol{\tau}$. For isotropic (i.e., the material properties are independent of direction) Newtonian fluids, the viscous stress tensor is related to the symmetric part of the velocity gradient tensor $\boldsymbol{\nabla}\mathbf{v}$, denoted as $\mathbf{D}$, by

$$\tau_{xx} = 2\mu D_{xx}, \quad \tau_{yy} = 2\mu D_{yy}, \quad \tau_{xy} = 2\mu D_{xy}, \tag{9.2.15}$$

where D_{xx}, D_{yy}, and D_{xy} are defined by

$$D_{xx} = \frac{\partial v_x}{\partial x}, \quad D_{yy} = \frac{\partial v_y}{\partial y}, \quad D_{xy} = \frac{1}{2}\left(\frac{\partial v_x}{\partial y} + \frac{\partial v_y}{\partial x}\right). \tag{9.2.16}$$

For an isotropic incompressible fluid, Eqs. (9.2.12) and (9.2.13) reduce to

$$\rho\left(v_x\frac{\partial v_x}{\partial x} + v_y\frac{\partial v_x}{\partial y}\right) - \frac{\partial}{\partial x}\left(2\mu\frac{\partial v_x}{\partial x}\right) + \frac{\partial}{\partial y}\left[\mu\left(\frac{\partial v_x}{\partial y} + \frac{\partial v_y}{\partial x}\right)\right] - \frac{\partial P}{\partial x} - f_x = 0, \tag{9.2.17}$$

$$\rho\left(v_x\frac{\partial v_y}{\partial x} + v_y\frac{\partial v_y}{\partial y}\right) - \frac{\partial}{\partial x}\left[\mu\left(\frac{\partial v_x}{\partial y} + \frac{\partial v_y}{\partial x}\right)\right] + \frac{\partial}{\partial x}\left(2\mu\frac{\partial v_y}{\partial y}\right) - \frac{\partial P}{\partial y} - f_y = 0. \tag{9.2.18}$$

9.2.2.4 Boundary conditions

The boundary conditions for the flow problem involve specifying one element of the duality pair $(\mathbf{v}, \mathbf{t})$ at every point on the boundary:

$$v_x = \hat{v}_x, \quad v_y = \hat{v}_y \quad \text{on} \quad \Gamma_v, \tag{9.2.19a}$$

$$t_x \equiv n_x\,\sigma_{xx} + n_y\,\sigma_{xy} = \hat{t}_x, \quad t_y \equiv n_x\,\sigma_{xy} + n_y\,\sigma_{yy} = \hat{t}_y, \quad \text{on} \quad \Gamma_\sigma, \tag{9.2.19b}$$

where $\hat{\mathbf{n}}$ is the unit normal to the boundary and Γ_v and Γ_σ are the boundary portions on which the velocity $\mathbf{v}$ and tractions (t_x, t_y), respectively, are specified.

9.3 Finite Element Model of Plane Elasticity

9.3.1 Weak Forms

The weak forms of the plane elasticity equations, Eqs. (9.2.9) and (9.2.10), over a typical finite element Ω^e with closed boundary Γ^e can be developed using the three-step procedure for each of the two differential equations: multiply the first equation with a weight function w_1 and the second equation with a weight function w_2 and integrate by parts to trade the differentiation equally between the weight functions and the dependent variables (u_x, u_y) in each case. The weight functions w_1 and w_2 have the meaning of first variations of u_x and u_y, respectively: $w_1 \sim \delta u_x$ and $w_2 \sim \delta u_y$. We obtain

$$\begin{aligned} 0 = h_e \int_{\Omega_e} \left[\frac{\partial w_1}{\partial x} \left(c_{11} \frac{\partial u_x}{\partial x} + c_{12} \frac{\partial u_y}{\partial y} \right) + \frac{\partial w_1}{\partial y} \left[c_{66} \left(\frac{\partial u_x}{\partial y} + \frac{\partial u_y}{\partial x} \right) \right] dxdy \\ - h_e \int_{\Omega_e} w_1 f_x \, dxdy - h_e \oint_{\Gamma_e} w_1 t_x \, ds \end{aligned} \quad (9.3.1)$$

$$\begin{aligned} 0 = h_e \int_{\Omega_e} \left[\frac{\partial w_2}{\partial x} \left[c_{66} \left(\frac{\partial u_x}{\partial y} + \frac{\partial u_y}{\partial x} \right) \right] + \frac{\partial w_2}{\partial y} \left(c_{12} \frac{\partial u_x}{\partial x} + c_{22} \frac{\partial u_y}{\partial y} \right) \right] dxdy \\ - h_e \int_{\Omega_e} w_2 f_y \, dxdy - h_e \oint_{\Gamma_e} w_2 t_y \, ds, \end{aligned} \quad (9.3.2)$$

where h_e is the thickness of the plane elastic body.

9.3.2 Finite Element Model

An examination of the weak forms in Eqs. (9.3.1) and (9.3.2) reveals that: (a) u_x and u_y are the primary variables, which must be carried as the primary nodal degrees of freedom; and (b) only first derivatives of u_x and u_y with respect to x and y appear in the weak forms. Therefore, u_x and u_y must be approximated by the Lagrange family of interpolation functions. The simplest elements that satisfy these requirements are the linear triangular and linear quadrilateral elements. Although u_x and u_y are independent of each other, they are the components of the displacement vector. Therefore, both components should be approximated using the same type and degree of interpolation.

Let $u \equiv u_x$ and $v \equiv u_y$ be approximated over the element domain by the finite element interpolations (the element label e is omitted in the interest of brevity)

$$u \equiv u_x \approx \sum_{j=1}^{n} u_j \, \psi_j^e(x, y), \qquad v \equiv u_y \approx \sum_{j=1}^{n} v_j \, \psi_j^e(x, y). \quad (9.3.3)$$

If a linear triangular element ($n = 3$) is used, we have two (u_i, v_i) $(i = 1, 2, 3)$ degrees of freedom per node and a total of six nodal displacements per element [see Fig. 9.3.1(a)]. For a bilinear quadrilateral element ($n = 4$), there are a total of eight nodal displacements per element [see Fig. 9.3.1(b)]. Since the first derivatives of ψ_i for a triangular element are element-wise constant, all strains

$(\varepsilon_{xx}, \varepsilon_{yy}, \varepsilon_{xy})$ computed for the linear triangular element are element-wise constant. Therefore, the linear triangular element for plane elasticity problems is known as the *constant-strain-triangular (CST) element.* For a quadrilateral element, the first derivatives of ψ_i are not constant: $\partial\psi_i^e/\partial\xi$ is linear in η and constant in ξ, and $\partial\psi_i^e/\partial\eta$ is linear in ξ and constant in η.

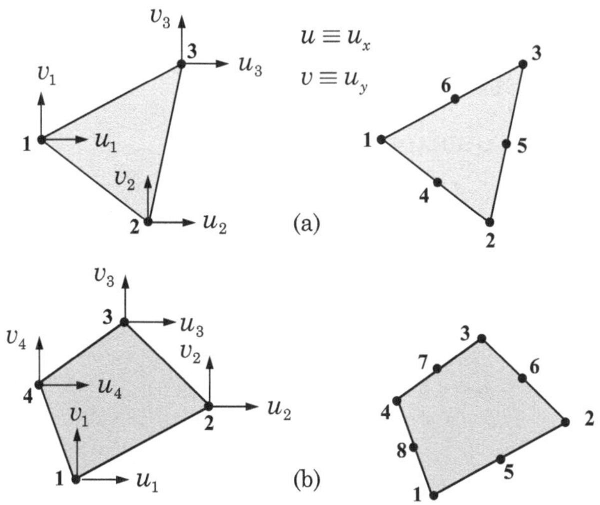

Fig. 9.3.1 Linear and quadratic (a) triangular and (b) quadrilateral elements for plane elasticity. Each node has two displacement degrees of freedom: (u, v).

Substituting Eq. (9.3.3) for u_x and u_y and $w_1 = \psi_i^e$ and $w_2 = \psi_i^e$, to obtain the ith algebraic equation associated with each of the weak statements in Eqs. (9.3.1) and (9.3.2), and writing the resulting algebraic equations in matrix form, we obtain

$$\begin{bmatrix} \mathbf{K}^{11} & \mathbf{K}^{12} \\ (\mathbf{K}^{12})^{\mathrm{T}} & \mathbf{K}^{22} \end{bmatrix} \begin{Bmatrix} \mathbf{u} \\ \mathbf{v} \end{Bmatrix} = \begin{Bmatrix} \mathbf{F}^1 \\ \mathbf{F}^2 \end{Bmatrix}, \tag{9.3.4}$$

where

$$\begin{aligned}
K_{ij}^{11} &= h_e \int_{\Omega_e} \left(c_{11} \frac{\partial \psi_i^e}{\partial x} \frac{\partial \psi_j^e}{\partial x} + c_{66} \frac{\partial \psi_i^e}{\partial y} \frac{\partial \psi_j^e}{\partial y} \right) dxdy \\
K_{ij}^{12} &= K_{ji}^{21} = h_e \int_{\Omega_e} \left(c_{12} \frac{\partial \psi_i^e}{\partial x} \frac{\partial \psi_j^e}{\partial y} + c_{66} \frac{\partial \psi_i^e}{\partial y} \frac{\partial \psi_j^e}{\partial x} \right) dxdy \\
K_{ij}^{22} &= h_e \int_{\Omega_e} \left(c_{66} \frac{\partial \psi_i^e}{\partial x} \frac{\partial \psi_j^e}{\partial x} + c_{22} \frac{\partial \psi_i^e}{\partial y} \frac{\partial \psi_j^e}{\partial y} \right) dxdy \\
F_i^1 &= h_e \int_{\Omega_e} \psi_i^e f_x \, dxdy + h_e \oint_{\Gamma_e} \psi_i^e t_x \, ds \equiv f_i^1 + Q_i^1, \\
F_i^2 &= h_e \int_{\Omega_e} \psi_i^e f_y \, dxdy + h_e \oint_{\Gamma_e} \psi_i^e t_y \, ds \equiv f_i^2 + Q_i^2
\end{aligned} \tag{9.3.5}$$

The body forces f_x and f_y are measured per unit volume, and the surface tractions t_x and t_y are measured per unit area. The coefficient matrix $\mathbf{K}^{\alpha\beta}$ corresponds to the coefficient of βth variable in the αth equation. Equations (9.3.1) and (9.3.2) are labelled as the first and second equations, respectively, and u and v are numbered as variables 1 and 2, respectively.

9.4 Dual Mesh Control Domain Model of Plane Elasticity

9.4.1 Governing Equations

The governing equations of plane elasticity, Eqs. (9.2.9) and (9.2.10), can be expressed in terms of the displacement components $u \equiv u_x$ and $v \equiv u_y$ as

$$\frac{\partial}{\partial x}\left(c_{11}\frac{\partial u}{\partial x} + c_{12}\frac{\partial v}{\partial y}\right) + \frac{\partial}{\partial y}\left[c_{66}\left(\frac{\partial u}{\partial y} + \frac{\partial v}{\partial x}\right)\right] + f_x = 0 \tag{9.4.1}$$

$$\frac{\partial}{\partial x}\left[c_{66}\left(\frac{\partial u}{\partial y} + \frac{\partial v}{\partial x}\right)\right] + \frac{\partial}{\partial y}\left(c_{12}\frac{\partial u}{\partial x} + c_{22}\frac{\partial v}{\partial y}\right) + f_y = 0 \tag{9.4.2}$$

where c_{ij} are the elasticity coefficients of the material, and (f_x, f_y) denote the x and y components, respectively, of the body force vector $\mathbf{f}$.

9.4.2 Control Domain Statements

To develop the discretized equations associated with Eqs. (9.4.1) and (9.4.2), first we identify the computational domain for the problem at hand. For the sake of simplicity, we assume that the domain of interest is rectangular in shape. The computational domain is represented with a primal mesh of rectangular finite elements, and the dual mesh is placed on the primal mesh such that it bisects the primal mesh, as shown in Fig. 9.4.1.

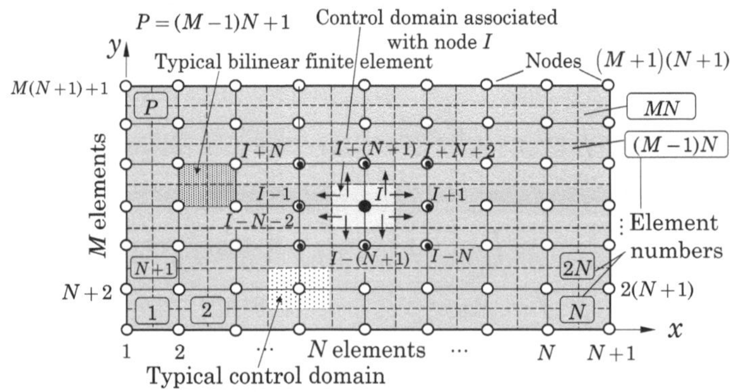

Fig. 9.4.1 Representation of a rectangular domain with a primal mesh (shown with solid lines) of bilinear rectangular finite elements and a dual mesh (shown with broken lines which bisect the solid lines) of rectangular control domains. For a uniform primal mesh, the dual mesh will be uniform and the nodes of the primal mesh will be at the center of the dual mesh.

We isolate a typical control domain, shown in Fig. 9.4.2, to develop the discretized equations associated with the governing equations, Eqs. (9.4.1) and (9.4.2). The control domain partially occupies the four elements of the primal mesh, and the nodes influencing the control domain are those coming from the finite element mesh. The node centered at the control domain is numbered as I, with the remaining nodes numbered based on a $N \times M$ mesh of finite elements (i.e., N finite elements along the x-axis and M finite elements along the y-axis; see Fig. 9.4.1).

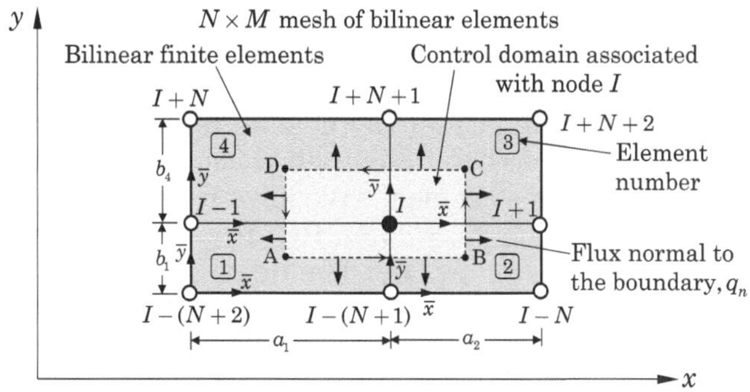

Fig. 9.4.2 The two-dimensional control domain associated with an interior node I. For a general $N \times M$ mesh of rectangular elements, $a_4 = a_1$, $a_3 = a_2$, $b_2 = b_1$, and $b_4 = b_3$; for a uniform mesh of $N \times M$ elements, $a_1 = a_2 = a$ and $b_1 = b_4 = b$.

Next, we satisfy the governing equations, Eqs. (9.4.1) and (9.4.2), over a typical control domain in an integral sense. This amounts to going back to the very derivation of the momentum equations in a linear elasticity course. The integral form is often known as the *global form* of the equilibrium equations. The local form is the differential equations, which is obtained by invoking the continuum assumption and shrinking an arbitrary volume to a point. Thus, writing the integral statement of the governing equations amounts to satisfying the global form of the differential equations. This is the most desirable feature of the FVM and DMCDM, which guarantees the satisfaction of the global form of the balance equations.

The integral statements of Eqs. (9.4.1) and (9.4.2) over a typical rectangular control domain Ω_R with boundary Γ_R (we note that Γ_R is defined piece-wise, i.e., it comprises of four line elements AB, BC, CD, and DA, enclosing Ω_R; see Fig. 9.4.2) are obtained as follows:

$$
\begin{aligned}
0 &= h \int_{\Omega_R} \left\{ \frac{\partial}{\partial x}\left(c_{11}\frac{\partial u}{\partial x} + c_{12}\frac{\partial v}{\partial y}\right) + \frac{\partial}{\partial y}\left[c_{66}\left(\frac{\partial u}{\partial y} + \frac{\partial v}{\partial x}\right)\right] + f_x \right\} dxdy \\
&= h \oint_{\Gamma_R} \left[\left(c_{11}\frac{\partial u}{\partial x} + c_{12}\frac{\partial v}{\partial y}\right) n_x + c_{66}\left(\frac{\partial u}{\partial y} + \frac{\partial v}{\partial x}\right) n_y \right] ds \\
&\quad + \int_{\Omega_R} f_x \, dxdy, \qquad (9.4.3)
\end{aligned}
$$

$$
\begin{aligned}
0 &= h\int_{\Omega_R}\left\{\frac{\partial}{\partial x}\left[c_{66}\left(\frac{\partial u}{\partial y}+\frac{\partial v}{\partial x}\right)\right]+\frac{\partial}{\partial y}\left(c_{12}\frac{\partial u}{\partial x}+c_{22}\frac{\partial v}{\partial y}\right)-f_y\right\}dxdy\\
&= h\oint_{\Gamma_R}\left[c_{66}\left(\frac{\partial u}{\partial y}+\frac{\partial v}{\partial x}\right)n_x+\left(c_{12}\frac{\partial u}{\partial x}+c_{22}\frac{\partial v}{\partial y}\right)n_y\right]ds\\
&\quad+\int_{\Omega_R} f_y\,dxdy, \qquad (9.4.4)
\end{aligned}
$$

where (n_x, n_y) are the direction cosines of the unit normal vector to the boundary of the control domain (see Fig. 9.4.2), and Γ_R is the boundary of the rectangular control domain $\Omega_R = [x_I - 0.5a, x_I + 0.5a] \times [y_I - 0.5b, y_I + 0.5b]$, with (x_I, y_I) being the global coordinates of the node labelled as I. The integration is taken all around the boundary of the control domain in the direction indicated. The boundary integrals can be simplified using the values of the direction cosines on each boundary line segment. We have $\hat{\mathbf{n}} = (n_x, n_y)$; $\hat{\mathbf{n}} = (0, -1)$ and $ds = dx$ on AB; $\hat{\mathbf{n}} = (1, 0)$ and $ds = dy$ on BC; $\hat{\mathbf{n}} = (0, 1)$ and $ds = -dx$ on CD; and $\hat{\mathbf{n}} = (-1, 0)$ and $ds = -dy$ on DA. The domain integrals involving f_x and f_y (measured per unit area) are evaluated using the one-third Simpson rule or the Gauss quadrature in each coordinate direction. For example, the integral of a function $g(x)$ is evaluated according to the one-third Simpson rule as

$$
\int_{x_1}^{x_3} g(x)\,dx = \frac{a}{3}\Big[g(x_1) + 4g(x_2) + g(x_3)\Big] \qquad (9.4.5)
$$

where $a = x_3 - x_2 = x_2 - x_1$ (i.e., $x_3 - x_1 = 2a$).

9.4.3 Discretized Equations

Over each element Ω^e of the primal mesh, each displacement component is approximated using the bilinear elements. Since the control domain partially occupies four bilinear elements (see Fig. 9.4.2), the line integrals in Eqs. (9.4.3) and (9.4.4) can be expressed as

$$
\begin{aligned}
\oint_{\Gamma_R}(\cdot)\,n_x\,ds &= -\int_{0.5b}^{b}(\cdot)\Big|_{\bar{x}=0.5a}^{(1)}d\bar{y} + \int_{0.5b}^{b}(\cdot)\Big|_{\bar{x}=0.5a}^{(2)}d\bar{y}\\
&\quad+\int_{0}^{0.5b}(\cdot)\Big|_{\bar{x}=0.5a}^{(3)}d\bar{y} - \int_{0}^{0.5b}(\cdot)\Big|_{\bar{x}=0.5a}^{(4)}d\bar{y} \qquad (9.4.6)
\end{aligned}
$$

Here the superscript (e) $(e = 1, 2, 3, 4)$ refers to the element number shown in Fig. 9.4.2 and $(\bar{x}, \bar{y})$ are the local coordinates with the origin at node 1 of each finite element. Similar relation holds for the closed contour (line) integral of $(\,\cdot\,)\,n_y$:

$$
\begin{aligned}
\oint_{\Gamma_R}(\cdot)\,n_y\,ds &= -\int_{0.5a}^{a}(\cdot)\Big|_{\bar{y}=0.5b}^{(1)}d\bar{x} - \int_{0}^{0.5a}(\cdot)\Big|_{\bar{y}=0.5b}^{(2)}d\bar{x}\\
&\quad+\int_{0}^{0.5a}(\cdot)\Big|_{\bar{y}=0.5b}^{(3)}d\bar{x} + \int_{0.5a}^{a}(\cdot)\Big|_{\bar{y}=0.5b}^{(4)}d\bar{x} \qquad (9.4.7)
\end{aligned}
$$

We have the following identities for the evaluation of the derivatives of a function $f(x, y)$ along the closed boundary Γ_R of an interior control domain Ω_R:

$$\begin{aligned}\oint_{\Gamma_R} \frac{\partial f}{\partial x} n_x\, ds = & \tfrac{1}{8}\frac{b_1}{a_1} F_{I-N-2} - \tfrac{1}{8}\left(\frac{b_1}{a_1} + \frac{b_2}{a_2}\right) F_{I-N-1} + \tfrac{1}{8}\frac{b_2}{a_2} F_{I-N} \\ & + \tfrac{3}{8}\left(\frac{b_1}{a_1} + \frac{b_4}{a_4}\right) F_{I-1} - \tfrac{3}{8}\left(\frac{b_1}{a_1} + \frac{b_2}{a_2} + \frac{b_3}{a_3} + \frac{b_4}{a_4}\right) F_I \\ & + \tfrac{3}{8}\left(\frac{b_2}{a_2} + \frac{b_3}{a_3}\right) F_{I+1} + \tfrac{1}{8}\frac{b_4}{a_4} F_{I+N} \\ & - \tfrac{1}{8}\left(\frac{b_3}{a_3} + \frac{b_4}{a_4}\right) F_{I+N+1} + \tfrac{1}{8}\frac{b_3}{a_3} F_{I+N+2},\end{aligned} \tag{9.4.8}$$

$$\begin{aligned}\oint_{\Gamma_R} \frac{\partial F}{\partial y} n_y\, ds = & \tfrac{1}{8}\frac{a_1}{b_1} F_{I-N-2} + \tfrac{3}{8}\left(\frac{a_1}{b_1} + \frac{a_2}{b_2}\right) F_{I-N-1} + \tfrac{1}{8}\frac{a_2}{b_2} F_{I-N} \\ & - \tfrac{1}{8}\left(\frac{a_1}{b_1} + \frac{a_4}{b_4}\right) F_{I-1} - \tfrac{3}{8}\left(\frac{a_1}{b_1} + \frac{a_2}{b_2} + \frac{a_3}{b_3} + \frac{a_4}{b_4}\right) F_I \\ & - \tfrac{1}{8}\left(\frac{a_2}{b_2} + \frac{a_3}{b_3}\right) F_{I+1} + \tfrac{1}{8}\frac{a_4}{b_4} F_{I+N} \\ & + \tfrac{3}{8}\left(\frac{a_3}{b_3} + \frac{a_4}{b_4}\right) F_{I+N+1} + \tfrac{1}{8}\frac{a_3}{b_3} F_{I+N+2},\end{aligned} \tag{9.4.9}$$

$$\oint_{\Gamma_R} \frac{\partial F}{\partial x} n_y\, ds = \oint_{\Gamma_R} \frac{\partial F}{\partial y} n_x\, ds = \frac{1}{4}\left(F_{I-N-2} - F_{I-N} + F_{I+N+2} - F_{I+N}\right), \tag{9.4.10}$$

where (a_i, b_i), $i = 1, 2, 3, 4$, denote the dimensions of the four rectangular elements associated with the control domain at node I. Due to the rectangular geometry and simplicity of the finite element interpolation functions, the integrals can be evaluated exactly.

Using the identities in Eqs. (9.4.8)–(9.4.10), Eqs. (9.4.3) and (9.4.4) can be expressed in terms of the values of the displacement components u and v at the nine nodes including the node I as follows (e.g., the values of u and v at node I are denoted with capital letters U_I and V_I, respectively; and symbol F is replaced by either symbol U or V, as indicated):

$$\begin{aligned}& h\, c_{11}\,[\text{Eq. (9.4.8) with } F \to U] + h\, c_{66}\,[\text{Eq. (9.4.9) with } F \to U] \\ & + \frac{h}{4}(c_{12} + c_{66})\,[\text{Eq. (9.4.10) with } F \to V] = F_I^x,\end{aligned} \tag{9.4.11}$$

$$\begin{aligned}& h\, c_{66}\,[\text{Eq. (9.4.8) with } F \to V] + h\, c_{22}\,[\text{Eq. (9.4.9) with } F \to V] \\ & + \frac{h}{4}(c_{12} + c_{66})\,[\text{Eq. (9.4.10) with } F \to U] = F_I^y,\end{aligned} \tag{9.4.12}$$

where h is the thickness of the body and

$$\begin{aligned} F_I^x = & \int_{x_I-0.5a}^{x_I} \int_{y_I-0.5b}^{y_I} f_x(x,y)\, dxdy + \int_{x_I}^{x_I+0.5a} \int_{y_I-0.5b}^{y_I} f_x(x,y)\, dxdy \\ & + \int_{x_I}^{x_I+0.5a} \int_{y_I}^{y_I+0.5b} f_x(x,y)\, dxdy + \int_{x_I-0.5a}^{x_I} \int_{y_I}^{y_I+0.5b} f_x(x,y)\, dxdy \end{aligned} \tag{9.4.13}$$

with a similar expression for F_I^y (f_x replaced with f_y).

For the nodes on the boundary, we must modify Eqs. (9.4.11) and (9.4.12), as explained in Chapter 5 [see Eqs. (5.4.13)–(5.4.18b)]. For a rectangular domain, the nodal point locations can be classified into several cases as shown in Fig. 5.4.4. Appendix 9A contains the evaluation of coefficients for interior as well as typical boundary nodes for plane elasticity problems.

This completes the derivation of the DMCDM equations for plane elasticity problems posed on rectangular domains with uniform meshes. Discrete models (either by the FEM or DMCDM) of finite elasticity are not considered here, as it requires the development of additional theoretical concepts (e.g., introduction of stress and strain measures).

Next, we consider the finite element model of creeping flows. The equations resulting from the omission of the convective terms in the Navier–Stokes equations, Eqs. (9.2.17) and (9.2.18), are known as the Stokes equations, which we term as the linearized Navier–Stokes equations. For the purpose of developing the discrete models of the Stokes or Navier–Stokes equations in terms of the velocity field only, we use the penalty function approach to eliminate the pressure variable. This approach does not affect the convective terms. Therefore, the pressure elimination is discussed as a part of the Stokes equations. The resulting equations (i.e., after the use of the penalty function method) are used to develop the discrete models using the FEM as well as DMCDM. The same equations are then modified to include the convective terms in developing the corresponding FEM and DMCDM models.

9.5 Finite Element Model of Creeping Flows

9.5.1 Penalty Function Formulation

In general, the penalty function method allows the inclusion of constraint conditions into variational statements or weak forms of the governing equations. In order to use the penalty function method for the problem at hand, we must first formulate the weak forms associated with Eqs. (9.2.12) and (9.2.13) [expressed in terms of the velocity components by using Eqs. (9.2.14)–(9.2.16)].

Suppose that the velocity field ($\mathbf{v}$) is such that the continuity equation, Eq. (9.2.11), is satisfied identically. Then, the variational problem can be stated as follows (see Reddy [8, 13] for details): among all vectors $\mathbf{v} = v_x\hat{\mathbf{e}}_x + v_y\hat{\mathbf{e}}_y$ that satisfy the continuity equation, Eq. (9.2.11), find the solution (v_x, v_y) that satisfy the variational problem (which is a combined form of the weak forms) defined by

$$0 = \int_{\Omega} \mu \left[2\frac{\partial w_x}{\partial x}\frac{\partial v_x}{\partial x} + \frac{\partial w_x}{\partial y}\left(\frac{\partial v_x}{\partial y} + \frac{\partial v_y}{\partial x} \right) \right] dxdy$$
$$- \int_{\Omega^e} \rho w_x f_x \, dxdy - \oint_{\Gamma} w_x t_x \, ds, \tag{9.5.1}$$

$$0 = \int_{\Omega} \mu \left[2\frac{\partial w_y}{\partial y}\frac{\partial v_y}{\partial y} + \frac{\partial w_y}{\partial x}\left(\frac{\partial v_x}{\partial y} + \frac{\partial v_y}{\partial x} \right) \right] dxdy$$
$$- \int_{\Omega} \rho w_y f_y \, dxdy - \oint_{\Gamma} w_y t_y \, ds, \tag{9.5.2}$$

where (w_x, w_y) are the weight functions such that they satisfy the condition

$$\frac{\partial w_x}{\partial x} + \frac{\partial w_y}{\partial y} = 0. \tag{9.5.3}$$

The variational problem in Eqs. (9.5.1) and (9.5.2) is a constrained variational problem, because the solution vector $\mathbf{v}$ is constrained to satisfy the continuity equation in Eq. (9.2.11).

The penalty function method is used to include the constraint condition back into the weak forms (to have an unconstrained problem) as explained in detail in [8, 10, 13] and is not repeated here. In essence, the penalty function approach applied to the equations at hand amounts to replacing the pressure variable in these equations with

$$P = -\gamma \left(\frac{\partial v_x}{\partial x} + \frac{\partial v_y}{\partial y} \right), \tag{9.5.4}$$

where γ is called the *penalty parameter*. The larger the value of γ, the more exactly the constraint is satisfied.

In view of Eq. (9.5.4), the weak forms of the Stokes equations [i.e., Eqs. (9.2.17) and (9.2.18) without the convective terms] are

$$0 = \int_{\Omega} \left[2\mu\frac{\partial w_x}{\partial x}\frac{\partial v_x}{\partial x} + \mu\frac{\partial w_x}{\partial y}\left(\frac{\partial v_x}{\partial y} + \frac{\partial v_y}{\partial x} \right) \right] dxdy - \int_{\Omega} \rho f_x \, w_x \, dxdy$$
$$- \oint_{\Gamma} w_x \, t_x \, ds + \int_{\Omega} \gamma \frac{\partial w_x}{\partial x}\left(\frac{\partial v_x}{\partial x} + \frac{\partial v_y}{\partial y} \right) dx\, dy \tag{9.5.5}$$

$$0 = \int_{\Omega} \left[2\mu\frac{\partial w_y}{\partial y}\frac{\partial v_y}{\partial y} + \mu\frac{\partial w_y}{\partial x}\left(\frac{\partial v_x}{\partial y} + \frac{\partial v_y}{\partial x} \right) \right] dxdy - \int_{\Omega} \rho f_y \, w_y \, dxdy$$
$$- \oint_{\Gamma} w_y \, t_y \, ds + \int_{\Omega} \gamma \frac{\partial w_y}{\partial y}\left(\frac{\partial v_x}{\partial x} + \frac{\partial v_y}{\partial y} \right) dxdy, \tag{9.5.6}$$

The two variational statements in Eqs. (9.5.5) and (9.5.6) are indeed the weak forms needed for the penalty finite element model. We note that the pressure does not appear explicitly in these weak forms, although it is a part of (t_x, t_y). An approximation for the pressure can be post-computed (at the reduced-order Gauss points) from Eq. (9.5.4) after the velocity field is determined from the discretized equations (by the FEM or the DMCDM).

It is useful in the sequel to rewrite Eqs. (9.2.17) and (9.2.18) by replacing the pressure P with Eq. (9.5.4):

$$\rho\left(v_x\frac{\partial v_x}{\partial x} + v_y\frac{\partial v_x}{\partial y}\right) - \frac{\partial}{\partial x}\left(2\mu\frac{\partial v_x}{\partial x}\right) + \frac{\partial}{\partial y}\left[\mu\left(\frac{\partial v_x}{\partial y} + \frac{\partial v_y}{\partial x}\right)\right]$$
$$-\frac{\partial}{\partial x}\left[\gamma\left(\frac{\partial v_x}{\partial x} + \frac{\partial v_y}{\partial y}\right)\right] - f_x = 0 \qquad (9.5.7)$$

$$\rho\left(v_x\frac{\partial v_y}{\partial x} + v_y\frac{\partial v_y}{\partial y}\right) - \frac{\partial}{\partial x}\left[\mu\left(\frac{\partial v_x}{\partial y} + \frac{\partial v_y}{\partial x}\right)\right] + \frac{\partial}{\partial x}\left(2\mu\frac{\partial v}{\partial y}\right)$$
$$-\frac{\partial}{\partial y}\left[\gamma\left(\frac{\partial v_x}{\partial x} + \frac{\partial v_y}{\partial y}\right)\right] - f_y = 0. \qquad (9.5.8)$$

9.5.2 Finite Element Model

The penalty finite element model of the linear equations is obtained using the weak forms in Eqs. (9.5.4) and (9.5.5) (with Ω replaced with Ω^e and Γ replaced with Γ^e; see [8]–[13] for details) by substituting the Lagrange interpolation of the velocity components

$$u \equiv v_x(x,y) \approx \sum_{j=1}^{m} u_j^e\,\psi_j^e(x,y), \quad v \equiv v_y(x,y) \approx \sum_{j=1}^{m} v_j^e\,\psi_j^e(x,y) \qquad (9.5.9)$$

where ψ_j^e are the Lagrange interpolation functions of element Ω^e and (u_j^e, v_j^e) are nodal values of the velocity components $(u \equiv v_x, v \equiv v_y)$, respectively. The weight functions (w_x, w_y) in the weak forms are replaced with $w_x = \psi_i^e$ and $w_y = \psi_i^e$. We obtain

$$\left(\mu\begin{bmatrix} 2\mathbf{S}^{11} + \mathbf{S}^{22} & (\mathbf{S}^{12})^{\mathrm{T}} \\ \mathbf{S}^{12} & \mathbf{S}^{11} + 2\mathbf{S}^{22}\end{bmatrix} + \gamma\begin{bmatrix}\mathbf{S}^{11} & \mathbf{S}^{12} \\ (\mathbf{S}^{12})^{\mathrm{T}} & \mathbf{S}^{22}\end{bmatrix}\right)\begin{Bmatrix}\mathbf{u}\\ \mathbf{v}\end{Bmatrix} = \begin{Bmatrix}\mathbf{F}^1\\ \mathbf{F}^2\end{Bmatrix} \qquad (9.5.10)$$

where $\mathbf{S}^{\alpha\beta}$ $(\alpha, \beta = 1, 2)$, $\mathbf{F}^1$, and $\mathbf{F}^2$ are given by

$$S_{ij}^{\alpha\beta} = \int_{\Omega^e} \frac{\partial \psi_i^e}{\partial x_\alpha}\frac{\partial \psi_j^e}{\partial x_\beta}\,dxdy \quad (\alpha, \beta = 1, 2;\ x_1 = x, x_2 = y),$$
$$F_i^1 = \int_{\Omega^e} \rho\,\psi_i f_x\,dxdy + \oint_{\Gamma^e} \psi_i t_x\,ds, \qquad (9.5.11)$$
$$F_i^2 = \int_{\Omega^e} \rho\,\psi_i f_y\,dxdy + \oint_{\Gamma^e} \psi_i t_y\,ds.$$

Figure 9.3.1 contains triangular and quadrilateral elements valid for the penalty finite element model with (u_i^e, v_i^e) denoting the nodal velocities.

The numerical evaluation of the coefficient matrices appearing in Eq. (9.5.10) requires special consideration. Equation (9.5.10) is of the form

$$(\mathbf{K}_\mu^e + \mathbf{K}_\gamma^e)\mathbf{v}^e = \mathbf{F}^e \tag{9.5.12}$$

where $\mathbf{K}_\mu^e$ is the contribution from the viscous terms and $\mathbf{K}_\gamma^e$ is from the penalty terms, which comes from the incompressibility constraint in Eq. (9.2.1). In theory, as we increase the value of γ, the conservation of mass is satisfied more exactly. However, in practice, for some large value of γ, the contribution from the viscous terms would be negligibly small compared to the penalty terms in a computer. Thus, if $\mathbf{K}_\gamma^e$ is a nonsingular (i.e., invertible) matrix, the solution of the global (i.e., assembled) equations associated with Eq. (9.5.12) for a large value of γ is trivial, $\{\mathbf{v}\} = \{\mathbf{0}\}$. While the solution satisfies the continuity equation, it does not satisfy the momentum equations. In this case the discrete problem in Eq. (9.5.12) is said to be overconstrained or "locked." If $\mathbf{K}_\gamma^e$ is singular, then the sum $\mu\mathbf{K}_\mu^e + \mathbf{K}_\gamma^e$ is nonsingular because $\mathbf{K}_\gamma^e$ is nonsingular for a problem without a "rigid-body mode" and a nontrivial solution to the problem is obtained.

It is found that if the coefficients of $\mathbf{K}_\gamma^e$ (i.e., penalty matrix coefficients) are evaluated using a numerical integration rule of an order less than that required to integrate them exactly, the finite element equations, Eq. (9.5.12), give acceptable solutions for the velocity field. This technique of underintegrating the penalty terms is known in the literature as reduced integration. For example, if a linear quadrilateral element is used to approximate the velocity field, the matrix coefficients $\mathbf{K}_\mu^e$ are evaluated using the 2×2 Gauss quadrature, and $\mathbf{K}_\gamma^e$ is evaluated using the one-point (1×1) Gauss quadrature. In addition, the value of the penalty parameter γ must be such that the contribution of the viscous terms to the coefficient matrix should be significant in a computer to be nonsingular. The value of the penalty parameter is found to be in the range of $\gamma = 10^6 - 10^{10}$ for best results (i.e., to satisfy the continuity equation within a meaningful error tolerance, say 10^{-3}; see Reddy [13]).

9.6 Dual Mesh Control Domain Model of Creeping Flows

9.6.1 Governing Equations

The governing equations of creeping flows of viscous incompressible fluids (i.e., the Stokes equations) using the penalty function method can be obtained from Eqs. (9.5.7) and (9.5.8) by omitting the convective terms (and using the notation $u \equiv v_x$ and $v \equiv v_y$)

$$\frac{\partial}{\partial x}\left(2\mu\frac{\partial u}{\partial x}\right) + \frac{\partial}{\partial y}\left[\mu\left(\frac{\partial u}{\partial y} + \frac{\partial v}{\partial x}\right)\right] + \frac{\partial}{\partial x}\left[\gamma\left(\frac{\partial u}{\partial x} + \frac{\partial v}{\partial y}\right)\right] + f_x = 0, \tag{9.6.1}$$

$$\frac{\partial}{\partial x}\left[\mu\left(\frac{\partial u}{\partial y} + \frac{\partial v}{\partial x}\right)\right] + \frac{\partial}{\partial x}\left(2\mu\frac{\partial v}{\partial y}\right) + \frac{\partial}{\partial y}\left[\gamma\left(\frac{\partial u}{\partial x} + \frac{\partial v}{\partial y}\right)\right] + f_y = 0, \tag{9.6.2}$$

where u and v denote the x and y components, respectively, of the velocity vector $\mathbf{v}$ referred to a rectangular Cartesian coordinates system (x, y), μ is the viscosity of the fluid, (f_x, f_y) denote the x and y components, respectively, of the body force vector $\mathbf{f}$, and γ is the penalty parameter whose magnitude dictates how closely the conservation of mass for incompressible fluids is satisfied:

$$\frac{\partial u}{\partial x} + \frac{\partial v}{\partial y} = 0. \tag{9.6.3}$$

Once the velocity field (u_γ, v_γ) is determined using Eqs. (9.6.1) and (9.6.2), the pressure variable P is post-computed from the velocity field using

$$P_\gamma = -\gamma \left(\frac{\partial u_\gamma}{\partial x} + \frac{\partial v_\gamma}{\partial y} \right). \tag{9.6.4}$$

9.6.2 Control Domain Statements

Due to the similarity of the governing equations of plane elasticity and those of the Stokes flow, all of the ideas already presented for the DMCDM model of plane elasticity are valid here, with the exception that the penalty terms have to be treated differently. To develop the discretized equations associated with Eqs. (9.6.1) and (9.6.2), we use the notation employed in the $M \times N$ primal mesh shown in Fig. 9.4.1. As in the case of plane elasticity, we isolate a typical control domain shown in Fig. 9.4.2 to develop the discretized equations associated with the governing equations (9.6.1) and (9.6.2).

Next, we set up integral statements of the governing equations, Eqs. (9.6.1) and (9.6.2), over a typical control domain. These statements are equivalent to the *global form* of the Stokes equations. The integral statements of Eqs. (9.6.1) and (9.6.2) over a typical rectangular control domain are obtained as [we note that (x_I, y_I) are the global coordinates of the node labeled as I, (n_x, n_y) are the direction cosines of the unit normal vector to the boundary of the control domain, as shown in Fig. 9.4.2, and Γ_R is the boundary of the rectangular control domain]:

$$\begin{aligned}
0 &= \int_{x_I-0.5a}^{x_I+0.5a} \int_{y_I-0.5b}^{y_I+0.5b} \left\{ \frac{\partial}{\partial x}\left(2\mu \frac{\partial u}{\partial x}\right) + \frac{\partial}{\partial y}\left[\mu\left(\frac{\partial u}{\partial y} + \frac{\partial v}{\partial x}\right)\right] \right. \\
&\qquad \left. + \frac{\partial}{\partial x}\left[\gamma\left(\frac{\partial u}{\partial x} + \frac{\partial v}{\partial y}\right)\right] + f_x \right\} dxdy \\
&= \oint_{\Gamma_R} \left[2\mu \frac{\partial u}{\partial x} n_x + \mu\left(\frac{\partial u}{\partial y} + \frac{\partial v}{\partial x}\right) n_y + \gamma\left(\frac{\partial u}{\partial x} + \frac{\partial v}{\partial y}\right) n_x \right] ds \\
&\quad + \int_{x_I-0.5a}^{x_I+0.5a} \int_{y_I-0.5b}^{y_I+0.5b} f_x \, dxdy,
\end{aligned} \tag{9.6.5}$$

$$
\begin{aligned}
0 = &\int_{x_I-0.5a}^{x_I+0.5a} \int_{y_I-0.5b}^{y_I+0.5b} \left\{ \frac{\partial}{\partial x}\left[\mu\left(\frac{\partial u}{\partial y}+\frac{\partial v}{\partial x}\right)\right] + \frac{\partial}{\partial y}\left(2\mu\frac{\partial v}{\partial y}\right) \right.\\
&\left. + \frac{\partial}{\partial y}\left[\gamma\left(\frac{\partial u}{\partial x}+\frac{\partial v}{\partial y}\right)\right] + f_y \right\} dxdy\\
= &\oint_{\Gamma_R}\left[\mu\left(\frac{\partial u}{\partial y}+\frac{\partial v}{\partial x}\right)n_x + 2\mu\frac{\partial v}{\partial y}n_y + \gamma\left(\frac{\partial u}{\partial x}+\frac{\partial v}{\partial y}\right)n_y\right] dxdy\\
&+ \int_{x_I-0.5a}^{x_I+0.5a} \int_{y_I-0.5b}^{y_I+0.5b} f_y\, dxdy,
\end{aligned} \tag{9.6.6}
$$

where the domain integrals (when f_x and f_y are nonzero) are evaluated using the one-third Simpson rule in each coordinate direction as given in Eq. (9.4.5).

9.6.3 Discretized Equations

First, a comment on the numerical evaluation of the penalty terms (i.e., terms involving γ) is in order. As discussed in Section 9.5.2, the penalty terms should be evaluated using "reduced" integration so that the coefficient matrix associated with the penalty terms is singular; otherwise, because of the large value of γ, the coefficient matrix will be dominated by the penalty terms over the viscous terms and yield erroneous solutions.

Over each element Ω^e of the primal mesh, each velocity component is approximated using the bilinear elements [see Eq. (9.5.9) with ψ_j^e being the bilinear interpolation functions]. Since the control domain partially occupies four bilinear elements (see Fig. 9.4.2), the line integrals in Eqs. (9.6.5) and (9.6.6) can be expressed as given in Eqs. (9.4.6) and (9.4.7). Using Eqs. (9.4.8)–(9.4.10) in Eqs. (9.6.5) and (9.6.6) (also see Appendix 9B), we obtain

$$
\begin{aligned}
&\left[-\mu\left(\frac{b_1}{4a_1}+\frac{a_1}{8b_1}\right)-\frac{\gamma}{4}\left(\frac{b_1}{a_1}\right)\right]U_{I-N-2}+\\
&\left[\mu\left(\frac{b_1}{4a_1}-\frac{3a_1}{8b_1}+\frac{b_2}{4a_2}-\frac{3a_2}{8b_2}\right)+\frac{\gamma}{4}\left(\frac{b_1}{a_1}+\frac{b_2}{a_2}\right)\right]U_{I-N-1}+\\
&\left[\mu\left(\frac{3b_1}{4a_1}+\frac{3a_1}{8b_1}+\frac{3b_2}{4a_2}+\frac{3a_2}{8b_2}+\frac{3b_3}{4a_3}+\frac{3a_3}{8b_3}+\frac{3b_4}{4a_4}+\frac{3a_4}{8b_4}\right)\right.\\
&\left.+\frac{\gamma}{4}\left(\frac{b_1}{a_1}+\frac{b_2}{a_2}+\frac{b_3}{a_3}+\frac{b_4}{a_4}\right)\right]U_I+\\
&\left[\mu\left(-\frac{3b_1}{4a_1}+\frac{a_1}{8b_1}-\frac{3b_4}{4a_4}+\frac{a_4}{8b_4}\right)-\frac{\gamma}{4}\left(\frac{b_1}{a_1}+\frac{b_4}{a_4}\right)\right]U_{I-1}+\\
&\left[-\mu\left(\frac{b_2}{4a_2}+\frac{a_2}{8b_2}\right)-\frac{\gamma}{4}\left(\frac{b_2}{a_2}\right)\right]U_{I-N}+\\
&\left[\mu\left(-\frac{3b_2}{4a_2}+\frac{a_2}{8b_2}-\frac{3b_3}{4a_3}+\frac{a_3}{8b_3}\right)-\frac{\gamma}{4}\left(\frac{b_2}{a_2}+\frac{b_3}{a_3}\right)\right]U_{I+1}+
\end{aligned}
$$

$$\left[-\mu\left(\frac{b_3}{4a_3}+\frac{a_3}{8b_3}\right)-\frac{\gamma}{4}\left(\frac{b_3}{a_3}\right)\right]U_{I+N+2}+$$
$$\left[\mu\left(\frac{b_3}{4a_3}-\frac{3a_3}{8b_3}+\frac{b_4}{4a_4}-\frac{3a_4}{8b_4}\right)+\frac{\gamma}{4}\left(\frac{b_3}{a_3}+\frac{b_4}{a_4}\right)\right]U_{I+N+1}+$$
$$\left[-\mu\left(\frac{b_4}{4a_4}+\frac{a_4}{8b_4}\right)-\frac{\gamma}{4}\left(\frac{b_4}{a_4}\right)\right]U_{I+N}+$$
$$\frac{1}{4}\left[-(\mu+\gamma)V_{I-N-2}+(\mu+\gamma)V_{I-N}-(\mu+\gamma)V_{I+N+2}+(\mu+\gamma)V_{I+N}\right]=0 \tag{9.6.7}$$

$$\left[-\mu\left(\frac{b_1}{8a_1}+\frac{a_1}{4b_1}\right)-\frac{\gamma}{4}\left(\frac{a_1}{b_1}\right)\right]V_{I-N-2}+$$
$$\left[\mu\left(\frac{b_1}{8a_1}-\frac{3a_1}{4b_1}+\frac{b_2}{8a_2}-\frac{3a_2}{4b_2}\right)-\frac{\gamma}{4}\left(\frac{a_1}{b_1}+\frac{a_2}{b_2}\right)\right]V_{I-N-1}+$$
$$\left[\mu\left(\frac{3b_1}{8a_1}+\frac{3a_1}{4b_1}+\frac{3b_2}{8a_2}+\frac{3a_2}{4b_2}+\frac{3b_3}{8a_3}+\frac{3a_3}{4b_3}+\frac{3b_4}{8a_4}+\frac{3a_4}{4b_4}\right)\right.$$
$$\left.+\frac{\gamma}{4}\left(\frac{a_1}{b_1}+\frac{a_2}{b_2}+\frac{a_3}{b_3}+\frac{a_4}{b_4}\right)\right]V_I+$$
$$\left[\mu\left(-\frac{3b_1}{8a_1}+\frac{a_1}{4b_1}-\frac{3b_4}{8a_4}+\frac{a_4}{4b_4}\right)+\frac{\gamma}{4}\left(\frac{a_1}{b_1}+\frac{a_4}{b_4}\right)\right]V_{I-1}+$$
$$\left[-\mu\left(\frac{b_2}{8a_2}+\frac{a_2}{4b_2}\right)-\frac{\gamma}{4}\left(\frac{a_2}{b_2}\right)\right]V_{I-N}+$$
$$\left[\mu\left(-\frac{3b_2}{8a_2}+\frac{a_2}{4b_2}-\frac{3b_3}{8a_3}+\frac{a_3}{4b_3}\right)+\frac{\gamma}{4}\left(\frac{a_2}{b_2}+\frac{a_3}{b_3}\right)\right]V_{I+1}+$$
$$\left[-\mu\left(\frac{b_3}{8a_3}+\frac{a_3}{4b_3}\right)-\frac{\gamma}{4}\left(\frac{a_3}{b_3}\right)\right]V_{I+N+2}+$$
$$\left[\mu\left(\frac{b_3}{8a_3}-\frac{3a_3}{4b_3}+\frac{b_4}{8a_4}-\frac{3a_4}{4b_4}\right)-\frac{\gamma}{4}\left(\frac{a_3}{b_3}+\frac{a_4}{b_4}\right)\right]V_{I+N+1}+$$
$$\left[-\mu\left(\frac{b_4}{8a_4}+\frac{a_4}{4b_4}\right)-\frac{\gamma}{4}\left(\frac{a_4}{b_4}\right)\right]V_{I+N}+$$
$$\frac{1}{4}\left[-(\mu+\gamma)U_{I-N-2}+(\mu+\gamma)U_{I-N}-(\mu+\gamma)U_{I+N+2}+(\mu+\gamma)U_{I+N}\right]=0 \tag{9.6.8}$$

For the nodes on the boundary, we must modify Eqs. (9.6.5) and (9.6.6), as explained in Section 5.4.3. Appendix 9B also contains the evaluation of coefficients for typical boundary nodes. This completes the derivation of DMCDM equations for the Stokes equations on rectangular domains and uniform meshes. Next, we extend the formulation to include the nonlinear (i.e., convective) terms of the Navier–Stokes equations.

9.7 Discrete Models of the Navier–Stokes Equations

9.7.1 Finite Element Model

The nonlinear terms can be added to Eqs. (9.6.5) and (9.6.6) to obtain the weak forms associated with the Navier–Stokes equations in Eqs. (9.2.17) and (9.2.18). We have the following weak forms over a typical element Ω^e with boundary Γ^e:

$$\begin{aligned}
0 = &\int_{\Omega^e} \left[w_x \,\rho \left(v_x \frac{\partial v_x}{\partial x} + v_y \frac{\partial v_x}{\partial y} \right) + 2\mu \frac{\partial w_x}{\partial x}\frac{\partial v_x}{\partial x} \right. \\
&\left. + \mu \frac{\partial w_x}{\partial y}\left(\frac{\partial v_x}{\partial y} + \frac{\partial v_y}{\partial x} \right) + \gamma \frac{\partial w_x}{\partial x}\left(\frac{\partial v_x}{\partial x} + \frac{\partial v_y}{\partial y} \right) \right] dxdy \\
&- \int_{\Omega^e} \rho f_x \, w_x \, dxdy - \oint_{\Gamma^e} w_x t_x \, ds,
\end{aligned} \tag{9.7.1}$$

$$\begin{aligned}
0 = &\int_{\Omega^e} \left[w_y \,\rho \left(v_x \frac{\partial v_y}{\partial x} + v_y \frac{\partial v_y}{\partial y} \right) + 2\mu \frac{\partial w_y}{\partial y}\frac{\partial v_y}{\partial y} \right. \\
&\left. + \mu \frac{\partial w_y}{\partial x}\left(\frac{\partial v_x}{\partial y} + \frac{\partial v_y}{\partial x} \right) + \gamma_e \frac{\partial w_y}{\partial y}\left(\frac{\partial v_x}{\partial x} + \frac{\partial v_y}{\partial y} \right) \right] dxdy \\
&- \int_{\Omega^e} \rho f_y \, w_y \, dx\,dy - \oint_{\Gamma^e} w_y \, t_y \, ds.
\end{aligned} \tag{9.7.2}$$

The penalty finite element model of the Navier–Stokes equations is obtained from Eqs. (9.7.1) and (9.7.2) by substituting the finite element interpolation from Eq. (9.5.9) for the velocity field, and $w_x = \psi_i^e$ and $w_y = \psi_i^e$:

$$\begin{aligned}
\left(\begin{bmatrix} \mathbf{C}(\mathbf{v}) & \mathbf{0} \\ \mathbf{0} & \mathbf{C}(\mathbf{v}) \end{bmatrix} + \mu \begin{bmatrix} 2\mathbf{S}^{11} + \mathbf{S}^{22} & (\mathbf{S}^{12})^{\mathrm{T}} \\ \mathbf{S}^{12} & \mathbf{S}^{11} + 2\mathbf{S}^{22} \end{bmatrix} + \gamma \begin{bmatrix} \mathbf{S}^{11} & \mathbf{S}^{12} \\ (\mathbf{S}^{12})^{\mathrm{T}} & \mathbf{S}^{22} \end{bmatrix} \right) \begin{Bmatrix} \mathbf{v}_x \\ \mathbf{v}_y \end{Bmatrix} \\
= \begin{Bmatrix} \mathbf{F}^1 \\ \mathbf{F}^2 \end{Bmatrix},
\end{aligned} \tag{9.7.3}$$

where $\mathbf{S}^{\alpha\beta}$ $(\alpha, \beta = 1, 2)$, $\mathbf{F}^1$, and $\mathbf{F}^2$ are the same as those defined in Eq. (9.5.11), and the nonlinear contribution due to the convective terms of the Navier–Stokes equations is included in the matrix coefficients

$$C_{ij} = \int_{\Omega^e} \rho \,\psi_i \left(v_x \frac{\partial \psi_j}{\partial x} + v_y \frac{\partial \psi_j}{\partial y} \right) dx\,dy. \tag{9.7.4}$$

As before, the linearization involves evaluating C_{ij} coefficients by assuming that v_x and v_y are known from the preceding iteration (r)

$$v_x(x,y) \approx \sum_{j=1}^{m} u_j^{(r)} \, \psi_j^e(x,y), \quad v_y(x,y) \approx \sum_{j=1}^{m} v_j^{(r)} \, \psi_j^e(x,y). \tag{9.7.5}$$

9.7.2 Dual Mesh Control Domain Model

By adding the convective (nonlinear) terms to Eqs. (9.6.1) and (9.6.2) (see [8]–[13] for details), we obtain (with $u \equiv v_x$ and $v \equiv v_y$)

$$\rho\left(u\frac{\partial u}{\partial x}+v\frac{\partial u}{\partial y}\right)-\frac{\partial}{\partial x}\left(2\mu\frac{\partial u}{\partial x}\right)-\frac{\partial}{\partial y}\left[\mu\left(\frac{\partial u}{\partial y}+\frac{\partial v}{\partial x}\right)\right]\\ -\frac{\partial}{\partial x}\left[\gamma\left(\frac{\partial u}{\partial x}+\frac{\partial v}{\partial y}\right)\right]-f_x=0, \tag{9.7.6}$$

$$\rho\left(u\frac{\partial v}{\partial x}+v\frac{\partial v}{\partial y}\right)-\frac{\partial}{\partial x}\left[\mu\left(\frac{\partial u}{\partial y}+\frac{\partial v}{\partial x}\right)\right]-\frac{\partial}{\partial x}\left(2\mu\frac{\partial v}{\partial y}\right)\\ -\frac{\partial}{\partial y}\left[\gamma\left(\frac{\partial u}{\partial x}+\frac{\partial v}{\partial y}\right)\right]-f_y=0, \tag{9.7.7}$$

where ρ denotes the density of the fluid.

The integral statements in Eqs. (9.6.5) and (9.6.6) are modified in this case to read

$$0=\int_{x_I-0.5a}^{x_I+0.5a}\int_{y_I-0.5b}^{y_I+0.5b}\left[\rho\left(u\frac{\partial u}{\partial x}+v\frac{\partial u}{\partial y}\right)+f_x\right]dxdy\\ -\oint_{\Gamma_R}\left[2\mu\frac{\partial u}{\partial x}n_x+\mu\left(\frac{\partial u}{\partial y}+\frac{\partial v}{\partial x}\right)n_y+\gamma\left(\frac{\partial u}{\partial x}+\frac{\partial v}{\partial y}\right)n_x\right]ds, \tag{9.7.8}$$

$$0=\int_{x_I-0.5a}^{x_I+0.5a}\int_{y_I-0.5b}^{y_I+0.5b}\left[\rho\left(u\frac{\partial v}{\partial x}+v\frac{\partial v}{\partial y}\right)+f_y\right]dxdy\\ -\oint_{\Gamma_R}\left[\mu\left(\frac{\partial u}{\partial y}+\frac{\partial v}{\partial x}\right)n_x+2\mu\frac{\partial v}{\partial y}n_y+\gamma\left(\frac{\partial u}{\partial x}+\frac{\partial v}{\partial y}\right)n_y\right]dxdy. \tag{9.7.9}$$

As in any numerical method, the nonlinear expressions (i.e., the convective terms) are linearized by assuming that the velocity field (u, v) is known from the previous iteration. Then the line integrals are evaluated as explained previously, while the area integrals in Eqs. (9.7.8) and (9.7.9) are numerically evaluated using an appropriate numerical integration scheme, such as the one-third Simpson rule given in Eq. (7.2.21a) but extended to two dimensions. In particular, for integration on a control domain of dimensions $0.5a \times 0.5b$, we have

$$\int_0^{0.5b}\int_0^{0.5a} f(x,y)\,dxdy\\ \approx \frac{ab}{144}\Big[f(x_1,y_1)+4f(x_2,y_1)+f(x_3,y_1)+4f(x_1,y_2)+16f(x_2,y_2)\\ +4f(x_3,y_2)+f(x_1,y_3)+4f(x_2,y_3)+f(x_3,y_3)\Big] \tag{9.7.10}$$

where (x_i, y_j) are evaluation points in the domain (i.e., $x = 0.0, 0.25a, 0.5a$ and $y = 0.0, 0.25b, 0.5b$).

9.8 Numerical Examples

Here we consider a number of examples to illustrate the ideas presented in the previous sections. The first two examples deal with linear elasticity followed by a Stokes flow example. Then, we solve the standard benchmark problem, namely the wall-driven cavity problem, using the Navier–Stokes equations. Numerical results obtained with the DMCDM are compared with the available numerical solutions obtained with the FEM and FVM.

Example 9.8.1

Consider an isotropic ($\nu = 0.25$ and $E = 207$ GPa) elastic plate of inplane dimensions a m and b m and thickness h m and subjected to a uniformly distributed edge stress of intensity t_0 N/m^2, as shown in Fig. 9.8.1. Assuming that the body forces are zero $f_x = f_y = 0$, determine the displacements and stresses using a 4×4 mesh of bilinear elements.

Solution: This problem exhibits biaxial symmetry. Hence, we can use one quadrant of the domain as the computational domain with suitable boundary conditions, as shown Fig. 9.8.1.

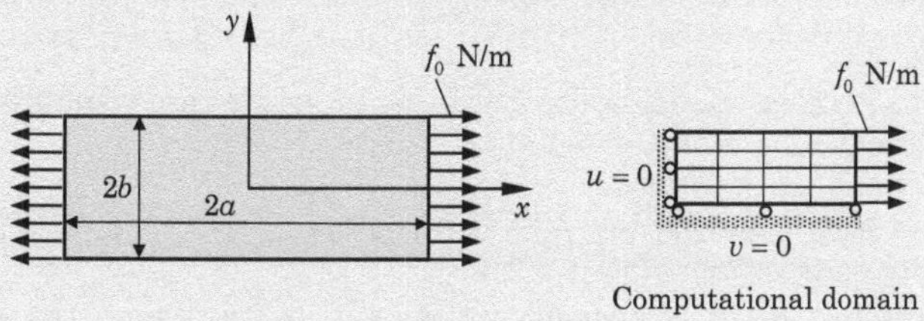

Fig. 9.8.1 Geometry and 4×4 mesh of bilinear elements for a plane elasticity problem.

The specified global displacement degrees of freedom and forces for the 4×4 mesh are

$$U_1 = U_6 = U_{11} = U_{16} = U_{21} = 0; \quad V_1 = V_2 = V_3 = V_4 = V_5 = 0. \tag{1}$$

The known nonzero forces in the x direction at global nodes 5, 10, 15, 20, and 25 are computed as follows ($\Delta y = a/4$ m):

$$F_{5x} = F_{25x} = 0.5\, t_0 h \Delta y \text{ N}, \quad F_{10x} = F_{15x} = F_{20x} = 0.5\,(t_0 h \Delta y + f_0 h \Delta y) \text{ N}. \tag{2}$$

For this problem, one can obtain the exact solution (see Reddy [23]) using the Airy stress function approach (it is easy to determine the stress field, but it is a bit involved to determine the displacement field). The exact stress field is given by $\sigma_{xx} = t_0$ N/m^2 and $\sigma_{yy} = \sigma_{xy} = 0$. Both the FEM and DMCDM predict the exact stress field. The x-component of displacement at $(x, y) = (a, y)$ is $u = t_0 a/E$. For $a = 1$ m, $E = 207 \times 10^9$ N/m^2, and $t_0 = 10^6$ N/m^2, we obtain $\sigma_{xx} = 1$ MN/m^2 and $u(a, y) = 0.4831 \times 10^{-5}$ m, which is what the FEM and DMCDM predicted. The results are independent of the mesh.

Example 9.8.2

Consider an isotropic thin elastic plate of inplane dimensions $a = 120$ in. (3.048 m) and $b = 160$ in. (4.064 m) and thickness $h = 0.036$ in. (0.9144 mm) and subjected to a uniformly distributed edge load of intensity $f_0 = 10$ lb/in. (1.75 kN/m), as shown in Fig. 9.8.2. Use $\nu = 0.25$, $E = 30 \times 10^6$ lb/in^2 (206.85 GPa), and $f_x = f_y = 0$. Determine the nodal displacements and stresses at the center of each element in the mesh using various meshes of bilinear elements.

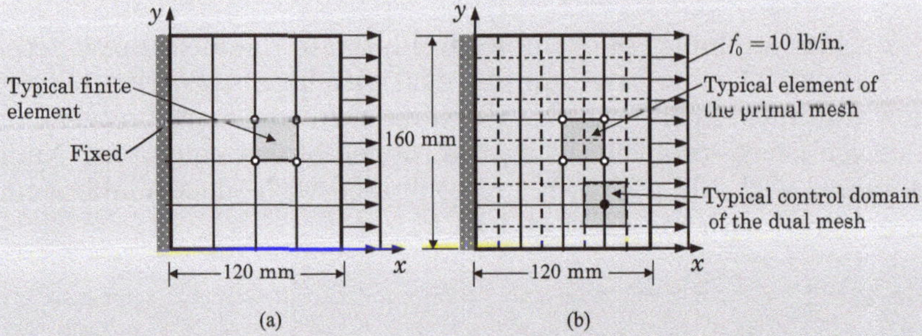

Fig. 9.8.2 Geometry and 4 × 4 mesh of bilinear elements (as an example) for (a) the FEM and (b) DMCDM of a plane elasticity problem.

Solution: The specified global displacement degrees of freedom and forces for the $m \times n$ mesh are obvious. For example, for a 4 × 4 mesh, we have

$$U_1 = V_1 = U_6 = V_6 = U_{11} = V_{11} = U_{16} = V_{16} = U_{21} = V_{21} = 0. \tag{1}$$

The known nonzero forces at global nodes 5, 10, 15, 20, and 25 are computed as follows (f_0 is the force per unit length; $h_y = 160/4 = 40$ in.):

$$F_{5x} = F_{25x} = \frac{f_0 h_y}{2} = 200\,\text{lb}, \quad F_{10x} = F_{15x} = F_{20x} = \frac{f_0 h_y}{2} + \frac{f_0 h_y}{2} = 400\,\text{lb}. \tag{2}$$

Table 9.8.1 contains the finite element solutions (deflections and stresses) obtained with various $n \times n$ meshes of FEM and DMCDM. The first line corresponds to FEM and the second line to DMCDM. We note that the FEM and DMCDM solutions are very close to each other.

Table 9.8.1 Deflections[a] and stresses obtained with various meshes of bilinear rectangular elements in an isotropic plate subjected to uniform edge load ($\bar{u} = u \times 10^4$ and $\bar{v} = v \times 10^4$).

Mesh	$\bar{u}(120,0)$	$\bar{v}(120,0)$	σ_{xx}	σ_{yy}	σ_{xy}
1 × 1	10.853	2.326	277.78	25.84	0.0
			(60, 80)	(60, 80)	(60, 80)[b]
	10.784	1.961	277.78	32.68	0.0
			(60, 80)	(60, 80)	(60, 80)
2 × 2	11.078	2.021	277.78	37.46	13.23
			(30, 40)	(30, 40)	(30, 40)
	11.058	1.946	277.78	39.13	13.09
			(30, 40)	(30, 40)	(30, 40)
4 × 4	11.150	2.009	288.09	49.75	27.73
			(15, 20)	(15, 60)	(15, 20)
	11.146	1.992	287.27	49.76	27.85
			(15, 20)	(15, 60)	(15, 20)
8 × 8	11.162	1.997	308.00	56.53	40.97
			(7.5, 10)	(7.5, 50)	(7.5, 10)
	11.161	1.995	306.57	56.52	41.53
			(7.5, 10)	(7.5, 50)	(7.5, 10)

[a]The first line for each mesh corresponds to the FEM solutions and the second corresponds to DMCDM. [b]Location of the stress.

Example 9.8.3

Consider the creeping flow of a viscous incompressible material squeezed between two long parallel plates [13]. Let V_0 be the velocity with which the two plates are moving toward each other in the y-direction (i.e., squeezing out the fluid), and let $2b$ and $2a$ denote, respectively, the distance between and the length of the plates at a fixed instant of time when the fluid is at rest. Assuming that the length of the plates (into the plane of the paper) is very large compared to both the width of and the distance between the plates, the problem can be treated as one of a plane flow. Exploiting the biaxial symmetry, use one quadrant of the domain as the computational domain, as shown in Fig. 9.8.3(a) and the 10×6 nonuniform mesh of linear elements shown in Figure 9.8.3(b) to determine the FEM and DMCDM solutions (note that the computational domain has singularities in the boundary conditions at all four corner points). Use the following element lengths to generate the 10×6 mesh:

$$\text{DX(I)} = \{1.0\ 1.0\ 1.0\ 1.0\ 0.5\ 0.5\ 0.25\ 0.25\ 0.25\ 0.25\},$$
$$\text{DY(I)} = \{0.25\ 0.25\ 0.5\ 0.5\ 0.25\ 0.25\}.$$

Also use a 20×16 nonuniform mesh of linear elements with

$$\text{DX(I)} = \{0.5\ 0.5\ 0.5\ 0.5\ 0.5\ 0.5\ 0.5\ 0.5\ 0.25\ 0.25\ 0.25\ 0.25\ 0.125\ 0.125\ 0.125\ 0.125\ 0.125\ 0.125\ 0.125\ 0.125\ 0.125\ 0.125\},$$
$$\text{DY(I)} = \{0.125\ 0.125\ 0.125\ 0.125\ 0.125\ 0.125\ 0.125\ 0.125\ 0.125\ 0.125\ 0.125\ 0.125\ 0.125\ 0.125\ 0.125\}$$

to obtain the numerical solutions and tabulate the results.

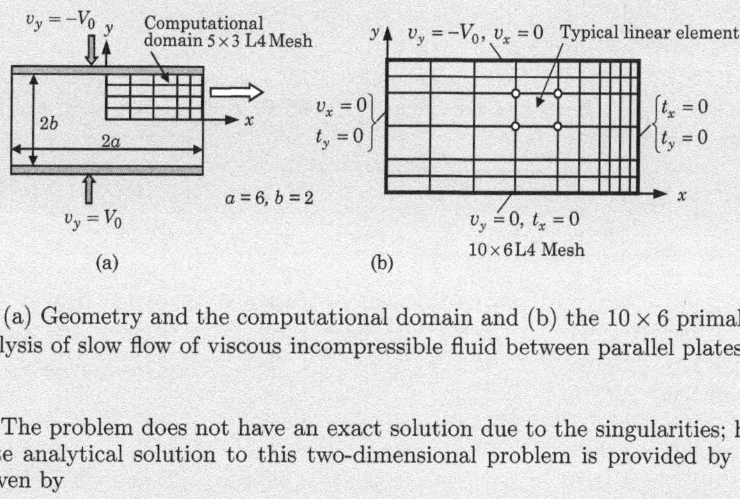

Fig. 9.8.3 (a) Geometry and the computational domain and (b) the 10×6 primal mesh used for the analysis of slow flow of viscous incompressible fluid between parallel plates.

Solution: The problem does not have an exact solution due to the singularities; however, an approximate analytical solution to this two-dimensional problem is provided by Nadai [47], and it is given by

$$u(x,y) \equiv v_x(x,y) = \frac{3V_0 x}{2b}\left(1 - \frac{y^2}{b^2}\right), \quad v(x,y) \equiv v_y(x,y) = -\frac{3V_0 y}{2b}\left(3 - \frac{y^2}{b^2}\right),$$
$$P(x,y) = \frac{3\mu V_0}{2b^3}\left(a^2 + y^2 - x^2\right). \tag{1}$$

Following the discussions presented in [8, 13], a penalty parameter value of $\gamma = 10^8$ is used in all cases.

A comparison of the results obtained with the FEM, DMCDM, and FVM are presented in Table 9.8.2. We note that the nodal locations in the FVM do not coincide with those in the FEM and DMCDM; the FVM solutions presented in Table 9.8.2 are the interpolated values at the control volume interfaces, which coincide with the nodal locations of the FEM and DMCDM. The solutions predicted by the FEM and DMCDM are close to each other (indistinguishable in the graphs) and agree well with the approximate analytical solutions for the velocity and pressure fields [obtained using Eq. (1)], as can be seen from Figs. 9.8.4 and 9.8.5, while the solution predicted by the FVM deviates slightly. Figure 9.8.6 shows the velocity contours in the computational domain as obtained using the DMCDM.

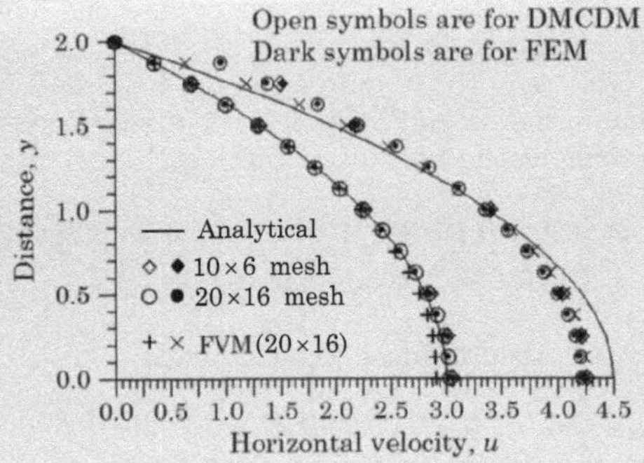

Fig. 9.8.4 Velocity component $v_x(x, y) = u(x, y)$ at $x = 4$ and $x = 6$ as a function of y for fluid squeezed between parallel plates. Results obtained using the FEM and DMCDM with 10×6 and 20×16 meshes and with the penalty parameter $\gamma = 10^8$ are presented. The solid lines denote approximate analytical solutions.

Table 9.8.2 FEM, DMCDM, and FVM solutions for the horizontal velocities $u(4, y)$ and $u(6, y)$ as functions y for the flow of viscous incompressible fluid squeezed between two parallel plates.

y	$x = 4$			$x = 6$		
	FEM	DMCDM	FVM	FEM	DMCDM	FVM
0.000	3.0301	3.0302	2.9073	4.2121	4.2150	4.2489
0.125	3.0179	3.0180	2.9000	4.1991	4.2019	4.2414
0.250	2.9803	2.9805	2.8744	4.1594	4.1622	4.2216
0.375	2.9185	2.9185	2.8260	4.0936	4.0961	4.1519
0.500	2.8315	2.8316	2.7543	4.0004	4.0030	4.0617
0.625	2.7207	2.7208	2.6589	3.8808	3.8830	3.9399
0.750	2.5853	2.5854	2.5397	3.7327	3.7348	3.7856
0.875	2.4269	2.4269	2.3964	3.5574	3.5589	3.5978
1.000	2.2445	2.2446	2.2288	3.3515	3.3531	3.3752
1.125	2.0402	2.0401	2.0366	3.1176	3.1181	3.1161
1.250	1.8128	1.8128	1.8197	2.8487	2.8497	2.8188
1.375	1.5644	1.5643	1.5779	2.5522	2.5506	2.4808
1.500	1.2937	1.2937	1.3114	2.2080	2.2088	2.0994
1.625	1.0026	1.0024	1.0201	1.8454	1.8382	1.6702
1.750	0.6897	0.6896	0.7041	1.3785	1.3828	1.1869
1.875	0.3558	0.3557	0.3641	0.9686	0.9513	0.6386

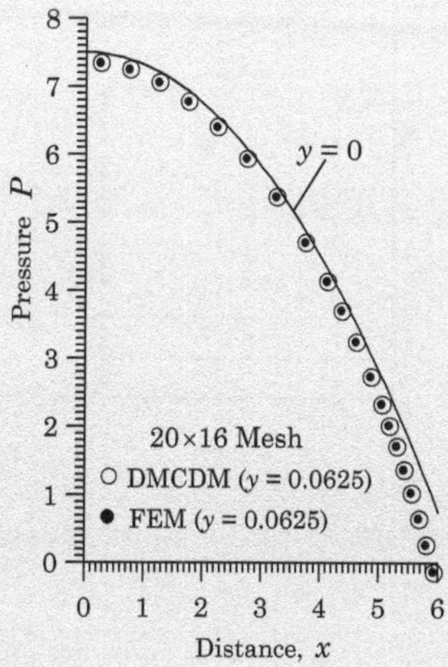

Fig. 9.8.5 Postcomputed pressure $P(x, y)$ versus x along the center line for fluid squeezed between parallel plates. The solid line denotes approximate analytical solution.

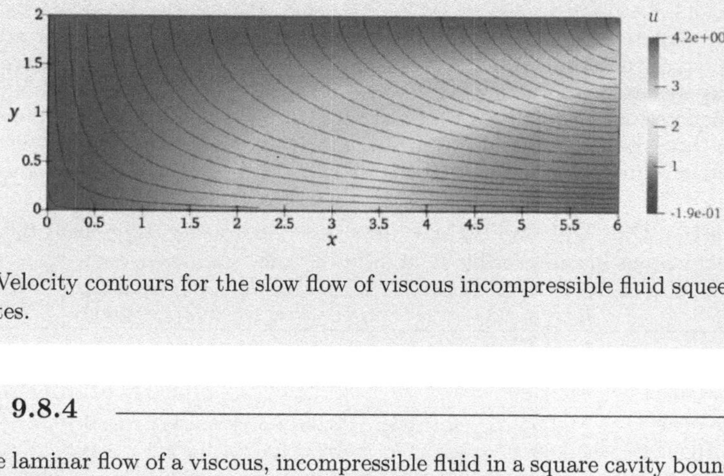

Fig. 9.8.6 Velocity contours for the slow flow of viscous incompressible fluid squeezed between parallel plates.

Example 9.8.4

Consider the laminar flow of a viscous, incompressible fluid in a square cavity bounded by three stationary walls (left, right, and bottom) and a lid at the top moving at a constant velocity $v_x \equiv u = V_0$ in its own plane, as shown in Fig. 9.8.7. Singularities exist at each corner where the moving lid meets a fixed wall (here we assume that $u(x, 1) = V_0$). Obtain the FEM and DMCDM solutions and plot the horizontal velocity $v_x = u$ along the vertical centerline (i.e., $u(0.5, y)$ versus y) for Reynolds numbers $Re = 0$ and $Re = 1000$ (take $\epsilon = 10^{-3}$ and $\gamma = 10^8$). Use two different meshes, a coarse 8×8 uniform mesh and a refined 16×20 nonuniform mesh with the following element lengths in the x and y directions:

DX(I) = {0.0625 0.0625 0.0625 0.0625 0.0625 0.0625 0.0625 0.0625
0.0625 0.0625 0.0625 0.0625 0.0625 0.0625 0.0625 0.0625}
DY(I) = {0.0625 0.0625 0.0625 0.0625 0.0625 0.0625 0.0625 0.0625
0.0625 0.0625 0.0625 0.0625 0.03125 0.03125 0.03125 0.03125
0.03125 0.03125 0.03125 0.03125}

This problem is often used as a benchmark problem to evaluate a computational scheme.

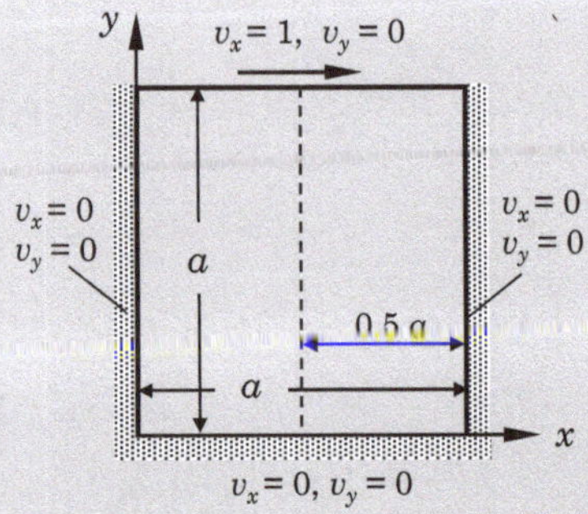

Fig. 9.8.7 Schematic of the wall-driven cavity problem.

Solution:

Table 9.8.3 contains a comparison of the velocity component, $u(0.5, y)$, obtained with various methods when the 16×20 mesh is used. In the FVM, various schemes are used to discretize the convective (nonlinear) terms. In the upwind scheme [3], the convective term of the Navier–Stokes equation is approximated by a backward derivative with respect to the velocity direction, which brings stability by adding an artificial numerical diffusion to the system. In the linear-upwind stabilized transport (LUST) scheme [48, 49], the linear-upwind is blended with linear interpolation to stabilize solutions while maintaining second-order accuracy. For this particular study, following Weller's example [48], a weight of 75% linear interpolation and 25% upwinding was used. In Table 9.8.3, FVM-U stands for the FVM with upwinding scheme and FVM-L denotes the FVM with LUST scheme. Clearly, the FVM results are significantly different from those of the FEM and DMCDM; in fact, the FVM requires many more elements to have the same degree of accuracy as the DMCDM.

Table 9.8.3 FEM, DMCDM, and FVM solutions for the velocity component $u(0.5, y)$ versus y of the flow of viscous incompressible fluid inside a wall-driven cavity.

	$Re = 0$			$Re = 1000$			
y	FEM	DMCDM	FVM	FEM	DMCDM	FVM-U	FVM-L
0.0625	−0.0369	−0.0369	−0.0388	−0.1224	−0.1430	−0.1247	−0.1802
0.1250	−0.0663	−0.0663	−0.0703	−0.2072	−0.2177	−0.1657	−0.2579
0.1875	−0.0920	−0.0919	−0.0976	−0.2616	−0.2502	−0.1705	−0.2709
0.2500	−0.1159	−0.1158	−0.1229	−0.2542	−0.2301	−0.1583	−0.2362
0.3125	−0.1389	−0.1386	−0.1470	−0.2001	−0.1776	−0.1354	−0.1785
0.3750	−0.1603	−0.1598	−0.1693	−0.1419	−0.1255	−0.1050	−0.1227
0.4375	−0.1782	−0.1775	−0.1881	−0.0928	−0.0803	−0.0701	−0.0754
0.5000	−0.1889	−01881	−0.1998	−0.0454	−0.0365	−0.0341	−0.0336
0.5625	−0.1865	−0.1857	−0.1989	0.0042	0.0091	0.0002	0.0071
0.6250	−0.1625	−0.1619	−0.1771	0.0566	0.0570	0.0315	0.0505
0.6875	−0.1057	−0.1057	−0.1237	0.1138	0.1097	0.0589	0.0986
0.7500	−0.0038	−0.0046	−0.0412	0.1715	0.1612	0.0794	0.1435
0.7813	0.0682	0.0673	0.0439	0.1994	0.1861	0.0938	0.1758
0.8125	0.1553	0.1542	0.1297	0.2245	0.2084	0.1071	0.2004
0.8438	0.2596	0.2584	0.2329	0.2455	0.2273	0.1215	0.2228
0.8750	0.3782	0.3772	0.3540	0.2577	0.2383	0.1403	0.2422
0.9063	0.5162	0.5149	0.4930	0.2680	0.2496	0.1753	0.2629
0.9375	0.6641	0.6634	0.6487	0.3088	0.2937	0.2603	0.3213
0.9688	0.8284	0.8279	0.8187	0.5406	0.5245	0.4849	0.5286

Figure 9.8.8 contains plots of the horizontal velocity component u along the vertical centerline (i.e., $u(0.5, y)$ versus y) for various Reynolds numbers and meshes ($\epsilon = 10^{-3}$ and $\gamma = 10^8$). The horizontal velocity contours obtained using the DMCDM for $Re = 0$ and $Re = 1000$ are included in Fig. 9.8.9. Both the FEM and DMCDM yield results that are very close to each other, as can be seen from the results presented in Fig. 9.8.9, although the DMCDM takes relatively less computational time because there are no element equations and their assembly in the DMCDM.

We close this example with a note that for higher Reynolds numbers, all numerical methods require refined meshes to resolve the viscous boundary layer (near the walls of the cavity) and satisfy the mass conservation. Further study of the problem for higher Reynolds numbers may be carried out by the interested readers.

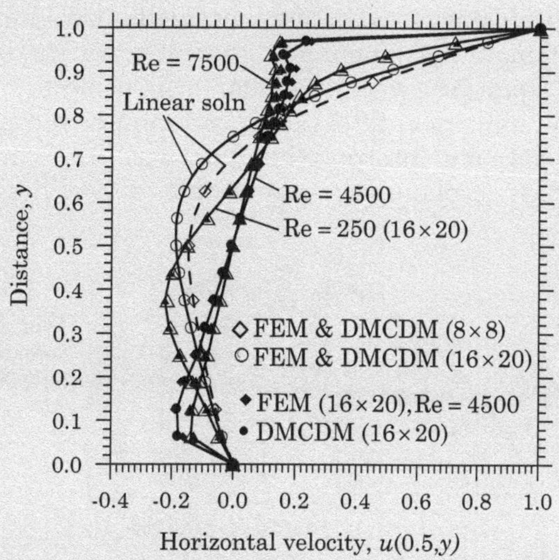

Fig. 9.8.8 Velocity component $u(0.5, y)$ versus y for Reynolds numbers $Re = 0, 250$, and 7,500 (obtained with the FEM and DMCDM models and two different meshes).

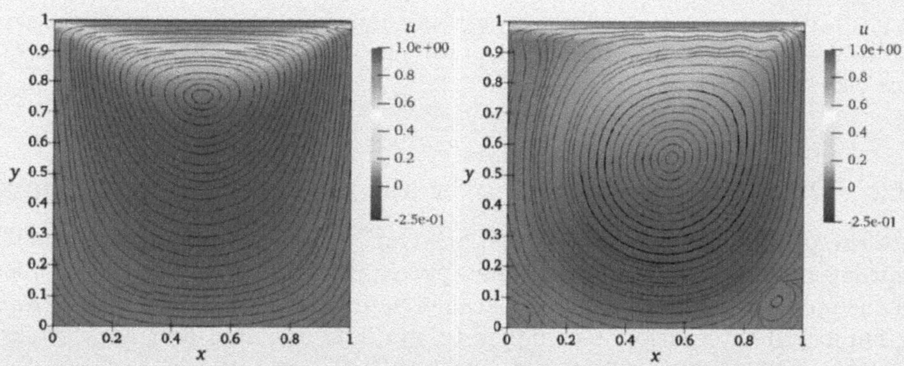

Fig. 9.8.9 Contours of the velocity component $u(x, y)$ for Reynolds numbers $Re = 0$ (left) and 1000 (right), obtained with the DMCDM model with the 16×20 mesh.

9.9 DMCDM with Arbitrary Meshes: 2D Elasticity

9.9.1 Preliminary Comments

So far, the DMCDM has been applied to solve two-dimensional problems in the fields of heat transfer, fluid mechanics, and solid mechanics with structured meshes. This section is devoted to the generalization of the DMCDM to irregular (i.e., non-rectangular) domains with unstructured primal meshes (e.g., arbitrary triangular and quadrilateral finite elements), as shown in Fig. 9.9.1. The resulting dual mesh of control domains will also be non-rectangular and irregular, making the DMCDM competitive with both the FEM and FVM in solving a variety of problems posed on nonrectangular two-dimensional domains. The evaluation of domain and boundary integrals over the control domains to obtain the discrete equations when irregular primal meshes are used requires, like in the FEM, the numerical integration techniques to evaluate the line and area integrals of the data and interpolation functions. These ideas are illustrated here through the study of plane elasticity problems (see Jiao et al. [50]).

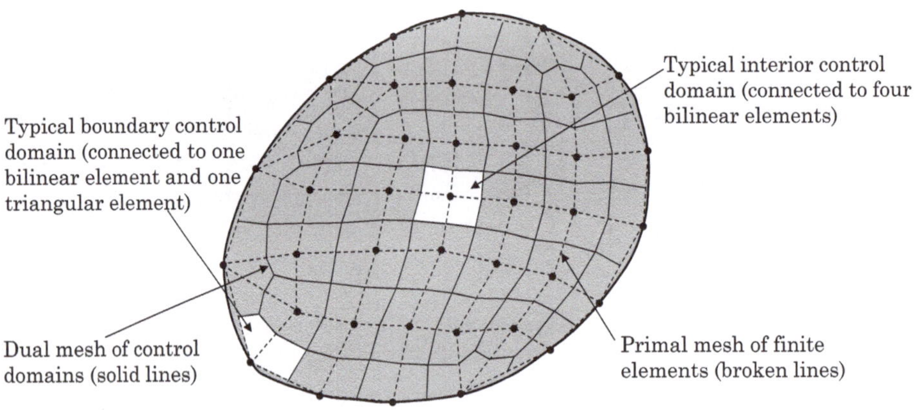

Fig. 9.9.1 Arbitrary primal mesh of linear and bilinear elements (broken lines) and associated dual mesh (solid lines). Depending on the primal mesh, the dual mesh consists of polygons with a different number of sides.

9.9.2 Discretized Equations over an Arbitrary Control Domain

Since the primal mesh is composed of arbitrarily shaped triangular elements or quadrilateral elements, and the same approximation is used for the geometry and the solution variables (i.e., isoparametric formulation), we need to employ the numerical integration techniques to evaluate the line and area integrals appearing in the integral statements in Eqs. (9.4.3) and (9.4.4). In this section, we limit our discussion to linear triangular elements and bilinear quadrilateral elements, whose interpolation functions are readily available for the so-called master elements from the books on the finite element method (see [8, 10, 12, 13]). It should be noted that when arbitrary meshes are used, the dual mesh of control domains will be quite irregular.

For example, for the isotropic case, Eqs. (9.4.3) and (9.4.4) over an arbitrary control domain $\Omega^{\rm cd}$ with boundary $\Gamma^{\rm cd}$ take the form

$$0 = \oint_{\Gamma^{\rm cd}} \left[\left(A\frac{\partial u}{\partial x} + B\frac{\partial v}{\partial y} \right) n_x + C\left(\frac{\partial u}{\partial y} + \frac{\partial v}{\partial x} \right) n_y \right] d\Gamma + \int_{\Omega^{\rm cd}} f_x \, d\Omega, \tag{9.9.1}$$

$$0 = \oint_{\Gamma^{\rm cd}} \left[C\left(\frac{\partial u}{\partial y} + \frac{\partial v}{\partial x} \right) n_x + \left(B\frac{\partial u}{\partial x} + A\frac{\partial v}{\partial y} \right) n_y \right] d\Gamma + \int_{\Omega^{\rm cd}} f_y \, d\Omega, \tag{9.9.2}$$

where $d\Omega$ is the area element, $d\Gamma$ is the line element, and

$$A = \frac{Eh}{1-\nu^2}, \quad B = \frac{\nu Eh}{1-\nu^2}, \quad C = \frac{Eh}{2(1+\nu)} = Gh \tag{9.9.3}$$

for the plane stress case and

$$A = \frac{(1-\nu)Eh}{(1+\nu)(1-2\nu)}, \quad B = \frac{\nu\, Eh}{(1+\nu)(1-2\nu)}, \quad C = \frac{Eh}{2(1+\nu)} = Gh \tag{9.9.4}$$

for the plane strain case.

The details of the evaluation of the integrals in Eqs. (9.9.1) and (9.9.2) for control domains associated with primal meshes of arbitrary triangular or bilinear elements (i.e., only one type of element is used in the mesh) are discussed next.

9.9.2.1 Primal mesh of triangles

Suppose that the computational domain is discretized using a primal mesh of linear triangular elements of arbitrary shape, and the dual mesh of control domains are constructed around the nodes of the primal mesh of triangular elements. The control domain boundary segments in each element of the primal mesh is selected here to be the line connecting the midpoint of the element edge to the element centroid, as shown in Fig. 9.9.2. Thus, an internal control domain in this case is a polygon of $2n$ sides, where n is the number of finite elements connected to the node under consideration. The solid lines in Fig. 9.9.2 indicate the primal mesh of triangular elements and the area (shaded) enclosed by the dotted lines (denoted by Ω^1 through Ω^6 in elements 1 through 6, respectively) is the control domain area. Here, I denotes the center of the control domain, C_i denotes the centroid of the ith triangular element, and P_1 through P_6 denote the midpoints of the sides of the each of six triangles that are connected to I. All of the labels are numbered counterclockwise.

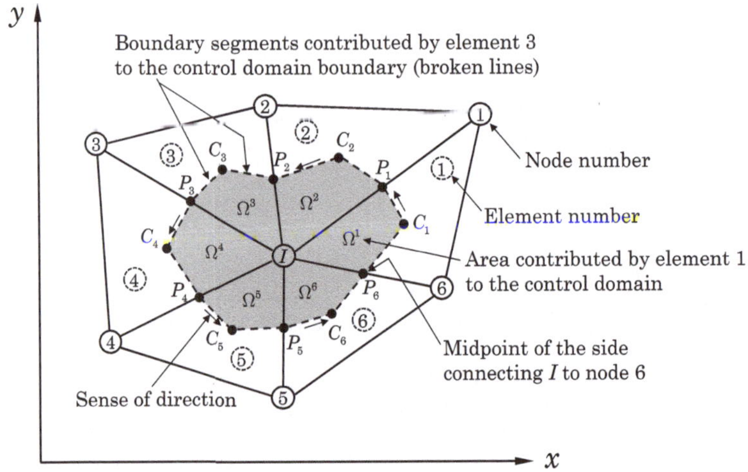

Fig. 9.9.2 Typical interior control domain formed by a primal mesh of arbitrary linear triangular elements.

The integral of Eqs. (9.9.1) and (9.9.2) can be expressed in the control domain of Fig. 9.9.2 as ($n_x\, d\Gamma = dy$ and $-n_y\, d\Gamma = dx$)

$$0 = \sum_{e=1}^{n=6} \int_{\Gamma^e} \left[\left(A\frac{\partial u}{\partial x} + B\frac{\partial v}{\partial y} \right) dy - C\left(\frac{\partial u}{\partial y} + \frac{\partial v}{\partial x} \right) dx \right] + \sum_{e=1}^{n=6} \int_{\Omega^e} f_x \, d\Omega, \tag{9.9.5}$$

$$0 = \sum_{e=1}^{n=6} \int_{\Gamma^e} \left[C\left(\frac{\partial u}{\partial y} + \frac{\partial v}{\partial x} \right) dy - \left(B\frac{\partial u}{\partial x} + A\frac{\partial v}{\partial y} \right) dx \right] + \sum_{e=1}^{n=6} \int_{\Omega^e} f_y \, d\Omega, \tag{9.9.6}$$

where Γ^e ($e = 1, 2, \ldots, n = 6$), denotes the boundary segment (Γ^e contains two line segments) of the control domain in the eth triangular element (i.e., $\Gamma^1 = P_6C_1P_1$, $\Gamma^2 = P_1C_2P_2$, $\Gamma^3 = P_2C_3P_3$, $\Gamma^4 = P_3C_4P_4$, $\Gamma^5 = P_4C_5P_5$, and $\Gamma^6 = P_5C_6P_6$).

As an example, consider the control domain area Ω^1 in element 1; to map a triangular mesh of arbitrary shape in the actual mesh to the master element, which is an isosceles right-triangle with side length 1 unit under the local normalized coordinate system, as shown in Fig. 9.9.3. According to the assumed isoparametric approximation of the displacements and the geometry, we have

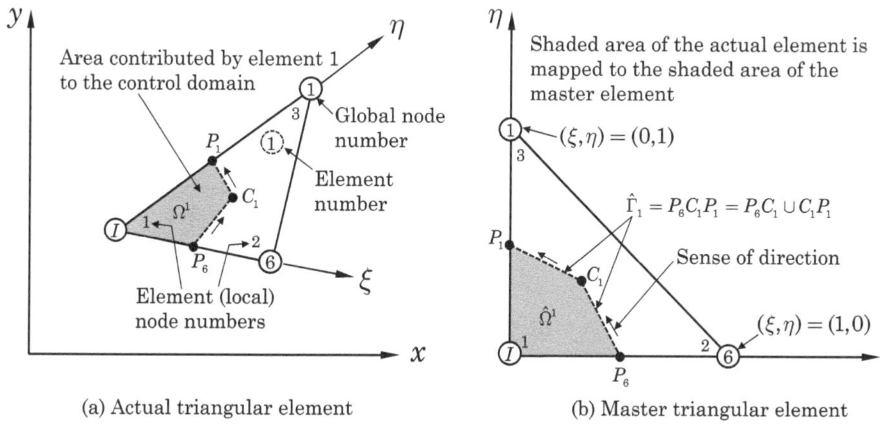

Fig. 9.9.3 Transformation of physical element area and boundary to those of the master triangular element.

$$u(x,y) = \sum_{j=1}^{m} u_j^e\, \psi_j^e(\xi,\eta), \quad v(x,y) = \sum_{j=1}^{m} v_j^e\, \psi_j^e(\xi,\eta), \tag{9.9.7}$$

$$x = \sum_{j=1}^{m} x_j^e\, \psi_j^e(\xi,\eta), \qquad y = \sum_{j=1}^{m} y_j^e\, \psi_j^e(\xi,\eta), \tag{9.9.8}$$

where (u_j^e, v_j^e) are the components of the displacement vector at node j of the eth element, (x_j, y_j) are the global coordinates of the jth (local) node, and $\psi_j^e(\xi,\eta)$ are the linear (i.e., $m = 3$) interpolation functions,

$$\psi_1^e(\xi,\eta) = 1 - \xi - \eta, \quad \psi_2^e(\xi,\eta) = \xi, \quad \psi_3^e(\xi,\eta) = \eta, \tag{9.9.9}$$

with derivatives with respect to the normalized local coordinates ξ and η

$$\begin{aligned} \frac{\partial \psi_1^e}{\partial \xi} &= -1, \quad \frac{\partial \psi_2^e}{\partial \xi} = 1, \quad \frac{\partial \psi_3^e}{\partial \xi} = 0, \\ \frac{\partial \psi_1^e}{\partial \eta} &= -1, \quad \frac{\partial \psi_2^e}{\partial \eta} = 0, \quad \frac{\partial \psi_3^e}{\partial \eta} = 1. \end{aligned} \tag{9.9.10}$$

The elemental lengths and areas in the coordinate systems are related by

$$dx = \frac{\partial x}{\partial \xi}\, d\xi + \frac{\partial x}{\partial \eta}\, d\eta, \quad dy = \frac{\partial y}{\partial \xi}\, d\xi + \frac{\partial y}{\partial \eta}\, d\eta, \quad dxdy = J\, d\xi d\eta, \tag{9.9.11}$$

where J is the determinant of the Jacobian transformation matrix $\mathbf{J}^e$(see Section 1.8.4)

$$\mathbf{J}^e = \begin{bmatrix} \frac{\partial x}{\partial \xi} & \frac{\partial y}{\partial \xi} \\ \frac{\partial x}{\partial \eta} & \frac{\partial y}{\partial \eta} \end{bmatrix}, \tag{9.9.12}$$

where the derivatives of (x, y) with respect to (ξ, η) can be determined using Eq. (9.9.8), and the Jacobian matrix and its determinant can be readily determined.

The derivatives of the interpolation functions with respect to the global coordinates [as required in Eqs. (9.9.5) and (9.9.6)] can be calculated using the following relations (see Reddy [8]):

$$\begin{Bmatrix} \frac{\partial \psi_i^e}{\partial \xi} \\ \frac{\partial \psi_i^e}{\partial \eta} \end{Bmatrix} = \mathbf{J}^e \begin{Bmatrix} \frac{\partial \psi_i^e}{\partial x} \\ \frac{\partial \psi_i^e}{\partial y} \end{Bmatrix} \rightarrow \begin{Bmatrix} \frac{\partial \psi_i^e}{\partial x} \\ \frac{\partial \psi_i^e}{\partial y} \end{Bmatrix} = (\mathbf{J}^e)^{-1} \begin{Bmatrix} \frac{\partial \psi_i^e}{\partial \xi} \\ \frac{\partial \psi_i^e}{\partial \eta} \end{Bmatrix}. \tag{9.9.13}$$

With the help of the relations in Eqs. (9.9.7)–(9.9.13), the boundary and domain integrals in Eqs. (9.9.5) and (9.9.6) can be evaluated (using numerical integration techniques; see [8]), and the discrete equations in terms of the nodal values of the elements connected can be determined.

Returning to the evaluation of integrals over the domain Ω^1 and its boundary, we substitute Eqs. (9.9.7)–(9.9.13) into Eq. (9.9.5) for $e = 1$ and obtain

$$\begin{aligned} 0 &= \int_{\Gamma^1} \left[\left(A\frac{\partial u}{\partial x} + B\frac{\partial v}{\partial y} \right) dy - C\left(\frac{\partial u}{\partial y} + \frac{\partial v}{\partial x} \right) dx \right] + \int_{\Omega^1} f_x \, d\Omega \\ &= \int_{\hat{\Gamma}^1} \left[\left(A\frac{\partial \psi_i^e}{\partial x}\frac{\partial y}{\partial \eta} - C\frac{\partial \psi_i^e}{\partial y}\frac{\partial x}{\partial \eta} \right) d\eta + \left(A\frac{\partial \psi_i^e}{\partial x}\frac{\partial y}{\partial \xi} - C\frac{\partial \psi_i^e}{\partial y}\frac{\partial x}{\partial \xi} \right) d\xi \right]^{e=1} u_i^{(1)} \\ &+ \int_{\hat{\Gamma}^1} \left[\left(B\frac{\partial \psi_i^e}{\partial y}\frac{\partial y}{\partial \eta} - C\frac{\partial \psi_i^e}{\partial x}\frac{\partial x}{\partial \eta} \right) d\eta + \left(B\frac{\partial \psi_i^e}{\partial y}\frac{\partial y}{\partial \xi} - C\frac{\partial \psi_i^e}{\partial x}\frac{\partial x}{\partial \xi} \right) d\xi \right]^{e=1} v_i^{(1)} \\ &+ \int_{\hat{\Omega}^1} f_x \, J \, d\xi d\eta, \end{aligned} \tag{9.9.14}$$

where $\hat{\Omega}^1$ denotes the control domain in the master triangular element, $\hat{\Gamma}^1$ denotes the pair of boundary line segments, P_6C_1 and C_1P_1, of $\hat{\Omega}^1$, and $u_i^{(1)}$ and $v_i^{(1)}$ are the displacement components at node i of element 1.

Using the following correspondence

$$\begin{aligned} &u_1^{(1)} = u_1^{(2)} = u_1^{(3)} = u_1^{(4)} = u_1^{(5)} = u_1^{(6)} = U_I, \\ &u_3^{(1)} = u_2^{(2)} = U_1, \quad u_3^{(2)} = u_2^{(3)} = U_2, \quad u_3^{(3)} = u_2^{(4)} = U_3, \\ &u_3^{(4)} = u_2^{(5)} = U_4, \quad u_3^{(5)} = u_2^{(6)} = U_5, \quad u_3^{(6)} = u_2^{(1)} = U_6, \end{aligned} \tag{9.9.15}$$

we obtain the pair of discrete equations associated with Eqs. (9.9.6) and (9.9.7):

$$\begin{aligned} &K_{Ix}^1 U_I + K_{Ix}^2 V_I + K_{1x}^1 U_1 + K_{1x}^2 V_1 + K_{2x}^1 U_2 + K_{2x}^2 V_2 + K_{3x}^1 U_3 + K_{3x}^2 V_3 \\ &\quad + K_{4x}^1 U_4 + K_{4x}^2 V_4 + K_{5x}^1 U_5 + K_{5x}^2 V_5 + K_{6x}^1 U_6 + K_{6x}^2 V_6 = F_{Ix}, \end{aligned} \tag{9.9.16a}$$

$$\begin{aligned} &K_{Iy}^1 U_I + K_{Iy}^2 V_I + K_{1y}^1 U_1 + K_{1y}^2 V_1 + K_{2y}^1 U_2 + K_{2y}^2 V_2 + K_{3y}^1 U_3 + K_{3y}^2 V_3 \\ &\quad + K_{4y}^1 U_4 + K_{4y}^2 V_4 + K_{5y}^1 U_5 + K_{5y}^2 V_5 + K_{6y}^1 U_6 + K_{6y}^2 V_6 = F_{Iy}, \end{aligned} \tag{9.9.16b}$$

where I represents the node number (around which the control domain is created), x and y represent the coordinate directions, and U_i and V_i are the nodal displacements in the x and y directions, respectively. Equations (9.9.16a) and (9.9.16b) can be extended to all interior nodes in the computational domain, and a system of algebraic equations with displacements as the nodal variables is obtained.

9.9.2.2 Primal mesh of quadrilaterals

Consider a typical control domain from the primal mesh of quadrilateral elements, as shown in Fig. 9.9.4, where I represents the center of the control domain, and C_1 through C_4 denote the centers of each of the four quadrilateral elements around node I. The centroid of each quadrilateral element of the primal mesh is connected to the middle of each edge, resulting in a polygon of eight sides as the control domain associated with an internal node I.

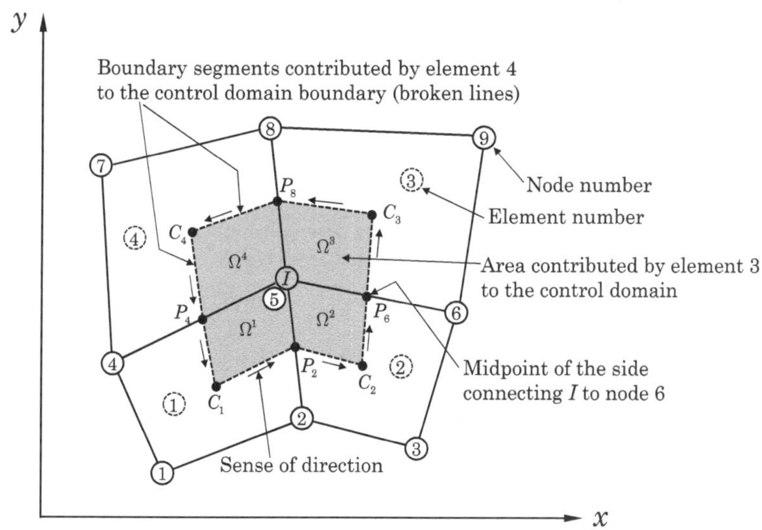

Fig. 9.9.4 Typical interior control domain formed by a primal mesh of arbitrary bilinear quadrilateral elements.

The integrals of Eqs. (9.9.1) and (9.9.2) can be expressed in the control domain of Fig. 9.9.4 as

$$0 = \sum_{e=1}^{n=4} \int_{\Gamma^e} \left[\left(A\frac{\partial u}{\partial x} + B\frac{\partial v}{\partial y} \right) dy - C \left(\frac{\partial u}{\partial y} + \frac{\partial v}{\partial x} \right) dx \right] + \sum_{e=1}^{n=4} \int_{\Omega^e} f_x \, d\Omega, \tag{9.9.17}$$

$$0 = \sum_{e=1}^{n=4} \int_{\Gamma^e} \left[C\left(\frac{\partial u}{\partial y} + \frac{\partial v}{\partial x}\right) dy - \left(B\frac{\partial u}{\partial x} + A\frac{\partial v}{\partial y}\right) dx \right] + \sum_{e=1}^{n=4} \int_{\Omega^e} f_y \, d\Omega, \tag{9.9.18}$$

where Γ^e is the pair of broken lines of the domain Ω^e (shaded area Ω^e of the eth quadrilateral element).

As discussed for a primal mesh of triangular elements, the evaluation of integrals is carried out using numerical integration, which requires the transformation of the integrands posed on the actual elements to the square master element of side 2. The finite element interpolation of the displacement components (u, v) and the geometry (x, y) is given by Eqs. (9.9.7) and (9.9.8), respectively, with $m = 4$ (i.e., bilinear elements). The bilinear interpolation functions are given by

$$\psi_j^e(\xi, \eta) = \tfrac{1}{4}(1 + \xi_j \xi)(1 + \eta_j \eta), \quad \frac{\partial \psi_j^e}{\partial \xi} = \tfrac{1}{4}\xi_j(1 + \eta_j \eta), \quad \frac{\partial \psi_j^e}{\partial \eta} = \tfrac{1}{4}(1 + \xi_j \xi)\eta_j, \tag{9.9.19}$$

where (ξ_j, η_j) denote the coordinates of the jth node of the master element $\hat{\Omega}$:

$$\begin{aligned} &(\xi_1, \eta_1) = (-1, -1), \quad (\xi_2, \eta_2) = (1, -1), \quad (\xi_3, \eta_3) = (1, 1) \\ &(\xi_4, \eta_4) = (-1, 1), \quad (\xi_5, \eta_5) = (0, -1), \quad (\xi_6, \eta_6) = (1, 0) \\ &(\xi_7, \eta_7) = (0, 1), \quad (\xi_8, \eta_8) = (-1, 0), \quad (\xi_9, \eta_9) = (0, 0). \end{aligned} \tag{9.9.20}$$

Considering the third element in Fig. 9.9.5(a) as an example, through the element transformation equations, the area and boundary integrals posed on the actual element can be transformed to the one posed on the master element shown in Fig. 9.9.5(b). For example, the x-momentum equation takes the form

$$\begin{aligned} &\int_{\Gamma^3} \left[\left(A\frac{\partial u}{\partial x} + B\frac{\partial v}{\partial y}\right) dy - C\left(\frac{\partial u}{\partial y} + \frac{\partial v}{\partial x}\right) dx \right] + \int_{\Omega^3} f_x \, d\Omega \\ &= \left[\int_{-1}^{0} \left(A\frac{\partial \psi_i^e}{\partial x}\frac{\partial y}{\partial \eta} - C\frac{\partial \psi_i^e}{\partial y}\frac{\partial x}{\partial \eta} \right) d\eta + \int_{0}^{-1} \left(A\frac{\partial \psi_i^e}{\partial x}\frac{\partial y}{\partial \xi} - C\frac{\partial \psi_i^e}{\partial y}\frac{\partial x}{\partial \xi} \right) d\xi \right]^{e=3} u_i^{(3)} \\ &+ \left[\int_{-1}^{0} \left(B\frac{\partial \psi_i^e}{\partial y}\frac{\partial y}{\partial \eta} - C\frac{\partial \psi_i^e}{\partial x}\frac{\partial x}{\partial \eta} \right) d\eta + \int_{0}^{-1} \left(B\frac{\partial \psi_i^e}{\partial y}\frac{\partial y}{\partial \xi} - C\frac{\partial \psi_i^e}{\partial x}\frac{\partial x}{\partial \xi} \right) d\xi \right]^{e=3} v_i^{(3)} \\ &+ \int_{-1}^{0}\int_{-1}^{0} f_x \, J \, d\xi d\eta, \end{aligned} \tag{9.9.21}$$

where the derivatives $\partial\psi_i^e/\partial x$, $\partial\psi_i^e/\partial y$, $\partial x/\partial\xi$, $\partial x/\partial\eta$, $\partial y/\partial\xi$, and $\partial y/\partial\eta$ can be expressed in terms of the global coordinates of the nodes and the local derivatives of the interpolation functions, as discussed in Eqs. (9.9.11)–(9.9.13).

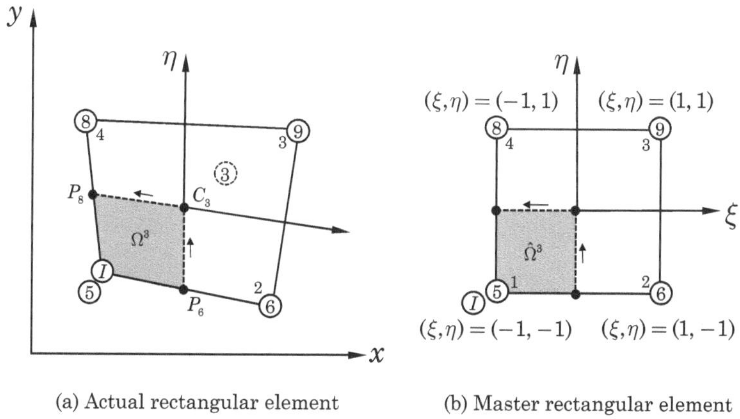

Fig. 9.9.5 Transformation of physical element area and boundary to those of the master rectangular element.

The pair of discrete equations associated with Eqs. (9.9.6) and (9.9.7) are

$$\begin{aligned} K^1_{1x}U_1 &+ K^2_{1x}V_1 + K^1_{2x}U_2 + K^2_{2x}V_2 + K^1_{3x}U_3 + K^2_{3x}V_3 + K^1_{4x}U_4 + K^2_{4x}V_4 \\ &+ K^1_{4x}U_4 + K^2_{4x}V_4 + K^1_{5x}U_5 + K^2_{5x}V_5 + K^1_{6x}U_6 + K^2_{6x}V_6 = F_{Ix}, \end{aligned} \tag{9.9.22a}$$

$$\begin{aligned} K^1_{1y}U_1 &+ K^2_{1y}V_1 + K^1_{2y}U_2 + K^2_{2y}V_2 + K^1_{3y}U_3 + K^2_{3y}V_3 + K^1_{4y}U_4 + K^2_{4y}V_4 \\ &+ K^1_{4y}U_4 + K^2_{4y}V_4 + K^1_{5y}U_5 + K^2_{5y}V_5 + K^1_{6y}U_6 + K^2_{6y}V_6 = F_{Iy}, \end{aligned} \tag{9.9.22b}$$

where I represents the node number (around which the control domain is created), x and y represent the global coordinate directions, and U_i and V_i are the nodal displacements in the x and y directions, respectively. Equations (9.9.22a) and (9.9.22b) can be extended to all interior nodes in the computational domain, and a system of algebraic equations with displacements as the nodal variables is obtained.

9.9.3 Control Domains at the Boundary

In the DMCDM, the components of forces at the boundary nodes are retained as a part of the discretized equations instead of replacing the displacement derivatives at the boundary with their interpolated values. When a displacement component at a node is known, we simply replace the nodal value with the known value. The corresponding dual variable (i.e., force component) is determined in the post-computation using the discretized equation at the node. The case when a force component is specified at a boundary node is discussed here.

Consider the case of a primal mesh of quadrilateral elements (the discussion is equally valid for meshes of triangular elements). In particular, we consider a node on the lower boundary of the computational domain, as shown in Fig.

9.9.6. For example, when a force component Q_{2x} (which is the integral of the distributed traction t_x on the boundary line segments) is specified, Eq. (9.9.21) takes the form

$$
\begin{aligned}
Q_{2x} =& \int_{\Gamma^2} \left[\left(A\frac{\partial u}{\partial x} + B\frac{\partial v}{\partial y} \right) dy - C\left(\frac{\partial u}{\partial y} + \frac{\partial v}{\partial x} \right) dx \right] + \int_{\Omega^2} f_x \, d\Omega \\
&+ \int_{\Gamma^1} \left[\left(A\frac{\partial u}{\partial x} + B\frac{\partial v}{\partial y} \right) dy - C\left(\frac{\partial u}{\partial y} + \frac{\partial v}{\partial x} \right) dx \right] + \int_{\Omega^1} f_x \, d\Omega \\
=& \left[\int_{-1}^{0} \left(A\frac{\partial \psi_i^e}{\partial x}\frac{\partial y}{\partial \eta} - C\frac{\partial \psi_i^e}{\partial y}\frac{\partial x}{\partial \eta} \right) d\eta + \int_{0}^{-1} \left(A\frac{\partial \psi_i^e}{\partial x}\frac{\partial y}{\partial \xi} - C\frac{\partial \psi_i^e}{\partial y}\frac{\partial x}{\partial \xi} \right) d\xi \right]^{e=2} u_i^{(2)} \\
&+ \left[\int_{-1}^{0} \left(B\frac{\partial \psi_i^e}{\partial y}\frac{\partial y}{\partial \eta} - C\frac{\partial \psi_i^e}{\partial x}\frac{\partial x}{\partial \eta} \right) d\eta + \int_{0}^{-1} \left(B\frac{\partial \psi_i^e}{\partial y}\frac{\partial y}{\partial \xi} - C\frac{\partial \psi_i^e}{\partial x}\frac{\partial x}{\partial \xi} \right) d\xi \right]^{e=2} v_i^{(2)} \\
&+ \left[\int_{0}^{-1} \left(A\frac{\partial \psi_i^e}{\partial x}\frac{\partial y}{\partial \eta} - C\frac{\partial \psi_i^e}{\partial y}\frac{\partial x}{\partial \eta} \right) d\eta + \int_{0}^{-1} \left(A\frac{\partial \psi_i^e}{\partial x}\frac{\partial y}{\partial \xi} - C\frac{\partial \psi_i^e}{\partial y}\frac{\partial x}{\partial \xi} \right) d\xi \right]^{e=1} u_i^{(1)} \\
&+ \left[\int_{0}^{-1} \left(B\frac{\partial \psi_i^e}{\partial y}\frac{\partial y}{\partial \eta} - C\frac{\partial \psi_i^e}{\partial x}\frac{\partial x}{\partial \eta} \right) d\eta + \int_{0}^{-1} \left(B\frac{\partial \psi_i^e}{\partial y}\frac{\partial y}{\partial \xi} - C\frac{\partial \psi_i^e}{\partial x}\frac{\partial x}{\partial \xi} \right) d\xi \right]^{e=1} v_i^{(1)} \\
&+ \int_{-1}^{0}\int_{0}^{-1} f_x \, J \, d\xi d\eta + \int_{0}^{-1}\int_{0}^{-1} f_x \, J \, d\xi d\eta. \qquad (9.9.23)
\end{aligned}
$$

where Γ^i denotes the boundary segment of the control domain in the ith element $(i = 1, 2)$. The boundary force component Q_{2y} can be treated in the same way.

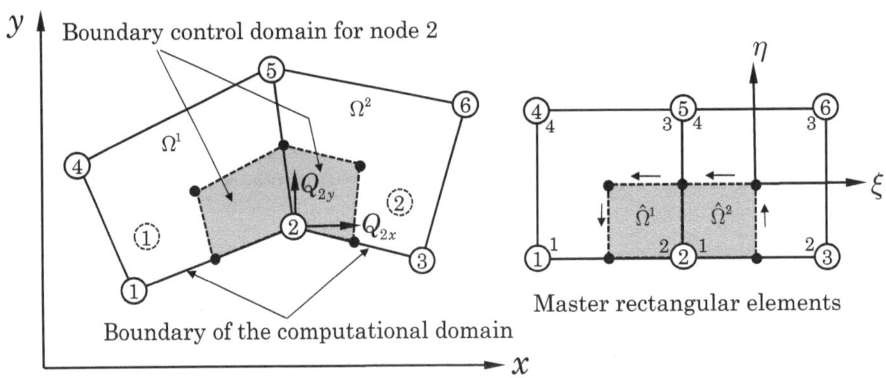

Fig. 9.9.6 Typical boundary control domain formed by a primal mesh of arbitrary bilinear quadrilateral elements.

9.9.4 Numerical Examples

9.9.4.1 Annular ring with internal pressure

Consider a long circular cylinder with internal radius $a = 0.05$ m and outside radius $b = 0.1$ m, held between two rigid walls, and subjected to an internal pressure $p = 120$ MPa; see Fig. 9.9.7(a). The elastic modulus of the material is $E = 200$ GPa and Poisson's ratio is $\nu = 0.3$. Exploiting the symmetry, the upper-right quarter of the domain is used as the computational domain, as shown in Fig. 9.9.7(b).

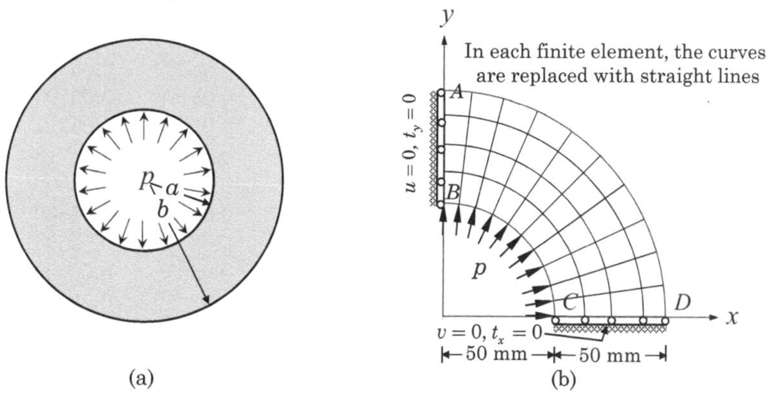

Fig. 9.9.7 Internally pressurized cylinder. (a) Problem domain and load. (b) Computational domain with a typical mesh of bilinear quadrilateral elements.

This plane strain problem has an analytical solution (see Reddy [23], pp. 295–297). The radial displacement and stress are given by

$$
\begin{aligned}
u_r(r) &= \frac{1}{E}\frac{a^2 b\, p}{b^2 - a^2}\left[(1+\nu)\frac{b}{r} + (1-2\nu)\frac{r}{b}\right] \\
\sigma_{rr}(r) &= \frac{a^2 p}{b^2 - a^2}\left(1 - \frac{b^2}{r^2}\right),
\end{aligned}
\tag{9.9.24}
$$

where r is the radial coordinate.

The area ABCD is divided into meshes of (a) triangular elements or (b) quadrilateral elements. Four different primal meshes are considered. The total number of elements in the meshes of linear triangles are: 80, 304, 696, 1216 (the number of elements along the sides AB or CD are 4, 8, 12, 16, respectively; the number of subdivisions along the circumference are obvious from the information given); the total number of elements in the meshes of bilinear quadrilaterals are 40, 152, 348, 608, respectively (the number of elements along the AB or CD sides are 4, 8, 12, 16, respectively). Note that for any given number of subdivisions along AB or CD, there are double the number of triangular elements compared to the quadrilateral elements.

The radial displacements at points C and D obtained with the DMCDM and FEM are compared with the analytical solution in Table 9.9.1. The stresses at the midpoint of the edge CD are compared in Table 9.9.2. It can be seen that when the mesh is refined, meshes of both triangles and quadrilaterals give good results. In comparison, the convergence of quadrilateral elements is significantly better than that of triangular elements. The results obtained with the DMCDM and FEM are quite close, with the DMCDM being a bit more accurate.

Table 9.9.1 Comparison of nodal displacements for the pressurized cylinder problem.

			Triangles		Quadrilaterals	
Node	Mesh[a]	Exact	FEM	DMCDM	FEM	DMCDM
C	4	0.05720	0.05729	0.05729	0.05596	0.05611
	8		0.05585	0.05585	0.05689	0.05692
	12		0.05735	0.05735	0.05706	0.05708
	24		0.05726	0.05726	0.05717	0.05717
	32		0.05724	0.05724	0.05718	0.05718
D	4	0.03640	0.03485	0.03485	0.03578	0.03603
	8		0.03631	0.03631	0.03624	0.03630
	12		0.03617	0.03617	0.03633	0.03636
	24		0.03633	0.03633	0.03638	0.03639
	32		0.03636	0.03636	0.03639	0.03639

[a]Number of subdivisions along AB or CD.

Table 9.9.2 Comparison of stresses at the midpoint of the edge CD for the pressurized cylinder problem.

			Triangles		Quadrilaterals	
Stress	Mesh[a]	Exact	FEM	DMCDM	FEM	DMCDM
σ_{xx}	4	-31.1111	-37.3499	-37.3499	-26.0511	-25.3032
	8		-28.2914	-28.2914	-29.8368	-29.6579
	12		-32.1435	-32.1435	-30.5439	-30.4651
	24		-31.4482	-31.4482	-30.9692	-30.9496
	32		-31.3314	-31.3314	-31.0313	-31.0202
σ_{yy}	4	111.1111	120.9427	120.9427	105.9847	106.9626
	8		104.9536	104.9536	109.8122	110.0541
	12		114.7269	114.7269	110.5324	110.6396
	24		112.9182	112.9182	110.9662	110.9930
	32		112.4662	112.4662	110.0296	110.0446

[a]Number of subdivisions along AB or CD.

9.9.4.2 Stretching of a square plate with a circular hole

In this example, we use the DMCDM to analyze a square plate with a circular hole stretched horizontally by a tensile stress applied at the edges, as shown in Fig. 9.9.8(a). The plate is of side length $b = 10$ mm, thickness $h = 1$ mm, and the radius of the concentric circular hole is $a = 1$ mm. The material properties are: elastic modulus $E = 210$ GPa and Poisson's ratio $\nu = 0.3$; tensile stress value is taken to be $\sigma_0 = 100$ MPa.

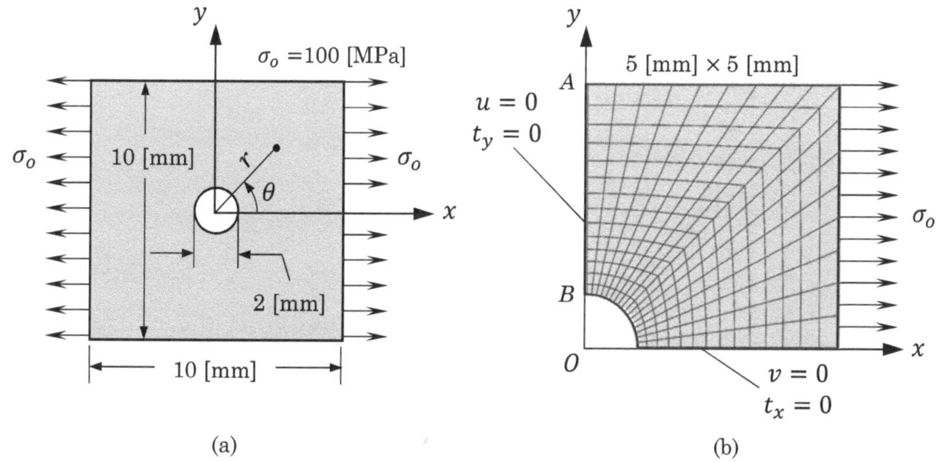

Fig. 9.9.8 Stretching of a square plate with a concentric hole. (a) Problem domain and applied stress. (b) Computational domain with a typical mesh of bilinear quadrilateral elements.

The problem is one of plane stress state. There is an analytical solution for the stresses for this plane stress problem (see Reddy [23], pp. 306–308) under the assumption that the plate is infinite in size:

$$\begin{aligned}
\sigma_{rr}(r,\theta) &= \frac{\sigma_0}{2}\left[\left(1-\frac{a^2}{r^2}\right)+\left(1+\frac{3a^4}{r^4}-\frac{4a^2}{r^2}\right)\cos 2\theta\right],\\
\sigma_{\theta\theta}(r,\theta) &= \frac{\sigma_0}{2}\left[\left(1+\frac{a^2}{r^2}\right)-\left(1+\frac{3a^4}{r^4}\right)\cos 2\theta\right],\\
\sigma_{r\theta}(r,\theta) &= -\frac{\sigma_0}{2}\left(1-\frac{3a^4}{r^4}+\frac{2a^2}{r^2}\right)\sin 2\theta.
\end{aligned} \tag{9.9.25}$$

The maximum normal stress occurs at $(r,\theta) = (a, \pm 90^\circ)$ and shear stress at $(r,\theta) = (\sqrt{3}a, -45^\circ)$:

$$\sigma_{\max} = \sigma_{\theta\theta}(a, \pm 90^\circ) = 3\sigma_0, \quad \sigma_{r\theta}(\sqrt{3}a, -45^\circ) = \frac{2}{3}\sigma_0. \tag{9.9.26}$$

Thus, the theoretical stress concentration factor is $K = \sigma_{\max}/\sigma_0 = 3$.

Using the biaxial symmetry, a quarter of the plate can be used as the computational domain for the analysis, as shown in Fig. 9.9.8(b). Two primal meshes are considered: one mesh with 1886 bilinear quadrilateral elements and the other with 3116 linear triangular elements. Both meshes yield results which are indistinguishable from each other when plotted. Figure 9.9.9 shows the variation of the circumferential stress along section AB (i.e., $x = 0$). Note that both the DMCDM and FEM results depart from the analytical solution, since the latter is applicable for a plate of infinite extent. For a plate of finite dimensions, the peak value of $\sigma_{\theta\theta}$ will be more than 300 MPa (as predicted by the DMCDM and FEM). The displacement and stress fields obtained with the DMCDM are compared with those of the FEM, both of which use the same mesh, in Figs. 9.9.10 and 9.9.11, respectively.

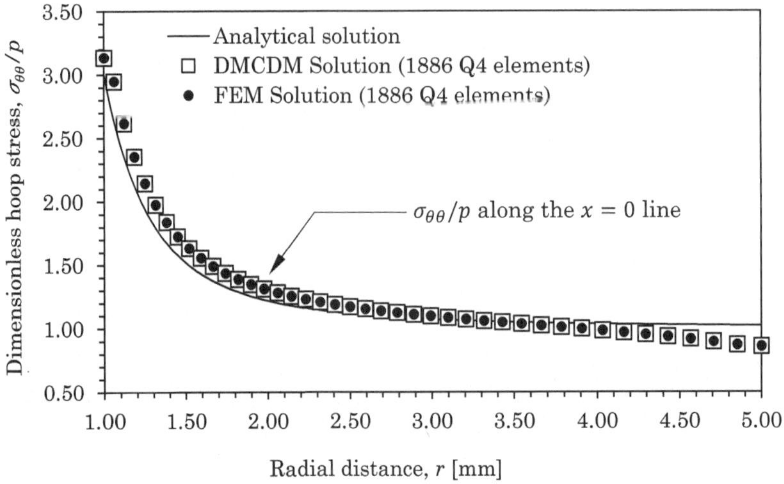

Fig. 9.9.9 Dimensionless circumferential stress distribution along the $x = 0$ edge.

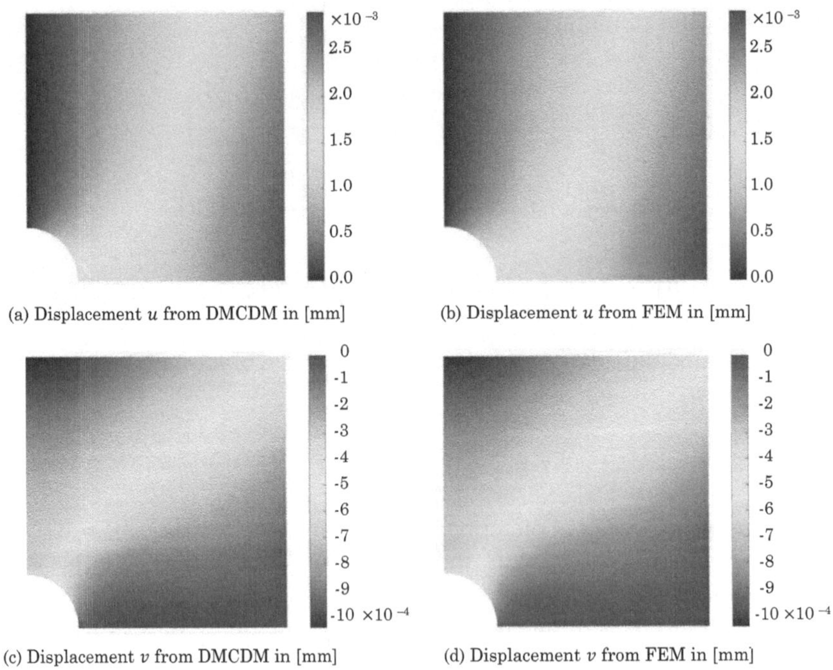

(a) Displacement u from DMCDM in [mm]

(b) Displacement u from FEM in [mm]

(c) Displacement v from DMCDM in [mm]

(d) Displacement v from FEM in [mm]

Fig. 9.9.10 Comparison of displacement fields obtained by the DMCDM and FEM for stretching of a square plate with a concentric hole.

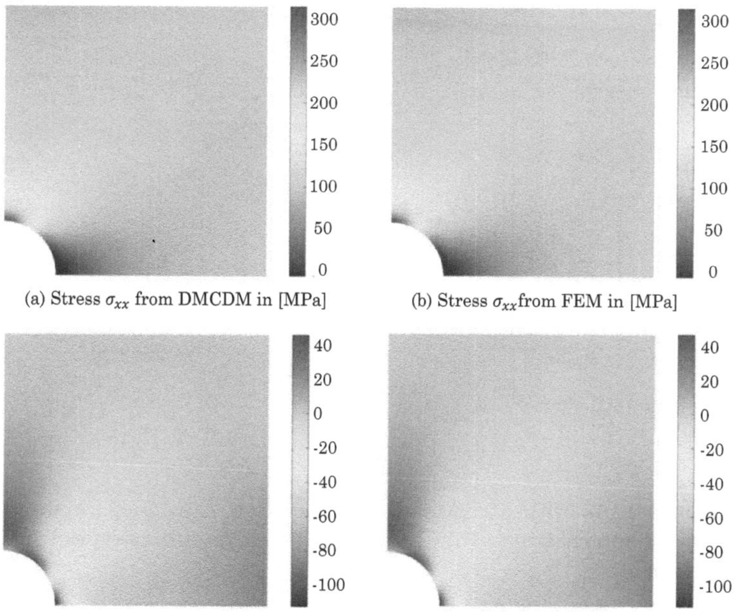

(a) Stress σ_{xx} from DMCDM in [MPa]

(b) Stress $\sigma_{xx}$from FEM in [MPa]

(c) Stress σ_{yy} from DMCDM in [MPa]

(d) Stress $\sigma_{yy}$from FEM in [MPa]

Fig. 9.9.11 Comparison of stress fields obtained by the DMCDM and FEM for stretching of a square plate with a concentric hole.

9.9.4.3 Pure shearing of a square plate with a concentric hole

Here, we consider a square plate with a small concentric circular hole, as in the previous example, except that the plate is subjected to pure shearing stress τ_0, as shown in Fig. 9.9.12(a). This problem also has an analytical solution (see Wang [51]). The stress components for this case are

$$\begin{aligned}\sigma_{rr}(r,\theta) &= \tau_0\left(1-\frac{a^2}{r^2}\right)\left(1-\frac{3a^2}{r^2}\right)\sin 2\theta,\\ \sigma_{\theta\theta}(r,\theta) &= -\tau_0\left(1+\frac{3a^4}{r^4}\right)\sin 2\theta,\\ \sigma_{r\theta}(r,\theta) &= \tau_0\left(1-\frac{a^2}{r^2}\right)\left(1+\frac{3a^2}{r^2}\right)\cos 2\theta.\end{aligned} \tag{9.9.27}$$

The minimum and maximum hoop stress occurs at $(r,\theta) = (a, \pm 45°)$:

$$(\sigma_{\theta\theta})_{\max} = \sigma_{\theta\theta}(a, \pm 45°) = \pm 4\,\tau_0. \tag{9.9.28}$$

This problem has symmetry about the two diagonals, allowing us to use a (rotated) quarter plate model, as shown in Fig. 9.9.12(b). Using a primal mesh of 528 linear quadrilateral elements, the hoop stress distribution along $x' = 0$ is shown in Fig. 9.9.13. The displacement and stress fields obtained with the DMCDM and FEM are shown in Figs. 9.9.14 and 9.9.15.

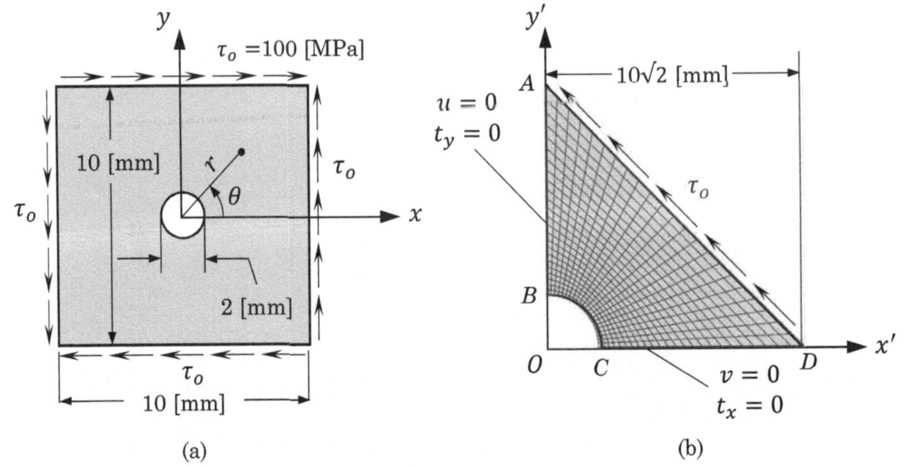

Fig. 9.9.12 Square plate with a concentric hole and subjected to pure shear stress. (a) Problem domain and applied stress. (b) Computational domain with a typical mesh of bilinear quadrilateral elements.

From Figs. 9.9.14 and 9.9.15, it can be seen that the distribution of displacement and stress fields are approximately the same for the two methods. Note that, akin to that in the previous example, the results of the DMCDM and FEM in Fig. 9.9.13 depart from the analytical solution. This is due to the fact that the analytical solution is pertinent to an infinite plate. For a plate of finite dimensions, the stresses will be comparably higher as predicted by the DMCDM and FEM.

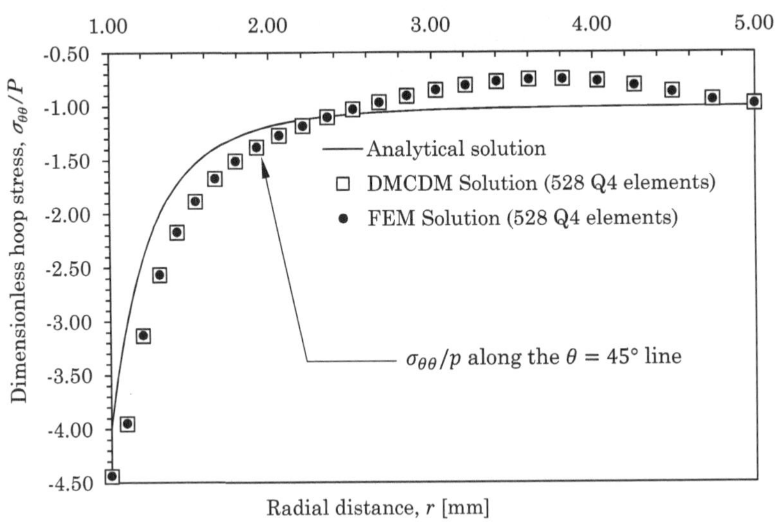

Fig. 9.9.13 Dimensionless hoop stress distribution along the $\theta = 45°$ line.

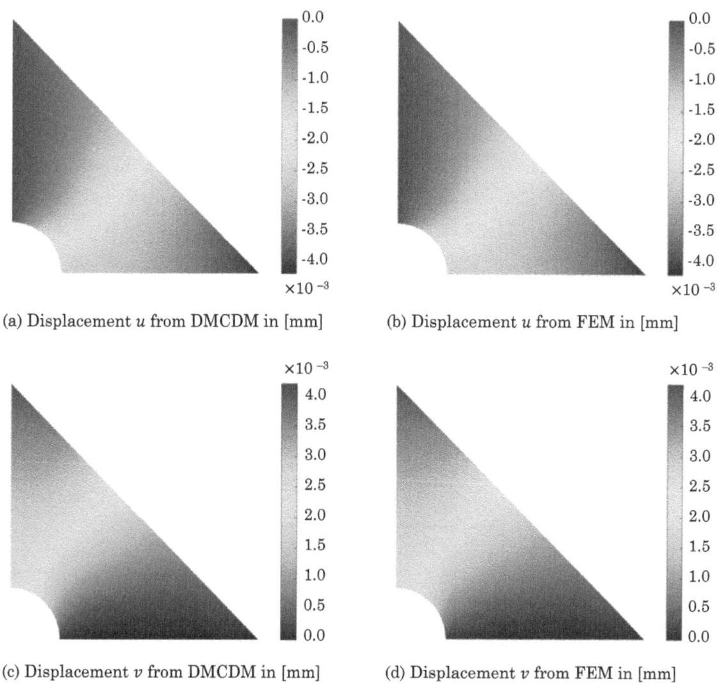

(a) Displacement u from DMCDM in [mm]

(b) Displacement u from FEM in [mm]

(c) Displacement v from DMCDM in [mm]

(d) Displacement v from FEM in [mm]

Fig. 9.9.14 Displacement fields obtained with the DMCDM and FEM using quadrilateral elements.

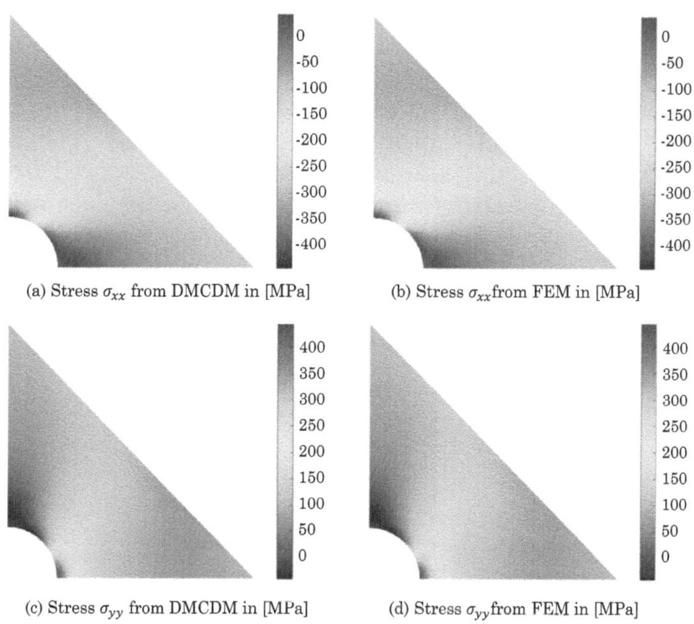

(a) Stress σ_{xx} from DMCDM in [MPa]

(b) Stress $\sigma_{xx}$from FEM in [MPa]

(c) Stress σ_{yy} from DMCDM in [MPa]

(d) Stress $\sigma_{yy}$from FEM in [MPa]

Fig. 9.9.15 Stress fields obtained with the DMCDM and FEM using quadrilateral elements.

9.9.4.4 Plate with a preexisting crack

This example deals with the study of a linear elastic fracture mechanics problem of an infinite plate with a preexisting crack. As shown in Fig. 9.9.16, a square plate of width $b = 2a$ with a crack along $y = 0$ and of length $a = 5$ is considered. Displacement boundary conditions are imposed on all four sides of the plate according to eq. (47) of [52] with $K_I = 1$. The problem is solved assuming plane stress condition and that the plate is made of a material with Young's modulus $E = 1$ and Poisson's ratio $\nu = 0.3$.

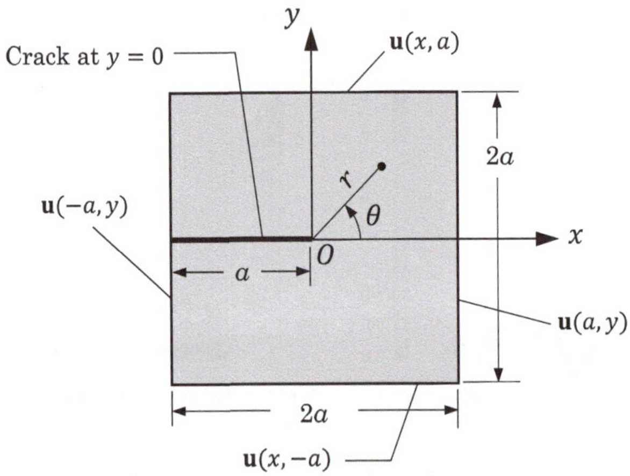

Plate with a horizontal crack.

Fig. 9.9.16 Problem domain for a plate with a horizontal crack. The problem domain is itself used as the computational domain for this problem.

The analytical solutions of the displacement field and stress field near the crack tip for the model-I crack problems are given in Eqs. (47) and (48), respectively, of [52], where (r, θ) are polar coordinates measured from the crack tip and K_I is the model-I stress intensity factor. The displacement and stress field equations are repeated below for ready reference:

$$\begin{aligned} u_x &= \frac{K_I}{2\mu}\sqrt{\frac{r}{2\pi}}\cos\tfrac{1}{2}\theta\left(\kappa - 1 + 2\sin^2\tfrac{1}{2}\theta\right) \\ u_y &= \frac{K_I}{2\mu}\sqrt{\frac{r}{2\pi}}\sin\tfrac{1}{2}\theta\left(\kappa + 1 - 2\cos^2\tfrac{1}{2}\theta\right) \end{aligned} \tag{9.9.29}$$

where $\kappa = \frac{3-\nu}{1+\nu}$ and $\mu = \frac{E}{2(1+\nu)}$

$$\begin{aligned} \sigma_{xx} &= \frac{K_I}{\sqrt{2\pi r}}\cos\tfrac{1}{2}\theta\left(1 - \sin\tfrac{1}{2}\theta\,\sin\tfrac{3}{2}\theta\right) \\ \sigma_{yy} &= \frac{K_I}{\sqrt{2\pi r}}\cos\tfrac{1}{2}\theta\left(1 + \sin\tfrac{1}{2}\theta\,\sin\tfrac{3}{2}\theta\right) \\ \sigma_{xy} &= \frac{K_I}{\sqrt{2\pi r}}\cos\tfrac{1}{2}\theta\left(\sin\tfrac{1}{2}\theta\,\cos\tfrac{3}{2}\theta\right) \end{aligned} \tag{9.9.30}$$

A mesh of 3364 bilinear quadrilateral elements is used to solve this problem. To simulate a preexisting crack in the DMCDM and FEM, all boundaries that coincide with cracks are defined as stress-free boundaries. The numerical solutions obtained using the DMCDM and FEM of u_y along $x = 0$ and σ_{yy} along $y = 0$ are compared with the exact solutions as shown in Fig. 9.9.17. It can be seen that the displacement as well as stress solutions are achieved with excellent accuracy. The stresses exhibit a singularity at the crack tip (i.e., at $r = 0$) and, hence, no amount of mesh refinement at that location will ever produce a converged value of stress.

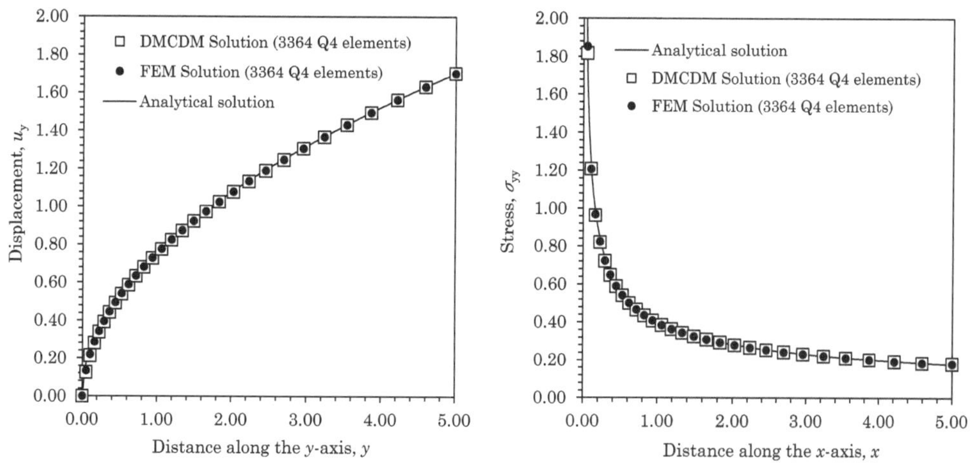

Fig. 9.9.17 Comparison of the displacement component u_y along $x = 0$ and stress component σ_{yy} along $y = 0$ obtained from the DMCDM and FEM with the analytical solution for the problem of a plate with a preexisting crack.

9.10 Summary

In this chapter, the FEM and DMCDM formulations are presented for the solution of problems involving plane elasticity and flows of viscous incompressible fluids. The penalty function formulation is employed for the solution of the Navier–Stokes equations governing viscous incompressible fluids. Through a number of numerical examples, it is shown that the DMCDM and FEM give results that are very close to each other.

In Section 9.9, the DMCDM is extended to unstructured grids in the context of linear plane elasticity problems. Details of the calculation of the discretized equations using isoparametric formulation (i.e., the same interpolation for the displacement components and the geometry are used) and numerical integration that requires the transformation of integrals posed on the actual element boundaries and domains to those on the master elements are presented. Both triangular and quadrilateral elements are discussed. Several numerical examples are presented to illustrate the accuracy of the DMCDM in comparison to the FEM. Numerical results show that the DMCDM with arbitrary quadrilateral and triangular elements has comparable accuracy to the FEM. The use of unstructured grids can be extended to other field problems discussed in this book. With the use of arbitrary meshes, the DMCDM is an attractive choice in lieu of the FEM because of its computational efficiency and the fact that the global form of the governing equations are satisfied in the DMCDM. Of course, application of the DMCDM to more practical and complex problems is the next major step.

The DMCDM ideas presented in this chapter may be extended to nonlinear elasticity problems. Extensions of the DMCDM to higher order elements as well as to three-dimensional problems and multi-field coupling problems are awaiting.

Problems

9.1 Verify the discrete equations in Eqs. (9.4.8)–(9.4.10) for a typical internal control volume shown in Fig. 9.4.2.

9.2 Derive the discrete equations associated with the nonlinear equations in Eqs. (9.7.8) and (9.7.9).

9.3 Verify the discrete equations in Eqs. (A9.4a)–(A9.5).

9.4 Verify the discrete equations in Eqs. (A9.6a)–(A9.7).

9.5 Verify the discrete equations in Eqs. (A9.8a) and (A9.8b).

Appendix 9A: Evaluation of line integrals in the DMCDM for plane elasticity

The bilinear finite element interpolation functions, defined in element coordinates $(\bar{x}, \bar{y})$, are

$$\begin{aligned} \psi_1^e(\bar{x},\bar{y}) &= \left(1-\frac{\bar{x}}{a}\right)\left(1-\frac{\bar{y}}{b}\right), & \psi_2^e(\bar{x},\bar{y}) &= \frac{\bar{x}}{a}\left(1-\frac{\bar{y}}{b}\right) \\ \psi_3^e(\bar{x},\bar{y}) &= \frac{\bar{x}}{a}\frac{\bar{y}}{b}, & \psi_4^e(\bar{x},\bar{y}) &= \left(1-\frac{\bar{x}}{a}\right)\frac{\bar{y}}{b} \end{aligned} \tag{A9.1}$$

Here $(\bar{x},\ \bar{y})$ denote the local coordinates with the origin located at node 1 of the element, and (a, b) denote the horizontal and vertical dimensions of the rectangle. The first derivatives of ψ_i^e with respect to x and y are

$$\begin{aligned} \frac{\partial\psi_i^e}{\partial x} &= \frac{\partial\psi_i^e}{\partial\bar{x}}, & \frac{\partial\psi_i^e}{\partial y} &= \frac{\partial\psi_i^e}{\partial\bar{y}} \quad (i=1,2,3,4) \\ \frac{\partial\psi_1^e}{\partial\bar{x}} &= -\frac{1}{a}\left(1-\frac{\bar{y}}{b}\right), & \frac{\partial\psi_1^e}{\partial\bar{y}} &= -\frac{1}{b}\left(1-\frac{\bar{x}}{a}\right) \\ \frac{\partial\psi_2^e}{\partial\bar{x}} &= \frac{1}{a}\left(1-\frac{\bar{y}}{b}\right), & \frac{\partial\psi_2^e}{\partial\bar{y}} &= -\frac{1}{b}\frac{\bar{x}}{a} \\ \frac{\partial\psi_3^e}{\partial\bar{x}} &= \frac{1}{a}\frac{\bar{y}}{b}, & \frac{\partial\psi_3^e}{\partial\bar{y}} &= \frac{1}{b}\frac{\bar{x}}{a} \\ \frac{\partial\psi_4^e}{\partial\bar{x}} &= -\frac{1}{a}\frac{\bar{y}}{b}, & \frac{\partial\psi_4^e}{\partial\bar{y}} &= \frac{1}{b}\left(1-\frac{\bar{x}}{a}\right) \end{aligned} \tag{A9.2}$$

We note that the derivatives $\partial\psi_i^e/\partial x$ are only (linear) functions of y while $\partial\psi_i^e/\partial y$ are only (linear) functions of x. In addition, we have the following integral identities:

$$\begin{aligned} \int_{0.5a}^{a}\left(1-\frac{\bar{x}}{a}\right)d\bar{x} &= \frac{a}{8}, & \int_{0}^{0.5a}\left(1-\frac{\bar{x}}{a}\right)d\bar{x} &= \frac{3a}{8} \\ \int_{0.5a}^{a}\frac{\bar{x}}{a}\,d\bar{x} &= \frac{3a}{8}, & \int_{0}^{0.5a}\frac{\bar{x}}{a}\,d\bar{x} &= \frac{a}{8} \end{aligned} \tag{A9.3}$$

Evaluation of integrals for control domains on the boundary

The statements in Eqs. (9.4.3) and (9.4.4) must be specialized to various boundary domains shown in Fig. 5.4.4. Here we present the discretized equations associated with Eqs. (9.4.3) and (9.4.4) for three representative control domains on the boundary of the rectangular domain shown in Fig. 9.4.1 (see Fig. 5.4.4 for representative boundary control domains).

Discretized equations for the boundary domain in Fig. A9.1. We have

$$
\begin{aligned}
& c_{11}\left[\tfrac{3}{8}\frac{b_1}{a_1}U_{I-1} - \tfrac{3}{8}\left(\frac{b_1}{a_1}+\frac{b_2}{a_2}\right)U_I + \tfrac{3}{8}\frac{b_1}{a_1}U_{I+1}\right.\\
&\qquad \left. + \tfrac{1}{8}\frac{b_1}{a_1}U_{I+N} - \tfrac{1}{8}\left(\frac{b_1}{a_1}+\frac{b_2}{a_2}\right)U_{I+N+1} + \tfrac{1}{8}\frac{b_2}{a_2}U_{I+N+2}\right] +\\
& c_{66}\left[-\tfrac{1}{8}\frac{a_1}{b_1}U_{I-1} - \tfrac{3}{8}\left(\frac{a_1}{b_1}+\frac{a_2}{b_2}\right)U_I - \tfrac{1}{8}\frac{a_2}{b_2}U_{I+1}\right.\\
&\qquad \left. + \tfrac{1}{8}\frac{a_1}{b_1}U_{I+N} + \tfrac{3}{8}\left(\frac{a_1}{b_1}+\frac{a_2}{b_2}\right)U_{I+N+1} + \tfrac{1}{8}\frac{a_2}{b_2}U_{I+N+2}\right] +\\
& \frac{c_{12}}{4}\left(V_{I-1} - V_{I+1} - V_{I+N} + V_{I+N+2}\right) +\\
& \frac{c_{66}}{4}\left(-V_{I-1} + V_{I+1} - V_{I+N} + V_{I+N+2}\right) = T_I^x + F_I^x
\end{aligned}
\tag{A9.4a}
$$

$$
\begin{aligned}
& c_{66}\left[\tfrac{3}{8}\frac{b_1}{a_1}V_{I-1} - \tfrac{3}{8}\left(\frac{b_1}{a_1}+\frac{b_2}{a_2}\right)U_I + \tfrac{3}{8}\frac{b_1}{a_1}V_{I+1}\right.\\
&\qquad \left. + \tfrac{1}{8}\frac{b_1}{a_1}V_{I+N} - \tfrac{1}{8}\left(\frac{b_1}{a_1}+\frac{b_2}{a_2}\right)V_{I+N+1} + \tfrac{1}{8}\frac{b_2}{a_2}V_{I+N+2}\right] +\\
& c_{22}\left[-\tfrac{1}{8}\frac{a_1}{b_1}V_{I-1} - \tfrac{3}{8}\left(\frac{a_1}{b_1}+\frac{a_2}{b_2}\right)U_I - \tfrac{1}{8}\frac{a_2}{b_2}V_{I+1}\right.\\
&\qquad \left. + \tfrac{1}{8}\frac{a_1}{b_1}V_{I+N} + \tfrac{3}{8}\left(\frac{a_1}{b_1}+\frac{a_2}{b_2}\right)V_{I+N+1} + \tfrac{1}{8}\frac{a_2}{b_2}V_{I+N+2}\right] +\\
& \frac{c_{66}}{4}\left(U_{I-1} - U_{I+1} - U_{I+N} + U_{I+N+2}\right) +\\
& \frac{c_{12}}{4}\left(-U_{I-1} + U_{I+1} - U_{I+N} + U_{I+N+2}\right) = T_I^y + F_I^y
\end{aligned}
\tag{A9.4b}
$$

where T_I^x and T_I^y are the x and y components of forces at node I (specified or to be determined) and F_I^x and F_I^y are the body forces at node I:

$$
\begin{aligned}
F_I^x &= \int_{x_I-0.5a_1}^{x_I}\int_{0.0}^{0.5b_1} f_x(x,y)\,dxdy + \int_{x_I}^{x_I+0.5a}\int_{0}^{0.5b_2} f_x(x,y)\,dxdy,\\
F_I^y &= \int_{x_I-0.5a_1}^{x_I}\int_{0.0}^{0.5b_1} f_y(x,y)\,dxdy + \int_{x_I}^{x_I+0.5a}\int_{0}^{0.5b_2} f_y(x,y)\,dxdy.
\end{aligned}
\tag{A9.5}
$$

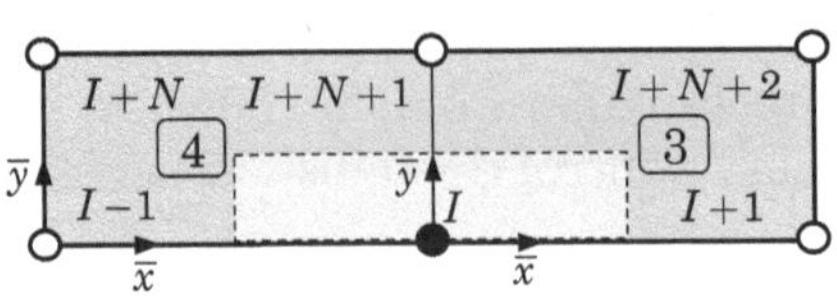

Fig. A9.1 Typical control domain on the bottom boundary of the domain.

Discretized equations for the boundary domain shown in Fig. A9.2. We have

$$
\begin{aligned}
&c_{11}\left[-\tfrac{3}{8}\frac{b_1}{a_1}U_1 - \tfrac{3}{8}\frac{b_1}{a_1}U_2 - \tfrac{1}{8}\frac{b_1}{a_1}U_{N+2} + \tfrac{1}{8}\frac{b_1}{a_1}U_{N+3}\right]+\\
&c_{66}\left[-\tfrac{3}{8}\frac{a_1}{b_1}U_1 - \tfrac{1}{8}\frac{a_1}{b_1}U_2 + \tfrac{3}{8}\frac{a_1}{b_1}U_{N+2} + \tfrac{1}{8}\frac{a_1}{b_1}U_{N+3}\right]+\\
&\frac{c_{12}}{4}\left(-V_1 + V_2 - V_{N+2} + V_{N+3}\right)+\\
&\frac{c_{12}}{4}\left(-V_1 - V_2 + V_{N+2} + V_{N+3}\right) = T_1^x + F_1^x, \qquad \text{(A9.6a)}\\
&c_{66}\left[-\tfrac{3}{8}\frac{b_1}{a_1}V_1 - \tfrac{3}{8}\frac{b_1}{a_1}V_2 - \tfrac{1}{8}\frac{b_1}{a_1}V_{N+2} + \tfrac{1}{8}\frac{b_1}{a_1}V_{N+3}\right]+\\
&c_{22}\left[-\tfrac{3}{8}\frac{a_1}{b_1}V_1 - \tfrac{1}{8}\frac{a_1}{b_1}V_2 + \tfrac{3}{8}\frac{a_1}{b_1}V_{N+2} + \tfrac{1}{8}\frac{a_1}{b_1}V_{N+3}\right]+\\
&\frac{c_{12}}{4}\left(-U_1 + U_2 - U_{N+2} + U_{N+3}\right)+\\
&\frac{c_{12}}{4}\left(-U_1 - U_2 + U_{N+2} + U_{N+3}\right) = T_1^y + F_1^y. \qquad \text{(A9.6b)}
\end{aligned}
$$

where

$$
F_1^x = \int_0^{0.5a_1}\int_{0.0}^{0.5b_1} f_x(x,y)\,dxdy, \quad F_1^y = \int_0^{0.5a_1}\int_{0.0}^{0.5b_1} f_y(x,y)\,dxdy. \qquad \text{(A9.7)}
$$

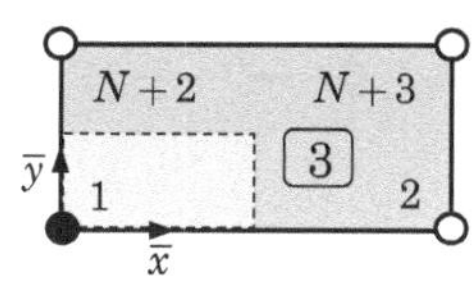

Fig. A9.2 Control domain on the bottom left corner of the domain.

Discretized equations for the boundary domain shown in Fig. A9.3. We have

$$
\begin{aligned}
&c_{11}\Bigg[-\tfrac{1}{8}\frac{b_1}{a_1}U_{I-N-1} - \tfrac{1}{8}\frac{b_1}{a_1}U_{I-N} - \tfrac{3}{8}\left(\frac{b_1}{a_1}+\frac{b_2}{a_2}\right)U_I\\
&\qquad + \tfrac{3}{8}\frac{b_2}{a_2}U_{I+1} - \tfrac{1}{8}\frac{b_2}{a_2}U_{I+N+1} + \tfrac{1}{8}\frac{b_2}{a_2}U_{I+N+2}\Bigg]+\\
&c_{66}\Bigg[\tfrac{3}{8}\frac{a_1}{b_1}U_{I-N-1} + \tfrac{1}{8}\frac{a_1}{b_1}U_{I-N} - \tfrac{3}{8}\left(\frac{a_1}{b_1}+\frac{a_2}{b_2}\right)U_I\\
&\qquad - \tfrac{1}{8}\frac{a_1}{b_1}U_{I+1} + \tfrac{3}{8}\frac{a_2}{b_2}U_{I+N+1} + \tfrac{1}{8}\frac{a_2}{b_2}U_{I+N+2}\Bigg]+
\end{aligned}
$$

$$
\begin{aligned}
&\frac{c_{12}}{4}\left(-V_{I-N-1} - V_{I-N} + V_{I+N+1} + V_{I+N+2}\right) + \\
&\frac{c_{66}}{4}\left(V_{I-N-1} - V_{I-N} - V_{I+N+1} + V_{I+N+2}\right) = T_I^x + F_I^x \qquad \text{(A9.8a)}
\end{aligned}
$$

$$
\begin{aligned}
&c_{66}\Bigg[-\tfrac{1}{8}\frac{b_1}{a_1}V_{I-N-1} - \tfrac{1}{8}\frac{b_1}{a_1}V_{I-N} - \tfrac{3}{8}\left(\frac{b_1}{a_1} + \frac{b_2}{a_2}\right)V_I \\
&\qquad + \tfrac{3}{8}\frac{b_2}{a_2}V_{I+1} - \tfrac{1}{8}\frac{b_2}{a_2}V_{I+N+1} + \tfrac{1}{8}\frac{b_2}{a_2}V_{I+N+2}\Bigg] + \\
&c_{22}\Bigg[\tfrac{3}{8}\frac{a_1}{b_1}V_{I-N-1} + \tfrac{1}{8}\frac{a_1}{b_1}V_{I-N} - \tfrac{3}{8}\left(\frac{a_1}{b_1} + \frac{a_2}{b_2}\right)V_I \\
&\qquad - \tfrac{1}{8}\frac{a_1}{b_1}U_{I+1} + \tfrac{3}{8}\frac{a_2}{b_2}V_{I+N+1} + \tfrac{1}{8}\frac{a_2}{b_2}V_{I+N+2}\Bigg] + \\
&\frac{c_{12}}{4}\left(-U_{I-N-1} - U_{I-N} + U_{I+N+1} + U_{I+N+2}\right) + \\
&\frac{c_{66}}{4}\left(U_{I-N-1} - U_{I-N} - U_{I+N+1} + U_{I+N+2}\right) = T_I^y + F_I^y \qquad \text{(A9.8b)}
\end{aligned}
$$

$$
\begin{aligned}
F_I^x &= \int_0^{0.5a_1}\int_{0.5b_1}^{b_1} f_x(x,y)\,dxdy + \int_0^{0.5a_2}\int_0^{0.5b_2} f_x(x,y)\,dxdy, \\
F_I^y &= \int_0^{0.5a_1}\int_{0.5b_1}^{b_1} f_y(x,y)\,dxdy + \int_0^{0.5a_2}\int_0^{0.5b_2} f_y(x,y)\,dxdy.
\end{aligned} \qquad \text{(A9.9)}
$$

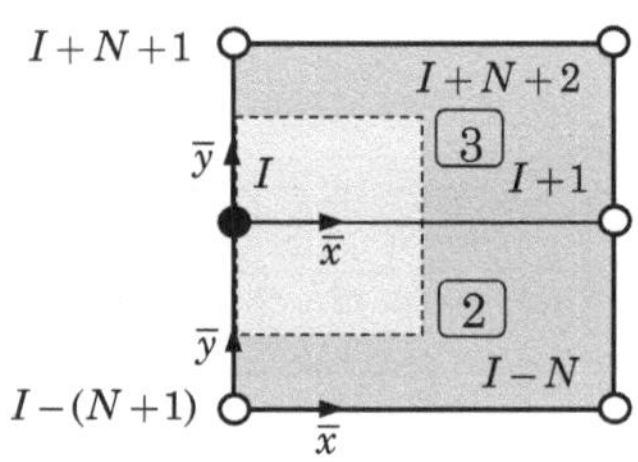

Fig. A9.3 Typical control domain on the left boundary of the domain.

Appendix 9B: Evaluation of line integrals in the DMCDM for fluid flow

Evaluation of integrals inside the domain

In the following simplifications, we omit the body force terms and determine the discrete matrix coefficients. Equations (9.6.5) and (9.6.6) over each of the four elements surrounding the node I in a typical internal control domain (see Fig. A9.4) are evaluated here. The penalty terms are evaluated using the "reduced" integration idea (i.e., the integrand is evaluated at the finite element center and multiplied by the area of the domain of the integration).

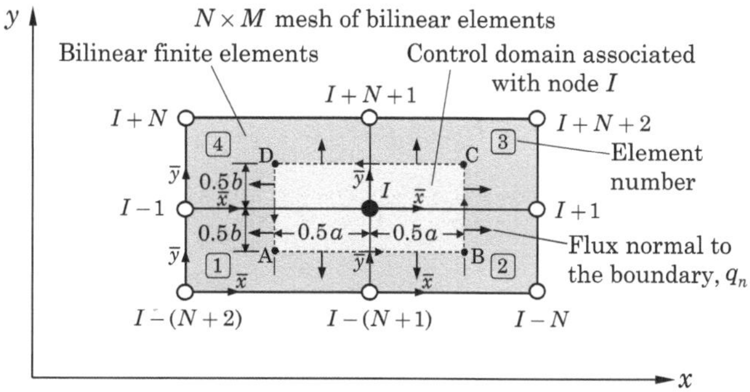

Fig. A9.4 Typical interior control domain.

Equation (9.6.5) *coefficients for element 1.* Equation (9.6.5) takes the form

$$0=\int_{0.5b_1}^{b_1}\left[2\mu\frac{\partial u}{\partial\bar{x}}+\gamma\left(\frac{\partial u}{\partial\bar{x}}+\frac{\partial v}{\partial\bar{y}}\right)\right]_{\bar{x}=0.5a_1}d\bar{y}+\int_{0.5a_1}^{a_1}\mu\left[\frac{\partial u}{\partial\bar{y}}+\frac{\partial v}{\partial\bar{x}}\right]_{\bar{y}=0.5b_1}d\bar{x}$$
$$=2\mu\left(-\tfrac{1}{8}\frac{b_1}{a_1}U_{I-N-2}+\tfrac{1}{8}\frac{b_1}{a_1}U_{I-N-1}+\tfrac{3}{8}\frac{b_1}{a_1}U_I-\tfrac{3}{8}\frac{b_1}{a_1}U_{I-1}\right)$$
$$+\mu\left(-\tfrac{1}{8}\frac{a_1}{b_1}U_{I-N-2}-\tfrac{3}{8}\frac{a_1}{b_1}U_{I-N-1}+\tfrac{3}{8}\frac{a_1}{b_1}U_I+\tfrac{1}{8}\frac{a_1}{b_1}U_{I-1}\right)$$
$$+\tfrac{1}{4}\frac{b_1}{a_1}\gamma\left(-U_{I-N-2}+U_{I-N-1}+U_I-U_{I-1}\right)$$
$$+\frac{\gamma}{4}\left(-V_{I-N-2}-V_{I-N-1}+V_I+V_{I-1}\right)$$
$$+\frac{\mu}{4}\left(-V_{I-N-2}+V_{I-N-1}+V_I-V_{I-1}\right) \tag{B9.1}$$

Equation (9.6.5) *coefficients for element 2.* Equation (9.6.5) takes the form

$$0=-\int_{0.5b_1}^{b_1}\left[2\mu\frac{\partial u}{\partial\bar{x}}+\gamma\left(\frac{\partial u}{\partial\bar{x}}+\frac{\partial v}{\partial\bar{y}}\right)\right]_{\bar{x}=0.5a_2}d\bar{y}+\int_{0}^{0.5a_2}\mu\left[\frac{\partial u}{\partial\bar{y}}+\frac{\partial v}{\partial\bar{x}}\right]_{\bar{y}=0.5b_1}d\bar{x}$$
$$=-2\mu\left(-\tfrac{1}{8}\frac{b_1}{a_2}U_{I-N-1}+\tfrac{1}{8}\frac{b_1}{a_2}U_{I-N}+\tfrac{3}{8}\frac{b_1}{a_2}U_{I+1}-\tfrac{3}{8}\frac{b_1}{a_2}U_I\right)$$
$$+\mu\left(-\tfrac{3}{8}\frac{a_2}{b_1}U_{I-N-1}-\tfrac{1}{8}\frac{a_2}{b_1}U_{I-N}+\tfrac{1}{8}\frac{a_2}{b_1}U_{I+1}+\tfrac{3}{8}\frac{a_2}{b_1}U_I\right)$$
$$-\tfrac{1}{4}\frac{b_1}{a_2}\gamma\left(-U_{I-N-1}+U_{I-N}+U_{I+1}-U_I\right)$$
$$-\frac{\gamma}{4}\left(-V_{I-N-1}-V_{I-N}+V_{I+1}+V_I\right)$$
$$+\frac{\mu}{4}\left(-V_{I-N-1}+V_{I-N}+V_{I+1}-V_I\right) \tag{B9.2}$$

Equation (9.6.5) *coefficients for element 3.* Equation (9.6.5) takes the form

$$
\begin{aligned}
0 = &-\int_0^{0.5b_2} \left[2\mu\frac{\partial u}{\partial \bar{x}} + \gamma\left(\frac{\partial u}{\partial \bar{x}} + \frac{\partial v}{\partial \bar{y}}\right)\right]_{\bar{x}=0.5a_2} d\bar{y} - \int_0^{0.5a_2} \mu\left[\frac{\partial u}{\partial \bar{y}} + \frac{\partial v}{\partial \bar{x}}\right]_{\bar{y}=0.5b_2} d\bar{x} \\
= &-2\mu\left(-\tfrac{3}{8}\frac{b_2}{a_2}U_I + \tfrac{3}{8}\frac{b_2}{a_2}U_{I+1} + \tfrac{1}{8}\frac{b_2}{a_2}U_{I+N+2} - \tfrac{1}{8}\frac{b_2}{a_2}U_{I+N+1}\right) \\
&- \mu\left(-\tfrac{3}{8}\frac{a_2}{b_2}U_I - \tfrac{1}{8}\frac{a_2}{b_2}U_{I+1} + \tfrac{1}{8}\frac{a_2}{b_2}U_{I+N+2} + \tfrac{3}{8}\frac{a_2}{b_2}U_{I+N+1}\right) \\
&- \tfrac{1}{4}\frac{b_2}{a_2}\gamma\left(-U_I + U_{I+1} + U_{I+N+2} - U_{I+N+1}\right) \\
&- \frac{\gamma}{4}\left(-V_I - V_{I+1} + V_{I+N+2} + V_{I+N+1}\right) \\
&- \frac{\mu}{4}\left(-V_I + V_{I+1} + V_{I+N+2} - V_{I+N+1}\right)
\end{aligned}
\tag{B9.3}
$$

Equation (9.6.5) *coefficients for element 4.* Equation (9.6.5) takes the form

$$
\begin{aligned}
0 = &\int_0^{0.5b_2} \left[2\mu\frac{\partial u}{\partial \bar{x}} + \gamma\left(\frac{\partial u}{\partial \bar{x}} + \frac{\partial v}{\partial \bar{y}}\right)\right]_{\bar{x}=0.5a_1} d\bar{y} - \int_{0.5a_1}^{a_1} \mu\left[\frac{\partial u}{\partial \bar{y}} + \frac{\partial v}{\partial \bar{x}}\right]_{\bar{y}=0.5b_2} d\bar{x} \\
= &\,2\mu\left(-\tfrac{3}{8}\frac{b_2}{a_1}U_{I-1} + \tfrac{3}{8}\frac{b_2}{a_1}U_I + \tfrac{1}{8}\frac{b_2}{a_1}U_{I+N+1} - \tfrac{1}{8}\frac{b_2}{a_1}U_{I+N}\right) \\
&- \mu\left(-\tfrac{1}{8}\frac{a_1}{b_2}U_{I-1} - \tfrac{3}{8}\frac{a_1}{b_2}U_I + \tfrac{3}{8}\frac{a_1}{b_2}U_{I+N+1} + \tfrac{1}{8}\frac{a_1}{b_2}U_{I+N}\right) \\
&+ \tfrac{1}{4}\frac{b_2}{a_1}\gamma\left(-U_{I-1} + U_I + U_{I+N+1} - U_{I+N}\right) \\
&+ \frac{\gamma}{4}\left(-V_{I-1} - V_I + V_{I+N+1} + V_{I+N}\right) \\
&- \frac{\mu}{4}\left(-V_{I-1} + V_I + V_{I+N+1} - V_{I+N}\right)
\end{aligned}
\tag{B9.4}
$$

Equation (9.6.6) *coefficients for element 1.* Equation (9.6.6) takes the form

$$
\begin{aligned}
0 = &\int_{0.5b_1}^{b_1} \left[\mu\left(\frac{\partial u}{\partial y} + \frac{\partial v}{\partial x}\right)\right]_{\bar{x}=0.5a_1} d\bar{y} + \int_{0.5a_1}^{a_1} \left[2\mu\frac{\partial v}{\partial y} + \gamma\left(\frac{\partial u}{\partial x} + \frac{\partial v}{\partial y}\right)\right]_{\bar{y}=0.5b_1} d\bar{x} \\
= &\,2\mu\left(-\tfrac{1}{8}\frac{a_1}{b_1}V_{I-N-2} - \tfrac{3}{8}\frac{a_1}{b_1}V_{I-N-1} + \tfrac{3}{8}\frac{a_1}{b_1}V_I + \tfrac{1}{8}\frac{a_1}{b_1}V_{I-1}\right) \\
&+ \mu\left(-\tfrac{1}{8}\frac{b_1}{a_1}V_{I-N-2} + \tfrac{1}{8}\frac{b_1}{a_1}V_{I-N-1} + \tfrac{3}{8}\frac{b_1}{a_1}V_I - \tfrac{3}{8}\frac{b_1}{a_1}V_{I-1}\right) \\
&+ \tfrac{1}{4}\frac{a_1}{b_1}\gamma\left(-V_{I-N-2} - V_{I-N-1} + V_I + V_{I-1}\right) \\
&+ \frac{\gamma}{4}\left(-U_{I-N-2} + U_{I-N-1} + U_I - U_{I-1}\right) \\
&+ \frac{\mu}{4}\left(-U_{I-N-2} - U_{I-N-1} + U_I + U_{I-1}\right)
\end{aligned}
\tag{B9.5}
$$

Equation (9.6.6) *coefficients for element 2.* Equation (9.6.6) takes the form

$$
\begin{aligned}
0 &= -\int_{0.5b_1}^{b_1} \left[\mu\left(\frac{\partial u}{\partial y}+\frac{\partial v}{\partial x}\right)\right]_{\bar{x}=0.5a_2} d\bar{y} + \int_0^{0.5a_2} \left[2\mu\frac{\partial v}{\partial y}+\gamma\left(\frac{\partial u}{\partial x}+\frac{\partial v}{\partial y}\right)\right]_{\bar{y}=0.5b_1} d\bar{x} \\
&= 2\mu\left(-\tfrac{3}{8}\frac{a_2}{b_1}V_{I-N-1}-\tfrac{1}{8}\frac{a_2}{b_1}V_{I-N}+\tfrac{1}{8}\frac{a_2}{b_1}V_{I+1}+\tfrac{3}{8}\frac{a_2}{b_1}V_I\right) \\
&\quad -\mu\left(-\tfrac{1}{8}\frac{b_1}{a_2}V_{I-N-1}+\tfrac{1}{8}\frac{b_1}{a_2}V_{I-N}+\tfrac{3}{8}\frac{b_1}{a_2}V_{I+1}-\tfrac{3}{8}\frac{b_1}{a_2}V_I\right) \\
&\quad +\tfrac{1}{4}\frac{a_2}{b_1}\gamma\left(-V_{I-N-1}-V_{I-N}+V_{I+1}+V_I\right) \\
&\quad +\frac{\gamma}{4}\left(-U_{I-N-1}+U_{I-N}+U_{I+1}-U_I\right) \\
&\quad -\frac{\mu}{4}\left(-U_{I-N-1}-U_{I-N}+U_{I+1}+U_I\right)
\end{aligned} \tag{B9.6}
$$

Equation (9.6.6) *coefficients for element 3.* Equation (9.6.6) takes the form

$$
\begin{aligned}
0 &= -\int_0^{0.5b_2} \left[\mu\left(\frac{\partial u}{\partial y}+\frac{\partial v}{\partial x}\right)\right]_{\bar{x}=0.5a_2} d\bar{y} - \int_0^{0.5a_2} \left[2\mu\frac{\partial v}{\partial y}+\gamma\left(\frac{\partial u}{\partial x}+\frac{\partial v}{\partial y}\right)\right]_{\bar{y}=0.5b_2} d\bar{x} \\
&= -2\mu\left(-\tfrac{3}{8}\frac{a_2}{b_2}V_I-\tfrac{1}{8}\frac{a_2}{b_2}V_{I+1}+\tfrac{1}{8}\frac{a_2}{b_2}V_{I+N+2}+\tfrac{3}{8}\frac{a_2}{b_2}V_{I+N+1}\right) \\
&\quad -\mu\left(-\tfrac{3}{8}\frac{b_2}{a_2}V_I+\tfrac{3}{8}\frac{b_2}{a_2}V_{I+1}+\tfrac{1}{8}\frac{b_2}{a_2}V_{I+N+2}-\tfrac{1}{8}\frac{b_2}{a_2}V_{I+N+1}\right) \\
&\quad -\tfrac{1}{4}\frac{a_2}{b_2}\gamma\left(-V_I-V_{I+1}+V_{I+N+2}+V_{I+N+1}\right) \\
&\quad -\frac{\gamma}{4}\left(-U_I+U_{I+1}+U_{I+N+2}-U_{I+N+1}\right) \\
&\quad -\frac{\mu}{4}\left(-U_I-U_{I+1}+U_{I+N+2}+U_{I+N+1}\right)
\end{aligned} \tag{B9.7}
$$

Equation (9.6.6) *coefficients for element 4.* Equation (9.6.6) takes the form

$$
\begin{aligned}
0 &= \int_0^{0.5b_2} \left[\mu\left(\frac{\partial u}{\partial y}+\frac{\partial v}{\partial x}\right)\right]_{\bar{x}=0.5a_1} d\bar{y} - \int_{0.5a_1}^{a_1} \left[2\mu\frac{\partial v}{\partial y}+\gamma\left(\frac{\partial u}{\partial x}+\frac{\partial v}{\partial y}\right)\right]_{\bar{y}=0.5b_2} d\bar{x} \\
&= -2\mu\left(-\tfrac{1}{8}\frac{a_1}{b_2}V_{I-1}-\tfrac{3}{8}\frac{a_1}{b_2}V_I+\tfrac{3}{8}\frac{a_1}{b_2}V_{I+N+1}+\tfrac{1}{8}\frac{a_1}{b_2}V_{I+N}\right) \\
&\quad +\mu\left(-\tfrac{3}{8}\frac{b_2}{a_1}V_{I-1}+\tfrac{3}{8}\frac{b_2}{a_1}V_I+\tfrac{1}{8}\frac{b_2}{a_1}V_{I+N+1}-\tfrac{1}{8}\frac{b_2}{a_1}V_{I+N}\right) \\
&\quad -\tfrac{1}{4}\frac{a_1}{b_2}\gamma\left(-V_{I-1}-V_I+V_{I+N+1}+V_{I+N}\right) \\
&\quad -\frac{\gamma}{4}\left(-U_{I-1}+U_I+U_{I+N+1}-U_{I+N}\right) \\
&\quad +\frac{\mu}{4}\left(-U_{I-1}-U_I+U_{I+N+1}+U_{I+N}\right)
\end{aligned} \tag{B9.8}
$$

Evaluation of integrals for control domains on the boundary

Here we present the discrete equations for a control domain associated with a boundary node. For a typical control domain on the bottom boundary (see Fig. A9.5), excluding the corner control domains [see Eqs. (A9.4a)–(A9.9) for other boundary control domains], we have the following two equations:

$$
\begin{aligned}
0 = & -2\mu\Big(-\tfrac{3}{8}\frac{b_2}{a_2}U_I + \tfrac{3}{8}\frac{b_2}{a_2}U_{I+1} + \tfrac{1}{8}\frac{b_2}{a_2}U_{I+N+2} - \tfrac{1}{8}\frac{b_2}{a_2}U_{I+N+1}\Big)\\
& - \mu\Big(-\tfrac{3}{8}\frac{a_2}{b_2}U_I - \tfrac{1}{8}\frac{a_2}{b_2}U_{I+1} + \tfrac{1}{8}\frac{a_2}{b_2}U_{I+N+2} + \tfrac{3}{8}\frac{a_2}{b_2}U_{I+N+1}\Big)\\
& - \tfrac{1}{4}\frac{b_2}{a_2}\gamma\left(-U_I + U_{I+1} + U_{I+N+2} - U_{I+N+1}\right)\\
& - \frac{\gamma}{4}\left(-V_I - V_{I+1} + V_{I+N+2} + V_{I+N+1}\right)\\
& - \frac{\mu}{4}\left(-V_I + V_{I+1} + V_{I+N+2} - V_{I+N+1}\right)\\
& + 2\mu\Big(-\tfrac{3}{8}\frac{b_2}{a_1}U_{I-1} + \tfrac{3}{8}\frac{b_2}{a_1}U_I + \tfrac{1}{8}\frac{b_2}{a_1}U_{I+N+1} - \tfrac{1}{8}\frac{b_2}{a_1}U_{I+N}\Big)\\
& - \mu\Big(-\tfrac{1}{8}\frac{a_1}{b_2}U_{I-1} - \tfrac{3}{8}\frac{a_1}{b_2}U_I + \tfrac{3}{8}\frac{a_1}{b_2}U_{I+N+1} + \tfrac{1}{8}\frac{a_1}{b_2}U_{I+N}\Big)\\
& + \tfrac{1}{4}\frac{b_2}{a_1}\gamma\left(-U_{I-1} + U_I + U_{I+N+1} - U_{I+N}\right)\\
& + \frac{\gamma}{4}\left(-V_{I-1} - V_I + V_{I+N+1} + V_{I+N}\right)\\
& - \frac{\mu}{4}\left(-V_{I-1} + V_I + V_{I+N+1} - V_{I+N}\right) + T_I^x
\end{aligned}
\tag{B9.9}
$$

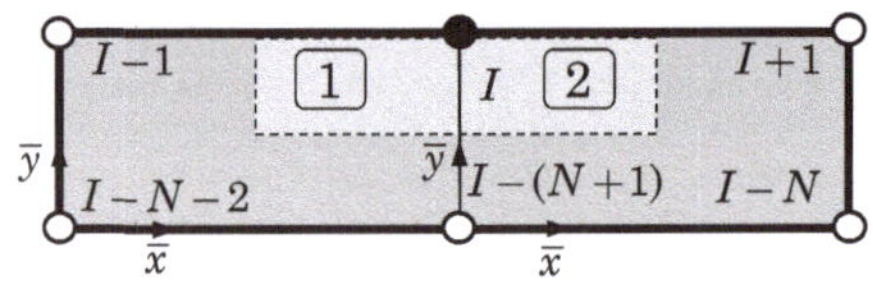

Fig. A9.5 Control domains associated with a typical node on the bottom boundary.

$$
\begin{aligned}
0 = & -2\mu\Big(-\tfrac{3}{8}\frac{a_2}{b_2}V_I - \tfrac{1}{8}\frac{a_2}{b_2}V_{I+1} + \tfrac{1}{8}\frac{a_2}{b_2}V_{I+N+2} + \tfrac{3}{8}\frac{a_2}{b_2}V_{I+N+1}\Big)\\
& - \mu\Big(-\tfrac{3}{8}\frac{b_2}{a_2}V_I + \tfrac{3}{8}\frac{b_2}{a_2}V_{I+1} + \tfrac{1}{8}\frac{b_2}{a_2}V_{I+N+2} - \tfrac{1}{8}\frac{b_2}{a_2}V_{I+N+1}\Big)\\
& - \tfrac{1}{4}\frac{a_2}{b_2}\gamma\left(-V_I - V_{I+1} + V_{I+N+2} + V_{I+N+1}\right)
\end{aligned}
$$

$$
\begin{aligned}
&-\frac{\gamma}{4}\left(-U_I+U_{I+1}+U_{I+N+2}-U_{I+N+1}\right)\\
&-\frac{\mu}{4}\left(-U_I-U_{I+1}+U_{I+N+2}+U_{I+N+1}\right)\\
&-2\mu\Big(-\tfrac{1}{8}\frac{a_1}{b_2}V_{I-1}-\tfrac{3}{8}\frac{a_1}{b_2}V_I+\tfrac{3}{8}\frac{a_1}{b_2}V_{I+N+1}+\tfrac{1}{8}\frac{a_1}{b_2}V_{I+N}\Big)\\
&+\mu\Big(-\tfrac{3}{8}\frac{b_2}{a_1}V_{I-1}+\tfrac{3}{8}\frac{b_2}{a_1}V_I+\tfrac{1}{8}\frac{b_2}{a_1}V_{I+N+1}-\tfrac{1}{8}\frac{b_2}{a_1}V_{I+N}\Big)\\
&-\tfrac{1}{4}\frac{a_1}{b_2}\gamma\left(-V_{I-1}-V_I+V_{I+N+1}+V_{I+N}\right)\\
&-\frac{\gamma}{4}\left(-U_{I-1}+U_I+U_{I+N+1}-U_{I+N}\right)\\
&+\frac{\mu}{4}\left(-U_{I-1}-U_I+U_{I+N+1}+U_{I+N}\right)+T_I^y
\end{aligned}
\tag{B9.10}
$$

Here T_I^x and T_I^y denote the x and y components of the stress vector at node I. On a free surface, $T_I^x = T_I^y = 0$ and Eqs. (B9.9) and (B9.10) provide the required equations. When U_I is specified, T_I^x is determined in the post-computation using Eq. (B9.9); similarly, when V_I is known, T_I^y is determined using Eq. (B9.10).

Appendix 9C: DMCDM equations for plane elasticity using primal mesh of arbitrary triangles

$$
\begin{aligned}
&\int_{P_1C_2P_2}\left[\left(A\frac{\partial u}{\partial x}+B\frac{\partial v}{\partial y}\right)dy-C\left(\frac{\partial u}{\partial y}+\frac{\partial v}{\partial x}\right)dx\right]+\int_{\Omega_2} f_x\,dxdy\\
&\quad=\left[\int_{\bar{l}_2}\left(A\frac{\partial\psi_i}{\partial x}\frac{\partial y}{\partial\eta}-C\frac{\partial\psi_i}{\partial y}\frac{\partial x}{\partial\eta}\right)^{(2)}d\eta+\int_{\bar{l}_2}\left(A\frac{\partial\psi_i}{\partial x}\frac{\partial y}{\partial\xi}-C\frac{\partial\psi_i}{\partial y}\frac{\partial x}{\partial\xi}\right)^{(2)}d\xi\right]u_i^{(2)}\\
&\quad+\left[\int_{\bar{l}_2}\left(B\frac{\partial\psi_i}{\partial y}\frac{\partial y}{\partial\eta}-C\frac{\partial\psi_i}{\partial x}\frac{\partial x}{\partial\eta}\right)^{(2)}d\eta+\int_{\bar{l}_2}\left(B\frac{\partial\psi_i}{\partial y}\frac{\partial y}{\partial\xi}-C\frac{\partial\psi_i}{\partial x}\frac{\partial x}{\partial\xi}\right)^{(2)}d\xi\right]v_i^{(2)}\\
&\quad+\int_{\hat{\Omega}_2} f_x J\,d\xi d\eta.
\end{aligned}
\tag{C9.1}
$$

$$
\begin{aligned}
&\int_{P_2C_3P_3}\left[\left(A\frac{\partial u}{\partial x}+B\frac{\partial v}{\partial y}\right)dy-C\left(\frac{\partial u}{\partial y}+\frac{\partial v}{\partial x}\right)dx\right]+\int_{\Omega_3} f_x\,dxdy\\
&\quad=\left[\int_{\bar{l}_3}\left(A\frac{\partial\psi_i}{\partial x}\frac{\partial y}{\partial\eta}-C\frac{\partial\psi_i}{\partial y}\frac{\partial x}{\partial\eta}\right)^{(3)}d\eta+\int_{\bar{l}_3}\left(A\frac{\partial\psi_i}{\partial x}\frac{\partial y}{\partial\xi}-C\frac{\partial\psi_i}{\partial y}\frac{\partial x}{\partial\xi}\right)^{(3)}d\xi\right]u_i^{(3)}\\
&\quad+\left[\int_{\bar{l}_3}\left(B\frac{\partial\psi_i}{\partial y}\frac{\partial y}{\partial\eta}-C\frac{\partial\psi_i}{\partial x}\frac{\partial x}{\partial\eta}\right)^{(3)}d\eta+\int_{\bar{l}_3}\left(B\frac{\partial\psi_i}{\partial y}\frac{\partial y}{\partial\xi}-C\frac{\partial\psi_i}{\partial x}\frac{\partial x}{\partial\xi}\right)^{(3)}d\xi\right]v_i^{(3)}\\
&\quad+\int_{\hat{\Omega}_3} f_x J\,d\xi d\eta
\end{aligned}
\tag{C9.2}
$$

$$\begin{aligned}
&\int_{P_3C_4P_4}\left[\left(A\frac{\partial u}{\partial x}+B\frac{\partial v}{\partial y}\right)dy-C\left(\frac{\partial u}{\partial y}+\frac{\partial v}{\partial x}\right)dx\right]+\int_{\Omega_4} f_x\,dxdy\\
&\quad=\left[\int_{\bar{l}_4}\left(A\frac{\partial\psi_i}{\partial x}\frac{\partial y}{\partial\eta}-C\frac{\partial\psi_i}{\partial y}\frac{\partial x}{\partial\eta}\right)^{(4)}d\eta+\int_{\bar{l}_4}\left(A\frac{\partial\psi_i}{\partial x}\frac{\partial y}{\partial\xi}-C\frac{\partial\psi_i}{\partial y}\frac{\partial x}{\partial\xi}\right)^{(4)}d\xi\right]u_i^{(4)}\\
&\quad+\left[\int_{\bar{l}_4}\left(B\frac{\partial\psi_i}{\partial y}\frac{\partial y}{\partial\eta}-C\frac{\partial\psi_i}{\partial x}\frac{\partial x}{\partial\eta}\right)^{(4)}d\eta+\int_{\bar{l}_4}\left(B\frac{\partial\psi_i}{\partial y}\frac{\partial y}{\partial\xi}-C\frac{\partial\psi_i}{\partial x}\frac{\partial x}{\partial\xi}\right)^{(4)}d\xi\right]v_i^{(4)}\\
&\quad+\int_{\hat{\Omega}_4} f_xJ\,d\xi d\eta
\end{aligned}\tag{C9.3}$$

$$\begin{aligned}
&\int_{P_4C_5P_5}\left[\left(A\frac{\partial u}{\partial x}+B\frac{\partial v}{\partial y}\right)dy-C\left(\frac{\partial u}{\partial y}+\frac{\partial v}{\partial x}\right)dx\right]+\int_{\Omega_5} f_x\,dxdy\\
&\quad=\left[\int_{\bar{l}_5}\left(A\frac{\partial\psi_i}{\partial x}\frac{\partial y}{\partial\eta}-C\frac{\partial\psi_i}{\partial y}\frac{\partial x}{\partial\eta}\right)^{(5)}d\eta+\int_{\bar{l}_5}\left(A\frac{\partial\psi_i}{\partial x}\frac{\partial y}{\partial\xi}-C\frac{\partial\psi_i}{\partial y}\frac{\partial x}{\partial\xi}\right)^{(5)}d\xi\right]u_i^{(5)}\\
&\quad+\left[\int_{\bar{l}_5}\left(B\frac{\partial\psi_i}{\partial y}\frac{\partial y}{\partial\eta}-C\frac{\partial\psi_i}{\partial x}\frac{\partial x}{\partial\eta}\right)^{(5)}d\eta+\int_{\bar{l}_5}\left(B\frac{\partial\psi_i}{\partial y}\frac{\partial y}{\partial\xi}-C\frac{\partial\psi_i}{\partial x}\frac{\partial x}{\partial\xi}\right)^{(5)}d\xi\right]v_i^{(5)}\\
&\quad+\int_{\hat{\Omega}_5} f_xJ\,d\xi d\eta
\end{aligned}\tag{C9.4}$$

$$\begin{aligned}
&\int_{P_5C_6P_6}\left[\left(A\frac{\partial u}{\partial x}+B\frac{\partial v}{\partial y}\right)dy-C\left(\frac{\partial u}{\partial y}+\frac{\partial v}{\partial x}\right)dx\right]+\int_{\Omega_6} f_x\,dxdy\\
&\quad=\left[\int_{\bar{l}_6}\left(A\frac{\partial\psi_i}{\partial x}\frac{\partial y}{\partial\eta}-C\frac{\partial\psi_i}{\partial y}\frac{\partial x}{\partial\eta}\right)^{(6)}d\eta+\int_{\bar{l}_6}\left(A\frac{\partial\psi_i}{\partial x}\frac{\partial y}{\partial\xi}-C\frac{\partial\psi_i}{\partial y}\frac{\partial x}{\partial\xi}\right)^{(6)}d\xi\right]u_i^{(6)}\\
&\quad+\left[\int_{\bar{l}_6}\left(B\frac{\partial\psi_i}{\partial y}\frac{\partial y}{\partial\eta}-C\frac{\partial\psi_i}{\partial x}\frac{\partial x}{\partial\eta}\right)^{(6)}d\eta+\int_{\bar{l}_6}\left(B\frac{\partial\psi_i}{\partial y}\frac{\partial y}{\partial\xi}-C\frac{\partial\psi_i}{\partial x}\frac{\partial x}{\partial\xi}\right)^{(6)}d\xi\right]v_i^{(6)}\\
&\quad+\int_{\bar{\Omega}_6} f_xJ\,d\xi d\eta
\end{aligned}\tag{C9.5}$$

The stiffness coefficients are as follows:

$$\begin{aligned}
K_{Ix}^1&=\int_{\bar{l}_1}\left(A\frac{\partial\psi_1}{\partial x}\frac{\partial y}{\partial\eta}-C\frac{\partial\psi_1}{\partial y}\frac{\partial x}{\partial\eta}\right)^{(1)}d\eta+\int_{\bar{l}_1}\left(A\frac{\partial\psi_1}{\partial x}\frac{\partial y}{\partial\xi}-C\frac{\partial\psi_1}{\partial y}\frac{\partial x}{\partial\xi}\right)^{(1)}d\xi\\
&+\int_{\bar{l}_2}\left(A\frac{\partial\psi_1}{\partial x}\frac{\partial y}{\partial\eta}-C\frac{\partial\psi_1}{\partial y}\frac{\partial x}{\partial\eta}\right)^{(2)}d\eta+\int_{\bar{l}_2}\left(A\frac{\partial\psi_1}{\partial x}\frac{\partial y}{\partial\xi}-C\frac{\partial\psi_1}{\partial y}\frac{\partial x}{\partial\xi}\right)^{(2)}d\xi\\
&+\int_{\bar{l}_3}\left(A\frac{\partial\psi_1}{\partial x}\frac{\partial y}{\partial\eta}-C\frac{\partial\psi_1}{\partial y}\frac{\partial x}{\partial\eta}\right)^{(3)}d\eta+\int_{\bar{l}_3}\left(A\frac{\partial\psi_1}{\partial x}\frac{\partial y}{\partial\xi}-C\frac{\partial\psi_1}{\partial y}\frac{\partial x}{\partial\xi}\right)^{(3)}d\xi
\end{aligned}$$

$$+ \int_{\bar{l}_4} \left(A \frac{\partial \psi_1}{\partial x} \frac{\partial y}{\partial \eta} - C \frac{\partial \psi_1}{\partial y} \frac{\partial x}{\partial \eta} \right)^{(4)} d\eta + \int_{\bar{l}_4} \left(A \frac{\partial \psi_1}{\partial x} \frac{\partial y}{\partial \xi} - C \frac{\partial \psi_1}{\partial y} \frac{\partial x}{\partial \xi} \right)^{(4)} d\xi$$
$$+ \int_{\bar{l}_5} \left(A \frac{\partial \psi_1}{\partial x} \frac{\partial y}{\partial \eta} - C \frac{\partial \psi_1}{\partial y} \frac{\partial x}{\partial \eta} \right)^{(5)} d\eta + \int_{\bar{l}_5} \left(A \frac{\partial \psi_1}{\partial x} \frac{\partial y}{\partial \xi} - C \frac{\partial \psi_1}{\partial y} \frac{\partial x}{\partial \xi} \right)^{(5)} d\xi$$
$$+ \int_{\bar{l}_6} \left(A \frac{\partial \psi_1}{\partial x} \frac{\partial y}{\partial \eta} - C \frac{\partial \psi_1}{\partial y} \frac{\partial x}{\partial \eta} \right)^{(6)} d\eta + \int_{\bar{l}_6} \left(A \frac{\partial \psi_1}{\partial x} \frac{\partial y}{\partial \xi} - C \frac{\partial \psi_1}{\partial y} \frac{\partial x}{\partial \xi} \right)^{(6)} d\xi \quad \text{(C9.6)}$$

$$K_{Ix}^2 = \int_{\bar{l}_1} \left(B \frac{\partial \psi_1}{\partial y} \frac{\partial y}{\partial \eta} - C \frac{\partial \psi_1}{\partial x} \frac{\partial x}{\partial \eta} \right)^{(1)} d\eta + \int_{\bar{l}_1} \left(B \frac{\partial \psi_1}{\partial y} \frac{\partial y}{\partial \xi} - C \frac{\partial \psi_1}{\partial x} \frac{\partial x}{\partial \xi} \right)^{(1)} d\xi$$
$$+ \int_{\bar{l}_2} \left(B \frac{\partial \psi_1}{\partial y} \frac{\partial y}{\partial \eta} - C \frac{\partial \psi_1}{\partial x} \frac{\partial x}{\partial \eta} \right)^{(2)} d\eta + \int_{\bar{l}_2} \left(B \frac{\partial \psi_1}{\partial y} \frac{\partial y}{\partial \xi} - C \frac{\partial \psi_1}{\partial x} \frac{\partial x}{\partial \xi} \right)^{(2)} d\xi$$
$$+ \int_{\bar{l}_3} \left(B \frac{\partial \psi_1}{\partial y} \frac{\partial y}{\partial \eta} - C \frac{\partial \psi_1}{\partial x} \frac{\partial x}{\partial \eta} \right)^{(3)} d\eta + \int_{\bar{l}_3} \left(B \frac{\partial \psi_1}{\partial y} \frac{\partial y}{\partial \xi} - C \frac{\partial \psi_1}{\partial x} \frac{\partial x}{\partial \xi} \right)^{(3)} d\xi$$
$$+ \int_{\bar{l}_4} \left(B \frac{\partial \psi_1}{\partial y} \frac{\partial y}{\partial \eta} - C \frac{\partial \psi_1}{\partial x} \frac{\partial x}{\partial \eta} \right)^{(4)} d\eta + \int_{\bar{l}_4} \left(B \frac{\partial \psi_1}{\partial y} \frac{\partial y}{\partial \xi} - C \frac{\partial \psi_1}{\partial x} \frac{\partial x}{\partial \xi} \right)^{(4)} d\xi$$
$$+ \int_{\bar{l}_5} \left(B \frac{\partial \psi_1}{\partial y} \frac{\partial y}{\partial \eta} - C \frac{\partial \psi_1}{\partial x} \frac{\partial x}{\partial \eta} \right)^{(5)} d\eta + \int_{\bar{l}_5} \left(B \frac{\partial \psi_1}{\partial y} \frac{\partial y}{\partial \xi} - C \frac{\partial \psi_1}{\partial x} \frac{\partial x}{\partial \xi} \right)^{(5)} d\xi$$
$$+ \int_{\bar{l}_6} \left(B \frac{\partial \psi_1}{\partial y} \frac{\partial y}{\partial \eta} - C \frac{\partial \psi_1}{\partial x} \frac{\partial x}{\partial \eta} \right)^{(6)} d\eta + \int_{\bar{l}_6} \left(B \frac{\partial \psi_1}{\partial y} \frac{\partial y}{\partial \xi} - C \frac{\partial \psi_1}{\partial x} \frac{\partial x}{\partial \xi} \right)^{(6)} d\xi \quad \text{(C9.7)}$$

$$K_{1x}^1 = \int_{\bar{l}_1} \left(A \frac{\partial \psi_3}{\partial x} \frac{\partial y}{\partial \eta} - C \frac{\partial \psi_3}{\partial y} \frac{\partial x}{\partial \eta} \right)^{(1)} d\eta + \int_{\bar{l}_1} \left(A \frac{\partial \psi_3}{\partial x} \frac{\partial y}{\partial \xi} - C \frac{\partial \psi_3}{\partial y} \frac{\partial x}{\partial \xi} \right)^{(1)} d\xi$$
$$+ \int_{\bar{l}_2} \left(A \frac{\partial \psi_2}{\partial x} \frac{\partial y}{\partial \eta} - C \frac{\partial \psi_2}{\partial y} \frac{\partial x}{\partial \eta} \right)^{(2)} d\eta + \int_{\bar{l}_2} \left(A \frac{\partial \psi_2}{\partial x} \frac{\partial y}{\partial \xi} - C \frac{\partial \psi_2}{\partial y} \frac{\partial x}{\partial \xi} \right)^{(2)} d\xi \quad \text{(C9.8)}$$

$$K_{1x}^2 = \int_{\bar{l}_1} \left(B \frac{\partial \psi_3}{\partial y} \frac{\partial y}{\partial \eta} - C \frac{\partial \psi_3}{\partial x} \frac{\partial x}{\partial \eta} \right)^{(1)} d\eta + \int_{\bar{l}_1} \left(B \frac{\partial \psi_3}{\partial y} \frac{\partial y}{\partial \xi} - C \frac{\partial \psi_3}{\partial x} \frac{\partial x}{\partial \xi} \right)^{(1)} d\xi$$
$$+ \int_{\bar{l}_2} \left(B \frac{\partial \psi_2}{\partial y} \frac{\partial y}{\partial \eta} - C \frac{\partial \psi_2}{\partial x} \frac{\partial x}{\partial \eta} \right)^{(2)} d\eta + \int_{\bar{l}_2} \left(B \frac{\partial \psi_2}{\partial y} \frac{\partial y}{\partial \xi} - C \frac{\partial \psi_2}{\partial x} \frac{\partial x}{\partial \xi} \right)^{(2)} d\xi \quad \text{(C9.9)}$$

$$K_{2x}^1 = \int_{\bar{l}_2} \left(A \frac{\partial \psi_3}{\partial x} \frac{\partial y}{\partial \eta} - C \frac{\partial \psi_3}{\partial y} \frac{\partial x}{\partial \eta} \right)^{(2)} d\eta + \int_{\bar{l}_2} \left(A \frac{\partial \psi_3}{\partial x} \frac{\partial y}{\partial \xi} - C \frac{\partial \psi_3}{\partial y} \frac{\partial x}{\partial \xi} \right)^{(2)} d\xi$$
$$+ \int_{\bar{l}_3} \left(A \frac{\partial \psi_2}{\partial x} \frac{\partial y}{\partial \eta} - C \frac{\partial \psi_2}{\partial y} \frac{\partial x}{\partial \eta} \right)^{(3)} d\eta + \int_{\bar{l}_3} \left(A \frac{\partial \psi_2}{\partial x} \frac{\partial y}{\partial \xi} - C \frac{\partial \psi_2}{\partial y} \frac{\partial x}{\partial \xi} \right)^{(3)} d\xi \quad \text{(C9.10)}$$

$$K_{2x}^2 = \int_{\bar{l}_2} \left(B \frac{\partial \psi_3}{\partial y} \frac{\partial y}{\partial \eta} - C \frac{\partial \psi_3}{\partial x} \frac{\partial x}{\partial \eta} \right)^{(2)} d\eta + \int_{\bar{l}_2} \left(B \frac{\partial \psi_3}{\partial y} \frac{\partial y}{\partial \xi} - C \frac{\partial \psi_3}{\partial x} \frac{\partial x}{\partial \xi} \right)^{(2)} d\xi$$
$$+ \int_{\bar{l}_3} \left(B \frac{\partial \psi_2}{\partial y} \frac{\partial y}{\partial \eta} - C \frac{\partial \psi_2}{\partial x} \frac{\partial x}{\partial \eta} \right)^{(3)} d\eta + \int_{\bar{l}_3} \left(B \frac{\partial \psi_2}{\partial y} \frac{\partial y}{\partial \xi} - C \frac{\partial \psi_2}{\partial x} \frac{\partial x}{\partial \xi} \right)^{(3)} d\xi \quad \text{(C9.11)}$$

$$K_{3x}^1 = \int_{\bar{l}_3} \left(A\frac{\partial \psi_3}{\partial x}\frac{\partial y}{\partial \eta} - C\frac{\partial \psi_3}{\partial y}\frac{\partial x}{\partial \eta} \right)^{(3)} d\eta + \int_{\bar{l}_3} \left(A\frac{\partial \psi_3}{\partial x}\frac{\partial y}{\partial \xi} - C\frac{\partial \psi_3}{\partial y}\frac{\partial x}{\partial \xi} \right)^{(3)} d\xi$$
$$+ \int_{\bar{l}_4} \left(A\frac{\partial \psi_2}{\partial x}\frac{\partial y}{\partial \eta} - C\frac{\partial \psi_2}{\partial y}\frac{\partial x}{\partial \eta} \right)^{(4)} d\eta + \int_{\bar{l}_4} \left(A\frac{\partial \psi_2}{\partial x}\frac{\partial y}{\partial \xi} - C\frac{\partial \psi_2}{\partial y}\frac{\partial x}{\partial \xi} \right)^{(4)} d\xi \quad \text{(C9.12)}$$

$$K_{3x}^2 = \int_{\bar{l}_3} \left(B\frac{\partial \psi_3}{\partial y}\frac{\partial y}{\partial \eta} - C\frac{\partial \psi_3}{\partial x}\frac{\partial x}{\partial \eta} \right)^{(3)} d\eta + \int_{\bar{l}_3} \left(B\frac{\partial \psi_3}{\partial y}\frac{\partial y}{\partial \xi} - C\frac{\partial \psi_3}{\partial x}\frac{\partial x}{\partial \xi} \right)^{(3)} d\xi$$
$$+ \int_{\bar{l}_4} \left(B\frac{\partial \psi_2}{\partial y}\frac{\partial y}{\partial \eta} - C\frac{\partial \psi_2}{\partial x}\frac{\partial x}{\partial \eta} \right)^{(4)} d\eta + \int_{\bar{l}_4} \left(B\frac{\partial \psi_2}{\partial y}\frac{\partial y}{\partial \xi} - C\frac{\partial \psi_2}{\partial x}\frac{\partial x}{\partial \xi} \right)^{(4)} d\xi \quad \text{(C9.13)}$$

$$K_{4x}^1 = \int_{\bar{l}_4} \left(A\frac{\partial \psi_3}{\partial x}\frac{\partial y}{\partial \eta} - C\frac{\partial \psi_3}{\partial y}\frac{\partial x}{\partial \eta} \right)^{(4)} d\eta + \int_{\bar{l}_4} \left(A\frac{\partial \psi_3}{\partial x}\frac{\partial y}{\partial \xi} - C\frac{\partial \psi_3}{\partial y}\frac{\partial x}{\partial \xi} \right)^{(4)} d\xi$$
$$+ \int_{\bar{l}_5} \left(A\frac{\partial \psi_2}{\partial x}\frac{\partial y}{\partial \eta} - C\frac{\partial \psi_2}{\partial y}\frac{\partial x}{\partial \eta} \right)^{(5)} d\eta + \int_{\bar{l}_5} \left(A\frac{\partial \psi_2}{\partial x}\frac{\partial y}{\partial \xi} - C\frac{\partial \psi_2}{\partial y}\frac{\partial x}{\partial \xi} \right)^{(5)} d\xi \quad \text{(C9.14)}$$

$$K_{4x}^2 = \int_{\bar{l}_4} \left(B\frac{\partial \psi_3}{\partial y}\frac{\partial y}{\partial \eta} - C\frac{\partial \psi_3}{\partial x}\frac{\partial x}{\partial \eta} \right)^{(4)} d\eta + \int_{\bar{l}_4} \left(B\frac{\partial \psi_3}{\partial y}\frac{\partial y}{\partial \xi} - C\frac{\partial \psi_3}{\partial x}\frac{\partial x}{\partial \xi} \right)^{(4)} d\xi$$
$$+ \int_{\bar{l}_5} \left(B\frac{\partial \psi_2}{\partial y}\frac{\partial y}{\partial \eta} - C\frac{\partial \psi_2}{\partial x}\frac{\partial x}{\partial \eta} \right)^{(5)} d\eta + \int_{\bar{l}_5} \left(B\frac{\partial \psi_2}{\partial y}\frac{\partial y}{\partial \xi} - C\frac{\partial \psi_2}{\partial x}\frac{\partial x}{\partial \xi} \right)^{(5)} d\xi \quad \text{(C9.15)}$$

$$K_{5x}^1 = \int_{\bar{l}_5} \left(A\frac{\partial \psi_3}{\partial x}\frac{\partial y}{\partial \eta} - C\frac{\partial \psi_3}{\partial y}\frac{\partial x}{\partial \eta} \right)^{(5)} d\eta + \int_{\bar{l}_5} \left(A\frac{\partial \psi_3}{\partial x}\frac{\partial y}{\partial \xi} - C\frac{\partial \psi_3}{\partial y}\frac{\partial x}{\partial \xi} \right)^{(5)} d\xi$$
$$+ \int_{\bar{l}_6} \left(A\frac{\partial \psi_2}{\partial x}\frac{\partial y}{\partial \eta} - C\frac{\partial \psi_2}{\partial y}\frac{\partial x}{\partial \eta} \right)^{(6)} d\eta + \int_{\bar{l}_6} \left(A\frac{\partial \psi_2}{\partial x}\frac{\partial y}{\partial \xi} - C\frac{\partial \psi_2}{\partial y}\frac{\partial x}{\partial \xi} \right)^{(6)} d\xi \quad \text{(C9.16)}$$

$$K_{5x}^2 = \int_{\bar{l}_5} \left(B\frac{\partial \psi_3}{\partial y}\frac{\partial y}{\partial \eta} - C\frac{\partial \psi_3}{\partial x}\frac{\partial x}{\partial \eta} \right)^{(5)} d\eta + \int_{\bar{l}_5} \left(B\frac{\partial \psi_3}{\partial y}\frac{\partial y}{\partial \xi} - C\frac{\partial \psi_3}{\partial x}\frac{\partial x}{\partial \xi} \right)^{(5)} d\xi$$
$$+ \int_{\bar{l}_6} \left(B\frac{\partial \psi_2}{\partial y}\frac{\partial y}{\partial \eta} - C\frac{\partial \psi_2}{\partial x}\frac{\partial x}{\partial \eta} \right)^{(6)} d\eta + \int_{\bar{l}_6} \left(B\frac{\partial \psi_2}{\partial y}\frac{\partial y}{\partial \xi} - C\frac{\partial \psi_2}{\partial x}\frac{\partial x}{\partial \xi} \right)^{(6)} d\xi \quad \text{(C9.17)}$$

$$K_{6x}^1 = \int_{\bar{l}_6} \left(A\frac{\partial \psi_3}{\partial x}\frac{\partial y}{\partial \eta} - C\frac{\partial \psi_3}{\partial y}\frac{\partial x}{\partial \eta} \right)^{(6)} d\eta + \int_{\bar{l}_6} \left(A\frac{\partial \psi_3}{\partial x}\frac{\partial y}{\partial \xi} - C\frac{\partial \psi_3}{\partial y}\frac{\partial x}{\partial \xi} \right)^{(6)} d\xi$$
$$+ \int_{\bar{l}_1} \left(A\frac{\partial \psi_2}{\partial x}\frac{\partial y}{\partial \eta} - C\frac{\partial \psi_2}{\partial y}\frac{\partial x}{\partial \eta} \right)^{(1)} d\eta + \int_{\bar{l}_1} \left(A\frac{\partial \psi_2}{\partial x}\frac{\partial y}{\partial \xi} - C\frac{\partial \psi_2}{\partial y}\frac{\partial x}{\partial \xi} \right)^{(1)} d\xi \quad \text{(C9.18)}$$

$$K_{6x}^2 = \int_{\bar{l}_6} \left(B\frac{\partial \psi_3}{\partial y}\frac{\partial y}{\partial \eta} - C\frac{\partial \psi_3}{\partial x}\frac{\partial x}{\partial \eta} \right)^{(6)} d\eta + \int_{\bar{l}_6} \left(B\frac{\partial \psi_3}{\partial y}\frac{\partial y}{\partial \xi} - C\frac{\partial \psi_3}{\partial x}\frac{\partial x}{\partial \xi} \right)^{(6)} d\xi$$
$$+ \int_{\bar{l}_1} \left(B\frac{\partial \psi_2}{\partial y}\frac{\partial y}{\partial \eta} - C\frac{\partial \psi_2}{\partial x}\frac{\partial x}{\partial \eta} \right)^{(1)} d\eta + \int_{\bar{l}_1} \left(B\frac{\partial \psi_2}{\partial y}\frac{\partial y}{\partial \xi} - C\frac{\partial \psi_2}{\partial x}\frac{\partial x}{\partial \xi} \right)^{(1)} d\xi \quad \text{(C9.19)}$$

$$F_{Ix} = \int_{\hat{\Omega}_1} f_x J \, d\xi d\eta + \int_{\hat{\Omega}_2} f_x J \, d\xi d\eta + \int_{\hat{\Omega}_3} f_x J \, d\xi d\eta$$
$$+ \int_{\hat{\Omega}_4} f_x J \, d\xi d\eta \int_{\hat{\Omega}_5} f_x J \, d\xi d\eta \int_{\hat{\Omega}_6} f_x J \, d\xi d\eta. \quad \text{(C9.20)}$$

The integral transformation in the y-direction and the value of the correlation coefficients can be obtained in the same way.

Appendix 9D: DMCDM equations for plane elasticity using primal mesh of arbitrary quadrilaterals

$$\int_{\partial\Omega_1} \left[\left(A\frac{\partial u}{\partial x} + B\frac{\partial v}{\partial y} \right) dy - C\left(\frac{\partial u}{\partial y} + \frac{\partial v}{\partial x} \right) dx \right] + \int_{\Omega_1} f_x \, dxdy$$
$$= \left[\int_1^0 \left(A\frac{\partial \psi_i}{\partial x}\frac{\partial y}{\partial \eta} - C\frac{\partial \psi_i}{\partial y}\frac{\partial x}{\partial \eta} \right)^{(1)} d\eta + \int_0^1 \left(A\frac{\partial \psi_i}{\partial x}\frac{\partial y}{\partial \xi} - C\frac{\partial \psi_i}{\partial y}\frac{\partial x}{\partial \xi} \right)^{(1)} d\xi \right] u_i^{(1)}$$
$$= \left[\int_1^0 \left(B\frac{\partial \psi_i}{\partial y}\frac{\partial y}{\partial \eta} - C\frac{\partial \psi_i}{\partial x}\frac{\partial x}{\partial \eta} \right)^{(1)} d\eta + \int_0^1 \left(B\frac{\partial \psi_i}{\partial y}\frac{\partial y}{\partial \xi} - C\frac{\partial \psi_i}{\partial x}\frac{\partial x}{\partial \xi} \right)^{(1)} d\xi \right] v_i^{(1)}$$
$$+ \int_0^1 \int_0^1 f_x J \, d\xi d\eta \quad \text{(D9.1)}$$

$$\int_{\partial\Omega_2} \left[\left(A\frac{\partial u}{\partial x} + B\frac{\partial v}{\partial y} \right) dy - C\left(\frac{\partial u}{\partial y} + \frac{\partial v}{\partial x} \right) dx \right] + \int_{\Omega_2} f_x \, dxdy$$
$$= \left[\int_0^1 \left(A\frac{\partial \psi_i}{\partial x}\frac{\partial y}{\partial \eta} - C\frac{\partial \psi_i}{\partial y}\frac{\partial x}{\partial \eta} \right)^{(2)} d\eta + \int_{-1}^0 \left(A\frac{\partial \psi_i}{\partial x}\frac{\partial y}{\partial \xi} - C\frac{\partial \psi_i}{\partial y}\frac{\partial x}{\partial \xi} \right)^{(2)} d\xi \right] u_i^{(2)}$$
$$= \left[\int_0^1 \left(B\frac{\partial \psi_i}{\partial y}\frac{\partial y}{\partial \eta} - C\frac{\partial \psi_i}{\partial x}\frac{\partial x}{\partial \eta} \right)^{(2)} d\eta + \int_{-1}^0 \left(B\frac{\partial \psi_i}{\partial y}\frac{\partial y}{\partial \xi} - C\frac{\partial \psi_i}{\partial x}\frac{\partial x}{\partial \xi} \right)^{(2)} d\xi \right] v_i^{(2)}$$
$$+ \int_{-1}^0 \int_0^1 f_x J \, d\xi d\eta \quad \text{(D9.2)}$$

$$\int_{\partial\Omega_4}\left[\left(A\frac{\partial u}{\partial x}+B\frac{\partial v}{\partial y}\right)dy-C\left(\frac{\partial u}{\partial y}+\frac{\partial v}{\partial x}\right)dx\right]+\int_{\Omega_4}f_x\,dxdy$$

$$=\left[\int_0^{-1}\left(A\frac{\partial\psi_i}{\partial x}\frac{\partial y}{\partial\eta}-C\frac{\partial\psi_i}{\partial y}\frac{\partial x}{\partial\eta}\right)^{(4)}d\eta+\int_1^{0}\left(A\frac{\partial\psi_i}{\partial x}\frac{\partial y}{\partial\xi}-C\frac{\partial\psi_i}{\partial y}\frac{\partial x}{\partial\xi}\right)^{(4)}d\xi\right]u_i^{(4)}$$

$$=\left[\int_0^{-1}\left(B\frac{\partial\psi_i}{\partial y}\frac{\partial y}{\partial\eta}-C\frac{\partial\psi_i}{\partial x}\frac{\partial x}{\partial\eta}\right)^{(4)}d\eta+\int_1^{0}\left(B\frac{\partial\psi_i}{\partial y}\frac{\partial y}{\partial\xi}-C\frac{\partial\psi_i}{\partial x}\frac{\partial x}{\partial\xi}\right)^{(4)}d\xi\right]v_i^{(4)}$$

$$+\int_0^1\int_{-1}^0 f_xJ\,d\xi d\eta \tag{D9.3}$$

The stiffness coefficients are as follows:

$$K_{1x}^1=\int_{-1}^{0}\left(A\frac{\partial\psi_1}{\partial x}\frac{\partial y}{\partial\eta}-C\frac{\partial\psi_1}{\partial y}\frac{\partial x}{\partial\eta}\right)^{(1)}d\eta+\int_0^{-1}\left(A\frac{\partial\psi_1}{\partial x}\frac{\partial y}{\partial\xi}-C\frac{\partial\psi_1}{\partial y}\frac{\partial x}{\partial\xi}\right)^{(1)}d\xi \tag{D9.4}$$

$$K_{1x}^1=\int_{-1}^{0}\left(B\frac{\partial\psi_1}{\partial y}\frac{\partial y}{\partial\eta}-C\frac{\partial\psi_1}{\partial x}\frac{\partial x}{\partial\eta}\right)^{(1)}d\eta+\int_0^{-1}\left(B\frac{\partial\psi_1}{\partial y}\frac{\partial y}{\partial\xi}-C\frac{\partial\psi_1}{\partial x}\frac{\partial x}{\partial\xi}\right)^{(1)}d\xi \tag{D9.5}$$

$$\begin{aligned}K_{2x}^1=&\int_0^{-1}\left(A\frac{\partial\psi_2}{\partial x}\frac{\partial y}{\partial\eta}-C\frac{\partial\psi_2}{\partial y}\frac{\partial x}{\partial\eta}\right)^{(1)}d\eta+\int_1^{0}\left(A\frac{\partial\psi_2}{\partial x}\frac{\partial y}{\partial\xi}-C\frac{\partial\psi_2}{\partial y}\frac{\partial x}{\partial\xi}\right)^{(1)}d\xi\\&+\int_{-1}^{0}\left(A\frac{\partial\psi_1}{\partial x}\frac{\partial y}{\partial\eta}-C\frac{\partial\psi_1}{\partial y}\frac{\partial x}{\partial\eta}\right)^{(2)}d\eta+\int_0^{-1}\left(A\frac{\partial\psi_1}{\partial x}\frac{\partial y}{\partial\xi}-C\frac{\partial\psi_1}{\partial y}\frac{\partial x}{\partial\xi}\right)^{(2)}d\xi\end{aligned} \tag{D9.6}$$

$$\begin{aligned}K_{2x}^2=&\int_0^{-1}\left(B\frac{\partial\psi_2}{\partial y}\frac{\partial y}{\partial\eta}-C\frac{\partial\psi_2}{\partial x}\frac{\partial x}{\partial\eta}\right)^{(1)}d\eta+\int_0^{1}\left(B\frac{\partial\psi_2}{\partial y}\frac{\partial y}{\partial\xi}-C\frac{\partial\psi_2}{\partial x}\frac{\partial x}{\partial\xi}\right)^{(1)}d\xi\\&+\int_{-1}^{0}\left(B\frac{\partial\psi_1}{\partial y}\frac{\partial y}{\partial\eta}-C\frac{\partial\psi_1}{\partial x}\frac{\partial x}{\partial\eta}\right)^{(2)}d\eta+\int_0^{-1}\left(B\frac{\partial\psi_1}{\partial y}\frac{\partial y}{\partial\xi}-C\frac{\partial\psi_1}{\partial x}\frac{\partial x}{\partial\xi}\right)^{(2)}d\xi\end{aligned} \tag{D9.7}$$

$$K_{3x}^1=\int_0^{-1}\left(A\frac{\partial\psi_2}{\partial x}\frac{\partial y}{\partial\eta}-C\frac{\partial\psi_2}{\partial y}\frac{\partial x}{\partial\eta}\right)^{(2)}d\eta+\int_1^{0}\left(A\frac{\partial\psi_2}{\partial x}\frac{\partial y}{\partial\xi}-C\frac{\partial\psi_2}{\partial y}\frac{\partial x}{\partial\xi}\right)^{(2)}d\xi \tag{D9.8}$$

$$K_{3x}^2=\int_0^{-1}\left(B\frac{\partial\psi_2}{\partial y}\frac{\partial y}{\partial\eta}-C\frac{\partial\psi_2}{\partial x}\frac{\partial x}{\partial\eta}\right)^{(2)}d\eta+\int_0^{1}\left(B\frac{\partial\psi_2}{\partial y}\frac{\partial y}{\partial\xi}-C\frac{\partial\psi_2}{\partial x}\frac{\partial x}{\partial\xi}\right)^{(2)}d\xi \tag{D9.9}$$

$$K_{4x}^1 = \int_0^1 \left(A\frac{\partial\psi_4}{\partial x}\frac{\partial y}{\partial\eta} - C\frac{\partial\psi_4}{\partial y}\frac{\partial x}{\partial\eta}\right)^{(1)} d\eta + \int_{-1}^0 \left(A\frac{\partial\psi_4}{\partial x}\frac{\partial y}{\partial\xi} - C\frac{\partial\psi_4}{\partial y}\frac{\partial x}{\partial\xi}\right)^{(1)} d\xi$$
$$+ \int_{-1}^0 \left(A\frac{\partial\psi_1}{\partial x}\frac{\partial y}{\partial\eta} - C\frac{\partial\psi_1}{\partial y}\frac{\partial x}{\partial\eta}\right)^{(4)} d\eta + \int_0^{-1} \left(A\frac{\partial\psi_1}{\partial x}\frac{\partial y}{\partial\xi} - C\frac{\partial\psi_1}{\partial y}\frac{\partial x}{\partial\xi}\right)^{(4)} d\xi \tag{D9.10}$$

$$K_{4x}^2 = \int_0^1 \left(B\frac{\partial\psi_4}{\partial y}\frac{\partial y}{\partial\eta} - C\frac{\partial\psi_4}{\partial x}\frac{\partial x}{\partial\eta}\right)^{(1)} d\eta + \int_{-1}^0 \left(B\frac{\partial\psi_4}{\partial y}\frac{\partial y}{\partial\xi} - C\frac{\partial\psi_4}{\partial x}\frac{\partial x}{\partial\xi}\right)^{(1)} d\xi$$
$$+ \int_{-1}^0 \left(B\frac{\partial\psi_1}{\partial y}\frac{\partial y}{\partial\eta} - C\frac{\partial\psi_1}{\partial x}\frac{\partial x}{\partial\eta}\right)^{(4)} d\eta + \int_0^{-1} \left(B\frac{\partial\psi_1}{\partial y}\frac{\partial y}{\partial\xi} - C\frac{\partial\psi_1}{\partial x}\frac{\partial x}{\partial\xi}\right)^{(4)} d\xi \tag{D9.11}$$

$$K_{5x}^1 = \int_1^0 \left(A\frac{\partial\psi_3}{\partial x}\frac{\partial y}{\partial\eta} - C\frac{\partial\psi_3}{\partial y}\frac{\partial x}{\partial\eta}\right)^{(1)} d\eta + \int_0^1 \left(A\frac{\partial\psi_3}{\partial x}\frac{\partial y}{\partial\xi} - C\frac{\partial\psi_3}{\partial y}\frac{\partial x}{\partial\xi}\right)^{(1)} d\xi$$
$$+ \int_0^1 \left(A\frac{\partial\psi_4}{\partial x}\frac{\partial y}{\partial\eta} - C\frac{\partial\psi_4}{\partial y}\frac{\partial x}{\partial\eta}\right)^{(2)} d\eta + \int_{-1}^0 \left(A\frac{\partial\psi_4}{\partial x}\frac{\partial y}{\partial\xi} - C\frac{\partial\psi_4}{\partial y}\frac{\partial x}{\partial\xi}\right)^{(2)} d\xi$$
$$+ \int_{-1}^0 \left(A\frac{\partial\psi_1}{\partial x}\frac{\partial y}{\partial\eta} - C\frac{\partial\psi_1}{\partial y}\frac{\partial x}{\partial\eta}\right)^{(3)} d\eta + \int_0^{-1} \left(A\frac{\partial\psi_1}{\partial x}\frac{\partial y}{\partial\xi} - C\frac{\partial\psi_1}{\partial y}\frac{\partial x}{\partial\xi}\right)^{(3)} d\xi$$
$$+ \int_{-1}^0 \left(A\frac{\partial\psi_2}{\partial x}\frac{\partial y}{\partial\eta} - C\frac{\partial\psi_2}{\partial y}\frac{\partial x}{\partial\eta}\right)^{(4)} d\eta + \int_1^0 \left(A\frac{\partial\psi_2}{\partial x}\frac{\partial y}{\partial\xi} - C\frac{\partial\psi_2}{\partial y}\frac{\partial x}{\partial\xi}\right)^{(4)} d\xi \tag{D9.12}$$

$$K_{5x}^2 = \int_1^0 \left(B\frac{\partial\psi_3}{\partial y}\frac{\partial y}{\partial\eta} - C\frac{\partial\psi_3}{\partial x}\frac{\partial x}{\partial\eta}\right)^{(1)} d\eta + \int_0^1 \left(B\frac{\partial\psi_3}{\partial y}\frac{\partial y}{\partial\xi} - C\frac{\partial\psi_3}{\partial x}\frac{\partial x}{\partial\xi}\right)^{(1)} d\xi$$
$$+ \int_0^1 \left(B\frac{\partial\psi_4}{\partial y}\frac{\partial y}{\partial\eta} - C\frac{\partial\psi_4}{\partial x}\frac{\partial x}{\partial\eta}\right)^{(2)} d\eta + \int_{-1}^0 \left(B\frac{\partial\psi_4}{\partial y}\frac{\partial y}{\partial\xi} - C\frac{\partial\psi_4}{\partial x}\frac{\partial x}{\partial\xi}\right)^{(2)} d\xi$$
$$+ \int_{-1}^0 \left(B\frac{\partial\psi_1}{\partial y}\frac{\partial y}{\partial\eta} - C\frac{\partial\psi_1}{\partial x}\frac{\partial x}{\partial\eta}\right)^{(3)} d\eta + \int_0^{-1} \left(B\frac{\partial\psi_1}{\partial y}\frac{\partial y}{\partial\xi} - C\frac{\partial\psi_1}{\partial x}\frac{\partial x}{\partial\xi}\right)^{(3)} d\xi$$
$$+ \int_{-1}^0 \left(B\frac{\partial\psi_2}{\partial y}\frac{\partial y}{\partial\eta} - C\frac{\partial\psi_2}{\partial x}\frac{\partial x}{\partial\eta}\right)^{(4)} d\eta + \int_1^0 \left(B\frac{\partial\psi_2}{\partial y}\frac{\partial y}{\partial\xi} - C\frac{\partial\psi_2}{\partial x}\frac{\partial x}{\partial\xi}\right)^{(4)} d\xi \tag{D9.13}$$

$$K_{6x}^1 = \int_1^0 \left(A\frac{\partial\psi_3}{\partial x}\frac{\partial y}{\partial\eta} - C\frac{\partial\psi_3}{\partial y}\frac{\partial x}{\partial\eta}\right)^{(2)} d\eta + \int_0^1 \left(A\frac{\partial\psi_3}{\partial x}\frac{\partial y}{\partial\xi} - C\frac{\partial\psi_3}{\partial y}\frac{\partial x}{\partial\xi}\right)^{(2)} d\xi$$
$$+ \int_0^{-1} \left(A\frac{\partial\psi_2}{\partial x}\frac{\partial y}{\partial\eta} - C\frac{\partial\psi_2}{\partial y}\frac{\partial x}{\partial\eta}\right)^{(3)} d\eta + \int_1^0 \left(A\frac{\partial\psi_2}{\partial x}\frac{\partial y}{\partial\xi} - C\frac{\partial\psi_2}{\partial y}\frac{\partial x}{\partial\xi}\right)^{(3)} d\xi \tag{D9.14}$$

$$K_{6x}^2 = \int_1^0 \left(B \frac{\partial \psi_3}{\partial y} \frac{\partial y}{\partial \eta} - C \frac{\partial \psi_3}{\partial x} \frac{\partial x}{\partial \eta} \right)^{(2)} d\eta + \int_0^1 \left(B \frac{\partial \psi_3}{\partial y} \frac{\partial y}{\partial \xi} - C \frac{\partial \psi_3}{\partial x} \frac{\partial x}{\partial \xi} \right)^{(2)} d\xi$$
$$+ \int_0^{-1} \left(B \frac{\partial \psi_2}{\partial y} \frac{\partial y}{\partial \eta} - C \frac{\partial \psi_2}{\partial x} \frac{\partial x}{\partial \eta} \right)^{(3)} d\eta + \int_1^0 \left(B \frac{\partial \psi_2}{\partial y} \frac{\partial y}{\partial \xi} - C \frac{\partial \psi_2}{\partial x} \frac{\partial x}{\partial \xi} \right)^{(3)} d\xi \tag{D9.15}$$

$$K_{7x}^1 = \int_0^1 \left(A \frac{\partial \psi_4}{\partial x} \frac{\partial y}{\partial \eta} - C \frac{\partial \psi_4}{\partial y} \frac{\partial x}{\partial \eta} \right)^{(4)} d\eta + \int_{-1}^0 \left(A \frac{\partial \psi_4}{\partial x} \frac{\partial y}{\partial \xi} - C \frac{\partial \psi_4}{\partial y} \frac{\partial x}{\partial \xi} \right)^{(4)} d\xi \tag{D9.16}$$

$$K_{7x}^2 = \int_0^1 \left(B \frac{\partial \psi_4}{\partial y} \frac{\partial y}{\partial \eta} - C \frac{\partial \psi_4}{\partial x} \frac{\partial x}{\partial \eta} \right)^{(4)} d\eta + \int_{-1}^0 \left(B \frac{\partial \psi_4}{\partial y} \frac{\partial y}{\partial \xi} - C \frac{\partial \psi_4}{\partial x} \frac{\partial x}{\partial \xi} \right)^{(4)} d\xi \tag{D9.17}$$

$$K_{8x}^1 = \int_1^0 \left(A \frac{\partial \psi_3}{\partial x} \frac{\partial y}{\partial \eta} - C \frac{\partial \psi_3}{\partial y} \frac{\partial x}{\partial \eta} \right)^{(4)} d\eta + \int_0^1 \left(A \frac{\partial \psi_3}{\partial x} \frac{\partial y}{\partial \xi} - C \frac{\partial \psi_3}{\partial y} \frac{\partial x}{\partial \xi} \right)^{(4)} d\xi$$
$$+ \int_0^1 \left(A \frac{\partial \psi_4}{\partial x} \frac{\partial y}{\partial \eta} - C \frac{\partial \psi_4}{\partial y} \frac{\partial x}{\partial \eta} \right)^{(3)} d\eta + \int_{-1}^0 \left(A \frac{\partial \psi_4}{\partial x} \frac{\partial y}{\partial \xi} - C \frac{\partial \psi_4}{\partial y} \frac{\partial x}{\partial \xi} \right)^{(3)} d\xi \tag{D9.18}$$

$$K_{8x}^2 = \int_1^0 \left(B \frac{\partial \psi_3}{\partial y} \frac{\partial y}{\partial \eta} - C \frac{\partial \psi_3}{\partial x} \frac{\partial x}{\partial \eta} \right)^{(4)} d\eta + \int_0^1 \left(B \frac{\partial \psi_3}{\partial y} \frac{\partial y}{\partial \xi} - C \frac{\partial \psi_3}{\partial x} \frac{\partial x}{\partial \xi} \right)^{(4)} d\xi$$
$$+ \int_0^1 \left(B \frac{\partial \psi_4}{\partial y} \frac{\partial y}{\partial \eta} - C \frac{\partial \psi_4}{\partial x} \frac{\partial x}{\partial \eta} \right)^{(3)} d\eta + \int_{-1}^0 \left(B \frac{\partial \psi_4}{\partial y} \frac{\partial y}{\partial \xi} - C \frac{\partial \psi_4}{\partial x} \frac{\partial x}{\partial \xi} \right)^{(3)} d\xi \tag{D9.19}$$

$$K_{9x}^1 = \int_1^0 \left(A \frac{\partial \psi_3}{\partial x} \frac{\partial y}{\partial \eta} - C \frac{\partial \psi_3}{\partial y} \frac{\partial x}{\partial \eta} \right)^{(3)} d\eta + \int_0^1 \left(A \frac{\partial \psi_3}{\partial x} \frac{\partial y}{\partial \xi} - C \frac{\partial \psi_3}{\partial y} \frac{\partial x}{\partial \xi} \right)^{(3)} d\xi \tag{D9.20}$$

$$K_{9x}^2 = \int_1^0 \left(B \frac{\partial \psi_3}{\partial y} \frac{\partial y}{\partial \eta} - C \frac{\partial \psi_3}{\partial x} \frac{\partial x}{\partial \eta} \right)^{(3)} d\eta + \int_0^1 \left(B \frac{\partial \psi_3}{\partial y} \frac{\partial y}{\partial \xi} - C \frac{\partial \psi_3}{\partial x} \frac{\partial x}{\partial \xi} \right)^{(3)} d\xi \tag{D9.21}$$

$$F_{Ix} = \int_0^1 \int_0^1 f_x J \, d\xi d\eta + \int_{\hat{\Omega}_2} f_x J \, d\xi d\eta + \iint_{\hat{\Omega}_3} f_x J \, d\xi d\eta$$
$$+ \iint_{\bar{\Omega}_4} f_x J \, d\xi d\eta \iint_{\bar{\Omega}_5} f_x J \, d\xi d\eta \iint_{\bar{\Omega}_6} f_x J \, d\xi d\eta \tag{D9.22}$$

10
Bending of Flat Plates

10.1 Introduction

A *plate* is a flat structural element with planform dimensions that are large compared to its thickness and is subjected to loads that cause bending deformation in addition to stretching in the plane. A plate can be viewed as a two-dimensional version of a beam. There exist two-dimensional plate theories, much like the beam theories considered in Chapter 7, which are used to study the deformation and stresses in plate structures. In this chapter, we consider the two-dimensional version of the Timoshenko beam theory, known as the *first-order shear deformation plate theory* (FST), with the assumptions of small strains, small to moderate rotations, and large displacements. Geometric nonlinearity is considered via the von Kármán strains, as in the case of beams.

First we review the governing equations of the FST. Much of the development follows what is discussed for the TBT, except accounting for the two-dimensional aspects of the theory. The FEM and DMCDM models of the FST presented in this chapter also closely resemble those presented for the TBT in Chapter 7.

10.2 Governing Equations

10.2.1 Displacement Field

The assumed form of the displacement field of the FST is

$$\begin{aligned} u_1(x,y,z) &= u(x,y) + z\phi_x(x,y), \\ u_2(x,y,z) &= v(x,y) + z\phi_y(x,y), \\ u_3(x,y,z) &= w(x,y), \end{aligned} \tag{10.2.1}$$

where $(u, v, w, \phi_x, \phi_y)$ are unknown functions to be determined. As before, (u, v, w) denote the displacements of a point on the plane $z = 0$ and ϕ_x and ϕ_y are the rotations of a transverse normal line about the $+y$ and $-x$ axes, respectively, as shown in Fig. 10.2.1 (i.e., the subscripts on ϕ_x and ϕ_y match the coordinate directions of the corresponding displacement components). The quantities $(u, v, w, \phi_x, \phi_y)$ are called the *generalized displacements*.

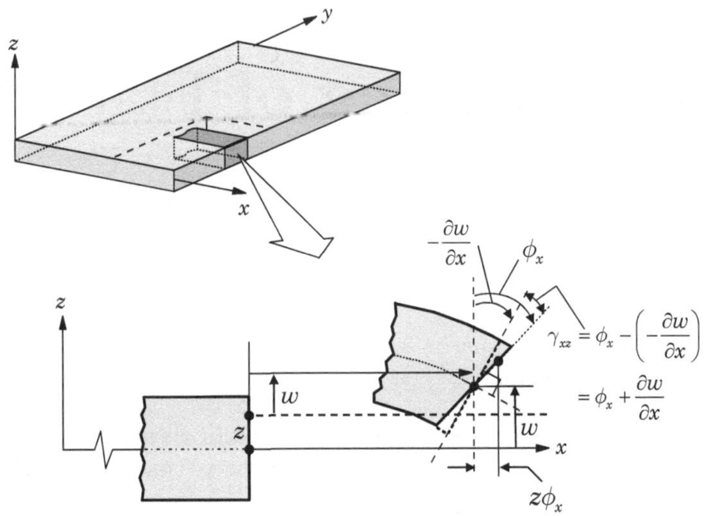

Fig. 10.2.1 Undeformed and deformed geometries of an edge of a plate under the assumptions of the FST.

For thin plates, that is, when the plate in-plane characteristic dimension to thickness ratio is on the order of 50 or greater, the rotation functions ϕ_x and ϕ_y should approach the respective slopes of the transverse deflection w:

$$\phi_x = -\frac{\partial w}{\partial x}, \quad \phi_y = -\frac{\partial w}{\partial y}. \tag{10.2.2}$$

However, this equality is not achieved in the discrete model, resulting in shear locking, as in the case of the TBT, when the same lower-order approximations are used for the transverse deflection w as well as the rotations ϕ_x and ϕ_y.

The von Kármán nonlinear strains associated with the displacement field in Eq. (10.2.1) are ($\varepsilon_{zz} = 0$)

$$\begin{Bmatrix} \varepsilon_{xx} \\ \varepsilon_{yy} \\ \gamma_{yz} \\ \gamma_{xz} \\ \gamma_{xy} \end{Bmatrix} = \begin{Bmatrix} \varepsilon_{xx}^{(0)} \\ \varepsilon_{yy}^{(0)} \\ \gamma_{yz}^{(0)} \\ \gamma_{xz}^{(0)} \\ \gamma_{xy}^{(0)} \end{Bmatrix} + z \begin{Bmatrix} \varepsilon_{xx}^{(1)} \\ \varepsilon_{yy}^{(1)} \\ 0 \\ 0 \\ \gamma_{xy}^{(1)} \end{Bmatrix} = \begin{Bmatrix} \frac{\partial u}{\partial x} + \frac{1}{2}\left(\frac{\partial w}{\partial x}\right)^2 \\ \frac{\partial v}{\partial y} + \frac{1}{2}\left(\frac{\partial w}{\partial y}\right)^2 \\ \frac{\partial w}{\partial y} + \phi_y \\ \frac{\partial w}{\partial x} + \phi_x \\ \frac{\partial u}{\partial y} + \frac{\partial v}{\partial x} + \frac{\partial w}{\partial x}\frac{\partial w}{\partial y} \end{Bmatrix} + z \begin{Bmatrix} \frac{\partial \phi_x}{\partial x} \\ \frac{\partial \phi_y}{\partial y} \\ 0 \\ 0 \\ \frac{\partial \phi_x}{\partial y} + \frac{\partial \phi_y}{\partial x} \end{Bmatrix} \tag{10.2.3}$$

Note that the strains (ε_{xx}, ε_{yy}, γ_{xy}) are linear through the plate thickness (i.e., they are linear functions of the thickness coordinate z), while the transverse shear strains (γ_{xz}, γ_{yz}) are constant through the thickness at any point (x, y) in the plane.

10.2.2 Principle of Virtual Displacements

The weak forms of the FST can be derived using the principle of virtual displacements

$$0 = \delta W \equiv \delta W_I + \delta W_E, \tag{10.2.4}$$

where the virtual strain energy δW_I and the virtual work done by applied forces δW_E in an arbitrary plate domain Ω are given by

$$\begin{aligned}\delta W_I = &\int_\Omega \left\{ \int_{-\frac{h}{2}}^{\frac{h}{2}} \left[\sigma_{xx}\left(\delta\varepsilon_{xx}^{(0)} + z\delta\varepsilon_{xx}^{(1)}\right) + \sigma_{yy}\left(\delta\varepsilon_{yy}^{(0)} + z\delta\varepsilon_{yy}^{(1)}\right)\right.\right.\\ &\left.\left. + \sigma_{xy}\left(\delta\gamma_{xy}^{(0)} + z\delta\gamma_{xy}^{(1)}\right) + K_s\sigma_{xz}\delta\gamma_{xz}^{(0)} + K_s\sigma_{yz}\delta\gamma_{yz}^{(0)}\right] dz \right\} dx\, dy\end{aligned} \tag{10.2.5}$$

$$\begin{aligned}\delta W_E = -&\left\{ \oint_\Gamma \int_{-\frac{h}{2}}^{\frac{h}{2}} \left[\sigma_{nn}\left(\delta u_n + z\delta\phi_n\right) + \sigma_{ns}\left(\delta u_s + z\delta\phi_s\right) + \sigma_{nz}\delta w\right] dz\, ds \right.\\ &\left. + \int_\Omega q\delta w \; dx\, dy \right\},\end{aligned} \tag{10.2.6}$$

where Ω denotes the undeformed midplane of a typical plate of total thickness h, $(\sigma_{nn}, \sigma_{ns}, \sigma_{nz})$ are the edge stresses along the (n, s, z) coordinates (see Fig. 10.2.2), and K_s is the shear correction coefficient ($K_s = 5/6$).

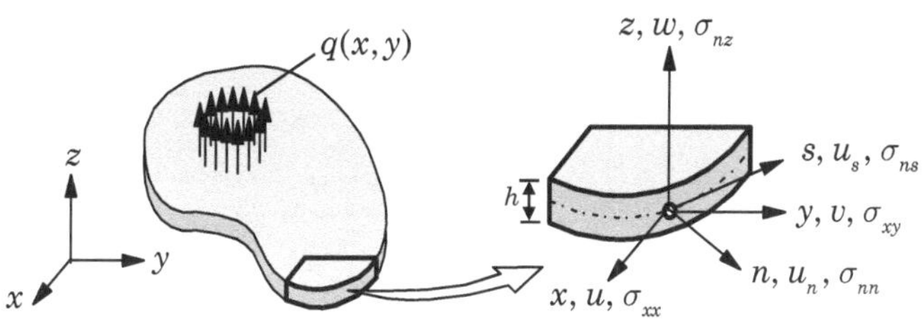

Fig. 10.2.2 Stresses and displacements at a point on the curved boundary of a plate element.

Substituting for δW_I and δW_E from Eqs. (10.2.5) and (10.2.6) into the virtual work statement in Eq. (10.2.4) and integrating through the thickness, we obtain

$$\begin{aligned}0 = &\int_\Omega \left[N_{xx}\delta\varepsilon_{xx}^{(0)} + M_{xx}\delta\varepsilon_{xx}^{(1)} + N_{yy}\delta\varepsilon_{yy}^{(0)} + M_{yy}\delta\varepsilon_{yy}^{(1)} + N_{xy}\delta\gamma_{xy}^{(0)} \right.\\ &\left. + M_{xy}\delta\gamma_{xy}^{(1)} + Q_x\delta\gamma_{xz}^{(0)} + Q_y\delta\gamma_{yz}^{(0)} - q\delta w \right] dx\, dy\\ &- \oint_\Gamma \left(N_{nn}\delta u_n + N_{ns}\delta u_s + M_{nn}\delta\phi_n + M_{ns}\delta\phi_s + Q_n\delta w\right) ds,\end{aligned} \tag{10.2.7}$$

where ϕ_n and ϕ_s are the rotations of a transverse normal about s and $-n$ coordinates, respectively, (N_{xx}, N_{yy}, N_{xy}) are the forces per unit length and (M_{xx}, M_{yy}, M_{xy}) are the moments per unit length, defined by

$$\begin{Bmatrix} N_{xx} \\ N_{yy} \\ N_{xy} \end{Bmatrix} = \int_{-\frac{h}{2}}^{\frac{h}{2}} \begin{Bmatrix} \sigma_{xx} \\ \sigma_{yy} \\ \sigma_{xy} \end{Bmatrix} dz \,, \quad \begin{Bmatrix} M_{xx} \\ M_{yy} \\ M_{xy} \end{Bmatrix} = \int_{-\frac{h}{2}}^{\frac{h}{2}} \begin{Bmatrix} \sigma_{xx} \\ \sigma_{yy} \\ \sigma_{xy} \end{Bmatrix} z\, dz \tag{10.2.8}$$

The transverse shear forces per unit length (Q_x, Q_y) are defined by

$$\begin{Bmatrix} Q_x \\ Q_y \end{Bmatrix} = K_s \int_{-\frac{h}{2}}^{\frac{h}{2}} \begin{Bmatrix} \sigma_{xz} \\ \sigma_{yz} \end{Bmatrix} dz. \tag{10.2.9}$$

Figure 10.2.3 shows various stress resultants (N_{xx},N_{yy},N_{xy},$Q_x \equiv N_{xz}$,$Q_y \equiv N_{yz}$,M_{xx},M_{yy},M_{xy}) on a plate edge with unit normal $\hat{\mathbf{n}} = n_x\hat{\mathbf{i}} + n_y\hat{\mathbf{j}}$; an alternative sign convention for moments, based on the right-hand screw rule, is displayed in Fig. 10.2.3. We note that the subscripts on N's and M's refer to the subscripts of stresses that produce them; u_n, u_s, and w are the displacements along the normal, tangential, and transverse directions (see Fig. 10.2.2), respectively, and

$$\begin{Bmatrix} N_{nn} \\ N_{ns} \end{Bmatrix} = \int_{-\frac{h}{2}}^{\frac{h}{2}} \begin{Bmatrix} \sigma_{nn} \\ \sigma_{ns} \end{Bmatrix} dz, \quad \begin{Bmatrix} M_{nn} \\ M_{ns} \end{Bmatrix} = \int_{-\frac{h}{2}}^{\frac{h}{2}} \begin{Bmatrix} \sigma_{nn} \\ \sigma_{ns} \end{Bmatrix} z\, dz, \quad Q_n = \int_{-\frac{h}{2}}^{\frac{h}{2}} \sigma_{nz}\, dz. \tag{10.2.10}$$

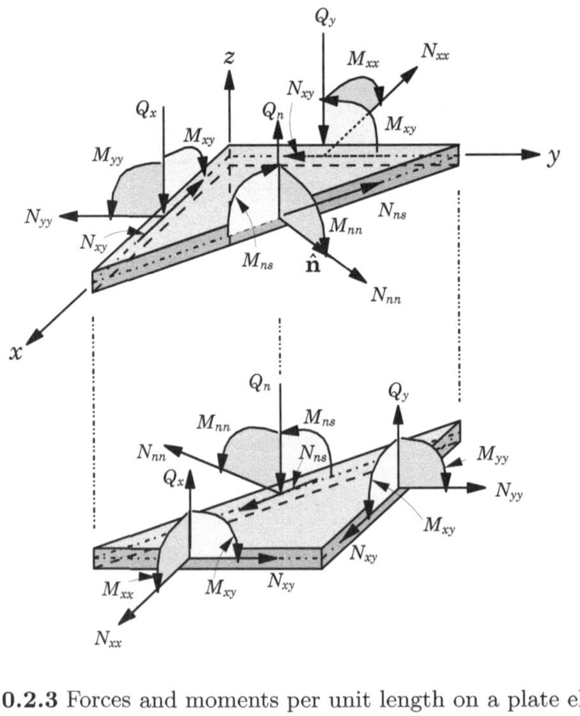

Fig. 10.2.3 Forces and moments per unit length on a plate element.

10.2.3 Governing Equations of Equilibrium

Equation (10.2.7) contains five different integral statements (i.e., weak forms) associated with the five virtual displacements $(\delta u, \delta v, \delta w, \delta\phi_x, \delta\phi_y)$, and they form the basis of the finite element model. The governing equations of equilibrium of the FST, although they are not needed to develop the finite element model, can be derived by first expressing the virtual strains in Eq. (10.2.7) in terms of the virtual displacements and then integrating-by-parts all expressions to relieve the virtual displacements $(\delta u, \delta v, \delta w, \delta\phi_x, \delta\phi_y)$ in Ω of any differentiation:

$$\begin{aligned}
0 = \int_\Omega \Big[&-(N_{xx,x} + N_{xy,y})\,\delta u - (N_{xy,x} + N_{yy,y})\,\delta v \\
&- (M_{xx,x} + M_{xy,y} - Q_x)\,\delta\phi_x - (M_{xy,x} + M_{yy,y} - Q_y)\,\delta\phi_y \\
&- (Q_{x,x} + Q_{y,y} + \mathcal{N} + q)\,\delta w\Big]\,dx\,dy \\
+ \oint_\Gamma \Big[&(N_{xx}n_x + N_{xy}n_y)\,\delta u + (N_{xy}n_x + N_{yy}n_y)\,\delta v \\
&+ (M_{xx}n_x + M_{xy}n_y)\,\delta\phi_x + (M_{xy}n_x + M_{yy}n_y)\,\delta\phi_y \\
&+ \Big(Q_x n_x + Q_y n_y + \mathcal{P}\Big)\delta w\Big]\,ds \\
- \oint_\Gamma &(N_{nn}\delta u_n + N_{ns}\delta u_s + M_{nn}\delta\phi_n + M_{ns}\delta\phi_s + Q_n\delta w)\,ds
\end{aligned} \tag{10.2.11}$$

where $\mathcal{N}$ and $\mathcal{P}$ are defined by

$$\begin{aligned}
\mathcal{N}(u,v,w) &= \frac{\partial}{\partial x}\left(N_{xx}\frac{\partial w}{\partial x} + N_{xy}\frac{\partial w}{\partial y}\right) + \frac{\partial}{\partial y}\left(N_{xy}\frac{\partial w}{\partial x} + N_{yy}\frac{\partial w}{\partial y}\right) \\
\mathcal{P}(u,v,w) &= \left(N_{xx}\frac{\partial w}{\partial x} + N_{xy}\frac{\partial w}{\partial y}\right)n_x + \left(N_{xy}\frac{\partial w}{\partial x} + N_{yy}\frac{\partial w}{\partial y}\right)n_y.
\end{aligned} \tag{10.2.12}$$

Then the fundamental lemma of calculus of variations is used to declare that the coefficients of $(\delta u, \delta v, \delta w, \delta\phi_x, \delta\phi_y)$ in Ω must be zero, giving the Euler equations

$$\delta u: \quad \frac{\partial N_{xx}}{\partial x} + \frac{\partial N_{xy}}{\partial y} = 0, \tag{10.2.13}$$

$$\delta v: \quad \frac{\partial N_{xy}}{\partial x} + \frac{\partial N_{yy}}{\partial y} = 0, \tag{10.2.14}$$

$$\delta w: \quad \frac{\partial Q_x}{\partial x} + \frac{\partial Q_y}{\partial y} + \mathcal{N}(u,v,w,\phi_x,\phi_y) + q = 0, \tag{10.2.15}$$

$$\delta\phi_x: \quad \frac{\partial M_{xx}}{\partial x} + \frac{\partial M_{xy}}{\partial y} - Q_x = 0, \tag{10.2.16}$$

$$\delta\phi_y: \quad \frac{\partial M_{xy}}{\partial x} + \frac{\partial M_{yy}}{\partial y} - Q_y = 0. \tag{10.2.17}$$

The boundary terms can be expressed in terms of the normal and tangential components u_n, u_s, ϕ_n, and ϕ_s using

$$u = n_x u_n - n_y u_s, \qquad v = n_y u_n + n_x u_s, \tag{10.2.18a}$$
$$\phi_x = n_x \phi_n - n_y \phi_s, \quad \phi_y = n_y \phi_n + n_x \phi_s. \tag{10.2.18b}$$

Thus, the primary and secondary variables of the theory are

$$\text{primary variables:} \qquad u_n,\ u_s,\ w,\ \phi_n,\ \phi_s \tag{10.2.19}$$
$$\text{secondary variables:} \quad N_{nn},\ N_{ns},\ Q_n,\ M_{nn},\ M_{ns} \tag{10.2.20}$$

10.2.4 Relations between Stress Resultants and Displacements

To express the forces (N_{xx},N_{yy},N_{xy},Q_x,Q_y) and moments (M_{xx},M_{yy},M_{xy}) in terms of the generalized displacements $(u, v, w, \phi_x, \phi_y)$, we must invoke appropriate stress–strain relations. Since $\varepsilon_{zz} = 0$, the transverse normal stress σ_{zz}, though not zero identically, does not appear in the virtual work statement and hence in the equations of equilibrium. Consequently, it amounts to neglecting the transverse normal stress. Thus, in theory, we have a case of both plane strain and plane stress. However, from practical considerations, a thin to moderately thick plate is in a state of plane stress because the thickness is small as compared to the in-plane dimensions. Hence, the plane stress-reduced constitutive relations are used.

For an orthotropic material with principal material axes (x_1, x_2, x_3) coinciding with the plate coordinates (x, y, z), the plane stress-reduced elastic constitutive equations can be expressed as

$$\begin{Bmatrix} \sigma_{xx} \\ \sigma_{yy} \\ \sigma_{xy} \end{Bmatrix} = \begin{bmatrix} Q_{11} & Q_{12} & 0 \\ Q_{12} & Q_{22} & 0 \\ 0 & 0 & Q_{66} \end{bmatrix} \begin{Bmatrix} \varepsilon_{xx} \\ \varepsilon_{yy} \\ \gamma_{xy} \end{Bmatrix}, \quad \begin{Bmatrix} \sigma_{yz} \\ \sigma_{xz} \end{Bmatrix} = \begin{bmatrix} Q_{44} & 0 \\ 0 & Q_{55} \end{bmatrix} \begin{Bmatrix} \gamma_{yz} \\ \gamma_{xz} \end{Bmatrix}, \tag{10.2.21}$$

where Q_{ij} are the plane stress-reduced elastic coefficients

$$\begin{aligned} Q_{11}(z) &= \frac{E_1(z)}{1-\nu_{12}\nu_{21}}, \quad Q_{12}(z) = \frac{\nu_{12}E_2(z)}{1-\nu_{12}\nu_{21}}, \quad Q_{22}(z) = \frac{E_2}{E_1}Q_{11}(z), \\ Q_{44}(z) &= G_{23}(z), \quad Q_{55}(z) = G_{13}(z), \quad Q_{66}(z) = G_{12}(z) \end{aligned} \tag{10.2.22}$$

and $(\sigma_{ij}, \varepsilon_{ij})$ are the stress and strain components, respectively. Here E_1 and E_2 denote Young's modulus in the x and y directions, respectively. For isotropic materials, we have $E_1(z) = E_2(z) = E(z)$ and $\nu_{12} = \nu_{23} = \nu_{13} = \nu$.

The plate constitutive equations relate the forces and moments per unit length in Eqs. (10.2.8) and (10.2.9) to the strains in Eq. (10.2.3) of the FST. For a single homogeneous orthotropic layer (i.e., Q_{ij} are constant), the plate constitutive relations are greatly simplified. They are

$$\begin{Bmatrix} N_{xx} \\ N_{yy} \\ N_{xy} \end{Bmatrix} = \begin{bmatrix} A_{11} & A_{12} & 0 \\ A_{12} & A_{22} & 0 \\ 0 & 0 & A_{66} \end{bmatrix} \begin{Bmatrix} \varepsilon_{xx}^{(0)} \\ \varepsilon_{yy}^{(0)} \\ \gamma_{xy}^{(0)} \end{Bmatrix}, \quad \begin{Bmatrix} M_{xx} \\ M_{yy} \\ M_{xy} \end{Bmatrix} = \begin{bmatrix} D_{11} & D_{12} & 0 \\ D_{12} & D_{22} & 0 \\ 0 & 0 & D_{66} \end{bmatrix} \begin{Bmatrix} \varepsilon_{xx}^{(1)} \\ \varepsilon_{yy}^{(1)} \\ \gamma_{xy}^{(1)} \end{Bmatrix}, \tag{10.2.23}$$

where A_{ij} are *extensional stiffness coefficients* and D_{ij} are *bending stiffness coefficients*, which are defined in terms of the elastic stiffness coefficients Q_{ij} as

$$(A_{ij}, D_{ij}) = \int_{-\frac{h}{2}}^{\frac{h}{2}} Q_{ij}\,(1, z^2)\,dz \quad \text{or} \quad A_{ij} = Q_{ij}\,h, \quad D_{ij} = Q_{ij}\,\frac{h^3}{12}. \tag{10.2.24}$$

The discrepancy between the actual stress state and the constant stress state predicted by the FST is often corrected in computing the transverse shear forces (Q_x, Q_y) by multiplying the transverse shear stiffness coefficients (Q_{44}, Q_{55}) with the shear correction coefficient K_s. Thus, the constitutive equations for transverse shear forces of an orthotropic plate are

$$\begin{Bmatrix} Q_y \\ Q_x \end{Bmatrix} = K_s \int_{-\frac{h}{2}}^{\frac{h}{2}} \begin{Bmatrix} \sigma_{yz} \\ \sigma_{xz} \end{Bmatrix} dz = \begin{bmatrix} A_{44} & 0 \\ 0 & A_{55} \end{bmatrix} \begin{Bmatrix} \gamma_{yz} \\ \gamma_{xz} \end{Bmatrix}, \tag{10.2.25}$$

where the transverse shear stiffness coefficients A_{44} and A_{55} are defined by

$$(A_{44}, A_{55}) = K_s \int_{-\frac{h}{2}}^{\frac{h}{2}} (Q_{44}, Q_{55})\,dz, \quad Q_{44} = G_{23}, \quad Q_{55} = G_{13}. \tag{10.2.26}$$

For functionally graded isotropic-material plates, we can express A_{ij}, B_{ij}, and D_{ij} in terms of the following stiffness coefficients:

$$\begin{aligned} A &= \frac{1}{1-\nu^2}\int_{-h/2}^{h/2} E(z)\,dz, \qquad & B &= \frac{1}{1-\nu^2}\int_{-h/2}^{h/2} E(z)z\,dz \\ D &= \frac{1}{1-\nu^2}\int_{-h/2}^{h/2} E(z)z^2\,dz, \qquad & S_{xz} &= \frac{K_s}{2(1+\nu)}\int_{-h/2}^{h/2} E(z)\,dz. \end{aligned} \tag{10.2.27}$$

Then we have

$$\begin{aligned} A_{11} = A_{22} = A, \quad A_{12} = \nu\,A, \quad A_{66} = 0.5A(1-\nu), \\ A_{44} = A_{55} = S_{xz} = 0.5K_sA(1-\nu), \end{aligned} \tag{10.2.28}$$

with similar expressions for B_{ij} and D_{ij}.

The explicit expressions of the above integrals for the power-law variation of $E(z)$ [see Eq. (8.1.1)]

$$E(z) = E_2\left[(M-1)f(z) + 1\right], \quad f(z) = \left(\frac{1}{2} + \frac{z}{h}\right)^n, \quad M = \frac{E_1}{E_2}, \tag{10.2.29}$$

are given as

$$\begin{aligned} A &= \frac{E_2 h}{1-\nu^2}\frac{(M+n)}{(n+1)}, \quad B = \frac{E_2 h^2}{2(1-\nu^2)}\frac{n(M-1)}{(1+n)(2+n)}, \\ D &= \frac{E_2 h^3}{12(1-\nu^2)}\left[\frac{(6+3n+3n^2)M + (8n+3n^2+n^3)}{(1+n)(2+n)(3+n)}\right], \\ S_{xz} &= K_s E_2 h(1-\nu)\frac{(M+n)}{2(n+1)}. \end{aligned} \tag{10.2.30}$$

Here E_1 and E_2 denote Young's moduli of the two isotropic material constituents being combined.

10.3 Finite Element Model Development

10.3.1 Weak Forms

Using the virtual work statement in Eq. (10.2.11), which is equivalent to five weak forms (each associated with δu, δv, δw, $\delta\phi_x$, and $\delta\phi_y$), we can construct the displacement finite element model of the equations governing the FST. The stress resultants in Eq. (10.2.11) are understood to be known in terms of the generalized displacements $(u, v, w, \phi_x, \phi_y)$ via Eqs. (10.2.23) and (10.2.3). The two-dimensional midplane of the plate is discretized into a set of finite elements. The five weak forms over a typical plate finite element Ω^e and its boundary Γ^e are

$$0 = \int_{\Omega^e} \left(\frac{\partial \delta u}{\partial x} N_{xx} + \frac{\partial \delta u}{\partial y} N_{xy} \right) dx\, dy - \oint_{\Gamma^e} (N_{xx} n_x + N_{xy} n_y)\, \delta u\, ds \tag{10.3.1}$$

$$0 = \int_{\Omega^e} \left(\frac{\partial \delta v}{\partial x} N_{xy} + \frac{\partial \delta v}{\partial y} N_{yy} \right) dx\, dy - \oint_{\Gamma^e} (N_{xy} n_x + N_{yy} n_y)\, \delta v\, ds \tag{10.3.2}$$

$$\begin{aligned} 0 = & \int_{\Omega^e} \left[\frac{\partial \delta w}{\partial x} Q_x + \frac{\partial \delta w}{\partial y} Q_y + \frac{\partial \delta w}{\partial x} \left(N_{xx} \frac{\partial w}{\partial x} + N_{xy} \frac{\partial w}{\partial y} \right) \right. \\ & \left. + \frac{\partial \delta w}{\partial y} \left(N_{xy} \frac{\partial w}{\partial x} + N_{yy} \frac{\partial w}{\partial y} \right) - \delta w q + k w \delta w \right] dx\, dy \\ & - \oint_{\Gamma^e} \left[\left(Q_x + N_{xx} \frac{\partial w}{\partial x} + N_{xy} \frac{\partial w}{\partial y} \right) n_x \right. \\ & \left. + \left(Q_y + N_{xy} \frac{\partial w}{\partial x} + N_{yy} \frac{\partial w}{\partial y} \right) n_y \right] \delta w\, ds \end{aligned} \tag{10.3.3}$$

$$\begin{aligned} 0 = & \int_{\Omega^e} \left(\frac{\partial \delta \phi_x}{\partial x} M_{xx} + \frac{\partial \delta \phi_x}{\partial y} M_{xy} + \delta \phi_x Q_x \right) dx\, dy \\ & - \oint_{\Gamma^e} (M_{xx} n_x + M_{xy} n_y)\, \delta \phi_x\, ds \end{aligned} \tag{10.3.4}$$

$$\begin{aligned} 0 = & \int_{\Omega^e} \left(\frac{\partial \delta \phi_y}{\partial x} M_{xy} + \frac{\partial \delta \phi_y}{\partial y} M_{yy} + \delta \phi_y Q_y \right) dx\, dy \\ & - \oint_{\Gamma^e} (M_{xy} n_x + M_{yy} n_y)\, \delta \phi_y\, ds. \end{aligned} \tag{10.3.5}$$

We note from the boundary terms in Eqs. (10.3.1)–(10.3.5) that (u_n, u_s, w, ϕ_n, ϕ_s) are the primary variables (or generalized displacements) and the associated secondary variables (or generalized forces) are, respectively (i.e., the duality pairs):

$$\begin{aligned} N_{nn} &\equiv N_{xx} n_x + N_{xy} n_y, \quad N_{ns} \equiv N_{xy} n_x + N_{yy} n_y \\ M_{nn} &\equiv M_{xx} n_x + M_{xy} n_y, \quad M_{ns} \equiv M_{xy} n_x + M_{yy} n_y \\ Q_n &\equiv \left(Q_x + N_{xx} \frac{\partial w}{\partial x} + N_{xy} \frac{\partial w}{\partial y} \right) n_x + \left(Q_y + N_{xy} \frac{\partial w}{\partial x} + N_{yy} \frac{\partial w}{\partial y} \right) n_y. \end{aligned} \tag{10.3.6}$$

10.3.2 Finite Element Model

The virtual work statements in Eqs. (10.3.1)–(10.3.5) contain at the most only the first derivatives of the dependent variables $(u, v, w, \phi_x, \phi_y)$. Since no derivatives of $(u,\ v,\ w,\ \phi_x,\ \phi_y)$ appear in the list of the primary variables, each generalized displacement may be interpolated using Lagrange interpolation functions of a chosen degree. Hence, the element is called C^0 element (i.e., the functions are continuous but not their derivatives across the elements) with respect to all dependent unknowns. Let

$$
\begin{aligned}
u(x,y) &= \sum_{j=1}^{m} u_j\psi_j^{(1)}(x,y), \quad v(x,y) = \sum_{j=1}^{m} v_j\psi_j^{(1)}(x,y)\\
w(x,y) &= \sum_{j=1}^{n} w_j\psi_j^{(2)}(x,y)\\
\phi_x(x,y) &= \sum_{j=1}^{p} s_j^1\psi_j^{(3)}(x,y), \quad \phi_y(x,y) = \sum_{j=1}^{p} s_j^2\psi_j^{(3)}(x,y)
\end{aligned}
\tag{10.3.7}
$$

where $\psi_j^{(\alpha)}$ $(\alpha = 1,2,3)$ are Lagrange interpolation functions. One can use linear, quadratic, or higher-order interpolations of these variables. Although the development is general, in the implementation of this element, we shall use equal interpolation of all variables.

Substituting Eq. (10.3.7) for $(u, v, w, \phi_x, \phi_y)$ into Eqs. (10.3.1)–(10.3.5), we obtain (the element label e is omitted):

$$
\begin{bmatrix}
\mathbf{K}^{11} & \mathbf{K}^{12} & \mathbf{K}^{13} & \mathbf{K}^{14} & \mathbf{K}^{15}\\
\mathbf{K}^{21} & \mathbf{K}^{22} & \mathbf{K}^{23} & \mathbf{K}^{24} & \mathbf{K}^{25}\\
\mathbf{K}^{31} & \mathbf{K}^{32} & \mathbf{K}^{33} & \mathbf{K}^{34} & \mathbf{K}^{35}\\
\mathbf{K}^{41} & \mathbf{K}^{42} & \mathbf{K}^{43} & \mathbf{K}^{44} & \mathbf{K}^{45}\\
\mathbf{K}^{51} & \mathbf{K}^{52} & \mathbf{K}^{53} & \mathbf{K}^{54} & \mathbf{K}^{55}
\end{bmatrix}
\begin{Bmatrix} \mathbf{u}\\ \mathbf{v}\\ \mathbf{w}\\ \mathbf{S}^1\\ \mathbf{S}^2 \end{Bmatrix}
=
\begin{Bmatrix} \mathbf{F}^1\\ \mathbf{F}^2\\ \mathbf{F}^3\\ \mathbf{F}^4\\ \mathbf{F}^5 \end{Bmatrix},
\tag{10.3.8}
$$

where the coefficients of the submatrices $\mathbf{K}^{\alpha\beta}$ and vectors $\mathbf{F}^{\alpha}$ are defined, for $\alpha, \beta = 1,2,3,4,5$, by the expressions

$$
\begin{aligned}
K_{ij}^{11} &= \int_{\Omega^e}\left(A_{11}\frac{\partial\psi_i^{(1)}}{\partial x}\frac{\partial\psi_j^{(1)}}{\partial x} + A_{66}\frac{\partial\psi_i^{(1)}}{\partial y}\frac{\partial\psi_j^{(1)}}{\partial y}\right)dx\,dy\\
K_{ij}^{12} &= \int_{\Omega^e}\left(A_{12}\frac{\partial\psi_i^{(1)}}{\partial x}\frac{\partial\psi_j^{(1)}}{\partial y} + A_{66}\frac{\partial\psi_i^{(1)}}{\partial y}\frac{\partial\psi_j^{(1)}}{\partial x}\right)dx\,dy\\
K_{ij}^{13} &= \frac{1}{2}\int_{\Omega^e}\left[\frac{\partial\psi_i^{(1)}}{\partial x}\left(A_{11}\frac{\partial w}{\partial x}\frac{\partial\psi_j^{(2)}}{\partial x} + A_{12}\frac{\partial w}{\partial y}\frac{\partial\psi_j^{(2)}}{\partial y}\right)\right.\\
&\qquad\left. + A_{66}\frac{\partial\psi_i^{(1)}}{\partial y}\left(\frac{\partial w}{\partial x}\frac{\partial\psi_j^{(2)}}{\partial y} + \frac{\partial w}{\partial y}\frac{\partial\psi_j^{(2)}}{\partial x}\right)\right]dx\,dy\\
K_{ij}^{22} &= \int_{\Omega^e}\left(A_{66}\frac{\partial\psi_i^{(1)}}{\partial x}\frac{\partial\psi_j^{(1)}}{\partial x} + A_{22}\frac{\partial\psi_i^{(1)}}{\partial y}\frac{\partial\psi_j^{(1)}}{\partial y}\right)dx\,dy
\end{aligned}
$$

$$
\begin{aligned}
K_{ij}^{23} &= \frac{1}{2}\int_{\Omega^e}\left[\frac{\partial \psi_i^{(1)}}{\partial y}\left(A_{12}\frac{\partial w}{\partial x}\frac{\partial \psi_j^{(2)}}{\partial x} + A_{22}\frac{\partial w}{\partial y}\frac{\partial \psi_j^{(2)}}{\partial y}\right)\right.\\
&\quad \left. + A_{66}\frac{\partial \psi_i^{(1)}}{\partial x}\left(\frac{\partial w}{\partial x}\frac{\partial \psi_j^{(2)}}{\partial y} + \frac{\partial w}{\partial y}\frac{\partial \psi_j^{(2)}}{\partial x}\right)\right] dx\, dy\\
K_{ij}^{31} &= \int_{\Omega^e}\left[\frac{\partial \psi_i^{(2)}}{\partial x}\left(A_{11}\frac{\partial w}{\partial x}\frac{\partial \psi_j^{(1)}}{\partial x} + A_{66}\frac{\partial w}{\partial y}\frac{\partial \psi_j^{(1)}}{\partial y}\right)\right.\\
&\quad \left. + \frac{\partial \psi_i^{(2)}}{\partial y}\left(A_{66}\frac{\partial w}{\partial x}\frac{\partial \psi_j^{(1)}}{\partial y} + A_{12}\frac{\partial w}{\partial y}\frac{\partial \psi_j^{(1)}}{\partial x}\right)\right] dx\, dy\\
K_{ij}^{32} &= \int_{\Omega^e}\left[\frac{\partial \psi_i^{(2)}}{\partial x}\left(A_{12}\frac{\partial w}{\partial x}\frac{\partial \psi_j^{(1)}}{\partial y} + A_{66}\frac{\partial w}{\partial y}\frac{\partial \psi_j^{(1)}}{\partial x}\right)\right.\\
&\quad \left. + \frac{\partial \psi_i^{(2)}}{\partial y}\left(A_{66}\frac{\partial w}{\partial x}\frac{\partial \psi_j^{(1)}}{\partial x} + A_{22}\frac{\partial w}{\partial y}\frac{\partial \psi_j^{(1)}}{\partial y}\right)\right] dx\, dy\\
K_{ij}^{33} &= \int_{\Omega^e}\left(A_{55}\frac{\partial \psi_i^{(2)}}{\partial x}\frac{\partial \psi_j^{(2)}}{\partial x} + A_{44}\frac{\partial \psi_i^{(2)}}{\partial y}\frac{\partial \psi_j^{(2)}}{\partial y}\right) dx\, dy\\
&\quad + \frac{1}{2}\int_{\Omega^e}\left\{\left[A_{11}\left(\frac{\partial w}{\partial x}\right)^2 + A_{66}\left(\frac{\partial w}{\partial y}\right)^2\right]\frac{\partial \psi_i^{(2)}}{\partial x}\frac{\partial \psi_j^{(2)}}{\partial x}\right.\\
&\quad + \left[A_{66}\left(\frac{\partial w}{\partial x}\right)^2 + A_{22}\left(\frac{\partial w}{\partial y}\right)^2\right]\frac{\partial \psi_i^{(2)}}{\partial y}\frac{\partial \psi_j^{(2)}}{\partial y}\\
&\quad \left. + (A_{12}+A_{66})\frac{\partial w}{\partial x}\frac{\partial w}{\partial y}\left(\frac{\partial \psi_i^{(2)}}{\partial x}\frac{\partial \psi_j^{(2)}}{\partial y} + \frac{\partial \psi_i^{(2)}}{\partial y}\frac{\partial \psi_j^{(2)}}{\partial x}\right)\right\} dx\, dy\\
K_{ij}^{34} &= \int_{\Omega^e} A_{55}\frac{\partial \psi_i^{(2)}}{\partial x}\psi_j^{(3)}\, dx\, dy, \quad K_{ij}^{35} = \int_{\Omega^e} A_{44}\frac{\partial \psi_i^{(2)}}{\partial y}\psi_j^{(3)}\, dx\, dy\\
K_{ij}^{44} &= \int_{\Omega^e}\left(D_{11}\frac{\partial \psi_i^{(3)}}{\partial x}\frac{\partial \psi_j^{(3)}}{\partial x} + D_{66}\frac{\partial \psi_i^{(3)}}{\partial y}\frac{\partial \psi_j^{(3)}}{\partial y} + A_{55}\psi_i^{(3)}\psi_j^{(3)}\right) dx\, dy\\
K_{ij}^{45} &= \int_{\Omega^e}\left(D_{12}\frac{\partial \psi_i^{(3)}}{\partial x}\frac{\partial \psi_j^{(3)}}{\partial y} + D_{66}\frac{\partial \psi_i^{(3)}}{\partial y}\frac{\partial \psi_j^{(3)}}{\partial x}\right) dx\, dy\\
K_{ij}^{55} &= \int_{\Omega^e}\left(D_{66}\frac{\partial \psi_i^{(3)}}{\partial x}\frac{\partial \psi_j^{(3)}}{\partial x} + D_{22}\frac{\partial \psi_i^{(3)}}{\partial y}\frac{\partial \psi_j^{(3)}}{\partial y} + A_{44}\psi_i^{(3)}\psi_j^{(3)}\right) dx\, dy\\
F_i^1 &= \oint_{\Gamma^e} N_{nn}\,\psi_i^{(1)}\, ds, \quad F_i^2 = \oint_{\Gamma^e} N_{ns}\,\psi_i^{(1)}\, ds\\
F_i^3 &= \int_{\Omega^e} q\psi_i^{(2)}\, dx\, dy + \oint_{\Gamma^e} Q_n\,\psi_i^{(2)}\, ds, \quad F_i^4 = \oint_{\Gamma^e} M_{nn}\,\psi_i^{(3)}\, ds\\
F_i^5 &= \oint_{\Gamma^e} M_{ns}\,\psi_i^{(3)}\, ds,
\end{aligned}
\tag{10.3.9}
$$

and [here $(\cdot)^{\mathrm{T}}$ denotes the transpose of the enclosed quantity]

$$\mathbf{K}^{21} = (\mathbf{K}^{12})^{\mathrm{T}}; \quad \mathbf{K}^{43} = (\mathbf{K}^{34})^{\mathrm{T}}; \quad \mathbf{K}^{53} = (\mathbf{K}^{35})^{\mathrm{T}}; \quad \mathbf{K}^{54} = (\mathbf{K}^{45})^{\mathrm{T}}$$
$$\mathbf{K}^{14} = \mathbf{K}^{15} = \mathbf{K}^{24} = \mathbf{K}^{25} = \mathbf{K}^{41} = \mathbf{K}^{42} = \mathbf{K}^{51} = \mathbf{K}^{52} = \mathbf{0}. \tag{10.3.10}$$

We note that the element stiffness matrix is *not* symmetric.

When the bilinear rectangular element is used for all generalized displacements $(u,\ v,\ w, \phi_x, \phi_y)$, the element stiffness matrices are of the order 20×20; and for the nine-node quadratic element they are 45×45.

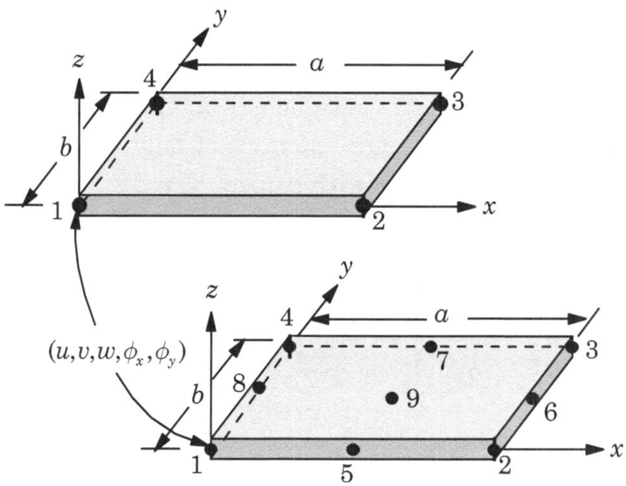

Fig. 10.3.1 Linear and nine-node quadratic rectangular elements for the FST.

10.3.3 Tangent Stiffness Coefficients

The tangent stiffness matrix coefficients for the FST (i.e., applying Newton's method of linearization) can be computed using the definition in Eq. (8.4.8) with $\gamma = 1, 2, \ldots, 5$. Suppose that the tangent stiffness matrix is of the same form as the direct stiffness matrix in Eq. (10.3.8). Then the coefficients of the submatrices $\mathbf{T}^{\alpha\beta}$ are defined by

$$T_{ij}^{\alpha\beta} = \frac{\partial R_i^\alpha}{\partial \Delta_j^\beta} \tag{10.3.11}$$

where the components of the residual vector $\mathbf{R}^\alpha$ are given by

$$R_i^\alpha = \sum_{\gamma=1}^{5} \sum_{k=1}^{n(\gamma)} K_{ik}^{\alpha\gamma} \Delta_k^\gamma - F_i^\alpha \tag{10.3.12}$$

$$\Delta_i^1 = u_i, \quad \Delta_i^2 = v_i, \quad \Delta_i^3 = w_i, \quad \Delta_i^4 = s_i^1, \quad \Delta_i^5 = s_i^2 \tag{10.3.13}$$

and $n(\gamma)$ denotes n, m, or p, depending on the nodal degree of freedom. Thus, we have (here $\mathbf{F}^\alpha$ is assumed to be independent of the solution vector)

$$\begin{aligned}
T_{ij}^{\alpha\beta} &= \frac{\partial}{\partial \Delta_j^\beta}\left(\sum_{\gamma=1}^{5}\sum_{k=1}^{n(\gamma)} K_{ik}^{\alpha\gamma}\Delta_k^\gamma - F_i^\alpha\right) = \sum_{\gamma=1}^{5}\sum_{k=1}^{n(\gamma)} \frac{\partial K_{ik}^{\alpha\gamma}}{\partial \Delta_j^\beta}\Delta_k^\gamma + K_{ij}^{\alpha\beta}\\
&= \sum_{k=1}^{m}\frac{\partial K_{ik}^{\alpha 1}}{\partial \Delta_j^\beta}u_k + \sum_{k=1}^{m}\frac{\partial K_{ik}^{\alpha 2}}{\partial \Delta_j^\beta}v_k + \sum_{k=1}^{n}\frac{\partial K_{ik}^{\alpha 3}}{\partial \Delta_j^\beta}w_k\\
&\quad + \sum_{k=1}^{p}\frac{\partial K_{ik}^{\alpha 4}}{\partial \Delta_j^\beta}s_k^1 + \sum_{k=1}^{p}\frac{\partial K_{ik}^{\alpha 5}}{\partial \Delta_j^\beta}s_k^2 + K_{ij}^{\alpha\beta}.
\end{aligned} \tag{10.3.14}$$

We note that the only coefficients that depend on the solution are K_{ij}^{13}, K_{ij}^{23}, K_{ij}^{31}, K_{ij}^{32}, and K_{ij}^{33}. Further, they are functions of only w (or functions of w_j). Hence, derivatives of all submatrices with respect to u_j, v_j, S_j^1, and S_j^2 are zero. Thus, we have (for $\alpha = 1, 2, 3, 4, 5$)

$$\begin{aligned}
&\mathbf{T}^{\alpha 1} = \mathbf{K}^{\alpha 1} = (\mathbf{K}^{1\alpha})^{\mathrm{T}}, \quad \mathbf{T}^{\alpha 2} = \mathbf{K}^{\alpha 2} = (\mathbf{K}^{2\alpha})^{\mathrm{T}}\\
&\mathbf{T}^{\alpha 4} = \mathbf{K}^{\alpha 4} = (\mathbf{K}^{4\alpha})^{\mathrm{T}}, \quad \mathbf{T}^{\alpha 5} = \mathbf{K}^{\alpha 5} = (\mathbf{K}^{5\alpha})^{\mathrm{T}},\\
&\mathbf{T}^{31} = \mathbf{K}^{31} = (\mathbf{T}^{13})^{\mathrm{T}}, \quad \mathbf{T}^{32} = \mathbf{K}^{32} = (\mathbf{T}^{23})^{\mathrm{T}},
\end{aligned} \tag{10.3.15}$$

and $\mathbf{T}^{\alpha 3}$, for $\alpha = 1, 2, 3$, are calculated as follows:

$$\begin{aligned}
T_{ij}^{13} &= \sum_{\gamma=1}^{5}\sum_{k=1}^{n(\gamma)} \frac{\partial K_{ik}^{1\gamma}}{\partial w_j}\Delta_k^\gamma + K_{ij}^{13} = \sum_{k=1}^{n}\frac{\partial K_{ik}^{13}}{\partial w_j}w_k + K_{ij}^{13}\\
&= \frac{1}{2}\int_{\Omega^e}\left[\frac{\partial \psi_i^{(1)}}{\partial x}\left(A_{11}\frac{\partial w}{\partial x}\frac{\partial \psi_j^{(2)}}{\partial x} + A_{12}\frac{\partial w}{\partial y}\frac{\partial \psi_j^{(2)}}{\partial y}\right)\right.\\
&\qquad\left. + A_{66}\frac{\partial \psi_i^{(1)}}{\partial y}\left(\frac{\partial w}{\partial x}\frac{\partial \psi_j^{(2)}}{\partial y} + \frac{\partial w}{\partial y}\frac{\partial \psi_j^{(2)}}{\partial x}\right)\right] dx\,dy + K_{ij}^{13}\\
&= K_{ij}^{13} + K_{ij}^{13} = 2K_{ij}^{13} = T_{ji}^{31},
\end{aligned} \tag{10.3.16}$$

$$\begin{aligned}
T_{ij}^{23} &= \sum_{\gamma=1}^{5}\sum_{k=1}^{n(\gamma)} \frac{\partial K_{ik}^{2\gamma}}{\partial w_j}\Delta_k^\gamma + K_{ij}^{23} = \sum_{k=1}^{n}\frac{\partial K_{ik}^{23}}{\partial w_j}w_k + K_{ij}^{23}\\
&= \frac{1}{2}\int_{\Omega^e}\left[\frac{\partial \psi_i^{(1)}}{\partial y}\left(A_{12}\frac{\partial w}{\partial x}\frac{\partial \psi_j^{(2)}}{\partial x} + A_{22}\frac{\partial w}{\partial y}\frac{\partial \psi_j^{(2)}}{\partial y}\right)\right.\\
&\qquad\left. + A_{66}\frac{\partial \psi_i^{(1)}}{\partial x}\left(\frac{\partial w}{\partial y}\frac{\partial \psi_j^{(2)}}{\partial x} + \frac{\partial w}{\partial x}\frac{\partial \psi_j^{(2)}}{\partial y}\right)\right] dx\,dy + K_{ij}^{23}\\
&= K_{ij}^{23} + K_{ij}^{23} = 2K_{ij}^{23} = T_{ji}^{32},
\end{aligned} \tag{10.3.17}$$

$$
\begin{aligned}
T_{ij}^{33} &= \sum_{\gamma=1}^{5}\sum_{k=1}^{n(\gamma)} \frac{\partial K_{ik}^{3\gamma}}{\partial w_j}\Delta_k^\gamma + K_{ij}^{33} = \sum_{k=1}^{n(\gamma)}\left(\frac{\partial K_{ik}^{31}}{\partial w_j}u_k + \frac{\partial K_{ik}^{32}}{\partial w_j}v_k + \frac{\partial K_{ik}^{33}}{\partial w_j}w_k\right) + K_{ij}^{33} \\
&= \int_{\Omega^e}\left(A_{55}\frac{\partial \psi_i^{(2)}}{\partial x}\frac{\partial \psi_j^{(2)}}{\partial x} + A_{44}\frac{\partial \psi_i^{(2)}}{\partial y}\frac{\partial \psi_j^{(2)}}{\partial y}\right)dx\,dy \\
&+ \int_{\Omega^e}\left\{N_{xx}\frac{\partial \psi_i^{(2)}}{\partial x}\frac{\partial \psi_j^{(2)}}{\partial x} + N_{yy}\frac{\partial \psi_i^{(2)}}{\partial y}\frac{\partial \psi_j^{(2)}}{\partial y} + N_{xy}\frac{\partial \psi_i^{(2)}}{\partial x}\frac{\partial \psi_j^{(2)}}{\partial y}\right. \\
&+ N_{xy}\frac{\partial \psi_i^{(2)}}{\partial y}\frac{\partial \psi_j^{(2)}}{\partial x} + \left(A_{12}+A_{66}\right)\frac{\partial w}{\partial x}\frac{\partial w}{\partial y}\left(\frac{\partial \psi_i^{(2)}}{\partial x}\frac{\partial \psi_j^{(2)}}{\partial y} + \frac{\partial \psi_i^{(2)}}{\partial y}\frac{\partial \psi_j^{(2)}}{\partial x}\right) \\
&+ \left[A_{11}\left(\frac{\partial w}{\partial x}\right)^2 + A_{66}\left(\frac{\partial w}{\partial y}\right)^2\right]\frac{\partial \psi_i^{(2)}}{\partial x}\frac{\partial \psi_j^{(2)}}{\partial x} \\
&+ \left.\left[A_{66}\left(\frac{\partial w}{\partial x}\right)^2 + A_{22}\left(\frac{\partial w}{\partial y}\right)^2\right]\frac{\partial \psi_i^{(2)}}{\partial y}\frac{\partial \psi_j^{(2)}}{\partial y}\right\}dx\,dy
\end{aligned} \tag{10.3.18}
$$

where (N_{xx}, N_{yy}, N_{xy}) are known in terms of u, v, and w through Eqs. (10.2.23) and (10.2.3). We note that the tangent stiffness matrix of the FST element is symmetric.

10.3.4 Shear and Membrane Locking

The C^0 plate bending elements based on the FST are among the simplest available in the literature. Unfortunately, when lower order (quadratic or less) equal interpolation of the generalized displacements is used, the elements become excessively stiff in the thin plate limit, yielding displacements that are too small compared to the true solution. Since bilinear Lagrange interpolation functions are used for all primary variables in the FEM, shear locking occurs. In the FEM, shear locking is eliminated using reduced integration technique in evaluating the integrals which correspond to the shear stress terms (called selective reduced integration). This is equivalent to treating the shear terms $\phi_x + \partial w/\partial x$ and $\phi_y + \partial w/\partial y$ to be constant when calculating the element stiffness coefficients. As discussed in Chapters 7 and 8 for beams and axisymmetric circular plates, we use full integration to evaluate all linear stiffness coefficients and reduced integration to evaluate the transverse shear stiffness coefficients, (i.e., all coefficients in $K_{ij}^{\alpha\beta}$ that contain A_{44} and A_{55}).

Membrane locking occurs due to the inconsistency of interpolation between (u,v) and w. When a plate undergoes pure bending without extension, the membrane strains should vanish:

$$
\varepsilon_{xx}^{(0)} = \frac{\partial u}{\partial x} + \left(\frac{\partial w}{\partial x}\right)^2 \to 0 \quad \text{and} \quad \varepsilon_{yy}^{(0)} = \frac{\partial v}{\partial y} + \left(\frac{\partial w}{\partial y}\right)^2 \to 0.
$$

Again, reduced integration is used to treat all nonlinear terms to be a constant and avoid membrane locking.

10.4 Dual Mesh Control Domain Model

10.4.1 Primal and Dual Meshes

In the DMCDM, the computational domain is discretized by two meshes, a primal mesh and a dual mesh, such that both meshes completely encompass the computational domain. The primal mesh connects the nodes of the computational domain and the primary variables are interpolated on the primal mesh of finite elements using linear Lagrange interpolation functions. The DMCDM seeks to establish linear algebraic relations among the values of the primary variables at the nodes of the computational domain with their corresponding dual variables. This is achieved by writing the integral statements of the governing equations on the elements of the dual mesh of control domains. The dual mesh is constructed such that the interior nodes are at the center of their corresponding control domain, while the nodes on the edges of the computational domain and the nodes on the corners of the computational domain are on the edges and corners of their corresponding control domains, respectively (see Fig. 10.4.1).

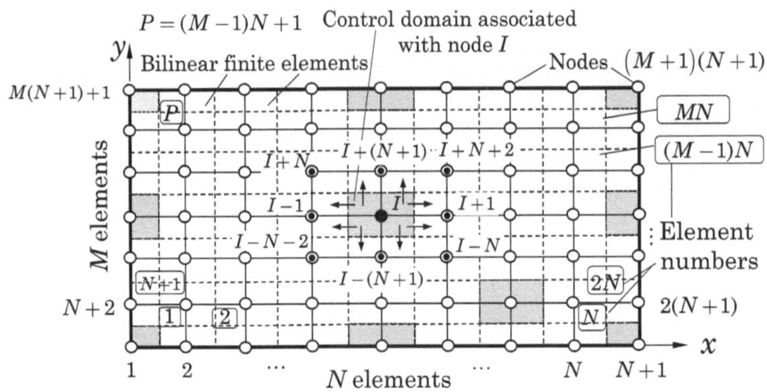

Fig. 10.4.1 Rectangular plate discretized using $N \times M$ primal mesh of linear finite elements (i.e., N elements along the x-axis and M elements along the y-axis). Dark lines represent the primal mesh, while dotted lines represent the dual mesh. Open circles are the nodes of the primal mesh. Typical control domains, including those on the boundary, are represented by shaded regions.

Figure 10.4.2(a) shows the control domain of typical interior node I along with the contributing primal mesh elements. On a typical primal mesh element $\Omega_p^{(e)}$ (here superscript e represents the number assigned to a particular primal mesh element) of the primal mesh, the dependent variables are approximated using bilinear Lagrange interpolation functions (see Nampally, Ruocco, and Reddy [53]). For a local coordinate system of $(\bar{x}, \bar{y})$ within each primal mesh element $\Omega_p^{(e)}$ [see Fig. 10.4.2(b)], a dependent variable, say u, can be approximated as follows:

$$u(x, y) \approx u_j^{(e)} \psi_j^{(e)}(\bar{x}, \bar{y}), \qquad \bar{x} = x - x_1^{(e)}, \qquad \bar{y} = y - y_1^{(e)} \tag{10.4.1}$$

where $u_j^{(e)}$ denote the unknown values of the variable u at the element nodes and $(x_1^{(e)}, y_1^{(e)})$ are the global coordinates of local node number 1 of element $\Omega_p^{(e)}$. Here $j = \{1, 2, 3, 4\}$ and summation convention is implied on the repeated index; $\psi_j^{(e)}$ are the bilinear Lagrange interpolation functions associated with the element and are given by

$$\begin{aligned} \psi_1^{(e)} &= \left(1 - \frac{\bar{x}}{a_e}\right)\left(1 - \frac{\bar{y}}{b_e}\right), & \psi_2^{(e)} &= \frac{\bar{x}}{a_e}\left(1 - \frac{\bar{y}}{b_e}\right), \\ \psi_3^{(e)} &= \frac{\bar{x}}{a_e}\frac{\bar{y}}{b_e}, & \psi_4^{(e)} &= \left(1 - \frac{\bar{x}}{a_e}\right)\frac{\bar{y}}{b_e}. \end{aligned} \tag{10.4.2}$$

Here a_e and b_e represent the horizontal and vertical dimensions of a typical rectangular primal mesh element $\Omega_p^{(e)}$, respectively. Other dependent variables are also approximated in a similar manner on each primal mesh element. Here, once again, we use the same set of bilinear Lagrange interpolation functions in approximating all the dependent variables on a given element (see [53]).

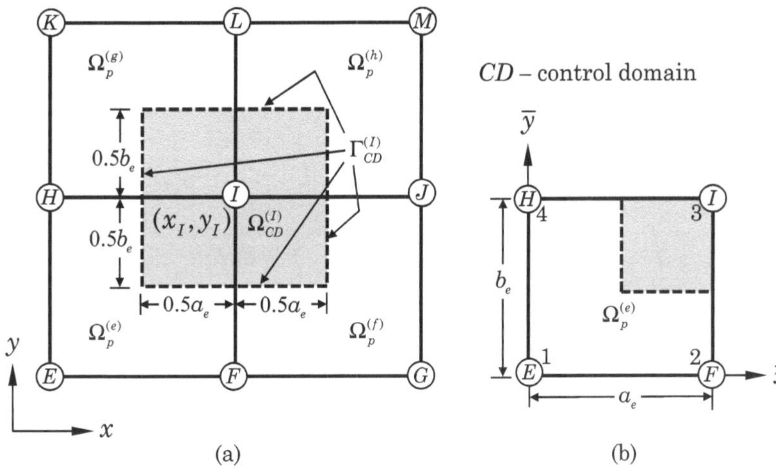

Fig. 10.4.2 (a) Control domain of typical interior node I along with the contributing primal mesh elements. (b) Local coordinate system for a typical primal mesh element $\Omega_p^{(e)}$. The node labels in circles are global, while node numbers outside circles are local to the element.

10.4.2 Discretized Equations

The displacement dual mesh control domain model of the FST governed by Eqs. (10.2.13)–(10.2.17) is developed for the linear case. The primary variables of such formulation are the displacements (u, v, w) and rotations (ϕ_x, ϕ_y), together called generalized displacements. The first step in developing the dual mesh control domain formulation is to write the integral statements of the governing equations (10.2.13)–(10.2.17) on a typical interior control domain, $\Omega_{CD}^{(I)}$, centered around node I:

$$0 = -\int_{\Omega_{CD}^{(I)}} \left(\frac{\partial N_{xx}}{\partial x} + \frac{\partial N_{xy}}{\partial y} + f_x \right) dxdy, \tag{10.4.3a}$$

$$0 = -\int_{\Omega_{CD}^{(I)}} \left(\frac{\partial N_{xy}}{\partial x} + \frac{\partial N_{yy}}{\partial y} + f_y \right) dxdy, \tag{10.4.3b}$$

$$0 = -\int_{\Omega_{CD}^{(I)}} \left(\frac{\partial Q_x}{\partial x} + \frac{\partial Q_y}{\partial y} + q \right) dxdy, \tag{10.4.3c}$$

$$0 = -\int_{\Omega_{CD}^{(I)}} \left(\frac{\partial M_{xx}}{\partial x} + \frac{\partial M_{xy}}{\partial y} - Q_x \right) dxdy, \tag{10.4.3d}$$

$$0 = -\int_{\Omega_{CD}^{(I)}} \left(\frac{\partial M_{xy}}{\partial x} + \frac{\partial M_{yy}}{\partial y} - Q_y \right) dxdy. \tag{10.4.3e}$$

The next step is to evaluate the integrals in the above equations such that the resulting boundary terms on the boundary of the control domain, $\Gamma_{CD}^{(I)}$, represent physically meaningful secondary variables which are dual to the primary variables. Thus we arrive at the following equations:

$$0 = -\oint_{\Gamma_{CD}^{(I)}} [N_{xx}\, n_x + N_{xy}\, n_y]\, ds - \int_{\Omega_{CD}^{(I)}} f_x\, dxdy, \tag{10.4.4}$$

$$0 = -\oint_{\Gamma_{CD}^{(I)}} [N_{xy}\, n_x + N_{yy}\, n_y]\, ds - \int_{\Omega_{CD}^{(I)}} f_y\, dxdy, \tag{10.4.5}$$

$$0 = -\oint_{\Gamma_{CD}^{(I)}} [Q_x\, n_x + Q_y\, n_y]\, ds - \int_{\Omega_{CD}^{(I)}} q\, dxdy, \tag{10.4.6}$$

$$0 = -\oint_{\Gamma_{CD}^{(I)}} [M_{xx}\, n_x + M_{xy}\, n_y]\, ds + \int_{\Omega_{CD}^{(I)}} Q_x\, dxdy, \tag{10.4.7}$$

$$0 = -\oint_{\Gamma_{CD}^{(I)}} [M_{xy}\, n_x + M_{yy}\, n_y]\, ds + \int_{\Omega_{CD}^{(I)}} Q_y\, dxdy. \tag{10.4.8}$$

We note that the stress and moment resultants are known in terms of the displacements through Eqs. (10.2.3), (10.2.22), and (10.2.23). Since the control domain spans four different primal mesh elements, and since the interpolation functions change from one primal mesh element to the other, each of the above integrals should be broken into four parts with each part corresponding to a different primal mesh element that contributes to the control domain. For example, consider Eq. (10.4.4) for a uniform primal mesh (i.e., all primal mesh elements are of same size $a_e \times b_e$). Equation (10.4.4), when evaluated on the part of the control domain $\Omega_{CD}^{(I)}$ which lies inside the primal mesh element $\Omega_p^{(e)}$ [see Fig. 10.4.2(a)], can be written as follows (for brevity, the element label 'e'

on a and b are omitted):

$$\begin{aligned}
&\left\{\int_{0.5a}^{a}\left[\frac{A(1-\nu)}{2}\frac{\partial\psi_i^{(e)}}{\partial\bar{y}}\right]_{\bar{y}=0.5b}d\bar{x}+\int_{0.5b}^{b}\left[A\frac{\partial\psi_i^{(e)}}{\partial\bar{x}}\right]_{\bar{x}=0.5a}d\bar{y}\right\}u_i^{(e)}\\
&+\left\{\int_{0.5a}^{a}\left[\frac{A(1-\nu)}{2}\frac{\partial\psi_i^{(e)}}{\partial\bar{x}}\right]_{\bar{y}=0.5b}d\bar{x}+\int_{0.5b}^{b}\left[A_{xx}\nu\frac{\partial\psi_i^{(e)}}{\partial\bar{y}}\right]_{\bar{x}=0.5a}d\bar{y}\right\}v_i^{(e)}\\
&+\left\{\int_{0.5a}^{a}\left[\frac{B(1-\nu)}{2}\frac{\partial\psi_i^{(e)}}{\partial\bar{y}}\right]_{\bar{y}=0.5b}d\bar{x}+\int_{0.5b}^{b}\left[B_{11}\frac{\partial\psi_i^{(e)}}{\partial\bar{x}}\right]_{\bar{x}=0.5a}d\bar{y}\right\}\phi_{x_i}^{(e)}\\
&+\left\{\int_{0.5a}^{a}\left[\frac{B(1-\nu)}{2}\frac{\partial\psi_i^{(e)}}{\partial\bar{x}}\right]_{\bar{y}=0.5b}d\bar{x}+\int_{0.5b}^{b}\left[B\nu\frac{\partial\psi_i^{(e)}}{\partial\bar{y}}\right]_{\bar{x}=0.5a}d\bar{y}\right\}\phi_{y_i}^{(e)}\\
&=\int_{0.5a}^{a}\int_{0.5b}^{b}f_x(\bar{x},\bar{y})d\bar{x}d\bar{y},
\end{aligned}\tag{10.4.9}$$

where $u_i^{(e)}, v_i^{(e)}, \phi_{x_i}^{(e)}, \phi_{y_i}^{(e)}$ represent the nodal values of u, v, ϕ_x, and ϕ_y, respectively, of primal mesh element $\Omega_p^{(e)}$ and $i = \{1, 2, 3, 4\}$, with summation on repeated index implied.

Equation (10.4.4) when evaluated on the part of the control domain which lies inside the primal mesh element $\Omega_p^{(f)}$ can be written as

$$\begin{aligned}
&\left\{\int_{0}^{0.5a}\left[\frac{A(1-\nu)}{2}\frac{\partial\psi_i^{(f)}}{\partial\bar{y}}\right]_{\bar{y}=0.5b}d\bar{x}-\int_{0.5b}^{b}\left[A\frac{\partial\psi_i^{(f)}}{\partial\bar{x}}\right]_{\bar{x}=0.5a}d\bar{y}\right\}u_i^{(f)}\\
&+\left\{\int_{0}^{0.5a}\left[\frac{A(1-\nu)}{2}\frac{\partial\psi_i^{(f)}}{\partial\bar{x}}\right]_{\bar{y}=0.5b}d\bar{x}-\int_{0.5b}^{b}\left[A\nu\frac{\partial\psi_i^{(f)}}{\partial\bar{y}}\right]_{\bar{x}=0.5a}d\bar{y}\right\}v_i^{(f)}\\
&+\left\{\int_{0.5a}^{a}\left[\frac{B(1-\nu)}{2}\frac{\partial\psi_i^{(f)}}{\partial\bar{y}}\right]_{\bar{y}=0.5b}d\bar{x}-\int_{0.5b}^{b}\left[B\frac{\partial\psi_i^{(f)}}{\partial\bar{x}}\right]_{\bar{x}=0.5a}d\bar{y}\right\}\phi_{x_i}^{(f)}\\
&+\left\{\int_{0}^{0.5a}\left[\frac{B(1-\nu)}{2}\frac{\partial\psi_i^{(f)}}{\partial\bar{x}}\right]_{\bar{y}=0.5b}d\bar{x}-\int_{0.5b}^{b}\left[B\nu\frac{\partial\psi_i^{(f)}}{\partial\bar{y}}\right]_{\bar{x}=0.5a}d\bar{y}\right\}\phi_{y_i}^{(f)}\\
&=\int_{0}^{0.5a}\int_{0.5b}^{b}f_x(\bar{x},\bar{y})d\bar{x}d\bar{y}.
\end{aligned}\tag{10.4.10}$$

For the part of the control domain which lies inside primal mesh element $\Omega_p^{(g)}$ we have

$$\begin{aligned}
&\left\{-\int_{0.5a}^{a}\left[\frac{A(1-\nu)}{2}\frac{\partial\psi_i^{(g)}}{\partial\bar{y}}\right]_{\bar{y}=0.5b}d\bar{x}+\int_{0}^{0.5b}\left[A\frac{\partial\psi_i^{(g)}}{\partial\bar{x}}\right]_{\bar{x}=0.5a}d\bar{y}\right\}u_i^{(g)}\\
&+\left\{-\int_{0.5a}^{a}\left[\frac{A(1-\nu)}{2}\frac{\partial\psi_i^{(g)}}{\partial\bar{x}}\right]_{\bar{y}=0.5b}d\bar{x}+\int_{0}^{0.5b}\left[A\nu\frac{\partial\psi_i^{(g)}}{\partial\bar{y}}\right]_{\bar{x}=0.5a}d\bar{y}\right\}v_i^{(g)}\\
&+\left\{-\int_{0.5a}^{a}\left[\frac{B(1-\nu)}{2}\frac{\partial\psi_i^{(g)}}{\partial\bar{y}}\right]_{\bar{y}=0.5b}d\bar{x}+\int_{0}^{0.5b}\left[B\frac{\partial\psi_i^{(g)}}{\partial\bar{x}}\right]_{\bar{x}=0.5a}d\bar{y}\right\}\phi_{x_i}^{(g)}\\
&+\left\{-\int_{0.5a}^{a}\left[\frac{B(1-\nu)}{2}\frac{\partial\psi_i^{(g)}}{\partial\bar{x}}\right]_{\bar{y}=0.5b}d\bar{x}+\int_{0}^{0.5b}\left[B\nu\frac{\partial\psi_i^{(g)}}{\partial\bar{y}}\right]_{\bar{x}=0.5a}d\bar{y}\right\}\phi_{y_i}^{(g)}\\
&=\int_{0.5a}^{a}\int_{0}^{0.5b}f_x(\bar{x},\bar{y})d\bar{x}d\bar{y}.
\end{aligned}\tag{10.4.11}$$

Finally, Eq. (10.4.4) when evaluated on the part of the control domain $\Omega_{CD}^{(I)}$ which lies inside the primal mesh element $\Omega_p^{(h)}$, will be

$$\begin{aligned}
&\left\{-\int_{0}^{0.5b}\left[A\frac{\partial\psi_i^{(h)}}{\partial\bar{x}}\right]_{\bar{x}=0.5a}d\bar{y}-\int_{0}^{0.5a}\left[\frac{A(1-\nu)}{2}\frac{\partial\psi_i^{(h)}}{\partial\bar{y}}\right]_{\bar{y}=0.5b}d\bar{x}\right\}u_i^{(h)}\\
&+\left\{-\int_{0}^{0.5b}\left[A\nu\frac{\partial\psi_i^{(h)}}{\partial\bar{y}}\right]_{\bar{x}=0.5a}d\bar{y}-\int_{0}^{0.5a}\left[\frac{A(1-\nu)}{2}\frac{\partial\psi_i^{(h)}}{\partial\bar{x}}\right]_{\bar{y}=0.5b}d\bar{x}\right\}v_i^{(h)}\\
&+\left\{-\int_{0}^{0.5b}\left[B\frac{\partial\psi_i^{(h)}}{\partial\bar{x}}\right]_{\bar{x}=0.5a}d\bar{y}-\int_{0}^{0.5a}\left[\frac{B(1-\nu)}{2}\frac{\partial\psi_i^{(h)}}{\partial\bar{y}}\right]_{\bar{y}=0.5b}d\bar{x}\right\}\phi_{x_i}^{(h)}\\
&+\left\{-\int_{0}^{0.5b}\left[A\nu\frac{\partial\psi_i^{(h)}}{\partial\bar{y}}\right]_{\bar{x}=0.5a}d\bar{y}-\int_{0}^{0.5a}\left[\frac{A(1-\nu)}{2}\frac{\partial\psi_i^{(h)}}{\partial\bar{x}}\right]_{\bar{y}=0.5b}d\bar{x}\right\}\phi_{y_i}^{(h)}\\
&=\int_{0}^{0.5a}\int_{0}^{0.5b}f_x(\bar{x},\bar{y})d\bar{x}d\bar{y}.
\end{aligned}\tag{10.4.12}$$

In all the above equations, $i=\{1,2,3,4\}$ and summation on repeated index is implied.

We note that

$$\begin{aligned}
&u_1^{(e)}=U_E,\quad u_2^{(e)}=u_1^{(f)}=U_F,\quad u_2^{(f)}=U_G,\quad u_4^{(e)}=u_1^{(g)}=U_H\\
&u_3^{(e)}=u_4^{(f)}=u_2^{(g)}=u_1^{(h)}=U_I,\quad u_3^{(f)}=u_2^{(h)}=U_J,\quad u_4^{(g)}=U_K\\
&u_3^{(g)}=u_4^{(h)}=U_L,\quad u_3^{(h)}=U_M,
\end{aligned}$$

where U_E, U_F, ... are the nodal values of the displacement u, identified with global node labels. Similar relations hold for other primary variables as well. Using these continuity relations, we can obtain the final discretized equation corresponding to Eq. (10.4.4) by adding Eq. (10.4.9) and Eqs. (10.4.10)–(10.4.12), which can be written in the following form:

$$\begin{aligned}
&K_{1(1)}U_E + K_{1(2)}V_E + K_{1(4)}\Phi_{x_E} + K_{1(5)}\Phi_{y_E} + K_{1(6)}U_F + K_{1(7)}V_F \\
&\quad + K_{1(9)}\Phi_{x_F} + K_{1(10)}\Phi_{y_F} + K_{1(11)}U_G + K_{1(12)}V_G + K_{1(14)}\Phi_{x_G} + K_{1(15)}\Phi_{y_G} \\
&\quad + K_{1(16)}U_H + K_{1(17)}V_H + K_{1(19)}\Phi_{x_H} + K_{1(20)}\Phi_{y_H} + K_{1(21)}U_I + K_{1(22)}V_I \\
&\quad + K_{1(24)}\Phi_{x_I} + K_{1(25)}\Phi_{y_I} + K_{1(26)}U_J + K_{1(27)}V_J + K_{1(29)}\Phi_{x_J} + K_{1(30)}\Phi_{y_J} \\
&\quad + K_{1(31)}U_K + K_{1(32)}V_K + K_{1(34)}\Phi_{x_K} + K_{1(35)}\Phi_{y_K} + K_{1(36)}U_L + K_{1(37)}V_L \\
&\quad + K_{1(39)}\Phi_{x_L} + K_{1(40)}\Phi_{y_L} + K_{1(41)}U_M + K_{1(42)}V_M + K_{1(44)}\Phi_{x_M} \\
&\quad + K_{1(45)}\Phi_{y_M} = F_{x_I}^{(e)} + F_{x_I}^{(f)} + F_{x_I}^{(g)} + F_{x_I}^{(h)}
\end{aligned} \tag{10.4.13}$$

The coefficients of K_{ij} are obtained by factoring out the nodal values. Similar discretized equations can be written for Eqs. (10.4.5)–(10.4.8).

Equation (10.4.13) was derived for a control domain corresponding to an interior node. Now we shall consider the discretization of Eq. (10.4.4) on a control domain corresponding to a boundary node. Consider the control domain corresponding to node 1 as shown in Fig. 10.4.3. Equation (10.4.4) on this control domain can be written as

$$\begin{aligned}
&\int_0^{0.5a} [N_{xy}]_{y=0}\, dx - \int_0^{0.5b} [N_{xx}]_{x=0.5a}\, dy - \int_0^{0.5a} [N_{xy}]_{y=0.5b}\, dx \\
&+ \int_0^{0.5b} [N_{xx}]_{x=0}\, dy = \int_0^{0.5a}\int_0^{0.5b} f_x\, dxdy.
\end{aligned} \tag{10.4.14}$$

However, we note that a part of the boundary of the control domain in Fig. 10.4.3 coincides with the boundary of the computational domain. Since on the boundary of the computational domain we either know the primary variable or the corresponding dual variable because of their duality, we shall not write the integrals on those boundaries of the control domain which coincide with the boundary of the computational domain in terms of the primary variables.

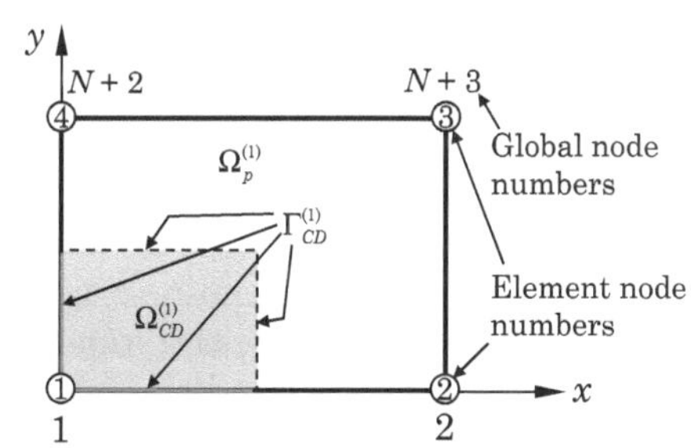

Fig. 10.4.3 Control domain corresponding to global node 1. The numbers in the circle are local node numbers of primal mesh element $\Omega_p^{(1)}$, while the numbers outside are global node numbers.

Hence, Eq. (10.4.4) will take the form

$$\begin{aligned}
&\left\{-\int_0^{0.5b}\left[A\frac{\partial\psi_i^{(1)}}{\partial\bar{x}}\right]_{\bar{x}=0.5a}d\bar{y}-\int_0^{0.5a}\left[\frac{A(1-\nu)}{2}\frac{\partial\psi_i^{(1)}}{\partial\bar{y}}\right]_{\bar{y}=0.5b}d\bar{x}\right\}u_i^{(1)}\\
&+\left\{-\int_0^{0.5b}\left[A\nu\frac{\partial\psi_i^{(1)}}{\partial\bar{y}}\right]_{\bar{x}=0.5a}d\bar{y}-\int_0^{0.5a}\left[\frac{A(1-\nu)}{2}\frac{\partial\psi_i^{(1)}}{\partial\bar{x}}\right]_{\bar{y}=0.5b}d\bar{x}\right\}v_i^{(1)}\\
&+\left\{-\int_0^{0.5b}\left[B\frac{\partial\psi_i^{(1)}}{\partial\bar{x}}\right]_{\bar{x}=0.5a}d\bar{y}-\int_0^{0.5a}\left[\frac{B(1-\nu)}{2}\frac{\partial\psi_i^{(1)}}{\partial\bar{y}}\right]_{\bar{y}=0.5b}d\bar{x}\right\}\phi_{x_i}^{(1)}\\
&+\left\{-\int_0^{0.5b}\left[B\nu\frac{\partial\psi_i^{(1)}}{\partial\bar{y}}\right]_{\bar{x}=0.5a}d\bar{y}-\int_0^{0.5a}\left[\frac{B(1-\nu)}{2}\frac{\partial\psi_i^{(1)}}{\partial\bar{x}}\right]_{\bar{y}=0.5b}d\bar{x}\right\}\phi_{y_i}^{(1)}\\
&=\int_0^{0.5a}\int_0^{0.5b}f_x(\bar{x},\bar{y})d\bar{x}d\bar{y}-\int_0^{0.5a}\left[N_{xy}\right]_{y=0}dx-\int_0^{0.5b}\left[N_{xx}\right]_{x=0}dy
\end{aligned}\tag{10.4.15}$$

Here $u_i^{(1)}, v_i^{(1)}, \phi_{x_i}^{(1)}, \phi_{y_i}^{(1)}$ represent the nodal values of u, v, ϕ_x and ϕ_y respectively of primal mesh element $\Omega_p^{(1)}$ and $i = \{1, 2, 3, 4\}$ with summation on repeated index implied.

The line integrals on the right-hand side of Eq. (10.4.15) represent the secondary variables dual to the primary variable u on the boundary. For a given boundary value problem, we either know the value of u on the boundary or its dual. Hence, when u is known on the boundary, the stress resultants on the boundary can be computed using Eq. (10.4.15). Conversely, when the stress resultant (secondary variable) is specified on the boundary, Eq. (10.4.15) is used to calculate u on the boundary. Similar procedure holds for all the control domains which share boundary with the boundary of the computational domain. Figure 10.4.2 shows such possible boundary control domains for a rectangular computational domain discretized with rectangular primal mesh elements.

10.4.3 Shear Locking

Since we are using bilinear Lagrange interpolation functions for all the primary variables in DMCDM, shear locking occurs even in the displacement-based DMCDM. We shall use the same technique to overcome shear locking in the DMCDM as well. Thus in evaluating the following integrals, the shear terms $\phi_x + \partial w/\partial x$ and $\phi_y + \partial w/\partial y$ are treated as constants within each primal mesh element

$$-\oint_{\Gamma_{CD}^{(I)}}\left[Q_x n_x + Q_y n_y\right]ds,\quad \int_{\Omega_{CD}^{(I)}} Q_x\, dxdy,\quad \int_{\Omega_{CD}^{(I)}} Q_y dxdy. \tag{10.4.16}$$

The value of the constant is taken to be the value of the shear term at the center of the primal mesh element.

For example, consider the integral (the element label 'e' on a and b are omitted for brevity)

$$\begin{aligned}
&\int_{\Omega_{CD}^{(I)}} Q_x\, dxdy \\
&= \int_{x_I-0.5a}^{x_I}\int_{y_I-0.5b}^{y_I} S_{xz}\left(\phi_x + \frac{\partial w}{\partial x}\right) dxdy + \int_{x_I}^{x_I+0.5a}\int_{y_I-0.5b}^{y_I} S_{xz}\left(\phi_x + \frac{\partial w}{\partial x}\right) dxdy \\
&+ \int_{x_I-0.5a}^{x_I}\int_{y_I}^{y_I+0.5b} S_{xz}\left(\phi_x + \frac{\partial w}{\partial x}\right) dxdy + \int_{x_I}^{x_I+0.5a}\int_{y_I}^{y_I+0.5b} S_{xz}\left(\phi_x + \frac{\partial w}{\partial x}\right) dxdy \\
&= \int_{0.5a}^{a}\int_{0.5b}^{b} \frac{S_{xz}}{4} d\bar{x}d\bar{y} \sum_{j=1}^{4} \phi_{x_j}^{(e)} + \left\{\int_{0.5a}^{a}\int_{0.5b}^{b} \left[S_{xz}\frac{\partial \psi_i^{(e)}}{\partial \bar{x}}\right]_{(0.5a,0.5b)} d\bar{x}d\bar{y}\right\} w_i^{(e)} \\
&+ \int_{0}^{0.5a}\int_{0.5b}^{b} \frac{S_{xz}}{4} d\bar{x}d\bar{y} \sum_{j=1}^{4} \phi_{x_j}^{(f)} + \left\{\int_{0}^{0.5a}\int_{0.5b}^{b} \left[S\frac{\partial \psi_i^{(f)}}{\partial \bar{x}}\right]_{(0.5a,0.5b)} d\bar{x}d\bar{y}\right\} w_i^{(f)} \\
&+ \int_{0.5a}^{a}\int_{0}^{0.5b} \frac{S_{xz}}{4} d\bar{x}d\bar{y} \sum_{j=1}^{4} \phi_{x_j}^{(g)} + \left\{\int_{0.5a}^{a}\int_{0}^{0.5b} \left[S_{xz}\frac{\partial \psi_i^{(g)}}{\partial \bar{x}}\right]_{(0.5a,0.5b)} d\bar{x}d\bar{y}\right\} w_i^{(g)} \\
&+ \int_{0}^{0.5a}\int_{0}^{0.5b} \frac{S_{xz}}{4} d\bar{x}d\bar{y} \sum_{j=1}^{4} \phi_{x_j}^{(h)} + \left\{\int_{0}^{0.5a}\int_{0}^{0.5b} \left[S_{xz}\frac{\partial \psi_i^{(h)}}{\partial \bar{x}}\right]_{(0.5a,0.5b)} d\bar{x}d\bar{y}\right\} w_i^{(h)},
\end{aligned} \tag{10.4.17}$$

where $i = \{1, 2, 3, 4\}$ and summation on repeated index is implied. Similarly, we have

$$\begin{aligned}
&-\oint_{\Gamma_{CD}^{(I)}} Q_x n_x ds \\
&= \int_{0.5b}^{b} \frac{S_{xz}}{4} d\bar{y} \sum_{j=1}^{4} \phi_{x_j}^{(e)} + \left\{\int_{0.5b}^{b} \left[S_{xz}\frac{\partial \psi_i^{(e)}}{\partial \bar{x}}\right]_{(0.5a,0.5b)} d\bar{y}\right\} w_i^{(e)} \\
&- \int_{0.5b}^{b} \frac{S_{xz}}{4} d\bar{y} \sum_{j=1}^{4} \phi_{x_j}^{(f)} - \left\{\int_{0.5b}^{b} \left[S_{xz}\frac{\partial \psi_i^{(f)}}{\partial \bar{x}}\right]_{(0.5a,0.5b)} d\bar{y}\right\} w_i^{(f)} \\
&+ \int_{0}^{0.5b} \frac{S_{xz}}{4} d\bar{y} \sum_{j=1}^{4} \phi_{x_j}^{(g)} + \left\{\int_{0}^{0.5b} \left[S_{xz}\frac{\partial \psi_i^{(g)}}{\partial \bar{x}}\right]_{(0.5a,0.5b)} d\bar{y}\right\} w_i^{(g)} \\
&- \int_{0}^{0.5b} \frac{S_{xz}}{4} d\bar{y} \sum_{j=1}^{4} \phi_{x_j}^{(h)} - \left\{\int_{0}^{0.5b} \left[S_{xz}\frac{\partial \psi_i^{(h)}}{\partial \bar{x}}\right]_{(0.5a,0.5b)} d\bar{y}\right\} w_i^{(h)}.
\end{aligned} \tag{10.4.18}$$

This completes the displacement dual mesh control domain formulation for first-order shear deformation theory of rectangular plates.

10.5 Numerical Examples

10.5.1 Linear Analysis

Example 10.5.1

Consider a functionally graded plate of length $a = 1$ in (25.4 mm), width $b = 1$ in (25.4 mm) and thickness $h = 0.01$ in (0.254 mm). The material properties of the functionally graded material are

$$E_1 = 3 \times 10^7 \text{ psi (206.84 GPa)}, \quad E_2 = 3 \times 10^6 \text{ psi (20.68 GPa)}, \quad \nu = 0.3 \tag{10.5.1}$$

The plate is subjected to a uniformly distributed load of intensity $q = 1$ lb/in^2 (0.00689 N/mm^2) on the top surface. Consider two different boundary conditions: (a) clamped and (b) simply supported. Since the loading and boundary conditions result in symmetry about the x- and y-axes, consider only the quarter plate in the positive quadrant as the computational domain (see Fig. 10.5.1) and obtain numerical results using the FEM and DMCDM.

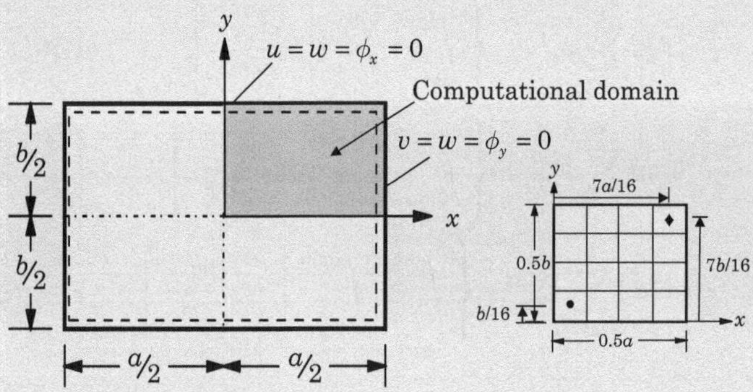

Fig. 10.5.1 Rectangular plate with the chosen coordinate system, computational domain, and the SS-1 (simply supported type 1) boundary conditions on the primary variables.

Solution: The boundary conditions on the computational domain are then given by

- *Clamped*

$$u = \phi_x = 0 \quad \text{at} \quad x = 0; \qquad v = \phi_y = 0 \quad at \quad \text{at} \quad y = 0$$
$$u = v = w = \phi_x = \phi_y = 0 \quad \text{at} \quad x = a/2 \quad \text{and} \quad y = b/2$$

- *Simply supported*

$$u = \phi_x = 0 \quad \text{at} \quad x = 0; \qquad v = \phi_y = 0 \quad at \quad \text{at} \quad y = 0$$
$$v = w = \phi_y = 0 \quad \text{at} \quad x = a/2; \qquad u = w = \phi_x = 0 \quad \text{at} \quad y = b/2$$

Further, we define the following dimensionless parameters:

$$\bar{w}(x,y) = w(x,y)\frac{E_1 h^3}{q a^4}, \quad \bar{u}(x,y) = u(x,y)\frac{E_1 h^2}{q a^3}$$
$$\bar{B} = \frac{B(1-\nu^2)}{E_1 h^2}, \quad \bar{D} = \frac{D(1-\nu^2)}{E_1 h^3}$$

$$\xi = \frac{2x}{a}, \quad \eta = \frac{z}{h}, \quad \chi = \frac{2\sqrt{x^2+y^2}}{\sqrt{a^2+b^2}}$$

$$\bar{\sigma}_{xx} = \sigma_{xx}\frac{h^2}{qa^2}, \quad \bar{\sigma}_{xy} = \sigma_{xy}\frac{h^2}{qa^2}, \quad \bar{\sigma}_{xz} = \sigma_{xz}\frac{h}{qa}$$

Table 10.5.1 presents the comparison between the dimensionless transverse center deflections of the functionally graded plate for various values of n, subjected to clamped boundary conditions under the action of a uniformly distributed transverse load $q = 1$ lb/in^2 as obtained from FEM and DMCDM for various mesh sizes. It can be seen that for a 32×32 uniform mesh there is no difference between the results from FEM and DMCDM up to the fifth decimal point. For all the further analysis, we use a 32×32 uniform primal mesh for DMCDM and finite element mesh for FEM. Once the displacement field is obtained from the dual mesh control domain method, the strains and stresses can be computed using Eqs. (10.2.3) and (10.2.21), respectively. The strains and stresses are evaluated at the center of each element of the primal mesh. This is similar to the evaluation of stresses at a single Gauss quadrature point of each element of the finite element mesh.

Table 10.5.1 Dimensionless center deflection, $\bar{w}(0,0)$ of a clamped square plate under uniformly distributed load of $q = 1$ lb/in^2 for various values of n.

Mesh	$n=0$		$n=1$		$n=2$		$n=3$		$n=4$		$n=5$	
	FEM	DMCDM	FEM	DMCDM	FEM	DMCDM	FEM	DMCDM	FEM	DMCDM	FEM	DMCDM
2×2	0.01326	0.01281	0.03102	0.02997	0.04548	0.04395	0.05370	0.05187	0.05796	0.05601	0.06039	0.05838
4×4	0.01368	0.01359	0.03201	0.03174	0.04692	0.04659	0.05544	0.05499	0.05985	0.05937	0.06237	0.06186
8×8	0.0138	0.01377	0.03231	0.03222	0.04737	0.04728	0.05592	0.05580	0.06036	0.06024	0.06291	0.06279
16×16	0.01383	0.01383	0.03237	0.03234	0.04749	0.04746	0.05604	0.05601	0.06051	0.06048	0.06306	0.06300
32×32	0.01384	0.01384	0.03237	0.03237	0.04749	0.04749	0.05607	0.05607	0.06054	0.06054	0.06309	0.06306

Note that the bending stiffness of the plate decreases with increase in n. This can be seen from the plot of dimensionless bending stiffness $\bar{D}$ as a function of n in Fig. 10.5.2(a). Thus we expect that the transverse deflection of the plate increases with an increase in the value of n. Figure 10.5.2(b) presents the dimensionless transverse deflection, $\bar{w}$ along the line $y = 0$ for various values of n of a simply supported functionally graded plate under the action of a uniformly distributed load of $q = 1$ lb/in^2. As expected, the magnitude of the dimensionless transverse deflection increases with increase in n. However, the bending-extension coupling stiffness B first increases and then decreases with increasing n. This is shown in Fig. 10.5.3(a), where the dimensionless extensional-bending coupling stiffness is plotted as function of n. Thus, we expect that the magnitude of the axial deflection of the plate will first increase with an increase in n and then decrease. This conjecture is verified in Fig. 10.5.3(b), where the dimensionless axial deflection, $\bar{u}$ of a simply supported functionally graded plate along the line $y = 0$ is plotted for various values of n. Further, the results from DMCDM are coincident with the results from FEM in both Fig. 10.5.2(b) and Fig. 10.5.3(b).

The dimensionless Young's modulus $\bar{E}$ of the functionally graded plate as a function of dimensionless thickness coordinate η for various values of n is presented in Fig. 10.5.4(a). It can be seen that as n increases the portion of the plate thickness on which the Young's modulus changes decreases. The first-order shear deformation theory predicts constant transverse shear strains along the thickness and, hence, for a homogeneous plate the transverse shear stresses will not be a function of the thickness coordinate z. However, for a functionally graded plate the Young's modulus changes with the thickness coordinate and hence, the shear modulus also changes in a similar fashion (since ν is assumed to be constant). Thus for a functionally graded plate although the first-order shear deformation theory predicts constant transverse shear strains along the thickness, the transverse shear stresses vary with the thickness coordinate in a similar fashion as Young's modulus. This is shown in Fig. 10.5.4(b) where the dimensionless transverse shear stress $\bar{\sigma}_{xz}$ is plotted as a function of the dimensionless thickness coordinate η at $(x, y) = (a/128, b/128)$ for clamped boundary conditions. For higher values of n, $\bar{\sigma}_{xz}$ is almost constant for the most part of the plate thickness and only changes toward the top surface in a similar fashion as Young's modulus, as expected. Also, the results from the DMCDM are compared with the results from FEM in Fig. 10.5.4(b).

Finally, the dimensionless stresses $\bar{\sigma}_{xx}$ and $\bar{\sigma}_{xy}$ are presented in Fig. 10.5.5(a) and Fig. 10.5.5(b), respectively, obtained from the DMCDM, on the top surface of a clamped functionally graded plate along the line $x = y$ for various values of n. It can be seen that the magnitude of both these dimensionless stresses increases with increasing n.

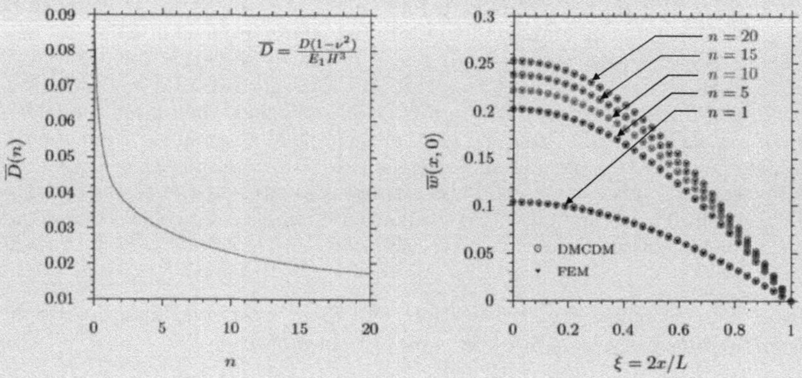

(a) (b)

Fig. 10.5.2 (a) Dimensionless bending stiffness as a function of n. (b) Dimensionless transverse deflection, $\bar{w}$ along $y = 0$ of a simply supported functionally graded plate under the action of a uniformly distributed load, $q = 1$ lb/in^2.

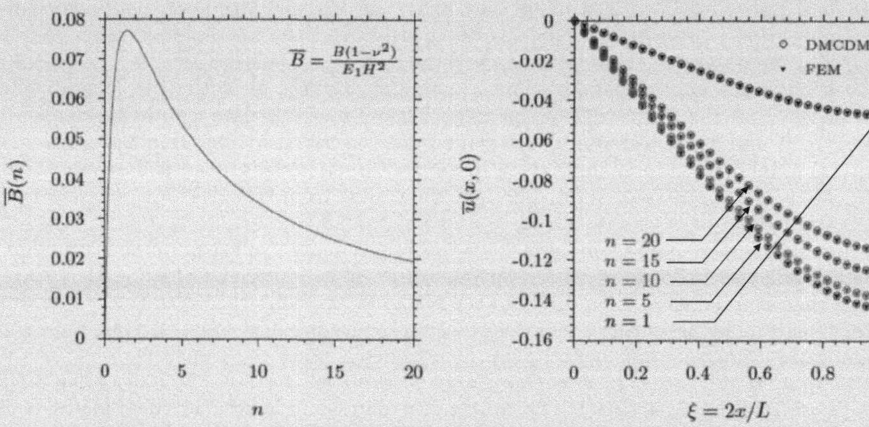

(a) (b)

Fig. 10.5.3 (a) Dimensionless extensional-bending coupling stiffness as a function of n. (b) Dimensionless axial deflection, $\bar{u}$ along $y = 0$ of a simply supported functionally graded plate under the action of a uniformly distributed load, $q = 1$ lb/in^2.

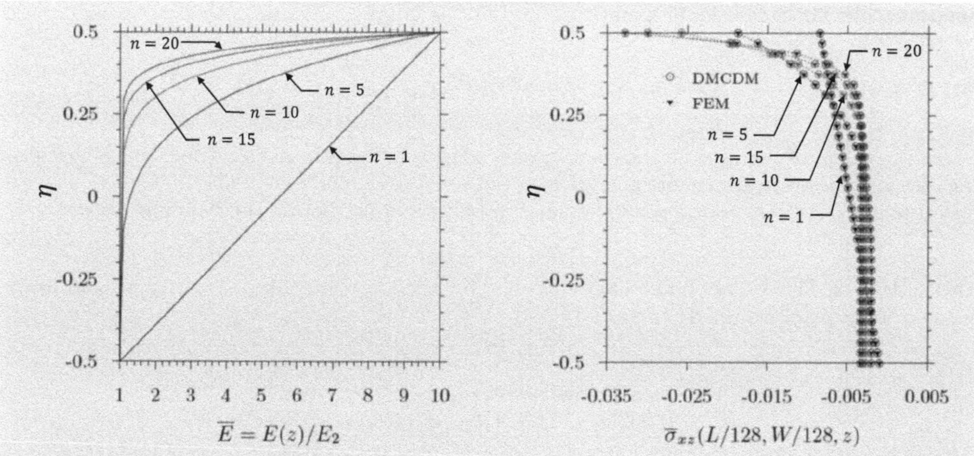

(a) (b)

Fig. 10.5.4 (a) Dimensionless Young's modulus $\bar{E}$ as a function of dimensionless thickness coordinate η for various values of n. (b) Dimensionless stress $\bar{\sigma}_{xz}$ along the thickness of a clamped functionally graded plate at $(x, y) = (L/128, W/128)$.

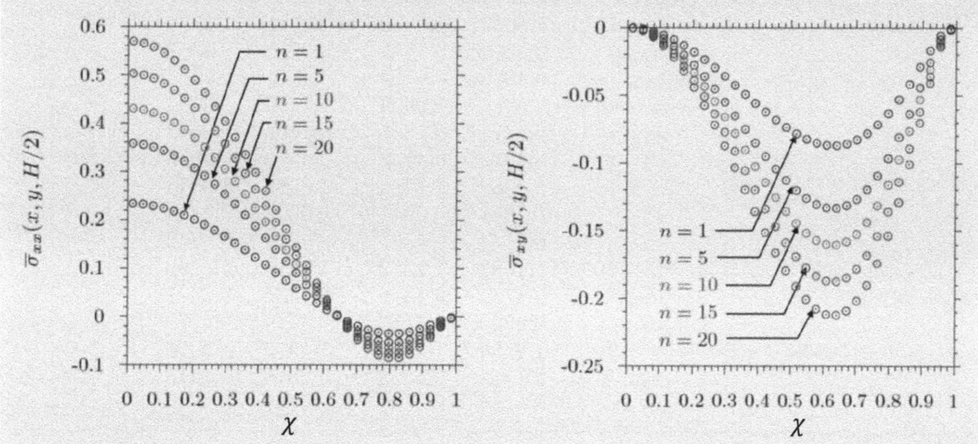

(a) (b)

Fig. 10.5.5 (a) Dimensionless stress $\bar{\sigma}_{xx}$ on the top surface along the line $x = y$ for clamped boundary conditions. (b) Dimensionless stress $\bar{\sigma}_{xy}$ on the top surface along the line $x = y$ for clamped boundary conditions.

Example 10.5.2

Analyze a clamped isotropic ($\nu = 0.25$ and $K_s = 5/6$) square plate under uniformly distributed transverse load of intensity q_0, using the FEM and various types of meshes.

Solution: The nondimensional quantities used are as follows:

$$\bar{w} = w(0,0)\frac{E_2 h^3}{a^4 q_0}, \quad \bar{\sigma}_{xx} = \sigma_{xx}(A, B, -h/2)(h^2/b^2 q_0), \quad \bar{\sigma}_{xz} = \sigma_{xz}(A, B, -h/2)(h/bq_0)$$

where the values of A and B for different meshes are given in Table 10.5.2. Table 10.5.3 contains the dimensionless displacements and stresses obtained using the FEM. The FST element with selective integration or reduced integration is accurate in predicting the bending response.

Table 10.5.2 The Gauss point locations at which the stresses are computed in the finite element analysis of clamped plates.

Point	4L	8L	2Q9	4Q9
A	$0.4375a$	$0.46875a$	$0.44717a$	$0.47358a$
B	$0.0625a$	$0.03125a$	$0.05283a$	$0.02642a$

Table 10.5.3 Effect of integration rule on the linear deflections $\bar{w}$ and stresses $\bar{\sigma}$ of clamped, isotropic, square plates under uniform transverse load q_0.

a/h	Integ.	Variable	4×4L	2×2Q9	8×8L	4×4Q9
10	F	$\bar{w} \times 10^2$	1.2593	1.5983	1.5447	1.6685
		$\bar{\sigma}_{xx}$	0.1190	0.1568	0.2054	0.2301
		$\bar{\sigma}_{xz}$	0.3890	0.4193	0.4463	0.4578
10	S	$\bar{w} \times 10^2$	1.6632	1.6880	1.6721	1.6758
		$\bar{\sigma}_{xx}$	0.1689	0.1813	0.2275	0.2357
		$\bar{\sigma}_{xz}$	0.4056	0.4118	0.4511	0.4566
10	R	$\bar{w} \times 10^2$	1.6854	1.6903	1.6776	1.6760
		$\bar{\sigma}_{xx}$	0.1718	0.1817	0.2284	0.2358
		$\bar{\sigma}_{xz}$	0.4045	0.4120	0.4509	0.4566
100	F	$\bar{w} \times 10^2$	0.0386	1.1222	1.3982	1.3546
		$\bar{\sigma}_{xx}$	0.0041	0.0947	0.0204	0.1825
		$\bar{\sigma}_{xz}$	0.3734	0.4732	0.4229	0.4862
100	S	$\bar{w} \times 10^2$	1.4093	1.4382	1.4219	1.4268
		$\bar{\sigma}_{xx}$	0.1731	0.1846	0.2337	0.2417
		$\bar{\sigma}_{xz}$	0.4236	0.4255	0.4729	0.4776
100	R	$\bar{w} \times 10^2$	1.4334	1.4417	1.4279	1.4271
		$\bar{\sigma}_{xx}$	0.1762	0.1853	0.2346	0.2418
		$\bar{\sigma}_{xz}$	0.4234	0.4284	0.4727	0.4779

10.5.2 Results of Nonlinear Analysis

Nonlinear analysis of plates using the DMCDM is yet to be developed (i.e., nonlinear results based on DMCDM are not available at this writing). Here we consider several examples of nonlinear bending of rectangular plates that have been solved by the FEM. Newton's iteration method is used to solve the nonlinear equations. Unless stated otherwise, a uniform mesh of 4×4 nine-node quadratic elements is used in a quarter plate for the FST. For this choice of mesh, full integration (F) means the 3×3 Gauss rule and reduced integration (R) means the 2×2 Gauss rule. In the examples, stresses were calculated at the center of the element.

Example 10.5.3

Consider a square isotropic plate with

$$a = b = 10 \text{ in}, \quad h = 1 \text{ in}, \quad E = 7.8 \times 10^6 \text{ psi}, \quad \nu = 0.3, \quad K_s = 5/6, \tag{1}$$

and subjected to a uniformly distributed transverse load of intensity q_0. Two types of simply supported boundary conditions, namely, SS-1 (see Fig 10.5.1) and SS-3 (see Fig. 10.5.6), are to be studied. The geometric boundary conditions for the two types of simply supported boundary conditions are as follows:

$$\text{SS-1: At } x = a/2: \quad v = w = \phi_y = 0; \quad \text{At } y = b/2: \quad u = w = \phi_x = 0. \tag{2}$$

$$\text{SS-3: } u = v = w = 0 \text{ on simply supported edges.} \tag{3}$$

The boundary conditions along the symmetry lines for both cases are given in Fig. 10.5.6.

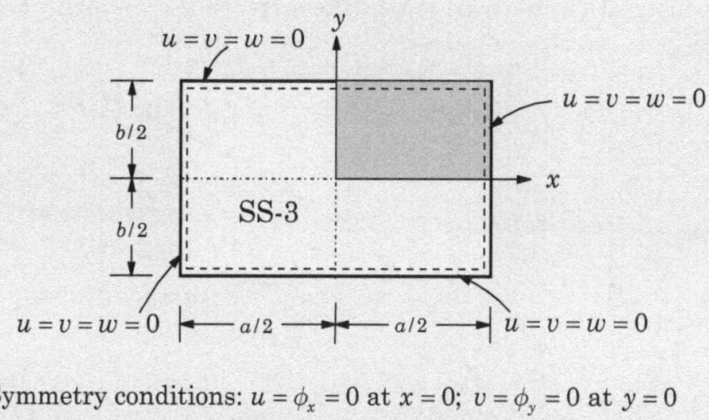

Fig. 10.5.6 Geometric boundary conditions used in the SS-3 type of simply supported rectangular plates.

Solution: Using the load parameter introduced earlier, $P \equiv q_0 a^4/Eh^4$, the incremental load vector is chosen to be

$$\{\Delta P\} = \{6.25, 6.25, 12.5, 25.0, 25.0, \ldots, 25.0\}. \tag{4}$$

A tolerance of $\epsilon = 10^{-2}$ is used in the Newton iteration scheme to check for convergence of the nodal displacements.

Table 10.5.4 contains the deflections $w(0,0)$ obtained with a uniform mesh of 4×4Q9 FST elements for various integration rules (also see Fig. 10.5.7). The number of iterations taken for convergence are listed in parentheses. The linear FST plate solution for load $q_0 = 4875$ psi (or $P = 6.25$) is $w = 0.2917$ in for SS-1 and $w = 0.3151$ in for SS-3 (for the R-R integration rule). The linear solution, being independent of the boundary conditions on u and v, of the SS-1 plate is smaller than that of the SS-3 plate. In the nonlinear case, the SS-3 type boundary conditions provide more edge restraint than the SS–1 boundary conditions and, therefore, would produce smaller transverse deflections. As discussed earlier, the 4×4Q9 meshes are not sensitive to shear or membrane locking, and therefore the results obtained with various integration rules are essentially the same. Table 10.5.5 contains the normal stresses $\bar{\sigma}_{xx} = \sigma_{xx}(a^2/Eh^2)$ obtained with a uniform mesh of 4×4Q9 FST elements for various integration rules (also see Fig. 10.5.8).

Table 10.5.4 Center deflection w of simply supported (SS-1 and SS-3) plates under uniformly distributed load q_0 ($4 \times 4Q9$ and $\epsilon = 10^{-2}$).

	SS-3			SS-1		
P	R–R*	F–R	F–F	R–R	F–R	F–F
6.25	0.2790 (4)	0.2790 (4)	0.2780 (3)	0.2813 (3)	0.2813 (3)	0.2812 (3)
12.5	0.4630 (3)	0.4630 (3)	0.4619 (3)	0.5186 (3)	0.5186 (3)	0.5185 (3)
25.0	0.6911 (3)	0.6911 (3)	0.6902 (3)	0.8673 (4)	0.8673 (4)	0.8672 (4)
50.0	0.9575 (3)	0.9575 (3)	0.9570 (3)	1.3149 (4)	1.3149 (4)	1.3147 (4)
75.0	1.1333 (3)	1.1333 (3)	1.1330 (3)	1.6241 (3)	1.6239 (3)	1.6237 (3)
100.0	1.2688 (3)	1.2688 (3)	1.2686 (3)	1.8687 (3)	1.8683 (3)	1.8679 (3)
125.0	1.3809 (2)	1.3809 (2)	1.3808 (2)	2.0758 (2)	2.0751 (2)	2.0746 (2)
150.0	1.4774 (2)	1.4774 (2)	1.4774 (2)	2.2567 (2)	2.2556 (2)	2.2549 (2)
175.0	1.5628 (2)	1.5629 (2)	1.5629 (2)	2.4194 (2)	2.4177 (2)	2.4168 (2)
200.0	1.6398 (2)	1.6399 (2)	1.6399 (2)	2.5681 (2)	2.5657 (2)	2.5645 (2)

* The first letter refers to the integration rule used for the nonlinear terms, while the second letter refers to the integration rule used for the shear terms; R = reduced; F = full integration.

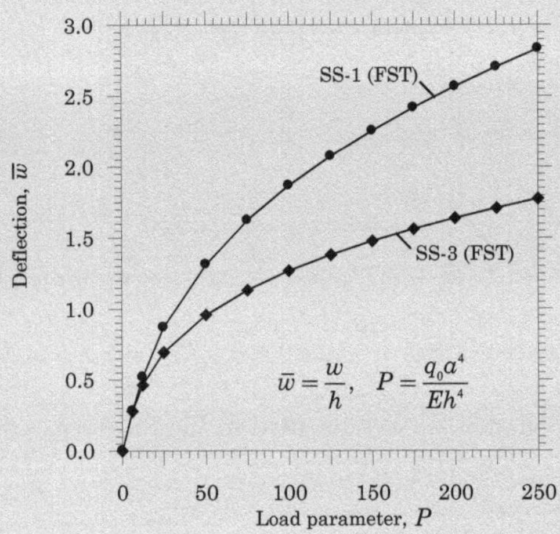

Fig. 10.5.7 Plots of normalized center deflection $\bar{w}$ versus load parameter P for isotropic ($\nu = 0.3$), simply supported square plates under uniformly distributed transverse load.

Table 10.5.5 Center stresses $\bar{\sigma}_{xx}$ of simply supported (SS-1 and SS-3) plates under uniformly distributed load q_0 (stresses were computed at the center of the element).

	SS-3			SS-1		
P	R–R	F–R	F–F	R–R	F–R	F–F
6.25	1.861	1.861	1.856	1.779	1.779	1.780
12.5	3.305	3.305	3.300	3.396	3.396	3.398
25.0	5.319	5.320	5.317	5.882	5.882	5.885
50.0	8.001	8.002	8.001	9.159	9.162	9.165
75.0	9.983	9.984	9.983	11.458	11.462	11.465
100.0	11.633	11.634	11.634	13.299	13.307	13.308
125.0	13.084	13.085	13.085	14.878	14.890	14.889
150.0	14.396	14.398	14.398	16.278	16.293	16.290
175.0	15.608	15.610	15.610	17.553	17.572	17.567
200.0	16.741	16.743	16.743	18.733	18.755	18.748

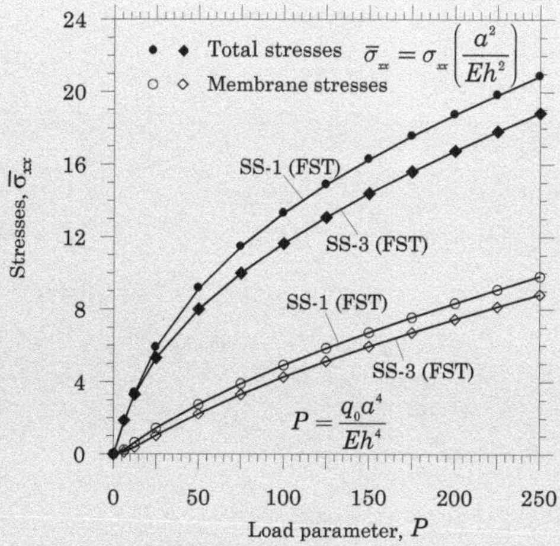

Fig. 10.5.8 Plots of center normal stress $\bar{\sigma}_{xx}$ versus load parameter P for isotropic ($\nu = 0.3$), simply supported square plates under uniformly distributed load.

Example 10.5.4

Consider an orthotropic square plate subjected to a uniformly distributed transverse load of intensity q_0. The geometric and material parameters to be used are

$$a = b = 12 \text{ in}, \quad h = 0.138 \text{ in}, \quad E_1 = 3 \times 10^6 \text{ psi}, \quad E_2 = 1.28 \times 10^6 \text{ psi}$$
$$G_{12} = G_{13} = G_{23} = 0.37 \times 10^6 \text{psi}, \quad \nu_{12} = 0.32, \quad K_s = 5/6. \tag{1}$$

Use a uniform mesh of 4×4Q9 elements in a quadrant, and analyze the problem for both SS-1 and SS-3 boundary conditions. Choose the vector of load increments to be $\{\Delta q_0\} = \{0.05, 0.05, 0.1, 0.2, 0.2, \ldots, 0.2\}$ with 12 load steps and error tolerance of $\epsilon = 0.01$.

Solution: Table 10.5.6 contains the center deflection and total normal stress as a function of the load for the two boundary conditions. Plots of load q_0 (psi) versus center deflection w (in) and q_0 versus normal stress (at the element center) $\bar{\sigma}_{xx} = \sigma_{xx}(a^2/E_2h^2)$ are shown in Figs. 10.5.9 and 10.5.10 for SS-1 and SS-3 plates, respectively.

Table 10.5.6 Center deflection w and normal stress $\bar{\sigma}_{xx}$ of simply supported orthotropic square plates under uniformly distributed load (4×4Q9).

	SS–1		SS–3	
q_0	w	$\bar{\sigma}_{xx}$	w	$\bar{\sigma}_{xx}$
0.05	0.0113	1.034	0.0113	1.056
0.10	0.0224	2.070	0.0218	2.116
0.20	0.0439	4.092	0.0397	4.058
0.40	0.0815	7.716	0.0650	7.103
0.60	0.1122	10.702	0.0824	9.406
0.80	0.1377	13.169	0.0959	11.284
1.00	0.1594	15.255	0.1069	12.894
1.20	0.1783	17.050	0.1162	14.316
1.40	0.1951	18.631	0.1244	15.602
1.60	0.2103	20.044	0.1318	16.783
1.80	0.2241	21.324	0.1384	17.880
2.00	0.2370	22.495	0.1445	18.909

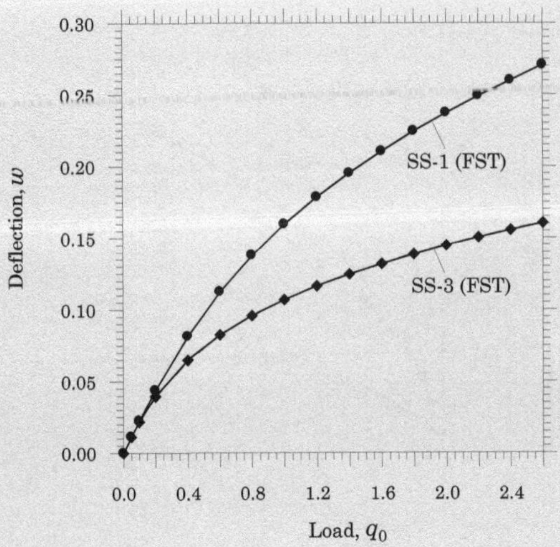

Fig. 10.5.9 Center deflection $w(0,0)$ as a function of the load q_0 for simply supported, orthotropic, square plates under uniformly distributed transverse load.

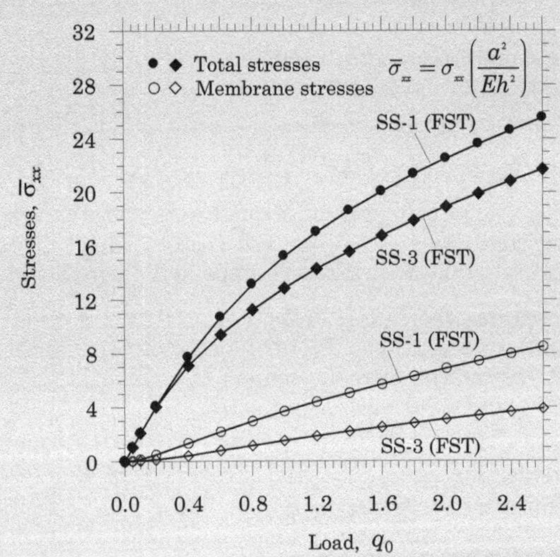

Fig. 10.5.10 Maximum stresses $\bar{\sigma}_{xx}$ as functions of the load q_0 for simply supported, orthotropic, square plates under uniformly distributed transverse load.

Example 10.5.5

Analyze an orthotropic plate with clamped edges (see Fig. 10.5.11). Assume that the boundary conditions of a clamped edge are

$$u = v = w = \phi_x = \phi_y = 0. \tag{1}$$

Use the geometric and material parameters given in Eq. (1) of **Example 10.5.4**, and assume a uniformly distributed transverse load of intensity q_0 on the plate. Use shear correction factor of $K_s = 5/6$.

Solution: The linear transverse deflection at the center of the plate is $w(0,0) = 0.0301$ for load $q_0 = 0.5$. Figure 10.5.12 contains a plot of load q_0 versus center deflection $w(0,0)$. Figure 10.5.12 also contains plots of the center deflection of an isotropic square plate ($h = 0.138$ in, $E = 1.28 \times 10^6$ psi, and $\nu = 0.3$). The results are predicted using the 4×4Q9 mesh of the FST elements. Table 10.5.7 contains numerical values of the center transverse deflection w and stress $\bar{\sigma}_{xx}$ (at the element center) as a function of q_0 for the same problem.

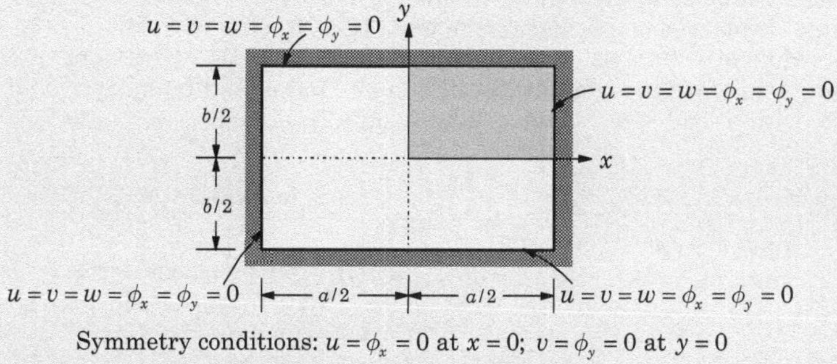

Fig. 10.5.11 Boundary conditions for a clamped rectangular plate.

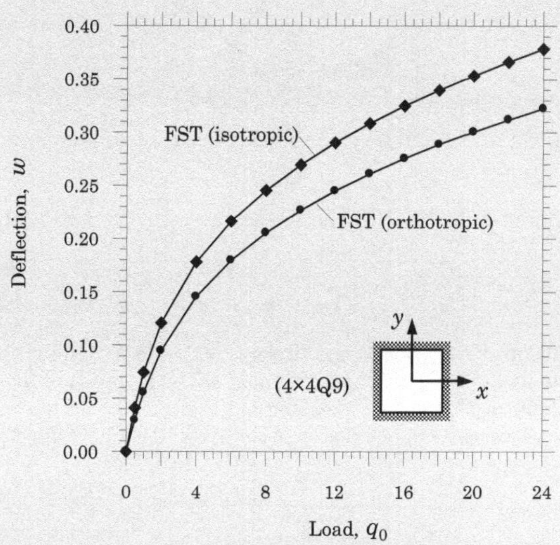

Fig. 10.5.12 Nonlinear center deflection $w(0,0)$ versus load parameter q_0 for clamped, isotropic and orthotropic, square plates under uniform load.

Table 10.5.7 Center deflection w and normal stress $\bar{\sigma}_{xx}$ for clamped orthotropic square plates under uniformly distributed load (4×4Q9 and $\epsilon = 10^{-2}$).

q_0	w	$\bar{\sigma}_{xx}$	q_0	w	$\bar{\sigma}_{xx}$
0.5	0.0294 (3)	4.317	12.0	0.2450 (2)	46.001
1.0	0.0552 (3)	8.467	14.0	0.2610 (2)	49.851
2.0	0.0948 (3)	15.309	16.0	0.2754 (2)	53.431
4.0	0.1456 (3)	24.811	18.0	0.2886 (2)	56.800
6.0	0.1795 (3)	31.599	20.0	0.3006 (2)	59.998
8.0	0.2054 (3)	37.078	22.0	0.3119 (2)	63.053
10.0	0.2267 (2)	41.793	24.0	0.3224 (2)	65.986

Example 10.5.6

Consider the nonlinear bending of a clamped, isotropic ($E = 10^6$ and $\nu = 0.3$), circular plate (radius $a = 100$ in and thickness $h = 10$ in) under uniformly distributed transverse load of intensity q_0. Due to the symmetry, the plate experiences axisymmetric deformation. However, to use the plate elements, one may use any sector of the plate. The geometric boundary condition along a radial line oriented at an angle other than 0° or 90° to the x- or y-axis is that the inplane displacement normal to the edge (i.e., u_n) be zero (due to the symmetry), which presents difficulty because u_n is not one of the degrees of freedom of the element. Analyze the problem using the computational domain, mesh, and boundary conditions shown in Fig. 10.5.13, with a load step of $\Delta q_0 = 2.5$ lb and a total load of $q_0 = 120$ lb.

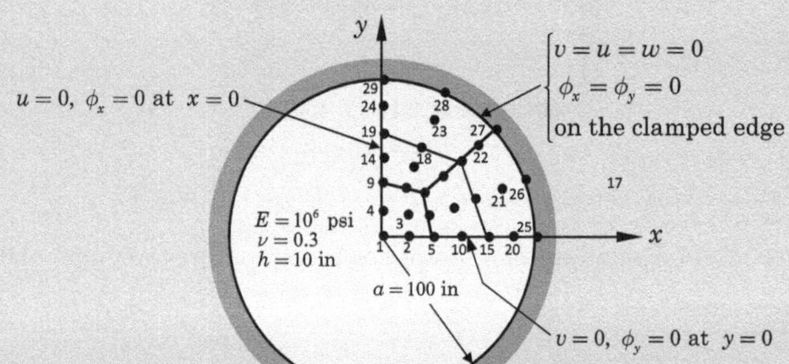

Fig. 10.5.13 Mesh and boundary conditions used for a clamped circular plate.

Solution: The exact center deflection for the linear case, for the two theories, is given by

$$w^{\text{FST}}(0) = \frac{q_0 a^4}{64D}\left(1 + \frac{8}{3(1-\nu)K_s}\frac{h^2}{a^2}\right), \quad w^{\text{CPT}}(0) = \frac{q_0 a^4}{64D}. \tag{1}$$

For $\nu = 0.3$, $K_s = 5/6$, and $a/h = 10$, the linear FST center deflection is given by $w(0) = 1.0457(q_0a^4/64D)$. The linear FST solution obtained with the mesh of five nine-node quadratic elements is $w(0) = 1.0659(q_0a^4/64D)$ (< 2% error). Figure 10.5.14 shows the nonlinear load-deflection curve and load versus $\bar{\sigma}_{xx} = \sigma_{xx}(A, A)(a/h)^2/10E$ curve for the problem.

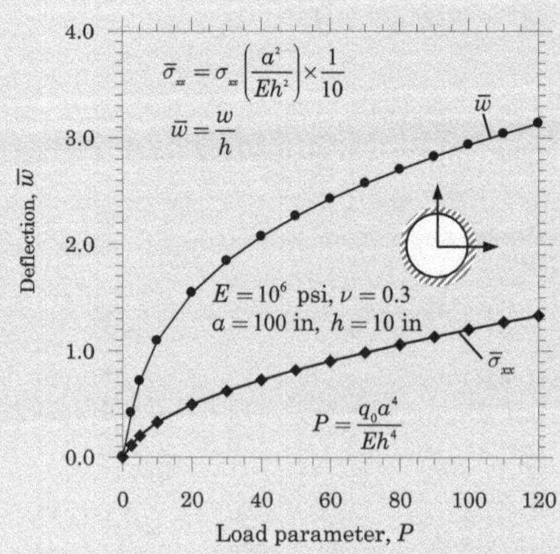

Fig. 10.5.14 Load-deflection curve for clamped isotropic circular plate under uniform load.

10.6 Summary

In this chapter, the governing equations of the first-order shear deformation theory (FST) of flat plates are derived using the principle of virtual displacements. Then the FEM and DMCDM formulations of the FST for rectangular plates are presented. In the case of the FEM, the von Kármán nonlinearity is also accounted for and the tangent stiffness coefficients are derived. The FST requires C^0-approximation (linear or quadratic) of all variables $(u, v, w, \phi_x, \phi_y)$ in the FEM; these variables are linearly interpolated in the DMCDM.

Like in the analysis of beams, both models (i.e., FEM and DMCDM) suffer from shear locking and membrane locking (when geometric nonlinearity is included). The shear locking is due to the use of the same low-order approximations of the rotations (ϕ_x, ϕ_y) and the transverse displacement w, while the membrane locking is a result of using the same low-order approximations of the in-plane displacements (u, v) and the transverse displacement w. Both types of locking are avoided here using reduced integration of shear and nonlinear terms of the stiffness coefficients.

Numerical results are presented to show the effect of boundary conditions and the material power-law index n on the plate deflection and stresses. The results also indicate that the DMCDM and FEM agree very well, although the DMCDM requires less formulation steps.

Although the present chapter discussed the methodology to implement DMCDM to functionally graded rectangular plates based on the FST, the same methodology can be used to extend DMCDM to higher-order plate theories, like the Reddy third-order plate theory. Such higher order theories can predict the variation of transverse shear stresses along the thickness of the plate and hence do not require shear correction factors, and they are very useful in the analysis of composite laminates [43]. Extension of the DMCDM to account for the geometric nonlinearity is also awaiting attention. Further, the DMCDM must be extended to nonlinear dynamics problems of fluid mechanics, elasticity, and plate bending. It is hoped that the present book would generate interest among the readers to embark on additional developments, extensions, and the study of the mathematical properties (e.g., existence and uniqueness as well error estimation) of the DMCDM.

Problems

WEAK FORMS AND FINITE ELEMENT MODELS

10.1 Develop the weak forms of the following linear equations governing the classical plate theory:

$$-\left(\frac{\partial^2 M_{xx}}{\partial x^2} + 2\frac{\partial^2 M_{xy}}{\partial x \partial y} + \frac{\partial^2 M_{yy}}{\partial y^2}\right) = q, \tag{1}$$

$$-\frac{\partial^2 w}{\partial x^2} - \left(\bar{D}_{22} M_{xx} - \bar{D}_{12} M_{yy}\right) = 0, \tag{2}$$

$$-\frac{\partial^2 w}{\partial y^2} - \left(\bar{D}_{12} M_{xx} - \bar{D}_{11} M_{yy}\right) = 0, \tag{3}$$

$$-2\frac{\partial^2 w}{\partial x \partial y} - (D_{66})^{-1} M_{xy} = 0, \tag{4}$$

where

$$\bar{D}_{ij} = \frac{D_{ij}}{D_0}, \quad D_0 = D_{11}D_{22} - D_{12}^2, \quad D_{ij} = Q_{ij}\frac{h^3}{12}, \quad (i = 1, 2, 6), \tag{5}$$

and Q_{ij} are the elastic stiffnesses defined in terms of the principal moduli (E_1, E_2), shear modulus G_{12}, and Poisson's ratio ν_{12} as

$$\begin{gathered} Q_{11} = \frac{E_1}{1 - \nu_{12}\nu_{21}}, \quad Q_{12} = \frac{\nu_{12}E_2}{1 - \nu_{12}\nu_{21}}, \quad Q_{22} = \frac{E_2}{1 - \nu_{12}\nu_{21}}, \\ Q_{66} = \frac{1}{S_{66}} = G_{12}, \quad \nu_{21} = \nu_{12}\frac{E_2}{E_1}. \end{gathered} \tag{6}$$

10.2 Develop the (mixed) finite element model associated with the equations in Problem 10.1. Assume approximation of the form

$$\begin{gathered} w = \sum_{i=1}^{r} w_i \psi_i^{(1)}, \quad M_{xx} = \sum_{i=1}^{s} M_{xi}\psi_i^{(2)}, \\ M_{yy} = \sum_{i=1}^{p} M_{yi}\psi_i^{(3)}, \quad M_{xy} = \sum_{i=1}^{q} M_{xyi}\psi_i^{(4)}, \end{gathered}$$

where $\psi_i^{(\alpha)}$, $(\alpha = 1, 2, 3, 4)$ are appropriate interpolation functions. Discuss the minimum requirements of the interpolation functions.

10.3 Develop the weak forms of the following nonlinear equations governing the classical plate theory:

$$-\left(\frac{\partial N_{xx}}{\partial x} + \frac{\partial N_{xy}}{\partial y}\right) = 0, \tag{1}$$

$$-\left(\frac{\partial N_{xy}}{\partial x} + \frac{\partial N_{yy}}{\partial y}\right) = 0, \tag{2}$$

$$-\left(\frac{\partial^2 M_{xx}}{\partial x^2} + 2\frac{\partial^2 M_{xy}}{\partial x \partial y} + \frac{\partial^2 M_{yy}}{\partial y^2}\right) - \mathcal{N}(u, v, w) = q, \tag{3}$$

$$-\frac{\partial^2 w}{\partial x^2} - (\bar{D}_{22}M_{xx} - \bar{D}_{12}M_{yy}) = 0, \tag{4}$$

$$-\frac{\partial^2 w}{\partial y^2} - (\bar{D}_{12}M_{xx} - \bar{D}_{11}M_{yy}) = 0, \tag{5}$$

$$-2\frac{\partial^2 w}{\partial x \partial y} - (D_{66})^{-1}M_{xy} = 0, \tag{6}$$

where $(N_{xx}, N_{yy}, N_{xy}, \mathcal{N})$ are known in terms of (u, v, w) and their derivatives by Eqs. (10.2.23) and (10.2.12). Note that the dependent unknowns are u, v, w, M_{xx}, M_{yy}, and M_{xy}.

10.4 Develop the (mixed) nonlinear finite element model of the equations in Problem 10.3. Assume interpolation of the form

$$u = \sum_{i=1}^{n} u_i \phi_i, \quad v = \sum_{i=1}^{n} v_i \phi_i, \quad w = \sum_{i=1}^{r} w_i \psi_i^{(1)}, \tag{1}$$

$$M_{xx} = \sum_{i=1}^{s} M_{xi}\psi_i^{(2)}, \quad M_{yy} = \sum_{i=1}^{p} M_{yi}\psi_i^{(3)}, \quad M_{xy} = \sum_{i=1}^{q} M_{xyi}\psi_i^{(4)}. \tag{2}$$

10.5 Develop the weak forms of the following nonlinear equations governing the classical plate theory:

$$-\left(\frac{\partial N_{xx}}{\partial x}+\frac{\partial N_{xy}}{\partial y}\right)=0, \tag{1}$$

$$-\left(\frac{\partial N_{xy}}{\partial x}+\frac{\partial N_{yy}}{\partial y}\right)=0, \tag{2}$$

$$-\frac{\partial^2 M_{xx}}{\partial x^2}+4D_{66}\frac{\partial^4 w}{\partial x^2\partial y^2}-\frac{\partial^2 M_{yy}}{\partial y^2}-\frac{\partial}{\partial x}\left(N_{xx}\frac{\partial w}{\partial x}+N_{xy}\frac{\partial w}{\partial y}\right)$$
$$-\frac{\partial}{\partial y}\left(N_{xy}\frac{\partial w}{\partial x}+N_{yy}\frac{\partial w}{\partial y}\right)=q, \tag{3}$$

$$-\frac{\partial^2 w}{\partial x^2}-(\bar{D}_{22}M_{xx}+\bar{D}_{12}M_{yy})=0, \tag{4}$$

$$-\frac{\partial^2 w}{\partial y^2}-(\bar{D}_{12}M_{xx}+\bar{D}_{11}M_{yy})=0, \tag{5}$$

where (N_{xx}, N_{yy}, N_{xy}) are known in terms of (u, v, w) and their derivatives by Eq. (10.2.23). Note that the dependent unknowns are u, v, w, M_{xx}, and M_{yy}.

10.6 Develop the (mixed) nonlinear finite element model of the equations in Problem 10.5 by assuming approximations of the form

$$u=\sum_{i=1}^{n}u_i\psi_i^{(1)},\quad v=\sum_{i=1}^{n}v_i\psi_i^{(1)},\quad w=\sum_{i=1}^{r}w_i\psi_i^{(2)}, \tag{1}$$

$$M_{xx}=\sum_{i=1}^{s}M_{xi}\psi_i^{(3)},\quad M_{yy}=\sum_{i=1}^{p}M_{yi}\psi_i^{(4)}. \tag{2}$$

10.7 Consider the following alternative form of nonlinear equations governing the first-order shear deformation plate theory:

$$-\left(\frac{\partial N_{xx}}{\partial x}+\frac{\partial N_{xy}}{\partial y}\right)=0, \tag{1}$$

$$-\left(\frac{\partial N_{xy}}{\partial x}+\frac{\partial N_{yy}}{\partial y}\right)=0, \tag{2}$$

$$-A_{55}\frac{\partial}{\partial x}\left(\frac{\partial w}{\partial x}+\phi_x\right)-A_{44}\frac{\partial}{\partial y}\left(\frac{\partial w}{\partial y}+\phi_y\right)$$
$$-\frac{\partial}{\partial x}\left(N_{xx}\frac{\partial w}{\partial x}+N_{xy}\frac{\partial w}{\partial y}\right)-\frac{\partial}{\partial y}\left(N_{xy}\frac{\partial w}{\partial x}+N_{yy}\frac{\partial w}{\partial y}\right)=q, \tag{3}$$

$$-\frac{\partial M_{xx}}{\partial x}-D_{66}\frac{\partial}{\partial y}\left(\frac{\partial \phi_x}{\partial y}+\frac{\partial \phi_y}{\partial x}\right)+K_sA_{55}\left(\frac{\partial w}{\partial x}+\phi_x\right)=0, \tag{4}$$

$$-\frac{\partial M_{yy}}{\partial y}-D_{66}\frac{\partial}{\partial x}\left(\frac{\partial \phi_x}{\partial y}+\frac{\partial \phi_y}{\partial x}\right)+K_sA_{44}\left(\frac{\partial w}{\partial y}+\phi_y\right)=0, \tag{5}$$

$$\frac{\partial \phi_x}{\partial x}-\left(\bar{D}_{11}M_{xx}+\bar{D}_{12}M_{yy}\right)=0, \tag{6}$$

$$\frac{\partial \phi_y}{\partial y}-\left(\bar{D}_{12}M_{xx}+\bar{D}_{22}M_{yy}\right)=0, \tag{7}$$

where (N_{xx}, N_{yy}, N_{xy}) are known in terms of (u, v, w) and their derivatives by [see Eq. (10.2.23)] and

$$\bar{D}_{ij}=\frac{D_{ij}}{D_0},\quad D_0=D_{11}D_{22}-D_{12}^2. \tag{8}$$

The dependent unknowns of the formulation (note the order) are $(u, v, w, \phi_x, \phi_y, M_{xx}, M_{yy})$. Develop the weak forms of Eqs. (4)–(7) such that they are symmetric in $(\phi_x, \phi_y, M_{xx}, M_{yy})$, and identify the primary and secondary variables.

10.8 Develop the (mixed) nonlinear finite element model of the equations in Problem 10.7 by assuming approximations of the form

$$u = \sum_{i=1}^{m} u_i \psi_i^{(1)}, \quad v = \sum_{i=1}^{m} v_i \psi_i^{(1)}, \quad w = \sum_{i=1}^{n} w_i \psi_i^{(2)}$$

$$\phi_x = \sum_{i=1}^{r} S_i^1 \psi_i^{(3)}, \quad \phi_y = \sum_{i=1}^{r} S_i^2 \psi_i^{(3)}, \quad M_{xx} = \sum_{i=1}^{s} M_{xi} \psi_i^{(4)}, \quad M_{yy} = \sum_{i=1}^{p} M_{yi} \psi_i^{(5)}.$$

10.9 Consider the governing equations of equilibrium of elastic plates according to the first-order shear deformation plate theory with the von Kármán nonlinearity. Assume that the material of the plate is isotropic and obeys the generalized Hooke's law. However, the elastic modulus E varies nonlinearly with the displacement gradients according to the equation

$$E = E_0 + E_1 \left(\frac{\partial u}{\partial x} + \frac{\partial v}{\partial y} \right). \tag{1}$$

Assuming that Poisson's ratio is constant, (a) develop the finite element model of the equations and (b) compute the tangent stiffness coefficients, T_{ij}^{11} and T_{ij}^{33}.

10.10 Consider the governing equations of equilibrium of elastic plates according to the first-order shear deformation plate theory with the von Kármán nonlinearity. Assume that the plate is made of two homogeneous isotropic materials that are combined according to the following power-law distribution:

$$E(z) = (E_1 - E_2)\, f(z) + E_2, \quad f(z) = \left(\frac{1}{2} + \frac{z}{h} \right)^n, \tag{1}$$

where E_1 and E_2 are moduli of the two materials that are being mixed, h is the total thickness of the plate, and n is the power-law index that dictates the volume fraction $f(z)$ of material 1. Develop the displacement finite element model of the functionally graded plate and determine the tangent stiffness coefficients.

References

[1] B. Carnahan, H. A. Luther, and J. O. Wilkes, *Applied Numerical Methods.* New York, NY, 1969; reprinted by Krieger Publishing, Malabar, FL: John Wiley & Sons, 1990.

[2] S. V. Patankar, *Numerical Heat Transfer and Fluid Flow.* Boca Raton, FL: CRC Press, 1980.

[3] H. K. Versteeg and W. Malalasekera, *Computational Fluid Dynamics, The Finite Volume Method,* 2nd ed. Harlow, England, UK: Pearson Education (Prentice-Hall), 2007.

[4] J. H. Ferziger and M. Perić, *Computational Methods for Fluid Dynamics,* 3rd ed. New York, NY: Springer-Verlag, 2002.

[5] F. Moukalled, L. Mangani, and M. Darwish, *The Finite Volume Method in Computational Fluid Dynamics.* New York, NY: Springer, 2016.

[6] S. Mazumder, *Numerical Methods for Partial Differential Equations. Finite Difference and Finite Volume Methods.* New York, NY: Elsevier, 2016.

[7] J. N. Reddy, N. K. Anand, and P. Roy, *Finite Element and Finite Volume Methods for Heat Transfer and Fluid Dynamics.* Cambridge, UK: Cambridge University Press, 2022.

[8] J. N. Reddy, *An Introduction to the Finite Element Method,* 4th ed. New York, NY: McGraw-Hill, 2019.

[9] P. M. Gresho and R. L. Sani, *Incompressible Flow and the Finite Element Method.* London, UK: John Wiley & Sons, 1998.

[10] J. N. Reddy and D. K. Gartling, *The Finite Element Method in Heat Transfer and Fluid Dynamics,* 3rd ed. Boca Raton, FL: CRC Press, 2010.

[11] B. Szabó and I. Babuška, *Introduction to Finite Element Analysis: Formulation, Verification and Validation.* New York, NY: John Wiley & Sons, 2011.

[12] K. J. Bathe, *Finite Element Procedures.* Prentice-Hall, Upper Saddle River, NJ, 1996.

[13] J. N. Reddy, *An Introduction to Nonlinear Finite Element Analysis*, 2nd ed. Oxford, U.K: Oxford University Press, 2015.

[14] K. S. Surana and J. N. Reddy, *The Finite Element Method for Boundary Value Problems, Mathematics and Computations.* Boca Raton, FL: CRC Press, 2017.

[15] K. S. Surana and J. N. Reddy, *The Finite Element Method for Initial Value Problems, Mathematics and Computations.* Boca Raton, FL: CRC Press, 2018.

[16] J. N. Reddy, "A dual mesh finite domain method for the numerical solution of differential equations," *Int. J. Comput. Methods Eng. Sci. Mech.*, vol. 20, no. 3, pp. 212–228, 2019.

[17] J. N. Reddy, *Energy Principles and Variational Methods in Applied Mechanics*, 3rd ed. New York, NY: John Wiley & Sons, 2018.

[18] E. Kreyszig, *Advanced Engineering Mathematics*, 10th ed. New York, NY: John Wiley & Sons, 2011.

[19] L. A. Pipes and L. R. Harvill, *Applied Mathematics for Engineers and Physicists*, 3rd ed. New York, NY: McGraw-Hill (Dover edition, 2014), 1970.

[20] J. N. Reddy and M. L. Rasmussen, *Advanced Engineering Analysis.* New York, NY: John Wiley & Sons, 1990.

[21] J. N. Reddy, *Applied Functional Analysis and Variational Methods in Engineering.* New York, NY: McGraw-Hill, 1991.

[22] M. McHale, J. Friedman, and J. Karian, "Standard for verification and validation in computational fluid dynamics and heat transfer," *The American Society of Mechanical Engineers, ASME V&V*, vol. 20, 2009.

[23] J. N. Reddy, *An Introduction to Continuum Mechanics*, 2nd ed. New York, NY: Cambridge University Press, 2013.

[24] P. J. Roache, *Verification and validation in computational science and engineering.* Hermosa Publishers, Socorro, NM, 1998.

[25] I. Babuška and J. T. Oden, "Verification and validation in computational engineering and science: basic concepts," *Computer Methods in Applied Mechanics and Engineering*, vol. 193, pp. 4057–4066, 2004.

[26] W. L. Oberkampf and C. J. Roy, *Verification and Validation in Scientific Computing.* New York, NY: Cambridge University Press, 2010.

[27] S. C. Chapra and R. P. Canale, *Numerical Methods for Engineers*, 8th ed. New York, NY: McGraw-Hill, 2021.

[28] E. Isaacson and H. B. Keller, *Analysis of Numerical Methods.* New York: John Wiley & Sons, 1966.

[29] F. B. Hildebrand, *Introduction to Numerical Analysis.* McGraw-Hill, New York, NY, 1956.

[30] T. J. Chung, *Computational Fluid Dynamics.* New York, NY: Cambridge University Press, 2002.

[31] P. D. Lax and R. D. Richtmyer, "Survey of the stability of linear finite difference equations," *Communications in Pure and Applied Mathematics*, vol. 9, pp. 267–293, 1956.

[32] R. D. Richtmyer, *Difference Methods for Initial Value Problems.* Interscience, New York, NY, 1957.

[33] G. O'Brien, M. Hyman, and S. Kaplan, "A study of the numerical solution of partial differential equations," *Journal of Mathematical Physics*, vol. 29, pp. 233–251, 1951.

[34] J. N. Reddy, M. Martinez, and P. Nampally, "A novel numerical method for the solution of nonlinear equations with applications to heat transfer," *International Journal of Numerical Methods in Heat Transfer and Fluid Flow*, vol. DOI 10.1108/HFF-07-2020-0397, 2021.

[35] V. R. Voller, *Basic Control Volume Finite Element Methods for Fluids and Solids.* World Scientific, Singapore, 2009, vol. IISc Research Monograph Series, Vol. 1.

[36] A. M. Winslow, "Numerical solution of the quasilinear poisson equation in a nonuniform triangle mesh," *Journal of Computational Physics*, vol. 2, pp. 149–172, 1967.

[37] B. R. Baliga and S. Patankar, "A new finite-element formulation for convection-diffusion problems," *Numerical Heat Transfer*, vol. 3(4), pp. 393–409, 1980.

[38] B. R. Baliga and S. V. Patankar, "A control volume finite-element method for two-dimensional fluid flow and heat transfer," *Numerical Heat Transfer*, vol. 6(3), pp. 245–261, 1983.

[39] C. Prakash and S. V. Patankar, "A control volume-based finite-element method for solving the navier-stokes equations using equal-order velocity-pressure interpolation," *Numerical Heat Transfer*, vol. 8, no. 3, pp. 259–280, 1985.

[40] S. Kattoura, A. Lamoureux, and B. R. Baliga, "A control-volume finite element method for the prediction of three-dimensional diffusion-type phenomena in anisotropic media," *Computational Thermal Sciences: An International Journal*, vol. 5, no. 3, pp. 249–260, 2013.

[41] J. N. Reddy and M. Martinez, "A dual mesh finite domain method for steady-state convection-diffusion problems," *Computers and Fluids*, vol. 251, p. 112648, 2020.

[42] J. N. Reddy, *Theories and Analyses of Beams and Axisymmetric Circular Plates.* Boca Raton, FL: CRC Press, 2002.

[43] J. N. Reddy, *Mechanics of Laminated Composite Plates and Shells: Theory and Analysis*, 2nd ed. CRC Press, Boca Raton, FL, 2004.

[44] J. N. Reddy, P. Nampally, and A. R. Srinivasa, "Nonlinear analysis of functionally graded beams using the dual mesh finite domain method and the finite element method," *International Journal of Non-Linear Mechanics*, vol. 127, p. 103575, 2020.

[45] P. Nampally and J. N. Reddy, "Bending analysis of functionally graded axisymmetric circular plates using the dual mesh finite domain method," *Lat. Am. J. Solids struct.*, vol. 17, 2020.

[46] J. N. Reddy, P. Nampally, and N. Phan, "Dual mesh control domain analysis of functionally graded circular plates accounting for moderate rotations," *Composite Structures*, p. 113153, 2020.

[47] A. Nadai, *Theory of Flow and Fracture of Solids.* New York, NY: McGraw–Hill, 1963.

[48] H. Weller, "Controlling the computational modes of the arbitrarily structured c grid," *Monthly Weather Review*, vol. 140, no. 19, pp. 3220–3234, 2012.

[49] OpenFOAM, *Controlling the computational modes of the arbitrarily structured C grid.* ESI Group, Paris, France, 2014.

[50] Z. Jiao, T. Heblekar, G. Wang, W. Xu, R. Chen, and J. N. Reddy, "Solution of plane elasticity problems using the dual mesh control domain method with arbitrary meshes," *Computational Methods in Applied Mechanics and Engineering*, vol. 416, p. 116342, 2023.

[51] G. N. Wang, "The elastic solutions of separable problems with the applications to multilayered structures," *Archive of Applied Mechanics*, vol. 88, pp. 1525–1543, 2018.

[52] T. L. Anderson, *Fracture Mechanics: Fundamentals and Applications*, 4th ed. Boca Raton, FL: CRC Press, 2017.

[53] P. Nampally, E. Ruocco, and J. N. Reddy, "Bending analysis of functionally graded rectangular plates using the dual mesh control domain method," *Computational Methods in Engineering Science and Mechanics*, vol. 22, no. 5, pp. 425–437, 2021.

Index